Basic mathematics
for the
physical sciences

Basic mathematics for the physical sciences

Edited by

Robert Lambourne

Department of Physics, Open University

and

Michael Tinker

J. J. Thompson Physical Laboratory, University of Reading

JOHN WILEY & SONS, LTD

Chichester · New York · Weinheim · Brisbane · Singapore · Toronto

Published by John Wiley & Sons, Ltd, The Atrium, Southern Gate,
Chichester, West Sussex PO19 8SQ, England
Telephone (+44) 1243 779777

Email (for orders and customer service enquiries): cs-books@wiley.co.uk
Visit our Home Page www.wileyeurope.com or www.wiley.com

Reprinted November 2001, January 2005, September 2006, December 2007

Other Wiley Editorial Offices

John Wiley & Sons Inc., 111 River Street, Hoboken, NJ 07030, USA

Jossey-Bass, 989 Market Street, San Francisco, CA 94103-1741, USA

Wiley-VCH Verlag GmbH, Boschstr. 12, D-69469 Weinheim, Germany

John Wiley & Sons Australia Ltd, 33 Park Road, Milton, Queensland 4064, Australia

John Wiley & Sons (Asia) Pte Ltd, 2 Clementi Loop #02-01, Jin Xing Distripark,
Singapore 129809

John Wiley & Sons (Canada) Ltd, 22 Worcester Road, Etobicoke, Ontario M9W 1L1

Wiley also publishes its books in a variety of electronic formats. Some content that appears in print may not be
available in electronic books.

Library of Congress Cataloging-in-Publication Data
Basic mathematics for the physical sciences / edited by Robert
Lambourne and Michael Tinker.
        p. cm.
    Includes index.
    ISBN 0–471–85206–6 (alk. paper). – ISBN 0–471–85207–4
  (pbk. : alk paper)
    1. Mathematics. I. Lambourne, Robert II. Tinker, Michael.
QA37.2.B378   2000
510–dc21   99–35925
                                            CIP

British Library Cataloguing in Publication Data
A catalogue record for this book is available from the British Library

ISBN 978-0–471–85206–3 (hbk)
ISBN 978-0–471–85207–0 (pbk)

Designed and typeset in TimesTen by The Florence Group, Stoodleigh, Devon

Contents

Preface
The FLAP project

This book is one of the many products to emerge from the *FLAP* project – the Flexible Learning Approach to Physics – a part of the Teaching and Learning Technology Programme funded by the four United Kingdom Higher Education Funding Bodies, HEFCE, SHEFC, HEFCW and DENI.

The main aim of *FLAP* was, and still remains, the provision of high quality teaching resources that will enable colleges and universities to broaden their intake in physical science and to ease the transition that students must make in going from school to university. In recent years particular problems have arisen within higher education, especially in engineering and the physical sciences, from the increasingly diverse prior learning of students within individual classes. *FLAP* aimed to meet this challenge by providing science, engineering and mathematics departments with a large number of compact, authoritative, self-study 'modules', that incorporated ready-to-study tests and fast track questions so that students could identify for themselves those parts of each module that they needed to study. Photocopy masters of the module texts, along with supporting question banks, study guides, AV and CAL materials were supplied to teachers and lecturers throughout the UK so that they could incorporate them into their own courses, as and when appropriate, with the greatest possible ease and efficiency. This approach was sufficiently successful, and the teaching materials sufficiently well-received, that this book has now been produced so that a wider audience, including schools and individual learners, can make use of the rich resource that *FLAP* provides.

Basic Mathematics for the Physical Sciences, together with its companion volume, *Further Mathematics for the Physical Sciences*, contains about 35% of the teaching text developed by the *FLAP* project. The two volumes include almost all of the material devoted specifically to applicable mathematics, but exclude all of the physics-based material. Further information about the full range of *FLAP* products, including this book and the related non-text teaching resources, may be obtained from the project's web site **http://yan.open.ac.uk/flap/FLAPHome.html**.

Like other *FLAP* products this book is a result of team effort. We, the editors, have created the chapters that follow by modifying the self-study modules produced by members of the original *FLAP* production team. We therefore take great pleasure in acknowledging the input of all the members of that team, and of the many others who assisted the project. In addition, and specifically in relation to this book, we should like to particularly acknowledge the input of Dr Rebecca Graham and the very substantial contribution of Dr Stuart Windsor. The advice and guidance of Andy Slade and Simon Plumtree of John Wiley and Sons, Ltd is also greatly appreciated.

FLAP Production Team

Management Team

Dr Robert Lambourne, Project Director, Open University
Dr Michael Tinker, General Editor, University of Reading
Dr Simone Pitman, Project Coordinator, Open University
Dr Stuart Windsor, Project Coordinator, University of Reading

Editorial Team

Dr Graham Farmelo, Science Museum, London
Dr Geoffrey Green, University of Reading
Dr David Grimes, University of Reading
Dr John Perring, formerly Harwell
Dr Graham Read, Open University
Dr Elizabeth Swinbank, University of York

Author Team

Dr David Betts, University of Sussex
Dr Derek Capper, formerly Queen Mary and Westfield College
Dr Anthony Evans, University of Sussex
Dr Raymond Flood, University of Oxford
Dr Frank Foster, University of Lancaster
Dr Caroline Fraser, Christchurch College, University of Oxford
Dr James Gotaas, University of Central Lancashire
Dr Geoffrey Green, University of Reading
Dr David Grimes, University of Reading
Dr Keith Higgins, Birkbeck College, University of London
Dr Susan Holmes, formerly Royal Holloway and Bedford New College
Dr Robert Lambourne, Open University
Dr David Martin, Coventry University
Dr Leslie Mustoe, University of Loughborough
Dr John Roche, Oxford Brookes University
Mrs Mary Stewart, University of Loughborough
Dr David Sumner, Tarragon Press
Dr Elizabeth Swinbank, University of York
Dr Michael Tinker, University of Reading
Mr Richard Walton, Sheffield Hallam University

OU Editorial Personnel

Dr Rebecca Graham
Mr David Tillotson
Mr Peter Twomey

The *FLAP* Team wish to acknowledge the work of the following Open University Course Teams, some of whose materials were used in the production of *FLAP*: S271, MS284, MS283, S272 and S102.

Thanks are also due to:
Dr John Bolton, Dr Andrew Norton, Dr Michael Bromilow, Dr Stuart Freake, Dr David Johnson, Dr Alan Durrant, Dr Dennis Dunne, Mrs Maureen Maybank, Dr Tom Smith, the *FLAP* Design Team (supported by GEC Marconi) and Dr Cheryl Newport.

Introduction
This book and how
to use it

This is not a conventional textbook. Unlike a normal book, where the reader is expected to start at the beginning and work through sequentially to the end, this book is designed to encourage flexible use, particularly by those who already have some knowledge of its subject matter but are either very rusty, or suffering from the kind of highly fragmented knowledge that many students exhibit. Whatever your individual strengths and weaknesses may be, this book is designed to let you make the most of any relevant prior learning you may have acquired, and to help you to overcome any problems you may be facing. At one extreme, for a reader who already possesses a good knowledge of basic applicable mathematics, but who is unfamiliar with just one or two topics, the book can act as a sort of academic aspirin; use it in the evening and your problem can be cured by the morning. At the other extreme, for those with little or no background in the subject, the book can provide a complete and coherent introduction. It may certainly be used to supplement a course, but it can also be used to replace a conventional course; providing the reader with all that is needed in the way of study advice, explanation, worked examples and self-assessment questions.

As its title implies, this book is mainly concerned with the most introductory parts of applicable mathematics; topics that are usually covered in school mathematics syllabuses. However, its treatment of these topics is unusually detailed since it is firmly aimed at those who are either currently pursuing, or intending to pursue, the study of physical science or engineering at university level. Signs of this orientation are to be found throughout the book, particularly in the widespread use of physical examples, and in the consistent inclusion of SI units in calculations. Combined study of this book and its companion volume – *Further Mathematics for the Physical Sciences* – should provide readers with all the maths they need to know to progress from compulsory level school mathematics to typical first year undergraduate courses in applied maths. This book will be particularly useful to those undertaking access or foundation courses, and to those finding their current undergraduate maths courses too taxing due to deficiencies in their prior learning. It should be of interest to school pupils who are wondering what university-level study holds in store for them, and for anyone who is planning to return to structured learning after a long break. Mature students and those involved in any form of distance education are likely to find its methods particularly well suited to their needs.

How to use this book

In order to get the most out of this book it is probably best to regard it as a collection of essentially free-standing chapters rather than a monolithic tome. All the chapters are structured in a similar way. Subsection 1.1 of each is an introduction that outlines the contents of the chapter. This is partly written for the benefit of those who already have some familiarity with the subject. If you recognise all of the (*italicised*) technical terms that arise in the introduction to a chapter there is a good chance that you will not need to read the chapter in detail, and you should give serious consideration to following the '*Fast track*' route outlined below.

Subsections 1.2 and 1.3 in each chapter are entitled *Fast track questions* and *Ready to study?*, respectively. Each of these subsections contains questions which you will need to work through. The *Fast track questions* should allow you to decide whether or not you *need to study* the chapter. If your answers to all of these questions are correct (you can check your answers against those given at the back of the book), then you probably don't need to study the chapter in detail. You can simply glance through the main text before looking carefully at the subsections entitled *Chapter summary* and *Achievements*. If you are familiar with all the summarised points and you believe you can meet all of the expected achievements then you should move on to the *End of chapter questions* that follow. If you can answer all these questions correctly, your task is complete!

If you are unfamiliar with some or all of the terms used in a chapter introduction, and if you find the *Fast track questions* hard or even meaningless then you should try to work through the whole chapter in detail. If you do decide that you must work through the whole chapter, Subsection 1.3, *Ready to study?*, will enable you to check that you have the right background knowledge to undertake such a task. If you find that the *Ready to study* section advises you to be familiar with some term or technique that you do not recognise then you can find out about that topic by looking it up in the index at the back of the book and consulting the references that you find there – especially references printed in **bold type**, which are generally major sources of information on the topic concerned. To avoid any ambiguity over the amount of detail that you need to know about particular topics, the *Ready to study?* subsection usually includes some questions, the answers to which are printed at the back of the book.

Once you embark on the main text of a chapter there are several features that are worth looking out for:

- All newly introduced terms are printed in **bold type** close to the place where they are defined. A few especially important terms are deliberately introduced more than once to increase the flexibility with which the book can be used. (Experience has shown that these repetitions are also effective in driving home the fundamental importance of certain topics.)

- Some other important scientific terms used in the text are printed in *italic type*. These are terms which are either covered in greater detail elsewhere, or else are too peripheral to deserve a full definition or a detailed discussion in this book.

- In all chapters the most important statements and equations are printed in open boxes to draw your attention to them.

- 'Helping hands' (☞) will sometimes appear in the text to direct you to read a helpful marginal note or comment. Less significant marginal notes may appear in the margin without warning – so keep your eyes open for them, they may help!

- Most chapters also include worked *Examples*. A problem is posed and its solution follows. These *Examples* are designed to help you develop a systematic approach to problem solving, so it is worth paying attention to how each problem is tackled.

- Two further sorts of questions appear in the main text:

  Practice questions (in boxes) are short questions that are immediately followed by their answers. They are intended to be treated as questions, but if you are finding things difficult you may prefer to treat some of them as examples.

  Text questions (marked by a capital T and a number) should always be treated as questions. You should make a serious effort to answer them before consulting the answers at the back of the book and certainly before moving on in the main text. Nonetheless, if you are having problems, these too can be treated as worked examples since the answers are always quite detailed and frequently contain good advice.

- The ends of *Text questions*, *Solutions* to *Examples* and solutions to *Practice questions* are marked by an open box symbol, so you can see where the normal text resumes. ☐

You might think that we are very keen on putting you through your paces and testing your understanding every few pages. If so, you'd be right! One of the best ways of making sure that you have understood a piece of mathematics is to answer a question about it. By thinking carefully about questions and getting into the habit of providing good answers to them you will enhance your problem solving skills, and deepen your understanding.

Chapter summaries are quite detailed and specific; they are designed to act as a reference list for the chapter and may also be useful for revision purposes. Each summary subsection is followed by two other subsections that deserve close attention; the *Achievements,* which lists all the things you should be able to do as a result of studying the chapter, and the *End of chapter questions* which test that you really *can* do them. *The End of chapter questions* are directly linked to the *Achievements,* with each question testing one or more of the *Achievements.* Answers to the *End of chapter questions* appear in an accompanying solutions book and contain references back to the text in case you

You will see, for example, that we encourage you always to substitute units as well as numbers once the time comes to evaluate quantities from an algebraic expression. This provides an excellent check on the calculation. To avoid an unholy mess, we also encourage you to make all these substitutions together just before a final evaluation, rather than piecemeal during the calculation.

had difficulty with the questions. When you have successfully completed the *End of chapter questions* for any particular chapter you should return to the corresponding set of *Fast track questions* in Subsection 1.2, and answer those if you have not already done so.

The *Maths handbook* is an extensive collection of useful formulae and mathematical results. This handy reference document also includes tables of common mathematical symbols, SI units (and some of the common non-SI ones too!) and the Greek alphabet. The *Maths handbook* is printed as an appendix, near the back of the book, just before the *Answers and comments*.

Part I

Basic arithmetic and algebra

Introducing arithmetic and algebra

Chapter

1

1.1 Opening items

1.1.1 Chapter introduction

This chapter is about basic algebra and arithmetic. It assumes that you already have some familiarity with these topics and concentrates on giving you an opportunity to practise the skills you have already acquired, and to sharpen up your use of the very precise terminology of mathematics.

As you work through this chapter remember that it has been designed as a tool to be used by *you* for your own benefit. You should not waste time on topics with which you are already fully familiar, but nor should you shy away from topics you have not completely mastered. Use the questions in the chapter to probe your own knowledge and ability, and use the text to rectify any deficiencies that you detect. By reinforcing your understanding of basic mathematics you will be preparing yourself for some of the more demanding mathematical challenges that you will inevitably meet as your studies progress.

The overall thrust of this chapter is towards *problem solving* – abstract principles and general definitions are certainly included, but their role is to support your development as a confident *user* of mathematics. Section 1.1 contains the various *Opening items* that are a common feature of all chapters. These are designed to give you a clear view of the chapter's content and the prior knowledge that it assumes, so that you can assess for yourself the extent to which you need to study the chapter and the degree to which you are prepared for such a study. Sections 1.2 and 1.3 contain the basic teaching material of the chapter, dealing with mathematical *expressions* and with relations between expressions, particularly *equality* and *inequality*. Section 1.2 starts with a review of the basic *operations* (addition, subtraction, multiplication and division) together with their symbolic representation and order of priority in written expressions. This topic is taken further in Subsection 1.2.2 which concentrates on *brackets* and their use in simplifying and expanding expressions. The manipulation of algebraic and arithmetic *fractions* (a common source of errors) is discussed in Subsection 1.2.3, and the topic of *powers*, *roots* and *reciprocals* is covered in Subsection 1.2.4. The writing and rearrangement of simple *equations* is dealt with in Subsections 1.3.1 and 1.3.2, while Subsection 1.3.3 looks at *proportionality*. (Both *direct* and *inverse* proportionality are considered, together with the equations that reflect such relationships.) Subsection 1.3.4 deals with the subject of *inequalities* and reviews the rules for their manipulation. The chapter ends with a section of *Closing items* (Section 1.4) that includes a summary, a list of the things that you should be able to do on completing the chapter, and some *End of chapter questions* designed to let you assess your achievements and alert you to any remaining difficulties. The answers to all the numbered questions included in the chapter can be found at the end of the book.

Study comment

Having read the introduction you may feel that you are already familiar with the material covered by this chapter and that you do not need to study it. If so, try the *Fast track questions* given in Subsection 1.1.2. If not, proceed directly to *Ready to study?* in Subsection 1.1.3.

1.1.2   Fast track questions

Study comment

Can you answer the following *Fast track questions*? The answers are given at the end of the book. If you answer the questions successfully you need only glance through the chapter before looking at the *Chapter summary* (Subsection 1.4.1) and the *Achievements* listed in Subsection 1.4.2. If you are sure that you can meet each of these achievements, try the *End of chapter questions* in Subsection 1.4.3. If you have difficulty with only one or two of the questions you should follow the guidance given in the answers and read the relevant parts of the chapter. However, if you have difficulty with more than two of the *End of chapter questions* you are strongly advised to study the whole chapter.

Question FI

$$z = [2xy - (x + y)^2]^2$$

(a)   Evaluate z when $x = 1$ and $y = 2$.

(b)   Expand the expression for z by removing all the brackets.

Question F2

Write each of these expressions as a single fraction expressed in as simple a form as possible:

(a) $2mu - \dfrac{2}{u}d$ (b) $\dfrac{2l^2}{3h} + \dfrac{h}{2}$

(c) $\dfrac{3l(h + f)^2}{2a^2} \div \dfrac{3a(h^2 - f^2)}{-2l^2}$

Question F3

A rocket of mass m_1 fired vertically at speed v from the surface of a planet of mass m_2 and radius r can escape from the planet provided its kinetic energy $m_1 v^2/2$ is greater than $Gm_1 m_2/r$. Derive an inequality, with v as its subject, that must be satisfied if the rocket is to escape.

1.1.3 Ready to study?

Study comment

Relatively little background knowledge is required to study this chapter, which deals with very basic topics in arithmetic and algebra. Nonetheless, it is assumed that you have previously performed a wide range of calculations involving the *addition*, *subtraction*, *multiplication* and *division* of *positive* and *negative quantities*, and that you have already encountered terms such as *fraction*, *power*, *square*, and *square root*. In addition it is assumed that you are accustomed to using a calculator and that you have one available while working on this chapter.

Formulae from many different branches of physics have deliberately been used to illustrate the mathematical points made in this chapter. As a result you may find that you are asked to deal with some unfamiliar physical quantities. If this should happen, don't worry. You will find that the physical properties of the quantities concerned are of no real importance in this chapter and that any unfamiliar quantity can be treated as a simple mathematical 'unknown'.

1.2 Expressions

1.2.1 Terms, operations and priority

Basic operations

The four basic **operations** of arithmetic are addition ($+$), subtraction ($-$), multiplication ($\times$) and division ($\div$ or $/$). Other operations include squaring (e.g. $x^2 = x \times x$) and taking square roots (e.g. $\sqrt{(x^2)} = x$). A mathematical **expression** is a combination of numbers and letters (which generally represent numbers) linked together by various operations. $5 \times 4 - 3$ and $3a - b/c$ are both examples of expressions (☞); the former is purely **arithmetic** (i.e. involving numbers only), the latter **algebraic** (i.e. involving numbers and letters).

Note: $3a = 3 \times a$

When working with expressions it is crucially important that they should be unambiguous. Thus, when dealing with the expression $5 \times 4 - 3$, for example, you need to know whether it means 'multiply 5 by 4 *then* subtract 3' or 'subtract 3 from 4 *then* multiply by 5' — the outcome is different in the two cases. The potential ambiguity is removed by insisting on a standard order of priority for carrying out the basic operations.

Standard order of priority for basic operations:

multiplication and division, then addition and subtraction

===

Question T1

Using the standard order of priority evaluate the following expressions:

(a) $3 + 6/2 - 4$

(b) $4.2 \times 3.0 - 1.4$

(c) $ab + c/2$, where $a = 2$, $b = 1/2$ and $c = -4$.   □

A note on terminology   For future reference, here are the definitions of some words that are often used in connection with the basic operations.

- The number that results from adding two (or more) numbers is their **sum**.
- The number that results from subtracting one number from another is their **difference**.
- The number that results from multiplying two numbers is their **product**.
- The number that results from dividing one number by another is their **quotient**.
- The numbers that are multiplied together in a product are **factors** of that product.
- The number by which a given number is divided in a quotient is called the **divisor**.

Each of these words may be applied to algebraic expressions in the same way that it is applied to arithmetic expressions.

Terms and ordering

It is often useful to regard a given expression as consisting of several distinct **terms** each of which may be a number, a letter or even another expression. Strictly speaking, an expression should be treated as a *sum* of terms, so $5 \times 4 - 3$ should be viewed as a positive term 5×4 *added* to a negative term -3. However, in practice the word *term* is often used more loosely and an expression such as $3ab$ is spoken of as a *product* of terms.

The terms making up an expression can usually be ordered in a variety of different ways. For instance, $5 \times 4 - 3$ could be equally well written as $-3 + 5 \times 4$, though it may *not* be written $3 - 5 \times 4$, and $3ab$ can be rewritten as $3ba$ or even $a3b$ or $ab3$, though these last two would be regarded as ugly and unusual. The conventional way of ordering terms within an expression is best picked up by seeing examples such as those contained in this chapter. The basic rules that govern reordering are explored in the following question.

Is the ordering of two terms, a and b, immaterial if they are (a) multiplied, (b) added, (c) subtracted, (d) divided?

(a) Yes, $ab = ba$ for every choice of a and b

(b) Yes, $a + b = b + a$ for every choice of a and b

(c) No, $a - b \neq b - a$ for every choice of a and b (☞)

(d) No, $a/b \neq b/a$ for every choice of a and b

The symbol $\neq$ means 'is *not* equal to'.

Using brackets to indicate priority

How would you write an expression corresponding to the instruction 'subtract 3 from 4 *then* multiply the result by 5'? You certainly can't write it as $4 - 3 \times 5$ because (according to the standard order of priority) that would mean 'multiply 3 by 5 and *then* subtract the result from 4'. One way to get round this problem is to use **brackets** to separate parts of an expression and to give the evaluation of one part priority over another. The standard order of priority still applies *within* the brackets but the bracketed part of the expression is treated as a single entity and its evaluation is given precedence over all other operations. Thus, the instruction 'subtract 3 from 4 *then* multiply the result by 5' can be written $(4 - 3) \times 5$, or more conventionally $5(4 - 3)$.

Quite often a bracketed expression includes a term that also involves brackets. In such a case one pair of brackets appears inside another pair as in $5(3 + 2(2 - 4))$. The two pairs of brackets are said to be **nested** and are dealt with by evaluating the contents of the innermost brackets first and then moving outwards. Thus

$$5(3 + 2(2 - 4)) = 5(3 + 2(-2)) = 5(3 - 4) = 5(-1) = -5$$

Sometimes brackets of different shapes (☞) are used to make it clearer which brackets are to be paired − the previous example might have been written $5[3 + 2(2 - 4)]$ − but this is not always done, even for complicated expressions.

The convention used for the order of brackets is as follows:

$$\{[(x)]\}$$

( ) are called parentheses
[ ] are called square brackets
{ } are called braces

Question T2

Evaluate the following expressions:

(a) $(6 + 3/(9 - 7))/2$

(b) $2[(3a - 4)b + 2a(3b + 1)]$, where $a = 1$ and $b = -1$

(c) $x\{[x - 2(x + 3)/(x - 1)] + [1 + (x + 3)/2]/x\}$,
     where $x = 2$ □

Powers, roots and reciprocals are the subject of Subsection 1.2.4.

When expressions involve *powers*, such as squares and square roots, as well as brackets the overall order of priority is as follows.

Overall order of priority for operations:
1.   First deal with anything in brackets
2.   then deal with powers (including roots and reciprocals)
3.   then multiply and divide
4.   then add and subtract.

Question T3

Evaluate the following expressions:

(a) $2 + 4/3^2$

(b) $2 + (4 - 3a)^2$, where $a = 1$

(c) $(2 + (2 + x)^2)^2$, where $x = 2$ □

Operations on your calculator

Many calculators are designed to carry out processes in the correct order if you key in a calculation exactly as it is written. Explore your own calculator to find out whether it does this — use the examples given above, or invent some for yourself and do the calculations 'by hand' as well as on your calculator.

You may think that if your calculator does it all for you, there is no need to take much notice of the order in which operations should be carried out. But when you are manipulating algebraic expressions, and rearranging equations, it is important to know in exactly what order you can do, or undo, processes. (Manipulating algebraic expressions features throughout this chapter.) Also, it is always possible to make a mistake when using a calculator, so it is often useful to carry out a rough calculation to make sure that the answer on your calculator is sensible.

1.2.2   Expanding and simplifying

In the previous subsection, brackets were used to separate one part of an expression from the rest. Sometimes it is useful either to rewrite an expression by removing brackets (in other words, to **expand** the expression), or to use brackets to group parts of an expression together (in other words to **simplify** the expression).

Expanding

The calculation $(3 + 4) \times 5$ means 'add 3 to 4, *then* multiply by 5'. In this case, 3 and 4 have been added, and their sum multiplied by 5, so the calculation could have been *expanded* as $3 \times 5 + 4 \times 5$. In cases like this, each and every term inside the brackets has to be multiplied by the factor outside. If the factor enclosed by brackets is a difference, rather than a sum, the expression can be expanded in exactly the same way. For example: $4 \times (7 - 5) = 4 \times 2 = 8$, and this could be rewritten as $4 \times 7 - 4 \times 5 = 28 - 20 = 8$.

If an expression is a product of two or more factors enclosed by brackets, it can still be expanded, but each term inside one pair of brackets must be multiplied by each term inside every other pair of brackets. For example: $(2 - 3)(4 - 5) = 2 \times 4 - 2 \times 5 - 3 \times 4 + 3 \times 5$. (☞)

Note that when writing out an expansion it is advisable to treat it systematically, by working along from left to right as above, otherwise you risk forgetting or duplicating some of the terms.

Numbers were used in the examples above because they make it easier to show what is happening, but in practice there is not often much point in expanding purely numerical expressions since it rarely makes the calculation any easier. Expanding algebraic expressions can, however, help to simplify an expression or equation.

> Expand the following expressions:
>
> (a) $-3(4x - 2)$ (b) $2(3x + 2y) - 4y$ (c) $3(x + 2(2x + 5))$
>
> ---
>
> (a) $-12x + 6$ (each term in the bracket is multiplied by -3)
>
> (b) $2(3x + 2y) - 4y = 6x + 4y - 4y$
>
> $$= 6x$$
>
> (c) $3(x + 2(2x + 5)) = 3(x + 4x + 10)$
>
> $$= 3(5x + 10)$$
>
> $$= 15x + 30 \quad \square$$

Brackets can be removed from a division calculation in the same way as for multiplication. Each term inside the brackets must be divided by the divisor outside. For example:

$$(6 + 9)/3 = 6/3 + 9/3 = 2 + 3 = 5$$

Question T4

Expand the following expressions so that in each case the resulting expression does not involve brackets:

(a) $3x(4 + 3y) + 2x$ (b) $(x + 2y)^2$ (c) $(x + y)(x - y)$

(d) $(3x + 6y)/2$ (e) $(p + q)/q$ (f) $(2xy)^2/2$ $\square$

Collecting terms together

Terms with a **common factor** (that is, a number or expression that is a factor of each) can be collected together using brackets. This is the reverse of the

When expanding brackets take particular care over negative terms. Even if you remember to multiply 4 by (-3) to get the third term you may still forget to multiply (-3) by (-5) to get the fourth term.

process shown above, and can often be used to simplify an expression, or at least to rewrite it in a more compact form. For example, all the terms in the expression $8x + 6y + 10$, have 2 as a common factor, so it could be written as $2(4x + 3y + 5)$. As another example: the expression $6x + 4x^2 + 8xy$ has $2x$ as a common factor, so could be rewritten as $2x(3 + 2x + 4y)$. The terms $2x$ and $4y$ also share a common factor of 2, so nested brackets could be used to collect these terms together, $2x(3 + 2(x + 2y))$. (☞)

You may think that using brackets to group terms together makes things look more complicated, rather than simpler but, as you will see in the rest of this chapter, it can be a useful step towards simplifying fractions and equations.

Question T5

By finding the common factors, simplify the following expressions:

(a) $3x^2y + 6xy$ (b) $8x^3 + 4x^2y + 2xy^2 + 2x$ ☐

The following expansions arise frequently and are worth remembering

$$(a + b)^2 = a^2 + 2ab + b^2 \tag{1}$$

$$(a - b)^2 = a^2 - 2ab + b^2 \tag{2}$$

$$(a + b)(a - b) = a^2 - b^2 \tag{3}$$

1.2.3 Fractions

The word **fraction** probably conjures up in your mind numbers like 1/2, 3/4 and 7/3. All these fractions are **ratios** of two whole numbers which can be rewritten as a single number by dividing the number on top (the **numerator**) by the one on the bottom (the **denominator**) to find their quotient. But in physics you also have to deal with fractions that are ratios of two algebraic expressions, such as Q/C, mv^2/r, and $q_1q_2/(4\pi\varepsilon_0 r^2)$. Algebraic fractions can be manipulated in exactly the same ways as arithmetic ones, and the terms numerator and denominator are still used for the expressions on top and underneath respectively.

Any number may be expressed as a fraction of any other number; 2, for instance, is 1/2 of 4, and 6 is 3/2 of 4. This observation provides the basis of the system of **percentages** in which one number is expressed as so many hundreths of another number. For example

$$2 = \frac{50}{100} \times 4 \quad \text{so we may say 2 is 50\% (read as 50 per cent) of 4}$$

More generally, if we want to express a as a percentage of b, that percentage is given by $(a/b) \times 100$.

Simplifying fractions by cancelling common factors

If the numerator and denominator of a fraction have a common factor, then that factor both multiplies and divides the rest of the fraction − and those two operations cancel each other out. **Cancelling** therefore allows common factors to be removed from both the numerator and denominator of a fraction without having any net effect. For example, in the case of 6/10, the numerator and denominator each have 2 as a factor and 6/10 can therefore be written as 3/5. The same principle applies to algebraic fractions such as $\pi r^2/(2\pi r)$, the ratio of the area of a circle to its circumference. The numerical constant π (☞) and the radius r are both common factors of the numerator and the denominator, so they may be cancelled from both, leaving $r/2$.

To three decimal places $\pi = 3.142$.

Question T6

Simplify these fractions by cancelling the common factors in their numerators and denominators:

(a) $\dfrac{7}{42}$ (b) $\dfrac{6x}{2x^3}$ (c) $\dfrac{8(x + 3)}{2x}$ ☐

Question T7

By first using brackets to group terms in the numerator and/or denominator, simplify the following fractions by cancelling common factors:

(a) $(x + 2)(x + 3)/(3x^2y + 6xy)$

(b) $(3xy + 3y^2)/(4x^2y + 4x^3)$ ☐

Notice that in Question T7 you were able to cancel factors that were not just single numbers or symbols, and that using brackets to write the various expressions as products made these factors much easier to spot.

Arithmetic operations with fractions

All the standard mathematical operations can be applied to fractions. *Multiplication* and *division* are especially easy − provided the problem is written out clearly in the first place. In general algebraic terms;

$$\frac{a}{b} \times \frac{c}{d} = \frac{ac}{bd} \tag{4}$$

$$\text{and } \frac{a}{b} \div \frac{c}{d} = \frac{ad}{bc} \tag{5}$$

So, dividing by (c/d) is the same as multiplying by (d/c).

Simplify the following fractions:

(a) $(3/5) \div (5/2)$ (b) $(mv^2/2)/(r/2)$.

(a) $\dfrac{3}{5} \div \dfrac{5}{2} = \dfrac{3 \times 2}{5 \times 5} = \dfrac{6}{25} = 0.24$

(b) $\dfrac{mv^2}{2} \div \dfrac{r}{2} = \dfrac{mv^2 \times 2}{2 \times r} = \dfrac{mv^2}{r}$ □

Adding and *subtracting* fractions is also straightforward provided they all have the same denominator — a so-called **common denominator**. Expressions such as 4/5 + 7/5, or 8/3 − 5/3 can be rewritten as (4 + 7)/5 or (8 − 5)/3. But if fractions that are to be added or subtracted do not share a common denominator, they can — and must — be rewritten in such a way as to give them one. For example, to add the fractions 4/5 and 8/3, you can write

$$\frac{4}{5} + \frac{8}{3} = \frac{(4 \times 3)}{(5 \times 3)} + \frac{(8 \times 5)}{(3 \times 5)}$$

Notice that the value of neither fraction has been changed, since all that has been done is to introduce a common factor into the numerator and denominator of each; also notice that the denominator of each fraction was used to rewrite the other. Now the addition can be written as

$$\frac{4}{5} + \frac{8}{3} = \frac{(4 \times 3) + (8 \times 5)}{(5 \times 3)}$$

This technique can, of course, also be used when fractions are to be subtracted, and algebraic fractions are treated in exactly the same way. Thus, in general algebraic terms:

$$\frac{a}{b} + \frac{c}{d} = \frac{ad + cb}{bd} \tag{6}$$

and $\dfrac{a}{b} - \dfrac{c}{d} = \dfrac{ad - cb}{bd}$ (7)

Rewrite the fractions $3x/5$ and $2/y$ so that they have a common denominator and then write $(3x/5) - (2/y)$ as a single fraction.

$$\frac{3x}{5} = \frac{3xy}{5y} \quad \text{and} \quad \frac{2}{y} = \frac{10}{5y}$$

Thus, $\dfrac{3x}{5} - \dfrac{2}{y} = \dfrac{3xy}{5y} - \dfrac{10}{5y} = \dfrac{3xy - 10}{5y}$ □

Avoiding common mistakes

The rules for manipulating fractions are fairly straightforward but under pressure some students panic when faced with fractions and make serious mistakes. Here is a short catalogue of common fallacies that you should try to avoid.

Forgetting the common denominator   It is not uncommon to see totally bogus additions such as:

$$\frac{a}{b} + \frac{c}{d} = \frac{a+c}{b+d}$$ it looks plausible, but it's *WRONG*.

Incomplete cancelling   It is also quite common to see half-hearted cancelling along the following lines:

$$\frac{1+ax}{x} = 1 + a$$ this too is *WRONG*. (☞)

Though $(1/x) + a$ would have been right.

Wrongly ordered divisions   If asked to divide $(2ax/b)$ by $(2a/b)$ most students will find the correct answer (x) without any difficulty. But different ways of presenting the same calculation can lead to mistakes. The following are all equal (if not equally attractive).

$$(2ax/b)/(2a/b) = \frac{2ax/b}{2a/b} = \frac{\dfrac{2ax}{b}}{\dfrac{2a}{b}} = \frac{2ax}{b(2a/b)} = \frac{2axb}{2ab} = x \quad ☞$$

Strictly speaking, the second, third and fourth of these expressions are ambiguous and need brackets to make them unambiguous. However, many would take the view that the style in which they are printed makes clear the presumed order of the various divisions.

Question T8

Simplify the following, if possible:

(a) $\dfrac{2}{3} + \dfrac{3}{5}$ (b) $\dfrac{x}{3} + \dfrac{x}{2}$ (c) $(1/u) + (1/v)$

(d) $\dfrac{1+a}{3+a}$ (e) $\dfrac{2(1+x)}{(1+x)^2}$ (f) $\dfrac{5/7}{2/3}$

(g) $\dfrac{3x}{x/3}$ (h) $\dfrac{3}{a} + \dfrac{7}{2a}$ (i) $(-7x^2) \div (-3x)$ ☐

1.2.4   Powers, roots and reciprocals

A **power** (often called an **index**) indicates repeated multiplication as in:

squares	$a^2 = a \times a$	(read as 'a squared')
cubes	$a^3 = a \times a \times a$	(read as 'a cubed')

or, generally

nth powers $a^n = \underbrace{a \times a \times \ldots \times a}_{n \text{ factors}}$ (read as 'a to the n')

(Note the use of the three dots (...), known as an **ellipsis**, to indicate that the sequence continues in a similar fashion.)

Powers (or indices) enable many expressions to be written neatly and compactly. For instance, Kepler's third law, which relates the time T that a planet requires to orbit the Sun to the mean distance R between that planet and the Sun, can be written

$$T^2 = kR^3 \tag{8}$$

where k is a constant. Relationships of this kind, that relate one quantity to a power of another, are often called **power laws**.

The use of powers is simple and straightforward as long as the power is a positive whole number (1, 2, 3, etc.), but you will encounter many situations in which the power is neither positive nor a whole number. Fortunately, the general rules for manipulating powers are simply extensions of the rules that apply to positive whole number powers, so let's start with those.

It follows directly from the definition of a^n that if m and n are positive whole numbers

$$a^m \times a^n = \underbrace{a \times a \times \ldots \times a}_{m \text{ factors}} \times \underbrace{a \times a \times \ldots \times a}_{n \text{ factors}}$$

Now, the right-hand side of this equation consists of $m + n$ factors of a, so it may be written a^{m+n}. Thus,

$$a^m \times a^n = a^{m+n} \tag{9}$$

Similarly, raising a^m to the power n gives

$$(a^m)^n = \overbrace{(a \times a \times \ldots \times a)}^{m \text{ factors } of \text{ a}} \underbrace{\times (a \times a \times \ldots \times a) \times \ldots \times (a \times a \times \ldots \times a)}_{n \text{ factors of } a^m}$$

The right-hand side now contains $m \times n$ factors of a. So,

$$(a^m)^n = a^{m \times n} = a^{mn} \tag{10}$$

Question T9

Using equations 9 and 10, simplify and then evaluate the following expressions:

(a) $2^2 \times 2^3$ (b) $2^3 \times 2^4 \times 2^5$ (c) $(-2)^4 \times (-2)^2$

(d) $(2^3)^2$ (e) $((-0.2)^2)^3$ (f) $((-1)^{17})^{23}$ □

Negative powers and reciprocals

We can make sense of negative powers by extending the pattern of behaviour that we find for a sequence of terms such as a^2, a^3, a^4 ... and so on. For the sake of having a concrete example we will assume that $a = 5$, but the general results that we deduce will be independent of this particular choice and would be equally true for any value of a other than zero. Given our chosen value of a, the first few terms in the sequence a^2, a^3, a^4 ... are

$$5 \times 5 \qquad 5 \times 5 \times 5 \qquad 5 \times 5 \times 5 \times 5 \qquad \ldots$$

To go from one term to the next in the sequence a^2, a^3, a^4 ... you simply *raise* the power by 1, that's to say you *multiply* the previous term by a, or 5 in this case.

Now suppose that the sequence is written in reverse order: ... a^4, a^3, a^2. What must you now do to go from one term to the next, working from left to right?

You have to *reduce* the power by 1, i.e. *divide* the previous number by a, or 5 in this case.   □

It is useful to continue this sequence of decreasing powers for a few steps:

$$\ldots a^3, a^2, a^1, a^0, a^{-1}, a^{-2} \ldots$$

and to evaluate some of the terms using our chosen value, $a = 5$.

Dividing 5^2 by 5 gives 5,   so $5^1 = 5$.

Dividing 5^1 by 5 gives 1,   so $5^0 = 1$.

Dividing 5^0 by 5 gives $\dfrac{1}{5}$,   so $5^{-1} = \dfrac{1}{5} = 0.2$.

Dividing 5^{-1} by 5 gives $\dfrac{1}{5^2}$,   so $5^{-2} = \dfrac{1}{5^2} = 0.04$.

You might like to repeat this process using some other value of a (especially a negative value such as $a = -2$) but as long as a is not zero the pattern should always be the same and you should always find the following:

The relation $a^0 = 1$ may look strange, but it is a logical step in the pattern.

$$a^1 = a \qquad a^0 = 1 \qquad a^{-1} = 1/a \qquad a^{-2} = 1/a^2 \qquad (\text{☏})$$

Given any non-zero quantity a, the quantity $1/a$ is called the **reciprocal** of a. So our deductions have shown that the reciprocal of a quantity a can be written a^{-1}.

Write down the reciprocal of a^2 in two different ways.

The reciprocal of a^2 may be written as $1/a^2$ or $(a^2)^{-1}$. □

If the power notation is to make sense we must require these two alternative ways of writing the reciprocal to be equal. But we already know that $1/a^2$ may be written as a^{-2}, so

$$(a^2)^{-1} = 1/a^2 = a^{-2}$$

This is an example of a more general result that works for any value of n.

$$(a^n)^{-1} = 1/a^n = a^{-n} \qquad (11)$$

Question T10

Evaluate the following:

(a) 2^{-4} (b) 10^{-3} (c) $\left(\dfrac{1}{2}\right)^{-3}$ (d) $(0.2)^{-2}$ □

The use of negative powers to indicate reciprocals is consistent with the two general rules that were introduced earlier (Equations 9 and 10). For example, we know that if m and n are positive whole numbers

$$a^m \times a^n = a^{m+n}$$

but even if m or n (or both) are negative, the rule still works

$$2^3 \times 2^{-5} = (2 \times 2 \times 2) \times \frac{1}{2 \times 2 \times 2 \times 2 \times 2} = \frac{1}{2 \times 2} = \frac{1}{2^2} = 2^{-2}$$

which is exactly the result we would expect from the general rule (Equation 9).

Similarly, for the second rule, $(a^m)^n = a^{mn}$

$$(2^2)^{-4} = \frac{1}{(2^2)^4} = \frac{1}{2^8} = 2^{-8}$$

which is also the result we would expect from the general rule (Equation 10).

Question T11

Express each of the following in the form a^n:

(a) $5^2 \times 5^{-5}$ (b) $(5^2)^{-3}$ (c) $(2^4)/(2^6)$

(d) $(3^{-1})^{-1}$ (e) $\left(\dfrac{1}{2}\right)^{-3}$ (f) $\left(\dfrac{2^5}{2^6}\right)^{-1} \times 2 \times 2^3$ □

Negative powers, or powers appearing in the denominator of an expression, are often referred to as **inverse powers**. So Newton's law of gravitation, which asserts that the strength of the gravitational force attracting a mass m_1 towards another mass m_2 a distance r away is

$$F = \frac{Gm_1m_2}{r^2}$$

is often described as an *inverse square law*.

Fractional powers and roots

Although Kepler's third law ($T^2 = kR^3$) is a valuable result, it is often even more valuable to know the way in which T itself (rather than T^2) is related to k and R. Such a relation is easily obtained from Kepler's law by taking *square roots*:

$$T = \sqrt{kR^3}$$

The root symbol ($\sqrt{\ }$) can be used to indicate a variety of **roots** such as

square roots $\qquad\qquad \sqrt{a}\,\sqrt{a} = a$

cube roots $\qquad\qquad \sqrt[3]{a}\,\sqrt[3]{a}\,\sqrt[3]{a} = a$

or, generally, nth roots $\qquad \underbrace{\sqrt[n]{a}\,\sqrt[n]{a}\ldots\sqrt[n]{a}}_{n\text{ factors}} = a$

In fact, it is somewhat unusual to see cube (or higher) roots written in this form since most physical scientists prefer to make use of another notation based on fractional powers. According to this alternative notation

$$\sqrt[n]{a} = a^{1/n} \tag{12}$$

Once again this new notation is consistent with the rules given earlier. For example, we know that $8^{1/3} = 2$ since $2 \times 2 \times 2 = 8$, but the second of our rules for powers ($(a^m)^n = a^{mn}$) would lead us to expect that

$(8^{1/3})^3 = 8^{(1/3)\times 3} = 8^1 = 8$ (☞)

which is indeed the case since $8^{1/3} = 2$ and $2^3 = 8$.

$8^{1/3}$ is the cube root of 8.

Basic arithmetic and algebra

All this may seem fairly straightforward, but, in fact, a certain degree of care is needed when dealing with roots and fractional powers. In particular, the following points should be noted.

(i) Positive numbers have two square roots It is true that $4 = 2 \times 2$, but it is equally true that $4 = (-2) \times (-2)$. So, both 2 *and* -2 are square roots of 4. Thus we may write $4^{1/2} = 2$ or $4^{1/2} = -2$. These two statements may be combined in the form

$$4^{1/2} = \pm 2$$

where the symbol $\pm$ is to be read as ' + or −'.

It is a common convention to insist that $a^{1/2}$ and $\sqrt{a}$ should always be used to represent the *positive* square root of a. The square roots of a will then be $\pm a^{1/2}$ or $\pm \sqrt{a}$. We shall follow this convention, though negative square roots will be ignored when there is a physical reason for doing so.

The need to remember that nth roots may be positive or negative is not restricted to the case of square roots where $n = 2$ − it applies to all cases where n is an even whole number.

(ii) Negative numbers do not have square roots Some negative numbers certainly have roots for instance:

$$-27 = (-3) \times (-3) \times (-3) \quad \text{so } (-27)^{1/3} = -3$$

But it is not possible to find any number a, positive or negative, such that

$$a^2 = -4 \quad \text{or} \quad a^2 = -0.1$$

Roots of negative quantities may be defined in the context of *complex numbers*, but we are ignoring that possibility in this chapter.

Indeed, it is generally said that negative quantities do not have square roots. Nor, in the same spirit, is it possible to find the nth root of any negative quantity unless n is an odd whole number.

Question T12

Evaluate the following:

(a) $9^{1/2}$ (b) $16^{1/2}$ (c) $16^{1/4}$

(d) $27^{1/3}$ (e) $\sqrt{64}$ (f) $\sqrt[4]{49}$ ☐

Question T13

Write down the following roots, or explain why they cannot be found:

(a) $(-8)^{1/3}$ (b) $\left(-\dfrac{1}{16}\right)^{1/2}$ (c) $(-2)^{1/2}$ ☐

Working with general powers

We have seen that, subject to certain cautionary notes, the rules for manipulating negative and fractional powers are similar to those for powers that are positive whole numbers. In fact the general rules for manipulating powers are:

$$a^p \times a^q = a^{p+q} \tag{13}$$

and $$(a^p)^q = a^{pq} \tag{14}$$

Although these equations are similar to Equations 9 and 10, p and q have been used to represent the powers so that you will no longer think of them as positive whole numbers. They may be negative, fractional or even numbers such as π that can be represented as infinite decimals but not as finite fractions.

Use the above rules to derive a similar rule expressing the *quotient* a^p/a^q as a power of a.

---

a^p divided by a^q is identical to a^p multiplied by a^{-q}.

So $a^p/a^q = a^p \times a^{-q} = a^{p-q}$

where we have used the rule for multiplication with q replaced by $-q$.   □

How would you interpret $a^{p/q}$ in words?

If we use the general rules we may equally well write

$a^{p/q} = (a^p)^{1/q}$ or $a^{p/q} = (a^{1/q})^p$

so $(a^p)^{1/q}$ is the qth root of the pth power a

and $(a^{1/q})^p$ is the pth power of the qth root of a.   □

Question T14

Evaluate the following:

(a) $7^{4/9}$

(b) $(2.7)^3/(2.7)^2$

(c) $\sqrt{2} \times \sqrt[3]{2}$

(d) $\dfrac{\sqrt{3}}{\sqrt[5]{3}}$ □

Quite apart from numerical evaluations, the rules for powers allow algebraic relations and physical laws to be written in convenient forms. For example, if you are dealing with a formula that involves a term $r\sqrt{r}$ you might well find it more convenient (and elegant) to rewrite the term as $r^{3/2}$ or even $r^{1.5}$. Such rewriting might also make the term easier to evaluate since many calculators have an x^y button but none are likely to have an $r\sqrt{r}$ button.

Question T15

Rewrite the following in the form r^p:

(a) $1/\sqrt{r}$

(b) $r^3\sqrt{r}$

(c) $r^5\sqrt{r}$ □

Question T16

Simplify the following expressions:

(a)　$4x^2y^2z^4$　　　(b)　$\dfrac{h^2n^2}{8m}\left(\sqrt[3]{L^3}\right)^{-2}$　　　(c)　$\left(\dfrac{8\pi^2 Z\sqrt{l^5}}{3^3\sqrt{Z}}\right)^{1/2}$　□

1.3　**Equations and inequalities**

In Section 1.2 we dealt with simplifying and rewriting expressions, but an expression on its own is not usually very useful. You are much more likely to deal with an **equation** – a statement that two expressions are equal, such as

$$s = ut + \frac{1}{2}at^2 \tag{15}$$

It is worth stressing that the = sign means 'has the same value as', so, while the expressions on either side of the equation may look very different, they actually represent *exactly the same* number. This holds true even for algebraic equations – the letters in such an equation simply stand for the numbers that make the two sides equal.

　　Equations generally involve **constants** (i.e. quantities with a fixed and pre-determined value) and **variables** (i.e. quantities that may take on any one of a range of values). Often, an equation will relate the value of one variable to the value (or values) of one or more other variables. Thus, in Equation 15, if the values of u, a and t are specified then the value of s may be determined. In this way, Equation 15 provides both a general relationship between the variables s, u, a and t, and a recipe for calculating the value of s when u, a and t have particular known values. In the case of Equation 15, s is said to be the **subject** of the equation since it is the single variable that is expressed in terms of the others.

1.3.1　*Simplifying equations*

In Section 1.2, you met various ways of using brackets and common factors to simplify expressions. These same techniques can be used to simplify equations, as the following example illustrates.

　　When a drop of water falls vertically through air three forces act on it: a downward gravitational force of strength F_g, an upward buoyancy force of strength F_b (due to air displacement), and another upward force (due to air resistance – or more properly viscosity) of strength F_v. The strengths of these forces are determined by the following equations:

ρ, σ and η are the Greek letters 'rho', 'sigma' and 'eta', respectively.

$$F_g = \frac{4}{3}\pi r^3 \rho g \qquad F_b = \frac{4}{3}\pi r^3 \sigma g \qquad F_v = 6\pi\eta r v$$

where r is the radius of the water drop, ρ is its density, g is a constant known as (the magnitude of) the *acceleration due to gravity*, σ is the density of air, η is a constant known as the *viscosity of air* and v is the speed of the drop.

If the directions of the forces are taken into account, the strength of the total force acting on the falling drop is $F = F_g - F_b - F_v$. Express F in terms of r, g, ρ, σ, η and v, and then simplify it as much as possible.

$$F = \frac{4}{3}\pi r^3 \rho g - \frac{4}{3}\pi r^3 \sigma g - 6\pi\eta r v$$

If a common factor of $\frac{4}{3}\pi r^3 g$ is extracted from the first two terms on the right

$$F = \frac{4}{3}\pi r^3 g(\rho - \sigma) - 6\pi\eta r v$$

As a last step it is possible to extract a common factor of $2\pi r$ from each term on the right-hand side (☞)

$$F = 2\pi r \left[\frac{2}{3}r^2 g(\rho - \sigma) - 3\eta v \right] \qquad (16) \quad \square$$

Many physicists would be reluctant to take this last step since it probably makes the physical interpretation of F less obvious.

Question T17

Show (justifying each step) that

$$\phi = \frac{b^2 V r \theta}{b^2 - a^2} - \frac{b^2 V a^2 \theta}{r(b^2 - a^2)} - \frac{a^2 U r \theta}{b^2 - a^2} + \frac{a^2 U b^2 \theta}{r(b^2 - a^2)}$$

may be simplified to give

$$\phi = \left[b^2 V \left(r - \frac{a^2}{r} \right) - a^2 U \left(r - \frac{b^2}{r} \right) \right] \frac{\theta}{b^2 - a^2} \quad \square$$

1.3.2 Rearranging equations

You will often need to rearrange equations, perhaps to simplify them, or to ensure that they express relationships in a more illuminating way, or simply because you want to calculate the value of a particular variable that is not already the subject of the equation. Whatever the reason, the **rearrangement** should provide a different but equivalent relationship between the relevant variables.

A given equation can usually be rearranged in a variety of ways. The important principle to remember while performing the rearrangement is that *both sides of the equation represent the same number.* So, if you add or subtract a term to one side of an equation you must add or subtract an identical term to the other side of the equation. The same principle applies to multiplication and division and to other operations such as squaring or taking square roots. (In the latter case you must remember that a positive number has both a positive and a negative square root.) When rearranging equations you should bear in mind two important points:

1. When one side of an equation consists of the sum of several terms, *each* of those terms must be treated in the same way. (It's no good just multiplying the first term on one side of an equation by some factor if you forget to do the same to all the other terms on that side.)

2. When dividing both sides of an equation by some divisor you must ensure that divisor is not equal to zero. (You wouldn't expect to get a meaningful result if you divided a number by zero, so you shouldn't try to do it to equations either.)

Although this is a bogus proof, note the way it is laid out. Clear layout is always helpful, especially if you have to find an error, as in this case.

The following 'proof' shows that $1 = 2$. It is, of course, quite wrong. Spot the error.

Let a and b represent two numbers and suppose $a = b$.

Multiply both sides by a,	$a^2 = ab$
Subtract b^2 from both sides,	$a^2 - b^2 = ab - b^2$
Extract a common factor of $(a - b)$ from both sides,	$(a - b)(a + b) = b(a - b)$
Divide both sides by $(a - b)$,	$(a + b) = b$
But, since $a = b$, this implies that	$2b = b$
Finally, divide both sides by b,	$2 = 1$

From a mathematical point of view this is a 'silly' proof from the outset, but there is not actually anything wrong with it until both sides are divided by $(a - b)$. Since $a = b$, this step amounts to dividing both sides by zero, which is not a legitimate procedure. The result of the division is quite bogus, as you can see by replacing a and b by some simple number such as 1 throughout the 'proof'. ☐

In general, if both sides of an equation are divided by an algebraic expression such as $(a - b)$ then the result of that division (and anything deduced from that result) will only be valid if $a \neq b$.

The following questions will give you some practice in rearranging equations. The first two make further use of the example of the falling water drop that was introduced in Subsection 1.3.1.

Question T18

Initially, as a drop of water falls through air its speed v will increase and F_v will become larger. Consequently the value of $F(= F_g - F_b - F_v)$ will decrease until v becomes so large that $F = 0$. The speed at which this happens is called the *terminal speed* and is denoted v_t. Find as simple an expression as possible for v_t in terms of r, g, ρ, σ and η. □

Question T19

The relative importance of F_g, F_b and F_v under various circumstances can be determined by studying the ratios F_b/F_g, F_v/F_g and F_b/F_v.

(a) Work out fully simplified equations relating each of these ratios to r, g, ρ, σ, η and v.

(b) If F_v is ten times greater than F_b when the drop reaches its terminal speed ($v = v_t$) find an equation that expresses the radius of the drop in terms of σ, g, η and v_t.

(c) If F_b is negligibly small when $v = v_t$ then $F_g = F_v$. Under these circumstances, find an equation that expresses r in terms of ρ, g, η and v_t. □

Question T20

Rearrange the following equations to make x the subject:

(a) $y = 2ax + b^2$ \qquad (b) $(y - b) + (x - a) = 5xh^2 + 3$

(c) $\dfrac{1}{x - a} + \dfrac{1}{y - b} = 3t^2$ \quad (d) $t + a = \sqrt{\dfrac{3b}{x - y}}$ □

Once you have completed the rearrangement of an equation it is often a good idea to check that the result you have obtained is consistent with the original equation. This can often be done quite easily by substituting some simple numbers for the various algebraic quantities that enter the equation. For instance, suppose you start out with the equation

$$v^2 = u^2 + 2as$$

and you rearrange it to find

$$s = \frac{v^2 - u^2}{2a}$$

As a simple check of the correctness of this rearrangement pick some simple values for u, a and s in the original equation and substitute them into the equation to find v^2.

Note that these values don't have to be realistic nor do they involve units of measurement — we're just checking for mathematical consistency. Letting $u = 2$ (i.e. $u^2 = 4$), $a = 3$ and $s = 4$, for example, gives $v^2 = 2^2 + 2 \times 3 \times 4 = 28$. Now, if these values are substituted into the rearranged equation we get $s = (28 - 4)/(2 \times 3) = 4$, which was exactly the value of s we chose. By showing that the same set of values for u^2, v^2, a and s satisfy both of the equations we have not *proved* that the rearrangement is correct but at least we haven't found any inconsistency, so there's a good chance that the rearrangement is right.

Use the method outlined above to show that $v^2 = u^2 + 2as$ may not be rearranged to give

$$\frac{v^2}{2s} = u^2 + a$$

Using the values chosen earlier

$$\frac{v^2}{2s} = \frac{28}{8} = 3.5, \quad \text{but} \quad u^2 + a = 4 + 3 = 7$$

Since the same set of values cannot satisfy the original equation and its supposed rearrangement there is clearly something amiss. ☐

Question T21

Rearrange the equation $F = Gm_1m_2/r^2$ to make r the subject, and check your result using small whole numbers before looking at the answer. ☐

1.3.3 Equations and proportionality

Many important relationships in science involve just two variables. For example, an object of fixed mass m travelling at speed v has momentum of magnitude p given by

$$p = mv \tag{17}$$

Similarly, light travelling at fixed speed c through a vacuum is characterized by a frequency f and a wavelength λ that are related by

$$f\lambda = c \tag{18}$$

In these two examples m and c are constants, so only two variables are involved in each case.

In the case of Equation 17, the variables p and v are said to be **directly proportional** (or just **proportional**). This indicates that if v is multiplied by some factor then p must be multiplied by the same factor. Or, to put it another way, the *ratio p/v* is constant.

In Equation 18 the relationship between f and λ is very different. Now, it is the *product* $f\lambda$ that is constant with the consequence that if f is multiplied by a given factor then λ must be divided by that same factor. This situation is described by saying that f and λ are **inversely proportional**.

Direct proportionality and inverse proportionality are closely related, as can be seen by dividing both sides of Equation 18 by λ to obtain

$$f = \frac{c}{\lambda}$$

This shows that if f and λ are inversely proportional then f and $(1/\lambda)$ are directly proportional.

The symbol $\propto$ is used to mean 'is directly proportional to' (☞) so the proportionalities embodied in Equations 17 and 18 can be written as

$$p \propto v \quad \text{and} \quad f \propto \frac{1}{\lambda}$$

Conversely, a proportional relationship may always be represented by an equation, as follows:

If $y \propto x$ then $y = Kx$,   where K is a constant.   (19)

A constant introduced in this way is generally referred to as a **constant of proportionality**. In Equation 17, m is the constant of proportionality relating p to v, while in Equation 18 c is the constant of proportionality relating f to $1/\lambda$.

Of course, physical relationships often involve more than two variables. For example, the pressure P, volume V and temperature T of a low-density sample of gas obey (at least approximately) an equation of the form

$$PV = NkT$$

where N is the number of particles in the sample and k is a constant (☞). Despite the relative complexity of this relationship the idea of proportionality is still very useful in this context, provided we keep the full relationship in mind. For instance, it is quite correct to say that V is proportional to T *provided P and N are kept constant*, and it is equally correct to say that P and V are inversely proportional *provided N and T are kept constant*.

Question T22

Write down equations that embody the following relationships. (You will have to introduce constants of proportionality for yourself; choose an appropriate letter and make a note to remind yourself that the newly introduced letter represents a constant.)

(a) $V \propto I$

(b) $R \propto L$ provided r is constant, and $R \propto 1/r^2$ provided L is constant

---

Note that the symbol $\propto$ is *not* the same as the Greek letter alpha α; take care to make them look different when you write them.

$k = 1.38 \times 10^{-23}\,\text{J\,K}^{-1}$ is known as *Boltzmann's constant.*

In reading this subsection you may be wondering what distinguishes a 'true' constant from a variable that is held constant? This is a deep question to which there is no simple answer. It has been suggested (notably by Paul Dirac (1902–1984)) that some of the fundamental constants of physics might actually change very gradually with time.

(c) $(E + \phi) \propto f$

(d) $F_{grav} \propto 1/r^2$ provided m_1 and m_2 are constant, $F_{grav} \propto m_1$ provided r and m_2 are constant and $F_{grav} \propto m_2$ provided r and m_1 are constant.   □

1.3.4   Inequalities

As you have seen, equations are often used to specify the precise value of a quantity, but it is sometimes useful to be able to express the *range* of values that a variable might take. For example, you might have a variable power-supply unit incorporating a circuit breaker that trips if the current reaches a certain maximum value I_{max}. The condition that the current must be less than I_{max} can be written as $I < I_{max}$: the symbol < means 'is less than'. The symbol ≤ means 'is less than or equal to'.

What do you think the symbols > and ⩾ mean?

> means 'is greater than' and

⩾ means 'is greater than or equal to'.   □

A statement that uses any of the symbols <, ≤, > or ⩾ to compare two values is called an **inequality**. Note that whichever symbol is used, the greater of the two quantities is always at the wider end of the symbol. So, you might correctly write $3 < 6$ or $6 > 3$ but *not* $6 < 3$ or $3 > 6$.

When using an inequality it is important to remember that any negative number is considered to be less than any positive number. In fact, numbers are treated as if they are laid out along a line, sometimes called the **number line**, with increasingly large positive numbers further and further to the right of zero, and increasingly large negative numbers further and further to the left of zero. Any given point on the number line then represents a number that is greater than the numbers represented by points to the left of the given point.

$$\text{lesser} \quad -4 \quad -3 \quad -2 \quad -1 \quad 0 \quad 1 \quad 2 \quad 3 \quad 4 \quad\quad \text{greater}$$

Question T23
Which of the following inequalities are correct?

(a) $3 > -4$ (b) $-4 > -16$ (c) $(-4)^2 > 8$

(d) $-10 \leqslant -8$ (e) $\dfrac{1}{2} \leqslant 0.5$ (f) $\dfrac{1}{2} < \dfrac{1}{8}$

(g) $\sqrt{2} > 1.40$ (h) $-1 > 0$ □

Inequality symbols can also be used to express both ends of a range of values. For example the statement $3 < x < 10$ is read as 'x is greater than 3 *and* less than 10', in other words 'x is in the range 3 to 10 − excluding the 'endpoints', 3 and 10.'

Use inequality symbols to represent the statement 'x is in the range 3 to 10 − *including* the endpoint values 3 and 10'.

$$3 \leqslant x \leqslant 10$$

(Note that in writing statements of this kind it is conventional to put the lesser value on the left − just like the number line.) □

Rearranging inequalities

In practice, purely arithmetic inequalities such as $2 > 1$ are rarely written down since they are 'obvious'. It is much more likely that you will have to deal with algebraic inequalities such as $x > b$ or $2x \geqslant b - y$. It is quite likely that you will be called on to rearrange such inequalities, either to simplify them or possibly to change their subject. The procedure is broadly similar to that for rearranging equations, but much greater care is needed when dealing with inequalities.

The basic rules for manipulating inequalities that involve the symbol $>$ are given below. Similar rules concerning the other inequality symbols may be obtained by replacing the $>$ by $\geqslant$, $<$ or $\leqslant$ throughout and simultaneously replacing the $<$ by $\leqslant$, $>$ or $\geqslant$ throughout.

Rules for manipulating inequalities:

1. If $x > y$ and a is any number, then $x + a > y + a$.

2. If $x > y$ and k is a positive number, then $kx > ky$.

3. If $x > y$ and k is a negative number, then $kx < ky$.

Pay particular attention to Rule 3: if both sides of an inequality are multiplied by the same *negative* quantity then the inequality must be reversed. (So, if $2 < 3$ then $(-2)(2) > (-2)(3)$, i.e. $-4 > -6$, which is correct.)

What problems might arise from the following actions?

(a) Divide both sides of an inequality by $(1 - a)$ when $a = 1$.

(b) Multiply both sides of an inequality by $(1 - a)$ when $a = 1$.

(c) Square both sides of an inequality.

(d) Take square roots of both sides of an inequality.

(a) Dividing by $(1 - a)$ when $a = 1$ is just another way of dividing by zero. It will produce meaningless results and should be avoided.

(b) Multiplying both sides of an inequality by zero is also dangerous. It's true that $3 < 6$, but is *not* true that $(0) \times (3) < (0) \times (6)$, i.e. $0 < 0$.

(c) Even if $a > b$ it is *not* generally true that $a^2 > b^2$. For instance $-3 > -4$ but it is *not* true that $(-3)^2 > (-4)^2$.

(d) Square roots also need care. Negative quantities don't have square roots and positive quantities have two square roots, so although $9 > 4$, it is *not* true that $-3 > -2$ though it *is* true that $3 > 2$. □

Obviously, the most important thing to do when manipulating inequalities is to *think carefully* about each step. Here are some questions that require you to do just that.

Question T24

Make x (alone) the subject of each of the following inequalities:

(a) $2x + 4 > 6$

(b) $2(a - x) \leqslant 7$

(c) $2(a - x) \leqslant 3x + b$. □

Question T25

The electrical resistance R of a piece of wire of length l and cross-sectional area A, made from a material of resistivity ρ is given by the expression $\rho l / A$. Suppose you need a sample of wire with resistance R such that $R > R_0$. Write down inequalities expressing, (a) the range of acceptable lengths for a sample of given ρ and A, and (b) the range of acceptable areas, for a sample of given l and ρ. □

1.4 Closing items

1.4.1 Chapter summary

1. The overall order of priority for *operations* is:
    - first deal with anything in brackets
    - then deal with powers (including roots and reciprocals)
    - then multiply and divide
    - then add and subtract.

2. *Expressions* containing a product of bracketed factors can be *expanded* by multiplying each term inside one pair of brackets by all the terms in the other pair of brackets, e.g.

$$(a + b)(c + d) = ac + ad + bc + bd$$

$$(a + b)^2 = a^2 + 2ab + b^2$$

$$(a + b)(a - b) = a^2 - b^2$$

$$(a - b)^2 = a^2 - 2ab + b^2$$

3. *Expressions* can be *simplified* by using brackets to group together terms with a *common factor*, e.g. $ab + ac = a(b + c)$.

4. *Fractions* may be multiplied and divided according to the following rules:

$$\frac{a}{b} \times \frac{c}{d} = \frac{ac}{bd} \tag{5}$$

$$\frac{a}{b} \div \frac{c}{d} = \frac{ad}{bc} \tag{6}$$

5. *Fractions* may be added or subtracted by rewriting them so that they have a *common denominator*:

$$\frac{a}{b} + \frac{c}{d} = \frac{ad + cb}{bd} \tag{7}$$

$$\frac{a}{b} - \frac{c}{d} = \frac{ad - cb}{bd} \tag{8}$$

6. *Powers* may be used to express *roots* and *reciprocals*. In particular, if x is greater than zero: $x^1 = x$, $x^0 = 1$, $x^{-1} = 1/x$ and $x^{1/n} = \sqrt[n]{x}$.

7. The rules for manipulating powers are:

$$a^p \times a^q = a^{p+q} \tag{13}$$

$$(a^p)^q = a^{pq} \tag{14}$$

$$a^p/a^q = a^{p-q}$$

$$a^p b^p = (ab)^p$$

8. An *equation* can be *rearranged* by performing exactly the same operations on the whole of both sides of the equation.

9. Two *variables* x and y are *directly proportional* ($x \propto y$) if their *ratio* can be expressed wholly in terms that do not include x and y. They are *inversely proportional* ($x \propto 1/y$) if their product can be expressed wholly in terms that do not include x and y.

    If $x \propto y$ then $x = Ky$ where K is a *constant of proportionality*.

10 Inequality symbols $<$, $\leqslant$, $>$ and $\geqslant$ may be used to express the range of values of a variable that fulfil particular conditions.

11 Inequalities can be manipulated in ways similar to equations, but more care is needed and the direction of an inequality is reversed if both sides are multiplied by the same negative quantity.

1.4.2 Achievements

Having completed this chapter, you should be able to:

A1. Define the terms that are emboldened in the text of the chapter.

A2. Decide the correct order of priority for operations in an arithmetic or algebraic expression and hence evaluate such expressions (using a calculator where appropriate).

A3. Use powers to express roots and reciprocals, and, where appropriate, combine powers in arithmetic and algebraic expressions.

A4. Manipulate and simplify fractions, both arithmetic and algebraic, including cases where there is initially no common denominator.

A5. Expand expressions that include brackets, and use brackets to simplify expressions.

A6. Rearrange equations in order to simplify them or to change their subject.

A7. Recognize examples of direct and inverse proportionality, and combine two or more proportional relationships.

A8. Rewrite proportional relationships as equations by introducing appropriate constants of proportionality.

A9. Interpret, use and rearrange inequalities.

1.4.3   End of chapter questions

Question E1   Write down the operations you would need to carry out, in the correct order, to evaluate the following expressions, e.g. for $(a - b)/3$ you subtract b from a and then divide by 3.

   (a) $5(x + 3y^2)$ (b) $E/(1 + r/R)$.

Question E2   Expand the following expressions and write them in as simple a form as possible:

   (a) $(x + 5)(x + 7)$ (b) $x(x + 2)(x^2 - 1) - x^2(2x - 1)$.

Question E3   Rewrite the following expression as a single fraction expressed in as simple a form as possible: $2p/q + 7pr/3$.

Question E4   The equation $v = \sqrt{E/\rho}$ relates the speed, v, of sound in a solid to the elastic modulus E and density ρ. Rearrange the equation so that ρ is the subject.

Question E5   The equation $(1/u) + (1/v) = 1/f$ describes a relationship between the distance u of an object from a lens of focal length f and the distance v from the lens to the image. Rearrange the equation so that f is the subject.

Question E6   The kinetic energy E of an object of mass m and speed v is given by $E = mv^2/2$. If the object increases its speed from an initial value v_1 to a final value v_2, its kinetic energy increases from E_1 to E_2. Derive expressions in terms of m, v_1 and v_2 for (a) the ratio of the final to the initial kinetic energy E_2/E_1, (b) the increase in kinetic energy $E_2 - E_1$, and (c) the fractional increase in kinetic energy $(E_2 - E_1)/E_1$. In each case, write the expression in as simple a form as possible.

Question E7   Simplify the following expressions:

   (a) $(l/\sqrt{l})/l^2$ (b) $z\left[\sqrt[5]{z^{-1}}\, x^{0.5}y^{0.1})/(x^{-1/2}y^{1/5})^2\right]^{1/3}$

Question E8   The strength F_{el} of the electrostatic force between two objects with positive charges q_1 and q_2, separated by a distance r, is given by $F_{el} = (q_1q_2)/(4\pi\varepsilon_0 r^2)$ where ε_0 is a constant known as the permittivity of free space. Express the relationships between the following variables as proportionalities (a) F_{el} and q_1, (b) F_{el} and r.

Question E9   A pendulum of length l, swinging under the influence of gravity, completes a full swing in a time T. Observations in many different places, each characterized by a local value for g (the magnitude of the acceleration due to gravity), show that $T \propto \sqrt{l}$ for fixed g and $T \propto 1/\sqrt{g}$ for fixed l. Write down an equation relating T, g and l.

Question E10

(a) Express in words the statement $3 < x \leq 20$.

(b) If $a^2x^2 - a^2x^2h \geq 2y$, find the condition under which

$$x^2 \leq \frac{2y}{a^2(1 - h)}.$$

Study comment

Now that you have completed the *End of chapter questions*, go back to Subsection 1.1.2 and try the *Fast track questions* if you have not already done so.

Numbers, units and physical quantities

Chapter 2

2.1 Opening items

2.1.1 Chapter introduction

The statements 'That car was travelling at about 80', and 'That car was travelling at 81.6 kilometres per hour', could both apply to the same situation. The first statement gives information in an approximate form and takes the units of measurement for granted. This is often appropriate in everyday life, but is seldom good enough for scientific purposes; not only is it imprecise, it is also open to misinterpretation since in the UK it would probably be taken to mean 80 *miles* per hour while in other European countries *kilometres* per hour would be presumed. The second statement is much more informative, the units of measurement are clearly stated and the speed is given with some precision; the car was not travelling at 81.5 kilometres per hour, nor at 81.7 kilometres per hour, but at 81.6 kilometres per hour.

This chapter is about using and interpreting numbers and units in the specification of physical quantities. Some very basic terms with which you may already be familiar are precisely defined, and some widely used conventions are explicitly spelt out. Throughout the chapter the main aim is to cause you to think more deeply and more incisively than usual about the exact meaning of numerical statements concerning physical quantities.

The main teaching text of this chapter is contained in Sections 2.2 and 2.3. The first of these sections is concerned mainly with numbers. Subsections 2.2.1 and 2.2.2 introduce *scientific notation* (also known as *powers of ten notation* or *standard form*) which is widely used to represent large and small numbers. Subsection 2.2.3 refines the definition of scientific notation by using *inequalities* and the *modulus notation* (e.g. $|a|$) for the *absolute value* of a number. The interpretation of quantities specified to a fixed number of *significant figures* or to a given number of *decimal places* is explained in Subsection 2.2.4, along with the value of using scientific notation to avoid spurious indications of precision. Subsection 2.2.5 sets out the procedure for *rounding* values to a given number of significant figures and describes the related issues of *rounding errors*, *approximations* and *order of magnitude* estimates. Section 2.2 ends with a brief introduction to *set notation*, an informal account of the set of *real numbers* and the definition of some of its major subsets such as *integer* and *rational numbers*. Section 2.3 is concerned with units and deals in sequence with the notational conventions of SI units, the correct usage of units in calculations and the investigation of quantities and equations by means of *dimensional analysis*.

2.1.2   Fast track questions

Question F1

Express the following numbers in scientific notation (this uses 'powers of ten' and is also known as standard form):

(a)   293.45   (b)   1 380   (c)   −2 804   (d)   0.005 67   □

Question F2

Select the numbers from the following list that have (a) the same number of significant figures, and (b) the same order of magnitude:

1.23×10^6 727 8.5×10^6 1.148×10^7 92.874   0.043 2   □

Question F3

Excited hydrogen atoms emit radiation at certain well defined frequencies. These frequencies are related to changes in the energy of electrons as they jump from one atomic energy level to another. The emissions are described by the equation $f = RcX$ where f is the frequency of the radiation, c the speed of light, X a dimensionless number that depends on the initial and final energy levels of the electron, and R a physical constant called Rydberg's constant. Use the rearrangement $R = f/cX$ to deduce the SI units and dimensions of R.   □

2.1.3   Ready to study?

Study comment

In order to study this chapter you will need to understand the following terms: *power*, *product*, *ratio*, *rearrangement* (*of an equation*), *reciprocal* and *root*. (These were all introduced in Chapter 1.) In addition, it is assumed that you are familiar with SI units such as metre and kilogram (though SI units in general are briefly reviewed in Section 2.3) and are accustomed to carrying out simple calculations involving the basic arithmetic operations (×, ÷, +, −), powers and brackets. The following questions will help you to check that you have the required level of familiarity with powers, roots, reciprocals and equation rearrangement.

Question R1

Rewrite each of the following expressions as a single power of x:

(a) $x^3 \times x^4$

(b) $x^5 \times x^{-2}$

(c) $1/x^2$

(d) $1/x^{-6}$

(e) x^8/x^3

(f) $(x^4)^{-6}$

(g) $\sqrt{x^3}\sqrt{1/x^5}$ $\square$

Question R2

Rearrange the equation $F = 6\pi\eta r v$ so that η is isolated on one side. $\square$

2.2 Numbers and physical quantities

Study comment

This section is designed to make you think carefully about the way numbers are written, discussed and interpreted. You should not use a calculator to answer any of the questions or exercises contained in this section unless you are specifically instructed to do so.

2.2.1 Scientific notation

Many physical quantities have numerical values that are much greater than or less than 1. For example, a non-SI unit of pressure (based on the pressure of the atmosphere) is 1 atm but the equivalent pressure expressed in SI units is 101 325 pascal; similarly, the charge of the electron $(-e)$ in SI units is $-0.000\,000\,000\,000\,000\,000\,160\,2$ coulomb. In physics it is customary to write such large or small numbers using **scientific notation** (☞): that is, as a product of a positive or negative *decimal number* (with one non-zero digit before the decimal point) and a suitable *power of ten*. Using scientific notation, 1 atm $= 1.013\,25 \times 10^5$ pascal, and $-e = -1.602 \times 10^{-19}$ coulomb. This subsection explores the numerical aspects of scientific notation and highlights its

Scientific notation is sometimes referred to as standard form, standard index form, floating point notation, or powers of ten notation.

Table 2.1 Powers of ten

Number written in full	Number written using a product of 10		Number written as a power of 10
100 000	$10 \times 10 \times 10 \times 10 \times 10$		10^5
10 000	$10 \times 10 \times 10 \times 10$		10^4
1 000			
100			
10			10^1
1			
0.1	1/10		10^{-1}
0.01	1/100	$1/(10 \times 10)$	$1/10^2$	10^{-2}
0.001			

advantages. Discussion of the role of units in scientific notation is mainly deferred to Section 2.3.

The **powers of ten** involved in scientific notation take the form 10^n where n can be any positive or negative whole number. Table 2.1 illustrates some features of powers of ten, and is the subject of the following questions.

Question T1

Using the printed entries in Table 2.1 as a guide, fill in the spaces in the three columns in the upper part of the table, and in all five columns in the lower part. ☐

Question T2

(a) If you look at each row of the *upper* part of the completed Table 2.1, what pattern links the number of zeros after the 1 in the left-hand column, the number of tens multiplied together, and the power of ten?

(b) If you look at each row of the *lower* part of the completed Table 2.1, what pattern links the position of the 1 after the decimal point in the left-hand column and the power of ten? ☐

The *reciprocal* of a quantity a is the quantity $1/a = a^{-1}$.

Notice that finding the *reciprocal* (☞) of any power of ten is equivalent to changing the sign of the power (e.g. $1/10^2 = 10^{-2}$), and that the relation $10^0 = 1$ forms part of the pattern you have been using.

A **decimal number** is any number expressed in **base** ten notation – the normal system for writing numbers, in which there are ten **digits** (0 to 9) and the position of each digit relative to the decimal point is related to a power of ten. For example, the number 345.6 in base ten notation denotes $3 \times 10^2 + 4 \times 10^1 + 5 \times 10^0 + 6 \times 10^{-1}$. A decimal number does not necessarily include a decimal point; 976, −4.05 and 0.038 are all decimal numbers.

Any number can be expressed as the product of a decimal number and a power of ten. Indeed, this can generally be done in many different ways:

Throughout this book we have used the convention of writing long strings of digits in blocks of three. Thus ten million is written 10 000 000. Some authors like to insert commas into the gaps as in 10,000,000 but we will not do so.

For example,	3 428	$= 3.428 \times 1\,000 = 3.428 \times 10^3$
but equally,	3 428	$= 34.28 \times 100 = 34.28 \times 10^2$
Similarly,	0.056 7	$= 5.67 \times 0.01 = 5.67 \times 10^{-2}$
and		0.056 7 $= 567 \times 0.000\,1 = 567 \times 10^{-4}$

Notice that, in principle, *any* power of ten can be chosen because the decimal number can always be adjusted to ensure that the value of the product remains unchanged. Thus, scientific notation is nothing more than an agreed convention that restricts the form of the decimal number and hence dictates the choice of power.

Question T3

Write the following numbers in full:

(a) 3.2×10^6 (b) 8.76×10^{-4} ☐

Question T4

Write the following numbers in scientific notation:

(a)   98 765   (b)   0.004 32   ☐

One advantage of scientific notation is that it is quite compact. Another advantage is the simplicity it gives to calculations that involve only multiplication and division; where powers of ten can be dealt with separately from the decimal numbers and the order in which multiplications are performed does not matter.

For example, $2 \times 10^4 \times 4 \times 10^5 \times 6 \times 10^{-10}$

can be written as $2 \times 4 \times 6 \times 10^4 \times 10^5 \times 10^{-10}$

If we combine the powers of ten we obtain:

$$2 \times 4 \times 6 \times 10^{-1} = 48 \times 10^{-1} = 4.8$$

Even in calculations that cannot so easily be simplified or reordered, it is often worth manipulating the powers of ten to produce a simpler expression before reaching for a calculator, since the fewer numbers you have to key in, the less likely you are to make a mistake.

Question T5

Rewrite the following expressions so that each involves a single power of ten:

(a) $1.40 \times 10^4 \times 5.50 \times 10^{13} \times 6.20 \times 10^{-15}$

(b) $2 \times 10^6 \times 3 \times 10^5/(8 \times 10^7)$ ☐

2.2.2   Scientific notation and calculators

All 'scientific' calculators handle scientific notation, though you may have to select a particular mode such as 'SCI' or 'ENG' on your calculator to activate this feature, or you might have to press a key marked 'EE' or 'EXP' (☞). The following exercises are aimed at helping you to avoid some common mistakes when using powers of ten on a calculator.

Calculators are not all identical, so the precise combination of key strokes that you need to use may differ slightly from those given here.

Key the number 10 000 000 into your calculator, and make your calculator display it in a more compact form. What do you get? (If it doesn't happen automatically and the keys mentioned above don't work, try '=' or '×' followed by 'EE' or 'EXP'. If all else fails, read the instructions!)

Your calculator should have displayed 10^7 or 1.00 E 7, 1 07, 1^7 or 1^{07} (or something similar). □

Clear the calculator, and repeat the previous exercise with 30 000 000.

The display should have read 3.00 E 7, 3 07, 3^7 or 3^{07}. □

The 'compact form' that calculators use to display numbers corresponds to scientific notation. Whichever form your calculator uses, you should *write* the numbers as 10^7 and 3×10^7. In particular, resist the temptation to write them as 1 07 and 3 07 (to avoid confusion with 107 and 307) or 1^7 and 3^7.

Why do you think it is important not to write, for example, 3×10^7 as 3^7 when writing down the number shown on your calculator? What does 3^7 really mean?

$$3^7 = 3 \times 3 \times 3 \times 3 \times 3 \times 3 \times 3 = 2\,187,$$

which is clearly *not* the same as 3×10^7. □

You have just illustrated one important point about using a calculator: *always be careful how you write down numbers from the calculator display.*

Now try entering first 10^7 and then 3×10^7 directly. Try keying 1 EE 7 or 1 × EE 7, and then 3 EE 7 or 3 × 1 EE 7. What do you get?

The numbers in the display should look exactly the same as in the previous exercises. □

A common mistake is to key 10 EE 7, or 1 × 10 EE 7, when entering 10^7, and 3 × 10 EE 7 (or similar) when entering 3×10^7. This is *WRONG*.

If you key 10 EE 7 (or similar) into a calculator, what number have you in fact entered? Why do you think this is a common mistake?

You have entered 10×10^7, i.e. 10^8. The mistake probably arises because a natural response to seeing the 10 written down as part of 10^7 is to key in the figure 10. □

The above *incorrect* attempt treats the EE key (or similar) as something that raises the first number (10 in this example) to the power of the second (7 in this example). *But that is not what the EE key does.* Remember that entering 10^7 involves keying in 1 EE 7 (or similar); you can think of the EE key as meaning 'times ten to the power'.

You have now illustrated another important point about using a calculator: *be careful to use the EE key correctly when entering powers of ten.*

Finally, look at the process of entering negative powers of ten.

Try entering 3×10^{-7} into your calculator first by keying 3 EE 7 +/− or 3 EE +/− 7 (or similar), then clear and try 3 EE − 7.

The first method should produce 3.00 E − 7 (or similar) but the second will probably give −4 (which may be displayed as −4 EE 00). □

The +/− key changes the sign of a number, whereas the − (minus) key subtracts the next number you enter. You have illustrated a third important point about using calculators: *always use the +/− key when entering negative powers of ten.*

Question T6

Expressed in scientific notation, what value does your calculator provide for the following?

(a) $1.40 \times 10^4 \times 5.50 \times 10^{13} \times 6.20 \times 10^{-15}$

(b) $2 \times 10^6 \times 3 \times 10^5/(8 \times 10^7)$ □

2.2.3   The modulus or absolute value of a number

As defined earlier, a number written in scientific notation is the product of a decimal and an appropriate power of ten, where the decimal may be positive or negative and has a single, non-zero, digit before the decimal point. In other words, if the decimal is positive it must be between 1 and 10 (but not actually equal to 10), and if the decimal is negative it must be between −1 and −10 (but not actually equal to −10). It is both useful and instructive to

Although *modulus* has been introduced in the context of scientific notation the idea is used widely and will be encountered in many different settings.

find a neater way of expressing this condition. This can partly be achieved by introducing the **modulus** or **absolute value** of a number. Given a number r, its *modulus* is denoted $|r|$ and represents the numerical value of r without its sign, e.g. $|3| = 3$ and $|-3| = 3$. More precisely:

If p is a positive number then $|p| = p$ and $|-p| = p$.

Question T7

Evaluate the following expressions:

(a) $|2.3|$ (b) $|-3.1|$ (c) $|(4 - 7)/2|$

(d) $|(4 - 5)(7 - 9)|$ (e) $|2|/|4|$ (f) $|-6||3|$ □

Using the modulus concept we can say that in scientific notation any number is represented as the product of a decimal r and an appropriate power of ten, where $|r|$ is between 1 and 10 (but not actually equal to 10).

This definition can be further refined by making use of an **inequality**. Inequalities are mathematical statements that compare numerical values using the symbols > (read as 'is greater than'), ≥ (read as 'is greater than or equal to'), < (read as 'is less than') and ≤ (read as 'is less than or equal to'). Equipped with these symbols we can say:

In scientific notation any number can be written as the product of a number r such that $1 \leqslant |r| < 10$, and an appropriate power of ten.

With experience you will probably find the following is a better interpretation 'the modulus of r is less than 10 but greater than or equal to 1'. It is $|r|$ which can vary, not 'one'.

How would you express '$1 \leqslant |r| < 10$' in words?

'One is less than or equal to the modulus of r, and the modulus of r is less than ten.'   □ (☜)

The three dots, technically known as an *ellipsis*, indicate that the number continues in the same way. If the length l was known to be 6.00 mm it would still be correct to write $l = 6.0$ mm, though to do so would be to deliberately ignore valid information about l. This may sometimes be desirable, but not usually.

2.2.4   Significant figures and decimal places

When a number is used to describe a physical quantity, each of its digits should carry some meaning, in other words, each should be a **significant figure**. For example, stating that a certain length l is 6.00 mm should mean that l *really is* 6.00 mm, rather than 6.01 mm or 5.99 mm. The use of three significant figures in this case implies that the length has been determined to the *nearest* 0.01 mm and is therefore known to be somewhere between 5.995 000 00 ... mm and 6.004 999 99 ... mm (☜). It would be quite wrong, for example, having determined a length as 6.00 mm to the nearest 0.01 mm to write that length as 6.000 mm. The presence of the last zero would imply that the length was in the range 5.999 500 00 ... mm to 6.000 499 99 ... mm, which would not be justified. Thus, the number of significant figures in a value should indicate the *precision* with which that value is known.

Even if a value is given to the correct level of precision it is not necessarily the case that all of its figures are significant. For example, without changing the information about the length l, you could write it with as many zeros as you wished at the front, for example, 06.00 mm or 0 006.00 mm. Such zeros carry no information and are not counted as significant figures. You could also express the same information using different units, metres for example, giving $l = 0.006\,00$ m. In this case the zeros at the start of the number *do* carry important information since they put the rest of the number in its rightful place relative to the decimal point, but they do no more than this. They do not indicate any increase in precision, so they are still not counted as significant figures. Generally then, zeros at the start of a number, even if important, are not counted as significant figures.

Zeros at the end of a number present a different problem. Suppose you went for a walk and measured its distance as 3.4 km to the nearest 0.1 km. If you were asked the length of the walk in metres you might be tempted to say 3 400 m. But that would not be correct; the two zeros on the right appear to be significant and falsely imply that you have measured the length of the walk to the nearest metre. Nonetheless, those zeros are playing an important role; how else could you express the measured length in metres? Fortunately, this question has a simple answer – use *scientific notation*. Just write the length as 3.4×10^3 m; this conveys the essential information without introducing misleading zeros. Avoiding spurious 'significant figures' is yet another advantage of using scientific notation. Despite this, data involving whole numbers (such as 3 400 m) are often written with unjustified zeros at the end, so you should treat such figures with care. If someone tells you the world's population is 6 000 000 000, ask them how they can be so precise.

The *significant figures* in a number are the meaningful digits that indicate its precision. They do not include any zeros to the left of the first non-zero digit. Using scientific notation avoids the need to write down any zeros that are not significant, either to the left or to the right of the significant figures, and thus avoids any ambiguity in writing or interpreting a number.

===

Decimal numbers are sometimes described in terms of their number of **decimal places** – that is, the number of digits after the decimal point. For example, the number 765.43 is specified to two decimal places. The number of decimal places may also be used to indicate the precision of a value. For example, you might say that the length $l = 6.00$ mm has been determined 'to two decimal places' but it's important to take the units into account. The same length, determined to the same level of precision, but expressed in metres would have to be given to five decimal places (i.e. 0.006 00 m).

Question T8

Describe each of the following values in terms of its number of significant figures and number of decimal places:

(a)   65.43   (b)   0.003 56   (c)   2 278   (d) 3.04×10^{-5} ☐

The number of significant figures or decimal places to which a physical quantity is known is generally limited by the precision of a measurement. This applies not only to quantities that are measured directly, but also to those derived from measured quantities via calculations. The following discussion illustrates the effect of combining quantities that are known only to a limited number of significant figures.

Suppose you have measured the length l and breadth b of a rectangle as $l = 3.4$ m and $b = 6.2$ m (i.e. to the nearest 0.1 m). What can you say about its area $A = l \times b$? You might be tempted to say 3.4 m $\times$ 6.2 m = 21.08 m^2, but that really isn't justified since l and b are only measured to the *nearest* 0.1 m, so their true values might be as much as 0.05 m larger or smaller.

Calculate the area A of the rectangle

(a)   if $l = 3.350\,0$ m and $b = 6.150\,0$ m, and

(b)   if $l = 3.450\,0$ m and $b = 6.250\,0$ m.

(a) $A = 20.602\,5$ m^2 (b) $A = 21.562\,5$ m^2 ☐

In view of these answers, it would be sensible to say that a rectangle of length 3.4 m and breadth 6.2 m has an area of 21 m^2, since that implies its area is between 20.500 0 ... m^2 and 21.499 9 ... m^2. It doesn't quite cover all the possibilities, but it is a reasonable compromise given the imprecision of the original measurements. Notice that this (sensible) answer has two significant figures, as did each of the values used to calculate it. (☜)

In this book, most calculations will be carried out using three significant figures. The answers will then also be quoted to three significant figures, though the last figure may not be entirely reliable.

If you use a calculator, you will often find that as many as eight digits are displayed. Do not be fooled into thinking that they are all significant figures. Unless you started with extremely precise initial values, most of the displayed figures will be completely useless. You should look back at the initial values that you used, find the one with the *fewest* significant figures, and then adjust your answer so that it has the same number of significant figures. For example, if your calculator displays a result as 8.362 114 5, but you know that only three figures could be significant (due to your initial input), you should write the answer as 8.36. If only two figures were significant, you should write 8.4, since 8.362 is closer to 8.4 than it is to 8.3. Generally, if the first 'insignificant' digit is 0, 1, 2, 3 or 4 it can just be dropped,

You can extend this discussion to cover division by finding the *ratio* of the sides of the rectangle using the same ranges of values for l and b.

but if it is 5, 6, 7, 8 or 9 then 1 must be added to the last significant digit. This process is called **rounding**. If the last significant digit is left unchanged the number is said to have been **rounded down**, and if the last significant digit is altered the number is said to have been **rounded up**. The errors that can arise in a final answer as a result of rounding are called **rounding errors**. Such errors will generally grow as the number of steps in a calculation increases.

In calculations where there are several steps it is generally wise to delay the process of *rounding* until the final step. This will minimise the chance of rounding errors introduced in one step amplifying rounding errors introduced in some other step. However, it is important to realise that allowing your calculator to carry all the given digits in the data through the interim steps of a calculation does not make those figures significant.

Of course, some quantities are known exactly or with very great precision. For example, the speed of light travelling through a vacuum is now *defined* to be exactly $2.997\,924\,58 \times 10^8$ m s^{-1}, and the mathematical constant π ($= 3.142$ to three decimal places) has been determined to an enormous number of decimal places. Similarly, the pure numbers that enter into formulae, such as the 2 in the formula $C = 2\pi r$ for the circumference C of a circle of radius r, are usually intended to represent exact values. Such exact or very precise values will not generally limit the precision of a calculation and should not, therefore, be allowed to limit the number of significant figures in a result.

> Note that rounding may have a knock-on effect on more than one digit, for example if an answer displayed as 4.98 has only two significant figures it must be written 5.0 as a consequence of adding 1 to the 9.

---

1. When multiplying or dividing two numbers, the result should not be quoted to more significant figures than the least precisely determined number. (So, $0.4 \times 1.21 = 0.5$.)
2. When adding or subtracting two numbers, the last significant figure in the result should be the last significant figure that appears in *both* the numbers when they are expressed in decimal form without powers of ten. (So, $0.4 + 1.21 = 1.6$ and $0.004 + 1.21 = 1.21$.)
3. When calculations are performed with a consistent number of significant figures throughout, *the final result may still not be reliable to that number of significant figures.*

Question T9

The dimensions of a thin metal sheet have been measured as 0.23 m, 6.26 m and 2.88×10^{-3} m. Someone has used these values to calculate the volume of the sheet and has obtained $4.146\,62 \times 10^{-3}$ m³. (a) How many figures in this answer are significant? What is a sensible value for the volume? (b) The mass has been measured as 18.17 kg. If the mass is divided by the volume to calculate the density, how many figures in that answer will be significant? ☐

2.2.5   Approximations and orders of magnitude

When performing calculations it is often useful to work with **approximations**. For example, if you know that a certain length lies somewhere between 3.1 m and 3.3 m you may wish to approximate it to 3.2 m even though the final '2' is not significant. By making such an approximation you might well obtain a better idea of the true outcome of the calculation than you would by saying that the length was 3 m and reducing the number of significant figures in the final answer. On the other hand the figures in your final answer will not all be significant and could mislead if taken too seriously. To avoid this danger the symbol ≈ (read as 'is approximately equal to') can be used to indicate that an approximation has been introduced.

Even when quantities are known to a certain level of precision it is sometimes useful to approximate them by 'simpler' values just to get a rough idea of the answer. For example

$$\frac{3.2 \times 5.9}{4.5} \approx \frac{3.0 \times 6.0}{4.5}$$

Sometimes it is only possible to obtain a very crude approximation to the value of a quantity. The use of such crude approximations is generally indicated by the symbol ~ which can usually be read as 'is very roughly equal to'. Sometimes a more precise meaning is attached to this symbol when it is used to indicate the value of a quantity to the nearest **order of magnitude** — that is, to the nearest power of ten. For example, $2.34 \times 10^7 \sim 10^7$ and $8.765 \times 10^7 \sim 10^8$. Notice that if the first digit of the decimal number is 5 or greater, then the original power of ten must be rounded up to give the order of magnitude.

Orders of magnitude give a quick way of making a rough comparison between two quantities. For example, the mass of the Earth is 5.98×10^{24} kg, and that of the Moon is 7.35×10^{22} kg, but for a quick comparison it is sufficient to say that the Earth's mass $\sim 10^{25}$ kg and that of the Moon $\sim 10^{23}$ kg, and that the mass of the Earth is two orders of magnitude greater than that of the Moon.

Estimating the order of magnitude of a calculated quantity often provides a useful way of checking that any value found with a calculator is reasonable. It won't guarantee the absence of a mistake, but it will help you to avoid silly mistakes.

Find the approximate value and the order of magnitude of the number y, where $y = 2\pi \times 3.21 \times 10^5 \times 864/(2.04 \times 10^2 \times 77)$.

If we write the numbers in scientific notation and collect together the powers of ten, then

$$y = \frac{2\pi \times 3.21 \times 8.64 \times 10^5 \times 10^2}{2.04 \times 7.7 \times 10^2 \times 10^1}$$

If we work to two significant figures,

$$y \approx \frac{2 \times 3.1 \times 3.2 \times 8.6 \times 10^4}{2.0 \times 7.7}$$

The crudity of the approximation in the following steps is a matter of taste. Personally, as a crude calculator, I would be happy to say that the 2s cancel, that 3.1×3.2 is about 10 and that $8.6/7.7$ is about 1.1, (☞) so $y \approx 1.1 \times 10^5$ and, as an order of magnitude estimate, $y \sim 10^5$. ☐

Notice that in estimating the order of magnitude of an expression you should *not* substitute order of magnitude approximations to each term. Doing so can easily produce an answer that is wrong by several orders of magnitude.

Question T10

Write down the order of magnitude of the following: (a) mass of the Sun = 1.99×10^{30} kg; (b) radius of the Sun = 6.96×10^8 m; (c) mass of a proton = 1.6726×10^{-27} kg; (d) mass of an electron = 9.11×10^{-31} kg. ☐

Question T11

Without using a calculator, find the order of magnitude of z, where $z = 7.3 \times 10^3 \times 869.3/(2.3 \times 10^4 \times 4.07 \times 10^5)$. ☐

2.2.6 Sets and types of number

This subsection introduces some mathematical terminology that is useful when describing different types of number. It is worth understanding these differences since the various types of number are interpreted in different ways.

Natural numbers and integers

For the purpose of counting, the only numbers we need are positive, whole, numbers 1, 2, 3 ... and so on. These 'counting' numbers are called **natural numbers**. Taken together they form a **set** – a collection of entities defined by some common characteristic. The fact that the natural numbers form a set is sometimes indicated by writing them in braces (curly brackets), though it is more usual to denote the whole set by the single symbol $\mathbb{N}$.

Thus, $\mathbb{N} = \{1, 2, 3, \ldots\}$

Each number belonging to the set $\mathbb{N}$ is said to be an **element** of $\mathbb{N}$. The fact that a number such as 57 is an element of $\mathbb{N}$ is shown by writing $57 \in \mathbb{N}$ where the symbol $\in$ is read as 'is an element of'.

Another important set includes negative whole numbers and zero as well as the counting numbers. This is the set of **integers** and is symbolised by $\mathbb{Z}$.

Thus, $\mathbb{Z} = \{\ldots -2, -1, 0, 1, 2, 3 \ldots\}$

The whole of the set $\mathbb{N}$ is included within the set $\mathbb{Z}$, a relationship that is recognised by saying that $\mathbb{N}$ is a **subset** of $\mathbb{Z}$ or by writing $\mathbb{N} \subset \mathbb{Z}$. Since integers are whole numbers it would be quite wrong to regard a given integer, such as 2, as meaning anything other than *exactly* 2. It is in this sense that the 2s in expressions such as $2\pi r$ or $mv^2/2$ are to be interpreted. Thus, *the presence of an integer in a calculation does not affect the number of figures that are significant in the final result (because it really has an infinite number of zeros after the decimal point).*

Real numbers, rational numbers and irrational numbers

All the numbers used in this chapter belong to the set of **real numbers**. This set is denoted by the symbol $\mathbb{R}$, and each of its elements can be represented as a positive or negative decimal number. Thus all 'ordinary' numbers such as 15.5, -0.009, 187 and 4.31×10^{-8} are real numbers (even the last of these *could* have been written as a decimal). The set of integers and the set of natural numbers are both subsets of the set of real numbers, and the numerical values of all measurable physical quantities *can* be expressed in terms of real numbers. (☞)

Notice that we are not saying that measurable quantities *must* be expressed in terms of real numbers, only that they *can* be.

You should note that when dealing with pure numbers in a mathematical context, they should normally be interpreted as exact values. This should be contrasted with the customary practice when dealing with the values of physical quantities, where a number generally indicates a range of values. On the whole, you will have to determine how a real number is supposed to be interpreted from the context in which it is used.

The set of real numbers contains two other important subsets; the *rational numbers* and the *irrational numbers*. The set of **rational numbers** includes every real number that can be expressed as a ratio in which one integer is divided by another. The set of **irrational numbers** contains every real number that cannot be expressed as a fraction. Every real number is either rational or irrational.

Real numbers such as 0.5 ($= 1/2$), -531.25 ($= -2\,125/4$), 53 ($= 53/1$) and $0.666\ldots$ ($= 2/3$) are all rational because each can be expressed as a ratio of integers. But there are many other real numbers that are not rational. Several real numbers which are commonly used in physical science are irrational, including many square roots. For example, it is easy to show that $\sqrt{2}$ cannot be expressed as a ratio of integers − there are *no* two integers with a ratio that is exactly equal to $\sqrt{2}$, i.e. there are no two integers m and n such that $\sqrt{2} = n/m$. The mathematical constant π ($= 3.142$ to three decimal places) is also irrational, as is another mathematical constant e ($= 2.718$ to three decimal places), which will be discussed later.

One characteristic of irrational numbers is that they extend indefinitely beyond the decimal point, without any repeating patterns of digits. Calculators generally use values of π, e and other irrationals (e.g. $\sqrt{2}$) that extend to at least eight figures, which for most purposes is sufficient to ensure that the presence of such numbers does not adversely affect the precision of any calculation.

Question T12

Which of the following statements are true?

(a) $\mathbb{Z} \subset \mathbb{R}$ (b) $\sqrt{8}$ is irrational

(c) $-4 \in \mathbb{N}$ (d) $2.00 \in \mathbb{Z}$. ☐

Question T13

The equation $T = 2\pi\sqrt{l/g}$ gives the time T that a pendulum of length l takes to complete a full swing (i.e. an oscillation) when the magnitude of the acceleration due to gravity is g. (a) Identify one integer and one irrational number used in the equation. (b) Explain why the number of significant figures in the answer will depend only on the values given for l and g, and not on the other numbers in the equation. ☐

2.3 Units and physical quantities

2.3.1 Symbols and units

Units of measurement play an essential role in the specification of most physical quantities. The *Système International d'Unités* (SI) is used throughout this book. The seven basic SI units, their standard abbreviations, the terminology of standard multiples, and a number of derived units are summarised in Tables 2.2, 2.3 and 2.4 for easy reference.

Physical quantities are often represented by algebraic symbols such as E for energy, l for length and so on. Such symbols should generally be presumed to represent the *whole* quantity − that is, the units as well as the numerical value. Thus, it would be correct to speak of 'a rod of length l', but it would be incorrect to say 'a rod of length l m' since, if l was 2.2 m the latter statement would mean 'a rod of length 2.2 m m' which would be misleading and wrong.

These seven basic units have been agreed by convention − others could have been chosen.

Table 2.2 The seven basic SI units

Physical quantity	Unit	Symbol for SI unit
Length	metre	m
Mass	kilogram	kg
Time	second	s
Electric current	ampere	A
Temperature	kelvin	K
Luminous intensity	candela	cd
Amount of substance	mole	mol

Table 2.3 Standard SI multiples

Multiple	Prefix	Symbol for prefix
10^{12}	tera	T
10^{9}	giga	G
10^{6}	mega	M
10^{3}	kilo	k
10^{0}		
10^{-3}	milli	m
10^{-6}	micro	μ
10^{-9}	nano	n
10^{-12}	pico	p
10^{-15}	femto	f

Table 2.4   A number of derived SI units that are given special names

Physical quantity	Unit	Symbol for derived SI unit	Definition
Energy	joule	J	$\text{kg m}^2\text{ s}^{-2}$
Force	newton	N	$\text{kg m s}^{-2} = \text{J m}^{-1}$
Power	watt	W	$\text{kg m}^2\text{ s}^{-3} = \text{J s}^{-1}$
Electric charge	coulomb	C	A s
Electric potential difference	volt	V	$\text{kg m}^2\text{ s}^{-3}\text{ A}^{-1} = \text{J A}^{-1}\text{ s}^{-1}$
Electric resistance	ohm	Ω	$\text{kg m}^2\text{ s}^{-3}\text{ A}^{-2} = \text{V A}^{-1}$
Electric capacitance	farad	F	$\text{A}^2\text{ s}^4\text{ kg}^{-1}\text{ m}^{-2} = \text{A s V}^{-1}$
Magnetic flux	weber	Wb	$\text{kg m}^2\text{ s}^{-2}\text{ A}^{-1} = \text{V s}$
Inductance	henry	H	$\text{kg m}^2\text{ s}^{-2}\text{ A}^{-2} = \text{V s A}^{-1}$
Magnetic field	tesla	T	$\text{kg s}^{-2}\text{ A}^{-1} = \text{Wb m}^{-2}$
Frequency	hertz	Hz	s^{-1}
Pressure	pascal	Pa	$\text{kg m}^{-1}\text{ s}^{-2} = \text{N m}^{-2} = \text{J m}^{-3}$

Units should generally be treated like algebraic quantities. Thus, if a runner travels a distance $d = 1\,000$ m in a time $t = 200$ s the runner's average speed v can be written

$$v = \frac{d}{t} = \frac{1000\text{ m}}{200\text{ s}} = 5.00\text{ m s}^{-1}$$

Notice that in accordance with convention the result of dividing m (for metre) by s (for second) has been written m s^{-1} rather than m/s or m per s. It is generally expected that combinations of units will be displayed in this way, using appropriate powers, as the right-hand column of Table 2.4 indicates.

Although oblique strokes (/) should be avoided when combining units, they may be used to rewrite a statement such as $\lambda = 567$ nm in the form $\lambda/\text{nm} = 567$ (☞). Notice that the unit is again being treated as an algebraic symbol, and has been used to divide both sides of the original statement. This way of writing a quantity is used mainly in column headings in tables, so that the units do not have to be written next to each entry (thus making the table less cluttered), and in labelling the axes on graphs where it ensures that the values plotted are, technically speaking, pure numbers (which pleases mathematical purists).

The abbreviation nm stands for nanometre. It arises from the combination of the prefix *nano* (Table 2.3) and the basic unit *metre* (Table 2.2). $1\text{ nm} = 10^{-9}$ m.

Why would it be inappropriate to write λ (nm) in a table column heading or on an axis of a graph?

λ (nm) could be interpreted as meaning $\lambda \times$ nm. So if, say, $\lambda = 567$ nm, then λ (nm) $= 567$ nm nm $= 567$ nm^2, which implies λ is an area rather than a length and is therefore clearly nonsense. This style of labelling used to be widely used but is now strongly disfavoured.   □

2.3.2 Units in equations

Since symbols represent *entire* quantities, you should *always include units when substituting numerical values into equations.* (☞) For example, when using the equation $T = 2\pi\sqrt{l/g}$ (☞) to find the period of oscillation of a pendulum, where $l = 1.2$ m and $g = 9.8$ m s^{-2}, it would be wrong to write $T = 2\pi\sqrt{1.2/9.8}$, it would be better to write:

This convention will be used throughout.

This equation was used earlier, in Question T13.

$$T = 2\pi\sqrt{\frac{1.2\,\text{m}}{9.8\,\text{m s}^{-1}}} = 2\pi\sqrt{\frac{1.2}{9.8}} \times \sqrt{\frac{\text{m}}{\text{m s}^{-2}}} = 2\pi\sqrt{\frac{1.2}{9.8}}\,\text{s}$$

This is obviously cumbersome and if you are confident about the use of units you might well cut out the intermediate steps, but you still shouldn't omit the units entirely, as though they have magically gone away while you deal with the numbers.

You should look upon units as a help rather than a hindrance. Their consistent inclusion is particularly advantageous in complicated and unfamiliar situations but even in straightforward cases they may help you to avoid elementary errors. For example, suppose you wrongly thought the equation for T was $T = 2\pi\sqrt{g/l}$. By including units in your calculation you would soon see that this equation would give a value for T of $2\pi\sqrt{(9.8\,\text{m s}^{-2})/(1.2\,\text{m})}$, so its units would be s^{-1}. Such an outcome should warn you that something has gone wrong. (Notice, though, that the units give no clue as to whether the numerical factor 2π is correct since it is a pure number and has no units.) Another common error that can be avoided by treating units carefully is that of forgetting to convert a quantity given in units other than base units (a distance in centimetres or kilometres, say, or a mass in grams) into the equivalent number of base units for the purposes of calculation.

Including the units in a calculation also allows the units of an unfamiliar quantity to be deduced. For example, Newton's law of gravitation $F_{grav} = Gm_1m_2/r^2$ which relates the magnitude F_{grav} of the gravitational force between two objects to their masses m_1 and m_2 and their separation r, can be used to deduce the units of G, Newton's gravitational constant. We can rearrange the equation to give $G = F_{grav}r^2/(m_1m_2)$. In SI units, F_{grav} could be expressed in newtons, r in metres, and m_1 and m_2 in kilograms. The units of G would then be N m^2/(kg $\times$ kg), or, more properly, N m^2 kg^{-2}.

Question T14

The equation $F_{el} = q_1q_2/(4\pi\varepsilon_0 r^2)$ relates the magnitude F_{el} of the electrostatic force between two particles of charges q_1 and q_2 to their separation r, where ε_0 is the permittivity of a vacuum. Use the equation to deduce the SI units of ε_0 when q_1 and q_2 are expressed in coulombs (C), F_{el} in newtons (N) and r in metres (m). □

2.3.3 Dimensional analysis

The discussion of units in Subsection 2.3.2 was based on the correct presumption that the quantities on both sides of an equation *could* be expressed in terms of the same units. Of course, the units do not *have* to be identical, though they must be equivalent. For example, it is quite true that 2.000 m = 2 000 mm even though the units differ, but it is obviously wrong to say 2.000 m = 2 000 kg. The plausibility of the first equation and the stupidity of the second can both be demonstrated by a process known as **dimensional analysis**.

In the context of dimensional analysis the term **dimension** relates to basic measurable quantities such as mass, length, time, temperature and electric current. These quantities differ fundamentally in that the units used to measure any of them cannot be entirely expressed in terms of the units used to measure the others. This basic incompatibility is recognised by saying that they have *different dimensions*. On the other hand, the fact that a quantity such as area can be measured in the same units as a product of two lengths is expressed by saying that area has the *same dimensions* as length[2].

When carrying out dimensional analysis it is useful to introduce single letters such as M, L and T to represent the dimensions of mass, length and time, respectively, and to enclose a quantity in square brackets when referring to its dimensional nature alone (☞). Thus, we may write

$$[\text{mass}] = M \qquad [\text{length}] = L \qquad [\text{time}] = T$$

It follows that $\qquad [\text{area}] = [\text{length}^2] = L^2$

Similarly, $\qquad [\text{acceleration}] = [\text{length/time}^2] = LT^{-2}$

and $\qquad [\text{force}] = [\text{mass} \times \text{acceleration}]$

$$= [\text{mass} \times \text{length/time}^2]$$

$$= MLT^{-2} \ (☞)$$

Any physical quantity may be dimensionally analysed in this way, though those which involve thermal quantities such as temperature, or electrical quantities such as electric currents will require the introduction of other basic dimensions in addition to M, L and T.

Question T15

What are the dimensions of the following quantities: (a) density (b) Newton's gravitational constant, G (c) speed? ☐

One of the main uses of dimensional analysis is in exposing dimensional inconsistencies in equations. Clearly, if $A = B$ then the dimensions of A and B must be the same (☞), i.e. $[A] = [B]$, it follows that:

An equation must be false if the quantities being equated, added or subtracted do not have the same dimensions.

[q] should be read as 'the dimensions of q.'

You can confirm the dimensions of force by examining the definition of the newton in Table 2.4.

Note that even if both sides of an equation have the same dimensions it does not follow that the equation is correct, only that the equation is *plausible* in dimensional terms.

For example, the claim that the volume of the outer third of a sphere of radius R is given by $V = \pi R^2/9$ is obviously bogus because it is dimensionally inconsistent; $[V] = L^3$ while $[\pi R^2/9] = [R^2] = L^2$.

Note that numerical factors such as $\pi/9$ are **dimensionless**, that is to say they have no dimensions and can be ignored in dimensional analysis (more formally we can write $[\pi/9] = 1$). The same is true of ratios such as length/breadth in which a quantity with certain dimensions is divided by another quantity with the same dimensions. Indeed, such ratios are called **dimensionless ratios**.

Dimensional analysis can also be used to help determine the form that a relationship should take. For example, suppose you have discovered that the period T of a pendulum depends on the length l of the pendulum and the magnitude of the acceleration due to gravity g. You might well arrive at the conclusion that the relationship between these quantities is of the general form

$$T = k \times l^p \times g^q$$

where k is a dimensionless constant and the powers p and q are unknown. If so dimensional analysis implies

Note that the *italic* letter T represents a physical quantity while the *roman* letter T stands for the dimension of time.

$$[T] = [k \times l^p \times g^q] = [k] \times [l^p] \times [g^q]$$

so, $\quad \text{T} = 1 \times \text{L}^p(\text{LT}^{-2})^q = \text{L}^p \times \text{L}^q\text{T}^{-2q}$

i.e. $\quad \text{T} = \text{L}^{p+q} \times \text{T}^{-2q}$

This equation can only hold true if $p + q = 0$ and $-2q = 1$, and this in turn requires that $q = -1/2$ and $p = 1/2$. Thus, on dimensional grounds

If $p + q = 0$ and $-2q = 1$, $\text{T} = \text{L}^0 \times \text{T}^1$ as consistency requires.

$$T = k \times l^{1/2} \times g^{-1/2} = k\sqrt{l/g}$$

This, of course, is the correct form of the relationship (you met it earlier), though the fact that the dimensionless constant k is actually 2π cannot be determined from dimensional considerations.

Question T16

If m is a mass, v a speed, h a height and F the magnitude of a force, determine which of the following expressions for the energy E of a system are implausible on dimensional grounds:

(a) $\quad mv^2$ $\qquad$ (b) $\quad mv^2/2$ $\qquad$ (c) $\quad mv/4$

(d) $\quad mh/2$ $\qquad$ (e) $\quad 2Fh$ $\qquad$ (f) $\quad Fh + mv^2$

(g) $\quad mv^2/2 + F^2h/(mv^2)$ $\quad \square$

2.4 Closing items

2.4.1 Chapter summary

1. In *scientific notation* a quantity is written as the product of a positive or negative *decimal number* (with one non-zero digit before the decimal point) and an appropriate *power of ten*, followed by the relevant physical units.

2. The *modulus* (or *absolute value*) of a number r, denoted by the symbol $|r|$, is its numerical value without its sign. Thus, if p is a positive number, $|p| = p$ and $|-p| = p$.

3. A number is given to n *significant figures* when it contains n meaningful digits (excluding zeros at the start of the number).

4. In calculations involving numbers known only to n significant figures, only n digits of the answer can be significant and the answer should be *rounded* accordingly. Even so, *rounding errors* may mean that the final answer is not reliable to n significant figures.

5. The fact that x is an *approximation* to y is indicated by writing $x \approx y$. Cruder approximations, particularly *order of magnitude* estimates (to the nearest power of ten), can be indicated by using $\sim$ in place of $\approx$. Thus $700 \sim 1\,000$ and $0.22 \sim 10^{-1}$.

6. Any number that can be expressed as a decimal is a *real number*. Every real number r is an *element* of the *set of real numbers* $\mathbb{R}$, as is indicated by writing $r \in \mathbb{R}$.

7. Important *subsets* (indicated by the symbol $\subset$) of the set of real numbers include the set of *natural numbers* $\mathbb{N} = \{1, 2, 3, \ldots\}$, the set of *integers* $\mathbb{Z} = \{\ldots, -2, -1, 0, 1, 2, 3, \ldots\}$, the set of *rational numbers* (each of which can be written as the ratio of two integers) and the set of *irrational numbers* (each of which cannot be written as a ratio of integers). The numbers π, e, and $\sqrt{2}$ are all irrational.

8. When an algebraic symbol is used to represent a physical quantity it represents both the numerical value *and* the units of that quantity. In calculations units may be treated as though they are algebraic quantities in their own right. (SI units and conventions are summarised in Tables 2.2, 2.3 and 2.4.)

9. *Dimensional analysis* assigns appropriate combinations of basic *dimensions* (mass, length, time, etc.) to each physical quantity and uses such assignments to investigate the plausibility of relationships between those quantities. Such analysis reveals nothing about the values of purely numerical factors or *dimensionless ratios*, but an equation must be false if the quantities being equated do not have the same dimensions or if it suggests adding or subtracting quantities with different dimensions.

2.4.2 Achievements

Having completed this chapter, you should be able to:

A1. Define the terms that are emboldened in the text of the chapter.

A2. Use and interpret scientific notation, and the mathematical symbols for modulus, approximation and order of magnitude.

A3. Perform calculations that involve quantities specified to different levels of precision and present appropriately rounded answers that do not contain spurious 'significant' figures.

A4. Carry out elementary approximations and order of magnitude calculations without using a calculator.

A5. Recognise, interpret and use elementary set notation.

A6. Manipulate SI units as part of a calculation involving physical quantities.

A7. Assign dimensions to appropriately defined physical quantities and use such assignments to investigate the plausibility of proposed relationships between such quantities.

2.4.3 End of chapter questions

Question E1   (a) Express the numbers 390 463 and 0.000 705 8 using scientific notation; (b) round your answers to three significant figures; and then (c) give your answers to the nearest order of magnitude.

Question E2   Evaluate the expression $3\,678 \times 2.45 \times 10^{12}/(9.43 \times 10^{-3})$ and express the answer in scientific notation with three significant figures.

Question E3   Without using a calculator, find the approximate value and order of magnitude of the strength $F_{grav} = Gm_1m_2/r^2$ of the gravitational force on a person of mass $m_1 = 74$ kg due to another person of mass $m_2 = 56$ kg when the two are separated by a distance $r = 0.6$ m. (Note that $G = 6.67 \times 10^{-11}$ N m^2 kg^{-2} to three significant figures.)

Question E4   A wire of length l carrying a current I at right-angles to a magnetic field experiences a force of magnitude F where $F = BIl$ and B is a measure of the strength of the magnetic field. Use the fact that, in SI units, I may be measured in amperes (A), l in metres (m) and F in newtons (N) to deduce an appropriate SI unit for B.

Question E5   The equation $T = 2\pi\sqrt{r/g}$ gives the orbital period T of a satellite in a circular orbit of radius r where g is the (local) magnitude of the acceleration due to gravity. (a) Calculate the period of a satellite in an orbit with $r = 6.6 \times 10^3$ km and $g = 9.8$ m s^{-2} and express your answer with an appropriate number of significant figures. (b) Use dimensional analysis to show that $T = 2\pi\sqrt{r/g}$, is a plausible relationship between T, r and g.

Question E6   If a and b are both integers, which of the following are always true?

(a) $|a - b| = |a| - |b|$ (b) $|ab| = |a| \, |b|$

(c) $a \in \mathbb{N}$ (d) $a, b \subset \mathbb{Z}$

(e) $\mathbb{Z} \in \mathbb{R}$ (f) $a/b \in \mathbb{R}$

Functions and graphs

Chapter 3

3.1 Opening items

3.1.1 Chapter introduction

What are *functions*? Dictionaries give the following definitions:

> Function *Math* a variable quantity regarded in relation to other(s) in terms of which it may be expressed or on which its value depends.
>
> (one of four meanings)
> *Concise Oxford Dictionary*, 8th Edition (1990)
> Oxford University Press.

> Function (*math*) a relation that associates with every ordered set of numbers $(x, y, z \ldots)$ a number $f(x, y, z \ldots)$ for all the permitted values of $x, y, z \ldots$
>
> (one of six meanings)
> *Longmans English Larousse*, (1968) Longmans.

Not very helpful, you may think. If the dictionaries cannot say anything clearer than that, how will you understand? Of course the purpose of a dictionary is to define rather than explain.

This chapter provides a first step into applicable mathematics beyond the most elementary arithmetic and algebra. The idea of a function is a very general one, more general than our purposes will require. In this chapter we shall see how functions are used in scientific contexts, and how they should be regarded, thought about and described.

One very useful way of looking at and thinking about a function is to use its *graph*; this is a pictorial representation of the function, and shows a substantial part of its behaviour at once. We shall devote a good deal of time in this chapter to the study of graphs and how to draw them.

Subsection 3.2.1 of this chapter defines and explains the concept of a function in very general terms. Subsections 3.2.2 and 3.2.3 examine functions from a more restricted mathematical point of view and introduce the terminology and notation used to describe them. The rest of the chapter is mainly concerned with the representation of functions. Subsection 3.2.4 deals with representation by *tables* and *equations*, and the whole of Section 3.3 is devoted to the important topic of graphical representation. Subsections 3.3.1

to 3.3.3 are concerned with the principles and conventions of graph drawing, while Subsections 3.3.4 to 3.3.7 present a catalogue of *polynomial functions* and also cover related matters such as the analysis of *straight-line graphs*, the nature of (*local*) *maxima*, (*local*) *minima* and *points of inflection*, and the behaviour of *reciprocal functions*. Section 3.4 deals briefly with the more advanced topics of *inverse functions* and *functions of functions*.

3.1.2   Fast track questions

Question F1

Sketch a graph of the function $f(x) = x^3 - 4x$, where x is any real number. ☐

Question F2

If $H(x) = x^2 - 4x + 6$, where x is any real number, rewrite $H(x)$ in completed square form and hence find the coordinates of the vertex. Does the graph of $H(x)$ intersect the x-axis? If so, where is the intersection located? Does $H(x)$ have an inverse function? ☐

Question F3

Find the asymptotes of the graph of the function $g(x) = (x - 2)^2/(x - 1)$, where $x \neq 1$ ☐

3.1.3   Ready to study?

Study comment

In order to study this chapter you will need to understand the following terms: *element (of a set)*, *equation*, *fraction*, *inequality*, *integer*, *modulus*, *powers*, *real number*, *reciprocal*, *set*, *roots*. (These are all explained in Chapters 1 and 2; see the index for precise references.) In addition, you will need to be familiar with *SI units*, and be able to *expand*, *simplify* and *evaluate algebraic expressions* that involve *brackets*. If you are uncertain about your ability to perform these operations, you should again refer to the index for specific references. The following questions will help you to check that you have the required level of skills and knowledge.

Question R1

Which of the following are integers, and which are real numbers: 2   7.0   7   3.1 π? ☐

Question R2

Which of the following inequalities are true:

(a) $1 < 4$ (b) $2.6 > -3.6$ (c) $|-2.6| > 1.2$

(d) $|x| \geqslant 0$ (e) $-1 \leqslant -2$ □

Question R3

Evaluate the following:

(a) 5^2 (b) $(-4)^3$ (c) $\sqrt{49}$

(d) $\sqrt[3]{-2\,000}$ (e) $\sqrt{0.09}$ (f) $|5|$

(g) $|-3.2|$ □

Question R4

Simplify the following expressions:

(a) $a^2 \times a$ (b) $a^3 \times a^2$ (c) b^3/b^2 (d) $c^4 \times c^2/c^6$ □

Question R5

Simplify the expression $a^2 + 3a - 4a + 7$ □

Question R6

Expand the following expressions:

(a) $(u + 3)(u - 3)$ (b) $(p + 2)(p - 4) + (p + 1)^2$ □

3.2   Functions

3.2.1   The concept of a function

In mathematics, especially in its applications to physical science, we are often interested in the relations and connections between different numbers or sets of numbers. A *function* is a way of expressing such a connection. If we have one set of numbers, values or items of any sort, and each of these is connected to a particular number, value or item in another set by some kind of rule, then the second set is said to be a function of the first. This is a very general statement – perhaps too general to be easily understood. So it is probably best to start with some specific examples. Here are some pairs of sets together with rules that relate a single element of the second set to each element of the first set.

1. *Set 1*: {All possible values for the area of a square kitchen floor.}

    Set 2: {All possible numbers of standard sized floor tiles that you might buy.}

    Rule: Given any value for the area of the kitchen floor, the associated number of floor tiles is the number that you would need to cover that floor.

2. *Set 1*: {All possible dates in June 1995.}

    Set 2: {All possible numbers of ice creams that a vendor might sell.}

    Rule: Given any date in June 1995, the associated number of ice creams is the number that a particular vendor sold on that day.

3. *Set 1*: {All possible temperatures measured in °C.}

    Set 2: {All possible temperatures measured in °F.}

    Rule: Given any temperature in °C, the associated temperature in °F is the physically equivalent temperature.

For each of the three examples given above, we can say that the second set is a *function* of the first; the number of tiles bought is a *function* of the floor area to be covered, the number of ice creams sold is a *function* of the date in June 1995, and so on. More generally:

A **function** is any combination of two sets and a rule that meets the following conditions:
- the rule may be applied to every element of the first set
- the rule associates a *single* element of the second set with each element of the first set.

Note that although the rule allows us to go from any element of the first set to a *single* element of the second set, the reverse need not be true. For instance, if 352 ice creams were sold on 10 June no other number could have been sold on that day, but it is quite possible that 352 ice creams were also sold on 14 June.

Is the time of sunrise at your home a function of the date?

Yes. There is only one sunrise time for each day (this assumes that you don't live within the Arctic or Antarctic Circle). □

Question T1

Is your height a function of the date? □

Question T2

Suppose you live within the Arctic Circle. Is the number of sunrises you can see on any date a function of that date? ☐

Question T3

Is the date a function of the time of day? ☐

3.2.2   Variables, constants and parameters

We can talk about numbers and quantities in various ways, depending on how they behave. A number or quantity that never changes is called a **constant**; one that takes on a range of values is called a **variable**; and one that remains constant throughout a particular discussion, but may change under different circumstances is called a **parameter**. For example, consider the equation that relates the pressure P, volume V, and temperature T of a sample of ideal gas:

$$PV = NkT \tag{1}$$

where N is the number of molecules in the sample and k is a constant (called *Boltzmann's constant*). If we were to study the behaviour of a given sample at a fixed temperature, then P and V would be variables, while N, k and T could be regarded as constants. However, if we were to study the relation between P and V in the same sample at different temperatures, then we might call T a parameter, while k and N would still be constants; if we were to use different samples then N also would become a parameter though k would remain constant.

Boltzmann's constant $k = 1.38 \times 10^{-23}$ J K^{-1}.

In a second series of investigations we might treat P and T as variables and V and N as parameters though k would have to remain constant since it is, in fact, one of the fundamental constants of nature. (Not every constant has such an exalted status.)

The important point about the ideal gas equation (Equation 1) is that once we know the values of all the constants and parameters concerned, it relates one variable to another. In the first series of investigations, those carried out with a single sample of gas at a fixed temperature, Equation 1 allows us to associate a single value of P with any given value of V. Thus, for a given sample at a given temperature we may say that pressure is a *function* of volume. Similarly, in the second series of investigations – those involving a fixed volume and a variable temperature – we may say that pressure is a *function* of temperature in a given sample of fixed volume.

Clearly, the distinction between a parameter and a variable is somewhat arbitrary, since any parameter is potentially a variable. For this reason it is often easiest to regard P, V, N and T as variables (even if some remain constant under certain circumstances) and to say that in an ideal gas any one of them is a function of all the others. Thus, we can say, for instance, that the pressure of a sample of ideal gas is a function of the volume, temperature and number of molecules in that sample. This perfectly reasonable view raises an interesting question.

A function always involves *two* sets and a rule. In a given sample of ideal gas (where the number of molecules N is constant) the pressure P is said to be a *function* of the volume V and the temperature T. What are the two sets and the rule that define this particular function?

The rule is obvious, it's the ideal gas equation $PV = NkT$. The second of the two sets is equally obvious, it's the set of all possible values for the pressure. The difficult item to identify is the first set: it can't be the set of all possible temperatures since the volume may vary, so a given temperature is no longer associated with a single pressure; nor, since the temperature may vary, can it be the set of all possible volumes. Instead, the first set is the set of all possible *pairs* of values for the volume and temperature taken together. The set of all possible pairs of values (V, T) may seem like a rather strange set to have to consider, but it's the price we pay for regarding P as a function of V and T in the given sample.   □ (✆)

We could, if we preferred, regard the first set as the set of all quotients T/V formed from pairs of values for V and T but that would hardly be an improvement.

3.2.3   Dependent and independent variables

When a function relates two sets, as described in Subsection 3.2.1, we call the first of those sets the **domain** of the function and the second the **codomain**. Thus, a function associates each element in its domain with a single element in its codomain. When the domain consists of the values of a variable we call that variable the **independent variable** of the function. (The area of a kitchen floor, for example.) If the codomain also consists of the values of a variable, we call that variable the **dependent variable**. (The number of tiles needed to cover the floor, say.) These names make sense, since any allowed value of the independent variable will determine a single value of the dependent variable via the rule of the function. A special notation is used to indicate this relationship between the two variables: if the independent variable is denoted by x, and the dependent variable by y then we say that 'y is a function of x' and we write $y = f(x)$ (read as 'y equals f of x'). (✆)

Note that $f(x)$ does *not* mean $f \times x$.

Of course, there is nothing sacred about the use of x and y to represent independent and dependent variables, or the use of f to denote a function that relates them. If dealing with the pressure and volume of a given sample of ideal gas at a fixed temperature, for example, you might well write $P = h(V)$ or $P = \phi(V)$. Any letter will do to indicate the existence of a functional relationship between P and V. Scientists, much to the horror of mathematicians, tend to use the same symbol to represent both a dependent variable *and* the function that relates that variable to some independent variable. Thus, a physicist might well write $P = P(V)$ to show that P is a function of V, or $y = y(x)$ to show that y is a function of x.

How would you interpret the equation $V = V(I, R)$?

The equation indicates that the dependent variable V is a function of the independent variables I and R. In other words, a given pair of values for I and R determine a single value of V. (We have followed the physicist's convention of representing the function by V, even though this symbol has already been used to represent the dependent variable. A mathematician would probably have written the same relationship as $V = f(I, R)$.) □

A useful way of thinking about an expression such as $y = f(x)$ is to regard the symbol $f(\ )$ as a sort of machine, something like an electronic calculator, awaiting some specific numerical input. When you put a particular number into the gap between the brackets, 2.6 or -497 say, the 'machine' processes that value and produces a specific numerical output, $f(2.6)$ or $f(-497)$. It is this specific numerical output that is the value of the dependent variable which the function associates with the input value of the independent variable. In fact, when using expressions such as $y = f(x)$ or $V = V(I, R)$ to show the existence of functional relationships, whatever appears inside the brackets of a particular function is called the **argument** of the function. The important thing to remember is that what we call the independent variable, x, X, z or whatever, makes no difference, it is the *value* of the argument that really counts since that is what determines the corresponding value of the dependent variable.

3.2.4 Representing functions by tables and equations

In order to have a detailed understanding of any particular function you really need to know the two sets concerned and the rule that relates them. In very simple cases when dealing with a **discrete variable** that can only take on certain isolated values, it may be possible to tabulate every possible value for the dependent variable alongside the corresponding value for the independent variable. An example of this kind is shown in Table 3.1. Such a **table of values** certainly defines the function concerned, but it is not a technique that can easily be applied to cases where the independent variable can take on a great many values.

In practice, most of the functions that you will meet in your studies will involve **continuous variables** that cover an unbroken range of real values and therefore *cannot be defined* by a table. Such functions are usually represented by *equations*. You have already seen (in Subsection 3.2.2) that an equation *can* provide the rule needed to relate two sets of values, so it shouldn't come as a shock to learn that this is how functions are *usually* defined. For example, a statement such as

$f(x) = x^2 + x + 1$ where the domain consists of all real numbers
                       and the codomain consists of all real numbers $\geqslant 3/4$

Table 3.1 Example of a function defined by a table

Independent variable	Dependent variable
1	3
2	7
3	13
4	21
5	31
6	43
7	57
8	73
9	91

is a perfectly good definition of a function; it identifies the two sets involved and explicitly states the rule that associates a single element in the codomain with each element in the domain. In practice, you are more likely to see functions defined by equations in the following way:

$$y = x^2 + x + 1 \qquad \text{where } x \text{ is any real number}$$

The domain is still indicated, but all reference to the codomain is omitted since it is usually taken to be the full range of y values that result from applying the rule to every allowed value of x.

In a similar spirit the specific form of the function $V = V(I, R)$ that we discussed earlier might have been defined by

$$V = IR \quad \text{or} \quad V(I, R) = IR$$

where V is the voltage across an electrical resistor, I the current flowing through the resistor and R the resistance. (You may recognise this as *Ohm's law*.) Of course, the function relating V to I and R didn't have to be Ohm's law, it might have been some other relation entirely, such as

$$V = \pi I R^2$$

where V is the volume of a cylinder, I is its height and R its radius.

Although functions that involve continuous variables cannot be *defined* by tables of values it is sometimes useful to compile such a table for some 'typical' or 'representative' values of the independent variable(s). Such a table can often provide more insight into the nature of a function than the equation itself. For example, Table 3.1 (which actually defines a function in its own right) can also be regarded as a representative table of values for the function $f(x) = x^2 + x + 1$ that was introduced above. Naturally, the table is restricted to just a few of the possible values of x, but it can be useful nonetheless, as you will see in the next section.

Suppose that m may be any of the whole numbers in the range $1 \leqslant m \leqslant 9$, and n may be any of the whole numbers belonging to the set $\{3, 7, 13, 21, 31, 43, 57, 73, 91\}$. Write down an equation relating n and m that defines the same function as Table 3.1.

$n = m^2 + m + 1$. This equation is the result of restricting the values of the real variable x in the equation $y = x^2 + x + 1$ (given above) to the given integer values of m (i.e. $1 \leqslant m \leqslant 9$). $\square$

Thus, functions that involve continuous variables cannot be defined by a table, but they can be defined by an equation, and representative values can be tabulated. Functions that involve a finite number of values of discrete variables can be defined by a table of values or an equation.

3.3 Graphs

'Every picture tells a story' and 'a picture is worth a thousand words'. These are sayings that are worth remembering in mathematics as well as in everyday life. Why? Because a picture is something which can be looked at as a whole at once; it is an excellent means of investigating and summarising the behaviour of a function in its entirety. Scientists will often plot data as they collect it during an experiment, so that they can see the general behaviour, and perhaps get early notice of any problems with the apparatus.

3.3.1 Cartesian coordinates

The framework used for drawing pictures of functions is that provided by **Cartesian coordinates**. Its basic ingredient is a pair of lines, called **coordinate axes**, at right angles to each other, as shown in Figure 3.1. The two lines may be regarded as infinitely long, extending as far as we like beyond the edges of the paper. By convention, the horizontal line is called the x-axis, and the vertical line the y-axis; the two lines intersect at a point called the **origin**. The lines are scaled to indicate the displacement from the origin. For the x-axis, displacements to the right of the origin are positive, and those to the left are negative; for the y-axis displacements above the origin are positive, and those below are negative. It is common to draw an arrowhead at the end of each axis to show the direction of increasing x or y.

Any pair of values for x and y can be represented by a point in the coordinate system. The pair (a, b) is represented by the point that is at a displacement a from the origin, measured parallel to the x-axis and a displacement b from the origin, measured parallel to the y-axis (☞). The point corresponding to the pair $(2, 1)$ is shown as the point A in Figure 3.1; a perpendicular line from A to the x-axis meets it at $x = 2$, and a perpendicular line from A to the y-axis meets it at $y = 1$. Similarly the number pairs $(1, 1)$ and $(-1, 2)$, represent, respectively, the points B and C in Figure 3.1.

Question T4

What values of (x, y) represent the points D, E, F, G and O in Figure 3.1? ☐

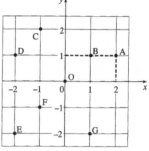

Figure 3.1 A Cartesian coordinate system and some points.

Note that the displacements referred to here are the 'scale' displacements shown on the axes.

3.3.2 Representing functions by graphs

Undoubtedly you will be familiar with the use of graphs to plot data; scientists use them all the time, and so do newspapers, but the idea of the graph of a function may be less familiar. Nonetheless, given a function $f(x)$, any allowed value of x together with the corresponding function value $y = f(x)$ forms a pair of numbers (x, y) that can be represented by a point in a Cartesian coordinate system. Figure 3.2 shows the result of plotting all such points for the function defined by Table 3.1. This plot is called the **graph** of the function.

It would not be correct to draw a curve through the points on the graph in Figure 3.2 as it is the graph of an equation with discrete variables (see Subsection 3.2.4).

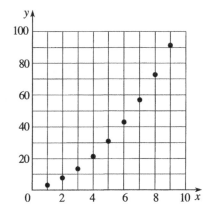

Figure 3.2 The graph of the function defined by Table 3.1.

In science we are more often interested in continuous variables. If x is a continuous variable then there will be an infinite number of points $(x, f(x))$ which can be plotted on the diagram. Consider, for instance, the function

$$f(x) = 2x + 1 \qquad\qquad (☎) \quad (2)$$

and let $y = f(x)$. For every value of x we have a corresponding value of y. For $x = 0$, $y = 1$; for $x = 1$, $y = 3$; for $x = -1$, $y = -1$ and so on; these pairs of numbers can be used to compile a representative table of values (Table 3.2) that can be used to help us plot the graph of the function $f(x)$. After plotting just a few points it soon becomes clear that all the points lie on a straight line as shown in Figure 3.3.

Similarly the function

$$g(x) = x^2 \qquad\qquad (3)$$

will give the curve shown in Figure 3.4. Here again we have taken $y = g(x)$; this is the customary way of proceeding, and in future we shall draw the graph of any function with an axis labelled as the y-axis.

From this we can see that any function may be characterised by its graph, and that a graph can show very simply many of the important

Throughout the rest of this section we will always assume that x may be any real number unless stated otherwise.

Table 3.2 Table of values for $f(x) = 2x + 1$

x	$f(x) = 2x + 1$
−2.0	−3.0
−1.5	−2.0
−1.0	−1.0
−0.5	0.0
0.0	1.0
0.5	2.0
1.0	3.0
1.5	4.0
2.0	5.0

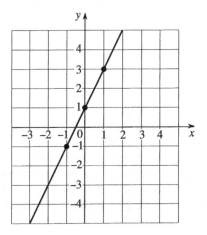

Figure 3.3 Graph of the function $f(x) = 2x + 1$.

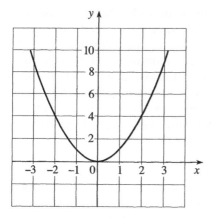

Figure 3.4 Graph of the function $g(x) = x^2$.

behavioural features of a function. It is, of course, for this reason that graphs are so commonly used both inside and outside science.

3.3.3   Drawing and sketching graphs

A graph should convey a visual message, either to you personally, or (more importantly) to anyone else who may look at it. You should therefore make sure that the information it presents is as clear and comprehensible as possible, just as you should do when writing something. Here are some points to pay attention to when drawing a graph.

Using graph paper

Graph paper is necessary if any degree of accuracy is to be maintained, but it may not be needed for a quick sketch. The grid provided by the paper will limit your choice of scale and hence the size of the graph. You should not try to use the paper completely at the expense of a sensible scale. A 1 cm square, for example, may conveniently represent an interval of 1, 2, or 5 scale units, but 4 and certainly 3 should be avoided. A single experience of reading intermediate points from such a scale should convince you of this.

We can call this the 'etiquette' of graph drawing, because it is good manners to make things easy for anyone you are communicating with – quite apart from the fact that it is always to your advantage to get your points over as clearly as possible.

Selecting the size, scale and orientation

The most important thing to do is to choose the right scales for both the horizontal and vertical axes. These should cover the whole range of interest, and a little more. The curves or points on the graph will then cover the whole area, with no unnecessary large blank areas at any of the edges. You can orientate your paper whichever way you want, but it is conventional to assign the horizontal axis to the independent variable and the vertical axis to the dependent variable.

Computer graphics programs, which plot graphs automatically, are useful but not usually very intelligent. They often produce unsuitable scales.

Marking and numbering the axes

Each axis should have the scale units indicated at reasonable intervals by graduation marks (small lines at right angles to the axis). There shouldn't be too many of these marks, but enough so that the viewer can estimate the location of intermediate points without difficulty. Similarly, the numbers written along the axes to show the scale values associated with some of the graduation marks should not be too frequent – though you will usually need at least one number for every five graduation marks. Take care to choose sensible sequences of numbers; choices such as 1, 2, 3, . . ., or 10, 20, 30, . . ., are obvious, but if the whole range is large you might prefer to use 2, 4, 6, . . ., or 5, 10, 15, ... Avoid using sequences like 3, 6, 9, . . .; even intervals of 4 may be found irritating. In brief, only display useful numbers, and try to avoid any appearance of 'fussiness' on the page.

Labelling the axes

Points that scientists need to pay particular attention to when drawing graphs are labelling axes and indicating any units that have been used in the measurement of physical quantities. If you are plotting a purely numerical variable (without physical units) you have nothing to worry about, just put the name of the variable or the symbol representing it (x or y or whatever) along the appropriate axis. However, if you are plotting values of variables such as mass or length that do require units it is usually best to label the axis as 'mass/kg' or 'length/m', as though dividing the variable by the relevant unit. It is then logical to write pure numbers along the axes rather than values that include units. If plotting very large or very small values it is generally a good idea to use multiples of units. For instance, if you have to plot masses in the range 2×10^6 kg to 2×10^7 kg it is probably best to label the axis 'mass/10^6 kg', so that the numbers are only in the range 2 to 20.

In bygone days you might have been advised to indicate units by brackets as in mass (kg). This is now regarded as potentially confusing, so you should use the mass/kg notation instead.

Joining the dots

When drawing a graph, you will only be able to plot a finite number of points through which the curve must pass, and you will then have to think about the best way of joining them up. As a general rule, the points should be joined with a pencil line in the smoothest manner possible, and there should be no kinks or discontinuities unless there are good mathematical or physical reasons for them. Knowing when to expect such kinks or discontinuities is a skill that comes with insight and experience; this chapter is only the starting point for the development of such a skill.

The figures in this chapter should give you an idea of how to set out a graph in the proper way.

So far we have talked about how to construct an accurate graph. This is usually called **plotting** the graph. However, very often this *plot* is not necessary. If you are asked to **sketch** the behaviour of a function, rather than to draw or plot it, all that is needed is a rough diagram showing the main features. It should always be possible to deduce some of the following from the defining equation of the function:

1. How does the function behave as x becomes large and positive?
2. How does the function behave as x becomes large and negative?
3. What is the value at $x = 0$ (i.e. where does it cross the y-axis)?

These characteristics are unlikely to be enough, and you will probably have to compute one or two more points; but on the whole you should try to do as few calculations as possible. Try to make use of the general characteristics of the particular type of function you are sketching; some of these are described in the following subsections, others are discussed elsewhere – particularly in the chapters devoted to differentiation (Part 4 of this book).

Question T5

Draw a graph showing the cost of posting a first class letter as a function of its mass. (In August 1993 the rates were: up to 60 g, 24p; 60 g to 100 g, 36p; 100 g to 150 g, 45p; 150 g to 200 g, 54p.) ☐

Question T6

Plot the graph of the function $f(x) = x^2 + x + 1$ for $x \geqslant 0$.

(*Hint*: remember the relationship between this function and the values in Table 3.1 or Figure 3.2.) ☐

3.3.4 Linear functions and straight-line graphs

In Subsection 3.2 you saw that the graph of the function $f(x) = 2x + 1$ was a straight line (Figure 3.3). Of course, this isn't the only function to have a graph that is a straight line. In fact, there is an entire class of functions, called *linear functions*, every member of which has a straight-line graph.

A **linear function** is any function that may be written in the form

$$f(x) = mx + c \qquad (4)$$

where m and c are constants, called the **gradient** and **intercept**, respectively.

Which of the following are linear functions? Give the values of the gradient m and intercept c for each of the linear functions.

The gradient of a line is sometimes called its slope.

(a) $f(x) = 1 + 2x$

(b) $f(t) = -6.0t - 3.8 \times 10^4$

(c) $f(x) = 6x + a$ where a is a constant

(d) $f(x) = cx + b$ where c and b are constant

(e) $f(x) = 3x^2 - 2.1$

(f) $f(x^2) = 3x^2 - 2.1$ (Pay attention to the argument of this function)

(a) $f(x)$ is linear, it is equal to $2x + 1$; $m = 2$, $c = 1$.

(b) $f(t)$ is linear; $m = -6.0$, $c = -3.8 \times 10^4$.

(c) $f(x)$ is linear; $m = 6$, $c = a$ (You may not know the value of a, but that makes no difference as long as you know it's constant.)

(d) $f(x)$ is linear; in this case the gradient is c and the intercept is b. Don't be confused by the fact that the given function includes a constant called c. That constant has no connection whatso-ever with the symbol c that is conventionally used to represent the intercept.

(e) $f(x)$ is *not* linear, so the terms gradient and intercept do not apply.

(f) $f(x^2)$ is linear, but note that it is a linear function of x^2 (though *not* a linear function of x). In other words if we let $y = f(x^2)$ and plot the graph of y against x^2 the result will be a straight line (though a graph of y against x would *not* be straight). Regarded as a function of x^2, the gradient is $m = 3$ and the intercept $c = -2.1$. □ (☜)

If you are confused by this answer just replace all occurrences of x^2 with some other symbol such as z.

The two constants that characterise any linear function are called the *gradient* and *intercept* for good reasons. To appreciate these reasons look again at Figure 3.3 − the graph of $y = 2x + 1$. Note that the straight line crosses the y-axis at $y = 1$, the value of the *intercept*. Also note that along the straight line the value of y increases twice as fast as the value of x; this factor of 2 corresponds to the *gradient* of the function which there-fore determines the steepness or inclination of the graph. These graphical interpretations of m and c are general properties, as we now show.

Given any linear function $f(x)$ the equation $y = f(x)$ will describe a straight line. Thus, we may define

---

the general **equation of a straight line**

$$y = mx + c \qquad\qquad (5)$$

where m and c are constants.

In this general case, shown graphically in Figure 3.5a, any point on the y-axis is at $x = 0$, and substituting this value of x into Equation 5 shows that the general straight line intersects the y-axis at $y = c$. Similarly, if x is increased by h (from x_1 to $x_2 = x_1 + h$, say) then Equation 5 shows that y increases by mh from

$$y_1 = mx_1 + c \quad \text{to} \quad y_2 = m(x_1 + h) + c = y_1 + mh$$

so $\dfrac{y_2 - y_1}{x_2 - x_1} = \dfrac{mh}{h} = m$

Note that Δx is a *single* quantity and *not* the product of some quantity Δ and some value of x.

This expression relating m to a change in y and to the corresponding change in x is very important. The symbols Δx and Δy, sometimes called the

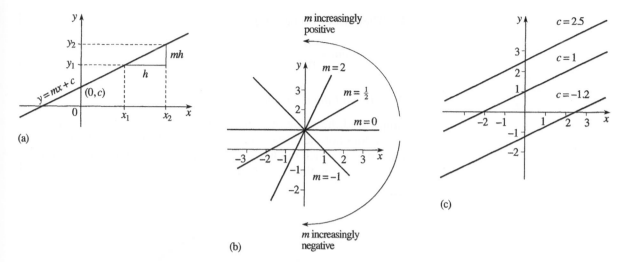

Figure 3.5 (a) The gradient−intercept form of a straight line, characterised by the constants m and c. (b) Changing the gradient m alters the steepness of the line. (c) Changing the intercept c alters the value of y at x = 0.

run and the *rise*, are used to represent the *changes* in x and y, so the gradient can be written in any of the following ways:

$$\text{gradient} = \frac{\text{rise}}{\text{run}} = \frac{\text{change in } y}{\text{change in } x} = \frac{y_2 - y_1}{x_2 - x_1} = \frac{\Delta y}{\Delta x} = m \qquad (6)$$

Figures 3.5b and c show, respectively, the effect on the graph of different values of m and c. Note that if m is zero the graph is horizontal whereas if m is negative the graph slopes downwards from left to right. The larger the value of m the 'steeper' the incline. Also note that if c is negative the intercept with the y-axis is below the x-axis.

Given a straight-line graph, how would you determine (a) the gradient and (b) intercept of the corresponding linear function?

(a) To find the gradient:

1. Pick two convenient points, (x_1, y_1) and (x_2, y_2), on the straight line.

2. Determine their horizontal separation $x_2 - x_1$. (This corresponds to the separation h in Figure 3.5a, and is often called the *run*.)

3. Determine their vertical separation $y_2 - y_1$, taking the points in the same order as before. (This corresponds to the separation mh in Figure 3.5a and is often called the *rise*.)

Remember, that all these separations are to be measured in the scale units appropriate to the graph. Thus the units of m (if there are any) will be the same as the units of the ratio y/x.

4. Then, from Equation 6,

$$\text{gradient} = \frac{\text{rise}}{\text{run}} = \frac{y_2 - y_1}{x_2 - x_1}$$

(b)   To find the intercept:

Simply find the value of y that corresponds to $x = 0$. (*Note*: if your graph does not include $x = 0$, do not make the mistake of taking the value of y at which the line crosses your vertical axis to be the intercept. It must be the y value at $x = 0$.)   □

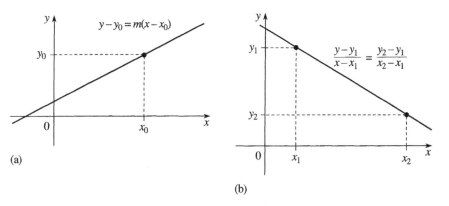

Figure 3.6 Alternative ways of representing a straight line. (a) The point–gradient and (b) two-point form of a straight line.

The **gradient–intercept form** (Equation 5) is one of the commonest ways of representing the equation of a straight line, but there are other useful representations. We shall briefly mention some of them here, but we defer their extensive discussion to Chapter 9. First, if we know the gradient m of a line and the fact that it passes through the point (x_0, y_0), then the equation of the line can be written

$$y - y_0 = m(x - x_0) \tag{7}$$

this is called the **point–gradient form** and is illustrated in Figure 3.6a.

Question T7

Rearrange Equation 7 to show that it is equivalent to Equation 5, and find the value of the intercept implied by Equation 7.   □

It is well known that given two points, (x_1, y_1) and (x_2, y_2) there is a unique straight line joining them; in other words two points determine a straight line. This fact provides the basis of another way of writing the equation of a straight line − the **two-point form** (illustrated in Figure 3.6b)

$$\frac{y - y_1}{x - x_1} = \frac{y_2 - y_1}{x_2 - x_1} \tag{8}$$

Question T8

Rearrange Equation 8 to show that it is also equivalent to Equation 5 and again find the implied value of the intercept. ☐

Yet another standard form for the equation of the straight line is

$$\frac{x}{a} + \frac{y}{b} = 1 \tag{9}$$

This is known as the **intercept form**.

Question T9

What is the graphical significance of the constants a and b in Equation 9? ☐

3.3.5 Quadratic functions and turning points

Linear functions and straight-line graphs are very common in mathematics and physical science, but there are many other functions which are also important. The *quadratic function*, a function of x which contains no higher power of x than x^2, is one such function.

A **quadratic function** is any function that may be written in the form

$$f(x) = ax^2 + bx + c \tag{10}$$

where a, b and c are constants, and $a \neq 0$.

A simple example of a quadratic function was shown in Figure 3.4; it was the function

$$g(x) = x^2 \tag{3}$$

which is obtained by setting $a = 1$, $b = 0$ and $c = 0$ in Equation 10. Two other quadratic functions, corresponding to different choices of a, b and c are shown in Figure 3.7. They are

$$h(x) = (x + 2)^2 \tag{☞) (11}$$

and $j(x) = -x^2 - 3 \tag{12}$

There are some important features that are common to all three of these examples of quadratic functions:

1. For each function there is a **turning point**, i.e. a point at which the value of the function ceases to increase or decrease and the graph turns back on itself. For quadratic functions the turning point is either a minimum or a maximum value of the function. (☞)

Figure 3.7 Graphs of the quadratic functions $h(x) = (x + 2)^2$ and $j(x) = -x^2 - 3$.

$h(x)$ may not look like a quadratic function, but if you expand the brackets you will see that it is

$$h(x) = x^2 + 4x + 4$$

Maxima, minima and turning points are defined more precisely and discussed more extensively in the chapters dealing with differentiation (particularly Chapter 16).

2.   For each function, the graph is symmetrical about a vertical line drawn through the turning point.

The graphical curve defined by a quadratic function is called a **parabola** and the turning point is called the **vertex** of the parabola.

What are the values of a, b and c (as given in Equation 10) for the functions $h(x)$ and $j(x)$?

_____

For $h(x)$; $a = 1$, $b = 4$, $c = 4$. For $j(x)$; $a = -1$, $b = 0$, $c = -3$. □

For any quadratic function, the values of the constants a, b and c determine the precise shape and location of the corresponding parabola.

For instance, if a is positive the vertex is at the 'bottom' of the parabola and thus represents the minimum value of y. If a is negative, the vertex is at the 'top' of the parabola and corresponds to a maximum value of y. Moreover, the precise value of a determines the 'width' of the parabola. The function

$$k(x) = \left(\frac{x}{2}\right)^2 \tag{13}$$

is twice as 'wide' as the function $g(x) = x^2$ for a given value of x.

Question T10

Sketch the graphs of $k(x)$ and $g(x)$ (Equations 13 and 3, respectively) and thus confirm the claim made above about the role of a in determining the 'width' of the parabola. □

The constants a, b and c also determine the location of the parabola's vertex.

Inspecting the graphs of Figure 3.7, you should be able to convince yourself that:

1.   the vertex of $h(x) = (x + 2)^2$ is at the point $(-2, 0)$; and
2.   the vertex of $j(x) = -x^2 - 3$ is at the point $(0, -3)$.

These are two special cases of a more general result, namely that the vertex of any quadratic function of the form $a(x - p)^2$ is at $(p, 0)$ and the vertex of any quadratic function of the form $ax^2 + q$ is at $(0, q)$.

Question T11

(a)   Write down a general expression for a quadratic function with a vertex at the point (p, q). (☎)

(b)   Write down the quadratic functions with $a = 1$ which have vertices at the points (i) $(1, 3)$, (ii) $(-2, 1)$ and (iii) $(2, -1)$. Sketch the graphs of these functions. □

This question is quite challenging; try to work it out for yourself but if you get stuck look at the answer.

Given any quadratic function $f(x)$, or the equation of the corresponding parabola $y = f(x)$, the items of information most often of interest are the position of the vertex and the locations of any points at which the parabola intersects the x and y axes. The position of the vertex can always be found by a process known as *completing the square*, whereas the intersections (if they exist) can be found by *factorisation*. We will deal with these processes in turn.

The **completed square form** of a quadratic function is:

$$f(x) = a(x - p)^2 + q \qquad (14)$$

where a, p and q are constant.

As indicated in the answer to Question T11, the graph of this function has its vertex located at the point (p, q).

The process of **completing the square** allows us to rewrite any given quadratic $f(x) = ax^2 + bx + c$ in the completed square form of Equation 14. To see that this is possible just note that

$$ax^2 + bx + c = a\left(x^2 + \frac{b}{a}x + \frac{c}{a}\right) = a\left[\left(x + \frac{b}{2a}\right)^2 - \frac{b^2}{4a^2} + \frac{c}{a}\right]$$

Expand the inner bracket to confirm that this is correct.

and thus,

$$ax^2 + bx + c = a\left(x + \frac{b}{2a}\right)^2 - \frac{b^2}{4a} + c \qquad (15)$$

We have now managed to isolate x within the brackets, just as in the completed square form (Equation 14). Comparing Equations 14 and 15, you should be able to see that they are the same if we make the following identifications

$$p = \frac{-b}{2a} \quad \text{and} \quad q = \frac{-b^2}{4a} + c$$

Thus, the vertex of the parabola $y = ax^2 + bx + c$ is located at the point

$$\left(\frac{-b}{2a}, \frac{-b^2}{4a} + c\right)$$

We now turn to the problem of determining the points at which the parabola $y = ax^2 + bx + c$ intersects the x and y-axes.

The intersection with the y-axis occurs when $x = 0$, so it follows from the equation of the parabola that at the point of intersection, $y = c$.

The parabola does not necessarily intersect the x-axis at all – the graph of $j(x)$ in Figure 3.7 has no such intersection, though $h(x)$ meets, rather than intersects, the axis once and one of the parabolas you drew in answering Question T11 has two such intersections. However, if the graph of $f(x) = ax^2 + bx + c$ does cross the x-axis at two points, let's say at $x = \alpha$ and $x = \beta$, then it is always possible to find those points by writing the quadratic in the so called **factorised form**:

$$f(x) = a(x - \alpha)(x - \beta) \qquad (16)$$

where a, α and β are constant.

Expanding this expression certainly gives a quadratic, since

$$f(x) = a[x^2 - (\alpha + \beta)x + \alpha\beta]$$

Moreover, if we substitute $x = \alpha$ into Equation 16 you can see that $f(\alpha) = 0$ and similarly if $x = \beta$, $f(\beta) = 0$. So the graph of this particular quadratic does indeed intersect the x-axis at $x = \alpha$ and $x = \beta$.

A general quadratic may always be written in the factorised form of Equation 16. The process by which the factorised form of a quadratic is determined is called **factorisation**. Sometimes, in very simple cases, it is possible to deduce the factorised form simply by 'inspecting' the original quadratic (this becomes easier with experience). More generally, it is always possible to use the following formula to find the values of x at which points of intersection occur, and hence the values of α and β.

If the parabola $y = ax^2 + bx + c$ intersects the x-axis, it does so at the points

$$x = \frac{-b \pm \sqrt{b^2 - 4ac}}{2a} \qquad (17)$$

The symbol $\pm$ is read as 'plus or minus' and reminds us that this single equation generally provides *two* values for x.

It is worth noting that this formula provides the values of α and β in the factorised form of a quadratic (Equation 16) even when α and β do not correspond to distinct points of intersection with the x-axis. The quantity $b^2 - 4ac$ that appears under the square root symbol in Equation 17 is called the **discriminant** and it is this that determines the number of times the graph of the quadratic function $ax^2 + bx + c$ meets or crosses (intersects) the x-axis.

1. If $b^2 - 4ac > 0$, there will be two crossing points with x-coordinates given by Equation 17.
2. If $b^2 - 4ac = 0$, there is only a single meeting point at $x = -b/(2a)$.
3. If $b^2 - 4ac < 0$, there is no real number equal to $\sqrt{b^2 - 4ac}$ and the parabola will not meet or cross the x-axis at all.

It is also worth noting that, since $y = 0$ when the parabola intersects the x-axis, the two values of x given by Equation 17 must satisfy an equation of the following form: (☞)

$$ax^2 + bx + c = 0 \qquad (18)$$

You can check this by substituting either of the values for x given in Equation 17 into Equation 18.

Equations of this kind are called **quadratic equations**. They are common in physical science and are dealt with in more detail in Chapter 4.

Question T12

For each of the following quadratic functions, determine the number of times its graph meets or crosses the horizontal axis, rewrite the function in completed square form and thus determine the location of the vertex:

(a) $f(x) = 3x^2 - 9x + 11$ (b) $f(t) = -t^2 - 2t - 6$

(c) $f(R) = -3R^2 + 15R - 18$ ☐

Question T13

For each of the following quadratic functions, determine the points (if there are any) at which its graph intersects the horizontal axis:

(a) $f(y) = y^2 + 5y - 1$ (b) $f(t) = 2t^2 - 3t + 4$

(c) $f(Z) = 3Z^2 + \dfrac{Z}{2} - \dfrac{1}{4}$ (d) $f(x) = x^2 - 5x + 6$ ☐

3.3.6 Polynomial functions and points of inflection

Linear and quadratic functions are simple examples of a wider class of functions that may involve higher powers of a variable.

A **polynomial function** of **degree** n is any function of the form

$$f(x) = a_0 + a_1 x + a_2 x^2 + \ldots + a_{n-2} x^{n-2} + a_{n-1} x^{n-1} + a_n x^n \qquad (19)$$

where n is an integer, and the $n + 1$ constants $a_0, a_1, a_2, \ldots a_{n-2}$, a_{n-1} and a_n are called the **coefficients** of the polynomial, with a_n not equal to zero.

The constant coefficients have been denoted by the subscripted symbols a_0, a_1, a_2, etc. to make it easier to associate them with the appropriate powers of x.

Thus, a polynomial function of x of degree n involves powers of x up to and including x^n but no higher powers. A linear function is a polynomial of degree 1, and a quadratic function is a polynomial of degree 2.

> Polynomial functions of degree 3 and degree 4 are called **cubic functions** and **quartic functions**, respectively. Write down general expressions for such functions similar to those given earlier for linear and quadratic functions, Equations 4 and 10, respectively.
>
> ---
>
> $f(x) = a + bx + cx^2 + dx^3$ is a cubic function
>
> $f(x) = a + bx + cx^2 + dx^3 + ex^4$ is a quartic function. □

Such functions, not surprisingly, are much more complicated, and show more varied behaviour, than linear and quadratic ones. The simplest cubic function:

$$f(x) = x^3 \tag{20}$$

has the graph shown in Figure 3.8. The two features to note about this are:

1. The behaviour of the graph far from the origin, where $f(x) \gg 0$ for $x \gg 0$, and $f(x) \ll 0$ for $x \ll 0$.
2. The behaviour near the origin, where the graph changes from a downward turning curve for $x < 0$, to an upward turning curve for $x > 0$. This behaviour is typical of cubic functions and is often seen in other polynomials; the point at which the change of curvature occurs (the origin in this case) is called the **point of inflection**.

Other cubic functions with different coefficients can show more complicated behaviour. For example, consider the cubic function

$$g(x) = x^3 - 4x \tag{21}$$

The graph of this, shown in Figure 3.9, has the extra features of a (local) minimum near $x = 1$ and a (local) maximum near $x = -1$, in addition to the

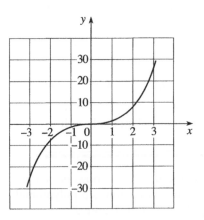

Figure 3.8 The cubic function $f(x) = x^3$.

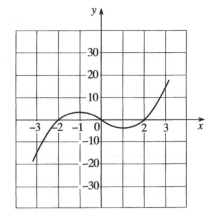

Figure 3.9 The cubic function $g(x) = x^3 - 4x$.

point of inflection at $x = 0$. This is the greatest degree of complication which arises with cubic functions; different choices for the coefficients a, b, c and d would alter the detailed shape of the curve and the locations of the point of inflection and the turning points (if there are any), but no choice of coefficients results in more than one point of inflection and two turning points.

Clearly when dealing with higher degree polynomials the complications will become even worse, so we shall not pursue individual cases any further here.

The only general points to be noted are that for a polynomial function of degree n:

- There will be at most a total of $(n - 1)$ local maxima and local minima.
- Between every local minimum and its neighbouring local maximum there will be a point of inflection.
- If n is even there will always be at least one maximum or minimum.
- If n (greater than 1) is odd there will always be at least one point of inflection.

3.3.7 Reciprocal functions and asymptotes

Although polynomial functions are important, they are not the only kind of function you are likely to meet. Another very common function is the **reciprocal function**

$$R(x) = 1/x \quad \text{where } x \neq 0 \qquad (\text{☞}) \ (22)$$

the behaviour of which is shown in Figure 3.10. The shape of this curve is known as a **hyperbola** and shows a number of interesting features:

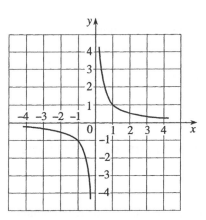

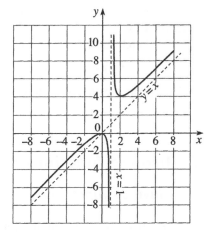

Figure 3.10 The hyperbola $R(x) = 1/x$.

Figure 3.11 The graph of $g(x) = x^2/(x - 1)$ and its asymptotes.

The symbol $\neq$ should be read as 'is not equal to'.

- The function $R(x)$ is not defined for $x = 0$. So in this case the argument x may be any real number *except* 0.
- The curve consists of two separate pieces, one for $x > 0$, and one for $x < 0$.
- As x approaches zero, $R(x)$ becomes large and positive if $x > 0$, but it becomes large and negative if $x < 0$.

The y-axis in Figure 3.10 forms a sort of limit which the curve approaches but never meets, no matter how far it is extended in either direction. Such a limiting line is called an **asymptote**, and the curve is said to approach it **asymptotically**. In this case the x-axis also forms an asymptote, since the curve never meets that either, though it approaches the asymptote more and more closely as x becomes increasingly positive or negative.

Asymptotes do not have to be vertical or horizontal. Consider the function

$$g(x) = x^2/(x - 1) \quad \text{where } x \neq 1 \tag{23}$$

and its graph in Figure 3.11. From what was said above it should be clear that $x = 1$ is a vertical asymptote. There is no horizontal asymptote in this case, but if we make x very large and positive, then the denominator $(x - 1)$ is very nearly equal to x, so that $g(x) \approx x$, and the larger x becomes, the closer the curve comes to the line $y = x$, which is an asymptote. We can use the same argument as x becomes large and negative.

When investigating the asymptotes of a function we are bound to be interested in some quantity (either x or y, usually) that is becoming either very large and positive or very large and negative. To aid such discussions it is useful to introduce the **infinity symbol** ∞ which is usually read as 'infinity'. This symbol should not be thought of as a number; rather it represents a quantity that is much larger than any other quantity that is likely to be considered. When discussing asymptotes we can then discuss the behaviour as x or y approaches ∞ or $-\infty$. For simplicity this is often written $x \to \infty$ or $x \to -\infty$. You should avoid writing $x = \pm\infty$ since, as already stated, ∞ is not a number.

Question T14

(a) Sketch the curves and asymptotes for the function

$$f(x) = x/(x + 1) \quad \text{where } x \neq -1 \tag{24}$$

(b) What are the asymptotes of $g(x) = 2x^2/(3x + 1)$? □

3.4   Inverse functions and functions of functions

3.4.1   Inverse functions

The statement $y = F(x)$ clearly indicates that y is a function of x, but sometimes it is very useful to look at things the other way round, and treat x as a function of y. As you will see shortly, it is not always possible to do this,

The symbol $\approx$ should be read as 'is approximately equal to'.

but if it can be done the process will define a new function called the *inverse function* of $F(x)$. Formally:

The **inverse function** of $F(x)$ is a function $G(y)$ such that if $y = F(x)$, then $x = G(y)$ for every value of x in the domain of $F(x)$.

===

Loosely speaking the effect of the function $F(x)$ is 'undone' by its inverse function $G(y)$.

Examples of inverse functions are easy to find. For instance in Subsection 3.3.2 we examined the function

$$f(x) = 2x + 1 \qquad (2)$$

the graph of which was shown in Figure 3.3. The inverse function of $f(x)$ is

$$g(y) = \frac{1}{2}(y - 1) \qquad (25)$$

Some authors indicate the inverse function of $F(x)$ by $F^{-1}(x)$. Due to the danger of misinterpreting this as $1/F(x)$ we prefer to use a totally different symbol such as $G(x)$.

Confirm the claim that $g(y)$ (Equation 25) is the inverse of $f(x)$ (Equation 2) for the specific values $x = -2$, $x = 0$ and $x = 3$.

When $x = -2$, $\quad y = f(-2) = 2(-2) + 1 = -3$,

and when $\quad y = -3$, $g(-3) = (-3 - 1)/2 = -2$.

When $x = 0$, $\quad y = f(0) = 2(0) + 1 = 1$,

and when $\quad y = 1$, $g(1) = (1 - 1)/2 = 0$.

When $x = 3$, $\quad y = f(3) = 2(3) + 1 = 7$,

and when $\quad y = 7$, $g(7) = (7 - 1)/2 = 3$.

So, for these three values of x at least, the functions $f(x)$ and $g(y)$ satisfy the requirement that $g(y)$ should 'undo' the effect of $f(x)$. $\quad\square$

The relationship between the functions $f(x)$ and $g(y)$ is perhaps more easily understood by noting that if

$$y = 2x + 1 \qquad (26)$$

then,

$$x = \frac{1}{2}(y - 1)$$

So, in this particular case the form of the inverse function can be found by simply rearranging Equation 26. (Not all cases are so simple.) More significantly, if you examine the graph of $g(y)$, shown in Figure 3.12, and compare it with the graph of $f(x)$, shown in Figure 3.3, you will see that the two graphs are related by a simple interchange of axes.

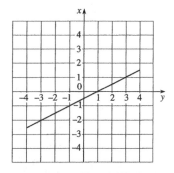

Figure 3.12 The graph of the function $g(y) = (y - 1)/2$, the inverse of the function $f(x) = 2x + 1$ shown in Figure 3.3. (Note the *axes*.) Of course, the domains of $g(y)$ and $g(x)$ also have to be the same.

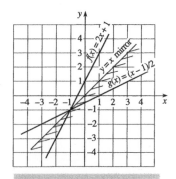

Figure 3.13 The function $f(x) = 2x + 1$ and its inverse $g(x) = (x - 1)/2$. These two lines are mirror images of each other in an imaginary mirror along the line $y = x$.

Although the inverse of $f(x)$ has been called $g(y)$ and its graph in Figure 3.12 has been plotted with y (the independent variable in this case) along the horizontal axis there is no need to show it in that way. Remember, given any function, its value is determined by the value of its argument — what you call the argument makes no difference. Thus, given

$$g(y) = \frac{1}{2}(y - 1) \qquad (25)$$

we can represent exactly the same function by

$$g(x) = \frac{1}{2}(x - 1)$$

Thanks to this, there is no need to introduce unconventional axes, such as those in Figure 3.12, when drawing the graph of an inverse function. In fact, given a function $f(x)$ and its inverse function $g(x)$ we can plot both functions on the same axes. Doing so, as in Figure 3.13, reveals an interesting phenomenon — the line $y = g(x)$ representing the inverse function can always be obtained by 'reflecting' the line $y = f(x)$ in an imaginary mirror placed along the line $y = x$. (This imaginary mirror is also indicated in Figure 3.13.) This 'reflection symmetry' between a function and its inverse is a general property, so given the graph of any function you can easily visualise the graph of its inverse — if that inverse exists.

The proviso 'if that inverse exists' is an important one. Any graph may be 'reflected' but not every such reflection is the graph of an inverse function. For example, the function

$$g(x) = x^2 \qquad (3)$$

was introduced in Subsection 3.3.2. (Do not confuse this function with the linear function $g(x)$ that was considered above. This is a different example.) The graph of Equation 3 is the parabola that was plotted in Figure 3.4. The reflection of that graph in an imaginary mirror along the line $y = x$ is shown in Figure 3.14a. Now you might be tempted to say that this is the graph of the function

$$f(x) = \sqrt{x} \qquad (27)$$

and that it is the inverse of $g(x)$ since it 'undoes' the effect of $g(x)$. However, a little more thought soon shows that the latter part of this claim cannot be true; $f(x)$ does not really 'undo' the effect of $g(x)$. If $x = -2$ then $g(-2) = (-2)^2 = 4$ but $f(4) = \sqrt{4} = 2$ (remember our convention that $\sqrt{x}$ is positive) so we are not unambiguously led back to the initial value of x. Moreover, the graph itself (Figure 3.14a) makes it clear that there are *two* values of y for each positive value of x and no values of y at all corresponding to negative values of x. The fact that two different values of y correspond to a single value of x means that Figure 14a cannot be the graph of any function (since a function relates each value of the independent variable to a *single* value of the dependent variable). In fact, the function $g(x) = x^2$ does not have an inverse.

Historically, the possibility of **multi-valued functions** of the kind represented by Figure 3.14a has been recognised and you may well encounter it. However, most modern definitions of 'function', including that given in Subsection 3.2.1, ensure that all functions are **single-valued functions**. Hence the term 'single-valued function' is actually a tautology and 'multi-valued function' is a misnomer.

Although the function $g(x) = x^2$ has no inverse when its domain consists of the set of all real numbers (positive and negative), it is possible to find closely related functions, with more restricted domains that do have well defined inverse functions. One such function involves the squares of non-negative numbers only and may be written

$$g_+(x) = x^2 \quad \text{where } x \geq 0$$

Another involves the squares of non-positive numbers and may be written

$$g_-(x) = x^2 \quad \text{where } x \leq 0$$

In either of these cases each value of the function (i.e. each element of the codomain) is associated with a different value of x (i.e. a unique element in the domain). As a result $g_+(x)$ and $g_-(x)$ have 'single valued' inverse functions that may be defined as follows

$$f_+(x) = \sqrt{x} \quad \text{for all } x \geq 0$$

$$\text{and} \quad f_-(x) = -\sqrt{x} \quad \text{for all } x \geq 0 \quad (\text{☞})$$

The graphs of these inverse functions are shown in Figures 3.14b and c.

The need to restrict the domain of a function in order to ensure that it has an inverse is quite common. It is also common for students to impose such a restriction part-way through a calculation, and then to forget that they have done so, leading to subsequent errors. Take care not to repeat this common mistake.

The general conclusion to be drawn from the above discussion is this:

In order for a function $F(x)$ to have an inverse, a necessary condition that must be satisfied is that each value of $F(x)$ must correspond to a unique value of x.

Question T15

Look at Figures 3.8 and 3.9, illustrating the cubic functions defined by Equations 20 and 21, respectively. Do these functions have inverses? If so, how might they be defined? □

Question T16

Would you expect a general quadratic function of the form $f(x) = ax^2 + bx + c$ to have an inverse? Explain your answer. □

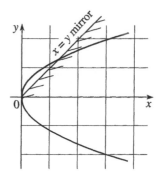

(a)

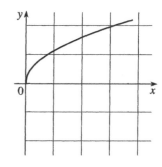

(b)

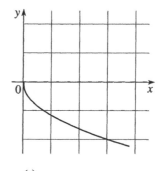

(c)

Figure 3.14
(a) The reflection of $y = x^2$.
(b) The curve $y = |\sqrt{x}|$ for $x \geq 0$. (c) The curve $y = -|\sqrt{x}|$ for $x \geq 0$.

3.4.2   Functions of functions

It is often convenient to combine functions together; that is to start with an independent variable x, find the value of some function $y = f(x)$ and then use that value of y as the argument of some other function $z = g(y)$. This two-step process allows us to relate a single value of the dependent variable z to each value of the independent variable x, so z is related to x by a function. We can indicate this by writing $z = h(x)$ where

$$h(x) = g(f(x))$$

The function $h(x)$ is said to be a **function of a function** or a **composite function**. The idea is more common and more straightforward than it sounds; for example,

if $f(x) = x^2$ and $g(y) = 1/y$ where $y \neq 0$

then $h(x) = g(f(x)) = g(x^2) = 1/x^2$ where $x \neq 0$

Note that in the final step we have simply taken the reciprocal of the argument of $g(x^2)$ – this is what the function $g(y) = 1/y$ tells us to do – it doesn't matter that the argument has been called x^2 rather than y, simply take its reciprocal!

Question T17
> Find $g(f(x))$ and $f(g(x))$ if $f(x) = x + 2$ and $g(x) = 1/x$. What is the largest possible domain for such a function, given that x is real?   ☐

Question T18
> Find $g(h(f(y)))$ for the same $f(x)$ and $g(x)$ as in Question T17, and $h(x) = x^2$.   ☐

Question T19
> Suppose $G(x)$ is the inverse of $F(x)$. Write down an explicit expression for the composite function $H(x) = G(F(x))$. (You do not need to know the explicit form of either function to answer this question.)   ☐

3.5   Closing items

3.5.1   Chapter summary

1. A *function* consists of two sets and a rule, related in such a way that each element of the first set (the *domain*) is associated with a single element of the second set (the *codomain*).

2. The rule involved in a particular function is often presented in the form of an equation that may involve a number of *variables*, *constants* and *parameters*.

3. The sets involved in a particular function are often the sets of values of specific variables which may be *continuous* or *discrete*. Under these circumstances the values belonging to the first set (the domain) are said to be values of the *independent variable(s)* while the associated values belonging to the second set (the codomain) are said to be values of the *dependent variable*. This sort of functional relationship may be indicated by the equation $y = f(x)$.

4. A *table of values* may be used to define a function of a discrete variable or to provide insight into the behaviour of a function of a continuous variable.

5. A *graph*, in which corresponding values of the independent and dependent variables are plotted as points on a pair of mutually perpendicular *coordinate axes*, provides a useful way of representing functions and equations. There are many standard conventions that apply to the drawing of graphs.

6. *Linear functions* of the form $y = mx + c$, where m is the *gradient* and c the *intercept*, have graphs that are *straight lines*. Graphically, m ($= rise/run$) determines the gradient of the line and c determines the point at which it intersects the y-axis. A straight line can be represented using a number of other forms, i.e. the *gradient−intercept*, *point− gradient*, *two-point* and *intercept forms*.

7. *Quadratic functions* of the form $f(x) = ax^2 + bx + c$ have *parabolic* graphs that each have a single *turning point* known as the *vertex* of the parabola. The constants a, b and c determine the location of the vertex,

$$\left(\frac{-b}{2a}, \frac{-b^2}{4a} + c\right),$$

and the precise shape of the graph, including any values of x at which the curve meets or crosses (intersects) the x-axis,

$$x = \frac{-b \pm \sqrt{b^2 - 4ac}}{2a}.$$

8. *Polynomial functions* of *degree n* have the form

$$f(x) = a_0 + a_1x + a_2x^2 + \ldots + a_{n-2}x^{n-2} + a_{n-1}x^{n-1} + a_nx^n$$

(with $a_n \neq 0$)

Their graphs generally exhibit (*local*) *maxima* and (*local*) *minima* separated by *points of inflection*.

9. The *reciprocal function*, $y = 1/x$, has a graph that is a *hyperbola* and which exhibits *asymptotes* – lines that are approached by the graph as x or y approach large positive or negative values.

10. Given a function $F(x)$, its *inverse function $G(x)$* 'undoes' the effect of $F(x)$. So, if $F(x) = y$ then $G(y) = x$ for every value of x in the domain of $F(x)$. For such a function to exist, each value of $F(x)$ must correspond to a different (unique) value of x.

11. A *function of a function* or a *composite function* is a function of the form $h(x) = g(f(x))$.

3.5.2 Achievements

Having completed this chapter, you should be able to:

A1. Define the terms that are emboldened in the text of the chapter.

A2. Describe the essential features of a function.

A3. Explain the roles of constants, parameters and variables in a function.

A4. Use equations to define functions and identify the independent and dependent variables in such definitions.

A5. Set up a Cartesian coordinate system and use it to represent the properties of a function.

A6. Plot an accurate graph of a given function.

A7. Draw the graph of a linear function given the gradient and intercept, and find the gradient and intercept of a given graph.

A8. Draw the graph of a linear function given either one point on the line and the gradient, or two points.

A9. Sketch the graph of a quadratic function, rewrite the function in completed square form and hence identify the coordinates of its vertex, and determine the number and location of any points at which the graph intersects the axes.

A10. Sketch the graph of a cubic polynomial function, identify its point of inflection and any local maximum, or local minimum, that it may have.

A11. Draw the graph of the reciprocal function, and identify asymptotes in simple cases.

A12. Recognise inverse functions (in simple cases), and distinguish between functions that may or may not have inverse functions.

A13. Describe and identify the domain and codomain of a function.

A14. Combine functions to produce a function of a function (in simple cases).

3.5.3 End of chapter questions

Question E1 A function $f(x)$ is defined by $f(x) = x$ for $x \geqslant 0$, and $f(x) = -x$ for $x < 0$. Plot a graph of $f(x)$ for values of x between -4 and $+4$.

Question E2 Sketch graphs of the following functions:

$$f(x) = 2x$$

$$f(x) = 2x + 3$$

$$f(x) = 2x - 1$$

$$f(x) = -\tfrac{1}{2}x$$

Question E3 What is the equation of the straight line (in the form $y = mx + c$) which passes through the point $(3, 5)$ with gradient 3?

Question E4 Give the equation (in gradient−intercept form) of the straight line which passes through the points $(-1, 5)$ and $(2, -1)$.

Question E5 Three quadratic functions are defined by the following:

$$f(x) = \tfrac{1}{2}x^2 - 3$$

$$g(x) = -\tfrac{1}{2}x^2 + 1$$

$$h(x) = x^2 + 2x + 2$$

What are the coordinates of their vertices? Which of the curves intersect the x-axis and where do such intersections occur? Sketch the graphs of the three functions on a single pair of axes.

Question E6 Sketch the graphs of $y = (x - 1)^3$ and $y = x^3 - 9x$.

In each case, how many times does the curve cross the x-axis? Where are the points of inflection located?

Question E7 The vertical velocity of a stone, thrown vertically upwards from the ground, is given by

$$v = u_0 - gt$$

where u_0 is the initial velocity of the stone and g is a constant. Assuming $g = 10$ m s^{-2} and $u_0 = 20$ m s^{-1} upwards, draw a properly labelled graph showing v as a function of t for the first 4 seconds of flight.

Question E8 At time t, the stone mentioned in Question E7 is at a height h above the ground, where

$$h = u_0 t - \tfrac{1}{2}gt^2$$

Draw a graph of h as a function of t and find (a) the maximum height the stone reaches, and (b) the time before it hits the ground. What features of the graph determined your answers?

Question E9   Sketch the graph of the function

$$f(x) = |\sqrt{x^2 + 1}|$$

Does this curve have asymptotes? Does $f(x)$ have an inverse function?

Question E10   If $f(x) = (x - 1)^2$ and $g(x) = x + 1$, what are $f(g(y))$ and $g(f(y))$? If y is real, what are the largest possible domains for these functions?

Solving equations

4.1 Opening items

4.1.1 Chapter introduction

When we do anything in physical science, there comes a point when we have to produce numbers; either as the result of an experimental measurement, or as the prediction of a theory, or both. Almost always this means having to do some calculations; not just adding numbers, or multiplying them, but manipulating them in some way so as to produce an answer; often this involves *solving* equations. In this chapter we will look at how to solve the simplest (and commonest) equations, and outline some of the more generally applicable ways in which we can solve more difficult problems.

Initially, we will look at solving *linear equations* (in Section 4.2), i.e. equations in which the independent variable appears only as the first power. In Section 4.3 the discussion is extended to the solution of *quadratic equations* by various methods, e.g. *factorising, completing the square* and using the quadratic equation formula. Finally, in Section 4.4 graphical and numerical procedures for solving *polynomial equations* are described.

The first thing you must do before starting this chapter is to **PUT YOUR CALCULATOR AWAY**. You will not need it for most of this chapter, and you will be told when you do need it. Calculators are very useful for doing calculations to great accuracy, but they can stop you thinking about the important features of calculations; they are often a good means of making mistakes quickly!

4.1.2 Fast track questions

Question F1

Find the value of x that satisfies the following equation:

$$\frac{x - 1}{x + 1} = 2$$

Question F2

Solve the following equations:

$$x + 2y = 4$$

$$2x - y = 3$$

Question F3

Find the roots of the following equation

$$x^2 + x - 1 = 0$$

leaving the square roots in the answer. How can you verify that your answers are right?

4.1.3   Ready to study?

Study comment

In order to study this chapter you will need to understand the following terms: *constant, equation, function, graph, powers, square root* and *variable*. In addition, you will need to be able to *expand, simplify* and *evaluate* simple *algebraic expressions* that involve *brackets*. You will also need to be able to *sketch* and *plot* the graphs of simple functions. The following *Ready to study questions* will help you to check that you have the required skills and knowledge before embarking on this chapter.

Question R1

Expand the following expressions:

(a) $(a - 2b)(a + 2b)$ (b) $(x + 3)(x - 4)$ (c) $(p - 3)^2$

Question R2

Simplify the expression $2(x^2 + 3x - 2) - (2x^2 - 6x - 4)$.

Question R3

If $f(x) = 3x + 2$, write down expressions for $f(2y)$ and $f(y - 1)$.

Question R4

What are the square roots of x^4?

Question R5

Sketch graphs of the following functions from $x = -1$ to $x = 3$:

$$f(x) = 2x - 1 \quad \text{and} \quad g(x) = x^2 - 3x + 2$$

Sketch means draw a rough graph, showing the salient features – it does not mean plot the graph accurately.

Question R6

Rewrite, with a single denominator, the expression

$$\frac{1}{1-x} - \frac{1}{1+x}$$

The *denominator* is the number or expression on the bottom of a fraction.

4.2 Solving linear equations

When we attempt to solve an equation for an **unknown variable**, x say, we aim to find an expression which looks something like this

$$x = ??? \tag{1}$$

where the right-hand side does not involve x and is something we can calculate immediately. When we have reached this point the mathematics is done and arithmetic takes over.

In this section we shall look at **linear equations**, which only involve the first power of the unknown variable, x, rather than x^2 or any higher power. In other words we will look at equations such as

$$3x + 2 = 5x - 2$$

which are said to be in **linear form**. We shall attempt to find the value (or values) of the unknown variable for which the equation is *true*. In the case of the above equation it is easy to see that the value $x = 2$ makes the equation true since for this value the left-hand side becomes $3 \times 2 + 2 = 8$, while the right-hand side is $5 \times 2 - 2 = 8$. We call this process **solving the equation**, and we speak of finding the **solution**, or finding the **root** of the equation. Later we will extend these notions to more than one equation. Equations that can be solved are sometimes said to be **soluble**, while those that can't are **insoluble**.

4.2.1 Simple linear equations; solving by rearrangement

Suppose you set your stop-watch to zero as you start cycling along a road at 10 mph. After you have travelled 15 miles you stop for a rest. How long has this taken you?

The distance you have travelled after a period of time t is

$$d = vt$$

distance = speed × time

where v is your speed. In this example we know what d and v are, and we have to find t. Dividing both sides of the above equation by v we see that

$$t = \frac{d}{v} = \left(\frac{15}{10}\right) \text{ hours}$$

Thus we have reached the point at which arithmetic takes over and the final answer is 1.5 hours. This simple problem was solved by rearranging the equation.

Now for a slightly more complicated problem. Forget for the moment that $d = 15$ miles and $v = 10$ mph, and think about the following problem in purely algebraic terms. Suppose that having cycled a distance d at speed v you then rested for a time t_R, and that following your rest you set out again on a second ride at speed v. If you kept cycling until your total journey time (including the rest) was T, what distance would you have covered in that time?

The time spent on the second ride is found by subtracting the rest time and the time of the first ride from the total journey time, so it is $(T - d/v - t_R)$; and the total distance D covered in both rides is just the sum of the distances covered in each ride, so

$$D = d + v(T - d/v - t_R) \tag{2}$$

It is easy to rearrange this equation to find an expression for T in terms of the total distance travelled, the speed, the distance d and the rest time. The form of this expression may already be obvious to you, but being systematic and spelling everything out in detail we can start by trying to isolate the bracket that contains T. To do this note that from Equation 2

$$D - d = v(T - d/v - t_R)$$

so that $(T - d/v - t_R) = (D - d)/v$

Thus, $T = d/v + t_R + (D - d)/v = d/v + t_R + D/v - d/v$

You can see that d/v cancels out on the right-hand side of the last equation, so

$$T = t_R + D/v \tag{3}$$

Now we can substitute in some numbers; suppose the total distance travelled is 35 miles and $v = 15$ mph; if the rest time was half an hour then $t_R = 0.5$ hours and

$$T = \left(\frac{1}{2} + \frac{35}{10}\right) \text{ hours} = (0.5 + 3.5) \text{ hours} = 4 \text{ hours}$$

You may be wondering why we didn't put these numbers in at the beginning. There are three reasons for this; first, if you look again at Equation 3 you will see that the variable d has disappeared, so less calculation is needed than if we had used numbers all the way through. Second, having worked out the general equation, we can change the distances, or the speed, and still find the time taken without having to go through the whole calculation again. Finally, it is often easier to check a calculation that involves algebraic variables rather than numerical values.

You might have realised that we could have moved directly to Equation 3 by equating the total time T to the sum of the rest time t_R and the total cycling time D/v.

Suppose you cycled non-stop for 36 miles at 12 mph. How long would this have taken?

We just need to substitute $t_R = 0$, $D = 36$ miles, $v = 12$ mph in Equation 3 to obtain

$$T = (0 + 36/12) \text{ hours} = 3 \text{ hours} \quad \square$$

This example provides us with a rule that applies throughout mathematics and physical science. To avoid unnecessary effort and to reduce the chance of mistakes:

Don't introduce numbers until it is essential; leave all the arithmetic until the end.

Let us now return to the business of solving equations. The following method applies to any equation in which the unknown variable appears to the first power only, and only in the numerator. The procedure in general is much like the one described above; for brevity let us suppose the unknown variable is x.

> The *numerator* is the number or expression on the top of a fraction.

To solve a linear equation, e.g. $3(x - 1) = 2x + 5(2 - x)$

1. if x appears in more than one bracket, expand the brackets; $3x - 3 = 2x + 10 - 5x$

2. collect together all the terms involving x on one side of the equation and the remaining terms on the other; $3x + 5x - 2x = 10 + 3$

3. combine the **coefficients** of x as far as possible (i.e. simplify all the terms multiplying x), and then simplify all the remaining terms; $6x = 13$

4. you should now have an equation that can be written in the form of Equation 1 with no more than one more step. $x = \dfrac{13}{6}$

With practice it is possible to condense this process by combining the steps but, be warned, doing too many steps in your head is likely to lead to mistakes. Try the next four questions, all of which can be solved by the above method.

Question TI

Solve the equation

$$2(x + 1) = x - 5 \qquad \square$$

Question T2

Solve the equation

$$(z - 2) + 3(2z + 1) - (5z - 7) = 0 \qquad \square$$

Question T3

If $a = 1$ and $b = 4$, find the value of p that satisfies

$$p - a = 3p + 2b \qquad \square$$

Question T4

If $a = 4$ and $b = -2$, find the value of p that satisfies

$$p - a = 3p + 2b \qquad \qquad \square$$

Always check that your answers satisfy the original equations and that they make sense.

It is sometimes possible to solve what appear to be more complicated problems in the way described above. The general principle is to reduce more complicated equations to a form that we know how to solve. For example,

$$\frac{x - 1}{x + 2} = 3 \tag{4}$$

can be rewritten, if both sides are multiplied by $(x + 2)$, as

$$x - 1 = 3(x + 2)$$

This is not a linear equation in x.

and then treated as before. However, it is not always possible to reduce such an equation to linear form.

Try to simplify the following:

$$\frac{x - 1}{x + 2} = \frac{3}{x}$$

If we multiply by $(x + 2)$ and by x, to obtain

$$x(x - 1) = 3(x + 2)$$

we are left with a term in x^2, which cannot be removed. We shall see later that such equations can be solved, but for the moment we stop here. $\square$

There are many equations that appear at first sight to be beyond the realm of linear equations, but which can be simplified to an unexpected degree. Look, for example, at the following

$$\frac{x - 1}{x + 2} = \frac{x + 3}{x - 1}$$

If we multiply both sides of the equation by $(x + 2)(x - 1)$ we have

$$(x - 1)^2 = (x + 3)(x + 2)$$

which implies

$$x^2 - 2x + 1 = x^2 + 5x + 6$$

i.e. $\quad 7x = -5$

so $\quad x = -5/7$

However, events like this should be regarded as just a lucky accident, and are not to be relied on − though it is legitimate to hope!

Question T5

Which of the following equations can be reduced to a linear form? Solve those which can.

(a) $\dfrac{1}{x-1} = 6$

(b) $x^2 - 1 = \dfrac{2}{x+1}$

(c) $x^2 + 2 = x(x-3)$

(d) $\dfrac{(x+2)(x-2)}{x^2-4} = x$ ☐

In real life it is not always obvious (as we can see from what we have just discussed) which equations can be reduced to the form of Equation 1. There is no general method which will allow us to take an arbitrary equation and say 'yes, this can be solved like this', and then write down the answer. Some equations can be solved easily, some with more difficulty, and some are impossible to solve. One of the advantages that comes from practice is the ability to recognise the category to which a particular equation belongs, so that you don't waste time trying to solve something insoluble, and you can go straight to the right method for something soluble.

4.2.2 Equations linear in a function of x

There is a slight and fairly obvious generalisation of the linear equation that arises if the variable is, say x^2, rather than x. For example,

$$x^2 + 4 = 2x^2 - 12$$

can be solved for the unknown variable x^2 to give $x^2 = 16$, i.e. $x = \pm 4$. A similar method can be used in the following example.

Solve the following equation $1/(x-1) + 2 = 2/(x-1) - 4$.

In this case we initially solve for the unknown variable $1/(x-1)$.

Thus $1/(x-1) - 2/(x-1) = -4 - 2$

so $1/(x-1) = 6$

We then solve this equation for x, as in Question T5(a), to obtain

$x = 7/6$ ☐

We could have reduced the chance of a mistake in copying, by substituting $y = 1/(x-1)$ at the start of this question. This trick is well worth remembering − try it on the next exercise.

Question T6

Solve the following equation:

$$2\left(\frac{x^2+1}{x^2-1}+1\right)=\frac{x^2+1}{x^2-1}+5 \quad \left(\text{hint: put } y=\frac{x^2+1}{x^2-1}\right) \quad \square$$

4.2.3 Simultaneous linear equations

Let us reconsider the cycling problem discussed in Subsection 4.2.1. Again suppose that you cycled along a road at a steady 10 mph. One hour after you left, a friend set off in a car at a steady 30 mph to catch you up. Where and when will you meet?

Let us take the distance to the meeting-point to be d_m, and the time on your stop-watch to be t_m when you meet. To keep things as general as possible, take the time at which the car starts to be T_0, and the speeds of the bicycle and the car to be v and V, respectively. At the instant you meet *you will each have travelled the same distance*, and so we have two equations analogous to Equation 2:

distance travelled by you = your speed × time spent cycling

i.e. $d_m = vt_m$ (5)

and

distance travelled by your friend = your friend's speed × time spent driving

i.e. $d_m = V(t_m - T_0)$

so $d_m = Vt_m - VT_0$ (6)

We know the values of v and V in these equations, and we can calculate VT_0; but we do not know the distance d_m or the time t_m, and we cannot work them out from either Equation 5 or Equation 6 alone (☞). Equations 5 and 6 are called **simultaneous linear equations** since they must be solved together if they are to be solved at all.

The basic method of solving a pair of equations of this kind involves **substitution**. This means that we find an expression for one of the variables using one equation, and we then substitute that expression for the selected variable in the other equation. For instance, Equation 5 tells us that we can replace d_m by vt_m; making this substitution in Equation 6 gives

$vt_m = Vt_m - VT_0$

i.e. $vt_m - Vt_m = -VT_0$

i.e. $(v - V)t_m = -VT_0$

so $t_m = \dfrac{-VT_0}{v - V} = \dfrac{VT_0}{V - v}$

Equations 5 and 6 each contain partial information about the problem; but we need all the information contained in both equations to solve the problem.

This is part of the solution we seek since it expresses one of the unknown quantities (t_m) in terms of known quantities. (It also makes good sense physically. Notice that t_m increases as T_0 increases, which we should expect, because the later your friend leaves, the longer it will be before you meet. Furthermore, if $V = v$ the bicycle and the car are travelling at the same speed, so the denominator becomes zero, t_m becomes infinite, and you never meet − which is again what we expect.) All that remains now is to find d_m, which we can do directly by substituting for t_m in Equation 5:

$$d_m = vVT_0/(V - v)$$

Now we can put in the numerical values $v = 10$ mph, $V = 30$ mph, $T_0 = 1$ h, and find

$$t_m = \left(\frac{30 \times 1}{30 - 10}\right) \text{hours} = 1.5 \text{ hours}$$

and $\qquad d_m = (10 \times 1.5)$ miles $= 15$ miles

Finally, we should check to see that these values do actually satisfy the original equations. We have used Equation 5 directly to find d_m, but Equation 6 correctly gives

$\qquad$ 15 miles $= 30$ mph $\times (1.5 - 1)$h $= 15$ miles

You will have noticed that we have put great emphasis on checking the answers; many wrong answers are due to bad arithmetic and algebraic slips. *Always check that your answers satisfy the original equations and that they make sense.* For example, an answer $t_m = -3$ hours would not be sensible.

Notice that the problem we have just solved had *two* equations and *two* unknown variables t_m and d_m. In general, a pair of simultaneous linear equations may look more like this:

$$x + 2y = 8 \tag{7}$$

$$2x - 3y = -5 \tag{8}$$

and we may be required to solve these equations for the two variables x and y.

We can express x in terms of y (using Equation 7)

$$x = -2y + 8 \tag{9}$$

and then substitute for x in Equation 8 to obtain

$$2(-2y + 8) - 3y = -5$$

so that $\qquad -4y + 16 - 3y = -5$

i.e. $\qquad 7y = 21$

giving $\qquad y = 3$

If we substitute this value for y in Equation 9 we obtain

$$x = (-2 \times 3) + 8 = 2$$

Now we should check to see that these answers satisfy the original equations. We have just used Equation 7 (in the form of Equation 9) to find x, so we check by substituting both values into the other equation, Equation 8, and we find

$$(2 \times 2) - (3 \times 3) = 4 - 9 = -5$$

which is correct. Notice that for checking we always use the original equation (before we have had a chance to make any mistakes), and we don't use the equation that has just been used to find one of the variables.

You can apply this same procedure to any pair of linear equations. However, if the coefficients are not the simple integers we have here, the algebra may soon start to look complicated, and it is useful to have some sort of drill for the process of solution. Look again at the equations we have just solved

$$x + 2y = 8 \tag{7}$$

$$2x - 3y = -5 \tag{8}$$

To get an equation involving just y, multiply the equations by suitable factors so that both have the same coefficient of x − in this case we multiply both sides of Equation 7 by 2 and leave Equation 8 unchanged:

$$2x + 4y = 16$$

$$2x - 3y = -5 \tag{8}$$

Now subtract the second equation from the first, to get

$$(2x + 4y) - (2x - 3y) = 16 - (-5)$$

i.e. $0x + 7y = 21$

so, as before

$$y = 3$$

This process is called **elimination**; we speak of *eliminating* x from the equations, though it's really just a systematic way of implementing substitution.

We now proceed as before and use Equation 7 to give $x = 2$ again. After this we should use Equation 8 to carry out the same check.

Alternatively, we could have eliminated y from Equations 7 and 8; and the simplest way to do that is to multiply Equation 7 by 3, and Equation 8 by 2 to obtain

$$3x + 6y = 24$$

$$4x - 6y = -10$$

Now we may *add* the above equations to obtain

$$7x + 0y = 14$$

which gives $x = 2$

We can then use either Equation 7 or Equation 8 to find y, and the other to check the answer.

So we have seen three alternative methods of solving Equations 7 and 8:

- substituting for x (found from one equation) in terms of y (in the other)
- eliminating x
- eliminating y.

In general we try to choose the method that is most convenient.

Question T7

Solve the following pairs of equations:

(a) $5x + 2y = 0$ (b) $4x + y\ = 10$
     $2x + 5y = 21$ $3x - 2y = 13$ ☐

We can also work out the general solution to a pair of simultaneous equations using symbols rather than numbers. For example, for the pair of equations

$$ax + by = p \tag{10}$$

$$cx + dy = q \tag{11}$$

the solution (provided $ad - bc \neq 0$) is

$$x = (dp - bq)/D \tag{12}$$

$$y = (aq - cp)/D \tag{13}$$

where $D = ad - bc$ (14)

This general solution is not particularly helpful, and certainly not worth remembering. However, you may meet it again when you study the theory of sets of three or more simultaneous equations.

Question T8

Show that Equations 12–14 provide the correct solution to Equations 7 and 8.   ☐

Before leaving the subject of simultaneous linear equations it's worth noting that such equations are not always soluble. Here are two situations where Equations 10 and 11 do not have a solution.

Case I

This is the case in which the equations are not **independent** — i.e. one equation is a multiple of the other, so the second equation provides no new information; an example would be the pair of equations

$$x + 2y = 8$$

and $2x + 4y = 16$

A fourth method of solving Equations 7 and 8 would have been to substitute for y in terms of x.

Try this for yourself by substituting for x (from one equation) in terms of y (in the other).

If we try to eliminate x from these equations we shall just end up with $0 = 0$, a statement which is true, but not much use.

Case 2

This is the case in which the equations are **inconsistent** − they cannot both be true at the same time. For example,

$$x + 2y = 8$$

and $$2x + 4y = 15$$

If we try to solve this pair of equations we shall find $1 = 0$, which is neither useful nor true.

The first case occurs when we don't have enough information to solve a problem, the second when some of the information is wrong or inconsistent. Note that in either case the general solution of Equations 12−14 does not apply since $D = 0$.

In this section we have been concerned with *two* linear equations, and *two* unknowns. There are many occasions, however, when one has to deal with three or more unknowns. In general:

If there are n unknowns, then n consistent independent linear equations are needed to solve for the unknowns.

===

If we are given more than n consistent linear equations, they cannot all be independent, so we are free to select n of them that *are* independent and use those to find the n unknowns. Finally, it should be noted that when seeking a set of three or more *independent* linear equations it is not enough to require that no equation should be a *multiple* of any other; rather we have to demand that none of the equations can be expressed as a *sum of multiples* of the others. It is actually fairly easy to determine whether or not a given set of equations satisfies this requirement, but it requires techniques beyond the scope of this chapter.

These techniques are discussed in the chapter on determinants in the companion volume, *Further Mathematics for the Physical Sciences.*

4.2.4   Graphical solutions

There is one other way of solving a pair of simultaneous equations that is worth mentioning − a graphical method. Let us consider Equations 7 and 8 yet again. If we write these in the equivalent form

$$y = \frac{8 - x}{2}$$

$$y = \frac{2x + 5}{3}$$

we can plot each of them as a straight-line graph, as in Figure 4.1. On the first line, $y = (8 - x)/2$, we have all the points (x, y) which make this equation a

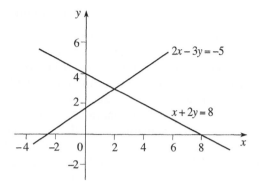

Figure 4.1 Graphical solution of Equations 7 and 8.

true statement. Similarly, on the second line, $y = (2x + 5)/3$, we have all the points (x, y) which make *this* equation a *true* statement. If *both* statements are true for a particular point (x, y) then this point must be where the two lines intersect, for it is only there that both equations are true simultaneously.

This method can be used to find an approximate solution to any pair of equations (linear or not) and it is often a good starting point, particularly if you have access to a graph-drawing computer or calculator.

Question T9

Use a graphical method to find approximate solutions of the simultaneous equations

$$x + 2y = k \quad \text{and} \quad 2x - 3y = -5$$

for $k = 2, 4, 6$ and 8. (You can make use of Figure 4.1.) ☐

4.3 **Solving quadratic equations**

Up to now we have dealt with *linear equations* and their roots, that is, equations in which the independent variable appears only as the first power. Let us now advance to *quadratic equations*, which involve the square of the independent variable. The general form of a **quadratic equation** is

$$ax^2 + bx + c = 0 \tag{15}$$

where a, b and c are constants, and $a \neq 0$. There is a general method of solving such equations which will be explained shortly, but it is instructive to examine the problem of solving quadratic equations from a number of other viewpoints, since it sheds a good deal of light on the general problem of solving equations.

4.3.1 *The solutions as zeros of a quadratic function*

As discussed in the last chapter, a **quadratic function** is any function $f(x)$ that may be written in the form

The right-hand side of Equation 16 is sometimes referred to as a *quadratic expression* or simply a *quadratic*.

Note the terminology; equations have *roots* (i.e. solutions) while functions have *zeros*.

The case $a > 0$ gives a 'happy' curve, and $a < 0$ gives a 'sad' curve. ('Happy' and 'sad' are not recognised mathematical terms.)

$$f(x) = ax^2 + bx + c \qquad (\text{☞}) \ (16)$$

where a, b and c are constants, and $a \neq 0$. Thus, when any given quadratic equation is written in the style of Equation 15, its left-hand side will define a corresponding quadratic function $f(x)$. Moreover, it follows from Equations 15 and 16 that $f(x) = 0$ for each value of x that satisfies the given quadratic equation. For this reason the roots (or solutions) of a quadratic equation are also referred to as the **zeros** of the corresponding quadratic function.

If we set $y = f(x)$ and plot the graph of a quadratic function the resulting curve is a **parabola**. Six different parabolas (corresponding to different values for a, b and c) are shown in Figure 4.2. If $a > 0$ the parabolas are as shown on the left of the figure, while the case $a < 0$ corresponds to those on the right.

The graph of any given quadratic function may meet the x-axis once or twice or not at all — it all depends on the values of a, b and c. However, if the graph does meet the x-axis it does so when $y = f(x) = 0$. It follows that each value of x at which the graph meets the x-axis is a zero of the quadratic function *and* therefore a solution (root) of the corresponding quadratic equation. The possibilities illustrated by the three parts of Figure 4.2 are as follows:

(a)   there are two unequal roots, and the parabola crosses the x-axis in two distinct points;

(b)   there is one root, at the point where the parabola just touches the x-axis;

(c)   there are no roots, because the parabola never reaches the x-axis.

Case (b), where there is just one root, is really a special limiting case of (a). If we move either of the parabolas in Figure 4.2a away from the x-axis its points of intersection with the x-axis (and hence the roots of the corresponding quadratic equation) will get closer and closer together, so that when the parabola just touches the x-axis, the two will have merged. For this reason we usually say that the root in (b) is a **repeated root** or that the equation has **coincident roots**.

These possibilities are the central feature of the process of solving quadratic equations, as we shall see.

4.3.2   Sums and products of roots: factorising the quadratic

Throughout this section our aim is to solve the quadratic equation $ax^2 + bx + c = 0$, by finding values for its roots, α and β say. But suppose for the moment that we know α and β and we ask instead 'what is the quadratic equation that has α and β as its roots?'

An answer to this question is given by the equation

$$(x - \alpha)(x - \beta) = 0 \qquad (17)$$

It is easy to see that if we put either $x = \alpha$ or $x = \beta$ in this equation then the left-hand side is zero (if α and β are equal, then we have a repeated root). Moreover, expanding the left-hand side of Equation 17 gives the quadratic equation

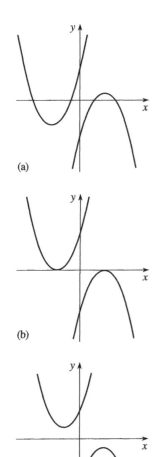

Figure 4.2   Graphs of six different quadratic functions.

$$x^2 - (\alpha + \beta)x + \alpha\beta = 0 \tag{18}$$

In other words, the quadratic equation $x^2 - (\alpha + \beta)x + \alpha\beta = 0$ has α and β as its roots.

Now, compare Equation 18 with the general quadratic equation we are trying to solve:

$$ax^2 + bx + c = 0 \tag{15}$$

We can make the general equation look more like Equation 18 by dividing both sides by a, which gives

$$x^2 + (b/a)x + (c/a) = 0 \tag{19}$$

It then becomes clear that Equations 18 and 19 are the same, provided that

$$\alpha + \beta = -b/a \quad \text{and} \quad \alpha\beta = c/a \tag{20}$$

Thus, we can make the following connections between the coefficients of a quadratic and its roots.

Given a general quadratic equation $ax^2 + bx + c = 0$ with roots α and β
-    the sum of its roots is $\alpha + \beta = -b/a$
-    and the product of its roots is $\alpha\beta = c/a$

Unfortunately it is not possible to use these equations directly to solve the equation; if you try to do so you will just end up with the original equation! However, they may help us to *factorise* the original equation, that is to re-express it in the form of Equation 17 and hence identify its solutions. This is often the quickest method of solution, especially if we know, or suspect, that the roots are integers (i.e. whole numbers).

Solve the equation $x^2 - 5x + 6 = 0$.

In this case $a = 1$, $b = -5$ and $c = 6$, so we wish to find roots, α and β, such that $\alpha + \beta = 5$ and $\alpha\beta = 6$. If the roots are integers then, apart from signs, they must either be 2 and 3, or 1 and 6, and both members of the pair must have the same sign, since their product is $+6$. In this case only the values 2 and 3 meet all the requirements. This means that the equation $x^2 - 5x + 6 = 0$ can be written in the alternative form $(x - 2)(x - 3) = 0$, and *now* we can see that the roots are 2 and 3. ☐

In the above solution we have used the process known as **factorisation** to find the roots of the equation (this was introduced in the last chapter). We can check that our answer is correct by expanding the brackets to give

$$(x - 2)(x - 3) = x^2 - 2x - 3x + 6 = x^2 - 5x + 6$$

For example, a quadratic equation with roots 2 and -5, is

$$(x - 2)(x + 5) = 0$$

which expands to $x^2 + 3x - 10 = 0$. Any constant multiple of this equation would also have roots 2 and -5.

The fact that there are very few choices of integers that multiply to give 6 helps us considerably in this case.

If we had to solve the almost identical equation

$$x^2 - 5x - 6 = 0$$

we would have $\alpha + \beta = 5$ and $\alpha\beta = -6$ and the roots would still involve 2 and 3, or 1 and 6, but this time the two numbers in each pair would have opposite signs, since the product $\alpha\beta$ is negative. We must choose the pair $+6$ and -1. We can check by expanding as before, and this time:

$$(x - 6)(x + 1) = x^2 - 6x + x - 6 = x^2 - 5x - 6$$

Question T10

Solve the following quadratic equations by factorisation:

(a) $x^2 - 3x + 2 = 0$ (b) $x^2 + x - 2 = 0$

(c) $x^2 + 7x + 12 = 0$ (d) $x^2 - x - 12 = 0$

(e) $x^2 - 2x - 15 = 0$ (f) $x^2 - 4x - 12 = 0$ □

The above equations all had the coefficient of x^2 equal to 1. It is a little more complicated if a is not 1; consider, for example, the expression

$$2x^2 + 5x + 3$$

If this can be factorised, then it must be possible to express it as

$$(2x + ?)(x + ?)$$

Trial and error shows us that the factorisation is

$$(2x + 3)(x + 1)$$

Thus, to solve the equation $2x^2 + 5x + 3 = 0$ we rewrite it as $(2x + 3) \times (x + 1) = 0$ and then it is clear that the roots are $-3/2$ and -1. (☞)

It is important to check your answer by substituting the values you have obtained back into the original equation to verify that they are indeed roots of the equation.

If $(2x + 3)(x + 1) = 0$ then either $(2x + 3) = 0$ or $(x + 1) = 0$. Thus the roots are $x = -3/2$ and $x = -1$.

Question T11

Solve the following equations by factorisation:

(a) $2x^2 + 9x - 5 = 0$ (b) $3x^2 - 11x - 4 = 0$

(c) $3x^2 - x - 2 = 0$ (d) $4x^2 - 3x - 1 = 0$

(e) $4x^2 - 4x + 1 = 0$ (f) $6x^2 + 5x - 6 = 0$ □

With practice this method can become very quick, but unfortunately it does not always work. The quadratic equation

$$x^2 - 3x + 1 = 0$$

cannot be factorised in the same way, because the roots are not integers. We must therefore find a more generally applicable method even if it is somewhat slower.

4.3.3 The quadratic equation formula: completing the square

The quadratic equation

$$x^2 - 7 = 0$$

is easily solved, for we may simply add 7 to both sides of the equation and then take square roots of both sides to obtain

$$x^2 = 7, \text{ so that } x = \pm\sqrt{7}$$

The quadratic equation

$$(x + 3)^2 - 5 = 0$$

is hardly more difficult, since we can add 5 to both sides, then take the square root of both sides to get $x + 3 = \pm\sqrt{5}$ so that $x = -3 \pm \sqrt{5}$.

If we could reduce the general quadratic equation to something like this form we should have another method of solution that might work more generally. Introducing such a method is the main purpose of this subsection, but first we must examine how one might write *any* quadratic equation in such a form. Another example may make the problem clear.

To solve the equation

$$3x^2 - 10x + 1 = 0$$

we first extract the factor 3 to give

$$3\left(x^2 - \frac{10}{3}x + \frac{1}{3}\right) = 0 \tag{21}$$

and we thus ensure that the coefficient of x^2 inside the brackets is 1. Next we arrange for the term $x^2 - \dfrac{10}{3}x$ to arise from a term of the form $(x + k)^2$ for some value k. (Such a term is called a **perfect square**.) Since $(x + k)^2 = x^2 + 2kx + k^2$ we will be forced to choose k so that $2k = -10/3$ in this case, in other words $k = -5/3$, which gives $k^2 = 25/9$. Now we add and subtract 25/9 to the term in the brackets in Equation 21 to get

$$3\left(x^2 - \frac{10}{3}x + \frac{25}{9} - \frac{25}{9} + \frac{1}{3}\right) = 0 \tag{22}$$

The advantage of this step is that the first three terms in the bracket can be written as a perfect square, so Equation 22 becomes

$$3\left[\left(x - \frac{5}{3}\right)^2 - \frac{25}{9} + \frac{1}{3}\right] = 0$$

and this can be simplified to

$$3\left[\left(x - \frac{5}{3}\right)^2 - \frac{22}{9}\right] = 0$$

so that $\left(x - \dfrac{5}{3}\right)^2 = \dfrac{22}{9}$ and therefore $x = \dfrac{5}{3} \pm \dfrac{\sqrt{22}}{3}$

The above process is known as **completing the square**; try the same method in the following exercises.

Question T12

Find the roots of the following equations by completing the square:

(a) $x^2 - x - 2 = 0$ (b) $x^2 - 2x - 8 = 0$

(c) $2x^2 + x - 10 = 0$ (d) $3x^2 - 18x + 24 = 0$ □

To see how this method can be applied in general, let us repeat the procedure for the general quadratic.

$$ax^2 + bx + c = a\left(x^2 + \frac{b}{a}x + \frac{c}{a}\right)$$

$$= a\left(x^2 + \frac{b}{a}x + \frac{b^2}{4a^2} - \frac{b^2}{4a^2} + \frac{c}{a}\right)$$

$$= a\left[\left(x + \frac{b}{2a}\right)^2 - \frac{b^2 - 4ac}{4a^2}\right]$$

Now if we put this expression equal to zero, we can divide both sides by the multiplying factor a, and get

$$\left(x + \frac{b}{2a}\right)^2 - \frac{b^2 - 4ac}{4a^2} = 0$$

so that $\left(x + \dfrac{b}{2a}\right)^2 = \dfrac{b^2 - 4ac}{4a^2}$

and $x + \dfrac{b}{2a} = \pm \sqrt{\dfrac{b^2 - 4ac}{4a^2}}$

---

thus $x = \dfrac{-b \pm \sqrt{b^2 - 4ac}}{2a}$ (23)

This is a very important formula, which is used so often that you should commit it to memory. In practice this formula provides the simplest and quickest way of solving most quadratic equations.

Question T13

Use Equation 23 to solve the problems in Question T12. ☐

Question T14

Use Equation 23 to solve the following:

(a) $x^2 - 6x + 1 = 0$ (b) $3x^2 - 9x + 2 = 0$

Do not use your calculator, but leave the answer in the simplest form possible, with the square roots not evaluated. Check that the sum and product of the roots are correct. ☐

Not only does Equation 23 simplify the process of solving quadratic equations, it also has a feature which enables us to say something about the general properties of the quadratic equation without actually having to solve it. This feature is known as the **discriminant**, and is given by the expression inside the square root, $b^2 - 4ac$.

The sign of the discriminant determines some useful properties of the roots of the equation:

1. If $b^2 - 4ac$ is positive, then we can find its square root, and the quadratic equation will have two different roots. This corresponds to the case (a) of Figure 4.2, where the parabola crosses the x-axis twice. (For reasons that will become clear in the next subsection, we also say that the equation has two distinct *real* roots.)

2. If $b^2 - 4ac$ is equal to zero, then the two roots of the equation will be repeated and the quadratic equation can be written as $a(x + k)^2 = 0$. This corresponds to the special case (b) of Figure 4.2 in which the parabola just touches the x-axis and we have just one (repeated) real root.

3. If $b^2 - 4ac$ is negative, then we cannot take the square root, and we say that there are no real roots to the equation. This case corresponds to case (c) of Figure 4.2, when the parabola does not intersect the x-axis at all.

Question T15

Use the discriminant to establish how many real roots the following equations have:

(a) $4x^2 - 20x + 25 = 0$ (b) $x^2 + x + 6 = 0$

(c) $x^2 + x - 6 = 0$ ☐

Although we now seem to have a recipe that will solve all our problems with quadratic equations, and make the whole business purely mechanical, this is not quite true. It often happens that when we have to substitute numbers in formulae, things are not quite as straightforward as appears at first sight. For example, consider the equation:

$$x^2 - 2x + 10^{-12} = 0$$

and notice that the constant term, c, is very small.

If the constant term 10^{-12} were not there, the equation would be $x^2 - 2x = 0$ which has roots at $x = 2$ and $x = 0$, so we would expect the given equation to have roots near these values.

Applying the formula (Equation 23) gives us the roots

$$x = (2 \pm \sqrt{4 - 4 \times 10^{-12}})/2$$

This shows the value of remembering the relations between the roots and coefficients of a quadratic equation as well as the formula.

which can only be evaluated as 2 or 0 to the accuracy of most calculators.

The first root will certainly be as close to 2 as most people could wish, but to get the second root we have to take the difference of two numbers that are very close together, which means that we have lost a lot of precision. However, there is an alternative. If we calculate the larger root (say α) using Equation 23, we find a value near to 2 as before. We can then use the fact that the product of the roots ($\alpha\beta$) is equal to c/a, i.e. 10^{-12}, to obtain the value $10^{-12}/2 = 5 \times 10^{-13}$ for the smaller root (say β).

4.3.4 Roots for all equations: complex numbers

Let us look again at what happens when the discriminant $b^2 - 4ac$ is negative. Consider the equation

$$x^2 - x + 1 = 0$$

for which $a = 1$, $b = -1$ and $c = 1$, so that $b^2 - 4ac = -3$.

If we try to solve this equation using Equation 23 we get

$$x = \frac{1}{2}(1 \pm \sqrt{-3})$$

which does not seem to make sense, since a negative number like -3 has no square root; yet if we substitute one of these 'roots' into the left-hand side of the equation we find:

$$x^2 - x + 1 = \left[\frac{1}{2}(1 - \sqrt{-3})\right]^2 - \frac{1}{2}(1 - \sqrt{-3}) + 1$$

$$= \frac{1}{4}\left[1 - 2\sqrt{-3} + (\sqrt{-3})^2\right] - \frac{1}{2}(1 - \sqrt{-3}) + 1$$

$$= \frac{1}{4}(1 - 2\sqrt{-3} - 3) - \frac{1}{2}(1 - \sqrt{-3}) + 1$$

$$= -\frac{1}{2} - \frac{1}{2}\sqrt{-3} - \frac{1}{2}(1 - \sqrt{-3}) + 1 = 0$$

So treating the quantity $\sqrt{-3}$ as an 'ordinary' number such that its square is -3 appears to give us a solution to the equation.

Question T16

Show that

$$x = \frac{1}{2}(1 + \sqrt{-3})$$

also satisfies the equation in the same way.　□

Both 'roots' satisfy the equation, even if they do not mean anything! Since there seems to be no contradiction, this is an invitation to mathematicians to invent something new. A quantity called i is introduced, which is used in place of $\sqrt{-1}$; it has the property that $i \times i = -1$. This enables us to deal with the square root of any negative number; for example

$$\sqrt{-3} = \sqrt{-1 \times 3} = \sqrt{-1} \times \sqrt{-3} = i\sqrt{3}$$

The number i can be given a perfectly sound mathematical definition, but that is not our concern here, we merely wish to use it. For example, we may write

$$i^4 = (i^2)^2 = (-1)^2 = 1$$

or　　$i^3 = i \times i \times i = -1 \times i = -i$

or　　$\dfrac{1}{i} = \dfrac{i}{i \times i} = \dfrac{i}{-1} = -i$

The quantity i and all its multiples, such as $i\sqrt{3}$, are examples of **imaginary numbers**. These should be contrasted with 'ordinary numbers' such as 35 or -2.6 that do not involve i and which are formally referred to as **real numbers**. By adding together real and imaginary numbers it is possible to create so-called **complex numbers** of the form

$a + bi$

where a and b are both real numbers. For example, $3 + 5i$ would be a typical complex number.

The real number a is called the **real part**, and the real number b the **imaginary part**, of the complex number. The real part of $3 + 5i$ is 3 while its imaginary part is 5. 'Real' and 'imaginary' are perhaps unfortunate names to have been chosen, since they suggest that a is more important than b, whereas we must treat both on an equal footing. However, the names are now too well established to be changed easily.

Question T17

Using complex numbers, find the solutions of the following equations:

(a)　$x^2 + 25 = 0$　　　　　(b)　$x^2 + x + 6 = 0$　□

Complex numbers play an important part in mathematics and physical science, and are considered in detail in the companion volume to this. We cannot go into all their applications here, but it is worth noting that if we allow roots to be complex, and if we count repeated roots twice, then *every* quadratic equation has *two* roots, the values of which are given by Equation 23.

Some engineers, and physical scientists who work with complex numbers use j instead of i, but all mathematicians, and most physical scientists, use i.

Notice that the imaginary part of $3 + 5i$ is 5 *not* $5i$.

In physical science and engineering, complex numbers are invaluable in calculations involving electrical circuits, where the two parts can be associated with the oscillatory and decay properties of the circuits. They are also useful in the study of waves and optics, and in many other areas.

4.4 **Solving more complicated equations**

The equations we have dealt with so far are a very small sample of the variety that you will meet. We mentioned simultaneous equations with more than two unknown quantities in Subsection 4.2.3, but you will also meet equations that involve higher powers of the unknown variable than the square, including **cubic equations** such as

$$x^3 + x^2 + x + 1 = 0$$

and **quartic equations** such as

$$x^4 + x^3 + x^2 + x + 1 = 0$$

Such equations, together with the linear and quadratic equations we considered earlier, are special cases of a class of **polynomial equations** which have the general form

$$a_0 + a_1 x + a_2 x^2 + \ldots + a_{n-1} x^{n-1} + a_n x^n = 0$$

where n is an integer called the **degree** of the polynomial, and the $n + 1$ constants $a_0, a_1, a_2, \ldots a_{n-1}$ and a_n are the *coefficients* of the polynomial. The left-hand side of such an equation constitutes a **polynomial expression** and may be used to define a corresponding **polynomial function**.

There is a general result for roots of polynomial equations:

The **fundamental theorem of algebra** states that *every polynomial equation of degree n has precisely n roots* as long as complex roots and repeated roots are counted.

Cubic and quartic equations can be solved using formulae similar to those used for solving quadratic equations, but these formulae are so complicated that they are rarely used. However, there are no general formulae for the roots of equations of the 5th and higher degrees, and in such cases we must usually be satisfied with numerical approximations. Many of the numerical methods that are used for finding these approximations can also be applied to non-polynomial equations and we shall examine such methods shortly. First, however, let us look at methods that allow us to solve *some* polynomial equations.

4.4.1   *Solving polynomial equations*

Sometimes, if we are fortunate, we are able to guess one, or more, solutions of a cubic, or higher order, equation and this makes the task of finding the remaining roots considerably easier. For example, if we substitute $x = 1$ into the left-hand side of the following equation:

$$x^3 - 6x^2 + 11x - 6 = 0 \tag{24}$$

we see immediately that this value of x makes the equation true. It follows that $(x - 1)$ is a factor of the left-hand side of Equation 24 so that

$$x^3 - 6x^2 + 11x - 6 = (x - 1) \times (\text{a quadratic expression in } x)$$

i.e. $\left(\dfrac{x^3 - 6x + 11x - 6}{x - 1} \right) = \text{a quadratic expression in } x$

We can find this quadratic expression by **algebraic division** as follows:

$$
\begin{array}{r}
x^2 - 5x + 6 \\
x - 1 \overline{) x^3 - 6x^2 + 11x - 6} \\
x^3 - x^2 \\
\hline
- 5x^2 + 11x \\
- 5x^2 + 5x \\
\hline
+ 6x - 6 \\
+ 6x - 6 \\
\hline
\cdots \cdots
\end{array}
$$

(☞)

Note the general form

$$\text{denominator} \overline{) \text{numerator}}^{\text{result}}$$

The result on the top is worked out term by term starting with x^2. We multiply the denominator by x^2 to get $x^3 - x^2$. This is written below the numerator and subtracted from it, leaving $-5x^2 + 11x - 6$. We then proceed to divide this by the dominator $x - 1$, thus finding that the second term in the result is $-5x$, and so on.

so that $x^3 - 6x^2 + 11x - 6 = (x - 1)(x^2 - 5x + 6)$.

The quadratic expression on the right is then easily factorised to give

$$x^3 - 6x^2 + 11x - 6 = (x - 1)(x - 2)(x - 3)$$

and now we can see that the roots of Equation 24 are 1, 2 and 3.

This method is quite restricted in its applications because it relies on our ability to guess a root, but it can still be very useful. (☞)

Some computer programs that perform algebraic manipulations will factorise expressions of this kind.

Question T18
Find the roots of $x^3 + x^2 - 4x - 4 = 0$. (*Hint:* try some integer values of x.) □

4.4.2 Graphical solutions

In Subsection 4.2.4 we saw how we could use graphs to solve a pair of simultaneous equations, and in Section 4.3 we saw how a graph of a quadratic function could give us clues about the roots of the corresponding quadratic equation. Similar methods can be applied to a wide variety of equations. Consider, for example, the function

$$f(x) = x^3 - 4x + 1 \tag{25}$$

the graph of which is shown in Figure 4.3.

The curve crosses the x-axis three times, so the equation $f(x) = 0$ has three roots, which are, respectively, near -2, between 0 and 1, and near $+2$. A larger scale graph would tell us that the values are nearer -2.1, 0.25 and $+1.9$. If this degree of precision is enough, then nothing further need be done. If greater accuracy is needed, then one has the choice of drawing a yet more accurate graph (or perhaps three separate graphs, one for the neighbourhood

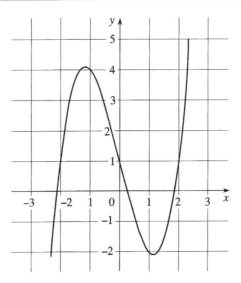

Figure 4.3 The graph of Equation 25.

of each root) or of using another method. Since more function values must be calculated to plot the new graphs, it is probably better to go straight to one of the *iterative* methods discussed in the next subsection.

Any pair of simultaneous equations can be solved approximately by graphical methods. For example, the equations

$$y = x^3 - 4x + 1 \tag{26}$$

and
$$4y = 3 - 2x - x^2 \tag{27}$$

can be solved by plotting the graphs of $y = x^3 - 4x + 1$ and $y = (3 - 2x - x^2)/4$. The points of intersection correspond to the solutions of this pair of equations. Figure 4.4 shows that there are three possible solutions, corresponding to the three intersections, at approximately

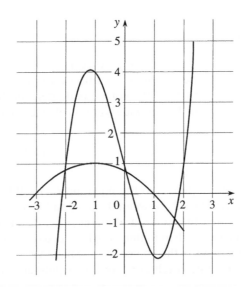

Figure 4.4 The graphs of Equations 26 and 27.

$$x = -2, \quad y = 0.7$$
$$x = 0, \quad y = 0.7$$
and
$$x = 1.7, \quad y = -0.8$$

which can be abbreviated to

$$(-2, 0.7), (0, 0.7), (1.7, -0.8)$$

More accurately, the points of intersection are

$$x = -2.0325 \quad y = 0.7336$$
$$x = 0.0719 \quad y = 0.7128$$
$$x = 1.7106 \quad y = -0.8369$$

(but these values were found by numerical, rather than graphical, methods).

4.4.3 Numerical procedures

Numerical procedures are methods of finding approximate solutions to a particular equation, in which all the constants are known numbers.

The bisection method

The simplest procedure, called the **bisection method**, is not much more than trial and error. If we want to solve the equation $f(x) = 0$, and we have some idea of the solution, perhaps from a graph, then we evaluate $f(x)$ on either side of our initial estimate so as to obtain a range of x values over which the sign of $f(x)$ changes. As an illustration, consider the equation

$$x^3 - 4x + 1 = 0$$

corresponding to the function $f(x) = x^3 - 4x + 1$. From Figure 4.3 we see that there is a root between $x = 1$ and $x = 2$, and this is confirmed by noticing that

$$f(1) = -2 \quad \text{and} \quad f(2) = 1$$

so that there is a sign change on the interval from $x = 1$ to $x = 2$.

At the mid-point, $x = 1.5$, we have $f(1.5) = -1.625$ so that the sign change (and therefore the root) must lie between $x = 1.5$ and $x = 2$.

Similarly, we can show that the root lies between $x = 1.75$ and $x = 2$; and we can continue in this fashion, bisecting the interval and evaluating the function at the mid-point, as far as we please and so obtain an approximation to the root which is as accurate as we need.

Regula falsi

The method of bisection is very reliable and simple to implement as a computer program. It is true that it is comparatively slow, but with the speed of modern computers this hardly matters for a single calculation. However, speed may become a critical factor if you need to perform a large number of evaluations of roots. A simple method of speeding up the method considerably is the variation known as *regula falsi*. Let us look in detail at a graph of the neighbourhood of the solution, in Figure 4.5.

If x_1 and x_2 are initial estimates for the root, and if $y_1 = f(x_1)$ and $y_2 = f(x_2)$ have opposite signs, then a straight line between the two points will intersect the x-axis at a point which is closer to the root. The equation of the straight line joining the two points (x_1, y_1) and (x_2, y_2) is

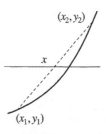

Figure 4.5 *Regula falsi.*

This is a version of the *two-point* form of the equation of a straight line.

$$\frac{y - y_1}{x - x_1} = \frac{y - y_2}{x - x_2} \tag{28}$$

If we put $y = 0$ in this equation, then the corresponding value of x will determine the point at which the line crosses the horizontal axis and will therefore give an improved estimate of the root. Thus,

$$\frac{-y_1}{x - x_1} = \frac{-y_2}{x - x_2}$$

which can be rearranged to give

$$x = \frac{x_2 y_1 - x_1 y_2}{y_1 - y_2}$$

This improved estimate can be used as the starting point for an even better estimate.

Finding a root by this method takes fewer steps, although more work is required in each step. In practice it can be a fine judgement as to which method is more efficient in a particular case.

Question T19

The function defined in Equation 25 has a zero near $x = 2$. Find this zero to an accuracy of five decimal places (i.e. so that the equation is satisfied to five decimal places) using one of the two numerical methods above. You may use your calculator. ☐

Iterative methods

An alternative class of methods that are often useful are the **iterative methods**; these are based on the idea that if one has an approximate solution to an equation, then it can be used to get a better value. You might consider either of the previous methods to be in this class, but we usually mean methods in which the next approximation is obtained by a special formula. Look again at the equation $x^3 - 4x + 1 = 0$, which we know, from Figure 4.3, has a solution near $x = 2$. We can rewrite the equation as

You may wish to verify that

$$x = \frac{2x^3 - 1}{3x^2 - 4}$$

is a rearrangement of the equation

$$x^3 - 4x + 1 = 0$$

Don't worry about how we found this rearrangement — that is dealt with in the companion volume, *Further Mathematics for the Physical Sciences*, under the heading *Newton–Raphson formula*.

$$x = \frac{2x^3 - 1}{3x^2 - 4} \qquad (☎)\ (29)$$

and we can use this to provide an **iteration formula** that relates an approximate root x_n to a more accurate approximate root x_{n+1}

$$x_{n+1} = \frac{2x_n^3 - 1}{3x_n^2 - 4} \tag{30}$$

So, if we set $n = 1$ in Equation 30 and take as our crude first approximation $x_1 = 2$, we can use Equation 30 to find a better approximation $x_2 = 15/8 = 1.875$. Setting $n = 2$ in Equation 30 we can now repeat this process (starting with $x_2 = 1.875$) to find an even better approximation x_3.

Continuing in this way we can obtain a succession of approximations x_2, x_3, x_4, x_5 and so on, with each step, or **iteration**, being an improvement on the last.

$$x_2 = 1.875$$

$$x_3 = 1.860\ 978\ 52$$

$$x_4 = 1.860\ 805\ 88$$

$$x_5 = 1.860\ 805\ 85$$

$$x_6 = 1.860\ 805\ 85$$

Notice that after three iterations the solution is consistent with the answer to Question T19 which was given to five decimal places.

We stop when two successive iterations are the same to the required number of decimal places, since we can expect no further improvement.

Question T20

The equation $x^3 + 2x - 1 = 0$, which has a root near $x = 0.45$, can be rearranged to give

$$x = \frac{1 - x^3}{2}$$

Using the first approximation $x_1 = 0.450\ 00$ and the iteration formula

$$x_{n+1} = \frac{1 - x_n^3}{2}$$

find this root to five decimal place accuracy. (You may use your calculator.) ☐

Unfortunately, this simple method must be used with great care, and careful choice of the iteration formula is often needed to produce an iteration that will result in a meaningful answer. However, if the method does work, then it has the great advantage that it is almost proof against a mistake, since any reasonable approximation will be improved; the worst that can happen is that the process is slowed down somewhat.

More advanced methods, beyond the scope of this chapter, enable one to produce very rapid convergence to the root.

Question T21

The following iterative formula was used in the early days of computers to find the *reciprocal* of a given number a:

$$x_{n+1} = x_n(2 - ax_n)$$

Use this iteration to find $1/0.8$ (i.e. let $a = 0.8$ in the above formula), starting with $x_1 = 1$. (You may use your calculator.) ☐

4.5 **Closing items**

4.5.1 *Chapter summary*

1. Given a function $f(x)$, a *solution* or *root* of the equation $f(x) = 0$ is a value of x that makes the equation true. Such a value is also a *zero* of the function and is represented graphically by any value of x at which the graph of $f(x)$ meets the x-axis.

2. A *linear equation* in a single variable x has the general form $ax + b = 0$ and may be *solved* by rearrangement to give $x = -b/a$.

3. A pair of *simultaneous linear equations* may be solved by *substitution*; the method of *elimination* provides a standard routine for carrying out the substitution.

4. A set of simultaneous linear equations involving n unknowns may be solved, provided that it contains at least n *independent* equations and provided that none of the equations are *inconsistent*.

5. Simultaneous linear equations may also be solved *graphically*, by drawing appropriate straight lines and determining their point of intersection.

6. A *quadratic equation* $ax^2 + bx + c = 0$ has roots

    $$x = \frac{-b \pm \sqrt{b^2 - 4ac}}{2a},$$

    as may be demonstrated by the process of *completing the square*.

7. If the *discriminant*, $b^2 - 4ac$, of a quadratic equation is positive there are two *real* roots, if it is zero there is a *repeated* root, and if the discriminant is negative there are no real roots. In the latter case it is possible to find two *complex* roots of the form $a + bi$ where $i = \sqrt{-1}$.

8. If the roots of $ax^2 + bx + c = 0$ are α and β then $\alpha + \beta = -b/a$ and $\alpha\beta = c/a$. Such a quadratic may be *factorised* and written in the form $(x - \alpha)(x - \beta) = 0$.

9. According to the *fundamental theorem of algebra*, a *polynomial equation* of *degree* n has n roots, provided we count complex roots and repeated roots.

10. Equations in general can be tackled by *graphical*, *numerical* and *iterative* techniques, but each case must be treated on its merits.

11. When solving equations that involve physical quantities, don't introduce numbers until it is essential; leave all the arithmetic until the end.

12. Always check that your answers satisfy the original equations and that they make sense.

4.5.2 Achievements

Having completed this chapter, you should be able to:

A1. Define the terms that are emboldened in the text of the chapter.

A2. Solve simple linear equations by rearrangement.

A3. Recognise an equation that is linear in a function of the unknown, and rearrange it for solution.

A4. Solve linear simultaneous equations by substitution, provided they are consistent and independent.

A5. Use graphical methods to solve simultaneous linear equations.

A6. Recall the general form of a quadratic equation and sketch the graph of any given quadratic function.

A7. Relate the coefficients of a given quadratic equation to the sum and product of the roots of that equation.

A8. Factorise a simple quadratic function, and so solve the corresponding quadratic equation.

A9. Explain the method of completing the square, and use it to solve a quadratic equation.

A10. Write down the general formula for the roots of a quadratic equation, and use it to solve any particular equation.

A11. Recall the definition of the discriminant, and explain how it determines the nature of the solutions of a quadratic equation.

A12. Recall the nature and significance of complex numbers, and use them in the solution of quadratic equations.

A13. Solve simple polynomial equations by factorisation.

A14. Describe the use of graphical methods in the solution of general equations, and use them to locate the roots of various equations.

A15. Use the methods of bisection and *regula falsi* in the solution of simple equations.

A16. Use iterative methods in the solution of equations given the iteration formula.

4.5.3 End of chapter questions

Question E1   Find x if $2(x - 1) + 3(x - 6) = 0$.

Question E2   Find x if

$$\frac{1}{x - 2} + 2\left(\frac{1}{x - 2} - 6\right) = 0$$

Question E3 Solve the following simultaneous equations:

$$2x + y = -4$$
$$3x + 5y = 1$$

Question E4 Solve the following simultaneous equations first by drawing suitable graphs, then algebraically

$$x + y = k$$
$$2x - y = 1 \qquad \text{for } k = 2, 5 \text{ and } 8.$$

Question E5 Sketch graphs to show quadratic functions with (a) two distinct zeros, (b) two coincident zeros, and (c) no real zeros.

Question E6 Solve the following equations by factorising the left-hand sides:

(a) $x^2 + 5x - 14 = 0$ $\qquad$ (b) $2x^2 - 21x + 27 = 0$

Check your answers by showing that the sum and product of the roots are correct in each case.

Question E7 Solve the following equations:

(a) $x^2 - 2x - 1 = 0$ $\qquad$ (b) $3x^2 + x - 2 = 0$

Question E8 State the conditions for the equation (in the variable x)

$$x^2 - 2rx + s = 0$$

to have (a) two real roots, and (b) repeated roots.

Question E9 Write down the roots of the equation in Question E8 above if neither of the conditions quoted there is fulfilled (so that the equation has complex roots).

Question E10 Guess one of the roots, and then solve the following equation by factorisation:

$$x^3 - 4x^2 + x + 6 = 0$$

Question E11 Sketch the graph of the function

$$g(x) = 3x^3 - 11x^2 + 5x + 8$$

and give approximate values for its zeros. (You may use your calculator.)

Question E12 Use a numerical method (and the results of Question E11 above) to find the largest root of the equation $3x^3 - 11x^2 + 5x + 8 = 0$ to two decimal places. (You may use your calculator.)

Trigonometric functions

Chapter 5

5.1 Opening items

5.1.1 Chapter introduction

Trigonometric functions have a wide range of application in physical science; examples include the addition and resolution of *vectors* (see Chapter 8), the description of *simple harmonic motion* and the formulation of quantum theories of the atom. Trigonometric functions are also important for solving certain *differential equations*, a topic explored in detail in *Further Mathematics for the Physical Sciences*.

In Section 5.2 of this chapter we begin by looking at the measurement of *angles* in *degrees* and in *radians*. We then discuss some basic ideas about *triangles*, including *Pythagoras's theorem*, and we use *right-angled triangles* to introduce the *trigonometric ratios* (*sin θ, cos θ* and *tan θ*) and the *reciprocal trigonometric ratios* (*sec θ, cosec θ* and *cot θ*). In Section 5.3 we extend this discussion to include the *trigonometric functions* (*sin (θ), cos (θ)* and *tan (θ)* and the *reciprocal trigonometric functions* (*cosec (θ), sec (θ)* and *cot (θ)*). These *periodic functions* generalise the corresponding ratios since the *argument* θ may take on values that are outside the range 0 to π/2. Subsection 5.3.2 discusses the related *inverse trigonometric functions* (*arcsin (x), arccos (x)* and *arctan (x)*), paying particular attention to the conditions needed to ensure they are defined. We end, in Section 5.4, by showing how the sides and angles of any triangle are related by the *sine rule* and the *cosine rule* and by listing some useful *identities* involving trigonometric functions.

5.1.2 Fast track questions

Question F1

State Pythagoras's theorem.

Question F2

Define the ratios sin θ, cos θ and tan θ in terms of the sides of the right-angled triangle. What are the exact values (i.e. don't use your calculator) of sin (45°) and tan (π/4)?

Question F3

Write down the sine and cosine rules for a triangle. Calculate the angles of a triangle which has sides of length 2 m, 2 m and 3 m.

Question F4

Sketch graphs of the functions, cosec (θ) and sec (θ), for $-2\pi \leqslant \theta \leqslant 2\pi$ and cot (θ) for $-3\pi/2 < \theta < 3\pi/2$.

Question F5

Sketch graphs of the functions, arccosec (x) and arccot (x) over the range $-10 < x < 10$.

5.1.3   Ready to study?

Study comment

In order to study this chapter you will need to understand the following terms: *constant*, *decimal places*, *function*, *power*, *reciprocal* and *variable*. An understanding of the terms *codomain*, *domain* and *inverse function* would also be useful but is not essential. You will need to be able to solve a pair of *simultaneous equations*, manipulate arithmetic and algebraic *expressions* – including those involving *squares*, *square roots*, *brackets* and *ratios* – and to plot and interpret *graphs*. The following *Ready to study questions* will help you to decide whether you need to review some of these topics before embarking on this chapter.

Question R1

Given that $x = a/h$, $y = b/h$, and $a^2 + b^2 = h^2$, rewrite the following expressions in terms of a, b and h, simplifying each as far as possible:

(a) $1/x$ (b) x/y (c) $x^2 + y^2$

Question R2

Solve the following simultaneous equations:

$$2x + 5y = 6 \quad \text{and} \quad 3x - 10y = 9$$

Question R3

What do we mean when we say that the position of an object is a *function* of time?

5.2 Triangles and trigonometric ratios

5.2.1 Angular measure: degrees and radians

When two straight lines pass through a common point, the **angle** between the lines provides a measure of the inclination of one line with respect to the other or, equivalently, of the amount one line must be rotated about the common point in order to make it coincide with the other line. The two units commonly used to measure angles are *degrees* and *radians* (discussed below) and we will use both throughout this chapter. Greek letters, α (alpha), β (beta), γ (gamma), ... θ (theta), ϕ (phi) ... are often used to represent the values of angles, but this is not invariably the case.

A **degree** is defined as the unit of angular measure corresponding to 1/360th of a circle and is written as 1°. In other words, a rotation through 360° is a complete revolution, and an object rotated through 360° about a fixed point is returned to its original position. Fractions of an angle measured in degrees are often expressed as decimals, as in 97.8°, but it is also possible to use subsidiary units usually referred to as *minutes* and *seconds* for subdivisions of a degree, with the definitions that sixty minutes are equivalent to one degree and sixty seconds are equivalent to one minute. To distinguish them from units of time, these angular units are called the **minute of arc** and **second of arc**, abbreviated to arcmin and arcsec, respectively. The symbols ' and " are often used for arcmin and arcsec, respectively. For example, 12' means 12 arcmin, and 35" means 35 arcsec.

(a) Express 6° 30' as a decimal angle in degrees.

(b) Express 7.2 arcmin in terms of arcsecs.

(a) $30' = 0.5°$, so $6° \, 30' = 6.5°$

(b) $7.2' = (7.2 \times 60)'' = 432''$ □

The angles 180° and 90° correspond to a rotation through half and one-quarter of a circle, respectively. An angle of 90° is known as a **right angle**. A line at 90° to a given line (or surface) is said to be **perpendicular** or **normal** to the original line (or surface).

It is conventional in mathematics and physical science to refer to rotations in an anti-clockwise direction as *positive* rotations. So, a positive rotation through an angle θ would correspond to the anticlockwise movement shown in Figure 5.1, while a negative rotation of similar size would correspond to a clockwise movement. The negative rotation may be described as a rotation through an angle $-\theta$.

An object which is rotated through an angle of 0° or 360° or 720° appears to remain unchanged and, in this sense, these rotations are equivalent. In the same sense rotations of 10°, 370°, 730°, −350° and so on are equivalent, since each can be obtained from the others by the addition of a

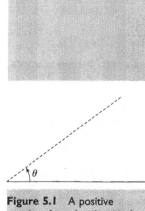

Figure 5.1 A positive rotation through a (positive) angle, θ.

The symbol < is read as 'is less than' and the symbol ≤ is read as 'is less than or equal to'.

multiple of 360°. When considering the orientational effect of a rotation through an angle θ it is only necessary to consider values of θ which lie in the range $0° \leq \theta < 360°$ since the orientational effect of every rotation is equivalent to a rotation lying in this range. For example, a rotation through $-1072°$ is equivalent to one through $-1072° + 3 \times 360° = 8°$. Notice that the range of non-equivalent rotations, $0° \leq \theta < 360°$, does *not* include 360°. This is because a rotation through 360° is equivalent to one through 0°, which *is* included.

Find a rotation angle θ in the range $0° \leq \theta < 360°$ that has the equivalent orientational effect to each of the following: 423.6°, $-3073.35°$ and 360°.

$423.6° - 360° = 63.6°$;

$-3073.35° + 9 \times 360° = 166.65°$;

$360° - 360° = 0°.$ □

Of course, angles that differ by a multiple of 360° are not equivalent in every way. For example, a wheel that rotates through 36 000° and thus completes 100 revolutions would have done something physically different from an identical wheel that only rotated through an angle of 360°, even if their final orientations were the same.

Despite the widespread use of degrees, a more natural (and important) unit of angular measure is the **radian**. As Figure 5.2 indicates, the radian may be defined as the angle subtended at the centre of a circle by an **arc** of the circle which has an **arc length** equal to the radius of the circle. As will be shown below, it follows from this definition that 1 radian (often abbreviated as 1 rad or sometimes 1^c) is equal to 57.30°, to two decimal places.

Radians are such a natural and widely used unit of angular measure that whenever you see an angular quantity quoted without any indication of the associated unit of measurement, you should assume that the missing unit is the radian.

An arc of a circle is a curve forming part of the circumference of that circle, and the arc length is the length of such a curve, measured along the curve. The meaning of 'subtended' can be seen from the figure below.

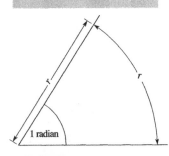

Figure 5.2   An angle of one radian.

Angular quantities such as ϕ generally represent the product of a number and a unit of angular measure such as a degree or a radian. Thus, the quotient ϕ/rad represents a pure number and may be read as 'the numerical value of ϕ measured in radians'.

In general, as indicated in Figure 5.3, if an arc length s at a distance r subtends an angle ϕ at the centre of the circle, then the value of ϕ, measured in radians, is:

$$\phi = \frac{s}{r}\,\text{rad} \tag{1}$$

i.e. ϕ/rad $= s/r$

This is a sensible definition of an angle since it is independent of the scale of Figure 5.3. For a given value of ϕ, a larger value of r would result in a larger value of s but the ratio s/r would be unchanged.

To determine the fixed relationship between radians and degrees it is probably easiest to consider the angle subtended at the centre of a circle by the complete circumference of that circle. A circle of radius r has a circumference of arc length $2\pi r$, where π represents the mathematical constant **pi**, an *irrational number* the value of which is 3.1416 to four decimal places. Consequently, the ratio of circumference to radius is 2π and the angle subtended at the centre of any circle by its circumference is 2π rad. But the complete angle at the centre of a circle is 360°, so we obtain the general relationship

$$2\pi \text{ radians} = 360°$$

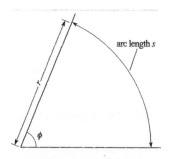

Figure 5.3 An angle measured in radians.

Note An *irrational number* is one that *cannot* be expressed as a quotient of two whole numbers, and hence extends to an infinite number of decimal places.

Since $2\pi = 6.2832$ (to four decimal places) it follows that 1 radian = 57.30°, as claimed earlier. Table 5.1 gives some angles measured in degrees and radians. As you can see from this table, many commonly-used angles are simple fractions or multiples of π radians, but note that angles expressed in radians are not *always* expressed in terms of π. Do not make the common mistake of thinking that π is some kind of angular unit; it is simply a number.

What is 10° expressed in radians?

$$\frac{10°}{180°} \times \pi \text{ rad} \approx \frac{\pi \text{ rad}}{18} \approx 0.175 \text{ rad} \quad \square$$

Question T1

The usual way of converting a quantity expressed in terms of one set of units into some other set of units is to multiply it by an appropriate *conversion factor*. For example, multiplying a distance measured in kilometres by the conversion factor 1000 m km^{-1} will provide the equivalent distance in metres. What are the conversion factors from radians to degrees and from degrees to radians? $\square$

Question T2

What is the non-negative rotation angle less than 2π rad that has the same orientational effect as a rotation through -201π rad? $\square$

Table 5.1 Equivalent angles in degrees and radians

Angle measured in degrees	Angle measured in radians
360	2π
180	π
90	$\pi/2$
60	$\pi/3$
45	$\pi/4$
30	$\pi/6$

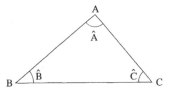

Figure 5.4 A general triangle.

AB̂C is spoken of as 'the angle ABC' and B̂ as 'the angle B'.

Another alternative notation, not used in this chapter, for the angle AB̂C is ∠ABC.

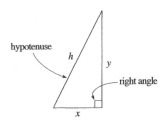

Figure 5.5 A right-angled triangle with sides of length x, y and h.

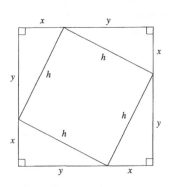

Figure 5.6 Pythagoras's theorem can be derived by writing expressions for the area of the large square.

5.2.2 Right-angled triangles

Figure 5.4 shows a general triangle – a geometrical figure made up of three straight lines. There are three angles formed by the intersections of each pair of lines. By convention, the three angles are labelled according to the corresponding vertices as Â, B̂ and Ĉ. An alternative notation is to use AB̂C instead of B̂, where AB and BC are the two lines defining the angle. In this chapter we use the AB̂C and B̂ notations interchangeably. (☜) The angles Â, B̂ and Ĉ in Figure 5.4 are known as the **interior angles** of the triangle.

The sum of the interior angles of *any* triangle is 180° (π radians). To see that this is so, imagine an arrow located at point A, superimposed on the line AB and pointing towards B. Rotate the arrow anticlockwise about the point A through an angle Â so that it is aligned with AC and points towards C. Then rotate the arrow through an additional angle Ĉ so that it is horizontal and points to the right, parallel with BC. Finally, rotate the arrow through an additional angle B̂ about point A. After this final rotation the arrow will again be aligned with the line AB, but will now point away from B having rotated through a total angle of 180°. So, rotating through Â + B̂ + Ĉ is equivalent to half a complete turn, i.e. 180°.

For the time being, we will be concerned only with the special class of triangles in which one interior angle is 90°. There are infinitely many such triangles, since the other two interior angles may have any values provided their sum is also 90°. Any such triangle is called a **right-angled triangle** (see Figure 5.5), and the side opposite the right angle is known as the **hypotenuse**. Notice the use of the box symbol, ∟, to denote a right angle. Right-angled triangles are important because:

- There is a special relationship between the lengths of the sides (Pythagoras's theorem).
- There is a strong link between right-angled triangles and the trigonometric ratios (sine, cosine and tangent).
- Applications involving right-angled triangles often occur in physics (and many other branches of science and engineering).

We shall consider each of these points in this chapter, but the rest of this subsection will be devoted to Pythagoras's theorem.

Pythagoras's theorem states that the square of the hypotenuse of a right-angled triangle is equal to the sum of the squares of the other two sides.

i.e. $h^2 = x^2 + y^2$

═══

Consideration of Figure 5.6 shows how this result comes about for a general right-angled triangle of sides x, y and h. One way of finding the area of the large outer square is by squaring the length of its sides, i.e.

$$\text{area} = (x + y)^2 = x^2 + y^2 + 2xy \tag{2}$$

However, the area of the large square can also be found by adding the area of the smaller square, h^2, to the areas of the four corner triangles. Each triangle has an area $xy/2$ (each is half a rectangle of sides x and y) so the area of the large square is

$$\text{area} = h^2 + 4xy/2 = h^2 + 2xy \tag{3}$$

Comparing the right-hand sides of Equations 2 and 3 shows that

$$h^2 = x^2 + y^2 \tag{4}$$

which is Pythagoras's theorem.

The *converse* of Pythagoras's theorem is also true; that is, if the sum of the squares of two sides of a triangle is equal to the square of the other side, then the triangle is right-angled. For the purpose of this chapter, we will accept the validity of the converse without proving it.

The angles of a triangle are 45°, 90° and 45° and two of the sides (that is, those sides opposite the 45° angles) have lengths of 10 m. What is the length of the hypotenuse?

By Pythagoras's theorem, the length of the hypotenuse is

$$\sqrt{10^2 + 10^2}\text{ m} = \sqrt{200}\text{ m}$$

$$= 14.14\text{ m (to two decimal places)}. \quad \square$$

The hypotenuse of a right-angled triangle is 7 m long while one of the other sides is of length 3 m. What is the length of the remaining side?

If we denote the length of the remaining side by x, then, from Pythagoras's theorem,

$$7^2\text{ m}^2 = 3^2\text{ m}^2 + x^2,$$

so that

$$x^2 = (7^2 - 3^2)\text{ m}^2 = (49 - 9)\text{ m}^2 = 40\text{ m}^2$$

and therefore

$$x = \sqrt{40}\text{ m} = 6.32\text{ m (to two decimal places)}. \quad \square$$

The sides of some right-angled triangles can be expressed entirely in terms of integers; probably the most famous is the $3:4:5$ triangle where the hypotenuse has length 5 units and the other two sides have lengths 3 and 4 units: $3^2 + 4^2 = 5^2$.

Note   Throughout the remainder of this chapter we will not usually express lengths in any particular units. This is because we are generally interested only in the *ratios* of lengths. Of course, when you are considering real physical situations, you must attach appropriate units to lengths.

Question T3

Show that triangles with sides in the ratios $5 : 12 : 13$ and $8 : 15 : 17$ are also right-angled triangles.   □

Question T4

Use Pythagoras's theorem to show that the hypotenuse is always the longest side of a right-angled triangle. (*Hint*: Consider $a^2 = b^2 + c^2$.)   □

5.2.3   The trigonometric ratios

In the previous subsection, we indicated that the ratios of the lengths of the sides of a triangle were often of more interest than the actual lengths themselves. Figure 5.7 shows some **similar triangles**, i.e. triangles that are the same shape but different sizes − in other words, triangles with corresponding angles that are equal but with corresponding sides of different lengths. Although the *lengths* of the sides of any one triangle may differ from those of any similar triangle, the *ratios* of the side lengths are the same in each triangle − for example, each triangle in Figure 5.7 has sides whose lengths are in the ratio $2 : 3 : 4$. In future, whenever we say that two or more triangles are similar we will mean it in the technical sense that they have the same interior angles and side lengths that are in the same ratio.

Figure 5.8 shows a right-angled triangle in which an angle θ has been marked for particular attention and the **opposite side** and **adjacent side** to this angle have been identified. Thus the three sides may be referred to as the opposite, the adjacent and the hypotenuse, and we may use these terms or the letters o, a and h to refer to their respective lengths. Thanks to the

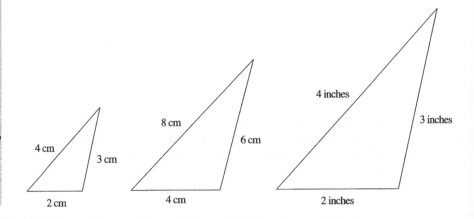

Figure 5.7 Some similar triangles. Each triangle has the same interior angles, and the lengths of the sides of any one triangle are in the same ratio to one another ($2 : 3 : 4$ in this case) as the lengths of the sides of any of the other similar triangles. (These are not to scale.)

special properties of right-angled triangles, the whole class of triangles that are similar to the triangle in Figure 5.8 can be characterised by the single angle θ, or, equivalently, by the ratio of the side lengths $o : a : h$. In fact, because of the Pythagorean relationship that exists between the sides of a right-angled triangle, the ratio of any *two* side lengths is sufficient to determine θ and identify the class of similar triangles to which a given right-angled triangle belongs. The ratios of the sides of right-angled triangles are therefore of particular importance. The study of right-angled triangles is known as **trigonometry**, and the three distinct ratios of pairs of sides are collectively known as the **trigonometric ratios**. They are called the **sine**, **cosine** and **tangent** of the angle θ – abbreviated to sin, cos and tan, respectively – and defined as follows:

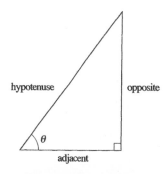

Figure 5.8 The labelling of sides in a right-angled triangle.

$$\sin \theta = \frac{\text{opposite}}{\text{hypotenuse}} \tag{5}$$

$$\cos \theta = \frac{\text{adjacent}}{\text{hypotenuse}} \tag{6}$$

$$\tan \theta = \frac{\text{opposite}}{\text{adjacent}} \tag{7}$$

The abbreviations used for trigonometric ratios are not entirely standard. Some texts, for example, abbreviate tangent as tg.

Given three side lengths o, a and h, it is possible to form *six* different ratios: o/h, a/h, o/a, h/o, h/a and a/o. However, the last three of these are merely the reciprocals of the first three and are therefore known as the *reciprocal trigonometric ratios*. Nonetheless, they are of some interest and are dealt with in the next subsection.

The angle θ that appears in these definitions must lie between 0 rad (0°) and $\pi/2$ rad (90°), but later in this chapter we will extend the definitions to all angles. It should be emphasised that the value of a particular trigonometric ratio depends only on the value of θ, so the sine, cosine and tangent are effectively *functions* of the angle θ. More will be made of this point in Section 5.3 when we actually define related quantities called the *trigonometric functions*.

It is extremely useful to remember the definitions of the trigonometric ratios. You may find it helpful to denote the sine, cosine and tangent by the letters s, c and t and then, using h, o and a to represent hypotenuse, opposite and adjacent, the three relations read, left to right and top to bottom, as *soh*, *cah* and *toa*.

You can use a calculator to find the sine, cosine or tangent of an angle expressed in either degrees or radians, provided you first switch it to the appropriate mode – this is usually done by pressing a key marked 'DRG' (or something similar) until either 'degrees' or 'radians' appears in the display. Then key in the angle followed by one of the function keys sin, cos or tan.

Note that we have used brackets to distinguish, for example, $\tan \dfrac{\pi}{3}$ from $\dfrac{\tan \pi}{3}$. Never hesitate to use brackets in this way if it will reduce the chance of confusion.

Use a calculator to find sin (15°); cos (72°); sin ($\pi/4$); tan ($\pi/3$); cos (0.63).

sin (15°) = 0.2588; cos (72°) = 0.3090; sin ($\pi/4$) = 0.7071; tan ($\pi/3$) = 1.732; cos (0.63) = 0.8080. (Note that the last three angles should be assumed to be in radians since no units are stated.)   □

The three trigonometric ratios are not all independent since we can write:

$$\tan \theta = \frac{\text{opposite}}{\text{adjacent}} = \frac{\text{opposite}}{\text{hypotenuse}} \times \frac{\text{hypotenuse}}{\text{adjacent}} = \sin \theta \times \frac{1}{\cos \theta}$$

and so we obtain the identity:

$$\tan \theta = \frac{\sin \theta}{\cos \theta} \qquad (8)$$

Furthermore, we can use Pythagoras's theorem to obtain a second relation. Starting with:

$$(\text{opposite})^2 + (\text{adjacent})^2 = (\text{hypotenuse})^2$$

we can divide both sides by $(\text{hypotenuse})^2$ to obtain

$$\frac{(\text{opposite})^2}{(\text{hypotenuse})^2} + \frac{(\text{adjacent})^2}{(\text{hypotenuse})^2} = 1$$

and therefore

$$(\sin \theta)^2 + (\cos \theta)^2 = 1$$

As you can see, writing powers of trigonometric functions can be rather cumbersome and so the convention that $\sin^n \theta$ means $(\sin \theta)^n$ (for *positive* values of n) is often used. Similar conventions are used for the other trigonometric functions. The notation cannot be used for negative values of n since $\sin^{-1} \theta$ is sometimes used for the inverse sine function, which we consider later in this chapter. The above relation can therefore be written as:

$$\sin^2 \theta + \cos^2 \theta = 1 \qquad (9)$$

Since there are two identities relating the trigonometric ratios, it follows that only one ratio is independent and therefore given one ratio we can find the other two. (This assumes that the trigonometric ratios are positive, which is true for $0° \leqslant \theta < 90°$.)

The angle ϕ in Figure 5.9 also has its sine, cosine and tangent. But the opposite and adjacent sides appropriate to θ are interchanged for ϕ and, as a consequence we can write

$$\tan \theta = \frac{1}{\tan \phi}, \sin \theta = \cos \phi, \cos \theta = \sin \phi$$

There are some special angles for which it is easy to write down the sine, cosine and tangent. As an example, consider the right-angled triangle with two sides of equal length, as shown in Figure 5.10.

Any triangle with two sides of equal length is called an **isosceles triangle**, and any isosceles triangle must contain two equal interior angles. The

Figure 5.9 A right-angled triangle with angles θ and ϕ.

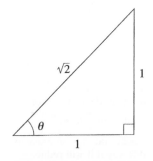

Figure 5.10 A right-angled triangle with two equal length sides, an isosceles triangle.

isosceles triangle of Figure 5.10 is special because it is also a right-angled triangle. Since the interior angles of any triangle add up to 180°, the angles of this particular triangle must be 45°, 90°, 45°. Also, since the two equal sides of this particular triangle are both of unit length it follows from Pythagoras's theorem that the length of the hypotenuse is $\sqrt{1^2 + 1^2} = \sqrt{2}$ and so we can write down the following results:

$$\sin(45°) = \frac{\text{opposite}}{\text{hypotenuse}} = \frac{1}{\sqrt{2}}$$

$$\cos(45°) = \frac{\text{adjacent}}{\text{hypotenuse}} = \frac{1}{\sqrt{2}}$$

$$\tan(45°) = \frac{\text{opposite}}{\text{adjacent}} = 1$$

Figure 5.11 shows an **equilateral triangle**, i.e. one with three sides of equal length and hence three equal interior angles which must be equal to 60°. A line has been drawn from one vertex (i.e. corner) to the middle of the opposite side, so that the angle between the line and the side is 90° (that is, the line is a *normal* to the side).

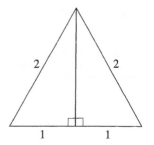

Figure 5.11 An equilateral triangle.

Question T5

By considering Figure 5.11, find the values of sin θ, cos θ and tan θ for θ equal to 30° ($\pi/6$ rad) and 60° ($\pi/3$ rad), and hence complete the trigonometric ratios in Table 5.2. ☐

Table 5.2 See Question T5.

θ/degrees	θ/radians	sin θ	cos θ	tan θ
0	0	0	1	0
30	$\pi/6$			
45	$\pi/4$	$1/\sqrt{2}$	$1/\sqrt{2}$	1
60	$\pi/3$			
90	$\pi/2$	1	0	undefined

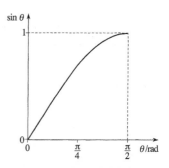

Figure 5.12 See Question T6.

Question T6

Figure 5.12 shows a graph of sin θ for $0 \leqslant \theta < \pi/2$. Using Table 5.2, your answer to Question T5, and any other relevant information given in this subsection, sketch corresponding graphs for cos θ and tan θ. ☐

In order to sketch the graphs in Question T6, we are assuming that the curve is smooth between the points where we have calculated its value.

The trigonometric functions can be evaluated directly using expressions involving θ. This subject is covered in the chapter on expansions in the companion volume.

One of the many reasons why trigonometric ratios are of interest to physical scientists is that they make it possible to determine the lengths of all the sides of a right-angled triangle from a knowledge of just one side length and one interior angle (other than the right angle). Here is an example of this procedure taken from the field of optics.

Question T7

Figure 5.13 represents a two-slit diffraction experiment. Derive a formula relating λ to d and θ. If $d = 2 \times 10^{-6}\,\text{m}$ and $\theta = 17.46°$, what is the value of λ?   □

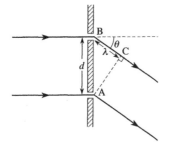

Figure 5.13   See Question T7.

5.2.4   The reciprocal trigonometric ratios

The ratios introduced in the previous subsection could all have been written the other way up. The resulting **reciprocal trigonometric ratios** occur so frequently that they too are given specific names; they are the **cosecant**, **secant**, and **cotangent** (abbreviated to cosec, sec and cot) and are defined by:

$$\operatorname{cosec} \theta = \frac{1}{\sin \theta} \quad \text{provided } \sin \theta \neq 0 \tag{10}$$

$$\sec \theta = \frac{1}{\cos \theta} \quad \text{provided } \cos \theta \neq 0 \tag{11}$$

$$\cot \theta = \frac{1}{\tan \theta} \quad \text{provided } \tan \theta \neq 0 \tag{12}$$

Notice that cosec is the reciprocal of sin, and sec the reciprocal of cos. This terminology may seem rather odd but it is easily remembered by recalling that each reciprocal pair − (sin, cosec), (cos, sec), (tan, cot) − involves the letters 'co' just once. In other words there is just one 'co' between each pair. Also notice that each reciprocal trigonometric function is undefined when its partner function is zero.

Throughout the domains on which they are defined, each of the reciprocal trigonometric ratios can also be written in terms of the sides of the triangle in Figure 5.8:

An alternative abbreviation for cosec is csc.

$$\operatorname{cosec} \theta = \frac{\text{hypotenuse}}{\text{opposite}} \tag{13}$$

$$\sec \theta = \frac{\text{hypotenuse}}{\text{adjacent}} \tag{14}$$

$$\cot \theta = \frac{\text{adjacent}}{\text{opposite}} \tag{15}$$

Two useful general relationships follow from Equations 13–15 and Pythagoras's theorem:

$$\cot \theta = \frac{\operatorname{cosec} \theta}{\sec \theta} \qquad (16)$$

and $1 + \tan^2 \theta = \sec^2 \theta$ \qquad (17)

Question T8

Using Figure 5.11, write down the values of cosec (30°), sec (30°), cot (30°) and cosec (60°), sec (60°), cot (60°). □

Calculators do not generally have keys that give the reciprocal trigonometric ratios directly, but the ratios can be found using the sin, cos and tan keys and the reciprocal (1/x) key.

Question T9

Use a calculator to find cosec (23°), sec (56°), cot (π/6), cot (1.5). □

Question T10

Figure 5.14 shows a graph of cosec θ for $0 < \theta < \pi/2$. Using values of reciprocal trigonometric ratios calculated above, and other information from this subsection, sketch graphs of sec θ and cot θ for $0 \leqslant \theta < \pi/2$. □

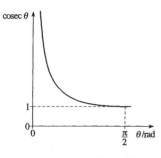

Figure 5.14 See Question T10.

5.2.5 Small angle approximations

We end this section with some useful approximations involving small angles. Figure 5.15 shows a right-angled triangle with one very small angle θ and the third angle almost a right angle. If θ is at the centre of a circle radius r, where r is the hypotenuse of the triangle, you can see from the diagram that the opposite side to θ is almost coincident with the arc length s and the adjacent side to θ is almost the same length as the hypotenuse. From Equation 1, s/r is the value of θ *in radians*. So, for the small angle θ, Equations 5 to 7 give $\sin \theta \approx s/r$, $\cos \theta \approx r/r$, $\tan \theta \approx s/r$ and hence:

for a small angle θ $\cos \theta \approx 1$

and $\sin \theta \approx \tan \theta \approx \theta/\text{rad}$

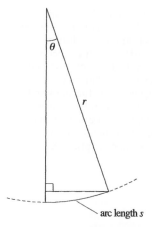

Figure 5.15 A right-angled triangle with a small angle θ.

Use a calculator to find sin θ, cos θ and tan θ for a few small angles, and hence show that the approximations expressed in the boxed equations above become increasingly good as θ becomes smaller. Try, for example, $\theta = 0.175\,00$ rad (i.e. $\theta \approx 10°$) and $\theta = 0.010\,00$ rad, and express the answers to five decimal places.

$$\sin(0.175\,00) = 0.174\,11; \qquad \cos(0.175\,00) = 0.984\,73;$$

$$\tan(0.175\,00) = 0.176\,81. \qquad \sin(0.010\,00) = 0.010\,00;$$

$$\cos(0.010\,00) = 0.999\,95; \qquad \tan(0.010\,00) = 0.010\,00. \quad \square$$

Write down approximate expressions for the reciprocal trigonometric ratios for a small angle θ.

$$\operatorname{cosec}\theta \approx \cot\theta \approx 1/(\theta/\mathrm{rad}); \qquad \sec\theta \approx 1. \quad \square$$

Question T11

Seen from Earth, the diameter of the Sun subtends an angle ϕ of about 0.5°. By expressing ϕ in radians, derive an expression for the Sun's diameter, s, in terms of its distance d from Earth. Your expression should *not* involve any trigonometric ratios. $\quad \square$

5.3 **Graphs and trigonometric functions**

5.3.1 *The trigonometric functions*

Throughout this section we will be careful to use brackets (as in sin (θ)) to distinguish the trigonometric functions from the trigonometric ratios (sin θ, etc.), but, since the trigonometric functions and ratios agree in those regions where they are both defined, this distinction is of little importance in practice.

The ratio definitions of the sine, cosine and tangent (i.e. Equations 5, 6 and 7) only make sense for angles in the range 0 to $\pi/2$ radians, since they involve the sides of a right-angled triangle. In this subsection we will define three **trigonometric functions**, also called sine, cosine and tangent, and denoted $\sin(\theta)$, $\cos(\theta)$ and $\tan(\theta)$, respectively. These functions will enable us to attach a meaning to the sine and cosine of *any* angle, and to the tangent of any angle that is not an odd multiple $\pi/2$. Like the trigonometric ratios that they generalise, these trigonometric functions are of great importance in physical science.

In defining the trigonometric functions we want to ensure that they will agree with the trigonometric ratios over the range 0 to $\pi/2$ radians. In order to do this, consider the point P shown in Figure 5.16 that moves on a circular path of unit radius around the origin O of a set of two-dimensional Cartesian coordinates (x, y). If we let θ be the angle between the line OP and the x-axis, it follows from the definition of the trigonometric ratios that the coordinates of P are

$$y = \sin \theta \quad \text{for } 0 \leqslant \theta < \pi/2$$

and $\quad x = \cos \theta \quad$ for $0 \leqslant \theta < \pi/2$

Although these trigonometric ratios are only defined over a very narrow range, it is easy to imagine the angle θ increasing in the positive (i.e. anticlockwise) direction to take up *any* positive value, with P crossing the positive x-axis whenever θ is equal to an integer multiple of 2π. And it is equally easy to imagine θ increasing in the negative (clockwise) direction to take up *any* negative value. Now, whatever the value of θ may be, large or small, positive or negative, the point P must still be located somewhere on the circle in Figure 5.16 and it must have a single x- and a single y-coordinate. We can use the particular values of x and y that correspond to a given value of θ to *define* the first two trigonometric functions

$$\sin (\theta) = y \quad \text{for any } \theta$$

and $\quad \cos (\theta) = x \quad$ for any θ

Defined in this way, it is inevitable that the trigonometric functions will agree with the trigonometric ratios when $0 \leqslant \theta \leqslant \pi/2$, but it is also clear that the functions, unlike the ratios, make sense for arbitrary values of θ.

Having defined the functions $\sin (\theta)$ and $\cos (\theta)$ we also want to define a function $\tan (\theta)$, but once again we want to ensure consistency with the behaviour of the trigonometric ratio $\tan \theta$ that was introduced earlier. We can do this by using a generalisation of Equation 8 (i.e. $\tan \theta = \sin \theta/\cos \theta$) as the basis of the definition. Thus, we define

$$\tan (\theta) = \frac{\sin (\theta)}{\cos (\theta)} \quad \text{for any } \theta \neq (2n + 1)\pi/2$$

Notice that unlike the sine and cosine functions, this function is not defined for values of θ that are odd integer multiples of $\pi/2$. This restriction is imposed because $\cos \theta = 0$ at those values of θ, and the quotient $\sin (\theta)/\cos (\theta)$ has no meaning. (☞)

Having defined the trigonometric functions it is only natural to enquire about their graphs, since graphs are usually a good way of gaining insight into the behaviour of functions. Perhaps the first thing to notice about the trigonometric functions is that they are not always positive. If θ is in the range $0 \leqslant \theta < \pi/2$ both the x- and y-coordinates of P will be positive, so both $\sin (\theta)$ and $\cos (\theta)$ will be positive as will their quotient $\tan (\theta)$. However, as θ enters the range $\pi/2 < \theta < \pi$ the x-coordinate of P becomes negative, so $\cos (\theta)$ and $\tan (\theta)$ will be negative, though $\sin (\theta)$ will remain positive. Similarly, when $\pi < \theta < 3\pi/2$, x and y are both negative so $\sin (\theta)$ and $\cos (\theta)$ are negative while $\tan (\theta)$ is positive; and if $3\pi/2 < \theta < 2\pi$, $\cos (\theta)$ is positive while $\sin (\theta)$ and $\tan (\theta)$ are negative. As θ increases beyond 2π (or when θ decreases below 0) the same pattern is repeated.

Figure 5.17 summarises the sign behaviour of the trigonometric functions. The positive function in each quadrant is indicated by its initial letter, or in the case of the first quadrant where all the functions are positive by the letter A. Most people who use the trigonometric functions find it helpful to memorise Figure 5.17 (or to remember Figure 5.16, so that they can work it

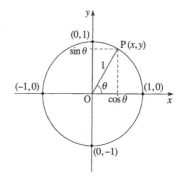

Figure 5.16 Defining the trigonometric functions for any angle. If $0 \leqslant \theta < \pi/2$, the coordinates of P are $x = \cos \theta$ and $y = \sin \theta$. For general values of θ we define $\sin (\theta) = y$ and $\cos (\theta) = x$.

If n may be any positive or negative whole number, $(2n + 1)$ represents an odd integer. So the condition on the right of the definition is a way of saying that the equation applies for any value of θ that is *not* an odd multiple of $\pi/2$.

1/0 has no meaning. It is not even correct to say that it is infinity.

A *quadrant* is a quarter of a circle.

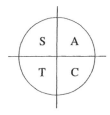

A all functions

S sine

T tangent

C cosine

Figure 5.17 Quadrants in which the trigonometric functions are positive.

out). A traditional mnemonic to help recall which letter goes in which quadrant is 'All Stations To Crewe', which gives the letters in positive (anticlockwise) order starting from the first quadrant.

Of course, it is not only the signs of the trigonometric functions that change as θ increases or decreases and P moves around the circle in Figure 5.16. The *values* of x and y, and consequently of sin (θ), cos (θ) and tan (θ) also vary.

If you were asked to draw a diagram similar to Figure 5.17, but showing which trigonometric function(s) increase as θ increases in each quadrant, how would you have to change the lettering on Figure 5.17?

A would become S, T (both sin (θ) and tan (θ) are increasing from zero in the first quadrant.)

S would become T (as sin (θ) decreases you might think that tan (θ) would also decrease, but cos (θ) is negative and decreasing in the second quadrant so tan (θ) becomes a smaller negative number as θ increases, i.e. the value of tan (θ) *increases*.)

T would become C, T (in this quadrant it is cos (θ) that is becoming less negative.)

C would become A (sin (θ) and tan (θ) are both becoming less negative and cos (θ) is increasing from zero in this quadrant.) □

The graphs of sin (θ), cos (θ) and tan (θ) are shown in Figures 5.18–5.20. As you can see, the values sin (θ) and cos (θ) are always in the range -1 to 1, and any given value is repeated each time θ increases or decreases by 2π.

Table 5.3   The trigonometric functions between 0 and 2π

θ/radians	sin (θ)	cos (θ)	tan (θ)
0	0	1	0
$\pi/6$	1/2	$\sqrt{3}/2$	$1/\sqrt{3}$
$\pi/4$	$1/\sqrt{2}$	$1/\sqrt{2}$	1
$\pi/3$	$\sqrt{3}/2$	1/2	$\sqrt{3}$
$\pi/2$	1	0	undefined
$2\pi/3$	$\sqrt{3}/2$	$-1/2$	$-\sqrt{3}$
$3\pi/4$	$1/\sqrt{2}$	$-1/\sqrt{2}$	-1
$5\pi/6$	1/2	$-\sqrt{3}/2$	$-1/\sqrt{3}$
π	0	-1	0
$7\pi/6$	$-1/2$	$-\sqrt{3}/2$	$1/\sqrt{3}$
$5\pi/4$	$-1/\sqrt{2}$	$-1/\sqrt{2}$	1
$4\pi/3$	$-\sqrt{3}/2$	$-1/2$	$\sqrt{3}$
$3\pi/2$	-1	0	undefined
$5\pi/3$	$-\sqrt{3}/2$	1/2	$-\sqrt{3}$
$7\pi/4$	$-1/\sqrt{2}$	$1/\sqrt{2}$	-1
$11\pi/6$	$-1/2$	$\sqrt{3}/2$	$-1/\sqrt{3}$
2π	0	1	0

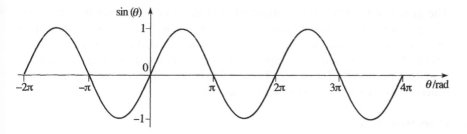

Figure 5.18   Graph of sin (θ).

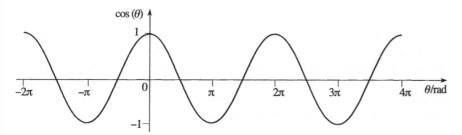

Figure 5.19   Graph of cos (θ).

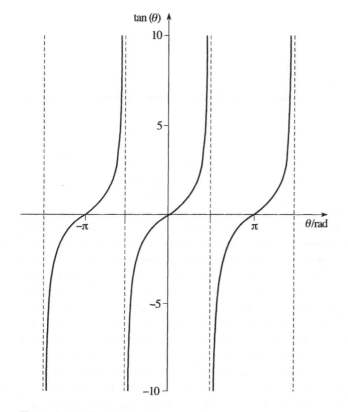

Figure 5.20   Graph of tan (θ).

The graph of tan (θ) is quite different. Values of tan (θ) cover the full range of real numbers, but tan (θ) tends towards $+\infty$ as θ approaches odd multiples of $\pi/2$ from below, and towards $-\infty$ as θ approaches odd multiples of $\pi/2$ from above. This emphasises the impossibility of assigning a meaningful value to tan (θ) at odd multiples of $\pi/2$. Table 5.3 lists the values of the trigonometric functions for some angles between 0 and 2π.

Question T12

Describe as many significant features as you can of the graphs in Figures 5.18 and 5.19. (Some of these features will be discussed in the next subsection.) □

Given the trigonometric functions, we can also define three **reciprocal trigonometric functions** cosec (θ), sec (θ) and cot (θ), that generalise the reciprocal trigonometric ratios defined in Equations 10–12. The definitions are straightforward, but a little care is needed in identifying the appropriate domain of definition in each case. (As usual we must choose the domain in such a way that we are not required to divide by zero at any value of θ.)

$$\operatorname{cosec}(\theta) = \frac{1}{\sin(\theta)} \qquad \theta \neq n\pi$$

$$\sec(\theta) = \frac{1}{\cos(\theta)} \qquad \theta \neq (2n+1)\pi/2$$

$$\cot(\theta) = \frac{\cos(\theta)}{\sin(\theta)} \qquad \theta \neq n\pi$$

Graphs of the reciprocal trigonometric functions are shown in Figures 5.21–5.23.

Throughout this subsection the argument θ of the various trigonometric and reciprocal trigonometric functions has always been an angle measured in radians. (This is true even though we have been conventionally careless about making sure that we always include the appropriate angular unit when assigning numerical values to θ.) However, the arguments of these functions do not *have* to be angles. If we regarded the numbers printed along the horizontal axes of Figures 5.18–5.23 as values of a purely numerical variable, x say, rather than values of θ in radians, we could regard the graphs as *defining* six functions of x; sin (x), cos (x), tan (x), etc. Strictly speaking these new functions are quite different from the trigonometric ratios and should be given different names to avoid confusion. But, given the tendency of physical scientists to be careless about domains and their habit of 'dropping' the explicit mention of radian from angular values, there is no practical difference between these new functions and true trigonometric ratios, so the confusion of names is harmless. Nonetheless, it is worth remembering that what appears as the argument of a trigonometric function is not necessarily an angle.

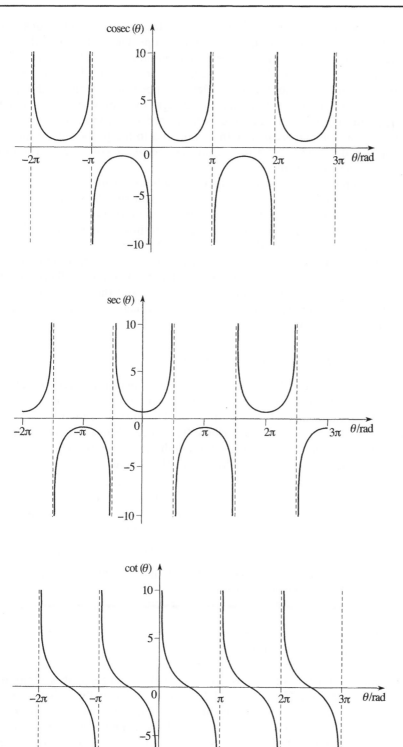

Figure 5.21 Graph of cosec (θ).

Figure 5.22 Graph of sec (θ).

Figure 5.23 Graph of cot (θ).

A common example of this arises in the study of oscillations where trigonometric functions are used to describe repeated back and forth motion along a straight line. In the simplest such motion, *simple harmonic motion*, the changing position x of a mass oscillating on the end of a spring may be represented by $x = A \cos(\omega t + \phi)$. Despite appearances none of the quantities inside the bracket is an angle (though they may be given angular interpretations); t is the time and is measured in seconds, ω is a constant known as the *angular frequency* that is related to the properties of the mass and spring and is measured in s^{-1}, and ϕ, the *phase constant*, is a number, usually in the range 0 to 2π.

One final point to note. As mentioned before, throughout this subsection we have been careful to use brackets (as in $\sin(\theta)$) to distinguish the trigonometric functions from the trigonometric ratios ($\sin\theta$, etc.), but since the trigonometric functions and ratios agree in those regions where they are both defined this distinction is also of little importance in practice. Consequently, as a matter of convenience, the brackets are usually omitted from the trigonometric functions unless such an omission is likely to cause confusion. In much of what follows we too will omit them and simply write the trigonometric and reciprocal trigonometric functions as $\sin x$, $\cos x$, $\tan x$, $\csc x$, $\sec x$ and $\cot x$.

5.3.2 Periodicity and symmetry

The trigonometric functions are all examples of **periodic functions**. That is, as θ increases steadily, the same sets of values are 'recycled' many times over, always repeating exactly the same pattern. The graphs in Figures 5.18−5.20 show this repetition, known as **periodicity**, clearly. More formally, a periodic function $f(x)$ is one which satisfies the condition $f(x) = f(x + nk)$ for every integer n, where k is a constant, known as the **period**.

Note that when we are dealing with a function of a sum of two angles the angle *must* be written in brackets to avoid ambiguity.

Adding or subtracting any multiple of 2π to an angle is equivalent to performing any number of complete rotations in Figure 5.16, and so does not change the value of the sine or cosine:

$$\sin(\theta) = \sin(\theta + 2n\pi) \quad \text{and} \quad \cos(\theta) = \cos(\theta + 2n\pi)$$

for any integer n. The functions $\sin(\theta)$ and $\cos(\theta)$, are therefore *periodic functions* with *period* 2π.

What is the period of $\tan(\theta)$?

Since $\tan(\theta) = \sin(\theta)/\cos(\theta)$ (if $\cos(\theta)$ is non-zero) it is tempting to say that $\tan(\theta)$ has period 2π, but we can actually do better than this. From Figure 5.20, you can see that it repeats with a period of π. You can see why this is so if you look at Figure 5.16. Rotating P through π radians leaves the sizes of x and y unchanged, but changes the sign of both of them, with the result that $\tan\theta \, (= y/x)$ will be unaffected. □

As noted in the answer to Question T12, the trigonometric functions have some symmetry either side of $\theta = 0$. From Figures 5.18–5.20 we can see the effect of changing the sign of θ:

$\sin(-\theta) = -\sin(\theta)$	(18)
$\cos(-\theta) = \cos(\theta)$	(19)
$\tan(-\theta) = -\tan(\theta)$	(20)

Any function $f(x)$ for which $f(-x) = f(x)$ is said to be **even** or **symmetric**, and will have a graph that is symmetrical about $x = 0$. Any function for which $f(-x) = -f(x)$ is said to be **odd** or **antisymmetric**, and will have a graph in which the portion of the curve in the region $x < 0$ appears to have been obtained by reflecting the curve for $x > 0$ in the vertical axis and then reflecting the resulting curve in the horizontal axis. It follows from Equations 5.18–5.20 that $\cos(\theta)$ is an even function, while $\sin(\theta)$ and $\tan(\theta)$ are both odd functions.

For each of the reciprocal trigonometric functions, state the period and determine whether the function is odd or even.

$\mathrm{cosec}(\theta)$ is an odd function with period 2π.

$\sec(\theta)$ is an even function with period 2π.

$\cot(\theta)$ is an odd function with period π. $\square$

An arbitrary function $f(x)$ is not necessarily even or odd, but any function can always be written as a sum of even and odd functions.

The combination of periodicity with symmetry or antisymmetry leads to further relationships between the trigonometric functions. For example, from Figures 5.18 and 5.19 you should be able to see that the following relationships hold true:

$$\sin(\theta) = \sin(\pi - \theta) = -\sin(\pi + \theta) = -\sin(2\pi - \theta) \tag{21}$$

and $\cos(\theta) = -\cos(\pi - \theta) = -\cos(\pi + \theta) = \cos(2\pi - \theta) \tag{22}$

Less apparent, but just as true, are the following relationships:

$$\cos(\theta) = \sin(\pi/2 - \theta) = \sin(\pi/2 + \theta)$$
$$= -\sin(3\pi/2 - \theta) = -\sin(3\pi/2 + \theta) \tag{23}$$

$$\sin(\theta) = \cos(\pi/2 - \theta) = -\cos(\pi/2 + \theta)$$
$$= -\cos(3\pi/2 - \theta) = \cos(3\pi/2 + \theta) \tag{24}$$

Thanks to periodicity, all of these relationships (Equations 21–24) remain true if we replace any of the occurrences of θ by $(\theta + 2n\pi)$, where n is any integer.

It is quite clear from Figures 5.18 and 5.19 that there must be a simple relationship between the functions $\sin(\theta)$ and $\cos(\theta)$; the graphs have exactly the same shape, one is just shifted horizontally relative to the other through a distance $\pi/2$. Equations 23 and 24 provide several equivalent ways of describing this relationship algebraically, but perhaps the simplest is that given by the first and third terms of Equation 23:

i.e. $\sin(\theta + \pi/2) = \cos(\theta)$

Clearly, adding a positive constant, $\pi/2$, to the argument of the function has the effect of shifting the graph to the left by $\pi/2$. In crude terms, the addition has boosted the argument and makes everything happen earlier (i.e. further to the left). Simple as it is, this is just one example of an important general principle that has many physical applications and deserves special emphasis.

Adding *any* positive constant ϕ to θ has the effect of shifting the graphs of $\sin\theta$ and $\cos\theta$ horizontally to the left by ϕ, leaving their overall shape unchanged. Similarly, subtracting ϕ shifts the graphs to the right. The constant ϕ is known as the **phase constant**.

Since the addition of a phase constant shifts a graph but does not change its shape, all graphs of $\sin(\theta + \phi)$ and $\cos(\theta + \phi)$ have the same 'wavy' shape, regardless of the value of ϕ: any function that gives a curve of this shape, or the curve itself, is said to be **sinusoidal**.

Sketch the graphs of $\sin(\theta)$ and $\sin(\theta + \phi)$ where $\phi = \pi/6$.

See Figure 5.24.   □

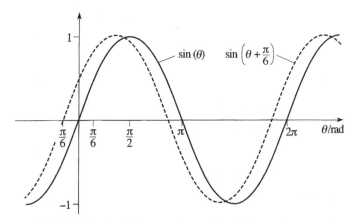

Figure 5.24   Graphs of $\sin(\theta)$ and $\sin(\theta + \pi/6)$.

Question T13

Use the terms periodic, symmetric or antisymmetric and sinusoidal, where appropriate, to describe the function tan (θ). What is the relationship between the graphs of tan (θ) and tan $(\phi + \theta)$?

☐

5.3.3   Inverse trigonometric functions

A problem that arises in physical science is that of finding an angle, θ, such that sin θ takes some particular numerical value. For example, given that sin $\theta = 0.5$, what is θ? You may know that the answer to this specific question is $\theta = 30°$ (i.e. $\pi/6$); but how would you write the answer to the general question, what is the angle θ such that sin $\theta = x$? The need to answer such questions leads us to define a set of **inverse trigonometric functions** that can 'undo' the effect of the trigonometric functions. These inverse functions are called **arcsine**, **arccosine** and **arctangent** (usually abbreviated to arcsin (x), arccos (x) and arctan (x)) and are defined so that:

$$\arcsin(\sin\theta) = \theta \quad -\pi/2 \leqslant \theta \leqslant \pi/2 \tag{25a}$$

$$\arccos(\cos\theta) = \theta \quad 0 \leqslant \theta \leqslant \pi \tag{25b}$$

$$\arctan(\tan\theta) = \theta \quad -\pi/2 < \theta < \pi/2 \tag{25c}$$

The abbreviations asin, acos and atan or alternatively $\sin^{-1}$, $\cos^{-1}$ and $\tan^{-1}$, are sometimes used for the inverse trigonometric functions.

Thus, since sin $(\pi/6) = 0.5$, we can write arcsin $(0.5) = \pi/6$ (i.e. $30°$), and since tan $(\pi/4) = 1$, we can write arctan $(1) = \pi/4$ (i.e. $45°$). Note that the *argument* of any inverse trigonometric function is just a number, whether we write it as x or sin θ or whatever, but the *value* of the inverse trigonometric function is always an angle. Indeed, an expression such as arcsin (x) can be crudely read as 'the angle whose sine is x.'

Notice that Equations 25a–c involve some very precise restrictions on the values of θ. These are necessary to avoid ambiguity and deserve further discussion. Looking back at Figures 5.18–5.20, you should be able to see that a single value of sin (θ), cos (θ) or tan (θ) will correspond to an infinite number of different values of θ. For instance, sin $(\theta) = 0.5$ corresponds to $\theta = \pi/6$, $5\pi/6$, $2\pi + (\pi/6)$, $2\pi + (5\pi/6)$, and any other value that may be obtained by adding an integer multiple of 2π to either of the first two values. To ensure that the inverse trigonometric functions are properly defined, we need to guarantee that each value of the function's argument gives rise to a *single* value of the function. The restrictions given in Equations 25a–c do ensure this, but they are a little too restrictive to allow those equations to be used as general definitions of the inverse trigonometric functions since they prevent us from attaching any meaning to an expression such as arcsin (sin $(7\pi/6)$).

In fact, the general definitions of the inverse trigonometric functions are as follows:

In each case, the range of allowed x values constitutes the *domain* of the inverse function and the range of allowed θ values constitutes the *codomain*.

If $\sin(\theta) = x$, where $-\pi/2 \leqslant \theta \leqslant \pi/2$ and $-1 \leqslant x \leqslant 1$	(26a)
then $\quad \arcsin(x) = \theta$	
If $\cos(\theta) = x$, where $0 \leqslant \theta \leqslant \pi$ and $-1 \leqslant x \leqslant 1$	(26b)
then $\quad \arccos(x) = \theta$	
If $\tan(\theta) = x$, where $-\pi/2 < \theta < \pi/2$	(26c)
then $\quad \arctan(x) = \theta$	

The graphs of these three functions are given in Figure 5.25.

Equations 26a–c look more intimidating than Equations 25a–c, but they embody the same ideas and they have the advantage of assigning meaning to expressions such as arcsin (sin ($7\pi/6$)).

What is $\arcsin(\sin(7\pi/6))$?

$\sin(7\pi/6) = -0.5$, so $\arcsin(\sin(7\pi/6)) = \arcsin(-0.5) = -\pi/6.$ □

Inverse trigonometric functions can be found on all 'scientific' calculators, often by using the trigonometric function keys in combination with the inverse key. The answer will be given in either radians or degrees, depending on the mode selected on the calculator, and will always be in the standard angular ranges given in Equations 26a–c.

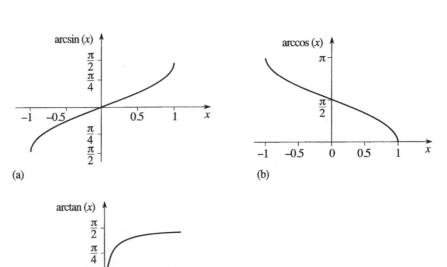

Figure 5.25 (a) The graph of arcsin (x). (b) The graph of arccos (x). (c) The graph of arctan (x).

When dealing with functions of any kind, physical scientists tradition- ally pay scant regard to such mathematical niceties as *domains* (ranges of allowed argument values) and *codomains* (ranges of allowed function values). This attitude persists because it rarely leads to error. However, care *is* needed when dealing with inverse trigonometric functions. If you know the sin, cos or tan of an angle and you want to find the angle itself, the inverse trigono- metric functions will give you *an* answer. But that answer will lie in one of the standard angular ranges and may not be the particular answer you are seeking. You will have to remember that there are other possible answers, your calculator is unlikely to warn you!

Earlier, arcsin (x) was crudely interpreted as 'the angle with a sine of x.' What would be a more accurate interpretation?

'The angle in the range $-\pi/2$ to $\pi/2$ with a sine of x.' $\square$

Question T14

(a) Use a calculator to find arcsin (0.65) both in radians and in degrees.

(b) Using information from earlier in this chapter, determine the values of:

arctan (1)   arcsin $(\sqrt{3}/2)$ arcos$(1/\sqrt{2})$ $\square$

It is possible to define a set of **inverse reciprocal trigonometric func- tions** in much the same way that we defined the inverse trigonometric functions. These functions are called **arccosecant**, **arcsecant** and **arccotangent**, and are usually abbreviated to arccosec, arcsec and arccot. The definitions are given below and the corresponding graphs are given in Figure 5.26.

If cosec $(\theta) = x$

where   either $0 < \theta \leqslant \pi/2$ and $x \geqslant 1$

or $-\pi/2 \leqslant \theta < 0$ and $x \leqslant -1$

then   arccosec $(x) = \theta$

If sec $(\theta) = x$

where   either $0 \leqslant \theta < \pi/2$ and $x \geqslant 1$

or $\pi/2 < \theta \leqslant \pi$ and $x \leqslant -1$

then   arcsec $(x) = \theta$

The codomains (i.e. ranges of θ) quoted here are not fully standardised. For example, some algebraic computing packages take the codomain of arccot (x) to be $0 < \theta < \pi$.

If $\cot(\theta) = x$

where $-\pi/2 < \theta < \pi/2$ and $-\infty < x < +\infty$

then $\text{arccot}(x) = \theta$

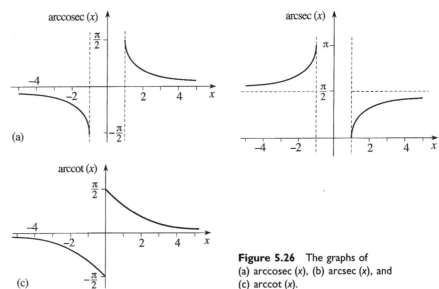

(a)

(c)

Figure 5.26 The graphs of (a) arccosec (x), (b) arcsec (x), and (c) arccot (x).

Finally, a few words of warning. The following notation is sometimes used to represent the inverse trigonometric functions:

$\sin^{-1}(x)$ for arcsin (x)

$\cos^{-1}(x)$ for arccos (x)

$\tan^{-1}(x)$ for arctan (x)

Notice that there is *no* connection with the *positive* index notation used to denote powers of the trigonometric functions (for example, using $\sin^2(\theta)$ to represent $(\sin(\theta))^2$. Also notice that although this notation might make it appear otherwise, there is still a clear distinction between the *inverse trigonometric functions* and the *reciprocal trigonometric functions*.

So $\sin^{-1}[\sin(\theta)] = \theta$

but $\text{cosec}(\theta) \times \sin(\theta) = 1$

Question T15

Figure 5.27 shows a ray of light travelling from glass to water. The angles θ_i and θ_r are related by Snell's law:

$$\mu_g \sin\theta_i = \mu_w \sin\theta_r$$

where μ_g and μ_w are constants.

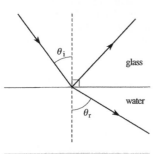

Figure 5.27 See Question T15.

(a) If the measured values of the angles are $\theta_i = 50.34°$ and $\theta_r = 60.00°$, what is the value of μ_g/μ_w?

(b) The *critical angle*, θ_c, is defined as the value of θ_i for which θ_r is 90°. Use your result for μ_g/μ_w to determine the critical angle for this glass–water interface. ☐

5.4 Trigonometric rules and identities

In this section we first extend the discussion in Section 5.2 to derive relationships between sides and angles of any arbitrary triangle, and then add to the discussion of Subsections 5.2.3 and 5.3.2 by introducing some further relationships involving trigonometric functions.

5.4.1 The sine and cosine rules

In Section 5.2, you saw that the sides and angles of a right-angled triangle are related by Pythagoras's theorem and by the trigonometric ratios. These relationships are in fact special cases of relationships that apply to *any* triangle. To derive these general relationships, consider Figure 5.28, where ABC is any triangle, with sides a, b and c. BD is drawn perpendicular to AC and has length p. We have called the length of AD, q, which means that the length of DC is $b - q$.

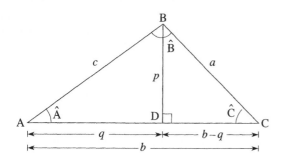

Figure 5.28
Construction used in deriving the sine and cosine rules.

From Figure 28,

$$\sin \hat{A} = \frac{p}{c}, \quad \text{so } p = c \sin \hat{A}$$

$$\sin \hat{C} = \frac{p}{a}, \quad \text{so } p = a \sin \hat{C}$$

and therefore $a \sin \hat{C} = c \sin \hat{A}$, or

$$\frac{a}{\sin \hat{A}} = \frac{c}{\sin \hat{C}}$$

Similarly, by drawing a perpendicular from C to AB,

$$\frac{a}{\sin \hat{A}} = \frac{b}{\sin \hat{B}}$$

By combining the two previous equations we obtain

the **sine rule**:

$$\frac{a}{\sin \hat{A}} = \frac{b}{\sin \hat{B}} = \frac{c}{\sin \hat{C}} \qquad (27)$$

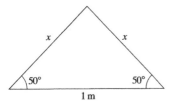

Figure 5.29 A triangle with two equal sides of unknown length x.

What is the length, x, in Figure 5.29?

The remaining angle is $180° - 2 \times 50° = 80°$, hence

$$\frac{x}{\sin (50°)} = \frac{1 \text{ m}}{\sin (80°)}$$

and

$$x = \frac{1 \text{ m} \times \sin (50°)}{\sin (80°)} \approx 0.778 \text{ m} \quad \square$$

It is possible for the value of one of the angles in a general triangle to be greater than 90° — the sine rule is still valid for such cases, since the trigonometric *function* sin (Â), for example, can be used instead of the trigonometric *ratio* sin Â.

There is also an important relationship between the three sides of a general triangle and the cosine of one of the angles. To derive this rule, consider Figure 5.28 again. ABD is a right-angled triangle so Pythagoras's theorem gives

$$c^2 = q^2 + p^2$$

In a similar way, since BCD is a right-angled triangle, we have:

$$a^2 = p^2 + (b - q)^2$$

Eliminating p^2 between these two *simultaneous equations* gives us:

$$a^2 = (c^2 - q^2) + (b - q)^2$$

which we can simplify by expanding the bracket on the right-hand side:

$$a^2 = c^2 - q^2 + b^2 - 2bq + q^2 = c^2 + b^2 - 2bq$$

But $q = c \cos \hat{A}$ and so:

$$a^2 = b^2 + c^2 - 2bc \cos \hat{A} \qquad (28)$$

which is called the **cosine rule**.

Question T16

Figure 5.30 shows the path of a ship that sailed 30 km due east, then turned through 120° and sailed a further 40 km. Calculate its distance x 'as the crow flies' from the starting point. ☐

Question T17

One of the interior angles of a triangle is 120°. If the sides adjacent to this angle are of length 4 m and 5 m, use the cosine rule to find the length of the side opposite the given angle. ☐

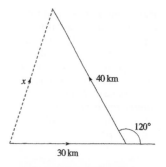

Figure 5.30 See Question T16.

5.4.2 Trigonometric identities

A good deal of applicable mathematics is concerned with equations. It is generally the case that these equations are only true when the variables they contain take on certain specific values; for example, $4x = 4$ is only true when $x = 1$. However, we sometimes write down equations that are true for *all* values of the variables, such as $(x + 1)^2 = x^2 + 2x + 1$. Equations of this latter type, i.e. ones that are true irrespective of the specific values of the variables they contain, are properly called **identities**. (☞)

There are a great many **trigonometric identities**, i.e. relationships between trigonometric functions that are independent of the specific values of the variables they involve. These have various applications and it is useful to have a list of them for easy reference. The most important are given below − you have already met the first seven (in slightly different forms) earlier in the chapter. Note that α and β may represent *any* numbers or angular values, unless their values are restricted by the definitions of the functions concerned.

The **symmetry relations**: (☞)

$$\sin(-\alpha) = -\sin(\alpha) \tag{18}$$

$$\cos(-\alpha) = \cos(\alpha) \tag{19}$$

$$\tan(-\alpha) = -\tan(\alpha) \tag{20}$$

The **basic identities**: (☞)

$$\tan(\alpha) = \frac{\sin(\alpha)}{\cos(\alpha)} \tag{8}$$

$$\sin^2(\alpha) + \cos^2(\alpha) = 1 \tag{9}$$

$$\cot(\alpha) = \frac{\cosec(\alpha)}{\sec(\alpha)} \tag{16}$$

$$1 + \tan^2(\alpha) = \sec^2(\alpha) \quad \alpha \neq (2n + 1)\frac{\pi}{2} \tag{17}$$

$$\cot^2(\alpha) + 1 = \cosec^2(\alpha)$$

Notice that the cosine rule reduces to Pythagoras's theorem for $\hat{A} = 90°$.

The cosine rule is also valid for $90° < \hat{A} < 180°$.

To show that two expressions are related by an identity we sometimes use the symbol ≡ which should be read as 'is identical to' rather than the more familiar = which should be read as 'is equal to.'

The symmetry relations show the oddness or evenness of the functions.

The basic identities arise from basic definitions and Pythagoras's theorem.

Replacing β by $-\beta$ and using the symmetry relations to replace $\sin(-\beta)$ by $-\sin(\beta)$, $\cos(-\beta)$ by $\cos(\beta)$, and $\tan(-\beta)$ by $-\tan(\beta)$ leads to further addition formulae (sometimes called the *subtraction formulae*) for $\sin(\alpha - \beta)$, $\cos(\alpha - \beta)$ and $\tan(\alpha - \beta)$.

The **addition formulae**: (☜)

$$\sin(\alpha + \beta) = \sin(\alpha)\cos(\beta) + \cos(\alpha)\sin(\beta) \tag{29}$$

$$\cos(\alpha + \beta) = \cos(\alpha)\cos(\beta) - \sin(\alpha)\sin(\beta) \tag{30}$$

$$\tan(\alpha + \beta) = \frac{\tan(\alpha) + \tan(\beta)}{1 - \tan(\alpha)\tan(\beta)} \tag{31}$$

for α, β and $(\alpha + \beta)$ not equal to $\dfrac{(2n + 1)\pi}{2}$

The double-angle formulae come from the addition formulae with $\alpha = \beta$. When deriving Equations 34 and 35, it is also necessary to make use of Equation 9.

The **double-angle formulae**: (☜)

$$\sin(2\alpha) = 2\sin(\alpha)\cos(\alpha) \tag{32}$$

$$\cos(2\alpha) = \cos^2(\alpha) - \sin^2(\alpha) \tag{33}$$

$$\cos(2\alpha) = 1 - 2\sin^2(\alpha) \tag{34}$$

$$\cos(2\alpha) = 2\cos^2(\alpha) - 1 \tag{35}$$

$$\tan(2\alpha) = \frac{2\tan(\alpha)}{[1 - \tan^2(\alpha)]} \tag{36}$$

for α and 2α not equal to $\dfrac{(2n + 1)\pi}{2}$

Equations 37 and 38 follow from Equations 34 and 35.

The **half-angle formulae**: (☜)

$$\cos^2(\alpha/2) = \tfrac{1}{2}[1 + \cos(\alpha)] \tag{37}$$

$$\sin^2(\alpha/2) = \tfrac{1}{2}[1 - \cos(\alpha)] \tag{38}$$

and if $t = \tan(\alpha/2)$ then, provided $\tan(\alpha/2)$ is defined:

$$\sin(\alpha) = \frac{2t}{1 + t^2} \tag{39}$$

$$\cos(\alpha) = \frac{1 - t^2}{1 + t^2} \tag{40}$$

$$\tan(\alpha) = \frac{2t}{1 - t^2} \tag{41}$$

for α not equal to $\dfrac{(2n + 1)\pi}{2}$

The sum formulae are used in the study of *oscillations* and waves, where we often need to combine two oscillations to determine their combined effect.

The **sum formulae**: (☜)

$$\sin(\alpha) + \sin(\beta) = 2\sin\left(\frac{\alpha + \beta}{2}\right)\cos\left(\frac{\alpha - \beta}{2}\right) \tag{42}$$

$$\sin(\alpha) - \sin(\beta) = 2\cos\left(\frac{\alpha + \beta}{2}\right)\sin\left(\frac{\alpha - \beta}{2}\right) \tag{43}$$

$$\cos(\alpha) + \cos(\beta) = 2\cos\left(\frac{\alpha + \beta}{2}\right)\cos\left(\frac{\alpha - \beta}{2}\right) \tag{44}$$

$$\cos(\alpha) - \cos(\beta) = -2\sin\left(\frac{\alpha + \beta}{2}\right)\sin\left(\frac{\alpha - \beta}{2}\right) \tag{45}$$

The **product formulae**: (☞)

$$2\sin(\alpha)\cos(\beta) = \sin(\alpha + \beta) + \sin(\alpha - \beta) \tag{46}$$

$$2\cos(\alpha)\sin(\beta) = \sin(\alpha + \beta) - \sin(\alpha - \beta) \tag{47}$$

$$2\cos(\alpha)\cos(\beta) = \cos(\alpha + \beta) + \cos(\alpha - \beta) \tag{48}$$

$$-2\sin(\alpha)\sin(\beta) = \cos(\alpha + \beta) - \cos(\alpha - \beta) \tag{49}$$

The product formulae can be of value in the procedure known as *integration*.

As indicated by the marginal notes, some of these identities can be derived straightforwardly from others. Rather than deriving the unexplained identities, one by one, we will consider just one identity in full and allow it to serve as an example.

To derive Equation 29, the addition formula for $\sin(\alpha + \beta)$, we use Figure 5.31. The line OB is at an angle, α, to the x-axis. OA has unit length and is at an angle, β, to OB. The angle AÔB is a right angle. (This is what fixes the length of OB, which is *not* of unit length.) It turns out to be useful to construct lines AN, BM and TB, such that AÑO, BM̂O and B̂TA are all right angles. If we let S represent the point of intersection of OB and AN, then OŜN, and consequently AŜB, are equal to $\alpha - 90°$. But SAB is a right-angled triangle, so TÂB is equal to α.

Since OA has unit length, the y-coordinate of the point, A, is $\sin(\alpha + \beta)$ and from the diagram we have:

$$\sin(\alpha + \beta) = NT + TA \tag{50}$$

To find TA, first consider the triangle, OAB. Since OA = 1, we have:

$$AB = \sin(\beta) \tag{51}$$

Next, considering triangle TAB, we have

$$\cos(\alpha) = \frac{TA}{AB} = \frac{TA}{\sin(\beta)} \tag{52}$$

Therefore,

$$TA = \cos(\alpha)\sin(\beta) \tag{53}$$

To find NT, we start by considering triangle OAB, obtaining

$$OB = \cos(\beta) \tag{54}$$

Next, consideration of triangle OBM gives

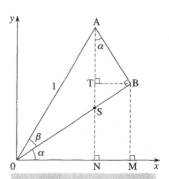

Figure 5.31 Construction used in deriving Equation 29.

$$\sin (\alpha) = \frac{MB}{OB} = \frac{NT}{\cos (\beta)} \tag{55}$$

Therefore

$$NT = \sin (\alpha) \cos (\beta) \tag{56}$$

We can now substitute in Equation 50 using Equations 53 and 56 to get the desired result:

$$\sin (\alpha + \beta) = \sin (\alpha) \cos (\beta) + \cos (\alpha) \sin (\beta) \tag{29}$$

The first of the double-angle identities, Equation 32, can be obtained by putting $\beta = \alpha$ in Equation 29:

$$\sin (\alpha + \alpha) = 2 \sin (\alpha) \cos (\alpha)$$

Use the addition identities and values from Table 5.3 to calculate the exact value of sin 75°.

From Equation 29,

$\sin 75° = \sin (45° + 30°)$

$\qquad = \sin 45° \cos 30° + \cos 45° \sin 30° \quad (\text{☞})$

$\qquad = \dfrac{1}{\sqrt{2}} \times \dfrac{\sqrt{3}}{2} + \dfrac{1}{\sqrt{2}} \times \dfrac{1}{2} = \dfrac{1 + \sqrt{3}}{2\sqrt{2}} \quad \square$

Question T18

Derive the identity given in Equation 33. $\square$

Question T19

Use Equation 34 to find the exact value of $\sin (15°)$. $\square$

Question T20

A feather trapped on the front of a loudspeaker moves back and forth whenever the speaker is in use. When a certain signal is played through the speaker, the position of the feather at time t is given by $x(t) = A \cos (\omega t + \phi_1)$, where A, ω and ϕ_1 are constants. A second similar signal causes a motion described by $x(t) = A \cos (\omega t + \phi_2)$, where ϕ_2 is another constant. Using an appropriate trigonometric identity, find a compact expression describing the motion $x(t) = A \cos (\omega t + \phi_1) + A \cos (\omega t + \phi_2)$ that results from relaying the two signals simultaneously. $\square$

The derivation given here assumes that $0 < (\alpha + \beta) < \pi/2$, but Equations 29 and 30 can be shown to be valid for all values of α and β.

Using Table 5.4 (see Answer T5).

5.5 Closing items

5.5.1 Chapter summary

1. Angles are usually measured in *degrees* or *radians*. An angle of 180° is equivalent to π rad.

2. The side opposite the *right angle* in a *right-angled triangle* is called the *hypotenuse*.

3. *Pythagoras's theorem* states that the square of the hypotenuse in a right-angled triangle is the sum of the squares of the other two sides. (The converse is also true.)

4. For a right-angled triangle, the *trigonometric ratios* are:

Useful mnemonics are *soh*, *cah* and *toa*.

$$\sin \theta = \frac{\text{opposite}}{\text{hypotenuse}} \tag{5}$$

$$\cos \theta = \frac{\text{adjacent}}{\text{hypotenuse}} \tag{6}$$

$$\tan \theta = \frac{\text{opposite}}{\text{adjacent}} \tag{7}$$

5. The *reciprocal trigonometric ratios* are:

$$\text{cosec } \theta = \frac{1}{\sin \theta} \quad \text{provided } \sin \theta \neq 0 \tag{10}$$

$$\sec \theta = \frac{1}{\cos \theta} \quad \text{provided } \cos \theta \neq 0 \tag{11}$$

$$\cot \theta = \frac{1}{\tan \theta} \quad \text{provided } \tan \theta \neq 0 \tag{12}$$

6. For a small angle θ, $\cos \theta \approx 1$

and $\sin \theta \approx \tan \theta \approx \theta/\text{rad}$

7. The sides and angles of a general triangle are related by the *sine rule*:

$$\frac{a}{\sin \hat{A}} = \frac{b}{\sin \hat{B}} = \frac{c}{\sin \hat{C}} \tag{27}$$

and the *cosine rule*:

$$a^2 = b^2 + c^2 - 2bc \cos \hat{A} \tag{28}$$

8. The *trigonometric functions* generalise the trigonometric ratios; $\sin(\theta)$ and $\cos(\theta)$ are *periodic functions* (with *period* 2π) and are defined for any value of θ. The function $\tan(\theta)$ is a periodic function (with period π) and is defined for all values of θ except odd multiples of $\pi/2$.

The *reciprocal trigonometric functions* generalise the reciprocal trigonometric ratios in a similar way.

9. The *inverse trigonometric functions*; arcsin (x), arccos (x) and arctan (x) are defined as follows:

If $\sin(\theta) = x$, where $-\pi/2 \leqslant \theta \leqslant \pi/2$ and $-1 \leqslant x \leqslant 1$ (26a)

    then $\arcsin(x) = \theta$

If $\cos(\theta) = x$, where $0 \leqslant \theta \leqslant \pi$ and $-1 \leqslant x \leqslant 1$ (26b)

    then $\arccos(x) = \theta$

If $\tan(\theta) = x$, where $-\pi/2 < \theta < \pi/2$ (26c)

    then $\arctan(x) = \theta$

The *inverse reciprocal trigonometric functions* may be defined in a similar way.

10. There are a large number of *trigonometric identities*, such as

$$\sin^2(\alpha) + \cos^2(\alpha) = 1 \tag{9}$$

The most important are listed in Subsection 5.4.2.

5.5.2 Achievements

Having completed this chapter, you should be able to:

A1. Define the terms that are emboldened in the text of the chapter.

A2. Given an angle measured in degrees, express its value in radians, and vice versa.

A3. Use Pythagoras's theorem.

A4. Use trigonometric ratios, and their reciprocals, to solve geometrical problems.

A5. Explain the sine and cosine rules and use these rules to solve problems involving general triangles.

A6. Explain how the sine, cosine and tangent functions can be defined for general angles and sketch their graphs.

A7. Recognise the most common identities for the trigonometric functions and apply them to solving mathematical and physical problems.

A8. Use the inverse trigonometric functions to solve mathematical and physical problems.

5.5.3 End of chapter questions

Question E1 Define the units commonly used in the measurement of angles.

Express the angles, $-10°$, $0°$, $275°$ in terms of radians.

Write the angles $2\pi/3$ and $\pi/4$ in terms of degrees.

Question E2 State Pythagoras's theorem.

If the two shorter sides of a right-angled triangle have lengths of 4961 m and 6480 m, what is the length of the third side?

Question E3 By drawing a suitable diagram, give definitions of the sine, cosine and tangent ratios. A right-angled triangle has a hypotenuse of length 7 mm, and one angle of $55°$. Determine the lengths of the other two sides.

Question E4 Two sides of a triangle have lengths 2 m and 3 m. If the angle between these two sides is $60°$, find the length of the third side and find the remaining angles.

Question E5 Sketch the graphs of $\sin(\theta)$, $\cos(\theta)$ and $\tan(\theta)$ for general values of θ. Why is $\tan(\theta)$ undefined when θ is an odd multiple of $\pi/2$?

Question E6 Use Figure 5.10 and suitable trigonometric identities to show that:

$$\sin(22.5°) = \sqrt{\frac{\sqrt{2}-1}{2\sqrt{2}}}$$

Question E7 Given that $\sin(60°) = \sqrt{3}/2$ and $\cos(60°) = 1/2$, use suitable trigonometric identities to find the exact values of $\sin(120°)$, $\cos(120°)$ and $\cot(2\pi/3)$ without using a calculator.

Question E8 Determine a general expression for all values of θ that satisfy the equation:

$\cot(\theta) = 4/3$

(*Hint:* Solve for $-\pi/2 < \theta < \pi/2$ and then consider how $\tan(\theta)$, and hence $\cot(\theta)$, behaves for other angles.)

Chapter 6   Exponential and logarithmic functions

6.1   Opening items

6.1.1   Chapter introduction

When the electric charge stored in a capacitor is discharged through a resistor, the rate of flow of charge through the resistor is proportional to the charge remaining on the capacitor. In a population of breeding organisms, the number of offspring produced in a given time, and hence the rate of population growth, is proportional to the size of the population. These processes of electrical discharge and population growth both provide examples of *exponential change*.

Exponential changes are the subject of Section 6.2. Subsection 6.2.1 introduces some more examples of exponential change and uncovers some of their common characteristics. Subsection 6.2.2 concerns the *rate of change* of a quantity and shows how this can be related to the gradient of the *tangent* to the graph of that quantity. In particular, by requiring that the rate of change of a quantity should always be equal to the instantaneous value of the quantity itself, we are led to define an *exponential function*, of the form $y(x) = e^x$, where e is an important mathematical constant, equal to 2.718 (to three decimal places). Subsection 6.2.3 examines the general mathematical properties of exponential functions, and in Subsection 6.2.4, exponential functions are used to describe various examples of exponential change, including the decay of radioactive nuclei. Section 6.2 ends with a more mathematical approach to the definition and evaluation of the number e that involves the concept of a *limit*.

In Section 6.3, *logarithmic functions* (logs) are introduced. We see how logarithms can be expressed in different *bases*, how the logs of products, quotients and powers can be expanded, and how the base of a logarithm can be changed. The *antilog* function is also introduced, and we look at how logs, antilogs and exponential functions can be handled on a calculator. The chapter ends with a brief look at how logarithmic functions are used in physical science to analyse data that obey an exponential law or a *power law*.

6.1.2   Fast track questions

Question F1

Plot a graph of $y = x^2$ by evaluating y when $x = 0$, ± 1, ± 2, ± 3. Estimate the gradient (i.e. slope) at $x = 2$ by drawing a tangent to the curve.

Question F2

Explain what is meant by:

$$\lim_{x \to \infty} \frac{1}{x}$$

Question F3

State the usual symbol for the following expression, and give its value to three decimal places:

$$\lim_{m \to \infty} (1 + 1/m)^m$$

Question F4

What is the gradient of the graph of $y = \exp(kx)$ at $x = 0$?

Question F5

Where possible, simplify the following expressions:

(a) $\log_e [(e^x)^y]$

(b) $\log_e (e^x + e^{2y})$

(c) $\exp [\log_e (x) + 2 \log_e (y)]$

(d) $\exp [2 \log_e (x)]$

(e) $a^{\log_a(x)}$

Question F6

If $P = kf^{-a}$, what are the gradient, and the intercept on the vertical axis, of the graph of $\log_{10}(P)$ (plotted vertically) against $\log_{10}(f)$?

6.1.3   Ready to study?

Study comment

In order to study this chapter you will need to understand the following terms: *constant of proportionality*, *decimal places*, *dependent variable*, *dimensions*, *function*, *independent variable*, *index*, *inverse function*, *power*, *proportional*, *reciprocal* and *root*. You will need to be able to use SI units, perform simple algebraic and numerical calculations (including using a calculator), plot the graphs of simple functions, and determine the *gradient* of a straight line that may be specified graphically or algebraically. The following *Ready to study questions* will allow you to establish whether or not you need to review some of the topics before embarking on this chapter.

Question R1

Write the following expressions in their simplest form:

(a) $\underbrace{a \times a \times a \times \ldots \times a}_{n \text{ factors}}$
(b) 5^0

Question R2

If a is any positive number, what is the value of x in the equation:

$$(a^4)^5 \times (a^2)^3 = a^x$$

Question R3

Write the following expressions in their simplest form:

(a) $16^{-1/4}$
(b) $16^{3/4}$
(c) $4^{5/2}$

(d) $27^{-2/3}$
(e) $1/(3^{-2})$

Question R4

If y is a function of x, given by $y = F(x)$, what is meant by saying that $G(x)$ is its inverse function? If $F(x) = x^3$, what is $G(x)$?

Question R5

Plot a graph of $y = x^2$ by evaluating the right-hand side of the equation at the points $x = 0, \pm1, \pm2, \pm3$. Use your graph to find solutions of the equation $x^2 = 2.72$.

Question R6

Which of the following expressions will give a straight line when y is plotted against x? For those that will give a straight line, state the value of its gradient. (All symbols except y and x represent non-zero constants.)

(a) $y = mx + c$
(b) $y = ax^2 + b$
(c) $y + x = k$

(d) $y/x = p$
(e) $y/x = qx + r$

Question R7

What is the gradient of the straight line joining the points with Cartesian coordinates $(1, 5)$ and $(3, 13)$?

6.2   Exponential functions

6.2.1   Exponential growth and decay

In physical science, and elsewhere, we are often concerned with how a quantity changes with time. The following examples all have an important feature in common. As you read, think what that feature might be.

Suppose that you were to invest £100 with a bank at an interest rate of 5% per year. At the end of the first year your money would have earned £5 in interest, and your total investment would be worth £105. To find the value of your investment after a further year, you would add 5% of £105 (i.e. £5.25) to obtain a total of £110.25 − and so on. Year after year, your total investment would increase, and so would the annual interest, since it would grow in proportion to your total investment. Thus, on an annual basis, the rate of growth of your investment (i.e. the interest gained per year) is proportional to your total investment.

When the electric charge Q stored in a capacitor is discharged through a resistor, the rate at which charge leaves the capacitor and flows through the resistor is described by the electric current I through the resistor. The size of this current is determined by the resistance R and the voltage V across the resistor: $I = V/R$. However, V itself depends on the capacitance C and the charge Q remaining in the capacitor: $V = Q/C$. It follows that $I = Q/(RC)$. So, at any moment, the rate at which charge is lost from the capacitor, I, is proportional to the charge, Q, remaining in the capacitor.

If you are unfamiliar with charge, current, voltage, resistance and capacitance, just regard Q, I and V as related variables, and R and C as constants.

The 'activity' of a radioactive sample is a measure of the number of atomic nuclei in that sample that disintegrate per second. Since each disintegration effectively removes one unstable nucleus from the sample it is also a measure of the rate at which unstable nuclei are lost from the sample. Now any individual nucleus is equally likely to decay in each second of its lifetime, so the number of disintegrations occurring in a sample in one second will be proportional to the number of unstable nuclei in that sample. Thus, at any time, the rate of decrease in the number of the unstable nuclei in a sample is proportional to the number of unstable nuclei that remain.

This example assumes that the decaying nuclei are all of the same type and that the nuclei produced by the decay are stable.

What do the above examples have in common?

Each describes a situation where the *rate of change* of some quantity at a given time is proportional to the value of that quantity at that time.   □

All the changes discussed above are examples of **exponential changes**. Such change may cause a quantity to increase (e.g. to grow) or to decrease (e.g. to decay), and may be characterised in the following way:

In an *exponential change* the rate of change of some quantity, y, at any time t, is directly proportional to the value of y itself at that time:

{rate of change of $y(t)$} $\propto y(t)$

i.e.   {rate of change of $y(t)$} $= ky(t)$ (1)

where k is a constant of proportionality.

If k is positive, y increases with time − this kind of change is called **exponential growth**.

If k is negative, y decreases with time − this kind of change is called **exponential decay**.

In our first example, the constant k was simply the interest rate, so, $k = 0.05$ year^{-1} (i.e. 5% per year). In the third example, the case of radioactive decay, the rate of change in the number of unstable nuclei was *negative* since the change *reduced* the number of such nuclei in the sample. (☞) In such cases the constant of proportionality is usually written as $-\lambda$, where λ is a positive quantity called the *decay constant* − for example, a certain isotope of polonium has a decay constant of $\lambda = 0.0133$ s^{-1}. Notice that the units of the decay constant reflect the units of the time interval used to define the rate.

Question TI
How many nuclear disintegrations per second would you expect from a sample containing 6.0×10^{18} polonium nuclei ($\lambda = 0.0133$ s^{-1})?   □

Question T2
In the second example above, what is the constant of proportionality relating the rate of discharge, I, of a capacitor to the charge Q remaining on the capacitor?   □

There are many examples of exponential change in physical science, some of which you will meet during this chapter. All exponential changes have an underlying mathematical similarity, and later in this section we will develop some powerful mathematical ideas and techniques relating to such changes. First, though, we need to have a more careful look at the idea of a *rate of change*.

6.2.2 Gradients and rates of change

Figure 6.1 illustrates how the volume, V, of water increases with time when a bath is filled. You can see from the graph that the water from the tap is running at a constant rate, because the volume of water increases by equal amounts in equal time intervals, i.e. by 2 litres in each second – the rate of change of volume is therefore 2 litres s^{-1}.

What is the gradient of the graph in Figure 6.1? What is the relationship between the gradient and the rate of change of volume?

The gradient is 2 litres s^{-1}. The gradient of the graph is *equal* to the rate of change of volume. ☐

What would the graph look like if the bath were *emptying* at a constant rate of 5 litres s^{-1}?

The graph would still be a straight line, but sloping in the opposite direction, and steeper. Its gradient would therefore be negative, -5 litres s^{-1}. ☐

We can generalise the above discussion to *any* constant rate of change:

For any quantity y that changes at a constant rate, the graph of y against time t is a straight line with a gradient equal to the rate of change of y.

In practice, a bath does not continue to empty, unaided, at a constant rate. As the water level falls, the water pressure also falls, and so the rate at which water flows out of the bath decreases. Figure 6.2 shows how the volume of water in an emptying bath might change with time. The graph drops steeply at first, corresponding to a rapid flow, and gradually becomes shallower as the flow rate diminishes. As before, we can relate the rate of change of volume to the steepness of the graph, even though the steepness is changing from moment to moment, but how do we do this? What feature of a *curved* graph such as that in Figure 6.2 will let us work out the rate of change of the plotted quantity at any time?

We could get a rough value for the rate of change of volume at, say, $t = 10$ s, by finding the change in the volume of water in the bath between $t = 6$ s and $t = 14$ s (about -7 litres) and then dividing that volume by the time interval of 8 s (14 s $- 6$ s) to get a rate of change of volume of about -0.9 litres s^{-1}. This is equivalent to finding the gradient of the straight line

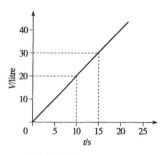

Figure 6.1 The volume of water in a bath that fills at a constant rate.

If conditions were such that the flow rate were proportional to the amount of water remaining, this would be another example of exponential change – but not otherwise.

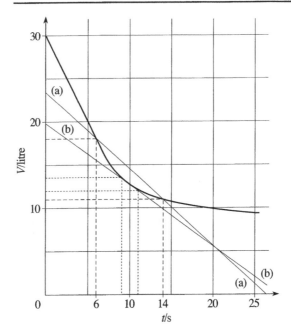

Figure 6.2
The volume of water in a bath emptying at a decreasing rate. The lines (a) and (b) are used to estimate the flow rate at $t = 10$ s.

joining the points on the graph at $t = 6$ s and $t = 14$ s. Or we could choose a shorter time interval, say 2 s, between $t = 9$ s and $t = 11$ s. You can see from the lines (a) and (b) drawn on Figure 6.2 that a different choice of time interval gives a straight line with a different gradient, and hence a different rate of change of volume. The difference arises because the flow rate is changing during the time intervals and what we have calculated is an average rate of change of volume over each of the specified periods. But how can we improve on this to find the rate of change at a particular time?

If we make the time interval *very small indeed*, we can hope that the flow rate hardly changes at all during that interval. Instead of a line joining two well-separated points on a curve, the situation is more like that shown in Figure 6.3; there will be a single straight line that just touches the curve at $t = 10$ s, the steepness of which matches exactly that of the curve at that point. Such a line is called a **tangent** to the curve. Using this idea we can define the gradient of the curve, at any particular point, to be the gradient of the tangent to the curve at that particular point. This gradient tells us the *instantaneous* rate of change at our chosen value of t, rather than the average rate of change over an interval. So, if we have a change in volume ΔV in a time interval Δt, the quantity $\Delta V/\Delta t$ will be equal to the average rate of change of volume during the interval Δt. But if we make Δt and ΔV smaller and smaller we can reasonably expect that $\Delta V/\Delta t$ will provide an increasingly good estimate of the gradient of the tangent. Indeed, if Δt and ΔV are small enough, we can expect $\Delta V/\Delta t$ to represent the (instantaneous) rate of change of V.

In the remainder of this chapter we will use the notation $\Delta V/\Delta t$ to represent the instantaneous rate of change of V with respect to t. In other words, no matter what value of t we are discussing, we will always assume that we can find suitable values Δt and ΔV to ensure that $\Delta V/\Delta t$ provides an accu-

ΔV (delta vee) means 'a small change in V'. It is *not* a factor Δ multiplied by the volume V, so the Δ *cannot be cancelled* in the expression $\Delta V/\Delta t$.

Exactly what is meant by 'small enough' is an important topic that occupies a good deal of the chapter that introduces differentiation.

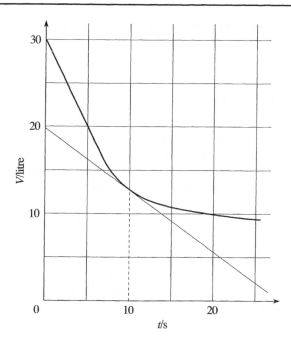

Figure 6.3 Finding the gradient at a point on a curve by drawing a tangent.

rate value for the gradient of the tangent at that value of t. From a strictly mathematical point of view this is not always justified, nor is it a particularly good use of notation, but we shall use it nonetheless. (Better methods will be introduced in Chapter 13.)

When trying to find an instantaneous rate of change from a graph you will probably have to make your best guess at an appropriate tangent, evaluate its gradient as accurately as possible, and accept that by working graphically you are limited to making *estimates* of rates of change. (Fortunately, there are algebraic techniques that enable us to work out rates of change accurately; these too will be introduced in Chapter 13.)

Question T3

Find the gradient of the tangent at $t = 10\,\text{s}$ shown in Figure 6.3. Draw a tangent to the curve at $t = 5\,\text{s}$ and hence estimate $\Delta V/\Delta t$ when $t = 5\,\text{s}$. ☐

The above discussion can be generalised to the rate of change of any quantity:

The rate of change of any quantity y, at a particular time, can be represented by the quotient $\Delta y/\Delta t$, provided the changes Δy and Δt are sufficiently small. The value of such a rate of change is given by the gradient of the tangent to the graph of y against t at the time in question.

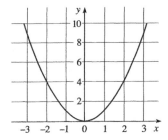

Figure 6.4 A graph of the equation $y = x^2$.

Furthermore, the idea of a gradient of a curved graph is not confined to graphs showing variation with time.

Figure 6.4 is a graph of the equation $y = x^2$. By drawing tangents to the curve, estimate $\Delta y/\Delta x$ when $x = 0$, $x = 2$, and $x = -1$.

At $x = 0$, you can see from the symmetry of the curve that the gradient is 0. The gradients of the other two tangents are 4 and -2, respectively, though your own results might be slightly different from these values since they will depend on how accurately you draw the tangents. □

Before we end this subsection, we briefly look at some more rates of change and their physical interpretations. One example is illustrated in Figure 6.5a, where the graph shows how the position coordinate, x, of an object moving along a straight line, changes with time. This is called *linear motion*. Where the graph is steep, the position changes rapidly with time – i.e. the object moves quickly – and shallower parts of the graph correspond to the object moving more slowly. A negative gradient corresponds to the object moving in the reverse direction. We can therefore say that, at any particular time, the gradient of this position–time graph is equal to the instantaneous velocity v_x of the object. Figure 6.5b shows how the velocity, v_x, of the object represented in Figure 6.5a changes with time.

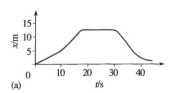

(a)

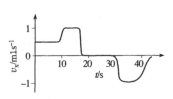

(b)

Figure 6.5 Graphs of (a) the position coordinate and (b) the instantaneous velocity of an object moving along a straight line.

Suggest a physical interpretation of the gradient of Figure 6.5b.

The gradient is $\Delta v_x/\Delta t$, the rate of change of velocity, which is equal to the acceleration. □

Using the notation introduced above, we can rewrite the condition for exponential change (Equation 1) as follows:

In an exponential change, at any time $\Delta y/\Delta t = ky$ (2)

Figure 6.6 shows an example of a quantity that is changing exponentially: N, the number of unstable nuclei in a radioactive sample, decays exponentially with time t. The gradient of the curve at any particular time is equal to the number of disintegrations per second occurring at that time. By drawing tangents to the graph at various times and measuring their gradients it is easy to see that the gradient is indeed proportional to N, as required. When $N = 10 \times 10^{10}$, 6×10^{10} and 2×10^{10}, the measured gradients are $-5 \times 10^4\,\text{s}^{-1}$, $-3 \times 10^4\,\text{s}^{-1}$ and $-1 \times 10^4\,\text{s}^{-1}$.

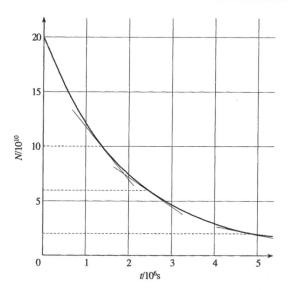

Figure 6.6 The number of unstable nuclei in a radioactive sample plotted against time, with tangents drawn at $N = 10 \times 10^{10}$, 6×10^{10} and 2×10^{10}. This is an example of exponential decay.

Figure 6.7 shows how the charge Q (measured in coulombs) stored in a capacitor changes with time. By measuring the gradients of tangents at various times, we can find the rate of flow of charge, i.e. the current, at these particular times.

Question T4

By drawing tangents to the curve, estimate the currents when $Q = 2$ C, 1 C and 0.5 C, and hence verify that Figure 6.7 shows exponential decay. □

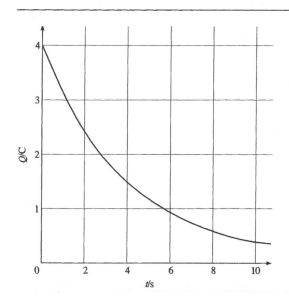

Figure 6.7 The charge remaining on a discharging capacitor. Another example of exponential decay.

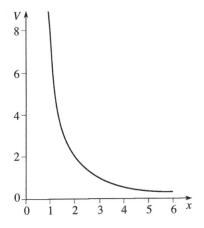

Figure 6.8   See Question T5.

Question T5

Figure 6.8 shows the graph of $V = 8/x^2$. Explain how you would show that this curve does not describe exponential decay.   □

6.2.3   Exponential functions and the number e

Remember that in exponential growth the constant k is positive, whereas in exponential decay k is negative.

In this subsection, you will learn how to write *functions* representing the values of quantities that change exponentially. We will begin with exponential growth, since we can then deal entirely with positive quantities. We will look for a function $y(x)$ such that, at any value of x, the rate of change of y is equal to the value of y itself, i.e. we will look for a function $y(x)$ such that $\Delta y/\Delta x = y$. For the sake of simplicity we will treat x as a purely numerical variable.

In Subsection 6.2.1, we saw that when there is a constant annual interest rate the value of an investment grows exponentially. With an interest rate of 5% per year, the value after one year is found by multiplying the initial sum by 1.05 and, after two years, by multiplying again by 1.05, i.e. in two years the value of the initial investment increases by a factor of $(1.05)^2$. If the investment is left for n years at the same interest rate, its initial value will be multiplied by a factor of $(1.05)^n$.

The above example suggests that we should look at functions of the form $y(x) = y_0 a^x$, where y_0 is the initial value of y and a is some (positive) constant number. It may therefore be helpful at this point to have a brief reminder of the properties of such functions. For any numbers a, x and y:

The relationships expressed in Equations 3 to 6 were derived and discussed in Chapter 1.

$$a^x a^y = a^{x+y} \tag{3}$$

$$(a^x)^y = a^{xy} \tag{4}$$

$$a^{-x} = \frac{1}{a^x} \tag{5}$$

$$a^{x/y} = (a^{1/y})^x \tag{6}$$

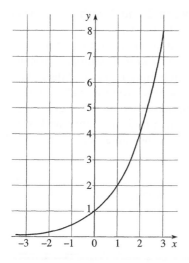

Figure 6.9 The graph of $y(x) = 2^x$.

Also, recall that $a^0 = 1$ and that by $a^{1/k}$ (for k a positive integer) we mean the kth root of a, i.e. a solution of the equation $x^k = a$.

Now let us look at a particular function: $y(x) = 2^x$. Figure 6.9 shows the graph of this function, and Table 6.1 shows its value for various values of x.

Table 6.1 Values of the function $y(x) = 2^x$

x	2^x	x	2^x
−3.5	0.09	0	1.00
−3.0	0.13	0.5	1.41
−2.5	0.18	1.0	2.00
−2.0	0.25	1.5	2.83
−1.5	0.35	2.0	4.00
−1.0	0.50	2.5	5.66
−0.5	0.71	3.0	8.00

Find the gradients of tangents to the graph of $y = 2^x$ at $x = -1, 0, 1$ and 2. Is this function growing exponentially?

At the given values of x the gradients are approximately 0.7, 1.4 and 2.8, respectively. From Table 6.1 the corresponding values of y are 1.0, 2.0 and 4.0. So, the gradient apparently increases in proportion to the value of the function. This indicates that $y(x) = 2^x$ increases exponentially with increasing x. ☐

In fact the gradient is consistent with $\Delta y / \Delta x \approx 0.7y$, not with $\Delta y / \Delta x = y$. If you were to plot the graph of $y(x) = 3^x$ you would find that it too grows exponentially, with a gradient approximately equal to 1.1×3^x. This suggests that there may be some number a between 2 and 3 such that if $y(x) = a^x$, then the gradient would be exactly equal to the value of y.

It turns out that there is indeed such a number. To three decimal places its value is 2.718, but like π and $\sqrt{2}$, it is an *irrational* number that cannot be accurately represented by any decimal with a finite number of decimal places. For this reason it is conventional to represent its accurate value by the letter e and simply substitute an appropriate numerical value whenever necessary. e is one of the most important numbers in physical science and in mathematics. (In Subsection 6.2.5, you will see how the value of e can be calculated to any desired degree of precision.)

An *irrational number* is one that cannot be written in the form p/q where p and q are whole numbers (i.e. integers).

$e = 2.718\ 281\ 828\ 459\ 05\ldots$ (to 14 decimal places)

If $y(x) = e^x$ (7)

then the rate of change $\Delta y/\Delta x = y(x)$.

Figure 6.10 is a graph of $y(x) = e^x$. Draw tangents at a few different points on the graph and measure their gradients. Confirm that at each point you examine the gradient is equal to the value of y.

You should have found that each gradient is close to the value of $y(x)$, thus confirming that the function satisfies $\Delta y/\Delta x = y$. □

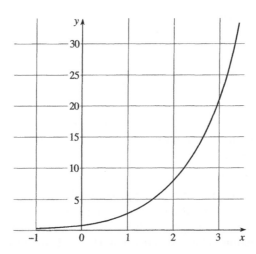

Figure 6.10 The graph of the function $y(x) = e^x$.

The function e^x is known as the **exponential function**, and to emphasise that it is a *function* of x it is often written as $\exp(x)$. So far, we have used it to describe how something varies with x, where x can be any variable without dimensions (since powers are always dimensionless numbers).

To evaluate exponential functions on a calculator, use the e^x key (possibly labelled $\exp(x)$) or its equivalent. If your calculator does not have

In this chapter we include parentheses in terms such as $\exp(x)$, but it is conventional to drop the parentheses unless they are required to avoid ambiguity.

such a key, you may be able to calculate e^x by using the y^x key with $y = 2.718$, but you are probably best advised to buy a new calculator.

Question T6

Evaluate e^2, e^3, $e^{1.43}$, e^{-1} and e^0. □

6.2.4 Exponential functions and exponential change

So far, you have seen that the function $y(x) = \exp(x)$ describes exponential changes in which the rate of change is equal to $y(x)$ at any given value of x. In this subsection, you will see how exponential functions can be used to describe exponential changes that correspond to any value of k in $\Delta y/\Delta x = ky$.

Table 6.2 gives some values for the function $y(x) = \exp(3x)$. Figure 6.11 shows the graph of this function for the range $x = 0$ to $x = 0.6$. Draw tangents at the points $y = 2$ and $y = 3$ and measure their gradients. Suggest a relationship between the values of y and the gradient.

At any value of y, the gradient is $3y$. Your measured values should have been close enough to the true values (6 and 9) to suggest this result. □

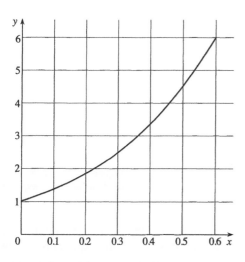

y-axis values: 6, 5, 4, 3, 2, 1
x-axis values: 0, 0.1, 0.2, 0.3, 0.4, 0.5, 0.6 x

Figure 6.11 The graph of $y(x) = \exp(3x)$.

In Section 6.3 you will see that there is another way to evaluate exponential functions on a calculator.

Table 6.2 A table of values (to three decimal places) for the function $y(x) = \exp(3x)$

x	$\exp(3x)$
-1.0	0.050
-0.5	0.223
0	1.000
0.5	4.482
1.0	20.086
1.5	90.017
2.0	403.429

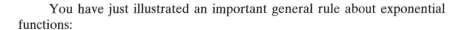

You have just illustrated an important general rule about exponential functions:

For any function $y(x) = \exp(kx)$, the gradient at any value of x is $ky(x)$, and so the rate of change $\Delta y/\Delta x = ky(x)$.

In other words, we seem to have found the function that satisfies the general condition for exponential change. So far, though, we have looked only at examples where k is positive (i.e. exponential growth), whereas we also need to be able to deal with exponential decay, where k is negative.

Figure 6.12 shows a graph of the function $y(x) = e^{-x}$ (i.e. $k = -1$). What is the sign of the gradient at any point on the curve? What is the relationship between the gradient and the value of y? (Draw tangents and measure their gradients if you have to, but you can probably guess the answer.)

The gradients are all negative. At any value of x the gradient is $-y(x)$. □

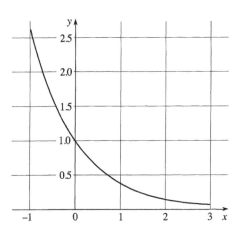

Figure 6.12 The graph of the function $y(x) = \exp(-x)$.

The above example illustrates that the exponential function $y(x) = \exp(kx)$, with a suitable value of k, can describe both exponential growth and exponential decay. So we now have an almost complete description of any exponential change. Why 'almost'? So far, we have neglected the initial value of y. If $y = \exp(kx)$, then when $x = 0$, $y = 1$. Clearly, this is not necessarily true in all physical situations. Also, since the start of Subsection 6.2.3 we have been treating x and y as a purely numerical variables, without any associated units. In practice we are certain to be interested in situations where y and x are physical quantities that involve units of measurement.

Dealing with these remaining conditions is actually very straight-forward. All we have to do to represent a general exponential change is to use a function of the form $y_0 \exp(kx)$. Thus:

For any exponential change, i.e. any change in which $\Delta y/\Delta x = ky$

$$y(x) = y_0 \exp(kx) \tag{8}$$

where y_0 is the value of y when $x = 0$.

As far as units are concerned, y will have the same units as y_0, and the product kt must be a pure number (i.e. dimensionless), so if x is a time in seconds, say, then the units of k should be seconds^{-1}. (☞)

Figure 6.13 shows the graphs of the exponential function $y = y_0 e^{kx}$ for various values of y_0 and k. If you measured the gradients of tangents to any of these graphs you would find that in each case they satisfy the equation $\Delta y/\Delta x = ky$ at every value of x. This illustrates the general rule that for exponential functions *the initial value of y does not alter the relationship between the value of y and the gradient*. So, we now have a complete 'recipe' for describing exponential change.

You may well be asked to deal with cases in which this last condition is not satisfied. For instance, x may be quoted in seconds, but k may be given in hours^{-1}. If so, you should convert one or other of the quantities, to ensure that kx is a pure number, before evaluating e^{kx}.

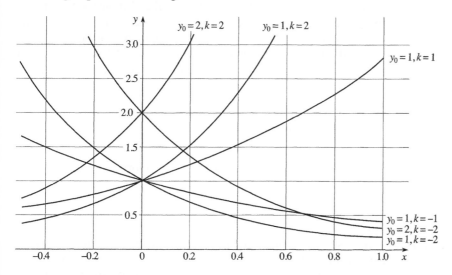

Figure 6.13 Graphs of the function $y(x) = y_0 e^{kx}$ for various values of y_0 and k.

Question T7

When a capacitor discharges, $\Delta Q/\Delta t = -Q/(RC)$. If the initial charge is Q_0, write an equation that describes how Q changes with time (i.e. express Q as a function of time). ☐

Question T8

Radioactive decay may be described by the equation $N = N_0 e^{kt}$. What are the values of N_0 and k for the decay shown in Figure 6.6? ☐

We will now consider some specific examples of exponential change.

Half-lives and decay absorption coefficients

Apart from the proportionality between the function and the value of its gradient, $y(t) = y_0 e^{kt}$ has another interesting property. To see this, consider the effect of increasing t in a series of equal steps, say from 0 to $1/k$, to $2/k$, to $3/k$, and so on. What are the corresponding values of $y(t)$? They are y_0, $y_0 e$, $y_0 e^2$, $y_0 e^3$, and so on. Clearly, equal additions to t cause $y(t)$ to change by equal multiplicative factors. This is true whatever the size of the additions to t may be (though different sized increments naturally correspond to different multiplicative factors), and this applies to exponential decay as well as exponential growth. In fact, the best known example of this is probably the use of **half-life** to characterise the decay of radioactive nuclei. Such decays are described by an exponential function of the form $N(t) = N_0 e^{-\lambda t}$, so in any time interval of length $1/\lambda$ the number of nuclei will decrease by a factor of e^{-1}. Similarly, in any interval of length $0.693/\lambda$ the number of nuclei is reduced by a factor of 0.5. That's why this is called the half-life. (☜)

So far we have been concentrating on changes with time, but we have already noted that an exponential function is not limited to such changes. For example, when a parallel beam of electromagnetic radiation passes through matter, its intensity I (☜) is related to the distance x that it has travelled through the medium by an exponential function: $I = I_0 \exp(-\mu x)$ where μ is called the *absorption coefficient*. It also depends on the nature of the radiation and the nature of the matter. For every additional distance $1/\mu$ that the beam travels through the matter, its intensity is reduced by a factor of e^{-1}.

What are the dimensions of μ?

Since μx must be dimensionless μ must be reciprocal length, $[L^{-1}]$. ☐

The function a^t

Throughout the last two subsections, we have been mainly concerned with functions of the form $y(x) = y_0 e^{kx}$. But we have also seen that functions such as 2^x and 3^x also satisfy the requirement that their gradients are proportional to their values at any given value of x. Clearly then, the exponential function $\exp(x)$, or e^x, is just one special example of the kind of function that can describe exponential change. More generally, any function of the type $y(x) = y_0 a^{kx}$ (where a is any positive number) will describe exponential change, though it will *not* satisfy the equation $\Delta y / \Delta x = ky$. For instance, in Subsection 6.2.3, you saw that the gradient of the graph of $y(x) = 2^x$ is approximately $\Delta y / \Delta x \approx 0.7y$ while that of $y(x) = 3^x$ is approximately $\Delta y / \Delta x \approx 1.1y$, so in both of these cases, where $k = 1$, it is certainly not the case that $\Delta y / \Delta x = ky$ even though $\Delta y / \Delta x$ is proportional to y. What really distinguishes the function $y(x) = y_0 e^{kx}$ is the fact that *for it alone* we can assert that $\Delta y / \Delta x = ky$ for all values of k.

$0.693/\lambda$ is accurate to three decimal places, but the precise relationship between the half-life and the characteristic decay constant λ involves the *logarithmic function* introduced in the next section. This relationship will be discussed in Question T17.

Intensity, I, is the rate at which energy is transferred across unit area.

Although the functions $y(x) = y_0 e^{kx}$ and $y(x) = y_0 a^{kx}$ are different (provided $a \neq e$), there is a simple relationship between them. Because it is always possible to find a number c such that $a = e^c$, for any positive value of a, it is always possible to write

$$y_0 a^{kx} = y_0 (e^c)^{kx} = y_0 e^{ckx} \quad (a > 0)$$

Thus, any function that describes exponential change can always be rewritten in terms of the exponential function; all we have to do is find the value of c that satisfies the equation $a = e^c$. We will return to this in Subsection 6.3.2, where you will see how it is done.

A note on terminology

In an expression such as q^p (where p and q are either constants or variables), the power p is sometimes called the **exponent** of q. Correspondingly, some authors use the term 'exponential function' to mean *all* functions of the type $y(x) = a^x$. Those authors still use exp (x) to represent e^x, but they sometimes call it the **natural exponential function** (since it arises from descriptions of naturally-occurring processes) in order to distinguish it from other functions of the type a^x. We shall generally use the term 'exponential function' to mean a function of the form $y(x) = e^x$ or $y(x) = e^{kx}$. Variables related by functions of the form $y(x) = y_0 e^{kx}$ are generally said to satisfy **exponential laws**. Such laws arise in many areas of physical science.

p is also known as the index, or the power, of q.

6.2.5 Evaluating the number e

In Subsection 6.2.3, the value of e was produced more or less out of a hat, and was then shown to have the necessary properties to describe exponential change. Now we will show how the value of e can be calculated from first principles.

We start from the requirement that we want a number e such that if $y(t) = y_0 e^{kt}$, then $\Delta y / \Delta t = ky$. In order to avoid having to deal with negative numbers, we will consider an example of continuous exponential growth. Our original example of the growth of an investment is now not a very good one, since the interest is added only once a year and so the value goes up in steps rather than continuously. A better example would be the growth of a large population of organisms (aphids, perhaps, or bacteria) where the population changes over even a very short time interval.

Suppose the number of organisms at time t is $N(t)$, and that at any time t the instantaneous rate of change is $\Delta N / \Delta t = kN(t)$. Even if we do not have an expression for $N(t)$ in terms of t, we can still use the given information to draw an *approximate* graph of N against t. To do this we note that the increase in the population over a small time interval Δt will be approximately $\Delta N = kN(t)\Delta t$. (This is only an approximation because the equation $\Delta N / \Delta t = kN(t)$ will not be completely accurate for any finite value of Δt, no matter how small.) So, if N_0 is the number of organisms at the *start* of a time interval Δt, the approximate number at the end of that interval will be $N_0 + \Delta N = N_0 + kN_0\Delta t$. This process can be repeated for the next interval Δt, by taking $N_1 = N_0 + kN_0\Delta t$ as the new starting value and $\Delta N = kN_1\Delta t$ as

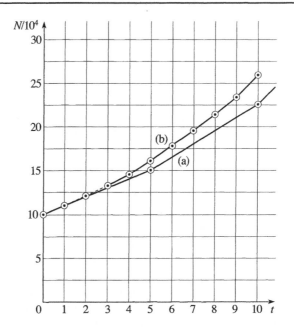

Figure 6.14
An approximate graph of N against t where $N_0 = 10^5$, $k = 0.1$ s^{-1} and (a) $\Delta t = 5$ s, (b) $\Delta t = 1$ s.

Table 6.3 Values of N calculated using (a) $\Delta t = 5$ s, and (b) $\Delta t = 1$ s. (Remember that $N/10^4$ in the column heading means that each entry in the column has been divided by 10^4)

t/s	(a) $N/10^4$ when $\Delta t = 5$ s	(b) $N/10^4$ when $\Delta t = 1$ s
0	10.000	10.000
1		11.000
2		12.100
3		13.310
4		14.641
5	15.000	16.105
6		17.716
7		19.487
8		21.436
9		23.579
10	22.500	25.937

the increment over that second interval. This can be continued until we have built-up a complete (though approximate) graph of $N(t)$ over any desired period of time. Figure 6.14a shows such a graph for $k = 0.1$ s^{-1} and $N_0 = 10^5$, drawn using a time interval of $\Delta t = 5$ s. If a time interval of $\Delta t = 1$ s were used (still with $k = 0.1$ s^{-1}), then the graph would not only be smoother but it would also climb more steeply, as shown in Figure 6.14b. The increased steepness arises because, instead of using the same rate of change for a whole 5 s, based on the value of N at the start of that interval, we calculate a new rate of change after only 1 s, based on a slightly larger N. Table 6.3 shows the values used to plot Figure 6.14.

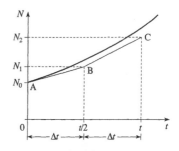

Figure 6.15 Estimating $N(t)$ by dividing the time t into small intervals. The curve represents the actual variation of N with t.

Which graph in Figure 6.14 is the better approximation to the real situation? How could the graphs be made even more realistic?

Figure 6.14b is a better approximation because it follows the changes in N and $\Delta N/\Delta t$ more closely. The graphs can be made even more realistic by choosing a smaller time interval. □

Let us now see what happens as we make Δt smaller and smaller. Figure 6.15 illustrates the procedure. N_0 is the initial number of organisms and t is some arbitrary period of time which can be subdivided into intervals. If we divide t into 2 equal intervals of length $\Delta t = t/2$, and if N_1 is the approximate number after the first interval of length $t/2$, then

$$N_1 = N_0 + \frac{kN_0t}{2} = N_0\left(1 + \frac{kt}{2}\right) \tag{9}$$

and after the second interval of length $t/2$

$$N_2 = N_1 + \frac{kN_1t}{2} = N_1\left(1 + \frac{kt}{2}\right) \tag{10}$$

Combining Equations 9 and 10, we find

$$N_2 = N_0\left(1 + \frac{kt}{2}\right)^2 \tag{11}$$

Now, if we had chosen to use 10 smaller intervals, each of length $t/10$, we would have obtained a somewhat different, and more accurate, estimate of the total number of organisms after time t:

$$N_{10} = N_0\left(1 + \frac{kt}{10}\right)^{10}$$

Similarly, if we had used n equal intervals, then we would have found

$$N_n = N_0\left(1 + \frac{kt}{n}\right)^n \tag{12}$$

If we now let the value of n become larger and larger we can expect the value of N_n to get closer and closer to the true final value (which we know to be N_0e^{kt}). Thus, as n becomes larger the following approximation becomes increasingly accurate:

$$N_0e^{kt} \approx N_0\left(1 + \frac{kt}{n}\right)^n$$

For reasons that will soon become apparent, it is useful to eliminate n from this relationship in favour of a new variable m defined by $m = n/kt$. In making this substitution it is important to remember that the final power of n must be replaced by mkt, and the statement that the approximation becomes increasingly accurate as n becomes larger should also be restated in terms of m. Thus, as m becomes larger the following approximation becomes increasingly accurate:

$$N_0 e^{kt} \approx N_0 \left(1 + \frac{1}{m}\right)^{mkt} = N_0 \left[\left(1 + \frac{1}{m}\right)^m\right]^{kt} \qquad (13)$$

It follows from Equation 13 that $e \approx (1 + 1/m)^m$, and that this approximation becomes increasingly accurate as m becomes larger and larger. This statement can be made more concise by using the mathematical idea of a **limit**. We say that e is the *limit* of $(1 + 1/m)^m$ as m tends to infinity. This is conventionally written as:

The symbol ∞ represents infinity.

$$e = \lim_{m \to \infty} (1 + 1/m)^m \qquad (14)$$

Table 6.4   See Question T9

m	a
2	2.2500
5	2.4883
10	2.5937
10^2	2.7048
10^3	
10^4	
10^5	

Inverse functions were discussed in Chapter 3.

Question T9

Given that $a = (1 + 1/m)^m$, complete Table 6.4 (using a calculator) and thus confirm that $(1 + 1/m)^m$ provides an increasingly good approximation to e as m becomes larger and larger.   □

6.3   Logarithmic functions

In Section 6.2, you saw how exponential changes can be described by a function of the general form $y(t) = y_0 \exp(kt)$. Given values for y_0 and k, this function provides a unique value of y for any given value of t. In this section our main aim is to investigate the *inverse function*, that tells us the value of t corresponding to any given value of y. However, we begin our investigation by examining a slightly different question: given that $x = 10^y$, what is the value of y that corresponds to a given value of x? In other words, given that $x = 10^y$, we want to know the inverse function that will enable us to write y as a function of x.

6.3.1   Logarithms to base 10: the inverse of 10^x

For some values of x, we can find y such that $x = 10^y$, without really thinking about inverse functions at all. For example, if $x = 100$ then, since $100 = 10 \times 10 = 10^2$, y must be equal to 2.

If $x = 10^y$, what is y when, (a) $x = 10\,000$, and (b) $x = 0.1$?

10 000 $= 10^4$, so $y = 4$; (b) $0.1 = 10^{-1}$, so $y = -1$. □

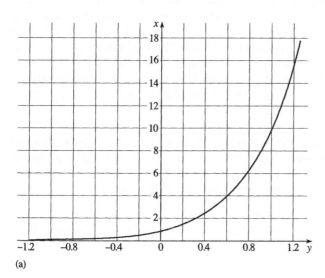

(a)

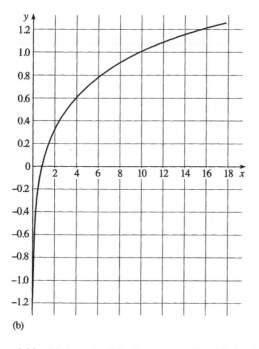

(b)

Figure 6.16 (a) A graph of the function $x = 10^y$ with the x-axis vertical and the y-axis horizontal. (b) A graph of the inverse function, $y = \log_{10}(x)$, plotted on axes with the conventional orientation.

Figure 6.16a is a graph of the equation $x = 10^y$, plotted with the x-axis vertical and the y-axis horizontal. This may look odd, but when plotting graphs it is conventional to plot values of the independent variable along the horizontal axis and values of the dependent variable along the vertical axis, and that is exactly what we have done. It is the choice of names for the variables that is unconventional in this case, not the graph plotting. In any event, it is clear from Figure 6.16a that each value of x corresponds to a different value of y, and that means it will be possible to define an inverse function to $x = 10^y$. The graph of this inverse function (with conventional x- and y-axes orientation) is shown in Figure 6.16b. It was obtained by re-plotting the data from Figure 6.16a so as to show y as a function of x. (This process is equivalent to reflecting the curve in Figure 6.16a in an imaginary mirror placed along the line $y = x$.) This shows that we can, in principle, find y for *any positive* value of x.

The function of x shown in Figure 6.16b is usually denoted by $y = \log_{10}(x)$ and its value for a given value of x is called the **logarithm to base 10** of x, or the **common logarithm** of x. This is an example of a **logarithmic function** − you will meet others in Subsection 6.3.2. As you can see from Figure 6.16b, it is only defined for positive values of x.

The logarithm to base 10 of x is the number y which satisfies the equation $x = 10^y$, i.e.

$$y = \log_{10}(x) \tag{15}$$

In other words, $\log_{10}(x)$ is the power to which we must raise 10 to obtain x.

===

We have written x in brackets in $\log_{10}(x)$, to emphasise that we are dealing with a *function*, but the brackets are often omitted in practice. Note, too, that $\log_{10}(x)$ is sometimes written as $\log(x)$.

From our definition we see that:

$\log_{10}(10) = 1$ since $10^1 = 10$

$\log_{10}(1) = 0$ since $10^0 = 1$

$\log_{10}\left(\dfrac{1}{10}\right) = -1$ since $10^{-1} = \dfrac{1}{10}$

Question T10

Without using your calculator, work out the values of:

(a) $\log_{10}(100)$ (b) $\log_{10}(1000)$ (c) $\log_{10}(0.1)$

(d) $\log_{10}(0.001)$ (e) $\log_{10}(10^{1/2})$ (f) $\log_{10}(10)$

(g) $\log_{10}(1)$ (h) $\log_{10}(10^{1.52})$ ☐

Use the x^y key on your calculator to find $10^{0.397\ 94}$ and $10^{-2.301\ 03}$, and hence find values for $\log_{10}(2.5)$ and $\log_{10}(0.005)$ to five decimal places.

$\log_{10}(2.5) = 0.397\ 94$, and $\log_{10}(0.005) = -2.301\ 03$. □

The values of $\log_{10}(x)$ are harder to calculate directly when x is not an obvious power of 10. However, they can be read from graphs such as that in Figure 6.16b, or found using the 'log' key on a calculator. For example, to find $\log_{10}(3.7)$, key in 3.7 then press the 'log' function key. You should obtain the answer 0.568 2017.

Use your calculator to find $\log_{10}(100)$, $\log_{10}(10)$, $\log_{10}(1)$, $\log_{10}(0.001)$, $\log_{10}(2)$ and $\log_{10}(3.16)$.

$\log_{10}(2) = 0.301$, and you can check your other answers with the answer to Question T10. Notice that, to two decimal places, $3.16 = 10^{1/2}$. □

Question T11

Use the log function key on your calculator to find x when

(a) $10^x = 6.8$ (b) $10^x = 537$ (c) $10^x = 0.34$.

(Give your answers to four significant figures.) □

Question T12

Use your calculator to find

(a) $\log_{10}(4.725)$ (b) $\log_{10}(47.25)$ (c) $\log_{10}(472.5)$

(d) Without further use of the calculator, find $\log_{10}(4725)$ and $\log_{10}(4.725 \times 10^7)$. □

We have already established that taking the logarithm to base 10 'undoes' the operation of raising 10 to a power (see part (e) of Question T10, for example). Likewise, raising 10 to a power 'undoes' the operation of finding the logarithm to base 10. These results, which follow from Equation 15, can be expressed as follows:

$$10^{\log_{10}(x)} = x$$

$$\text{and}\quad \log_{10}(10^x) = x \tag{16}$$

Remember that 10^x and antilog (x) are identical.

The function 10^x is sometimes called the **antilog** or **antilogarithmic function**. Antilogs can be found on a calculator using the 'inverse' and 'log' function keys.

Use the function keys on your calculator to find antilog (x) for $x = 2$, $x = -1$, $x = 0.397\,94$ and $x = -2.301\,03$ — key in the number, then press 'inv' and 'log'.

---

antilog $(2) = 100$	antilog $(-1) = 0.1$
antilog $(0.397\,94) = 2.5$	antilog $(-2.301\,03) = 0.005$

Notice that you have seen some of these results before, in Question T10 and the subsequent boxed question. □

6.3.2 Logarithms to base e and other bases

In Subsection 6.3.1, we looked at logarithms based on powers of 10. It is possible to use numbers other than 10 as the **base** for logarithms — in fact, it is possible to have logarithms to *any* base a that is greater than zero:

The **logarithm to base a** of x is the number y such that $x = a^y$.

In other words, the logarithm to base a of x is the power to which we must raise a to obtain x, i.e. if $x = a^y$ then

$$y = \log_a(x) \tag{17}$$

Just as the functions 10^x and $\log_{10}(x)$ are the inverse of each other, so the functions a^x and $\log_a(x)$ are the inverse of each other. This can be expressed by eliminating y from Equation 17 to obtain $x = a^{\log_a(x)}$ or alternatively by eliminating x to obtain $y = \log_a(a^y)$. Replacing y by x throughout the second of these results (which we are free to do since we can always rename a variable) we see that for any base a:

---

if $\qquad a^{\log_a(x)} = x$

then $\quad \log_a(a^x) = x \tag{18}$

Given that $N/N_0 = 2^{kT}$, find an expression for T.

(*Hint*: Start by taking appropriate logs of both sides of the given equation.) (☞)

Taking $\log_2$ of both sides, we find $\log_2 (N/N_0) = \log_2 (2^{kT})$

thus, from Equation 18, $\log_2 (N/N_0) = kT$

so, $T = (1/k) \log_2 (N/N_0)$ □

The most widely-used logarithms and logarithmic functions are those based on powers of 10 or on powers of the number e.

The **logarithm to base e** of x is the number y such that $x = e^y$, i.e.

if $\quad x = e^y$

then $\quad y = \log_e (x)$ $\hspace{4cm}$ (19)

and hence

$\quad \exp [\log_e (x)] = x$

and $\quad \log_e [\exp (x)] = x$ $\hspace{3cm}$ (20)

Since the number e arises from the description of naturally-occurring processes (radioactive decay), the logarithmic function based on powers of e is often known as the **natural logarithm** and $\log_e (x)$ is sometimes denoted by $\ln (x)$. Note that $\log_e (x)$ can be found on a calculator, in the same way as finding $\log_{10} (x)$. The key is usually marked $\log_e (x)$ or $\ln (x)$. (You may have to press 'inv' followed by 'e^x' if your calculator doesn't have a '$\log_e$' or '$\ln$' key.)

The process of 'taking logs of both sides' is the standard method of extracting a variable from an exponent. Similarly, you can often extract a variable from the argument of a logarithm by 'exponentiating both sides' of an equation.

Natural logarithms are also known as Naperian logarithms or hyperbolic logarithms.

Given that e = 2.718 (to three decimal places), use the y^x key on a calculator to find $e^{1.5041}$, and hence write down an approximate value for $\log_e (4.5)$.

$e^{1.5041} = 4.500$ (to three decimal places), so $\log_e (4.5) \approx 1.5041$. You can check the answer by using the $\log_e (x)$ key on a calculator. □

Question T13

Without using your calculator, what are the values of $\log_e (1)$ and $\log_e (e)$? □

The relationship between logs of different bases is discussed further in Subsection 6.3.3.

Question T14

Using your answers to Question T13, and the value of $\log_e (10)$ found using your calculator, sketch an approximate graph of the function $\log_e (x)$ on the same axes as a graph (such as that in Figure 6.16b) of $\log_{10} (x)$. □

Natural logarithms can, as you have just seen, be handled on a calculator in a similar way to common logarithms. Their inverse function, too, can be dealt with similarly.

The term 'antilog' is normally used *only* to denote the inverse of the logarithm in base 10.

On some calculators you must press the function keys *before* keying in the numbers.

Suggest how to find $\exp (x)$ on a calculator using the 'inv' (i.e. inverse) and 'log' keys. Use this method to find $e^{1.5041}$.

Key in 1.5041, then press the inverse and $\log_e$ function keys. You should obtain a value very close to 4.5. □

We are now able to solve the problem posed at the start of Section 6.3 – namely, to write t as a function of y given that $y(t) = y_0 \exp (kt)$. If we divide both sides of this equation by y_0 we obtain $y/y_0 = \exp (kt) = e^{kt}$. We can then use Equation 21 (or take $\log_e$ of both sides) to deduce that:

$$\text{if} \qquad y = y_0 \exp (kt)$$

$$\text{then} \qquad \log_e (y/y_0) = kt \qquad\qquad (21)$$

$$\text{so that} \qquad t = (1/k) \log_e (y/y_0)$$

When dealing with exponential decay, the constant k will be negative, and care will need to be taken when dealing with signs. For example, when a capacitor discharges through a resistor, the charge Q on the capacitor is given as a function of time by $Q = Q_0 \exp (-t/RC)$. To find the time at which Q reaches a certain value, divide both sides by Q_0 to obtain $Q/Q_0 = \exp (-t/RC)$, and then take $\log_e$ of both sides to see that $\log_e(Q/Q_0) = -t/RC$. So, $t = -RC \log_e (Q/Q_0)$.

We can also now solve the problem encountered at the end of Subsection 6.2.4, where we wanted to find a number c such that $a^{kt} = \exp (ckt)$. In other words, since $e^{ckt} = (e^c)^{kt}$, we wanted to find the value of c such that $a = e^c$. If we take $\log_e$ of both sides of this equation we find $c = \log_e (a)$. You will recall that this is an important result since it enables us to rewrite any function of the form $y(t) = y_0 a^{kt}$ in terms of the exponential function $y(t) = y_0 e^{ckt}$.

Question T15

Use your calculator to find x when (a) $e^x = 4.8$, (b) $e^x = 10$, and (c) $e^x = 0.56$. ☐

Question T16

The number of organisms, N, in a certain population increases exponentially: $N = N_0 \exp(kt)$ where $k = 0.02\,\text{h}^{-1}$. If $N_0 = 1000$, how long does it take for the population to double, i.e. what is t when $N = 2000$? Without doing any further calculation, write down the time taken for N to increase from 2000 to 4000. ☐

Question T17

In a sample of radioactive material, the number, N, of polonium nuclei decays exponentially: $N = N_0 \exp(-\lambda t)$ where $\lambda = 0.0133$ s^{-1}. How long does it take for the number of polonium nuclei to halve? ☐

Question T18

By finding a suitable value of k, rewrite the function $y(x) = 3^x$ in the form $y(x) = e^{kx}$. Check your answer by finding 3^x and $\exp(kx)$ when $x = 2$. ☐

6.3.3 Properties of logarithms

In this subsection we examine some further properties of logarithms that can be deduced from the relationships we have already discussed together with the rules for manipulating indices that were summarised in Equations 3–6.

Products and quotients

First, let us look at the result of multiplying two numbers together where each is expressed as 10 raised to a power. Suppose that $x = 10^p$ and $y = 10^q$. By definition then, $\log_{10}(x) = p$ and $\log_{10}(y) = q$ (see Equation 15).

Now, $\qquad xy = 10^p \times 10^q = 10^{p+q}$

so $\qquad \log_{10}(xy) = \log_{10}(10^{p+q}) = p + q = \log_{10}(x) + \log_{10}(y)$

thus $\qquad \log_{10}(xy) = \log_{10}(x) + \log_{10}(y)$

Of course, instead of working to base 10, we could equally well have written x and y as powers of some other number a, and then taken logs to base a. If we had done so we would have found the following general result:

In Subsections 6.2.4 and 6.3.2 you saw that any positive number could be expressed as e raised to some power. In fact, any positive number can be expressed in terms of *any* given positive number (such as 10 or e) raised to a suitable power.

$$\log_a(xy) = \log_a(x) + \log_a(y) \qquad\qquad (22)$$

You can illustrate this rule with the aid of a calculator. For example, $\log_{10}(3) = 0.477$, $\log_{10}(7) = 0.845$ and $\log_{10}(21) = 1.322$ (all to three decimal places), i.e. $\log_{10}(21) = \log_{10}(3) + \log_{10}(7)$.

Confirm that $\log_e(21) = \log_e(3) + \log_e(7)$.

$\log_e(3) = 1.099$, $\log_e(7) = 1.946$, $\log_e(21) = 3.045$

$(= 1.099 + 1.946)$. $\square$

You have in fact already met another example of this rule. In Question T12, you saw that $\log_{10}(4.725 \times 10^n) = n + \log_{10}(4.725)$ — which follows from Equation 22 because $\log_{10}(10^n) = n$.

The next example demonstrates some properties of logs of quotients and reciprocals. (✎)

Reciprocals: recall that

$(a/b)^{-1} = 1/(a/b) = b/a$

$(a^x)^{-1} = 1/a^x = a^{-x}$

$(a^{-x})^{-1} = a^x.$

(a) By writing $x = 10^p$ and $y = 10^q$, find an expression for $\log_{10}(x/y)$ in terms of $\log_{10}(x)$ and $\log_{10}(y)$.

(b) Use your answer to express $\log_{10}(1/x)$ in terms of $\log_{10}(x)$.

(a) $\log_{10}(x/y) = \log_{10}(10^{p-q}) = p - q = \log_{10}(x) - \log_{10}(y)$.

(b) It follows that $\log_{10}(1/x) = \log_{10}(1) - \log_{10}(x)$.

But, $\log_{10}(1) = 0$. So, $\log_{10}(1/x) = -\log_{10}(x)$. $\square$

The example above suggests the following general rules, which apply to logs to any base a:

$$\log_a(x/y) = \log_a(x) - \log_a(y) \tag{23}$$

$$\log_a(1/x) = -\log_a(x) \tag{24}$$

Check Equations 23 and 24 with the aid of your calculator. For example, try $x = 24$, $y = 8$, using (a) logs to base 10, and (b) logs to base e.

(a) $\log_{10}(24) = 1.380$, $\log_{10}(8) = 0.903$;

$\log_{10}(3) = 0.477 = \log_{10}(24) - \log_{10}(8)$

$\log_{10}(1/24) = \log_{10}(0.0417) = -1.380 = -\log_{10}(24)$.

(b) $\log_e(24) = 3.178$, $\log_e(8) = 2.079$;

$\log_e(3) = 1.099 = \log_e(24) - \log_e(8)$

$\log_e(1/24) = \log_e(0.0417) = -3.178 = -\log_e(24)$. $\square$

Powers and roots

Suppose we have some number $y = x^n$ where n is an integer, then

$$y = \underbrace{x \times x \times x \ldots \times x}_{n \text{ factors}}$$

It follows from Equation 24 that

$$\log_a (y) = \underbrace{\log_a (x) + \log_a (x) + \ldots + \log_a (x)}_{n \text{ terms}}$$

thus, $\log_a (x^n) = n \log_a (x)$.

This is indicative of a more general result that applies to any power of x, not just integer powers:

$$\log_a (x^b) = b \log_a (x) \tag{25}$$

You can check Equation 25 with the aid of a calculator. For example, you can show that $\log_{10} (3^2) = \log_{10} (9) = 2 \log_{10} (3)$; that $\log_{10} (49^{1/2}) = \log_{10} (7) = [\log_{10} (49)]/2$ and that $\log_{10} (4^{1.73}) = \log_{10} (11.004) = 1.73 \times \log_{10} (4)$.

> By taking logs to base 10 of both sides of the equation, find the value of x when $2^x = 3 \times 5^x$.
>
> ---
>
> $\quad\quad\quad x \log_{10} (2) = \log_{10} (3) + x \log_{10} (5).$
> So $\quad\quad x[\log_{10} (2) - \log_{10} (5)] = \log_{10} (3),$
> and hence $\quad x \log_{10} (2/5) = \log_{10} (3),$
> $\quad\quad\quad\quad x = \log_{10} (3)/\log_{10} (2/5) = -1.199. \quad \square$

As you will see in Subsection 6.3.4, Equations 22–25 enable logs to be used in tackling a wide variety of problems in physical science. But before we end this subsection, we will extend our discussion and show how to convert between different bases for logarithmic functions.

Changing the base of logarithms

In Subsection 6.3.2 we derived the following expression involving $\log_a (x)$:

$$a^{\log_a(x)} = x \tag{18}$$

Now suppose we take logs to *another base*, b, of both sides:

$$\log_b (a^{\log_a (x)}) = \log_b (x) \tag{26}$$

From Equation 25,

$$\log_a (x) \times \log_b (a) = \log_b (x) \tag{27}$$

Note that we are using logs to a general base a since Equation 24 applies for any positive value of a.

You have already met some other examples of this rule. From Equation 24,

$$\log_a (1/x) = \log_a (x^{-1})$$
$$= -\log_a (x)$$

i.e. $b = -1$ in Equation 25. Also, if $y/y_0 = e^{kt}$, then taking natural logs of both sides gives us

$$\log_e (y/y_0) = \log_e (e^{kt})$$
$$= kt \log_e (e)$$
$$= kt$$

(remember that $\log_e (e) = 1$) which is the expression we found in Subsection 6.3.2.

Note that we now know how to express a^x as e^{cx}, and $\log_a (x)$ as $\log_e (x)/c$, where $c = \log_e (a)$.

So we have a relationship between the log of x to two different bases — most usefully, it enables us to convert between the two most widely-used bases:

$$\log_{10} (x) \times \log_e (10) = \log_e (x) \tag{28}$$

To three decimal places, $\log_e (10) = 2.303$, and so $\log_e (x) = 2.303 \times \log_{10} (x)$. Using your calculator, you can verify that Equation 28 describes the relationships between $\log_e (x)$ and $\log_{10} (x)$ for various values of x. But if you look at the answer to Question T14 you will see that the graph of $\log_e (x)$ can be obtained from the graph of $\log_{10} (x)$ by re-scaling the y-axis by a factor of $\log_e (10)$, just as Equation 28 implies.

Question T19

Use logs to base ten to find the value of x when $6 = 2^x$. ☐

Question T20

By finding $\log_{10} (5)$ and $\log_{10} (2)$ on a calculator, calculate $\log_2 (5)$. ☐

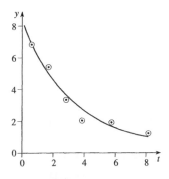

Figure 6.17 Experimental measurements of a quantity y that varies with time.

6.3.4 Using logarithms in physics

We have already seen one use of logarithms in physics — given that $y = y_0 \exp (kt)$ we can find t in terms of y, i.e. $t = [\log_e (y/y_0)]/k$. (See Subsection 6.3.2, in particular Questions T16 and T17.) Now consider another problem related to exponential change. Suppose an experiment gives the results shown in Figure 6.17. Does this graph follow an exponential law? And if so, what is the value of k? It is hard to see whether or not Figure 6.17 really does represent an exponential function — you can measure gradients at various points, as we did in Subsection 6.2.2 (e.g. in Question T4), but this is time-consuming, and there are always uncertainties involved with drawing tangents.

Let us consider an easier approach. If y does follow an exponential law, we must have $y = y_0 \exp (kt)$, and hence $\log_e (y/y_0) = kt$ (Equation 21). We can use Equation 23 to write $\log_e (y) - \log_e (y_0) = kt$, and hence

$$\log_e (y) = kt + \log_e (y_0) \tag{29}$$

Note that we could get the same information by plotting $\log_e (y/y_0)$ — the choice of which to plot will in practice depend on whether it is easier to measure y itself, or as a fraction of its initial value.

If we recognise $\log_e (y)$ as a simple variable (call it Y if you like) then we see that Equation 29 has the general form of the equation of a straight line ($y = mx + c$), so if we plot $\log_e (y)$ against t, we should expect to get a straight line with gradient k that intersects the vertical axis at $\log_e (y_0)$. If the results do not follow an exponential law, the graph will not be a straight line.

Question T21

Figure 6.18 shows graphs of the results from two (hypothetical) experiments. Which graph shows that the relationship between y and t follows an exponential law, and what is the value of k in that case? ☐

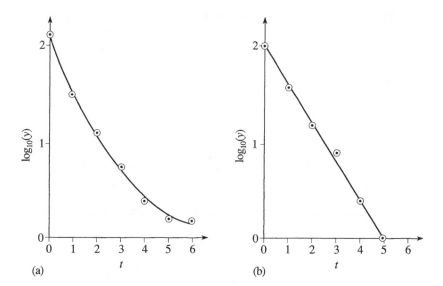

(a)

(b)

Similar techniques can be used for variables that obey a **power law**, i.e. a relation of the form $y = kx^p$ (or $y = kt^p$) where the power, p, is some constant number. For example, suppose you think that variables F and r might obey a power law of the form

$$F = \frac{k}{r^2}$$

(☞) (30)

where k is an unknown constant. How can you test your hypothesis experimentally? And how can you find the value of k? The properties of logarithms, as summarised in Equations 22–25, provide a way. First, write the expression $F = k/r^2$ in terms of $\log_{10}(F)$ and $\log_{10}(r)$, using Equation 23:

$$\log_{10}(F) = \log_{10}\left(\frac{k}{r^2}\right) = \log_{10}(k) - \log_{10}(r^2)$$

(31)

Then, using Equation 25 we find:

$$\log_{10}(F) = -2\log_{10}(r) + \log_{10}(k)$$

(32)

Such a power law is known as an *inverse square law*. Coulomb's law for the electrostatic force between two charged particles, and Newton's law for the gravitational attraction of two masses, are both examples of an inverse square law.

Alternatively, we could write $F = kr^{-2}$ and use Equation 22 to write

$$\log_{10}(F) = \log_{10}(k) +$$
$$\log_{10}(r^{-2})$$

then use Equation 25 to produce the same result.

The intercept is sometimes called the y-intercept.

Suppose you were to plot a graph of $\log_{10}(F)$ against $\log_{10}(r)$. What would be its gradient and its intercept (☞) on the vertical axis?

Equation 32 is an equation of the form $y = mx + c$, so the gradient would be -2 and the intercept $\log_{10}(k)$. □

Plotting a graph using logarithms thus enables you to test the data to see whether they follow the expected exponential law, and to find the value of the constant k. Note that you do not have to use logs to base 10, since Equations 22–25 apply to logs to any base a. This technique can be summarised as follows:

---

if $y = kx^p$

then $\log_a (y) = p \log_a (x) + \log_a (k)$ (33)

so a graph of $\log_a (y)$ against $\log_a (x)$ has gradient p and intercept $\log_a (k)$. The logs may have any base, though bases 10 or e are generally used.

In Kepler's law, $k = (4\pi^2/GM)$ where G is the universal gravitational constant and M the mass of the Sun.

Question T22

According to one of Kepler's laws of planetary motion, the orbital period, T, of a planet is related to its (average) distance, R, from the Sun: $T^2 = kR^3$. What would be the gradient of a graph of $\log_{10} (T)$ against $\log_{10} (R)$? What would be the gradient if log to base e was used rather than base 10?   □

Question T23

An experimenter measures the period of oscillation, T, of various masses m supported by a spring and suggests, on theoretical grounds, that $T \propto \sqrt{m}$. How could the experimental data be tested to see whether $T \propto \sqrt{m}$?   □

6.4   Closing items

6.4.1   Chapter summary

1. If the graph of some function $y(t)$ is a straight line, then the *rate of change* of y is constant, and is equal to the gradient of the graph. If the graph is a curve, then the rate of change at any point is defined by the gradient of the *tangent* to the curve at that point.

2. If the rate of change of a variable is always proportional to the current value of that variable, then we can say that variable changes *exponentially*. If we represent the variable by y, and its rate of change by $\Delta y/\Delta t$, we can write $\Delta y/\Delta t = ky$, where a positive value of k characterises *exponential growth*, and a negative value of k characterises *exponential decay*.

3. If $\Delta y/\Delta t = ky$, then $y(t) = y_0 \exp (kt)$, where the *exponential function* $\exp (kt) = e^{kt}$. The units of k are such that kt is a dimensionless quantity.

4. The following relationships are a consequence of the general properties of powers:

$$e^x e^y = e^{x+y}$$

$$(e^x)^y = e^{xy}$$

$$e^{-x} = \frac{1}{e^x}$$

$$e^{x/y} = (e^{1/y})^x$$

5. Exponential functions are not restricted to describing changes with time. For example, the function $\exp(kx)$ (with k such that kx is dimensionless) could describe how a quantity varies with position or any other quantity.

6. The constant, e, is given in terms of a *limit* by:

$$e = \lim_{m \to \infty} (1 + 1/m)^m = 2.718 \quad \text{(to three decimal places)}$$

7. The *logarithm to base a* of x is the power to which we must raise a to obtain x. So, if $x = a^y$, then $y = \log_a(x)$. The function $\log_a(x)$ is thus the inverse of the function a^x, hence $a^{\log_a(x)} = x$ and $\log_a(a^x) = x$.

8. The most widely-used logarithmic functions have $a = 10$ and $a = e$.

9. Any function a^x can be written in terms of the exponential function:

$a^x = \exp(cx)$, where $c = \log_e(a)$.

10. The following relationships arise from the definition of a logarithmic function and the rules for combining indices:

$$\log_a(xy) = \log_a(x) + \log_a(y) \tag{22}$$

$$\log_a(x/y) = \log_a(x) - \log_a(y) \tag{23}$$

$$\log_a(1/x) = -\log_a(x) \tag{24}$$

$$\log_a(x^b) = b\log_a(x) \tag{25}$$

11. The base (a or b) of any logarithmic function can be changed using the relationship

$$\log_a(x) \times \log_b(a) = \log_b(x) \tag{27}$$

So $\log_a(x) = \log_e(x)/c$, where $c = \log_e(a)$.

12. If $y = y_0 \exp(kt)$, then $\log_e(y) = \log_e(y_0) + kt$, and a graph of $\log_e(y)$ against t has gradient k and intercept $\log_e(y_0)$.

13. If y and x are related via a *power law*, $y = kx^p$, then

$$\log_a(y) = p\log_a(x) + \log_a(k) \tag{33}$$

and a graph of $\log_a(y)$ against $\log_a(x)$ has gradient p and intercept $\log_a(k)$.

6.4.2 Achievements

Having completed this chapter, you should be able to:

A1. Define the terms that are emboldened in the text of the chapter.

A2. Recognise examples of exponential change.

A3. Estimate the gradient at a point on a curved graph by drawing a tangent.

A4. Sketch a graph of $y = y_0 \exp(kt)$ and describe its significant features.

A5. Write down an expression for the rate of change of y, where $y = y_0 \exp(kt)$, for any given value of t or y.

A6. Recognise and interpret the common expression for the number e written in terms of a limit.

A7. Explain the relationships between the functions $\log_a(x)$, a^x and antilog (x) and know that many common applications of such functions use $a = e$ or $a = 10$.

A8. Sketch graphs of $y = \log_e x$ and $y = \log_{10} x$ and describe their significant features.

A9. Find the value of c such that $a^t = e^{ct}$.

A10. Manipulate and simplify expressions involving logarithmic and exponential functions of products, quotients and powers.

A11. Change the base of a logarithmic function and find the value of c such that $\log_a(x) = \log_e(x)/c$.

A12. Use a calculator to evaluate exponential functions, and logarithmic functions in base e and base 10.

A13. Use logarithmic functions to test whether experimental data obey an exponential law or a power law, and to find the values of unknown constants in such relationships.

6.4.3 End of chapter questions

Question E1 Figure 6.19 shows how a certain variable y changes with x. By drawing tangents, estimate the gradient $\Delta y/\Delta x$ at $y = 20$, $y = 40$ and $y = 60$. Are your answers consistent with y being an exponential function of x?

Question E2 For the function $y = y_0 \exp(kt)$ with $y_0 = 3$ and $k = 0.1$, (a) state a general expression for the gradient of the graph at any value of t, (b) calculate the gradient of the graph when $t = 0$ and when $t = 2$, and (c) sketch a graph of the function.

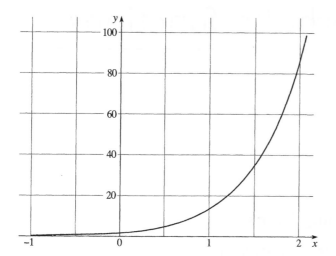

Figure 6.19
See Question E1.

Question E3 Rewrite the equation $y = 10^{bx}$ in terms of the exponential function exp (x).

Question E4 Given that $\log_{10}(3) = 0.477$ and $\log_{10}(2) = 0.301$, evaluate the following without using a calculator: (a) $\log_{10}(300)$, (b) $\log_{10}(6)$, (c) $\log_{10}(60)$.

Question E5 Without using your calculator, simplify

(a) $\exp[\log_e(\pi)]$ (b) $\log_e(e^\pi)$ (c) $\log_\pi(\pi^{\sqrt{2}})$.

Question E6 Find $\log_e(\pi)$ and $\log_e(2)$ on your calculator and hence evaluate $\log_2(\pi)$.

Question E7 The results of an (hypothetical) experiment are given in Table 6.5. Suppose the data obey a law of the form $T = kA^\alpha$. Find k and α by plotting a suitable graph.

Table 6.5
See Question E7

A	T
1	0.70
3	1.46
6	2.31
10	3.25
20	5.16
30	6.76
45	8.86

Chapter 7 | Hyperbolic functions

7.1 Opening items

7.1.1 Chapter introduction

Figure 7.1 shows a situation that you have certainly seen countless times, a heavy cable suspended between two pylons. We all know that the cable sags in the middle, but is there a mathematical function which describes the shape of this curve? It turns out that there is such a function, and it is particularly relevant to this chapter because it is one of the so-called *hyperbolic functions*. The hyperbolic functions sinh (x), cosh (x), tanh (x), sech (x), arctanh (x) and so on, which have many important applications in mathematics, physics and engineering, correspond to the familiar trigonometric functions: sin (x), cos (x), tan (x), sec (x), arctan (x), etc.

Section 7.2 begins with the definitions of the basic hyperbolic functions sinh (x), cosh (x) and tanh (x). It goes on to consider more advanced aspects of hyperbolic functions, including the reciprocal and inverse hyperbolic functions. Section 7.3 lists some useful identities which are analogous to those listed in Chapter 5 for the trigonometric functions.

This is a short chapter that is mainly intended to provide a repository of useful results. More will be said about the hyperbolic functions in the companion volume, following the introduction to complex numbers differentiation and integration. It is in the context of calculus that they are most frequently encountered in the physical sciences, and it is via complex numbers that their close relationship to the trigonometric functions is most easily explained.

Figure 7.1 A heavy cable suspended between two pylons.

7.1.2 Fast track questions

Question F1

Define sinh (x), cosh (x) and tanh (x) in terms of the exponential function, exp $(x) = e^x$.

Question F2

Show that

$$\text{arccoth}(x) = \frac{1}{2}\log_e\left(\frac{x+1}{x-1}\right) \quad \text{for } |x| > 1$$

7.1.3 Ready to study?

Study comment

To begin the study of this chapter you need to be familiar with the following: the general formula for the solution of a *quadratic equation* and the general definition of a *function*, particularly the *exponential function*, e^x (including the main features of its *graph*) and the *trigonometric functions* (including their *graphs*, *identities*, *reciprocals* and *inverse functions*). The following *Ready to study questions* will help you to establish whether you need to review some of the above topics before embarking on this chapter.

Question R1

(a) Evaluate the function e^x when $x = 2$, and when $x = -2$.

(b) Write down the solutions of the quadratic equation $x^2 - ax + 1 = 0$ where $a \geqslant 2$. Hence write down the solutions of the equation $e^{2w} - ae^w + 1 = 0$.

Question R2

Use your calculator to find the values of:

(a) $\arcsin(1/3)$ and $\dfrac{1}{\sin(1/3)}$,

(b) $\sin[\sin(1/3)]$ and $\sin^2(1/3)$.

Many authors use the notation sin^{-1} in place of arcsin.

Question R3

Explain the meaning of the term inverse function. What condition must a function satisfy if it is to have an inverse?

7.2 Some hyperbolic functions

7.2.1 Defining sinh, cosh and tanh

The **hyperbolic functions**, **sinh**(x) and **cosh**(x), are defined in terms of the exponential function by

$$\sinh(x) = \frac{e^x - e^{-x}}{2} \qquad \text{for all } x \qquad (1)$$

$$\cosh(x) = \frac{e^x + e^{-x}}{2} \qquad \text{for all } x \qquad (2)$$

These functions are given their particular names because they have much in common with the corresponding trigonometric functions, but we leave a discussion of this until later.

Write down (without using your calculator) the values of $\sinh(0)$ and $\cosh(0)$.

Since $e^0 = 1$ we have $\sinh(0) = 0$ and $\cosh(0) = 1$. □

Question T1

Use the basic definitions of $\cosh(x)$ and $\sinh(x)$ in terms of exponentials to show that:

(a) $\sinh(x)$ is an odd function while $\cosh(x)$ is an even function

(b) $e^x = \cosh(x) + \sinh(x)$

(c) $\cosh^2(x) - \sinh^2(x) = 1$ □

A function $f(x)$ is

odd if $f(x) = -f(-x)$

even if $f(x) = f(-x)$ for all values of x.

It is also convenient to define a third hyperbolic function, **tanh** (x), by

$$\tanh(x) = \frac{\sinh(x)}{\cosh(x)} \qquad \text{for all } x \qquad (3)$$

What is the value of $\tanh(0)$?

$\tanh(0) = 0$ since (as we saw before) $\sinh(0) = 0$ and $\cosh(0) = 1$.   □

The graphs of these three basic hyperbolic functions are given in Figures 7.2–7.4. Notice in Figure 7.3 that $\cosh(x) > 1$. In Figures 7.2 and 7.3 we see that $\sinh(x)$ and $\cosh(x)$ are very close in value when x is large and positive, but $\sinh(x) < \cosh(x)$ for other values of x. From Figure 7.4 we see that $\tanh(x)$ is approximately 1 when x is large and positive, and approximately -1 when x is large and negative, and that $-1 < \tanh(x) < 1$.

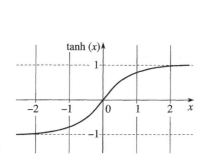

Figure 7.4 Graph of tanh (x).

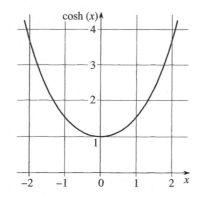

Figure 7.3 Graph of cosh (x).

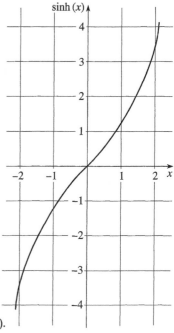

Figure 7.2 Graph of sinh (x).

Why are sinh (x) and cosh (x) called *hyperbolic* functions?

To answer this question, we first define variables x and y by

$$x = a \cosh (s) \tag{4}$$

$$y = b \sinh (s) \tag{5}$$

Then we use the result given in Question T1(c) to obtain

$$\frac{x^2}{a^2} - \frac{y^2}{b^2} = \cosh^2 (s) - \sinh^2 (s) = 1 \tag{6}$$

Now, when all the points (x, y) that satisfy this equation are plotted, the result is the curve shown in Figure 7.5. This curve is known as a *hyperbola*. Equation 6 is said to be the *equation of a hyperbola*, and Equations 4 and 5 are the *parametric equations* of a hyperbola. This is the reason why sinh and cosh are known as hyperbolic functions. (Hyperbolae and parametric equations are discussed in more detail in Chapter 10.)

As we shall see, there are many other functions (such as tanh) which can be defined in terms of sinh and cosh and these are all collectively known as *hyperbolic functions*.

Notice that, unlike the corresponding trigonometric functions, the sinh, cosh and tanh functions are not periodic. Moreover, sinh (x) and cosh (x) may take very large values (unlike sin (x) and cos (x)).

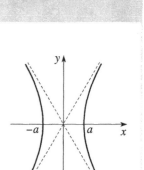

Figure 7.5 The hyperbola corresponding to $x^2/a^2 - y^2/b^2 = 1$.

Use the exponential function on your calculator to find sinh (1), cosh (1) and tanh (1). From these results calculate $\cosh^2(1) - \sinh^2(1)$.

$$\sinh(1) = \frac{e - e^{-1}}{2} \approx \frac{2.718\,28 - 0.367\,88}{2} \approx 1.175$$

$$\cosh(1) = \frac{e + e^{-1}}{2} \approx \frac{2.718\,28 + 0.367\,88}{2} \approx 1.543$$

$$\tanh(1) = \frac{\sinh(1)}{\cosh(1)} \approx \frac{1.175}{1.543} \approx 0.762$$

and $\cosh^2(1) - \sinh^2(1) \approx (1.543)^2 - (1.175)^2 \approx 1.000\,22$ $\square$

Note   A more accurate result for this calculation (and also for the calculation of tanh (1)) could be obtained by retaining as many decimal places as possible in the intermediate steps. In fact, from Question T1(c) we know that the result for $\cosh^2(1) - \sinh^2(1)$ should be exactly 1.

Question T2
The speed of waves in shallow water is given by the following equation

$$v^2 = A\lambda \tanh\left(\frac{6.3d}{\lambda}\right)$$

where λ is the wavelength and d is the depth of the water. Find the speed, v when $d = 6$ m, $\lambda = 18.9$ m and $A = 1.8$ m s^{-2}. $\square$

7.2.2   Reciprocal hyperbolic functions

The **reciprocals** of the basic hyperbolic functions occur fairly frequently and so are given special names, **cosech**, **sech** and **coth**:

Some computer algebra packages use csch in place of cosech.

$$\text{cosech}(x) = \frac{1}{\sinh(x)} \quad \text{provided } x \neq 0 \tag{7}$$

$$\text{sech}(x) = \frac{1}{\cosh(x)} \quad \text{for all } x \tag{8}$$

$$\text{coth}(x) = \frac{1}{\tanh(x)} \quad \text{provided } x \neq 0 \tag{9}$$

Notice that cosech (x) is the reciprocal of sinh (x) and that sech (x) is the reciprocal of cosh (x). As in the case of trigonometric functions, this terminology may seem rather odd, but it is easily remembered by recalling that each reciprocal pair − (sinh, cosech), (cosh, sech), (tanh, coth) − involves the letters 'co' just once. In other words, there is just one 'co' between each pair. Also notice that the cosech and coth functions are undefined when their partner functions are zero. By contrast, sech (x) is defined for all real x since cosh (x) is never zero, as Figure 7.3 shows.

The reciprocal functions can be written in terms of e^x as

$$\text{cosech}\,(x) = \frac{2}{e^x - e^{-x}} \tag{10}$$

$$\text{sech}\,(x) = \frac{2}{e^x + e^{-x}} \tag{11}$$

$$\text{coth}\,(x) = \frac{e^x + e^{-x}}{e^x - e^{-x}} \tag{12}$$

and graphs of these functions are given in Figures 7.6–7.8, respectively.

Question T3
Show that $\text{coth}^2(x) - \text{cosech}^2(x) = 1$. □

7.2.3 Inverse hyperbolic functions (and logarithmic forms)

A common problem is to find what argument gives rise to a particular value of some hyperbolic function. For example, if sinh $(x) = 3.6$, what is the value of x? We can get an approximate value of x by looking at Figure 7.2.

Use Figure 7.2 to find the (approximate) value of x that corresponds to sinh $(x) = 3.6$.

x is approximately 2.0. □

To answer such questions more precisely we need to know the **inverse hyperbolic functions** which 'undo' the effect of the hyperbolic functions. These new functions are known as **arcsinh**, **arctanh** and **arccosh**, and the first two functions are defined by

arcsinh $[\sinh(x)] = x$ (13)

arctanh $[\tanh(x)] = x$ (14)

Graphs of these functions are given in Figures 7.9 and 7.10, respectively. Notice that the domain of arctanh (x) is $-1 < x < 1$.

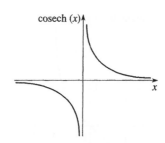

Figure 7.6 Graph of cosech (x).

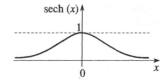

Figure 7.7 Graph of sech (x).

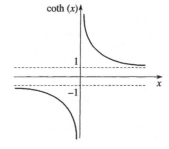

Figure 7.8 Graph of coth (x).

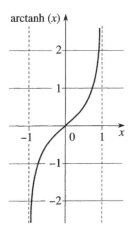

Figure 7.10 Graph of arctanh (x).

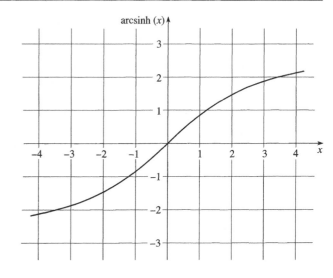

Figure 7.9 Graph of arcsinh (x).

From Figure 7.3, estimate values of x that correspond to cosh $(x) = 1, 2$ and 3.

The values of x are approximately 0, ± 1.3 and ± 1.8. □

This is precisely the same problem that we have when defining the inverse of the function $F(x) = x^2$. We solve it by assigning the unique positive value to the square root in order to make $\sqrt{x}$ a function. Remember that a function must give rise to a unique value $f(x)$ for each value of x.

From Figure 7.3, we see that for any given value of cosh $(x) > 1$ there are *two* corresponding values of x; in other words, the equation cosh $(x) = a > 1$ has two solutions. Consequently, in order for arccosh to be a function, we must impose a condition which picks out just one value (i.e. $x \geqslant 0$). Figure 7.11 shows the graph of the inverse function (the continuous line) and, as you can see, we choose the positive value for arccosh (x), which gives

$$\cosh [\text{arccosh} (x)] = x \quad \text{for } x \geqslant 1 \qquad (15a)$$

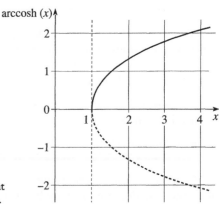

Figure 7.11 Graph of arccosh (x). Note that only the solid line is involved in the definition.

(a) Simplify $\sqrt{x^2}$ and $(\sqrt{x})^2$. Do you need to place any restrictions on the values of x in either case?

(b) Simplify arccosh [cosh (x)] and cosh [arccosh (x)]. Do you need to place any restrictions on the values of x in either case?

(a) In this chapter we always use the positive square root, so that $\sqrt{x^2} = |x|$ and there are no restrictions on the value of x. On the other hand, $\sqrt{x}$ only makes sense if $x \geqslant 0$, in which case $(\sqrt{x})^2 = x$.

(b) arccosh always gives rise to a positive value, and cosh (x) is defined for all values of x, so that arccosh [cosh (x)] = $|x|$ for all x. On the other hand, arccosh (x) only makes sense if $x \geqslant 1$, in which case cosh [arccosh (x)] = x. □

If we agree to restrict ourselves to positive values of x, so that we can be sure that $|x| = x$, there is no problem with the inverse of cosh (just as there is no problem with the inverse of 'squaring' if we restrict ourselves to positive numbers) for then we have

$$\text{arccosh } [\cosh (x)] = x \quad \text{for } x > 0 \tag{15b}$$

From Figure 7.3, what are the approximate values of arccosh (2) and arccosh (3)?

We find arccosh (2) ≈ 1.3 and arccosh (3) ≈ 1.8. These values are consistent with Figure 7.11. □

Find all solutions of the equation cosh $(x) = 2$.

Our first thought might be that we have just shown that arccosh (2) ≈ 1.3 and therefore *the* solution is approximately 1.3. However, from Figure 7.3 we know that there are two solutions of the equation cosh $(x) = 2$. These solutions are given by

$$x = \pm\text{arccosh } (2) \approx \pm 1.3 \quad □$$

As in the case of trigonometric functions, an alternative notation is sometimes used to represent the inverse hyperbolic functions:

$\sinh^{-1}(x)$ for $\text{arcsinh}(x)$

$\cosh^{-1}(x)$ for $\text{arccosh}(x)$

$\tanh^{-1}(x)$ for $\text{arctanh}(x)$

Notice that there is *no* connection with the *positive* index notation used to denote powers of the hyperbolic functions, for example, using $\sinh^2(x)$ to represent $[\sinh(x)]^2$. Also notice that, although this notation might make it appear otherwise, there is still a clear distinction between the inverse hyperbolic functions and the reciprocal hyperbolic functions. So

$$\sinh^{-1}[\sinh(x)] = \text{arcsinh}[\sinh(x)] = x$$

but $\quad \text{cosech}(x) \times \sinh(x) = \dfrac{1}{\sinh(x)} \times \sinh(x) = 1$

More inverse hyperbolic functions

There are three more inverse hyperbolic functions which we have not mentioned yet; these are **arccosech** (x), **arcsech** (x) and **arccoth** (x). As you might expect, they are defined so that

$$\text{arccosech}[\text{cosech}(x)] = x \qquad (16)$$

$$\text{arcsech}[\text{sech}(x)] = x \qquad (17)$$

$$\text{arccoth}[\text{coth}(x)] = x \qquad (18)$$

Graphs of these functions are given in Figures 7.12–7.14 and the following exercise leads to some useful identities.

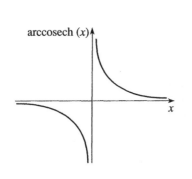

Figure 7.12 Graph of arccosech (x).

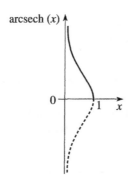

Figure 7.13 Graph of arcsech (x).

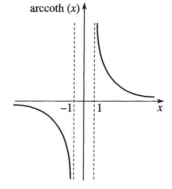

Figure 7.14 Graph of arccoth (x).

Question T4

Show that (a) arccosech (x) = arcsinh $(1/x)$, (b) arcsech (x) = arccosh $(1/x)$, and (c) arccoth (x) = arctanh $(1/x)$. ☐

Connection with logarithmic function

Given the relationship between the exponential and logarithmic functions, and the fact that the hyperbolic functions are defined in terms of the exponential function, it might be expected that the inverse hyperbolic functions can be expressed in terms of the logarithmic function. This is indeed the case, and it can be shown that the three basic hyperbolic functions can be written as

$$\text{arcsinh}\,(x) = \log_e\,(x + \sqrt{x^2 + 1}) \tag{19}$$

$$\text{arccosh}\,(x) = \log_e\,(x + \sqrt{x^2 - 1}) \quad \text{for } x \geqslant 1 \tag{20}$$

$$\text{arctanh}\,(x) = \frac{1}{2}\log_e\left(\frac{1 + x}{1 - x}\right) \quad \text{for } -1 < x < 1 \tag{21}$$

As an example, we now prove the second of these identities.

We start by setting $\quad y = \text{arccosh}\,(x)$

which implies that $\quad x = \cosh\,(y) = \dfrac{e^y + e^{-y}}{2}$, and we see that $x \geqslant 1$. (☞)

Multiplying the right- and left-hand sides of this last relationship by $2e^y$ and rearranging, we obtain

$$e^{2y} - 2xe^y + 1 = 0,$$

a quadratic equation in e^y which we can solve to obtain

$$e^y = \frac{2x \pm \sqrt{4x^2 - 4}}{2} = x \pm \sqrt{x^2 - 1}$$

The question now is whether either (or both) of the signs on the right-hand side of this equation are allowable. First we note that the right-hand side must be greater than or equal to one, because $y = \text{arcosh}\,(x) > 0$, so $e^y \geqslant 1$; then we can eliminate the negative sign as follows. (☞)

We consider the graphs of

$$Y_1 = (x + \sqrt{x^2 - 1}) \quad \text{and} \quad Y_2 = (x - \sqrt{x^2 - 1})$$

for $x \geqslant 1$ (shown in Figure 7.15), and ask if it is possible for either Y_1 or Y_2 to be greater than or equal to 1 as required.

From the graphs it is clear that $Y_2 \leqslant 1$, which makes this choice impossible, and therefore the only acceptable solution is

$$e^y = x + \sqrt{x^2 - 1}$$

If $y = \text{arccosh}\,(x)$ then $\cosh\,(y) = x$. See Figure 7.3.

Alternatively, if we suppose that

$$e^y = x - \sqrt{x^2 - 1}\,,$$

then we would have

$$x - \sqrt{x^2 - 1} \geqslant 1\,,$$

which would imply

$$(x - 1)^2 \geqslant x^2 - 1\,,$$

which would give

$$-2x + 1 \geqslant -1, \text{ or } x \leqslant 1.$$

But we know that $x \geqslant 1$, so this is a contradiction.

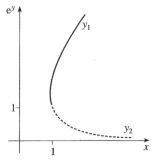

Figure 7.15   Graphs of
$Y_1 = (x + \sqrt{x^2 - 1})$ (——)
and $Y_2 = (x - \sqrt{x^2 - 1})$ (- - - -)
for $x \geqslant 1$.

Taking the logarithm of this result gives us

$$y = \log_e (x + \sqrt{x^2 - 1})$$

The remaining two identities can be proven in a similar way, as the following question demonstrates.

Question T5
Show that

$$\text{arctanh}\,(x) = \frac{1}{2} \log_e \left(\frac{1 + x}{1 - x}\right) \quad \text{for } -1 < x < 1. \quad \Box$$

Expressing the inverse hyperbolic functions in terms of the logarithmic function can be a useful way of solving certain algebraic equations. For example, suppose we want the values of x which satisfy

$$\sinh^2 (x) - 3 \cosh (x) + 3 = 0$$

then we can use the identity

$$\cosh^2 (x) - \sinh^2 (x) = 1$$

to obtain an equation involving just $\cosh (x)$

$$\cosh^2 (x) - 3 \cosh (x) + 2 = 0$$

This can be factorised to give

$$(\cosh (x) - 2)(\cosh (x) - 1) = 0$$

Don't forget that in general there are two solutions to equations such as $\cosh (x) = 2$.

so, either $x = \pm\, \text{arccosh}\,(2)$ (☜)

or $x = \pm\, \text{arccosh}\,(1)$

Using $\text{arccosh}\,(x) = \log_e (x + \sqrt{x^2 - 1})$

gives $x = \pm\, \text{arccosh}\,(2) = \pm \log_e (2 + \sqrt{3})$

or $x = \pm\, \text{arccosh}\,(1) = \pm \log_e (1) = 0$

(One advantage of this approach is that the logarithmic function is more likely to be available on a calculator.)

Question T6
Solve $\cosh^2 (x) + \sinh (x) = 3$, in terms of the logarithmic function. $\Box$

7.3 Identities

There are many identities involving trigonometric functions, and there are an equally large number of identities involving hyperbolic functions. In fact, since the trigonometric and hyperbolic functions are related (by, as you will learn in the companion volume, $\sinh (ix) = i \sin (x)$, where $i^2 = -1$), every trigonometric identity has an analogue for hyperbolic functions. The most important **hyperbolic function identities** are listed below, and it is worth comparing them with their trigonometric partners which are given in Chapter 5.

The **symmetry relations:** (☞)

$$\sinh (-x) = -\sinh (x) \tag{22}$$

$$\cosh (-x) = \cosh (x) \tag{23}$$

$$\tanh (-x) = -\tanh (x) \tag{24}$$

It is not suggested that you should be able to quote all of these identities. Most people just learn the few that they use most frequently and look the rest up in tables of such identities. The really important point is to know what kind of identities exist.

The symmetry relations show the oddness and evenness of the function.

The **basic identities:**

$$\tanh (x) = \frac{\sinh (x)}{\cosh (x)} \tag{25}$$

$$\cosh^2 (x) - \sinh^2 (x) = 1 \tag{26}$$

$$\coth (x) = \frac{\operatorname{cosech} (x)}{\operatorname{sech} (x)} \quad \text{for } x \neq 0 \tag{27}$$

$$1 - \tanh^2 (x) = \operatorname{sech}^2 (x) \tag{28}$$

The **addition identities:**

$$\sinh (x + y) = \sinh (x) \cosh (y) + \cosh (x) \sinh (y) \tag{29}$$

$$\cosh (x + y) = \cosh (x) \cosh (y) + \sinh (x) \sinh (y) \tag{30}$$

$$\tanh (x + y) = \frac{\tanh (x) + \tanh (y)}{1 + \tanh (x) \tanh (y)} \tag{31}$$

The corresponding formulae for trigonometric functions are usually known as the 'addition formulae'.

The **double-argument identities:**

$$\sinh (2x) = 2 \sinh (x) \cosh (x) \tag{32}$$

$$\cosh (2x) = \cosh^2 (x) + \sinh^2 (x) \tag{33}$$

$$\cosh (2x) = 1 + 2 \sinh^2 (x) \tag{34}$$

$$\cosh (2x) = 2 \cosh^2 (x) - 1 \tag{35}$$

$$\tanh (2x) = \frac{2 \tanh (x)}{1 + \tanh^2 (x)} \tag{36}$$

The corresponding formulae for trigonometric functions are usually known as the 'double-argument formulae'.

The corresponding formulae for trigonometric functions are usually known as the 'half-argument formulae'.

The **half-argument identities**:

$$\cosh^2\left(\frac{x}{2}\right) = \frac{1}{2}[1 + \cosh(x)] \tag{37}$$

$$\cosh^2\left(\frac{x}{2}\right) = \frac{1}{2}[-1 + \cosh(x)] \tag{38}$$

and if $t = \tanh(x/2)$ then

$$\sinh(x) = \frac{2t}{1 - t^2} \tag{39}$$

$$\cosh(x) = \frac{1 + t^2}{1 - t^2} \tag{40}$$

$$\tanh(x) = \frac{2t}{1 + t^2} \tag{41}$$

The **sum identities**:

$$\sinh(x) + \sinh(y) = 2\sinh\left(\frac{x+y}{2}\right)\cosh\left(\frac{x-y}{2}\right) \tag{42}$$

$$\sinh(x) - \sinh(y) = 2\cosh\left(\frac{x+y}{2}\right)\sinh\left(\frac{x-y}{2}\right) \tag{43}$$

$$\cosh(x) + \cosh(y) = 2\cosh\left(\frac{x+y}{2}\right)\cosh\left(\frac{x-y}{2}\right) \tag{44}$$

$$\cosh(x) - \cosh(y) = 2\sinh\left(\frac{x+y}{2}\right)\sinh\left(\frac{x-y}{2}\right) \tag{45}$$

The **product identities:**

$$2\sinh(x)\cosh(y) = \sinh(x+y) + \sinh(x-y) \tag{46}$$

$$2\cosh(x)\cosh(y) = \cosh(x+y) + \cosh(x-y) \tag{47}$$

$$2\sinh(x)\sinh(y) = \cosh(x+y) - \cosh(x-y) \tag{48}$$

The identities can be 'derived' quite simply from their trigonometric counterparts using Osborne's rules: change sin to sinh, cos to cosh, etc. in the trigonometric identities and then whenever sin × sin occurs change the sign (i.e. in $\sin^2$, $\tan^2$, cosec^2 and $\cot^2$).

All of these identities can be derived from the definitions of $\sinh(x)$ and $\cosh(x)$ in terms of e^x as the following questions demonstrate.

Use the definitions of $\sinh(x)$ and $\cosh(x)$ in terms of e^x to show that $\sinh(x + y) = \sinh(x)\cosh(y) + \cosh(x)\sinh(y)$.

The left-hand side can be written as

$$\frac{1}{2}\left(e^{x+y} - e^{-x-y}\right)$$

The right-hand side seems complicated but reduces to the left-hand side after some algebraic manipulation:

$$\left(\frac{e^x - e^{-x}}{2}\right)\left(\frac{e^y + e^{-y}}{2}\right) + \left(\frac{e^x + e^{-x}}{2}\right)\left(\frac{e^y - e^{-y}}{2}\right)$$

$$= \frac{1}{4}\left[\left(e^{x+y} + e^{x-y} - e^{-x+y} - e^{-x-y}\right) + \left(e^{x+y} - e^{x-y} + e^{-x+y} - e^{-x-y}\right)\right]$$

$$= \frac{1}{2}\left(e^{x+y} - e^{-x-y}\right) \quad \square$$

Question T7

Use the definition of the hyperbolic functions in terms of the exponential function to show that

$$\sinh(2x) = 2\sinh(x)\cosh(x). \quad \square$$

In many cases it is more convenient to derive identities from other identities rather than from the basic definitions. For example, by putting $x = y$ in

$$\sinh(x + y) = \sinh(x)\cosh(y) + \cosh(x)\sinh(y)$$

we obtain

$$\sinh(2x) = 2\sinh(x)\cosh(x)$$

This is certainly more straightforward than using the basic definitions; the other double-argument identities follow from the addition identities in a similar fashion.

Question T8

Use the addition identities to show that

$$\cosh(2x) = 2\cosh^2(x) - 1$$

and

$$\tanh(2x) = \frac{2\tanh(x)}{1 + \tanh^2(x)}.$$

7.4 Closing items

7.4.1 Chapter summary

1. There are *hyperbolic functions* corresponding to all the trigonometric functions, and though they are similar in some respects they are very different in others. In particular, the hyperbolic functions are not periodic, and, sinh (x) and cosh (x) are not constrained to lie between -1 and 1. Hyperbolic functions are named by appending 'h' to the corresponding trigonometric function.

2. sinh (x) and cosh (x) are defined by

$$\sinh (x) = \frac{e^x - e^{-x}}{2} \tag{1}$$

$$\cosh (x) = \frac{e^x + e^{-x}}{2} \tag{2}$$

3. tanh (x) is defined by

$$\tanh (x) = \frac{\sinh (x)}{\cosh (x)} = \frac{e^x - e^{-x}}{e^x + e^{-x}} \tag{3}$$

4. A hyperbola can be defined by means of the parametric equations

$$x = a \cosh (s) \tag{4}$$

$$y = b \sinh (s) \tag{5}$$

5. *Reciprocal hyperbolic functions* are defined by

$$\text{cosech } (x) = \frac{1}{\sinh (x)} \quad \text{provided } x \neq 0 \tag{7}$$

$$\text{sech } (x) = \frac{1}{\cosh (x)} \quad \text{for all } x \tag{8}$$

$$\coth (x) = \frac{1}{\tanh (x)} \quad \text{provided } x \neq 0 \tag{9}$$

6. *Inverse hyperbolic functions* are defined by

$$\text{arcsinh } (\sinh (x)) = x \tag{13}$$

$$\text{arccosh } (\cosh (x)) = x \quad \text{for } x \geqslant 0 \tag{15b}$$

$$\text{arctanh } (\tanh (x)) = x \tag{14}$$

7. The inverse hyperbolic functions can be written in terms of the logarithmic function

$$\text{arcsinh}\,(x) = \log_e (x + \sqrt{x^2 + 1}) \tag{19}$$

$$\text{arccosh}\,(x) = \log_e (x + \sqrt{x^2 - 1}) \quad \text{for } x \geqslant 1 \tag{20}$$

$$\text{arctanh}\,(x) = \frac{1}{2}\log_e \left(\frac{1 + x}{1 - x}\right) \quad \text{for } -1 < x < 1 \tag{21}$$

8. There are a large number of *hyperbolic function identities*, such as

$$\cosh^2(x) - \sinh^2(x) = 1 \tag{26}$$

The most important are listed in Section 7.3.

7.4.2   Achievements

Having completed this chapter, you should be able to:

A1. Define the terms that are emboldened in the text of this chapter.

A2. Recognise the similarities that exist between hyperbolic and trigonometric functions, and the limitations to those similarities.

A3. Sketch graphs for the more common hyperbolic functions.

A4. Recognise the more common identities for the hyperbolic functions and use them to solve mathematical and physical problems.

7.4.3   End of chapter questions

Question E1   Use the double-argument identities to show that

(a) $\cosh^2\left(\dfrac{x}{2}\right) = \dfrac{1}{2}[1 + \cosh(x)]$

(b) $\sinh^2\left(\dfrac{x}{2}\right) = \dfrac{1}{2}[-1 + \cosh(x)]$

Question E2   Suppose that we are given the values of A and B. Use the identities given in Section 7.3 to find values of C and y (in terms of A and B) such that

$$A \sinh(x) + B \cosh(x) = C \sinh(x + y)$$

Explain any restrictions on the possible values for A and B.

Question E3   From the graphs given in this chapter, which hyperbolic functions would you expect to have a domain which excludes the point $x = 0$? Confirm your suspicions by looking at algebraic expressions for the functions that you have picked out. (You should include the reciprocal and inverse hyperbolic functions in your discussion.)

Question E4   Starting from the definition of sinh (x) in terms of the exponential function, show that

$$\text{archsinh } (x) = \log_e (x + \sqrt{x^2 + 1})$$

Part 2

Basic geometry

Introducing geometry

8.1 Opening items

8.1.1 Chapter introduction

You may very well have met some of the concepts discussed in this chapter before; in fact it is hardly conceivable that you have not. Nevertheless there are many geometric terms which are in general use among physical scientists but which are commonly omitted from the standard school syllabus. The main aim of this chapter is to remedy that deficiency.

Geometry is of importance to all scientists and engineers since it deals with the relationships between points and lines, and with a variety of shapes in two and three dimensions.

In Section 8.2 we examine straight lines and the angles between them, and then the shapes, such as triangles and rectangles, which can be formed by straight lines. Such concepts are fundamental to the study of physical science and engineering, for example:

- It is well known in the study of optics that 'the angle of incidence is equal to the angle of reflection'.
- Parallel lines drawn on a flat surface do not meet, and, if another line is drawn to intersect a pair of parallel lines, certain angles are formed which are equal.
- The simplest shape formed by straight lines is a triangle, but, in spite of being simple, such shapes are very important to the engineer because they form the simplest rigid structure.

In Subsection 8.3.1 we discuss *congruence*. Roughly speaking, if you cut two *congruent* shapes from paper, then one will fit exactly over the other. In Subsection 8.3.2 we introduce *similarity*, and again speaking roughly, this means that one shape is a scaled version of the other (for example, the same shape but twice as big). In Section 8.4 we examine circles, and, in particular, the important concepts of *radian measure* and *tangent* lines. Finally in Section 8.5 we discuss the *areas* and *volumes* of various standard shapes.

Before we begin the chapter we should give you a word of warning. If you glance at Figure 8.1 you will see that the lines HF and DF appear to be approximately the same length. However in this topic equality means 'exactly equal' and not 'approximately equal'. It is very easy to assume that because

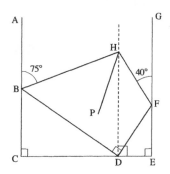

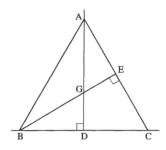

Figure 8.1 See Questions F1 and F2.

Figure 8.2 See Question F3.

two lines in a figure *look* the same length they *are* the same length, and similarly for angles. In this subject you are either told that two lengths are equal or you deduce that they are equal from the given information. There is no other way!

8.1.2 Fast track questions

Question F1

In Figure 8.1 the line HP bisects the angle $B\hat{H}F$. Find the value of angle $P\hat{H}D$.

Question F2

Again in Figure 8.1, $B\hat{C}D = B\hat{D}F = D\hat{E}F = H\hat{D}E = 90°$, and $BD = 15\,\text{m}$, $DF = 5\,\text{m}$, $CE = 15\,\text{m}$, $DE = 3\,\text{m}$. Show that the triangles BCD and DEF are similar, and hence find the lengths of EF and BC.

Question F3

In Figure 8.2, $AC = BC$ and $B\hat{E}C = A\hat{D}C = 90°$. By calculating the area of the triangle ABC using two different methods, or otherwise, show that $BE = AD$. Hence show that the triangles BEC and ADC are congruent.

Question F4

A steel pipe is $10\,\text{m}$ long. Its inside diameter is $40\,\text{mm}$ and its outside diameter is $50\,\text{mm}$. Calculate the volume of the steel in m^3.

8.1.3 Ready to study?

Study comment

In order to study this chapter you will need to be familiar with the following terms: *angle*, *line*, *two dimensions* and *three dimensions*. In addition, you will need to be familiar with *SI units*. It is also assumed that you can carry out basic algebraic and arithmetical manipulations, in particular calculations using *indices*. You will also need a ruler (measuring centimetres), a protractor and a pair of compasses. The following *Ready to study questions* will allow you to establish whether you need to review some of the topics before embarking on this chapter.

Question R1

(a) If $\dfrac{x}{y} = \dfrac{z}{w}$ and $y = 6$, $z = 15$, $w = 10$, find x.

(b) Given that $\dfrac{x}{y} = \dfrac{y}{z}$, $x = 6$ and $z = 24$,
find the possible values of y.

(c) Given that $x^3 = 5$, calculate the value of x^2.

Question R2

Draw a line AB of length 7 cm, then, using a pair of compasses, draw a circle centre A of radius AB and a circle centre B of radius AB. Label the points at which the circles intersect as P and Q, then draw the line PQ and label as M the point where it meets the line AB. Use a protractor to measure the angle between the two lines (PQ and AB), then use a ruler to measure AM and MB in cm. Use your protractor to measure the angle between the lines PA and PB, and then the angle between the lines PA and QA.

8.2 Euclidean plane geometry

8.2.1 Lines and angles

The most fundamental geometric concept is that of a **point**, which occupies a position but has no size or dimension. Initially we will be concerned with points that lie on a flat surface (like this sheet of paper), usually called a plane surface or, more simply, a **plane**. We will also investigate the properties of lines drawn on the plane. A line has length as its sole dimension, but it has no thickness. Of course, these are theoretical concepts; in practice, when we plot a point or draw a line we have to give it thickness in order to be able to see it.

A **straight line** is the shortest distance between two points. It is the path traced out by moving from one point to the other in a constant direction. If we wish to be precise then we would have to allow the straight line to extend infinitely away from any point on it and define that part of it between two particular points as a **straight line segment**. Any two points on the plane can be joined by a unique straight line segment which lies entirely in the plane. (☞) We say that two lines **intersect** if they have a point in common, known as the **intercept** (if they have more than one point in common then they must be identical because of the uniqueness property of straight lines). Two lines that do not intersect are said to be **parallel**. That part of a straight line which extends from a given point in one direction only is called a **ray** – you may have heard of the term 'a ray of light'.

Figure 8.3 is intended to illustrate a number of important concepts. First, the two lines YD and RV intersect at the point Z; of course the lines are not depicted in full (they are infinite in extent, after all) but by segments. The points C and X also lie on the line YD, and we say that they lie on the line DY **produced**. (☞)

There are four angles formed by the two lines YD and RV: $\hat{YZV}$, $\hat{VZD}$, $\hat{DZR}$ and $\hat{RZY}$. The notation is fairly obvious; the caret (ˆ) sign is placed

Mathematicians have developed many different kinds of geometry, but the geometry that we discuss here is the one that you probably met in school. It is often known as *Euclidean geometry*. Euclid was a Greek mathematician (c. 330– c. 275 BC), born in Alexandria, author of a famous treatise on geometry known as the Elements (*Stoicheia*) and of whom little else is known.

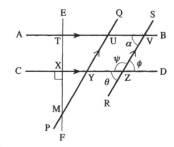

Figure 8.3 Properties of parallel lines.

The order of the points DY or YD does not really matter when defining the line, but the phrase 'DY produced' indicates that you move in the direction of D to Y when extending the line.

over the letter representing the point of the angle. It is also common practice to denote angles by letters (often Greek letters), so that in this case $Y\hat{Z}R = \theta$ (the Greek letter theta).

An angle is a measure of the rotation from one ray to another, for example from ZV to ZD. To quantify this rotation we measure a full rotation (i.e. one that gets you back to where you started) by 360 **degrees**, written 360°. Hence turning the ray ZV first to ZD, then ZR, then ZY and finally on to its starting point, is a rotation of 360°. When the four angles are equal to each other, and therefore each is equal to 90°, the lines are said to be **perpendicular** to each other or to intersect at **right angles**. This is the case for the lines EF and CD intersecting at X. Notice the symbol (a box) used at X to denote a right angle.

The angles θ and ψ (the Greek letter psi) together form half a complete turn about the point Z, and so their sum must be 180°. This is also the case for the angles ψ and ϕ (phi), and it follows that $\theta = \phi$. The angles θ and ϕ are said to be **vertically opposite**, and we have just shown that vertically opposite angles must always be equal.

In Figure 8.3, if $U\hat{V}Z = 50°$ what are the values of the three other angles at V?

$S\hat{V}B = 50°$ ($U\hat{V}Z$ and $S\hat{V}B$ are vertically opposite angles and therefore equal.)

$U\hat{V}S = B\hat{V}Z = 130°$ ($U\hat{V}Z$ and $U\hat{V}S$ form half a complete turn, while $U\hat{V}S$ and $B\hat{V}Z$ are vertically opposite.) $\square$

Angles less than 90° are called **acute angles**, those greater than 90° and less than 180° are **obtuse angles**, and those greater than 180° and less than 360° are **reflex angles**.

Classify the angles 135°, 60°, 250°, 310°, 90°.

Obtuse, acute, reflex, reflex, right angle. $\square$

When two angles add up to 180° they are said to be **supplementary**; when they add up to 90° they are **complementary**.

What are the angles that are supplementary to 130°, 60°, 90°? What are the angles that are complementary to 30°, 45°, 60°, 90°?

The supplementary angles are 50°, 120°, 90°.

The complementary angles are 60°, 45°, 30°, 0°. $\square$

8.2.2 Parallel lines

Parallel lines are indicated by arrowheads, as in Figure 8.3, and notice that different pairs of parallel lines are indicated by arrowheads of different styles. A line which intersects two parallel lines is called a **transversal**, so that (in Figure 8.3) RS is a transversal to the pair of parallel lines AB and CD. When two parallel lines are intersected by a transversal two sets of equal angles are created.

The angles θ and α (the Greek letter alpha) in Figure 8.3 are equal; they are **corresponding angles** – they occur on the same side of the transversal.

The angles ϕ and α are equal; they are **alternate angles** – they occur on opposite sides of the transversal.

In Figure 8.3 PQ and RS are parallel lines and AB is a transversal.

(a)   Identify the corresponding angle to α that occurs at point U.
(b)   Identify which angle at V is vertically opposite to α.
(c)   Give an example of two angles at U that are supplementary.
(d)   Identify which angle at U is alternate to α.

(a)   TÛY   (b)   SV̂B

(c)   TÛY and TÛQ (or TÛQ and QÛV, or QÛV and VÛY,

or VÛY and TÛY)   (d)   QÛV   □

8.2.3 Triangles

A **triangle** is a closed figure which has three sides, each of which is a straight line segment. A triangle is usually labelled with a capital letter at each of the corners which are more properly known as the vertices (the plural of vertex). We speak of the triangle ABC, for example.

Figure 8.4 shows three types of triangle (for the moment you can ignore the dotted lines and labelled angles).

In each of the three triangles the angles BÂC, AĈB and CB̂A are known as the **interior angles** of the triangle (and sometimes referred to as the angles Â, B̂ and Ĉ, or even more briefly as the angles A, B and C).

- In Figure 8.4a all the interior angles are acute.
- In Figure 8.4b one of the angles, A, is obtuse.
- In Figure 8.4c one of the angles, C, is 90°, i.e. a right angle.

Notice that we have also introduced the notation of a lower-case letter to represent each side of a triangle; the side a is opposite the angle A, and so on.

In the case of Figures 8.4a and 8.4c it is natural to refer to the side BC as the base of the triangle, however any side may be regarded as the *base* – it just depends on how you want to view it.

A complementary discussion of triangles, particularly right angled triangles, is given in Section 5.2.

Results of this kind are often known as *theorems*, meaning that the result can be deduced from the initial assumptions (or *axioms*) and any previous theorems.

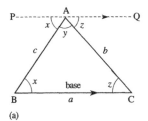

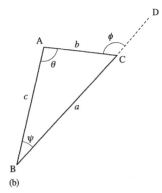

(a)

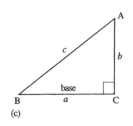

(b)

(c)

Figure 8.4 Various types of triangle.

Note that here we are just using the letters A, B, C to represent the interior angle at corresponding vertices.

If all three sides of a triangle are equal (i.e. their lengths are equal) then the triangle is known as an **equilateral triangle** and it is a consequence of this definition that the angles are equal. It is also the case that if the angles are equal then so are the sides.

If two sides of the triangle are equal the triangle is said to be **isosceles**, and the interior angles at either end of the third side are equal. A triangle for which one of the interior angles is a right angle is called a **right-angled triangle** and the side of the triangle opposite to the right angle (i.e. the longest side) is called the **hypotenuse**. A triangle whose three sides are unequal is called **scalene**.

An important result is that the sum of the interior angles of *any* triangle is 180°. (☎)

===

In Figure 8.4a the dotted line PQ is drawn through the vertex A and parallel to the opposite side BC. In the figure we have marked a pair of equal alternate angles x and a pair of equal alternate angles z. The sum of the angles (at A) on a straight line is 180° and hence $x + y + z = 180°$, which proves the required result.

The angle ϕ in Figure 8.4b is called an **exterior angle** of the triangle.

By drawing a ray from C in Figure 8.4b, parallel to BA, show that $\phi = \theta + \psi$.

The angle between AC and the ray is an alternate angle to θ, and the angle between the ray and CD is a corresponding angle to ψ. It follows that the exterior angle (ϕ) is equal to the sum of the opposite interior angles (θ and ψ). This result is true in general, and worth remembering. □

There is an alternative proof of the fact that the interior angles of a triangle sum to 180° which is worth mentioning since the same method can be applied very easily to figures with more than three sides.

Imagine that you are standing at B in Figure 8.4b and that you walk along the side BC until you reach C. You now turn anticlockwise through the angle ϕ until you are facing A; in other words you turn through an angle (180° − C). Now continue along CA until you reach A, where you turn anticlockwise through an angle (180° − A). Finally you proceed along AB until you reach B, then turn anticlockwise through an angle (180° − B) so that you have returned to your starting point and you are facing in your original direction, which means that you have made a complete turn of 360°.

However this complete turn was composed of three exterior angles, so that

$$(180° − A) + (180° − B) + (180° − C) = 360°$$

and therefore A + B + C = 180°.

Question T1

Use an argument similar to the above to show that the sum of the interior angles of a four-sided figure (such as a square) is 360°. □

Pythagoras's theorem

The lengths of the sides of a right-angled triangle are related by **Pythagoras's theorem**, which, using the notation of Figure 8.4c, my be written as

$$c^2 = a^2 + b^2 \tag{1}$$

where c is the longest side, i.e. the hypotenuse. We state the theorem without proof here since an outline proof was given in Section 5.2.2.

Given that $a = 0.8$ cm and $c = 1.7$ cm in Figure 8.4c, calculate the value of b.

Since $b^2 = c^2 - a^2$ it follows that $b^2 = 2.25$ cm² and therefore $b = 1.5$ cm. □

It should be noted that any triangle with sides of lengths a, b and c that satisfies Pythagoras's theorem *must* be a right-angled triangle. Thus, even if the only thing you know about a triangle is that its sides are of lengths 3 units, 4 units and 5 units, you can still be sure that it is a right-angled triangle simply because $3^2 + 4^2 = 5^2$. Other well known sets of numbers that satisfy $a^2 + b^2 = c^2$ are 5, 12, 13 and 6, 8, 10.

Is a triangle with sides 9 m, 12 m, and 15 m a right-angled triangle?

Yes. You can show this by checking that the values satisfy $a^2 + b^2 = c^2$, or more simply by noting that if you express the given lengths as multiples of 3 m 'units' the sides are of length 3 units, 4 units and 5 units. □

8.2.4 Polygons

Quadrilaterals

Any closed geometric shape whose sides are straight line segments is called a **polygon**. We have already met a three-sided polygon – a triangle. Four-sided polygons are known as **quadrilaterals**. When all the sides of a polygon are equal it is, not surprisingly, called an **equilateral polygon**; if all the internal angles are equal it is **equiangular**; and a polygon that is both equilateral and

You are not likely to meet often the terms *equiangular* and *equilateral* used to describe polygons; however, *regular* is a term that you should remember.

equiangular is said to be **regular**. Thus, for example, a regular quadrilateral is just a **square**. The **perimeter** of a polygon is the sum of the lengths of the sides.

A **parallelogram** is a special case of a quadrilateral in which the opposite sides are parallel (as, for example, UVZY in Figure 8.3).

Imagine a line drawn from U to Z in Figure 8.3. Is the triangle UVZ necessarily isosceles?

No. We cannot be sure that UV = VZ.   □

A parallelogram with all sides equal in length is called a **rhombus**, popularly known as diamond-shaped.

If the internal angles of a parallelogram are each 90° then the quadrilateral is a **rectangle**, and if the sides are also equal in length then the rectangle is a *square*.

Is every rectangle a square? Is every square a rectangle? Is every rectangle a rhombus? Is every square a rhombus?

Every square is a rectangle, but not all rectangles are squares. Not every rectangle is a rhombus, but every square is a rhombus.   □

A quadrilateral in which one pair of opposite sides is parallel is known as a **trapezium** (see Figure 8.5).

General polygons

Although triangles and quadrilaterals are the most commonly occurring polygons, the following result is sometimes useful since it can be applied to polygons with any number of sides. (☞) As one would expect, the corners of a polygon are known as **vertices** and the angle formed at a vertex which lies in the interior of the polygon is called an *interior angle*. The method that we used to show that the sum of the interior angles of a quadrilateral is 360° can easily be adapted to show that:

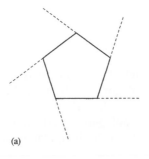
(a)

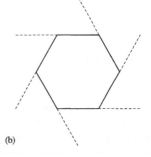
(b)

Figure 8.5   A trapezium.

pentagon (5)
hexagon (6)
heptagon (7)
octagon (8)
enneagon (9)
decagon (10)
hendecagon (11)
dodecagon (12)

Figure 8.6   (a) A regular pentagon, and (b) a regular hexagon.

The sum of the interior angles of an n sided polygon is $(2n - 4) \times 90°$, i.e. $(2n - 4)$ right angles.

Figure 8.6 shows a regular pentagon and a regular hexagon.

What is the size of an interior angle in (a) a regular pentagon, and (b) a regular hexagon?

(a) The sum of the interior angles in a pentagon is $(10 - 4) \times 90° = 540°$. Since there are five sides each interior angle in a regular pentagon is $1/5 (540°) = 108°$.

(b) The sum of the interior angles in a regular hexagon is $(12 - 4) \times 90° = 720°$ so that each interior angle is $1/6 (720°) = 120°$. □

8.3 Congruence and similarity

8.3.1 Congruent shapes

Geometric figures which are the same shape and size are said to be **congruent**. If we were to cut congruent triangles out of a piece of paper, we would be able to fit one exactly on top of the other, though we might have to turn one over or rotate it to do so. It is very important to be able to decide when triangles are congruent and the main aim of this section is to discuss this question.

In order to decide whether triangles are congruent, we need to know what is the minimum information to specify completely a triangle of unique size and shape. All triangles drawn with such a specification will then be copies of each other, and so congruent. There are four possible ways to specify a triangle of unique size and shape. These are as follows:

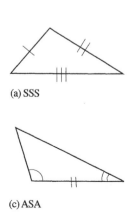

(a) SSS

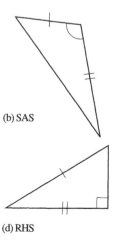

(b) SAS

(c) ASA

(d) RHS

Figure 8.7 Minimal specifications for a triangle. The cross bars and angles indicate the sides and angles referred to in the specifications.

You should memorise these.

Minimal specifications of a triangle
(a) State the length of all three sides. (SSS)
(b) State the lengths of two sides and the angle between them. (SAS)
(c) State the length of one side and the angles at each end of it. (ASA) (☎)
(d) Specify a right angle, the length of the hypotenuse and the length of one other side. (RHS)

Note that if the angle at one end of a side and the angle opposite the given side are known, the angle at the other end of the side can be found because the three angles must add up to 180°.

These specifications are shown diagramatically in Figure 8.7 together with abbreviations which may help you to remember them. The bars across the lines are often a useful device for indicating lines of equal length and in Figure 8.7 they indicate the lines referred to in the specifications above. We can now try to construct a triangle from each of these four specifications.

Circles will be discussed later in this chapter.

(a) Suppose that we are told that the lengths of the sides of a triangle are 7.4 cm, 6.0 cm and 3.8 cm. We first draw a line AB with the length of one of the sides, say 7.4 cm, as in Figure 8.8a. Then we draw a *circle*, with its centre at A, with the length of another side, say 6.0 cm, as its radius. Finally we draw a second circle with B as centre and the length of the third side, 3.8 cm, as its radius. If the circles intersects at a point C_1, then the triangle ABC_1 has sides of the required lengths. (The other intersection, C_2, would do equally well: triangle ABC_1 is a mirror-image in the line AB of triangle ABC_2.)

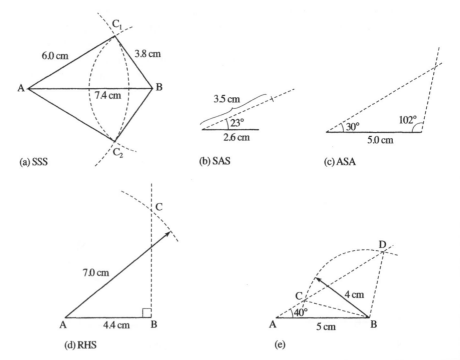

Figure 8.8 Constructing triangles using the minimal specifications. (These figures are not to scale.)

(b)   Suppose that we are told that the lengths of two of the sides of a triangle are 2.6 cm and 3.5 cm, and that the angle between them is 23°. First we draw a line segment of one of the known lengths, say 2.6 cm, as in Figure 8.8b. Next we use a protractor to draw a straight line through one end of this line segment and at the known angle, 23°, to it. We make the length of the second line the second known length 3.5 cm. Finally we join the end of the second line to the end of the first, giving the required triangle.

(c)   Suppose that we are given the length of one side, say 5.0 cm, and the angles at either end of this line, say 30° and 102°. We first draw a line segment of the known length, 5.0 cm, which immediately gives us two vertices of the triangle, as in Figure 8.8c. Then from each vertex we draw a line at the specified angle to the line segment. The intersection of these two lines gives us the third vertex.

(d)   Suppose that we are told that one of the angles in the triangle is a right angle and we are given the lengths of the hypotenuse and one other side, say 7.0 cm and 4.4 cm, respectively. We first draw a line segment the length of the given side other than the hypotenuse, AB in Figure 8.8d. Then we draw a line at right angles to it at one end, B say; finally, we draw a circle with its centre at the other end, A, and of radius equal to the length of the hypotenuse. The point at which this circle intersects the second line, C, is the third vertex of the triangle.

Three pieces of information are not always sufficient to specify a triangle completely. If we are given the lengths of two sides and an angle other than the one between the sides, it is possible that we will be able to draw two (non-congruent) triangles, each meeting the given specification. For example, suppose we are told that a triangle has two sides of length 4 cm and 5 cm, and that the angle at the end of the 5 cm line (not between it and the 4 cm line) is 40°. We can construct two possible triangles (ABC and ABD), but they are not congruent with each other, as illustrated in Figure 8e.

Would the three interior angles completely specify a unique triangle?

No. There would be a free choice of the length of any one side.   □

Triangles like these which are of the same shape but of different sizes are called *similar triangles* and we will be discussing them in detail in the next subsection. Of course once the length of a particular side is chosen, you then have a side and the angle at each end of it specified, and so you can draw a unique triangle as in case (c) above.

Having determined the minimal amount of information needed to uniquely specify a triangle (the conditions SSS, SAS, ASA and RHS of the previous box), we can now present a simple test for the congruency of two triangles.

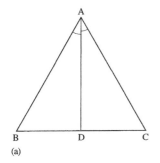

(a)

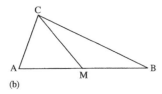

(b)

Figure 8.9 (a) See Question T2. (b) See Question T4.

Test for congruent triangles
Two triangles must be congruent if the information contained in any of the conditions SSS, SAS, ASA or RHS is the same for both.

It should be noted that although the above test only requires *one* of the minimal specifications to be common to both triangles it must follow that if the triangles are congruent then common values can be found for all three of the general specifications (SSS, SAS, ASA) and for the fourth specification (RHS) if it applies.

Bisectors
Isosceles triangles have some useful properties arising from their symmetry. In the triangle ABC of Figure 8a, AB = AC. The line AD is drawn so that $B\hat{A}D = C\hat{A}D = \frac{1}{2}(B\hat{A}C)$. This line is called the **bisector** of the angle $B\hat{A}C$. The term is also applied to line segments, so that a point lying at the mid-point of a line segment is sometimes said to bisect the line segment.

Question T2
By finding a pair of congruent triangles in Figure 8.9a, show that:

(a) $A\hat{B}C = A\hat{C}B$, (b) $A\hat{D}B = A\hat{D}C = 90°$ (c) BD = CD. ☐

The properties established in Question T2 are quite general for isosceles triangles and may be summarised as follows.

In any isosceles triangle:
(a) The angles opposite the equal sides are themselves equal.
(b) The line bisecting the angle between the equal sides also bisects the third side, and is perpendicular to it.

Question T3
Show that the angles in an equilateral triangle are of equal size, and find their magnitude. ☐

A line which bisects a line segment and intersects it at right angles is called a **perpendicular bisector** of the line segment. Thus the line AD in Figure 8.9a is a perpendicular bisector of the side BC.

A **median** is a line drawn from one vertex of a triangle to the mid-point of the opposite side. Although it is not the case for all triangles, for an isosceles triangle the median, perpendicular bisector and the angle bisector are identical.

Question T4
In Figure 8.9b the line CM = AM and CM is a median so that AM = MB. Show that the angle $A\hat{C}B$ is 90°. ☐

8.3.2 Similarity, ratios and scales

If two roads meet at an angle of 30°, we expect the same angle to occur on an Ordnance Survey map of the area, in other words we expect maps and plans to preserve the true angles. However, it would be ridiculous to construct a road map with the same dimensions as the original site, so map-makers and designers use the concept of **scaling**. Often a short distance on the map or plan represents a much longer distance in reality, but sometimes the reverse is true and a relatively large distance on the plan may represent a minute length in the real object (as, for example, in the plan of a circuit in a microchip). On one popular series of Ordnance Survey maps, 1 cm on the map represents 501 000 cm or 0.5 km on the ground so that 2 cm represents 1 km.

The essence of scaling is that it preserves shape. You have already come across this idea in Section 8.3.1. When two (or more) triangles are the same shape they are said to be **similar triangles**. The triangles shown in Figures 8.7b and c are actually similar (although it requires a rotation and a reflection of one of them to put them into the same orientation). If you look at Figure 8.3 you should be able to see that triangles MXY and MTU are also similar. You should note the convention used when discussing similar triangles; the order of the letters indicates the correspondence of the vertices. It is also true that any two *congruent* triangles are also *similar*. (☞)

An important property of similar triangles is that corresponding sides are in the same ratio, i.e. they are *scaled* by the same factor. (☞)

Suppose that in a map the scale is 2 cm to represent 1 km on the ground. We can express this scale as the ratio of the distance on the map to the corresponding distance on the ground: this ratio is then 1 : 501 000. In other words, actual distances are reduced by a factor of 50 000 before they are drawn on the map. The essential point is that only one scale is used for *all* distances in *all* directions (even curved paths). If A, B and C in Figure 8.10 represent the positions of three towns, and P, Q and R are their respective positions on the map, then

$$PQ = \frac{AB}{50\,000} \quad QR = \frac{BC}{50\,000} \quad RP = \frac{CA}{50\,000} \qquad (☞)$$

It follows that

$$\frac{PQ}{AB} = \frac{QR}{BC} = \frac{RP}{CA}$$

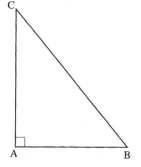

Figure 8.10 Similar triangles.

If corresponding angles in two triangles are equal, they will be similar triangles, i.e. the same shape. However, this does not apply to a figure with more than three sides; for example, a square and a rectangle have all four equal angles but they are not necessarily the same shape.

This is also true of any linear property of the triangles; for example, the perimeters are in the same ratio as the corresponding sides. However, this is not the same as the ratio of their areas.

The triangles in Figure 8.10 are right-angled; however, these results are true for any pair of similar triangles.

In any two similar triangles the ratios of corresponding sides are equal, and this result is true whatever the scale.

Question T5

Drivers are expected to be able to read car number plates at a distance of about 25 metres. The letters on the number plates are 8 cm high. You wish to put up a road sign which drivers travelling at 30 m s^{-1} will be able to read with ease. If it takes 3 s to read what is on the sign, and the driver must have completed reading it 100 m before the car reaches the sign (so that the driver can respond to what she/he reads), how large should you make the lettering on the road sign? This is illustrated in Figure 8.11. ☐

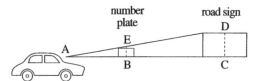

Figure 8.11 See Question T5.

Similar triangles and shadows

Now suppose that in Figure 8.10, AC represents a tree while AB represents the shadow cast by the tree; RP is a measuring pole of length 1 m and PQ represents its shadow. Suppose that we measure the length of the shadows and find them to be AB = 3.0 m and PQ = 0.75 m, we can then use the fact that ABC and PQR are similar triangles to calculate the height of the tree. Since the triangles are similar,

$$\frac{CA}{RP} = \frac{AB}{PQ}$$

Hence $CA = \dfrac{AB}{PQ} \times RP = \dfrac{(3\ m)}{(0.75\ m)} \times (1\ m) = 4\ m$

Question T6

Suppose that the pole RP (in Figure 8.10) is 2 m long, its shadow is 1 m long and the shadow of the tree is 31m long. How tall is the tree? ☐

Question T7

A vertical tree is growing on a hill that slopes upwards away from the sun at an inclination of 10° to the horizontal. The shadow of a pole 2 m long (held vertically on the hill) at the time of measurement is 1.5 m long and the shadow of the tree is 4.5 m long. Draw a diagram representing the situation and show that the properties of similar triangles still apply. What is the height of the tree?

(Assume that the tops of the shadows from the pole and the tree fall in the same place.) ☐

Sometimes it is not obvious when two triangles are similar. In Figure 8.12 ABC is a right-angled triangle with the right angle at B. We have drawn BD perpendicular to AC, and labelled the angles at A and C as α and β, so $\alpha + \beta = 90°$.

How many similar triangles are there in Figure 8.12?

In fact there are three. Look first at triangle ABD. One angle (AD̂B) is a right angle; one (DÂB) is α. So the third angle, AB̂D, must be $(180° - 90° - \alpha)$, and hence must be equal to β. Triangle ABD is therefore similar to triangle ABC, since it has the same three interior angles. To emphasize which sides are in the same ratio we can be more precise in the order in which we write the letters representing corresponding vertices, and say that the triangles ADB and ABC of Figure 8.12 are similar. This means that

$$\frac{AD}{AB} = \frac{DB}{BC} = \frac{AB}{AC}.$$

Now look at triangle BDC. Here, two of the angles are known: BD̂C = 90° and BĈD = β ; so the third angle (DB̂C) will be equal to α, and so this triangle will be similar to both ABC and ADB. Hence triangles ADB, ABC and BDC are similar. ☐

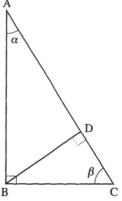

Figure 8.12 Similar triangles within a triangle.

8.4 Circles

8.4.1 Parts of a circle

Various terms are used to describe the geometry of circles, and it is necessary to commit them to memory since they occur frequently.

A **circle** is the set of points in a plane which are the same distance (i.e. **equidistant**) from a given point (in the same plane) called the **centre** of the circle and the distance is called the **radius** of the circle. The set of points with a given property is sometimes called the **locus**, so that in this case we might say that a circle is the locus of all points equidistant from a given point (i.e. the centre). The perimeter of the circle is known as the **circumference** of the circle. A straight line segment which passes through the centre of the circle and divides the interior of the circle into two equal parts is called a **diameter** (which is equal in length to twice the radius).

In Figure 8.13a C is the centre of the circle, ACB is a diameter and CP is a radius. The circumference of a circle plus its interior is sometimes known as a **disc**. The two halves of the circumference of the circle on either side of a diameter are called **semicircles**; two diameters which intersect at right angles

Many terms in geometry such as *hypotenuse, radius* and *diameter* are habitually used both to describe the geometric properties of a line segment and to mean the length of the line segment. For example, we often write 'the diameter of a circle' when strictly we should write 'the length of the diameter of a circle'.

(at the centre of the circle) divide the disc into four **quadrants** and when the two diameters are vertical and horizontal, as in Figure 8.13b, the quadrants are labelled as shown.

A portion of the circumference of the circle between two points, such as D and E in Figure 8.13a, is called an **arc** of the circle. When DE is not a diameter, the longer route along the circumference from D to E is called a **major arc** and the shorter route is called a **minor arc**. The straight line segment DE which connects two points on the circumference is called a **chord**. When the chord is not a diameter, it divides the disc into two unequal regions, called **segments**. The smaller region (the shaded portion from D to E) is called a **minor segment**, the larger region a **major segment**.

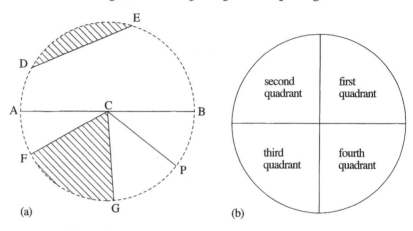

Figure 8.13 Features of a circle.

The region inside a disc cut off by two radii, such as the shaded region FCG in Figure 8.13a, is called a **sector**, as is the remainder of the disc when this shaded region is removed. The minor arc FG of the circle is said to **subtend** an acute or obtuse angle FĈG at the centre of the circle, whereas the major arc FG of the circle *subtends* a reflex angle FĈG at the centre.

Two or more circles with the same centre but different radii are said to be **concentric** (see Figure 8.14b) and the region bounded by two concentric circles is called an **annulus**.

What term is used for a *segment* formed by an *arc* of a *circle* that *subtends* a right angle at the *centre* of a circle?

——————————————————————

This *minor segment* of the circle is known as a *quadrant*. □

8.4.2 Properties of circles

The ratio of the circumference of a circle to its diameter is the well-known constant π (π is the Greek letter pi), which has an approximate value of 3.141 159 (though the less precise fractional value of 22/7 is often used).

So we can write $c = \pi d$

where c and d are the circumference and diameter of the circle, respectively.

The circumference of a circle of radius r is $2\pi r$ (2)

---

since the diameter is twice the radius.

So far in this chapter we have measured angles in degrees, a method which goes back to ancient times. In spite of its age, and its common use, this is not in fact a 'natural' system of measurement since there is nothing natural about choosing 360 rather than some other number.

An alternative scale, based on the lengths of arcs formed on a circle of unit radius (i.e. a unit circle), turns out to be far more appropriate in many circumstances (in particular in the context of *calculus*). Figure 8.14a represents a circle of radius 1 m, and the arc AB has been chosen so that it is exactly 1 m in length. The angle subtended by the arc at C is defined to be one **radian**. Since angles are preserved by scaling, there is nothing special about the choice of metres as our units of length, and in fact it is more usual to define a *radian* in terms of a circle of unit radius, without specifying what that unit of length might be.

In Figure 8.14b we have not specified the units of length, we have simply referred to them as 'units'. Thus in Figure 8.14b the circle is of radius 1 unit. An arc, PQ say in Figure 8.14b, of length α units subtends an angle α radians at the centre O of the circle, as does an arc RS of length αr units on a circle of radius r. Thus we have RS $= \alpha r$ so that $\alpha = $ RS/r, and since both RS and r have dimensions of length, it follows that angles measured in radians are dimensionless quantities. You should remember that:

An arc of a circle of radius r, subtending an angle α (in radians) at the centre of the circle, has length αr so that, in Figure 8.14b,

$$\text{RS} = \alpha r \qquad\qquad (3)$$

The number 360 is of course a good choice since it has so many factors.

Radians were introduced in a similar way in Section 5.2.1. You may find it instructive to compare the two treatments if you have not alreay studied the earlier one.

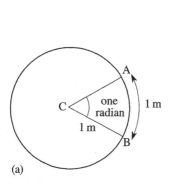

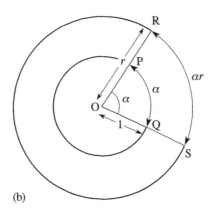

(a)   (b)

Figure 8.14   The radian measure.

If we extend the arc in Figure 8.14b so that RS is actually the circumference of the circle (and R coincides with S) we saw, in Equation 2, that its length will be $2\pi r$. Therefore $RS = \alpha r = 2\pi r$ and α for this 'arc' is a full turn, i.e. 360°. Therefore a complete turn of 360° corresponds to 2π radians, which we write as

---

360 degrees = 2π radians

From this we can deduce other relationships, for example,

$180° = \pi$ radians, $90° = (\pi/2)$ radians,

$60° = (\pi/3)$ radians, $45° = (\pi/4)$ radians, $30° = (\pi/6)$ radians.

1 radian is equivalent to $\left(\dfrac{180}{\pi}\right)°$ which is approximately 57.3°.

Express $3\pi/2$ radians in degrees and 135° in radians.

2π radians = 360°, therefore 1 radian $= \dfrac{360°}{2\pi}$,

thus $\dfrac{3\pi}{2}$ radians $= \dfrac{360°}{2\pi} \times \dfrac{3\pi}{2} = 270°$

$360° = 2\pi$ radians, therefore $1° = \dfrac{2\pi \text{ radians}}{360°}$,

thus $135° = \dfrac{2\pi \text{ radians}}{360°} \times \dfrac{135°}{1} = \dfrac{3}{4}\pi$ radians   □

Figure 8.15 See Question T8.

Question T8

In Figure 8.15 a belt ABCDA passes round two pulleys as shown. Find the length of the belt.   □

8.4.3   Tangents

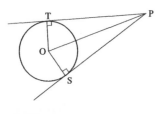

Figure 8.16 Tangents to a circle.

In Figure 16 the line TP intersects the circle at just one point. Such a line is called a **tangent** to the circle and T is the point of contact where the tangent touches the circle. The tangent is perpendicular to the radius at T. The radius OT joins the point of contact to the centre of the circle. In Figure 8.15 the line AB is a tangent to both circles, in which case we say that it is a **common tangent**.

From any point outside the circle, say P in Figure 8.16, it is possible to draw two tangents to the circle PT and PS, and the lengths of PT and PS are equal.

Show that the lengths of PT and PS are equal.

OT = OS (since both are the radii of the circle)

OT̂P = OŜP = 90° (the angle between radius and tangent) and OP (the hypotenuse) is common to both triangles. The triangles are therefore *congruent* by case (d) in our test for congruence in Section 8.3.1. It follows, therefore, that the lengths PT and PS are equal. ☐

8.5 Areas and volumes

8.5.1 Areas of standard shapes

The areas of quadrilaterals and triangles

The problem of finding areas of plane shapes is a common one. The area of the rectangle shown in Figure 8.17a is the product ab of the lengths of two adjacent (i.e. neighbouring) sides. Provided that we are given the sizes, it is easy to calculate the area of a shape such as that shown in Figure 8.17b by dividing it into suitable rectangles.

Figure 8.17c shows a rectangle ABCD with area ah. The triangles AED and BFC are congruent and so of equal area, and thus the area of the parallelogram EFCD is also ah. (☞) With DC as the base of the parallelogram, the length h is known as the **perpendicular height**, so that:

Imagine the triangle AED being 'cut' from one side of the rectangle and stuck on the other side.

The area of a parallelogram = base × perpendicular height

These statements can be proved using an argument based on congruent triangles.

Figure 8.17d shows the same parallelogram as in Figure 8.17c, and the diagonal of the parallelogram EC divides the parallelogram into two parts of equal area (as does the diagonal DF). It follows that

area of triangle EDC = area of triangle DFC = $ah/2$

and in general (☞):

Be careful when calculating the area of a triangle such as DFC in Figure 8.17d. It is important to distinguish between FC and the perpendicular height of the triangle.

The area of a triangle = $\frac{1}{2}$ base × perpendicular height

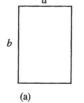 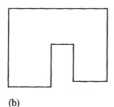 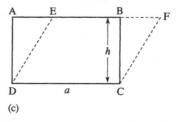 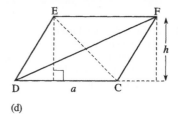

(a)　　　　(b)　　　　(c)　　　　(d)

Figure 8.17　Rectangles, parallelograms and triangles.

Another result, that is sometimes of use, relates the area of a triangle to the lengths of its sides. We quote this result without proof

The area of a triangle $= \sqrt{s(s-a)(s-b)(s-c)}$

where $s = \frac{1}{2}(a+b+c)$

A trapezium may be regarded as the sum of a parallelogram and a triangle. It follows that its area is given by

The area of a trapezium $= \frac{1}{2}$ sum of parallel side lengths $\times$ perpendicular height

Area enclosed by a circle

We state without proof that the area enclosed by a circle of radius r is πr^2.

This area is commonly (loosely) referred to as 'the area of the circle', though strictly we should say 'the area enclosed by the circle' or 'the area of the disc'.

The area of a circle of (radius r) $= \pi r^2$ (4)

Area of a sector

Consider the shaded sector FCG shown in Figure 8.13a and suppose that the radius of the circle is r and that $\hat{FCG} = \theta$ (radians). The ratio of the area of the sector to the area of the circle is the same as the ratio of the angle θ to the angle of a complete turn, i.e. 2π. In other words,

$$\frac{\text{area of sector}}{\pi r^2} = \frac{\theta}{2\pi}$$ (Remember that θ *must* be measured in radians.)

Hence:

The area of sector $= \frac{1}{2}\theta\pi r^2$ (5)

Area of a segment

We can find the area of a segment by finding the area of a sector and then subtracting the area of a triangle. (☞)

Be careful not to confuse a segment (which is bounded by an arc and a chord) with a sector (which is bounded by an arc and two radii).

Question T9

In Figure 8.13a suppose that a minor arc DE is of length 2π metres and that the radius of the circle is 4 metres.

(a) Find the angle θ subtended by the arc DE at C in both degrees and radians.

(b) Find the area of the triangle DEC.
(c) Find the area of the sector DEC.
(d) Find the area of the (larger) segment DAFGPBE.
(e) Find the area of the segment from D to E (the hatched area in Figure 8.13a).

Leave your answers in terms of π where appropriate. □

Area of the surface of a cylinder

The ends of a cylinder are circles. The curved surface of a cylinder can be opened out to form a rectangle as shown in Figure 8.18 so its area is the height of the cylinder times its circumference: that is, $2\pi R \times L$.

8.5.2 Volumes of standard solids

Prisms

Just as it is useful to be able to find the area of geometrical shapes, so it is useful to know the volume of various solid shapes. One of the simplest solid shapes is a box, known more formally as a **cuboid** or a **rectangular block**.

If the sides of the cuboid are of length a, b and c, its volume is $a \times b \times c$. In fact, a cuboid is a special case of a class of solids called *prisms*. A triangular **prism** is a solid for which the cross sections cut parallel to a certain direction are congruent triangles (oriented the same way), as shown by the hatched area in Figure 8.19. Every slice through the prism parallel to the base will produce a triangle identical to the shaded triangle. A *cuboid* is a prism for which the cross sections parallel to the base are identical quadrilaterals, and a cylinder is a prism for which the cross sections perpendicular to its axis are identical circles (see Figure 8.18). The *perpendicular height* of the prism is the length labelled b in Figure 8.19, and the hatched triangle is known as the **cross-sectional area** of the prism. These terms can be easily generalised to prisms of any cross-sectional shape, and in general:

The volume of a prism = cross-sectional area × perpendicular height

---

Find the volume of a cylinder with perpendicular height 1.5 m and whose cross section is a circle of radius 10 cm.

The cross-sectional area is $\pi r^2 = \pi(0.1 \text{ m})^2 = 0.01\pi \text{ m}^2$, and the volume is therefore $(0.01\pi \text{ m}^2) \times (1.5 \text{ m}) = 0.015\pi \text{ m}^3$. □

Question T10

A prism has perpendicular height 5 m and its cross section is an annulus formed by circles of radii 1 m and 2 m. What is the volume of the prism? □

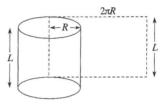

Figure 8.18 Surface area of a cylinder.

Note that Figure 8.18 is not drawn to scale.

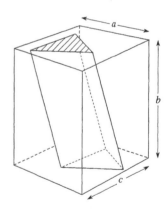

Figure 8.19 A cuboid and a prism.

The prism in Question T10 would more often be described as a hollow cylinder or pipe.

Spheres

A **sphere** is a three-dimensional surface whose points are at an equal distance (the radius) from a fixed point (the centre of the sphere). *Calculus* provides the easiest way of finding the area and volume of a sphere, and we will simply quote the results:

Calculus is discussed in Part 4 of this book.

The surface area of a sphere of radius r is $4\pi r^2$ and its volume is $\frac{4}{3}\pi r^3$ (6)

The surface area of a sphere is 10 m². What is its volume?

We have $4\pi r^2 = 101$ m² and therefore

$$r = \sqrt{\frac{10}{4\pi}}\ \text{m} \approx 0.892\ \text{m}$$

Hence the volume is $\frac{4}{3}\pi r^3 \approx 2.974$ m³. □

Question T11

A given sphere has twice the volume of another. What is the ratio of their surface areas? □

8.6   Closing items

8.6.1   Chapter summary

1. Much of this chapter is devoted to the various terms that are used to describe *geometric figures*.

2. When a line *intersects* a pair of *parallel lines* various angles between the lines are equal, in particular *corresponding* and *alternate* angles.

3. The sum of the *interior angles* of a *polygon* with n sides is $(2n - 4)$ *right angles*.

4. *Similar* figures (in particular triangles) are the same shape, and corresponding lengths in two similar figures are in the same ratio. *Congruent* figures are the same shape and the same size. The four minimal specifications of a unique triangle are as follows:

    (a)   State the length of all three sides. (SSS)
    (b)   State the lengths of two sides and the angle between them. (SAS)
    (c)   State the length of one side and the angles at each end of it. (ASA)
    (d)   Specify a right angle, the length of the hypotenuse and the length of one other side. (RHS)

If any of these specifications is common to two triangles, then those triangles must be congruent.

5. The sides a, b and c of a right-angled triangle are related by *Pythagoras's theorem*

$$c^2 = a^2 + b^2 \qquad (1)$$

where c is the hypotenuse of the triangle.

6. The *circumference* of a circle of radius r is $2\pi r$.

7. One *radian* is the angle subtended at the centre of a unit circle by an *arc* of unit length. The length of an arc of a circle of *radius r* which *subtends* an angle θ (in radians) at the centre is $r\theta$.

$$360 \text{ degrees} = 2\pi \text{ radians}$$

8. A *tangent* is a line that meets a circle at one point. A right angle is formed by the tangent and a radius drawn to the point of contact.

9. The area of a *parallelogram* is the base times the *perpendicular height*, and the area of a triangle $= \frac{1}{2}$ base $\times$ perpendicular height.

10. The area of a *circle* of radius r is πr^2 and the area of a sector *subtending* an angle θ at the centre is $\frac{1}{2}\theta r^2$.

11. The volume of a *prism* is the *cross-sectional area* times the *perpendicular height*.

12. The surface area of a *sphere* of radius r is $4\pi r^2$ and its volume is $\frac{4}{3}\pi r^3$.

8.6.2 Achievements

Having completed this chapter, you should be able to:

A1. Define the terms that are emboldened in the text of the chapter.

A2. Identify the different types of angles formed when lines intersect and calculate their values.

A3. Identify which angles formed by a line intersecting two parallel lines are equal.

A4. Classify the different types of quadrilateral.

A5. Understand and apply the simple properties of polygons.

A6. Classify the different types of triangle and apply the test for the congruence of two triangles using the four minimal specifications of a triangle.

A7. Recognise when two triangles are similar and compare the lengths of corresponding sides.

A8. Recognise the main features of a circle, including tangents, arcs, segments and sectors.

A9. Evaluate areas and volumes of standard shapes.

A10. Apply Pythagoras's theorem and use it to identify right-angled triangles.

8.6.3   End of chapter questions

Question E1   Find the angle $\hat{BCD}$ in Figures 8.20a and b.

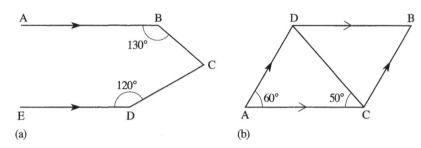

Figure 8.20   See Question E1.

Question E2   Suppose that you are told that in Figure 8.9b $\hat{ACB} = 90°$ and that AM = MC. Prove that M is the mid-point of AB.

Question E3   In Figure 8.21 ABCD is a parallelogram. A line through A meets DC at E and meets BC produced at F. Show that triangles ADE and FCE are similar. If DE = 15 mm, EC = 10 mm and FC = 5 mm find AD.

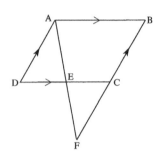

Figure 8.21   See Question E3.

Question E4   In Figure 8.22 AB is a tangent to the circle of radius a and centre C. DB represents the perpendicular height of the triangle ABC with AC as its base. The lengths CD and DA are x and y, respectively. Use an argument based on similar triangles to show that:

$$\frac{x}{a} = \frac{a}{x + y} \quad \text{and} \quad \frac{y}{c} = \frac{c}{x + y}$$

Show that $(AC)^2 = a^2 + c^2$.

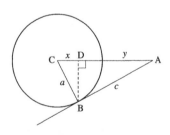

Figure 8.22   See Question E4.

Question E5   A washer is in the form of a cuboid with sides of length 1.5 cm and thickness 4 mm with a circular hole of radius 0.6 cm drilled through its centre. Calculate the volume of the washer.

Coordinate geometry

Chapter 9

9.1 Opening items

9.1.1 Chapter introduction

You probably think of geometry as a subject that involves pictures of points, lines, curves, surfaces and so on; and indeed the origins of the topic are intimately related to the figures that one might draw on a piece of paper. However, if you stick to this classical approach to the subject, you very soon find that the pictures become exceedingly complicated and difficult to visualise; and particularly so in three dimensions. A breakthrough came in the 17th century when René Descartes (1596–1650) showed that algebraic methods could be applied to geometry, and formulated the subject that we know today as coordinate (or Cartesian) geometry. Of course we continue to use diagrams, but we are no longer bound by them, and most of our arguments are entirely algebraic.

You have almost certainly met the idea of a *Cartesian coordinate system* in school (although you may not have called it that), and so our first task in this chapter will be to review some of the fundamental ideas. In Sections 9.2 and 9.3 we are concerned solely with geometry in two dimensions, but in Section 9.4 we discuss briefly how some of the ideas can be extended to three dimensions.

The essence of the Cartesian approach to geometry is to specify the position of a point by distances, in mutually perpendicular directions from a given fixed point. In a Cartesian coordinate system two distances are needed to specify the location of a point on a flat surface (known as a *plane*) and these distances are called *coordinates*. The given fixed point is called the *origin* (of coordinates) and the two directions are called *axes*: one of these is known as the x-axis and the other as the y-axis and the coordinates are written as an *ordered pair* of numbers. Thus the point represented by the pair (2, 3) has x-coordinate equal to 2 and y-coordinate equal to 3.

Having specified the points in a plane by a Cartesian coordinate system, it is natural to ask how we can determine the distance between two points in terms of their coordinates. We answer this question in Subsection 9.2.6.

Apart from a point, a straight line is the simplest geometric entity, and we show, in Subsection 9.2.2 that it can be represented by a simple equation relating values of y for points on the line to corresponding values of x. There is just one equation for a given line, but in Subsections 9.2.3 and 9.2.4 we will see that this equation can be manipulated into various equivalent forms in order to provide the information that we require.

In Section 9.3 we take an introductory look at the equation of a circle. In Subsection 9.3.1 we first derive the equation for a circle with its centre at the origin; then we generalise this equation to cover the cases where the centre is at any point in the plane. In Subsection 9.3.2 we derive equations for the tangents to a circle, in cases where the tangent passes through a given point outside the circle and in cases where the point of contact on the circle is specified. Subsection 9.3.3 deals briefly with an important alternative coordinate system – the *polar coordinate* system.

Section 9.4 extends some of the earlier results to three dimensions. Subsection 9.4.1 deals with points in three dimensions and the calculation of the distance between any two of them. Subsection 9.4.2 presents the equation of a plane surface which has similarities with the general form of the equation of a line in two dimensions. Finally, Subsection 9.4.3 considers the specification of a line in three dimensions.

9.1.2   Fast track questions

Question F1

Find the equations of the following straight lines:

(a)   the line with gradient -2 and y-intercept 3;

(b)   the line with gradient 3 passing through the point $(-1, 5)$;

(c)   the line through the points $(-2, -3)$ and $(-3, 1)$;

(d)   the line with intercepts -1 and 2 on the x- and y-axes, respectively.

Question F2

Find the equations of the circles:

(a)   centred at the origin and of radius 5;

(b)   centred at $(-1, -1)$ and of radius 2.

(c)   Find the equations of the tangents to the first circle which pass through the point $(4, 4)$.

(d)   Find the equation of the tangent to the first circle at the point $(-3, 4)$ on the circle.

Question F3

(a)   What is the distance between the points $(1, 2, 3)$ and $(-1, 3, -2)$?

(b)   Write down the equations which determine the line joining these points.

(c)   Does this line lie in the plane $x - 3y - z = -8$?

9.1.3   Ready to study?

Study comment

In order to study this chapter you will need to be familiar with the following terms: *circle, dimensions, distance, equation, function, intersection, inverse trigonometric functions (arcsin* (x), *arccos* (x) *and arctan* (x)), *line, modulus* | x |, *plane, point, simultaneous equations, tangent line, trigonometric ratios* and *variable.* You will need to be proficient in manipulating algebra and be able to apply *Pythagoras's theorem*; and you will also need to be able to find the area of a triangle, and to solve *quadratic equations* and state the condition that ensures that such equations have two real roots. In addition, you will need to be familiar with SI units. One question requires the use of certain *trigonometric identities*, but these are provided in the text. The following *Ready to study questions* will allow you to establish whether you need to review some of the topics before embarking on this chapter.

$\sin^{-1}(x)$, $\cos^{-1}(x)$ and $\tan^{-1}(x)$ are used by some authors in place of arcsin (x), arccos (x) and arctan (x).

Question R1

Which of the following are correct statements?

(a) $(x - 1)(y - 2) - 3(x - 5)(y - 4) - 12 = -2(x - 7)(y - 5)$
for all values of x and y.

(b) $\dfrac{3a + b}{a} = 3 + b$ for all values of a and b.

(c)   The quadratic equation $x^2 + 2x + 1 = 5$ has just one real root.

(d) $(m^2 + 3m + 2)^2 = m^4 + 9m^2 + 4$ for all values of m.

Question R2

If two sides of a right-angle triangle are 5 m and 12 m, does it follow that the third side is 13 m?

Question R3

What is the definition of a circle?

Question R4

What does it mean to say that 'a line is a tangent to a circle'?

9.2 Straight lines

9.2.1 Cartesian coordinates

The framework that we use when drawing pictures of curves is provided by a **Cartesian coordinate system**, of the sort introduced in Chapter 3. It consists of a pair of straight lines at right angles, called **coordinate axes** (Figure 9.1), which may be supposed to be infinitely long, even though we usually draw only a portion of each. The point where they cross is known as the **origin**. The horizontal line is called the x-axis and the vertical line the y-axis, and both lines are scaled to indicate distance from the origin along the line. Distances to the right of the origin along the x-axis are positive while those to the left are negative; along the y-axis, distances above the origin are positive while those below are negative. It is usual to place arrowheads on the axes to show the direction of increasing x and y (since occasionally we may wish to change the orientation of the axes).

    Any pair of values for x and y can be used to determine a point in the coordinate system. Thus, for example, the **ordered pair** (2, 1) is shown as point A in Figure 9.1; a vertical line from A to the x-axis meets it at $x = 2$, while a horizontal line from A to the y-axis meets it at $y = 1$. Similarly, the points B and C in Figure 9.1 represent the ordered pairs of numbers (1, 1) and (−1, 2), respectively.

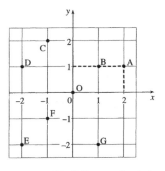

Figure 9.1 A Cartesian coordinate system and some points.

What can you say about the coordinates of a point, (a) on the x-axis, (b) on the y-axis?

(a) y is zero on the x-axis, and the point is of the form $(x, 0)$;

(b) x is zero on the y-axis, and the point is of the form $(0, y)$. ☐

Question T1
> Which values of (x, y) correspond to the points D, E, F, G, O in Figure 9.1? ☐

The points where a given line crosses the x- and y-axes are known as the x- and y-**intercepts**, respectively (although some authors refer to the y-intercept as *the* intercept).

9.2.2 Equation of a straight line – gradient and y-intercept

If you have already worked through the similar discussion of straight lines in Chapter 3, you should be able to jump to page 243.

A *linear function* is a *function* of the form $f(x) = mx + c$ where m and c are constants. For example, $f(x) = 2x + 1$ when $m = 2$ and $c = 1$, and, if we let y denote the values of $f(x)$, we may plot the graph of $y = 2x + 1$ as in Figure 9.2. It should come as no surprise that the graph is a straight line; this is after all why such functions are called *linear*.

You may have used a *table of values* (as in Table 9.1) to construct such a graph.

In fact, once you know that the graph must be a straight line, it is quite unnecessary to tabulate all these values. You just need to know two well spaced points, such as $(-3, -5)$ and $(3, 7)$, then join them with a straight line (although it is wise to calculate a third point and check that it lies on the line that you have drawn).

It is essential to realise that the equation $y = 2x + 1$ is a true statement if, and only if, we choose a point (x, y) that lies on the line. For example, the values $x = 2.5$ and $y = 6$ correspond to the point $(2.5, 6)$ which lies on the line because $6 = 2 \times 2.5 + 1$; on the other hand, the point $(3.5, 7)$ does *not* lie on the line because $7 \neq 2 \times 3.5 + 1$. (Notice that we do not need to draw the graph in order to reach these conclusions.)

Table 9.1 A table of values for $f(x) = 2x + 1$

x	y
-3	-5
-2	-3
-1	-1
0	1
1	3
2	5
3	7

Do the following points lie on the line $y = 2x + 1$:
(a) $(0.5, 2)$,   (b) $(989.13, 1979.26)$,   (c) $(0.0013, 1.0025)$?

(a) From Figure 9.2 we can see that the point $(0.5, 2)$ lies on the graph. Alternatively we can be sure that this is so because
$$\underbrace{2}_{y} = 2 \times \underbrace{0.5}_{x} + 1.$$

(b) $\underbrace{1979.26}_{y} = 2 \times \underbrace{989.13}_{x} + 1$ and therefore the point $(989.13, 1979.26)$ lies on the line.

(c) $\underbrace{1.0025}_{y} \neq 2 \times \underbrace{0.0013}_{x} + 1$ so that the point $(0.0013, 1.0025)$ does *not* lie on the line. □

The two constants, m and c, which characterise a linear function are called the **gradient** (or sometimes the **slope**) and the **y-intercept**, respectively.

Look again at Figure 9.2. Note that the straight line $y = 2x + 1$ crosses the y-axis at the point where $y = 1$, which is the value of c (i.e. the y-intercept). Also, from Table 9.1, note that a unit increase in the value of x gives rise to an increase of two units in the value of y. This factor of 2 corresponds to the value of m, the gradient of the linear function, and, in graphical terms, it is this value that determines how steep the line will be. The general result to draw from this, as described in Chapter 3, is that:

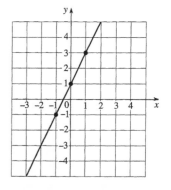

Figure 9.2 The graph of $y = 2x + 1$.

The **gradient-intercept form** of the equation of a straight line is

$$y = \underbrace{m}_{\text{gradient}} x + \underbrace{c}_{\text{y-intercept}} \tag{1}$$

where m and c are constants.

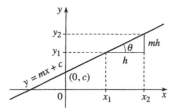

Figure 9.3 The gradient–intercept form of the straight line $y = mx + c$.

Sketch the straight lines corresponding to the following choices:
(a) $m = 1$ and $c = 0$, (b) $m = 0$ and $c = 1$.

(a)   This corresponds to the equation $y = x$, and is a straight line through the points E, F and B in Figure 9.1. (Note that the line goes through the origin since $c = 0$.)

(b)   This corresponds to the equation $y = 1$, and is a straight line through the points D, B and A in Figure 9.1. (Note that the line is horizontal since the gradient is zero.) □

Figure 9.3 represents an arbitrary line with equation $y = mx + c$. The constant c is known as the y-intercept because the point $(0, c)$ on the y-axis lies on this line.

Suppose now that (x_1, y_1) and (x_2, y_2) are any two points lying on the line, then

$$y_1 = mx_1 + c$$

and $y_2 = mx_2 + c$

Subtracting these equations we have

$$y_2 - y_1 = mx_2 - mx_1$$

---

and it follows that $$m = \frac{y_2 - y_1}{x_2 - x_1} \tag{2}$$

If we let $x_2 - x_1 = h$, as in Figure 9.3, then $y_2 - y_1 = mh$ and $\tan \theta = m$.

Figures 9.4a and b show, respectively, the effects on the line of different values of m and of c.

Figure 9.4 (a) Varying the gradient m. (b) Varying the intercept c.

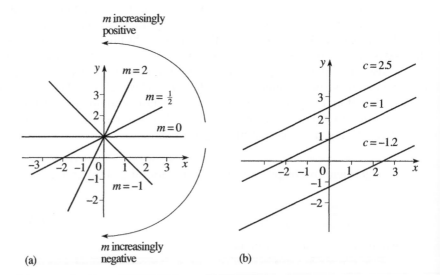

Varying the value of m, and keeping c fixed, will rotate the line about a fixed point (the intercept) on the y-axis. If m is positive the graph slopes upwards from left to right, whereas if m is negative the graph slopes downwards from left to right. Note that the graph is horizontal when $m = 0$.

Varying the value of c, and keeping m fixed, produces lines with the same gradient (they are parallel to one another). If $c = 0$ the line passes through the origin; increasing c raises the line, while decreasing c lowers it.

What is the gradient of the line that passes through the points $(-1.3, 2.8)$ and $(-3.4, 1.2)$?

Let $x_1 = -1.3$, $y_1 = 2.8$ and $x_2 = -3.4$, $y_2 = 1.2$ then (using Equation 2) we have

$$m = \frac{y_2 - y_1}{x_2 - x_1} = \frac{1.2 - 2.8}{-3.4 - (-1.3)} = \frac{-1.6}{-2.1} \approx 0.762 \quad \square \qquad (\text{☞})$$

Here we have chosen $(-1.3, 2.8)$ to be the first point and $(-3.4, 1.2)$ to be the second. We would have obtained the same answer had we chosen the points the other way round.

What is the gradient-intercept equation of the line that passes through the points $(-1.3, 2.8)$ and $(-3.4, 1.2)$?

The equation of the line is $y = mx + c$, and from the previous question we know that $m \approx 0.762$, but we also know that the point $(-1.3, 2.8)$ lies on the line, so that $2.8 = m(-1.3) + c$ and substituting the known value for m we have $2.8 \approx 0.762\ 3 \times (-1.3) + c$ which gives $c \approx 3.791$; thus the equation of the line is (approximately) $y = 0.762x + 3.791$. $\square$

Equations of lines need not be in gradient-intercept form; for example the equation $x = 5y + 3$ represents a line, but its gradient and y-intercept are not immediately obvious. However, if we rearrange this equation into the form

$$y = \frac{1}{5}x - \frac{3}{5}$$

it becomes clear that the gradient is 1/5 while the y-intercept is $-3/5$.

Question T2

What are the gradients and y-intercepts of the following lines?

(a) $y = 2 - x/3$ (b) $2y = 4x - 5$ (c) $2x + 3y = 5$

(d) $y = 3(x - 1)$ (e) $\dfrac{y - 3}{x + 2} = 1$ (f) $\dfrac{2}{x - 1} = \dfrac{1}{y + 3}$ $\square$

9.2.3 Straight line with a given gradient through a given point

Suppose that we wish to find the equation of the line through the point $(1, -2)$ with gradient 0.5. We know the gradient, so the equation must be of the form

$$y = 0.5x + c$$

and it remains to find the value of c. Using the fact that the point $(1, -2)$ lies on the line we have $\qquad -2 = 0.5 + c$

from which it follows that $\qquad c = -2.5$

and so the required equation is $\qquad y = 0.5x - 2.5.$

Alternatively, we could use Equation 2, letting the known point $(1, -2)$ be the first point on the line, and an arbitrary point (x, y) on the line be the second, then

$$0.5 = \frac{y - (-2)}{x - 1}$$

which can be rearranged to give $y = 0.5x - 2.5$ as before.

To obtain a general formula we can let the given point have coordinates (x_1, y_1) and let the gradient of the line be m, then for an arbitrary point (x, y) on the line, using Equation 2, we have

$$m = \frac{y - y_1}{x - x_1}$$

which can be rearranged into

The **point-gradient form** of the equation of a straight line

$$y - y_1 = m(x - x_1) \tag{3}$$

---

i.e. $\quad y = mx + y_1 - mx_1$

from which it is easy to see that the y-intercept is $y_1 - mx_1$. (☞)

You should try to remember Equation 2, but it is hardly worth remembering Equation 3 since it is easy to derive. A useful way of recalling the definition of the gradient is to remember that it is the *rise* divided by the *run*, i.e. $y_2 - y_1$ divided by $x_2 - x_1$.

Given that a line has gradient -5 and passes through the point $(1, 3)$ use the following methods to find its equation:

(a) by finding the value of c in the equation $y = mx + c$,

(b) and then by using Equation 2.

(a) We know that $y = -5x + c$ and, since $(1, 3)$ lies on the line, $3 = (-5) \times 1 + c$ so that $c = 8$, and the equation is $y = -5x + 8$.

(b) From Equation 2 we have $-5 = \dfrac{y - 3}{x - 1}$

which can be rearranged to give $y = -5x + 8$. ☐

While it is quite acceptable to find the equation of the line by calculating the value of c, it is perhaps slightly easier to use Equation 2.

What is the equation of the straight line through the point $(-1, -2)$ with gradient 3?

From Equation 2 we have

$$3 = \frac{y - (-2)}{x - (-1)}$$

which can be rearranged to give $y = 3x + 1$.   □

Question T3

The equation $5 = \dfrac{y + 2}{x - 1}$ is the equation of a line.

(a)   What is the gradient of the line?

(b)   Does the line pass through the point $(-1, 2)$?

(c)   What is the equation of this line in gradient-intercept form?

(d)   What is the y-intercept of this line?   □

9.2.4   Other forms of the equation for a straight line

Given two points on the line
Suppose that we are given two points, say $(2, -5)$ and $(7, 3)$, and that we are asked to construct the equation of the line joining them. A straightforward method would be to first use Equation 2 to find the gradient, and then to find c, the y-intercept. From Equation 2 we have

$$m = \frac{y_2 - y_1}{x_2 - x_1} = \frac{3 - (-5)}{7 - 2} = \frac{8}{5}$$

If the equation of the line is $y = mx + c$, and using either one of the points on the line, $(7, 3)$ say, we have

$$3 = \frac{8}{5} \times 7 + c$$

so that $c = -41/5$. Finally we can write the equation of the line is

$$y = \frac{8}{5}x - \frac{41}{5}.$$

The above method is quite acceptable, but it is slightly easier to use Equation 2 twice. First we find the gradient of the line to be 8/5 as before, but then we choose an arbitrary point (x, y) on the line, and use Equation 2 again with say $(7, 3)$ as the first point and (x, y) as the second, which gives

$$m = \frac{y - 3}{x - 7}$$

Since we know the value of m to be 8/5, this gives

$$\frac{8}{5} = \frac{y-3}{x-7} \quad \text{which can be rearranged to give} \quad y = \frac{8}{5}x - \frac{41}{5}.$$

We will now generalise this result, using the second of these methods.

Let the coordinates of two given points A and B be (x_1, y_1) and (x_2, y_2) respectively, and let P be (x, y), a general point on the line, as in Figure 9.5.

Using Equation 2 and the coordinates of points A and B we have

$$m = \frac{y_2 - y_1}{x_2 - x_1}$$

but using Equation 2 with the coordinates of points A and P gives us

$$m = \frac{y - y_1}{x - x_1}$$

Therefore equating these two expressions for m gives us

The **two-point form** of the equation of a straight line

$$\frac{y - y_1}{x - x_1} = \frac{y_2 - y_1}{x_2 - x_1} \tag{4}$$

===

Equation 4 merely says that the gradient of the line from A to P is the same as the gradient of the line from A to B.

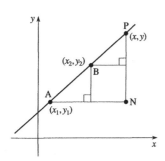

Figure 9.5 Graph of a straight line through two given points.

Find the equation of the straight line passing through the points A (2, 5), and B (−1, 3).

$$\frac{y - 5}{x - 2} = \frac{3 - 5}{-1 - 2} = \frac{-2}{-3} = \frac{2}{3} \quad \text{so that} \quad y - 5 = \frac{2}{3}(x - 2)$$

i.e. $$y = \frac{2}{3}x - \frac{4}{3} + 5 \quad \text{so} \quad y = \frac{2}{3}x + \frac{11}{3} \qquad \square$$

In physical science we are often confronted with relationships which are certainly not linear, for example the periodic time T for small vibrations of a simple pendulum is given by the formula $T = 2\pi\sqrt{l/g}$ where l is the length of the pendulum and g is the magnitude of the acceleration due to gravity.

Such equations can sometimes be simplified by a change of variable, and in this case if we let $y = T$ and $x = \sqrt{l}$ then the equation becomes $y = (2\pi/\sqrt{g})x$ which is the equation of a line through the origin with gradient $2\pi/\sqrt{g}$.

Such an expression could form the basis for an experiment to determine the value of g. We could perhaps vary the length l of the pendulum and measure the corresponding times T, then plot y against x and measure the gradient. Once we have estimated this value it would be then a simple matter to estimate the value of g, as the following exercise shows.

In the experiment described above only two measurements were made (allowing 10 cycles of the pendulum in each case) giving the following values.

$T = 1.4$ s when $l = 0.5$ m, and $T = 1.7$ s when $l = 0.75$ m.

Fit a straight line to the data in the manner described above, and hence estimate the value of g.

If we let $\quad x_1 = \sqrt{0.5}\ \text{m}^{1/2} \qquad y_1 = 1.4$ s

and $\qquad\quad x_2 = \sqrt{0.75}\ \text{m}^{1/2} \qquad y_2 = 1.7$ s

then using Equation 4 we have

$$\frac{y - (1.4\ \text{s})}{x - (\sqrt{0.5}\ \text{m}^{1/2})} = \frac{(1.7 - 1.4)\ \text{s}}{(\sqrt{0.75} - \sqrt{0.5})\ \text{m}^{1/2}}$$

which can be rearranged to give

$$y \approx (1.89\ \text{s m}^{-1/2})x + 0.065\ \text{s} \quad \Box$$

The gradient of this line is $1.89\ \text{s m}^{-1/2}$ and therefore $2\pi/\sqrt{g} \approx (1.89\ \text{s m}^{-1/2})$ so that the estimate for the magnitude of the acceleration due to gravity is

$$g \approx \left(\frac{2\pi}{1.89\ \text{s m}^{-1/2}}\right)^2 = 11.05\ \text{m s}^{-2} \quad \Box$$

The correct value of g is approximately $9.81\ \text{m s}^{-2}$.

Given the intercepts on the axes

Where does the line $\dfrac{x}{5} + \dfrac{y}{3} = 1$ meet the axes?

If we put $x = 0$ in the equation we see that the y-intercept is at $(0, 3)$, and if we put $y = 0$ in the equation we see that the x-intercept is at $(5, 0)$. $\quad\Box$

The form of the above equation makes it particularly easy to find the two intercepts; but the reverse is also true. If the intercepts are given it is straightforward to write down the equation of the line.

Which of the following equations is the equation of the line with intercepts $(0, 2)$ and $(7,0)$?

(a) $\dfrac{x}{2} + \dfrac{y}{7} = 1$ $\qquad$ or $\quad$ (b) $\dfrac{x}{7} + \dfrac{y}{2} = 1$

The correct answer is (b), as we can see if we substitute first $x = 0$ then $y = 0$ into the equation. $\quad\Box$

In Figure 9.6 the intercepts of the line on the x- and y-axes are A $(a, 0)$ and B $(0, b)$, respectively, and the equation of the line may be written using

---

The **intercept form** of the equation of a straight line

$$\frac{x}{a} + \frac{y}{b} = 1$$

(5)

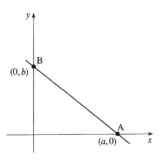

Figure 9.6 Graph of the straight line with given intercepts on the axes.

What is the gradient of the line $\dfrac{x}{a} + \dfrac{y}{b} = 1$

The equation may be rearranged to give $y = \dfrac{-bx}{a} + b$ and therefore the gradient is $-b/a$. ☐

Find the equation of the line with intercepts A $(1/2, 0)$ and B $(0, -6)$. What is its gradient?

Here $a = 1/2$, $b = -6$, so that the equation of the line is

$$\frac{x}{1/2} + \frac{y}{-6} = 1 \text{ which simplifies to } 2x - \frac{y}{6} = 1 .$$

This equation can be rearranged into the form $y = 12x - 6$ and we see that the gradient is 12. ☐

We can illustrate the use of Equation 5 with the following example.

At time t, the velocity v_x of an object moving in a straight line with constant acceleration a_x is given by $v_x = u_0 + a_x t$, where u_0 is its initial velocity in the x-direction.

As the velocity is along a straight line, say the x-direction, we need only consider the x-component of the velocity vector v_x in this example.

A particle is moving in a straight line with constant acceleration, and its velocity is found to be $v_{x1} = 10\ \text{m s}^{-1}$ at time $t_1 = 3\ \text{s}$, and $v_{x2} = 2.5\ \text{m s}^{-1}$ at time $t_2 = 5\ \text{s}$. Find the acceleration of the particle, estimate the time when its velocity will be zero, and also the initial velocity of the particle.

The equation $v_x = u_0 + a_x t$ represents the equation of a straight line (in the form of Equation 1) where the variables have been changed, so that

x has been replaced by t

y has been replaced by v_x

c has been replaced by u_0

m has been replaced by a_x

We are told that the points (t_1, v_{x1}) and (t_2, v_{x2}) lie on this line, so we may use Equation 4 (with appropriate change of variables) to give

$$\frac{v_x - v_{x1}}{t - t_1} = \frac{v_{x2} - v_{x1}}{t_2 - t_1}$$

If we rearrange this equation and substitute in the values, we obtain

$$v_x = (21.25 \text{ m s}^{-1}) - (3.75 \text{ m s}^{-2})t$$

from which we see that the acceleration is -3.75 m s^{-2} and the initial velocity is 21.25 m s^{-1}. Also the velocity will be zero when the line intercepts the t-axis,

i.e. when $t = \dfrac{21.25 \text{ m s}^{-1}}{3.75 \text{ m s}^{-2}} \approx 5.67$ s   □

General form of the equation of a straight line
The previous forms of the equation of a straight line have a deficiency which is not immediately apparent, as the following example is intended to show.

Example 1
Find the equation of the straight line passing through the points (1, 2) and (1, 4). Draw the line.

Solution   If we attempt to use Equation 4 we obtain

$$\frac{y - 2}{x - 1} = \frac{4 - 2}{1 - 1} = \frac{2}{0}$$

but 2 divided by zero is undefined!

On the other hand, we might try to find the constants m and c in the general form $y = mx + c$, in which case the point (1, 2) on the line gives

$$2 = m + c$$

while the point (1, 4) on the line gives the equation

$$4 = m + c$$

Clearly it is impossible for $m + c$ to be equal to both 2 and 4, and we conclude that the line cannot be written in the form $y = mx + c$.

The form of the equation of the line becomes clear if we list some points that lie on it. Each of the points (1, 0), (1, −2), (1, 2), (1, 2.3), (1, 99) lies on the line, and for each of them the x value is 1. This is the property which characterises such points and the equation of the line is $x = 1$, and the line is parallel to the y-axis.   □

Any line parallel to the y-axis has an equation $x = b$ for some constant b.

It is often useful to have a general form for the equation of a line that includes *all* possible lines. Thus

The general equation of a line in two dimensions is

$$ax + by + c = 0 \qquad\qquad (6)$$

Note that by putting $b = 0$ we obtain the equation $ax + c = 0$ from which we find that

$$x = -\frac{c}{a} \quad \text{which is a line parallel to the } y\text{-axis.}$$

If $b \neq 0$ find the gradient and intercept of the line $ax + by + c = 0$.

First subtract $ax + c$ from both sides of Equation 6 so that

$$by = -ax - c$$

Then divide both sides by b so that

$$y = \frac{-a}{b}x - \frac{c}{b}$$

Hence the gradient is $(-a/b)$ and the intercept is $(-c/b)$. $\square$

Question T4

Write the equation $y = -3x - 2$ in each of the three forms described in this subsection, corresponding to Equations 4, 5 and 6.

(For Equation 4 you will need to identify two points on the line; choose these to be the points with x-coordinates 1 and -1, respectively.) $\square$

9.2.5   Intersection of two lines

Parallel lines and perpendicular lines

If two lines have the *same gradient* then they are *parallel*. For example, the lines with equations $y = 2x + 3$ and $y = 2x - 1$ are parallel.

Are the lines $y = x + 1$, $y = -x + 1$ parallel?

The first line has gradient 1, whereas the second equation has gradient -1; the lines are therefore not parallel. □

Example 2
What is the relationship between the lines OA and OC in Figure 9.7?

Solution The line OA in Figure 9.7 joins the points $(0, 0)$ and $(1, m)$, and its equation is

$$y = mx$$

The line OC joins the points $(0, 0)$ and $(1, -1/m)$ and its equation is

$$y = -x/m$$

Notice that applying Pythagoras's theorem to triangle OBA gives us

$$OA^2 = OB^2 + BA^2 = 1 + m^2$$

and applying the same theorem to triangle OBC gives us

$$OC^2 = OB^2 + BC^2 = 1 + \frac{1}{m^2}$$

It follows that

$$AC^2 = (AB + BC)^2 = \left(m + \frac{1}{m}\right)^2 = (1 + m^2) + \left(1 + \frac{1}{m^2}\right)$$

$$= OA^2 + OC^2$$

and therefore (from the converse of Pythagoras's theorem) the angle AOC is a right angle. □

Figure 9.7 Two perpendicular lines through the origin.

The converse of Pythagoras's theorem is the statement that 'If the sides a, b, c of a triangle satisfy a relation of the form $a^2 + b^2 = c^2$ then the angle opposite side c is a right angle'.

The essential point to notice from Example 2 is that we have two lines through the origin, $y = mx$ and $y = -x/m$, which must be perpendicular to each other for any value of m. This means, for example, that we can see instantly that the two lines $y = 2x$ and $y = -0.5x$ are perpendicular because the product of their gradients is -1.

We quote without proof the general result that if the equations of two lines are

$$y = m_1x + c_1 \text{ and } y = m_2x + c_2 \text{ then:}$$

Two lines of gradient m_1 and m_2 are *perpendicular* if

$$m_2m_1 = -1 \tag{7}$$

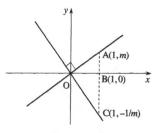

Are the lines $y = 2x + 3$ and $x + 2y = 4$ perpendicular?

The first line has gradient $m_1 = 2$. The equation of the second line can be rearranged to $2y = -x + 4$ or $y = -x/2 + 2$ from which we obtain its gradient $m_2 = -\frac{1}{2}$. Then $m_2 m_1 = -\frac{1}{2} \times 2 = -1$. The lines are therefore perpendicular. ☐

Note that the lines $x = a$, $y = b$, where a and b are constants, are perpendicular. We cannot use Equation 7 in this case (but we do not need to).

Intersection of two lines

Two lines that are not parallel intersect at a point. In order to find the coordinates of the point of intersection we could certainly draw the graphs of the lines on the same set of axes, and find where they cross. The lines $y = x - 1$ and $x + y = 3$, which appear to intersect at $(2, 1)$, are shown in Figure 9.8.

A graphical approach is sometimes useful, but it can be laborious and rather inaccurate, so we would prefer an algebraic method.

The equation $y = x - 1$ is a true statement only for points (x, y) lying on the first line, while the equation $x + y = 3$ is a true statement only for points (x, y) lying on the second line. If we are told that *both* equations are true, then we must be referring to the particular point (x, y) that lies on both lines.

A pair of equations such as

$$\left. \begin{array}{l} y = x - 1 \\ \text{and } x + y = 3 \end{array} \right\}$$

which are true for the same values of x and y are known as *simultaneous equations*, and they can be solved to find these values of x and y. In this instance we already suspect (from the graph) that the values $x = 2$ and $y = 1$ are the solution, and we have merely to verify that this is so by substituting these values into the equations. We do indeed find that both equations are satisfied, because $1 = 2 - 1$ (the first equation) and $2 + 1 = 3$ (the second equation).

Algebraic solution

There are several methods of solving a pair of simultaneous equations, but they all reduce to essentially the same idea – we use one of the equations to eliminate one of the variables, then solve the resulting equation in the remaining variable.

We will use the equations

$$y = x + 1 \tag{i}$$

$$2x + y = 3 \tag{ii}$$

to illustrate the general methods. Notice that we have numbered the equations in order to clarify the subsequent algebraic steps, and you should do likewise in your own work.

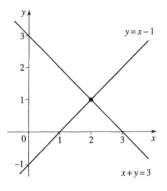

Figure 9.8 Intersection of two lines.

Method 1 – **solving by direct substitution**   We substitute for y using Equation (i) into Equation (ii) to obtain

$$2x + (x + 1) = 3$$

so that $3x + 1 = 3$

i.e. $3x = 2$

and therefore $x = \dfrac{2}{3}$

Substituting this value for x into Equation (i) we find that

$$y = \dfrac{5}{3}$$

We then check that these values also satisfy Equation (ii),

and we find $2x + y = 2 \times \dfrac{2}{3} + \dfrac{5}{3} = \dfrac{4}{3} + \dfrac{5}{3} = \dfrac{9}{3} = 3$.

Method 2 – **solving by indirect substitution**   We re-write the two equations as

$$-x + y = 1 \tag{i}$$
$$2x + y = 3 \tag{ii}$$

We eliminate y by subtracting Equation (i) from Equation (ii) to obtain

$$2x - (-x) + y - y = 3 - 1$$

i.e. $3x = 2$

as before. Then we can use either Equation (i) or Equation (ii) to calculate y and use the other equation as a check.

Alternatively, we could begin by eliminating x, and multiply Equation (i) by 2 and add the result to Equation (ii). (☞)

$$-2x + 2x + 2y + y = 2 \times 1 + 3$$

i.e. $3y = 2 + 3 = 5$

so that $y = \dfrac{5}{3}$

Either Equation (i) or (ii) yields $x = 2/3$ and the other equation can be used as a check.

A note such as $2 \times$ (i) + (ii) is sufficient in your written work.

Use the value $y = 5/3$ in Equation (ii) to obtain the value $x = 2/3$ and check the result using Equation (i).

$2x + y = 3$ becomes

$$2x + \frac{5}{3} = \frac{9}{3},$$

so that $\quad 2x = \dfrac{4}{3} \quad$ and hence $\quad x = \dfrac{2}{3}.$

Substituting these values for x and y into Equation (i) we obtain

$$-\frac{2}{3} + \frac{5}{3} = \frac{3}{3} = 1, \text{ as required.} \quad \square$$

The choice of method is often a matter of taste, although sometimes the numbers involved ensure that one method is quicker than another. It can be helpful to sketch the graphs of the two lines in order to get a rough idea of where they intersect, and perhaps to give an extra check on the results.

Strange cases

There are some particular cases that may lead you to some strange results. Suppose we were asked to find the point of intersection of the lines

$$x - 2y = 3 \text{ and } 2x - 4y = 6$$

and let us suppose that inspiration strikes and you guess the solution $x = 5$ and $y = 1$. You try these values in the equations, and you find that they work; so everything is fine and you have found the solution. Right?

However, the values $x = -1$, $y = -2$ also satisfy both equations.

How can this be? How can a pair of lines intersect at more than one point?

In fact, the two equations both represent the *same* straight line. (You can obtain the second equation by multiplying the first equation by 2.) *Any* point on the common line has values of x and y which satisfy both equations.

Question T5

Find where the following pairs of lines intersect:

(a) $y = 3x - 1$, $y = -2x + 2$

(b) $x + 2y = -1$, $y - 3x = 4$

(c) $2x + 5y = 10$, $-x + 4y = 2$

(d) $2x + 4y = 8$, $y = 2 - (1/2)x$ $\quad \square$

Ill-conditioned equations

Example 3

Solve the (rather unpleasant) pair of simultaneous equations

$$x - \pi y = 1 \tag{i}$$

and $$2000x - 6283y = 1995 \tag{ii}$$

Solution As a first step simplify Equation (i) a little by replacing π by 3.14 say, so that Equation (i) becomes

$$x - 3.14y = 1 \tag{iii}$$

now solving Equations (ii) and (iii) (and we won't bore you with the details) you would find a solution

$$x = \frac{187}{30} \approx 6.23 \quad \text{and} \quad y = \frac{5}{3} \approx 1.67$$

and everything appears to be fine. However, what if you were to choose a slightly better approximation for π, will this make much difference to the final result?

Suppose that you replace π by 3.141 59, then Equation (i) becomes

$$x - 3.141\ 59y = 1 \tag{iv}$$

and solving Equations (ii) and (iv) (and again we won't bother with the details) you would find

$$x = -\frac{30\ 559}{3600} \approx -86.27 \quad \text{and} \quad y = -\frac{250}{9} \approx -27.78.$$

This appears to be a disaster! A minor difference in the approximation for π gives startlingly different results. Why has it happened? ☐

Perhaps another example will make things clearer.

Example 4

Consider the equations

$$x + 2y = 4 \tag{i}$$
$$20x + 40.1y = 80.1 \tag{ii}$$

Solution If we subtract 20 times Equation (i) from Equation (ii) we obtain

$$0.1y = 0.1$$

so that $y = 1$ and from Equation (i) we obtain $x = 2$. This is the (exact) unique solution. The lines intersect at the point (2, 1). However, if you look at Figure 9.9 you will see that the two lines are very close together and it is hard to detect precisely where the lines meet.

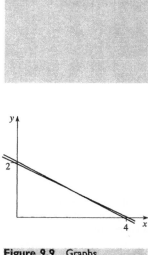

Figure 9.9 Graphs corresponding to ill-conditioned equations.

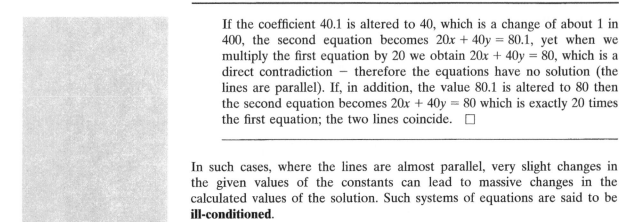

If the coefficient 40.1 is altered to 40, which is a change of about 1 in 400, the second equation becomes $20x + 40y = 80.1$, yet when we multiply the first equation by 20 we obtain $20x + 40y = 80$, which is a direct contradiction − therefore the equations have no solution (the lines are parallel). If, in addition, the value 80.1 is altered to 80 then the second equation becomes $20x + 40y = 80$ which is exactly 20 times the first equation; the two lines coincide. □

In such cases, where the lines are almost parallel, very slight changes in the given values of the constants can lead to massive changes in the calculated values of the solution. Such systems of equations are said to be **ill-conditioned**.

Question T6

Solve the following pairs of equations:

(a) $2x + y = 3$ and $x + 0.501y = 6$

(b) $2x + y = 3$ and $x + 0.499y = 6$ □

9.2.6   Distance between two points

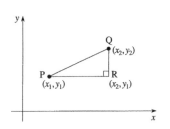

Figure 9.10 Distance between two points.

We have already seen in Example 2 how Pythagoras's theorem can be used to find distances, but here we discuss the distance between two arbitrary points P and Q, as shown in Figure 9.10. The point R has the same y-coordinate as that of P and the same x-coordinate as that of Q.

The distance PR is the difference in the x-coordinates of P and R, i.e. $(x_2 - x_1)$. The distance RQ is the difference in the y-coordinates of R and Q, i.e. $(y_2 - y_1)$. Applying Pythagoras's theorem to the triangle PQR gives the equation $(PQ)^2 = (PR)^2 + (RQ)^2$, i.e. $(PQ)^2 = (x_2 - x_1)^2 + (y_2 - y_1)^2$.

Hence the distance we require is given by

$$PQ = \sqrt{(x_2 - x_1)^2 + (y_2 - y_1)^2} \qquad (8)$$

Find the distance between the points P $(2, -1)$ and Q $(3, 4)$.

Here we can put $(x_1, y_1) = (2, -1)$ and $(x_2, y_2) = (3, 4)$. It follows that

$$PQ = \sqrt{(3 - 2)^2 + (4 - (-1))^2} = \sqrt{1^2 + 5^2} = \sqrt{26} .$$

Note that it does not matter which way we label the points. If we let $(x_1, y_1) = (3, 4)$ and $(x_2, y_2) = (2, -1)$ the distance is still $\sqrt{26}$. □

Since the square root is positive, such distances are always positive.

Question T7

Which two of the points (2, 3), (−1, 2), (3, −1) and (−2, −2) are
(a) closest to each other and (b) furthest apart? □

9.3 Circles

9.3.1 Equation of a circle, given its centre and radius

Our objective in this section is to determine the general equation of a circle,
but first we consider a simple case, when the centre of the circle is at the
origin. The point P lies on the circumference of the circle as shown in Figure
9.11a. Since OP is the radius of the circle its length will be the same whatever the choice of P (x, y) on the circumference, i.e. OP = R where R is the
radius of the circle. But, using Equation 8, $(OP)^2 = (x - 0)^2 + (y - 0)^2$
$= x^2 + y^2$ and therefore the equation of the circle is

$$x^2 + y^2 = R^2 \tag{9}$$

What is the radius of the circle $x^2 + y^2 = 4$?

From Equation 9 we see that $R^2 = 4$, so that the radius $R = 2$. □

It is a common error for
students to suppose that the
radius is 4 rather than $\sqrt{4}$.

In the general case where the centre of the circle, C, is the point (x_0, y_0) in
Figure 9.11b, it is the distance CP which is constant and equal to R. Therefore,
for an arbitrary point P(x, y) on the circle, $(CP)^2 = R^2$ and so:

The standard **equation of a circle** with centre (x_0, y_0) and radius
R is

$$(x - x_0)^2 + (y - y_0)^2 = R^2 \tag{10}$$

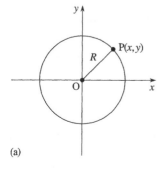

(a)

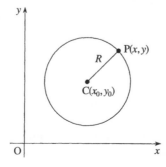

(b)

Figure 9.11 Circles.

Write down the equations of the following circles:
(a) with centre $(1, 2)$ and radius 5; (b) with centre $(-3, -2)$ and radius 3.

(a) $(x - 1)^2 + (y - 2)^2 = 5^2 = 25$

(b) $(x - (-3))^2 + (y - (-2))^2 = 3^2$, i.e. $(x + 3)^2 + (y + 2)^2 = 9$. □

Where does the line $y = x + 1$ meet the circle $(x - 1)^2 + (y - 2)^2 = 4$?

If we suppose that both equations are true, then we must be referring to the point (or points) where the line meets the circle. Substituting $y = x + 1$ into the equation of the circle we obtain

$$(x - 1)^2 + (x + 1 - 2)^2 = 4$$

which simplifies to give the quadratic equation $(x - 1)^2 = 2$, so that $x = 1 \pm \sqrt{2}$. Substituting these values into the equation of the line we obtain $y = 2 \pm \sqrt{2}$, and the points of intersection are $(1 + \sqrt{2}, 2 + \sqrt{2})$ and $(1 - \sqrt{2}, 2 - \sqrt{2})$. (☜)   □

Notice that the value $x = 1 + \sqrt{2}$ is associated with the value $y = 2 + \sqrt{2}$, while the value $x = 1 - \sqrt{2}$ is associated with the value $y = 2 - \sqrt{2}$.

Suppose that we wish to find the points of intersection of the circles

$$x^2 + y^2 = 1 \tag{i}$$

and $(x - 1)^2 + (y + 1)^2 = 4$ (ii)

As a first step we may expand Equation (ii) to give

$$x^2 - 2x + y^2 + 2y = 2 \tag{iii}$$

We can now subtract (i) from (iii) to give

$$-2x + 2y = 1 \tag{iv}$$

which is the equation of a line. *These equations are all true provided that we are referring to the points (x, y) where the circles meet.* This means that the points where the circles meet must also lie on the line (iv). Instead of completing the calculation by finding these points we now invite you to carry out some similar calculations for yourself.

Question T8
Where do the circles $(x - 1)^2 + (y - 2)^2 = 25$ and $(x + 3)^2 + (y + 2)^2 = 9$ meet?   □

9.3.2   Tangents to a circle

In this subsection we are mainly concerned with circles whose centres are at the origin.

We consider the intersections of a straight line QP with such a circle. As depicted in Figure 9.12, the line could

- cut the circle at two points A and B as in diagram (a),
- not intersect the circle as in diagram (b), or
- touch the circle at one point T as in diagram (c).

In this last case the line is a **tangent** to the circle.

Let the equation of the straight line be

$$y = mx + c \qquad\qquad (i)$$

and the equation of the circle be

$$x^2 + y^2 = R^2 \qquad\qquad (ii)$$

To find where the line meets the circle we solve Equations (i) and (ii) simultaneously.

Substituting the expression for y from Equation (i) into (ii)

$$x^2 + (mx + c)^2 = R^2$$

which can be rearranged to give a quadratic equation in x:

$$(1 + m^2)x^2 + 2mcx + (c^2 - R^2) = 0 \qquad\qquad (iii)$$

Now, we know that the general quadratic equation

$$\alpha x^2 + \beta x + \gamma = 0$$

- has two real roots if $\beta^2 > 4\alpha\gamma$
- has no real roots if $\beta < 4\alpha\gamma$
- has one repeated real root if $\beta^2 = 4\alpha\gamma$.

So, for the particular case of Equation (iii) there will be one root and therefore one point of contact if $(2\,cm)^2 = 4(1 + m^2)(c^2 - R^2)$

i.e. $4c^2m^2 = 4c^2 + 4m^2c^2 - 4R^2(1 + m^2)$

i.e. $c^2 = R^2(1 + m^2) \qquad\qquad (11)$

which can be rearranged to give

$$m^2 = \frac{c^2 - R^2}{R^2} \qquad\qquad (12)$$

From Equation 11 we see that for any given value of m there are two values of c;

$$c = R\sqrt{1 + m^2} \quad \text{and} \quad c = -R\sqrt{1 + m^2}\,.$$

These values of c together with the given value of m determine, via Equation (i), the two straight lines of gradient m that are tangential to a circle of radius R centred on the origin.

From Equation 12 we see that there are two values of m for any value of c provided that $c < -R$ or $c > R$. These values,

$$m = \sqrt{c^2 - R^2}/R \quad \text{and} \quad m = -\sqrt{c^2 - R^2}/R\,,$$

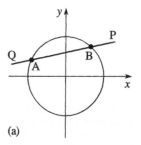

(a)

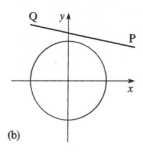

(b)

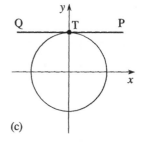

(c)

Figure 9.12 The intersection of a line and a circle.

together with the given value of c, determine, via Equation (i), the two straight lines passing through $y = (0, c)$ that are tangential to a circle of radius R centred on the origin. (Note that the condition $c < -R$ or $c > R$ ensures that the point $(0, c)$ is outside the circle and thus guarantees that there *are* two tangents passing through that point.)

You can see why there is a restriction on the values of c if you imagine a tangent line rolling around the circle of radius R. The value of c corresponds to the y-intercept on this line, and this intercept can never lie inside the circle if the circle is centred on the origin.

Given the circle $x^2 + y^2 = 1$ and the line $y = 2x + c$ find the values of c for which the line is a tangent to the circle and interpret your answer geometrically.

Figure 9.13a shows two tangents to the circle with the same slope.

With $R = 1$ and $m = 2$ the condition for tangency, Equation 11, is $1(1 + 4) = c^2$ so that $c = \pm\sqrt{5}$. The tangent lines are therefore

$$y = 2x + \sqrt{5} \quad \text{and} \quad y = 2x - \sqrt{5}$$

Figure 9.13a illustrates the general fact that in each direction there are two tangents to the circle. □

Given the circle $x^2 + y^2 = 25$ and the point P $(0, -10)$, find the equations of the lines through P that are tangents to the circle.

Refer to Figure 13b. Here $R = 5$, $c = -10$. Hence, from Equation 11, $25(1 + m^2) = 100$ so that $1 + m^2 = 4$ and $m = \pm\sqrt{3}$.

Hence the equations of the tangent lines are

$$y = \sqrt{3}\,x - 10 \quad \text{and} \quad y = -\sqrt{3}\,x - 10 \quad □$$

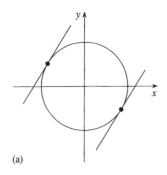

(a)

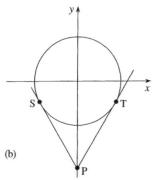

(b)

Figure 9.13 Pairs of tangents.

Note that x_1, y_1, and R are constants so Equation (i) is the equation for a straight line *not* the equation of a circle.

Equation of the tangent at a given point on a circle

We can obtain an interesting result if we consider an arbitrary point (x_1, y_1) and the points of intersection of the line

$$x_1 x + y_1 y = R^2 \qquad (☜)\ \text{(i)}$$

with the circle $\qquad x^2 + y^2 = R^2 \qquad \text{(ii)}$

The equation of the line (i) can be rewritten as

$$y = -\frac{x_1}{y_1} x + \frac{R^2}{y_1}$$

(so that $\quad m = -\dfrac{x_1}{y_1} \quad$ and $\quad c = \dfrac{R^2}{y_1}$)

and this line will be a tangent to the circle if Equation 11 is satisfied, i.e. if $\left(\dfrac{R^2}{y_1}\right)^2 = R^2\left[1 + \left(-\dfrac{x_1}{y_1}\right)^2\right]$

which reduces to $x_1^2 + y_1^2 + R^2$

In other words:

---

the line $x_1 x + y_1 y = R^2$ (13)

is a tangent to the circle $x^2 + y^2 = R^2$ provided that the point (x_1, y_1) lies on the circle.

What is the equation of the tangent to the circle $x^2 + y^2 = 25$ at the point $(4, 3)$ on the circle?

Here $(x_1, y_1) = (4, 3)$. The equation is $4x + 3y = 25$. □

Question T9

(a) Show that the line $y = 33.8 - 2.4x$ is a tangent to $x^2 + y^2 = 169$.

(b) Where is the point of contact?

(c) Write down the equation of the parallel tangent and find its point of contact. □

So far we have only considered circles centred on the origin. However, by introducing a new system of coordinates, with axes X and Y and a different origin from that of the usual (x, y) system, we can investigate the equation of the tangent to a circle that is not centred on the origin. This is what we do in the next example.

Example 5

Given a circle of radius 5, centred on the point $(-3, -4)$, what is the equation of a tangent to the circle through a given point on the circle?

Solution If we make a change of coordinates to $X = x - 3$ and $Y = y - 4$ then the origin in the (x, y) coordinate system becomes the point $(-3, -4)$ in the new (X, Y) coordinate system. If we rearrange the coordinate relations to give $x = X + 3$ and $y = Y + 4$, the circle $x^2 + y^2 = 25$, with radius 5 and centre at the origin in the (x, y) coordinate system, becomes the circle $(X + 3)^2 + (Y + 4)^2 = 25$ in the new coordinate system. In other words, its radius remains unchanged, but its centre is now at the point $(-3, -4)$ in the (X, Y) coordinate system.

We can similarly transform the tangent $xx_1 + yy_1 = 25$, where (x_1, y_1) is a point on the circle in the (x, y) coordinate system, to obtain

$$(X + 3)x_1 + (Y + 4)y_1 = 25$$

then if we put $x_1 = X_1 + 3$ and $y_1 = Y_1 + 4$ we obtain the equation of the tangent

$$(X + 3)(X_1 + 3) + (Y + 4)(Y_1 + 4) = 25$$

to the circle $(X + 3)^2 + (Y + 4)^2 = 25$, where (X_1, Y_1) is a point on the circle in the (X, Y) coordinate system (because $(X_1 + 3)^2 + (Y_1 + 4)^2 = x_1^2 + y_1^2 = 25$).

This method of obtaining results by making a simple transformation is often very convenient (see Question E5). ☐

Question T10

In this question we assume that the Earth is a perfect sphere. (You will need to use the trigonometric identities $\cos^2 A + \sin^2 A = 1$ and $\cos A \cos B + \sin A \sin B = \cos (A - B)$.)

(a) At time t, a point P has coordinates $x = a \cos (\omega t)$ and $y = a \sin (\omega t)$, where $a > 0$. Show that the point lies on a circle of radius a.

(b) Imagine now that you are looking down at the Earth from a point high above the North Pole, so that the Earth is rotating beneath you. You imagine the x- and y-axes to be two fixed directions (each perpendicular to the axis of the Earth). If P is a particular point on the Earth's equator (so that it rotates with the Earth), how long does it take the point P to move through a complete circle, and what is the value of its *angular speed* ω (in radians s^{-1})?

(c) A satellite S is placed in a circular orbit above the Earth's equator so that at time t its position is given by the coordinates

$$x = R \cos (\Omega t) \text{ and } y = R \sin (\Omega t)$$

Take a to be the radius of the Earth. What condition involving a tangent to the Earth's equator must be satisfied whenever P is moving directly towards or away from the satellite? At what time t is this condition satisfied? ☐

9.3.3   Polar coordinates

Cartesian coordinates provide a valuable method of specifying the location of a point relative to a given origin, but they are not the only way in which such information can be specified. Other kinds of coordinate system exist and may be better suited to specific problems. A case in point is that of two-dimensional **polar coordinates**, which are particularly useful when distance from a fixed origin is of paramount importance.

The polar coordinate system is illustrated in Figure 9.14. The origin is represented by the point O and the point of interest by P. The line segment from O to P (sometimes called the **radius vector** of P) is of length r and is inclined at an angle θ, measured in the anticlockwise direction, from an arbitrarily chosen ray, called the **polar axis**, that emanates from O. The length r is called the **radial coordinate** and the θ the **polar angle** of P.

The *polar coordinates* r and θ of the point P are usually combined so that P is represented by the ordered pair (r, θ).

Figure 9.15 shows four points P_1, P_2, P_3 and P_4 together with their polar coordinates.

Notice that an *anticlockwise* rotation is taken to be positive so that, for example, the point P_4 has polar coordinates $(4, 315°)$. Since clockwise rotation is negative, and we could reach P_4 by a clockwise rotation of $45°$ from the polar axis, its polar coordinates could be written $(4, -45°)$.

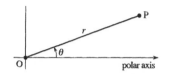

Figure 9.14 Polar coordinate system.

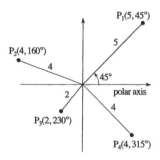

What is an alternative polar representation of P_3?

Since $230° = 360° - 130°$, P_3 can be represented as $(2, -130°)$. ☐

The polar coordinates of a given point are not uniquely determined, because we can add or subtract any multiple of $360°$ to the angle. We sometimes prefer polar coordinates to be completely specified, in which case the angle θ may be restricted to lie in the range $-180° < \theta \leqslant 180°$ (although there are occasions when the range $0 \leqslant \theta < 360°$ is more convenient).

Figure 9.15 Some points and their polar coordinates.

Linking Cartesian and polar coordinates

If the origin of polar coordinates O coincides with the origin of Cartesian coordinates, and if the polar axis is chosen as the positive x-axis, we may link the Cartesian and polar coordinates as in Figure 9.16.

Since r is the hypotenuse of a right-angled triangle, it is clear that

---

$$x = r \cos \theta \quad \text{and} \quad y = r \sin \theta \tag{14}$$

---

and these equations enable us to convert from polar to Cartesian coordinates.

The polar coordinates of a point P are $(5, 30°)$; what are its Cartesian coordinates?

Since $r = 5$ and $\theta = 30°$, we have

$$x = 5 \cos 30° = 5 \times \frac{\sqrt{3}}{2} \quad \text{and} \quad y = 5 \sin 30° = 5 \times \frac{1}{2} = 2.5 \quad ☐$$

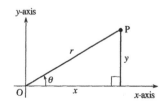

Figure 9.16 Polar and Cartesian coordinates.

It is also possible to deduce from Figure 9.16 that

$$r^2 = x^2 + y^2, \quad \sin\theta = y/r \quad \text{and} \quad \cos\theta = x/r \tag{15}$$

and these equations enable us to convert from Cartesian to polar coordinates. Since r is a positive quantity, its determination presents no difficulty; the same is not true for θ.

Example 6

Find the polar coordinates of the point P with Cartesian coordinates (3, 4).

Solution Since $r^2 = 3^2 + 4^2 = 25$, we have $r = 5$. It remains to find an angle θ such that $\sin\theta = 4/5 = 0.8$ and $\cos\theta = 3/5 = 0.6$.

If you use a calculator to find arcsin (0.8) and arccos (0.6) you will see that both give the same result $\theta \approx 53.13°$. Therefore the polar coordinates of P are approximately (5, 53.13°). □

Example 7

Find the polar coordinates of the point P with Cartesian coordinates $(-3, 4)$.

Solution Since $r^2 = (-3)^2 + 4^2 = 25$, we have $r = 5$, as in Example 6. Also we have $\sin\theta = 4/5 = 0.8$ and $\cos\theta = 3/5 = -0.6$, and now when you use your calculator to find the angle you will find arcsin (0.8) $\approx 53.13°$ as before, but arccos $(-0.6) \approx 126.87°$. This illustrates the limitations of your calculator, rather than a real problem in the mathematics.

The correct answer is $\theta \approx 126.87°$ as you can very easily verify. If you use your calculator to evaluate sin (126.87°) and cos (126.87°) you will soon see that sin (126.87°) ≈ 0.8 and cos (126.87°) ≈ -0.6.

So the polar coordinates are (5, 126.87°). □

You may have noticed that the two equations

$$\sin\theta = y/r \quad \text{and} \quad \cos\theta = x/r$$

can be combined to give $\tan\theta = y/x$, but specifying the tangent of an angle does not uniquely determine the angle, so this does not resolve the difficulty.

Example 8

Determine the polar coordinates of the points P_1 and P_2 with Cartesian coordinates (2, −2) and (−2, 2), respectively.

Solution In each case $r^2 = 2^2 + 2^2 = 8$ so that $r = 2\sqrt{2}$.

In each case $\tan\theta = -1$, and my calculator gives arctan $(-1) = -45°$, but P_1 and P_2 are clearly different points. The answer to our problems is to construct a sketch and plot the points; see Figure 9.17.

If we plot the point $(2, -2)$ we can see at once that the angle θ_1 is $45°$, and, since it is measured clockwise from OX the polar coordinates of P_1 are $(2\sqrt{2}, -45°)$, or equivalently $(2\sqrt{2}, 315°)$. The angle θ_2 is $180° - 45° = 135°$ and the polar coordinates of P_2 are $(2\sqrt{2}, 135°)$. $\square$

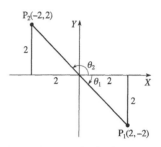

Figure 9.17 Points whose polar angles differ by $180°$.

The message is clear:

When determining the polar angle θ it is wise to *sketch* a diagram to show the position of the point.

9.4 Three-dimensional geometry

9.4.1 Points in three dimensions

We need two pieces of information to determine the position of a point in a plane, but in three dimensions we need to provide three *independent* pieces of information.

Figure 9.18 shows a Cartesian coordinate system in three dimensions. The point N is 'vertically' below the point P so that the line NP is parallel to the z-axis.

A point O is chosen as the origin of coordinates and three mutually perpendicular rays x, y and z are drawn as shown. The position of P is specified by three coordinates: x, y and z as indicated; and we refer to the point as $P(x, y, z)$. Having decided on the x- and y-axes, there are two possible directions, opposite to one another, in which the z-axis can be directed, so that it is normal (i.e. perpendicular) to both the x- and y-axes. To remove any ambiguity about the choice of direction of the z-axis, a **right-handed Cartesian coordinate system** is almost invariably used. There are various ways of describing this system, but one simple method only is included here. If the thumb and first two fingers of the right hand are arranged so that they are approximately mutually perpendicular as in Figure 9.19, then, if the first and second finger point along the x- and y-axes, respectively, the thumb points along the z-axis in a right-handed system.

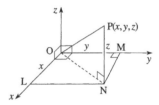

Figure 9.18 Three-dimensional Cartesian coordinate system.

Which of the following is a right-handed coordinate system:

(a) x points west, y points north and z points vertically upward,

(b) x is vertically downward, y points south and z points west?

Case (b) is right-handed, case (a) is not. $\square$

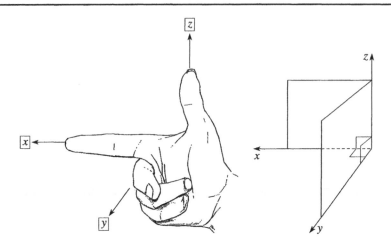

Figure 9.19 Picture of a hand showing the directions of the axes in a right-handed Cartesian coordinate system.

Distance between two points

The distance OP in Figure 9.18 can be found by applying Pythagoras's theorem twice. In the horizontal plane which contains the x- and y-axes,

$$(ON)^2 = (OL)^2 + (LN)^2$$

In the vertical plane which contains the lines ON, NP and OP,

$$(OP)^2 = (ON)^2 + (NP)^2 = (OL)^2 + (LN)^2 + (NP)^2$$

Hence $(OP)^2 = x^2 + y^2 + z^2$

Now we consider the distance between two general points.

Let the two points be $P_1(x_1, y_1, z_1)$ and $P_2(x_2, y_2, z_2)$. We can 'move' the point P_1 to the origin by subtracting x_1 from its first coordinate, y_1 from its second coordinate and z_1 from its third coordinate (this has the same effect as moving the origin to the point P_1). The same transformation will move P_2 to a point whose coordinates are $(x_2 - x_1, y_2 - y_1, z_2 - z_1)$. If we call this point P we can apply the result just obtained to determine the distance between P_1 and P_2. Thus the distance d between the points (x_1, y_1, z_1) and (x_2, y_2, z_2) is given by

$$d^2 = (x_2 - x_1)^2 + (y_2 - y_1)^2 + (z_2 - z_1)^2 \tag{16}$$

What is the distance between the points $(1, 2, -3)$ and $(-2, 0, 4)$?

Let $x_1 = 1$, $y_1 = 2$, $z_1 = -3$, $x_2 = -2$, $y_2 = 0$, $z_2 = 4$.

The distance d is given by

$$d^2 = (-2 - 1)^2 + (0 - 2)^2 + (4 - (-3))^2$$
$$= (-3)^2 + (-2)^2 + (7)^2 = 9 + 4 + 49 = 62.$$

Hence $d = \sqrt{62}$.   □

It is worth noting that the formula for the distance between two points in two dimensions (Equation 8) is the special case of the formula for three dimensions with z_1 and z_2 both put equal to zero.

Question T11

Which of the following points are (a) closest together, (b) furthest apart

$(1, 2, 3)$, $(-1, -2, -1)$, $(2, 2, -2)$, $(3, 0, 1)$? □

9.4.2   Planes in three dimensions

A plane is a flat surface which has the property that a straight line joining any two points in the plane lies entirely in the plane itself.

The general **equation of a plane** in three dimensions is often written as

$$ax + by + cz = d \tag{17}$$

where a, b, c and d are constants. Note that if we choose $a = 1$ with $b = c = d = 0$ the equation becomes $x = 0$ which is the (y, z) plane.

What values of a, b, c, d correspond to (a) the (x, y) plane, (b) the (z, x) plane?

(a) $a = b = d = 0$ gives $cz = 0$ and if $c \neq 0$ then $z = 0$ which is the equation of the (x, y) plane.

(b) $a = c = d = 0$, gives $by = 0$ and if $b \neq 0$ then $y = 0$ which is the equation of the (z, x) plane. □

Other planes can be described by choosing suitable combinations of the constants a, b, c and d. The plane $z = 2$ is parallel to the plane $z = 0$ (the (x, y) plane) and cuts the z-axis at $z = 2$, i.e. the point $(0, 0, 2)$. The equation $z = 2$ corresponds to Equation 17 with $a = 0$, $b = 0$, $c = 1$ and $d = 2$.

The three points $(1, 3, 5)$, $(1, -2, 7)$, $(1, -1, 0)$ lie in a plane. What is the equation of the plane?

The x-coordinate of all three points is 1, and therefore the equation for the plane is $x = 1$. □

Note that if $d = 0$ in Equation 17 then the plane will pass through the origin, since $x = y = z = 0$ will satisfy the equation of the plane.

Intersection of two planes

If two planes are not parallel then they meet in a line. Imagine two consecutive pages of a book. Hold the pages taut and you see that they meet in the line of the binding connecting them. For example, the (x, y) plane meets the (y, z) plane on the y-axis, this means that the y-axis corresponds to the points where both x and z are zero; in other words, we can see instantly that a point such as $(0, 5, 0)$ must lie on the y-axis.

Which axis corresponds to the intersection of the two planes $x = 0$ and $y = 0$?

The z-axis. □

Where does the plane $x + y + 2z = 6$ meet the x-, y- and z-axes?

The plane meets the x-axis when $y = z = 0$, so that $x = 6$, i.e. at the point $(6, 0, 0)$. Similarly the plane meets the y-axis at $(0, 6, 0)$, and the z-axis at $(0, 0, 3)$. □

If we keep the values of a, b and c fixed in Equation 17, but double the value of d, we obtain a plane that is parallel to the original plane (but twice as far from the origin). Compare the results of the following exercise with those of the previous exercise.

Where does the plane $x + y + 2z = 12$ meet the x-, y- and z-axes?

The plane meets the x-axis when $y = z = 0$, so that $x = 12$, i.e. at the point $(12, 0, 0)$. Similarly the plane meets the y-axis at $(0, 12, 0)$, and the z-axis at $(0, 0, 6)$. □

Question T12

Find the equation of a plane that is parallel to the plane $3x - 4y + z = 2$ and passes through the point $(1, 2, 3)$. □

9.4.3   Lines in three dimensions

In three dimensions, straight lines are often specified by the intersection of two planes. Of course we try to choose the planes so that the specification is as neat as possible; for example, we could specify a line as the intersection of the planes

Equations of planes and lines are best dealt with in the context of *vectors*; for example, the equation of a plane can be written very neatly in vector form as $a \cdot r = d$. Also, note in passing that the vector (a, b, c) is perpendicular to the plane $ax + by + cz = d$. Vectors are introduced in Part 3 of this book, and further developed in the companion volume.

$$3x - 2y = 5$$

and $5y - 3z = -11$

but this specification is not as neat as it might be.

The first of these equations can be rewritten in the form

$$\frac{x-1}{2} = \frac{y+1}{3}$$

while the second can be rewritten in the form

$$\frac{y+1}{3} = \frac{z-2}{5},$$

so that the two equations can be combined into the particularly neat form

$$\frac{x-1}{2} = \frac{y+1}{3} = \frac{z-2}{5} \qquad (18)$$

In this form we can see instantly that the point $(1, -1, 2)$ must lie on the line. The values 2, 3 and 5 are also significant, since they can be used to determine the direction of the line (but it would take us beyond the scope of this chapter to explain how this is done).

    If we put each of the fractions in Equations 18 equal to s, then they can be written in the equivalent form

$$\left. \begin{array}{l} x = 1 + 2s \\ y = -1 + 3s \\ z = 2 + 5s \end{array} \right\} \qquad (19)$$

The variable s is then a parameter that moves the point (x, y, z) up and down the line as we change its value.

    It is often this form of the equations of a line which is most useful, as the following exercise illustrates.

> Where does the line $\dfrac{x-1}{2} = \dfrac{y+1}{3} = \dfrac{z-2}{5}$ meet the plane
> $x + 2y + 3z = 1$?
>
> ---
>
> First we write the equations of the line in the form of Equation 19 $x = 1 + 2s$, $y = -1 + 3s$ and $z = 2 + 5s$, then substitute for x, y and z into the equation of the plane. This gives the following equation for s
>
> $$(1 + 2s) + 2(-1 + 3s) + 3(2 + 5s) = 1$$
>
> which simplifies to give $s = -4/23 \approx -0.17$. We can now find the values of x, y and z by substituting this value of s into the equations of the line, and we find the point of intersection $(0.66, -1.51, 1.15)$. ☐

We started with the neat specification and worked backwards.

In three dimensions only lines that lie in a plane will meet in a point. Indeed, in three dimensions, it is more likely that a pair of lines chosen at random will not meet, in which case they are said to be **skew** lines.

In general, the **equations of a line in three dimensions**

$$\frac{x - a}{l} = \frac{y - b}{m} = \frac{z - c}{n} \tag{20}$$

specify a line through the point (a, b, c) and the constants l, m, n are known as the **direction ratios** (or **direction cosines**) of the line.

Find the equations of a line through the points $(1, 2, 3)$ and $(6, 5, 4)$.

We can see at once that the equations can be written in the form of Equation 20.

$$\frac{x - 1}{l} = \frac{y - 2}{m} = \frac{z - 3}{n}$$

But the point $(6, 5, 4)$ lies on the line, and therefore

$$\frac{6 - 1}{l} = \frac{5 - 2}{m} = \frac{4 - 3}{n}$$

i.e. $\quad \dfrac{5}{l} = \dfrac{3}{m} = \dfrac{1}{n}$.

and we can see that $l = 5$, $m = 3$ and $n = 1$ provide a solution to these equations. The required equations are therefore

$$\frac{x - 1}{5} = \frac{y - 2}{3} = \frac{z - 3}{1} \quad \square$$

Question T13

Given the equations of a line in the form

$$x = 1 + 2s, \ y = -2 + s, \ z = 2 - 3s$$

(where s can take any value), show that the point $P(1, -2, 2)$ lies on the line. Find the points on the line that are at a distance $\sqrt{14}$ from P. Make s the subject of each equation and hence find the direction ratios for the line. $\square$

'Make s the subject of the equation' means that you have to write it in the form $s = \ldots$

9.5 Closing items

9.5.1 Chapter summary

1. In two dimensions the *Cartesian system* of coordinates consists of two perpendicular lines, called axes, which meet at the *origin* of coordinates; one of these lines, usually drawn horizontally, is designated the x-axis, the other, which is usually drawn vertically, is the y-axis. The x-coordinate of a point is the perpendicular distance of that point from the y-axis, with a negative sign attached if the point lies to the left of the y-axis. The y-coordinate of the point is its perpendicular distance from the x-axis, with a negative sign attached if the point lies below the x-axis.

2. The equation of a straight line can appear in a variety of forms; the one chosen depends on the information available about the line and the information likely to be required about the line.

3. If the line has a known gradient m and a known *intercept* (on the y-axis) c then the equation relating the x- and y-coordinates of a point on it is

$$y = mx + c \tag{1}$$

4. If the line has a known gradient m and passes through the point (x_1, y_1) then the equation of the line is

$$y - y_1 = m(x - x_1) \tag{3}$$

5. If the line meets the x-axis at $x = a$ and the y-axis at $y = b$ then the equation of the line is

$$\frac{x}{a} + \frac{y}{b} = 1 \tag{5}$$

6. The so-called general form of the equation of a line is

$$ax + by + c = 0 \tag{6}$$

7. If two lines are parallel then their gradients are equal, and if two lines are perpendicular then the product of their gradients is -1.

8. If two lines intersect then the coordinates of the point of intersection can be found by solving simultaneously the equations of the lines.

9. The distance between the points $P(x_1, y_1)$ and $Q(x_2, y_2)$ is given by

$$PQ = \sqrt{(x_2 - x_1)^2 + (y_2 - y_1)^2} \tag{8}$$

10. The equation of a circle centred at the origin and of radius R is

$$x^2 + y^2 = R^2 \tag{9}$$

11. The equation of a circle centred at the point (x_0, y_0) and of radius R is

$$(x - x_0)^2 + (y - y_0)^2 = R^2 \tag{10}$$

12. The equation of the tangent to the circle $x^2 + y^2 = R^2$ at a point (x_1, y_1) on the circle is given by

$$x_1 x + y_1 y = R^2 \tag{13}$$

13. In the *polar coordinate system* a point is specified by its distance r from the origin and the *polar angle* θ which it makes with a given line through the origin. The link between the two systems is provided by the pair of equations

$$x = r \cos \theta \text{ and } y = r \sin \theta \tag{14}$$

and $r^2 = x^2 + y^2, \quad \sin \theta = y/r, \quad \cos \theta = x/r \tag{15}$

where we must take care to identify the correct value of θ.

(This can often be done by plotting the point on a diagram, and then using the fact that $\tan \theta = y/x$.)

14. The distance d between the points (x_1, y_1, z_1) and (x_2, y_2, z_2) in three dimensions is given by

$$d^2 = (x_2 - x_1)^2 + (y_2 - y_1)^2 + (z_2 - z_1)^2 \tag{16}$$

15. The general equation of a plane in three dimensions is

$$ax + by + cz = d \tag{17}$$

If two planes intersect, they do so in a line.

16. The general equations of a line in three dimensions are

$$\frac{x - a}{l} = \frac{y - b}{m} = \frac{z - c}{n} \tag{20}$$

where (a, b, c) is a point on the line and l, m, n are constants, known as *direction ratios* (or *direction cosines*) of the line.

9.5.2 Achievements

Having completed this chapter, you should be able to:

A1. Define the terms that are emboldened in the text of the chapter.

A2. Set up a Cartesian coordinate system and plot points in it.

A3. Recognise the different forms of equation of a straight line and produce the appropriate equation from the information given about the line, then derive the required information from it.

A4. Convert one form of the equation of a line into another.

A5. Write down the equation of a circle, given its centre and radius, and deduce the centre and radius of a circle from its equation.

A6. Produce the equation of tangents to a circle, both from a given point outside the circle and at a given point of contact on the circle.

A7. Set up the polar coordinate system, plot points in it and convert from Cartesian coordinates to polar coordinates and vice versa.

A8. Calculate the distance between two points in three dimensions.

A9. Recognise the equation of a plane, and use it to indentify points that lie in the plane.

A10. Recognise the equations of a line in three dimensions and use them to identify points that lie on the line and the point at which a line intersects a given plane.

9.5.3   End of chapter questions

Question E1   Plot the following points on a Cartesian coordinate system: A(-1, 4), B(3, -1), C(-4, -1), D(0, 2), E(0, -3).

Question E2   Find the equations of the following lines:

(a)   with gradient -5 and y-intercept -2;

(b)   with gradient -2, passing through the point $(-1, 4)$;

(c)   passing through the points $(2, -3)$ and $(-1, -1)$;

(d)   with intercepts at -2 on both axes.

Question E3   Rewrite each of the equations in Question E2 in the general form $ax + by = c$.

Question E4   Find the gradient and y-intercept for the line of Question E2(c) and the gradient of the line in Question E2(d).

Question E5   A line $ax + by - c = 0$, with $a \neq 0$ and $b \neq 0$ meets the x- and y-axes at the points A and B, respectively.

(a)   Find the area of the triangle AOB (where O is the origin).

(b)   Find the length AB.

(c)   Find the shortest length h from the line to the origin O.

   (*Hint*: The shortest line will be perpendicular to AB, and what is the area of the triangle AOB when AB is the base?)

(d)   Find the coordinates of the point N where the line $y = 2$ meets the line $x = 5$.

(e)   If we make a change to the coordinate system by putting

$$X = x - 5 \text{ and } Y = y - 2$$

what are the new coordinates of the point N (in the XY system)?

(f)   We can find the equation of the line AB in the new coordinate system if we put

$$x = X + 5 \text{ and } y = Y + 2$$

into the equation $ax + by - c = 0$. What does this equation become?

(g)   What is the shortest distance from the point $(5, 2)$ to the line $ax + by - c = 0$?

Question E6   Which of the lines $2x + 3y = 2$ or $2x - 3y = 2$ is perpendicular to the line $6x + 4y = 5$? Find the equation of the line perpendicular to the line $6x + 4y = 5$ which passes through the point $(4, 4)$.

Question E7   Write down the equations of the circles of radius 3 which have centres at $(1, -2)$ and $(2, 1)$, respectively, and find where they meet.

Question E8   Find the point of intersection of the tangents to the circle $x^2 + y^2 = 4$ at the points $(1, \sqrt{3})$ and $(0, -2)$.

Question E9   Show that the points $(1, 2, 1)$, $(2, -3, -2)$ and $(-1, 0, -5)$ lie on the plane $2x + y - z = 3$. Which two are the closest together? Where does the line

$$\frac{x - 1}{3} = \frac{y - 1}{5} = \frac{z}{2}$$

meet the plane $2x + y - z = 3$?

Conic sections

10.1 Opening items

10.1.1 Chapter introduction

If you throw a stone, then to a good approximation, it follows a parabolic path. As a planet orbits the Sun it traces out a closed path in a plane, and this closed path is an ellipse. When an α-particle approaches the nucleus of an atom it undergoes a deflection, and the particle follows a hyperbolic path.

Parabolas, *ellipses* and *hyperbolas* are particular examples of a family of curves known as *conic sections*, for the very good reason that they can be obtained by taking a slice through a cone (or more precisely a *double cone*). The edge of the slice is called a conic section. We pursue this idea in Subsection 10.2.1, and then examine each of the major classes of conic section in turn in the subsequent subsections.

In Subsections 10.2.2 to 10.2.5 we study the circle, the parabola, the ellipse and finally the hyperbola. In each case our approach is very similar. First we introduce the equation of the curve in so-called *standard form*, which simply means that we choose the axes so that the equation of the curve turns out to be particularly simple. We investigate the tangents to the curve, and we consider parametric representations of the curve (this is particularly relevant to applications in physical science). At the end of each subsection we consider particular properties of the curve under discussion; for example, an ellipse is a closed curve, (☞) whereas, at a great distance from the origin, the points on a hyperbola approach one of two straight lines (called *asymptotes*). In spite of their very different shapes, it turns out that the various forms of conic section have a great deal in common. In each case we introduce the polar form of the equation, and in Subsection 10.2.5 we show how this form of the hyperbola may be applied to a well known experiment in nuclear physics.

For the equation of a conic in standard form one or both of the coordinate axes are aligned with an axis of symmetry of the curve, and the origin of coordinates is placed at a point which ensures that the equation of the curve is particularly simple. In Subsections 10.3.1 and 10.3.2 we introduce the general equation of a conic, and consider the effect of moving the coordinate axes from the 'standard position'. Finally in Subsection 10.3.3 we consider an application of the general form of the equation of a conic to electrostatics.

A closed curve returns to its starting point.

10.1.2   Fast track questions

Question F1

(a) Find the radius and the coordinates of the centre of the circle $x^2 + y^2 - 2x - 6y + 6 = 0$.

(b) Find the coordinates of the points P and Q at which the line $y = x/2$ meets the circle $x^2 + y^2 - 8x + 6y - 15 = 0$. Also find the equation of the circle passing through P, Q and the point $(1, 1)$.

(c) Find the equations of the tangents to the circle $x^2 + y^2 = 25$ which are parallel to the line $3x + 4y = 0$.

Question F2

(a) Find the equations of the tangents and normals to the parabola $y^2 = 16x$ at the points $(16, 16)$ and $(1, -4)$. If these tangents intersect at T and the normals intersect at R show that the line TR is parallel to the x-axis.

(b) Write down the parametric representation of a point on a parabola. Show that the tangents to a parabola at the ends of a chord which passes through the focus are perpendicular to each other.

Question F3

(a) Show that the tangent to the ellipse $x^2/16 + y^2/9 = 1$ at the point $(16/5, 9/5)$ makes equal intercepts with the coordinate axes.

(b) Find the equation of the tangent to the rectangular hyperbola $xy = 12$ at the point $(3, 4)$.

10.1.3   Ready to study?

Study comment

In order to study this chapter you will need to be familiar with the following terms: *chord, coordinate axes, graph, normal, repeated roots* of *quadratic equations, straight lines, tangent.* You should also be familiar with *trigonometric identities* (although the relevant formulae are repeated here). In addition, you will need to be familiar with *SI units.*

In this chapter, somewhat unusually, we have used the techniques of differential calculus (introduced in Part 4) to prove certain standard results. In all cases these proofs may be regarded as optional and should be omitted on a first reading if you are not familiar with the techniques of calculus.

The following *Ready to study questions* will allow you to establish whether you need to review some of the topics before embarking on this chapter.

Question R1

Calculate the distance between the points P(0, -2) and Q(3, 1) and the gradient of the line PQ.

Question R2

Find the equation of the line through the point (-2, 1) which has gradient 3. Find also the equation of the line through the same point which is perpendicular to the first line.

Question R3

Find the equation of the line passing through the points (2, -1) and (-1, 4) and find its point of intersection with the line $y = 3x$.

Question R4

(a) Write down the condition that the quadratic equation $Ax^2 + Bx + C = 0$ has a repeated root at $x = -B/(2A)$.

(b) Show that the quadratic equation $x^2 + (mx + c)^2 = a^2$ has a repeated root if the constants a, c and m satisfy the equation $c^2 = (1 + m^2)a^2$.

10.2 Conic sections

10.2.1 What are conic sections?

Figure 10.1 shows a **double cone**; it looks like one cone placed upside down on top of another. It is to be understood that this shape extends infinitely upwards and downwards.

To generate this figure we take the line of which AOB is a part and rotate it about a vertical axis through O in such a way that the point B traces out a circle which is perpendicular to the axis POQ and which has Q at its centre. The point A traces out a similar circle with P as its centre. The line of which POQ is a part is called the axis of symmetry of the double cone. P and Q are the centres of the circles traced out by A and B, respectively. The line through A, O and B is a **generator** of the double cone.

The curves we produce by cutting the double cone with a plane are called **conic sections** or, more simply **conics**. These curves are the *boundaries* of the sections exposed by the cuts. Figure 10.2 shows the possibilities.

(a) When the plane making the cut is perpendicular to the axis, the curve produced is a **circle**.

(b) When the cutting plane is inclined at an angle between the horizontal and that of a generator the boundary curve is an **ellipse**.

(c) When the cutting plane is parallel to a generator the resulting curve is a **parabola**.

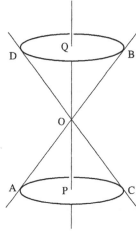

Figure 10.1 Double cone.

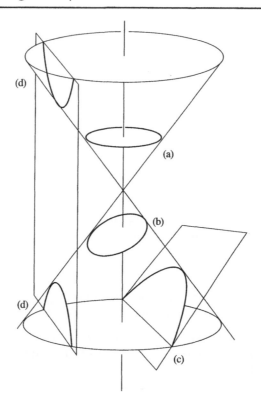

Figure 10.2 Conic sections. Note that the section representing the hyperbola (d) does *not* have to be parallel to the vertical axis — any angle of inclination steeper than (c) will suffice.

(d)   If the cutting plane meets both parts of the cone at any angle the curve produced is in two parts called **branches** and is a **hyperbola**.

In Subsections 10.2.2 to 10.2.5 we derive the equations of these four curves from quite different starting points. We shall not *prove* that the two approaches lead to the same curves in each case; rest assured that they do.

The conics of various kinds have a common property that we have illustrated in Figure 10.3. There is a fixed point called the **focus**, and a fixed line called the **directrix**, such that for any point P on the conic, the distance to the focus and the distance to the directrix have a constant ratio, called the **eccentricity** and denoted by e.

In other words,

$$\frac{FP}{PN} = e \qquad\qquad (1)$$

Figure 10.3 A conic with its focus and directrix.

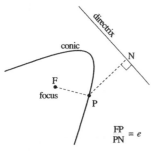

We shall see later that an ellipse has two foci. (Foci is the plural of focus.)

It is the value of the eccentricity that determines the nature of the conic, and Table 10.1 contains a short summary of some important results, which you may find useful as a quick reference.

The circle can be regarded as the limiting case of an ellipse as the directrix moves further from the origin, e approaches 0, and the foci (✍) approach the centre of the ellipse. It corresponds to putting $b = a$ in the equation $x^2/a^2 + y^2/b^2 = 1$. In this chapter when we refer to ellipses we mean to exclude circles, unless we say otherwise. The equations given in Table 10.1 will be known as the **standard equations** or **standard forms** of the conics. (We will

Table 10.1 Summary of the properties of conic sections

Conic	Circle	Parabola	Ellipse	Hyperbola
Eccentricity	$e = 0$	$e = 1$	$0 \leqslant e < 1$	$e > 1$
Standard equation	$x^2 + y^2 = a^2$	$y^2 = 4ax$	$\dfrac{x^2}{a^2} + \dfrac{y^2}{b^2} = 1$	$\dfrac{x^2}{a^2} - \dfrac{y^2}{b^2} = 1$
Graph	see Figure 10.4a	see Figure 10.4b	see Figure 10.4c	see Figure 10.4d

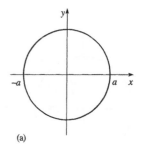

(a)

see later that it is sometimes convenient to represent the conics in various other forms.)

> Loosely speaking, a curve is symmetric about a line if it can be folded onto itself with a fold along the line. Which of the standard equations in Table 10.1 represent conics which are symmetric about (a) the x-axis, (b) the y-axis?
>
> The symmetries of the curves are clear from the figures in Figure 10.4, however we could use an algebraic approach. A curve is symmetric about the x-axis if the equation of the curve is unchanged when y is replaced by $-y$; it is symmetric about the y-axis if the equation of the curve is unchanged when x is replaced by $-x$. The equation $y^2 = 4ax$ represents a parabola that is symmetric about the x-axis only, and the other equations represent conics that are symmetric about both axes. In general ellipses and hyperbolas have two axes of symmetry whereas parabolas have one axis of symmetry. $\square$

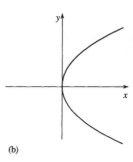

(b)

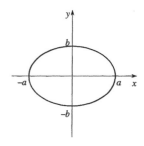

(c)

10.2.2 The circle

As described in the last chapter, a circle consists of all the points in a plane that are at some given distance from a fixed point. The fixed point is called the **centre** of the circle and the distance from the centre to any point on the circle is called the **radius** of the circle. If the fixed point is the origin of coordinates, O, and if P is an arbitrary point on the circumference with coordinates (x, y), then the distances ON and PN shown in Figure 10.5 are equal to the absolute values of x and y, the distance OP is equal to the radius of the circle, which we shall call a in this case and we may use Pythagoras's theorem to write

i.e. $\quad x^2 + y^2 = a^2$ $\qquad\qquad$ (2)

This is the standard equation of a circle.

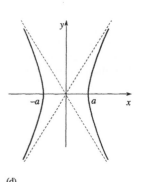

(d)

Figure 10.4 (a) Circle; (b) parabola; (c) ellipse; (d) hyperbola.

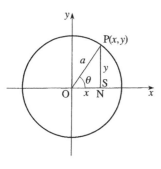

Figure 10.5 The circle and the angle θ.

The intersection of a line with a circle

We can discover the x-coordinates of the points where a line

$$y = mx + c$$

meets the circle

$$x^2 + y^2 = a^2$$

by substituting for y from the first equation into the second equation to obtain

$$x^2 + (mx + c)^2 = a^2$$

which may be written in the form

$$(1 + m^2)x^2 + 2mcx + (c^2 - a^2) = 0$$

As we saw in the last chapter (Subsection 9.3.2), this implies that the line will intersect the circle in two points, miss it entirely, or meet it in just one point according to whether the quadratic equation has:

two real roots, in which case $(2mc)^2 > 4(1 + m^2)(c^2 - a^2)$

no real roots in which case $(2mc)^2 < 4(1 + m^2)(c^2 - a^2)$

one real root in which case $(2mc)^2 = 4(1 + m^2)(c^2 - a^2)$

(See Question R4.) It is the third of the three cases that interests us here, since this is the case in which the line is a **tangent** to the circle. Rearranging the third condition, we see that a necessary condition for the line $y = mx + c$ to be a tangent to the circle $x^2 + y^2 = a^2$ is that

$$c^2 = (1 + m^2)a^2 \qquad\qquad (3)$$

Question T1

Show that the quadratic equation $x^2 + (mx + c)^2 = a^2$ has equal roots if $m = 3$, $a = 5$ and $c = 5\sqrt{10}$, and then show that Equation 3 is satisfied. What can you say about the line $y = 3x + 5\sqrt{10}$ and the circle $x^2 + y^2 = 25$? ☐

Tangents and normals to the circle

As a rule of thumb, 'half' of the x's in the equation of the circle have become x_1, while 'half' of the y's have become y_1.

We may use Equation 3 to show that the equation of the tangent to the circle $x^2 + y^2 = a^2$ at the point (x_1, y_1) on the circle is

$$x_1x + y_1y = a^2 \qquad\qquad (4)\ (\text{☞})$$

this was proved in Subsection 9.3.2.

You may omit the following question if you are not familiar with calculus.

Find the gradient of the tangent at the point $P(x_1, y_1)$ on the circle by differentiating Equation 2, and hence show that Equation 4 represents the tangent at P.

We may differentiate Equation 2 implicitly to obtain

$$2x + 2y\frac{dy}{dx} = 0 \quad \text{and hence} \quad \frac{dy}{dx} = -\frac{x}{y}$$

Thus, at the point $P = (x_1, y_1)$ on the circle, the derivative is

$$-\frac{x_1}{y_1}.$$

The tangent at P has the same slope as this derivative, and so the equation of the tangent at P is

$$y - y_1 = -\frac{x_1}{y_1}(x - x_1)$$

so that $y_1y - y_1^2 = -x_1x + x_1^2$

i.e. $\quad y_1y + x_1x = x_1^2 + y_1^2 = a^2$

because (x_1, y_1) is a point on the circle. $\quad\square$

From Equation 4 we can write

$$y = -\left(\frac{x_1}{y_1}\right)x + \frac{a^2}{y_1}.$$

Comparing this with $y = mx + c$ we can see that the gradient of the tangent at a point (x_1, y_1) on the circle is

$$-\frac{x_1}{y_1}.$$

The **normal** at (x_1, y_1) is by definition a line perpendicular to the tangent, and so its gradient must be y_1/x_1. (☞)

It follows that the equation of the normal to the circle at a point $P(x_1, y_1)$ on the circle is

$$y - y_1 = \frac{y_1}{x_1}(x - x_1)$$

which simplifies to

$$y = \left(\frac{y_1}{x_1}\right)x \tag{5}$$

This is the equation of a line through the origin, as you would expect.

If two lines are perpendicular then the product of their gradients is -1.

Question T2

What are the equations of the tangents and the normals to the circle $x^2 + y^2 = 4$ at the following points?

(a) $(\sqrt{2}, \sqrt{2})$ (b) $(0, 2)$ (c) $(2, 0)$

(d) $(-1, -\sqrt{3})$ (e) $(1.2, -1.6)$ □

Parametric equations for the circle

The equation $x^2 + y^2 = a^2$ defines a set of points (x, y) that lie on a circle, but in many applications this is not quite enough. We may also need to indicate that the points on the curve are traced in a particular order or that a point is moving along the curve in an anticlockwise direction starting at the point $(a, 0)$, for example. In Figure 10.5 the point P on the circle is determined by the angle θ, and from triangle OPN we see that

$$x = a \cos \theta \quad \text{and} \quad y = a \sin \theta \tag{6}$$

As θ takes values in the range $0 \leqslant \theta < 360°$ then P moves round the circle in an anticlockwise direction from S$(a, 10)$. A variable such as θ, which is used to determine the position of a point on a curve is known as a **parameter**. In many applications to physics the natural parameter to choose would be time, so that the position of P is determined by the value of the time.

Figure 10.6 Points on a circle of radius 2.

Mark on the circle $x^2 + y^2 = 4$ the points $P_1 = (2 \cos 60°, 2 \sin 60°)$, $P_2 = (2 \cos 240°, 2 \sin 240°)$.

See Figure 10.6. □

Note that the points P_1 and P_2 are at the opposite ends of a diameter. This will be true of any two points corresponding to two values of the parameter that differ by 180°.

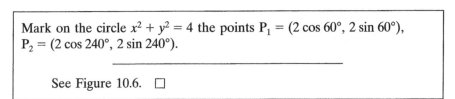

Suppose that the angle θ in Figure 10.5 is given by $\theta = \omega t$ where ω is a positive quantity and t is the time.

(a) Does the point P move clockwise or anticlockwise round the circle as t increases? (b) What are x and y of Figure 10.5 in terms of ω and t?

(a) θ increases as t increases, so P moves anticlockwise round the circle.

(b) $x = a \cos(\omega t)$, $y = a \sin(\omega t)$. □

Suppose the position of a point P on a circle is determined by a parameter t representing time in seconds, and $x = 4\cos(\omega t)$, $y = 4\sin(\omega t)$. What is the radius of the circle? In which direction does P traverse the circle if ω is negative; and if $\omega = -\pi$ rad s^{-1}, how long does it take P to trace out the circle?

The value of a in Figure 10.5 is 4 units, and $\theta = \omega t$ at time t. Since ω is negative, the value of θ decreases as t increases and so the point P moves in a clockwise direction around a circle of radius 4. When P returns to its starting point (for the first time), it will have traversed 2π radians (in the negative direction) and therefore putting $\theta = -2\pi$ radians and $\omega = -\pi$ rad s^{-1} into the formula $\theta = \omega t$ we obtain $t = 2$ s. $\square$

What is the equation of the tangent to the circle $x^2 + y^2 = 16$ at the point P($4\cos\theta$, $4\sin\theta$) on the circle?

From Equation 4, the equation of the tangent is $(4\cos\theta)x + (4\sin\theta)y = 16$ which simplifies to $x\cos\theta + y\sin\theta = 4$. $\square$

Question T3

What is the equation of the normal to the circle $x^2 + y^2 = 16$ at the point P($4\cos\theta$, $4\sin\theta$) on the circle? $\square$

Alternative forms of the circle

We normally regard Equations 2 or 6 as the simplest forms of the equations of a circle, but we will mention several other forms because they sometimes arise naturally in physical science.

The first form is simply a consequence of choosing the centre of the circle to be at a point other than the origin. Using Equation 10 from Subsection 10.3.1, the equation of a circle of radius a, centre (p, q) will be

$$(x - p)^2 + (y - q)^2 = a^2 \tag{7}$$

Alternatively you may encounter the equation of a circle in the general form

$$x^2 + y^2 + 2Gx + 2Fy + C = 0 \tag{8}$$

in which case the centre of the circle is at the point $(-G, -F)$ and its radius is $\sqrt{G^2 + F^2 - C}$.

Expand the brackets in Equation 7 and verify that Equation 8 represents a circle of radius $\sqrt{G^2 + F^2 - C}$ with centre at $(-G, -F)$.

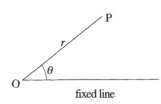

Figure 10.7 Polar coordinates.

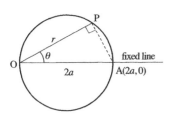

Figure 10.8 The polar form of a circle.

Table 10.2 The values of $r = 4 \cos\theta$ for various values of θ

$\theta/°$	r/cm	$\theta/°$	r/cm
−90	0.0	10	3.9
−80	0.7	20	3.8
−60	2.0	40	3.1
−40	3.1	60	2.0
−20	3.8	80	0.7
−10	3.9	90	0.0
0	4.0		

Equation 7 gives $x^2 - 2px + p^2 + y^2 - 2qy + q^2 = a^2$

or $x^2 + y^2 - 2px - 2qy + (p^2 + q^2 - a^2) = 0$

Then comparing the coefficients of this equation with those of Equation 8 we see that $p = -G, q = -F$, and $a = \sqrt{G^2 + F^2 - C}$. ☐

You are quite likely to encounter situations in which circles are described in terms of **polar coordinates** as in Figure 10.7 (these too were introduced in Chapter 9). The equation of a circle of radius a with its centre at the origin is particularly simple in such a coordinate system, it's just

$$r = a \quad (\text{with } a \geqslant 0) \tag{9}$$

You should be familiar with the representation of curves in terms of (x, y) coordinates, but you may not be aware that we can do something very similar with (r, θ) coordinates. For example the equation

$$r = 2a \cos \theta \tag{10}$$

represents a curve; but which curve? It's actually a circle, a fact that may be demonstrated as follows.

Suppose that the point P in Figure 10.7 has coordinates (x, y), then

$$r^2 = x^2 + y^2 \quad \text{and} \quad \cos \theta = \frac{x}{r} = \frac{x}{\sqrt{x^2 + y^2}}.$$

Substituting for r and $\cos \theta$ in Equation 10 we obtain $x^2 + y^2 = 2ax$ which gives $x^2 - 2ax + y^2 = 0$ implying $x^2 - 2ax + y^2 + a^2 = a^2$ so that

$$(x - a)^2 + y^2 = a^2$$

Comparing this with Equation 7 we see that it represents a circle of radius a with its centre at $(a, 0)$. Equation 10 therefore represents the circle shown in Figure 10.8.

In Figure 10.9 we provide you with graph paper marked in polar coordinates, and Table 10.2 is obtained from the formula $r = 4 \cos\theta$ cm. Plot the points and convince yourself that they do indeed lie on a circle. (Use a *pencil* as this graph paper will be needed again later on.)

The points lie on a circle of radius 2 cm with its centre at the point (2 cm, 0) with respect to the x- and y-axes shown in the figure. ☐

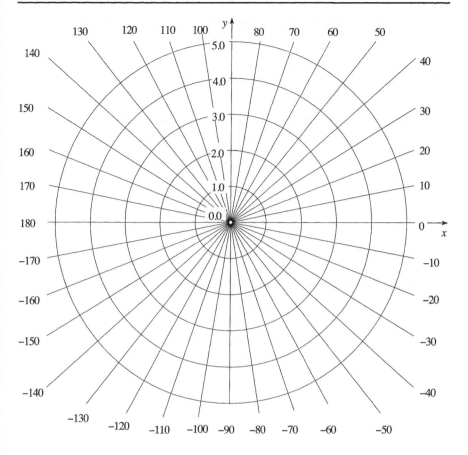

Figure 10.9 Polar graph paper.

10.2.3 The parabola

You may be aware that the graph of a quadratic function

$$y = Ax^2 + Bx + C \tag{11}$$

is a geometric shape known as a parabola, but here we begin our investigation of parabolas with the definition of a conic, Equation 1 with $e = 1$. Such shapes occur in many areas of physics; in particular, a simple model of the motion of a *projectile* (☞) under gravity near the Earth predicts that it will move along a parabolic path.

> A *projectile* is a body that moves in unpowered flight.

We will choose the position of the focus and the directrix in such a way that the equation of the parabola becomes particularly simple. In Figure 10.10 you will see that we have placed the focus on the x-axis at the point $(a, 0)$, and we have arranged for the parabola to pass through the origin by choosing the directrix to be the line $x = -a$. The point on the parabola closest to the directrix is called the **vertex** of the parabola.

The distance between two points (x_1, y_1) and (x_2, y_2) is

$$\sqrt{(x_1 - x_2)^2 + (y_1 - y_2)^2}$$

Notice that we have effectively replaced the product y^2 by $(y_1 y)$ and $4ax = 2a(x + x)$ by $2a(x + x_1)$ to obtain Equation 13. In other words, our rule of thumb: 'half' the x's in the equation have become x_1, while 'half' the y's have become y_1, works for the parabola as well as for the circle.

Let P be a point on the parabola. Then by Equation 1 with $e = 1$, we have FP = PN where PN is perpendicular to the directrix.

> Given that P is the point (x, y), calculate $(PN)^2$ and $(FP)^2$.
>
> ---
>
> PN = $x + a$ so that $(PN)^2 = (x + a)^2$; and, using the expression for the distance between two points (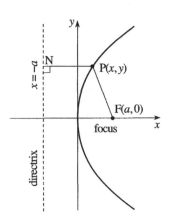), we have
>
> $$(FP)^2 = (x - a)^2 + (y - 0)^2 = (x - a)^2 + y^2 \quad \square$$

From the definition of a parabola we know that FP = PN so that $(FP)^2 = (PN)^2$ and therefore $(x - a)^2 + y^2 = (x + a)^2$ which simplifies to

---

$$y^2 = 4ax \tag{12}$$

This is the standard equation of the parabola.

Tangents and normals to the parabola

We can find the equation of the tangent to a parabola in much the same way as we did for the circle, by constructing a line which touches the curve at one point. We will spare you the details and quote the result.

The tangent at a point $P(x_1, y_1)$ on the parabola (Equation 12) has the equation ( )

$$y_1 y = 2a(x + x_1) \tag{13}$$

You may omit the following question if you are not familiar with calculus.

> Differentiate Equation 12 and hence verify that Equation 13 represents the tangent at a point $P(x_1, y_1)$ on the parabola.
>
> ---
>
> We may differentiate Equation 12 implicitly with respect to x to obtain
>
> $$2y \frac{dy}{dx} = 4a \quad \text{which can be rearranged to give} \quad \frac{dy}{dx} = \frac{2a}{y},$$
>
> and at the point $P(x_1, y_1)$ the gradient of the parabola is therefore
>
> $$\frac{2a}{y_1}.$$
>
> The tangent line at P has the same slope as this gradient and so its equation is

Figure 10.10 The parabola.

$$y - y_1 = \frac{2a}{y_1}(x - x_1).$$

Multiplying by y_1 we obtain $y_1 y - y_1^2 = 2ax - 2ax_1$.

Since P lies on the parabola we have $y_1^2 = 4ax_1$ so that the above equation becomes $y_1 y - 4ax_1 = 2a(x - x_1)$ which simplifies to $y_1 y = 2a(x + x_1)$ as required. □

You do not need to use calculus to answer the following question.

Question T4

(a) What are the equations of the tangent and the normal to the parabola $8y^2 = x$ at the points $(\frac{1}{2}, \frac{1}{4})$ and $(\frac{1}{2} - \frac{1}{4})$? (*Hint*: Use Equation 13.)

(b) Show that the point M $(a/m^2, 2a/m)$ lies on the parabola $y^2 = 4ax$ for any value of $m \neq 0$, and then use Equation 13 to show that the line

$$y = mx + \frac{a}{m} \qquad (14)$$

is a tangent to the parabola $y^2 = 4ax$, and hence the gradient of the parabola at M is m. □

Which of the following lines are tangents to the parabola $y^2 = 8x$?

(a) $y = 2x + 2$　(b) $y = 3x + 2/3$　(c) $y = kx + k^2$

The parameter a in Equation 12 is 2 since $4a = 8$.

(a) $m = 2$, $c = 2$ and $a/m = 1$, is not a tangent because $c \neq a/m$.

(b) $m = 3$, $c = \frac{2}{3}$ and $a/m = \frac{2}{3}$, is a tangent because $c = a/m$.

(c) $m = k, c = k^2$ and $a/m = 2/k$ and $c = a/m$ provided that $k^2 = 2/k$ which will be so if $k^3 = 2$ so that $k = \sqrt[3]{2}$. For any other values of k the line will not be a tangent. □

Parametric equations for the parabola

You may recall that the use of a parameter enabled us to trace out the points of a circle in a particular order, and we may do something very similar for a parabola. If we introduce the **parametric equations**

$$x = at^2 \quad \text{and} \quad y = 2at \qquad (15)$$

then both x and y are determined by the value of the parameter t. If we use the second of these equations to write $t = y/(2a)$ and then substitute for t in the first equation, we obtain $x = a(y/(2a))^2$, which reduces to $y^2 = 4ax$. This means that any point $(x, y) = (at^2, 2at)$ specified by Equation 15 must lie on the parabola defined by Equation 12. Moreover, each point on the parabola corresponds to one and only one value of t.

Many results relating to the parabola can be derived via the parametric representation. For example, consider two points P and Q corresponding to two values t_1 and t_2 of the parameter. The line passing through the two points P $(at_1^2, 2at_1)$ and Q $(at_2^2, 2at_2)$ has the equation

$$\frac{y - 2at_1}{2at_2 - 2at_1} = \frac{x - at_1^2}{at_2^2 - at_1^2}$$

which can be simplified to

$$2x - (t_1 + t_2)y + 2at_1t_2 = 0$$

If the two points P and Q coincide, i.e. if $t_1 = t_2 = t$ say, then this line becomes a tangent to the parabola at the point $(at^2, 2at)$ with equation

$$2x - 2ty + 2at^2 = 0 \quad \text{which simplifies to} \quad y = \frac{x}{t} + at$$

so that the gradient of the parabola at $(at^2, 2at)$ is $1/t$.

(This final equation is identical to Equation 14 except that m has been replaced by $1/t$.)

You may omit the following question if you are unsure of calculus.

What is the value of dy/dx in terms of t, at a general point $(at^2, 2at)$ on the parabola?

Since $dx/dt = 2at$ and $dy/dt = 2a$ it follows (from the chain rule) that

$$\frac{dy}{dx} = \frac{dy}{dt} \Big/ \frac{dx}{dt} = \frac{2a}{2at} = \frac{1}{t}. \quad \square$$

You do not need to use calculus to answer the following question.

Find the equation of the normal to the parabola at the point $(at^2, 2at)$.

The normal is perpendicular to the tangent at the point of contact and so the product of their gradients is -1. Hence the gradient of the normal is $-t$; the equation of the normal is therefore $y - 2at = -t(x - at^2)$ which simplifies to

---

$$y + tx = 2at + at^3 \qquad\qquad \square \quad (16)$$

Question T5

A line passing through the focus of the parabola $y^2 = 4ax$ cuts the parabola at the points $P(at_1^2, 2at_1)$ and $Q(at_2^2, 2at_2)$. Find a relationship between t_1 and t_2. ☐

Question T6

The tangents to the parabola $y^2 = 4ax$ at $P(at_1^2, 2at_1)$ and $Q(at_2^2, 2at_2)$ meet at a point R. What are the coordinates of R? ☐

Reflection property of the parabola

Parabolas have the interesting and useful property, which is illustrated in Figure 10.11, that parallel rays of light striking a parabolic mirrored surface are all reflected through a fixed point − the focus of the parabola.

A distant source of light will produce a nearly parallel beam, and a receiver placed at the focus of the parabolic reflector can collect a strong signal. Receiving dishes, such as those used in radio astronomy or to recover television signals from a satellite, are usually in the shape of a **paraboloid**, which is the surface obtained by rotating a parabola about its axis of symmetry.

The property can also be used in reverse. Domestic electric fires are often provided with parabolic reflectors and the heating element is placed at the focus of the reflecting surface. Heat rays (mainly infra-red radiation) then emerge in parallel and the arrows in Figure 10.11 would be reversed.

Figure 10.11 Reflection property of a parabola.

Give an example in which this geometric property is used to produce parallel rays of light.

Examples include a motor car headlamp, a torch, a theatre spotlight or a searchlight. ☐

Alternative forms of the parabola

Just as for the circle, the equation of a parabola that arises from our work in physical science may not be exactly like Equation 12. As a trivial example, the x and y may be swapped over to give $y = x^2/4a$, and a change of origin will then lead to Equation 11 (which is probably the most common form of the equation of a parabola).

The polar form of a parabola

$$L/r = 1 + \cos \theta \quad \text{for } -\pi < \theta < \pi \tag{17}$$

where L is a constant length, occurs quite often in the discussion of central orbits. We will not discuss this form in detail, but the following exercise is intended to convince you that the equation does indeed represent a parabola.

Table 10.3 See Question T7

$\theta/°$	r/cm
0	2
±20	2.1
±40	2.3
±60	2.7
±80	3.4
±90	4.0
±100	4.8

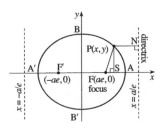

Figure 10.12 The ellipse.

Question T7

Table 10.3 shows the values of r for various values of θ (measured in degrees) obtained from the equation $L/r = 1 + \cos \theta$ where $L = 4$ cm. Plot these values on the polar graph paper provided in Figure 10.9. Use your graph to decide which of the following equations is equivalent to Equation 17 in this case.

(a) $(y - 4)^2 = -x/2$ (b) $(y - 4)^2 = x/2$

(c) $y^2 + 16 = x/2$ (d) $8y = -(x - 4)^2$

(e) $y^2 - 16 = -8x$ (f) $y = -2(x + 4)^2$ □

10.2.4   The ellipse

Our discussion for the ellipse is very similar to that for the parabola. Again we begin with Equation 1, but this time with $0 \leqslant e < 1$. Once again we choose the axes and the origin so as to produce the simplest form of the equation, and our choice is shown in Figure 10.12.

Let us suppose that a is an arbitrary positive number, then the focus is chosen at the point $(ae, 0)$ while the directrix is the line $x = a/e$.

From Equation 1 we have

$$\frac{FN}{PN} = e \tag{1}$$

and from the right-angled triangle FPS we have

$$FP^2 = FS^2 + PS^2$$

so that $(e \times PN)^2 = (x - ae)^2 + y^2$

But $PN = a/e - x$ and so $e \times PN = a - ex$,

therefore $(a - ex)^2 = (x - ae)^2 + y^2$

which gives $a^2 - 2aex + e^2x^2 = x^2 - 2aex + a^2e^2 + y^2$

or $a^2 - a^2e^2 = a^2 (1 - e^2) = (1 - e^2) x^2 + y^2$

Therefore $(1 - e^2)x^2 + y^2 = a^2(1 - e^2)$

Dividing both sides of this final equation by $a^2 (1 - e^2)$ we obtain

$$\frac{x^2}{a^2} + \frac{y^2}{a^2(1 - e^2)} = 1$$

Noting that $a^2 (1 - e^2) > 0$ we may replace it by b^2

so that $b^2 = a^2(1 - e^2)$

and $b = a\sqrt{1 - e^2} \tag{18}$

to produce the standard equation of an ellipse

$$\frac{x^2}{a^2} + \frac{y^2}{b^2} = 1 \tag{19}$$

What is the equation of the standard ellipse which passes through the points $(-2, 0)$ and $(1, \sqrt{3}/2)$?

Substituting $(-2, 0)$ into Equation 19 we obtain $4/a^2 = 1$ i.e. $a = 2$. Substituting

$(1, \sqrt{3}/2)$ into Equation 19 gives us

$$\frac{1}{a^2} + \frac{3}{4b^2} = 1 \quad \text{i.e.} \quad \frac{1}{4} + \frac{3}{4b^2} = 1$$

so that $b = 1$.

If we put $y = 0$ in Equation 19 we obtain $x^2/a^2 = 1$, so that $x = \pm a$ and the ellipse meets the x-axis at the points $A'(-a, 0)$ and $A(a, 0)$ of Figure 10.12.

If we put $x = 0$ in Equation 19 we obtain $y^2/b^2 = 1$, so that $y = \pm b$ and the ellipse meets the y-axis at the points $B'(0, -b)$ and $B(0, b)$ of Figure 10.12.

With this choice of axes, the ellipse is symmetric both about the x-axis and about the y-axis; and because of this latter symmetry there is a second focus $F'(-ae, 0)$ and a second directrix $x = -a/e$ which could have been used to draw the same ellipse.

Any chord passing through O is called a diameter, but, unlike those of a circle, the diameters of the ellipse are not all of the same length. The **major axis** is defined to be the longest diameter AA', while the **minor axis** is defined to be the shortest diameter BB'. If we put $y = 0$ in Equation 19 we obtain $x = \pm a$, so that the major axis is $2a$. Similarly putting $x = 0$ in Equation 19 we have $y = \pm b$ and the minor axis is $2b$.

For an ellipse we require $0 \leqslant e < 1$, but how does the shape of the ellipse change as we vary e within this range?

Remember that $b^2 = a^2(1 - e^2)$. When $e = 0$ then $b = a$ and we get a circle. Notice that as e decreases to zero the directrix $x = a/e$ moves further and further from the origin (until finally it is undefined when $e = 0$). Also as e decreases to zero the foci $(\pm ae, 0)$ get closer to the origin, which is then the centre of the circle. The closer e is to 1 then the smaller is $1 - e^2$ and so the ratio b/a is smaller, and the ellipse departs more from a circle to a squashed oval shape. □

You may well encounter situations in physical science where the relationships between the focus, eccentricity and directrix are all-important, as for example when you discuss the motion of the planets about the Sun, or the path of an α-particle near the nucleus of an atom.

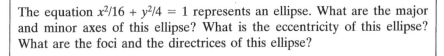

The equation $x^2/16 + y^2/4 = 1$ represents an ellipse. What are the major and minor axes of this ellipse? What is the eccentricity of this ellipse? What are the foci and the directrices of this ellipse?

Comparing the given equation with Equation 19 we see that $a = 4$ and $b = 2$ so the major axis is 8 and the minor axis is 4. Substituting these values for a and b into Equation 18 we obtain

$$2 = 4\sqrt{1 - e^2}$$

which gives $e = \sqrt{3}/2$. The foci are at $(\pm ae, 0) = (\pm 2\sqrt{3}, 0)$ while the directrices are $x = \pm a/e = \pm 8/\sqrt{3}$. □

Aside   From Equation 18 and the fact that $0 \leqslant e < 1$ it is clear that $b \leqslant a$ in the standard equation of the ellipse. Actually, Equation 19 with this condition $b \leqslant a$ would perhaps be more accurately described as 'the standard equation of an ellipse with its foci on the x-axis'. If we throw away the condition $b \leqslant a$ then Equation 19 is still the equation of an ellipse, but the foci may lie on the y-axis. For example, $x^2/4 + y^2/16 = 1$ is certainly the equation of an ellipse, with major axis 8 (in the direction of the y-axis) and minor axis 4 (in the direction of the x-axis) but its foci lie on the y-axis, rather than on the x-axis. In Equation 19 the greater of a and b is known as the **semi-major axis** while the lesser of a and b is known as the **semi-minor axis**.

In Figure 10.12 we have shown an ellipse with its major axis along the x-axis. You would need to turn Figure 10.12 through 90° in order to find the foci and directrices of an ellipse for which $b > a$.

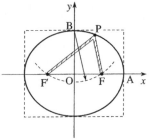

Figure 10.13 Drawing an ellipse with pins, string and pencil.

It is clear from Figure 10.13 that an ellipse is a bounded closed figure lying in a rectangle $-a \leqslant x \leqslant a$ and $-b \leqslant y \leqslant b$. This figure shows a simple method of constructing an ellipse inside a given rectangle. First, measure the length OA (the semi-major axis) then find where a circle centred at B with this length as radius meets the major axis. These two points are the foci F′ and F. Now place a loop of thread round pins through F′ and F and, keeping the thread taught, trace out the curve (so that the length F′PF is fixed).

Tangents and normals to the ellipse

The equation of the tangent to the ellipse defined by Equation 19, at a point P(x_1, y_1) on the ellipse, is

$$\frac{x_1 x}{a^2} + \frac{y_1 y}{b^2} = 1 \qquad\qquad (☜)\ (20)$$

Notice that we have effectively replaced x^2 and y^2 in Equation 19 by $x_1 x$ and $y_1 y$ respectively, and our rule of thumb still works. (See the margin note in Subsection 10.2.3)

This result is established in the next exercise, which you may omit if you are unsure of calculus.

Show that Equation 20 is the equation of the tangent to the ellipse at a point $P(x_1, y_1)$ on the ellipse.

Let $P(x_1, y_1)$ be an arbitrary point on the ellipse, then we differentiate Equation 19 implicitly to obtain

$$\frac{2x}{a^2} + \frac{2y}{b^2}\frac{dy}{dx} = 0 \quad \text{which gives us} \quad \frac{dy}{dx} = \frac{-b^2x}{a^2y}.$$

It follows that at $P(x_1, y_1)$ the gradient of the ellipse is

$$\frac{-b^2x_1}{a^2y_1},$$

and therefore the equation of the tangent at P is

$$y - y_1 = \frac{-b^2x_1}{a^2y_1}(x - x_1)$$

so that $\quad \dfrac{y_1y}{b^2} - \dfrac{y_1^2}{b^2} = \dfrac{-x_1x}{a^2} + \dfrac{x_1^2}{a^2}$

and hence $\quad \dfrac{y_1y}{b^2} + \dfrac{x_1x}{a^2} = \dfrac{x_1^2}{a^2} + \dfrac{y_1^2}{b^2}.$

However

$$\frac{x_1^2}{a^2} + \frac{y_1^2}{b^2} = 1,$$

because P lies on the ellipse, and this gives Equation 20. $\quad\square$

Write down the equation of the normal to the ellipse

$$\frac{x^2}{a^2} + \frac{y^2}{b^2} = 1$$

at the point (x_1, y_1) on the ellipse.

From Equation 20 the slope of the tangent at (x_1, y_1) is $-b^2x_1/a^2y_1$. So the slope of the normal is a^2y_1/b^2x_1. The equation of the normal at (x_1, y_1) is therefore

$$y - y_1 = \frac{a^2y_1}{b^2x_1}(x - x_1). \quad\square$$

Basic geometry

Notice that this is an example of an ellipse with its major axis along the *y*-axis.

Question T8

What are the equations of the tangent and the normal to the ellipse $x^2/4 + y^2/16 = 1$ at the points (☞)

(a) $(1, 2\sqrt{3})$ (b) $(1, -2\sqrt{3})$ (c) $(-1, 2\sqrt{3})$

(d) $(2, 0)$ (e) $(0, -4)$ ☐

Parametric equations for an ellipse

We introduce parametric equations

$$x = a \cos \theta \quad \text{and} \quad y = b \sin \theta \tag{21}$$

for the ellipse for the same reasons as for the circle and the parabola. It is clear that the point $P(a \cos\theta, b \sin\theta)$ lies on the ellipse (Equation 19) for any value of θ, since

$$\frac{(a \cos \theta)^2}{a^2} + \frac{(b \sin \theta)^2}{b^2} = \cos^2 \theta + \sin^2 \theta = 1$$

Each value of θ in the range $0 \leqslant \theta < 360°$ (or $-180° < \theta \leqslant 180°$) corresponds to one and only one point on the ellipse: see Figure 10.14. If we choose an arbitrary point $x_1 = a \cos \theta_1$, $y_1 = b \sin \theta_1$ on the ellipse, then Equation 20 becomes

$$\frac{\cos \theta_1}{a} x + \frac{\sin \theta_1}{b} y = 1 \tag{22}$$

and this is very often the most useful form of the equation of a tangent to the ellipse.

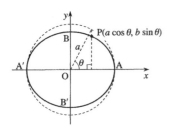

Figure 10.14 Parametric representation of an ellipse.

Equation 22 gives the equation of the tangent to the ellipse at the point corresponding to choosing $\theta = \theta_1$. Find the equation of the tangent line at the point corresponding to $\theta = \theta_1 + 90°$.

The equation of the tangent at the point corresponding to $\theta = \theta_1 + 90°$ is obtained by replacing θ_1 by $\theta_1 + 90°$ in Equation 22 which gives

$$\frac{\cos (\theta_1 + 90°)}{a} x + \frac{\sin (\theta_1 + 90°)}{b} y = 1$$

But $\cos(\theta_1 + 90°) = -\sin \theta_1$ and $\sin(\theta_1 + 90°) = \cos \theta_1$ so the equation of the tangent line becomes

$$\frac{-\sin \theta_1}{a} x + \frac{\cos \theta_1}{b} y = 1 . \quad ☐$$

Question T9

Show that the tangents to the ellipse

$$\frac{x^2}{a^2} + \frac{y^2}{b^2} = 1$$

at the points with parameters θ_1 and $\theta_1 + 90°$ meet at the point $x = a(\cos \theta_1 - \sin \theta_1)$, $y = b(\cos \theta_1 + \sin \theta_1)$ and show that this point lies on the ellipse

$$\frac{x^2}{a^2} + \frac{y^2}{b^2} = 2 . \quad \square$$

Alternative forms of the ellipse

The parabola, ellipse and hyperbola are of special interest because they are the paths which are followed by bodies which move under the so-called **inverse square law**. This occurs when the net force on a body is directed towards a fixed point, and is inversely proportional to the square of the distance between the body and the fixed point. Projectiles (in the absence of air resistance) travel in parabolic paths, planets move around the Sun in ellipses, and, in a famous experiment by Geiger and Marsden, α-particles were deflected by nuclei along hyperbolae.

The polar equation of a conic is

$$L/r = 1 + e \cos \theta \tag{23}$$

where L is a fixed length, and for $0 \leqslant e < 1$ the conic is an ellipse (we have already seen the case $e = 1$ in Subsection 2.3 when we discussed the parabola). The following exercise is intended to convince you that Equation 23 is indeed the equation of an ellipse.

Johannes Kepler (1571–1630) discovered that the planets move round the Sun in elliptical orbits with the Sun at one focus, a result which formed the basis for Newton's mathematical model of planetary motion.

The old British £1 note had a picture of an ellipse on the reverse side, with the Sun in the wrong place! It was at the origin rather than at one of the foci.

By using the range $\theta \leqslant e < 1$ we are including the circle $(e = d)$ as a special case

Question T10

Table 10.4 shows the values of r for various values of θ obtained from the equation $L/r = 1 + e \cos \theta$ where $L = 2.5$ cm and $e = 0.5$. Plot these values on the polar graph paper provided in Figure 10.9. Is the major axis of the ellipse parallel to the x-axis or to the y-axis? $\square$

Table 10.4 See Question T10

$\theta/°$	r/cm	$\theta/°$	r/cm
0	1.7	±110	3.0
±30	1.7	±130	3.7
±50	1.9	±150	4.4
±70	2.1	±175	5.0
±90	2.5	±180	5.0

10.2.5   The hyperbola

Our discussion for the hyperbola is very similar to that for the ellipse. Again we begin with Equation 1, but this time with $e > 1$. Once again we choose the axes and the origin so as to produce the simplest form of the equation, and our choice is shown in Figure 10.15.

Let us suppose that a is an arbitrary positive number, then the focus is chosen at the point $(ae, 0)$ while the directrix is the line $x = a/e$.

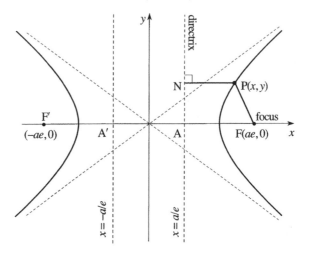

Figure 10.15
The hyperbola.

From Equation 1 we have $(FP)^2 = e^2(PN)^2$

so that $\quad (x - ae)^2 + (y - 0)^2 = e^2 \left(x - \dfrac{a}{e} \right)^2$

expanding gives us $\quad x^2 - 2aex + a^2e^2 + y^2 = e^2x^2 - 2aex + a^2$

factorising gives us $\quad (e^2 - 1)\, x^2 - y^2 = a^2\, (e^2 - 1)$

Hence $\quad\quad\quad \dfrac{x^2}{a^2} - \dfrac{y^2}{a^2(e^2 - 1)} = 1$

Since $a^2(e^2 - 1) > 0$ we may replace it by b^2, where b is a constant, to obtain the standard form of the hyperbola:

$$\frac{x^2}{a^2} - \frac{y^2}{b^2} = 1 \qquad\qquad (\text{✆})\ (24)$$

As with the ellipse it follows from the symmetry that there is a second focus and a second directrix which will produce the same curve. The two foci are the points $F(ae, 0)$ and $F'(-ae, 0)$ and the two directrices have equations $x = a/e$ and $x = -a/e$.

Notice from Figures 10.4d and 10.15 that the hyperbola has two distinct branches: this is quite different from an ellipse which is always a closed curve.

We can readily verify for Equation 24 that when $y = 0$, $x = \pm a$ (giving the points A' and A in Figure 10.15) but there is no value of y for which $x = 0$; indeed, since Equation 24 can be rearranged to give the equation

$$\frac{y^2}{b^2} = \frac{x^2 - a^2}{a^2}$$

in which the left-hand side is positive; it follows that $x^2 \geqslant a^2$ so there is no part of the curve in the band $-a < x < a$.

Notice the similarity to Equation 19, and the vital difference: the minus sign on the left-hand side of Equation 24.

Asymptotes

There is one feature of the hyperbola that is not present in any of the other curves we have considered. The dashed lines in Figure 10.15 correspond to the equations

$$y = \pm \frac{b}{a} x \qquad (25)$$

The branches of the hyperbola are 'hemmed in' by these lines as x and y both become large in magnitude. The hyperbola approaches one of these lines ever more closely but does not touch or cut it. These lines are called **asymptotes**, and the hyperbola is the only conic section which possesses them. (The general concept of an asymptote was introduced in Chapter 3, using the hyperbola as an example.)

The asymptotes are a useful aid in sketching a hyperbola.

What is the equation of the hyperbola which passes through the points $(-4, 0)$ and $(2\sqrt{5}, 1)$? What are the equations of its asymptotes?

Substituting $(-4, 0)$ into Equation 24 we obtain $16/a^2 = 1$, so that $a = 4$.

For $(2\sqrt{5}, 1)$ the resulting equation is

$$\frac{20}{16} - \frac{1}{b^2} = 1 \quad \text{so that} \quad \frac{1}{b^2} = \frac{1}{4} \quad \text{and} \quad b = 2 .$$

Hence the equation is

$$\frac{x^2}{16} - \frac{y^2}{4} = 1 ,$$

and from Equation 25 the asymptotes have equations

$$y = \pm \frac{2}{4} x , \quad \text{i.e.} \quad y = \pm x/2 . \quad \square$$

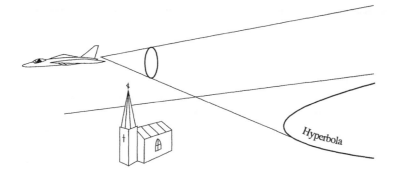

Figure 10.16 The cause of a sonic boom.

The 'boom' from a supersonic jet is the result of a cone shaped 'shock wave' that trails behind the jet. At the surface of this cone there is a sudden change of pressure which causes the sonic boom. The intersection of the cone with the ground is (usually) a branch of a hyperbola, and the strength of the bang decreases as the distance from the aircraft increases. See Figure 10.16.

Tangents and normals to the hyperbola

The derivation of the equation of the tangent is almost identical to that for the ellipse. You may omit the following question if you are unsure of calculus, but notice that Equation 26 is the equation of a tangent to the hyperbola of Equation 24 at the point (x_1, y_1).

Find the equation of the tangent line at a point $P(x_1, y_1)$ on the hyperbola defined by Equation 24.

Differentiating Equation 24 implicitly we obtain

$$\frac{2x}{a^2} - \frac{2y}{b^2}\frac{dy}{dx} = 0 \quad \text{so that} \quad \frac{dy}{dx} = \frac{b^2x}{a^2y}$$

To find the tangent at $P(x_1, y_1)$ on the hyperbola we first note that the gradient at this point is b^2x_1/a^2y_1. The equation of the tangent at P is therefore

$$y - y_1 = \frac{b^2x_1}{a^2y_1}(x - x_1)$$

which gives us

$$\frac{x_1x}{a^2} - \frac{y_1y}{b^2} = \frac{x_1^2}{a^2} - \frac{y_1^2}{b^2} = 1$$

The equation of the tangent line at a point $P(x_1, y_1)$ on the hyperbola is therefore

---

$$\frac{x_1x}{a^2} - \frac{y_1y}{b^2} = 1 \quad \square$$

(✎) (26)

Notice that once again x^2 and y^2 in Equation 24 have effectively been replaced by x_1x and y_1y, respectively, to produce Equation 26. Our rule of thumb applies in this case also.

Question T11

Find the equations of the tangent and normal to the hyperbola $9x^2 - 4y^2 = 36$ at the point $(4, 3\sqrt{3})$. $\square$

Parametric equations for a hyperbola

The *hyperbolic functions* cosh and sinh are defined by

$$\cosh x = \frac{e^x + e^{-x}}{2} \quad \text{and} \quad \sinh x = \frac{e^x - e^{-x}}{2} \tag{27}$$

where e^x is the exponential function.

We need to know very little about these functions in this chapter, except for the fact that they satisfy the following useful identity.

Show that $\cosh^2 \theta - \sinh^2 \theta = 1$

$$\cosh^2 \theta - \sinh^2 \theta = \left(\frac{e^x + e^{-x}}{2}\right)^2 - \left(\frac{e^x - e^{-x}}{2}\right)^2$$

$$= \left(\frac{e^{2x} + 2 + e^{-2x}}{4}\right) - \left(\frac{e^{2x} - 2 + e^{-2x}}{4}\right) = 1 \quad \square$$

Show that the point $x = a \cosh \theta$, $y = b \sinh \theta$ lies on the hyperbola of Equation 24.

$$\frac{x^2}{a^2} - \frac{y^2}{b^2} = \cosh^2 \theta - \sinh^2 \theta = 1 \quad \square$$

We may now use the equations

$$x = a \cosh \theta \quad \text{and} \quad y = b \sinh \theta \tag{28}$$

to parametrise the hyperbola because Equation 24 is satisfied for any value of θ. The functions cosh and sinh are known as hyperbolic functions precisely because of this association with the hyperbola. Just as for the ellipse, we can show that the tangent to the hyperbola at $\theta = \theta_1$ has equation

$$\frac{\cosh \theta_1}{a} x - \frac{\sinh \theta_1}{b} y = 1 \tag{29}$$

The polar equation of a hyperbola is Equation 23

$$L/r = 1 + e \cos \theta \tag{23}$$

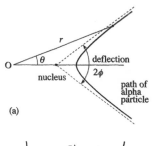

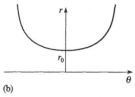

(a)

(b)

Figure 10.17 (a) The path of an alpha particle near the nucleus of an atom. (b) An upper estimate for the size of the nucleus.

where L is a fixed length, and $e > 1$ (we have already seen the cases $e = 1$ and $0 < e < 1$ when we discussed the parabola and the ellipse).

A famous example from physics

When an α-particle passes close to the nucleus of an atom (of gold, say) a certain mathematical model based on the inverse square law predicts that its path will be a hyperbola. We imagine an alpha particle heading along a straight line towards the nucleus, and then being deflected through an angle, 2ϕ say, which we can measure, as in Figure 10.17a. The paths along which the particle approaches and retreats from the nucleus lie close to the asymptotes of a hyperbola, and when r is very large we have $\theta \approx \pm \phi$ so that, from Equation 23, we have

$$1 + e \cos \phi = 0 \quad \text{and therefore} \quad e = -\frac{1}{\cos \phi}$$

and Equation 23 becomes $\quad r = \dfrac{L \cos \phi}{\cos \phi - \cos \theta}$

where ϕ and L are constants to be determined by experiment.

Figure 10.17b shows a typical graph of r plotted against θ (not a conic) but the significant point to notice is that there is a minimum value for r, say r_0. We conclude that this minimum value of r provides an upper estimate for the size of the nucleus. A similar calculation was first performed by Lord Rutherford (1871–1937), the discoverer of the nucleus, in the early years of the 20th century.

Rectangular hyperbola

A special case arises when $a = b$, for then Equation 24 reduces to

$$x^2 - y^2 = a^2 \tag{30}$$

and the asymptotes (from Equation 25) are $y = x$ and $y = -x$ which are perpendicular to each other. Such a curve is known as a **rectangular hyperbola**.

The following alternative form of the rectangular hyperbola arises in many applications:

$$y = \frac{c^2}{x} \tag{31}$$

Such a rectangular hyperbola has the x- and y-axes as its asymptotes. Equation 31 can be obtained from Equation 30 by rotating the axes about the origin through 45°, but the details are beyond the scope of this chapter.

Show that $x = ct$, $y = c/t$ is a parametrisation of the rectangular hyperbola $xy = c^2$.

If $x = ct$ and $y = c/t$ then $xy = (ct)(c/t) = c^2$. $\quad\square$

Find the equation of the line joining the points on the rectangular hyperbola $(ct_1, c/t_1)$ and $(ct_2, c/t_2)$.

The slope of the line is $\dfrac{(c/t_2) - (c/t_1)}{ct_2 - ct_1} = -\dfrac{1}{t_1 t_2}$

The equation of the line is

$$y - \frac{c}{t_2} = -\frac{1}{t_1 t_2}(x - ct_2)$$

which gives us $\quad t_1 t_2 y + x - c(t_1 + t_2) = 0$.

The result of the previous exercise can be used to find the equation of the tangent to the hyperbola at the point $x = ct$, $y = c/t$. Putting $t_1 = t_2 = t$, so that the two points coincide, we obtain

$$t^2 y + x - 2tc = 0 \quad \square \tag{32}$$

Question T12

Find the equations of the tangents to the hyperbola

$$\frac{x^2}{a^2} - \frac{y^2}{b^2} = 1$$

that pass through the point $S(a/2, 0)$. $\quad \square$

10.3 **The general form of a conic**

10.3.1 *Classifying conics*

In this short section we mention very briefly some topics which do not arise very often in physics, but nevertheless might cause you some difficulties on the rare occasions they do occur if you have never seen them before.

It can be shown that the general equation

$$Ax^2 + 2Hxy + By^2 + 2Gx + 2Fy + C = 0 \tag{33}$$

includes all the different types of conics: circles, ellipses, parabolas and hyperbolas. We state without proof that the following conditions determine the nature of the conic:

(a) $H = 0$, $A = B \neq 0$ is a circle (☞)   See Equation 8 with $H = 0$.

(b) $H^2 = AB$ is a parabola

(c) $H^2 < AB$ is an ellipse

(d) $H^2 > AB$ is a hyperbola

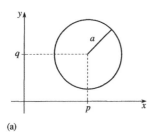

(a)

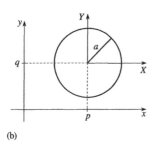

(b)

Figure 10.18 Translation of axes.

What type of conic is defined by the equation

$$3x^2 + 4xy + 3y^2 - 8x + 2y + 4 = 0 \ ?$$ (34)

In this example $H = 2$, $A = B = 3$ so that $H^2 = 4$ and $AB = 9$. Since $H^2 < AB$ the curve is an ellipse. □

Equation 34 represents an ellipse, but not the sort of ellipse with which we are familiar. Very roughly speaking, the xy term indicates that the ellipse has undergone a *rotation* so that its major and minor axes no longer coincide with the coordinate axes. The linear x and y terms indicate that the centre of the ellipse is not at the origin of coordinates so that the ellipse has undergone a *translation*.

Question T13

To which class of conic does each of the following belong?

(a) $5x^2 + 6xy - 4y^2 + 2x + 4y - 3 = 0$

(b) $2x^2 + 3xy + 2y^2 + 2 = 0$ □

10.3.2 Changing the axes

Let us consider the equation $(x - p)^2 + (y - q)^2 = a^2$. If we draw a graph of this for particular values of p, q, and a, we get a circle, with centre (p, q) and radius a (Figure 10.18a). Now if we make the substitutions $X = x - p$, and $Y = y - q$ we get $X^2 + Y^2 = a^2$, and this amounts to drawing the figure in a new coordinate system referred to as X- and Y-axes, whose origin is at the point (p, q) (Figure 10.18b).

What is the equation of an ellipse centred at the point (2, 4) with major axis 10 parallel to the x-axis and minor axis 6 parallel to the y-axis?

If this shape of ellipse was centred at (0, 0) its equation would be

$$\frac{x^2}{5^2} + \frac{y^2}{3^2} = 1$$

since $a = \frac{1}{2} \times 10$ and $b = \frac{1}{2} \times 6$. To effect the transformation we replace x by $x - 2$ and y by $y - 4$. The resulting equation is

$$\frac{(x - 2)^2}{25} + \frac{(y - 4)^2}{9} = 1$$

or written out in full $9x^2 + 25y^2 - 36x - 200y + 211 = 0$. □

Question T14

What is the effect on the standard equations of the ellipse and the hyperbola if they are moved so that they are centred at the point $(-1, 2)$ with the directions of the major and minor axes unchanged? □

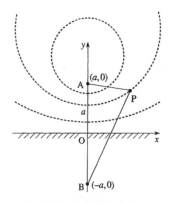

Figure 10.19 The equipotentials due to a line charge at A and a plane surface at zero potential along the x-axis.

10.3.3 An example from electrostatics: equipotentials

You do not need to understand the physics in this example, but we will outline the physical background in order to set it in context. You have probably encountered static electricity when you comb your hair, or after walking on a nylon carpet. Lightning is probably the best known, and certainly the most impressive, manifestation of the phenomenon. Figure 10.19 represents an electrical charge distributed along a very long straight line at A$(0, a)$ above a very large flat copper plate represented by the x-axis. Since we are viewing 'edge on', all we see of the line charge is a point (namely A) and similarly we see only the edge of the metal plate (along the x-axis). We suppose that the plate is earthed or, more technically, we say that it is at zero potential. The line charge and the metallic surface combine to create an electric field, and a small point charge will experience a force due to this field. The magnitude and direction of this force depend on where the point charge is placed. Through each chosen point there is a curve, called an *equipotential*, with the property that no work is required to move a point charge along the curve. In many respects the equipotentials are similar to the contour curves on a map showing points of equal height above sea level, in that relatively little effort is required to walk along such paths. The force experienced by a small charge is always perpendicular to the equipotential through the location of the charge.

In Figure 10.19 the dashed lines represent the equipotentials, and B$(0, -a)$ is the image of the point A$(0, a)$ in the flat surface (as in a mirror). It is known that the equipotentials have the property that AP/PB $= \lambda$, a constant. Different choices of the value of λ will produce the various equipotentials. In this example we shall investigate the nature of these curves.

First we suppose that P has coordinates (x, y). Then, since $(AP)^2 = \lambda^2(PB)^2$ we have, for a particular value of λ,

$$x^2 + (y - a)^2 = \lambda^2 [x^2 + (y + a)^2]$$

which can be rearranged to give

$$x^2 + y^2 - 2ay \frac{1 + \lambda^2}{1 - \lambda^2} + a^2\lambda^2 = 0 \tag{35}$$

This equation corresponds to putting $H = 0$ and $A = B = 1$ in Equation 33, and we see immediately that the equipotentials are circles. If we compare Equation 35 with Equation 8 we see that $G = 0$ and $F = -a(1 + \lambda^2)/(1 - \lambda^2)$, so that the centre of the circle is at $(0, a(1 + \lambda^2)/(1 - \lambda^2))$ on the y-axis. For $\lambda = 1$ we have AP $=$ PB and the equipotential lies along the x-axis (see Figure 10.19). For $\lambda < 1$ the equipotentials are circles in the upper half-plane. For small values of λ, P is close to A, so that the equipotential is a circle which is close to A. In the case $\lambda = 0$ the point P is actually at A and the equipotential reduces to a single point. (The case $\lambda > 1$ is of no interest since it gives rise to circles which lie below the x-axis, and these do not correspond to equipotentials.)

10.4 **Closing items**

10.4.1 *Chapter summary*

1. *Conic sections*, or *conics*, are the curves which arise when a *double cone* is cut by a plane. The main classes are *parabolas*, *ellipses* and *hyperbolas*; the circle is a special case of the ellipse.

2. Table 10.5 (see p. 299) summarises the main results.

3. The general equation of a conic has the form

    $$Ax^2 + 2Hxy + By^2 + 2Gx + 2Fy + C = 0.$$

    (a) $H = 0, A = B \neq 0$ a circle

    (b) $H^2 = AB$ a parabola

    (c) $H^2 < AB$ an ellipse

    (d) $H^2 > AB$ a hyperbola

    When the axes are translated (or rotated) then it is possible to obtain the equation of a conic relative to the new coordinates.

10.4.2 *Achievements*

Having completed this chapter, you should be able to:

A1. Define the terms that are emboldened in the margins of the chapter.

A2. Recognise the graphs of the main types of conic sections, write down their equations in standard form, and explain how the value of the eccentricity governs the nature of the conic.

A3. Appreciate that a tangent to a curve is a line that touches the curve at one point, and use the equations of the tangents to the conics.

A4. Use the parametric representations of the conics.

A5. Appreciate that a circle is a particular case of an ellipse, and that ellipses are the only bounded conics.

Table 10.5 Conic sections and their properties

Property	Circle	Parabola	Ellipse	Hyperbola
Eccentricity	$e = 0$ see Figure 4a	$e = 1$ see Figure 4b	$0 \leqslant e < 1$ see Figure 4c	$e > 1$ see Figure 4d
Focus Directrix		$(a, 0)$ $x = -a$	$(\pm ae, 0)$ $x = \pm a/e$	$(\pm ae, 0)$ $x = \pm a/e$
Standard equation	$x^2 + y^2 = a^2$ centre $(0, 0)$ radius a	$y^2 = 4ax$	$\dfrac{x^2}{a^2} + \dfrac{y^2}{b^2} = 1$ $b = a\sqrt{1 - e^2}$	$\dfrac{x^2}{a^2} - \dfrac{y^2}{b^2} = 1$ $b = a\sqrt{e^2 - 1}$
Asymptotes	none	none	none	$y = \pm \dfrac{b}{a} x$
Parametric form	$x = a \cos \theta$ $y = a \sin \theta$	$x = at^2$ $y = 2at$	$x = a \cos \theta$ $y = b \sin \theta$	$x = a \cosh \theta$ $y = b \sinh \theta$
Tangent at (x_1, y_1)	$x_1 x + y_1 y = a^2$	$y_1 y = 2a(x + x_1)$	$\dfrac{x_1 x}{a^2} + \dfrac{y_1 y}{b^2} = 1$	$\dfrac{x_1 x}{a^2} - \dfrac{y_1 y}{b^2} = 1$
Polar form	$r = a$, centre $(0, 0)$, radius a $r = 2a \cos \theta$, centre $(a, 0)$, radius a	$\dfrac{L}{r} = 1 + \cos \theta$	$\dfrac{L}{r} = 1 + e \cos \theta$ $0 \leqslant e < 1$	$\dfrac{L}{r} = 1 + e \cos \theta$ $e > 1$
Some other forms and special cases	$(x - p)^2 + (x - q)^2$ $= a^2$ centre (p, q) radius a $x^2 + y^2 + 2Gx + 2Fy$ $+ C = 0$ centre $(-G, -F)$ radius $\sqrt{G^2 + F^2 - C}$	$y = Ax^2 + Bx + C$		rectangular hyperbola $x^2 - y^2 = a^2$ rectangular hyperbola $y = \dfrac{k}{x}$

A6. Determine the equations of the asymptotes of a hyperbola in standard form.

A7. Recognise the various forms of the equations of the conics, in particular a circle centred at the point (p, q), the general equation of a circle, a rectangular hyperbola and especially a rectangular hyperbola whose asymptotes are the coordinate axes.

A8. Recognise the polar forms of the conics.

A9. Recognise the general form of a conic.

10.4.3   End of chapter questions

Question E1   Classify the following conics as circles, ellipses, parabolas or hyperbolas:

(a) $y^2 + (x - 1)^2 = 4$ (b) $x(x + y) = 2$

(c) $4x^2 + 9y^2 = 1$ (d) $r = 2\cos\theta$

(e) $\dfrac{2}{r} = 1 + 5\cos\theta$ (f) $r = \dfrac{1}{1 + \cos\theta}$

(g) $xy = 4$ (h) $x^2 + y^2 + x + y = 0$

Question E2

(a)   Show that the line $x + 3y = 1$ is a tangent to the circle $x^2 + y^2 - 3x - 3y + 2 = 0$ and find the coordinates of the point of contact. Use Equation 8 to find the centre and radius of this circle.

(b)   Prove, by calculation, that the point P(3, 2.5) lies outside the circle and calculate the length of a tangent to the circle from P.

Question E3   The normal to the parabola $y^2 = 4ax$ at the point $P(at^2, 2at)$ meets the x-axis at the point G, and GP is extended, beyond P, to the point Q so that PQ = GP. Show that Q lies on the curve $y^2 = 16a(x + 2a)$. What kind of curve is this?

Question E4   Find the equations of the tangents to the ellipse $x^2 + 2y^2 = 8$ which are parallel to the asymptotes of the hyperbola $4x^2 - y^2 = 1$.

Part 3

Basic vector algebra

Introducing scalars and vectors

Chapter 11

11.1 **Opening items**

11.1.1 *Chapter introduction*

Many of the quantities that physical scientists use are easy to specify; measurements of mass, length, time, area, volume and temperature can all be expressed as simple numbers together with appropriate units of measurement. Such quantities are known as *scalar quantities* or *scalars*. However, some quantities are more tricky to deal with. For instance, if you want to travel from one place to another you will not only want to know how far apart the two places are, you will also need to know the direction that leads from one to the other. The physical quantity that combines distance and direction is called *displacement* and is clearly more complicated than distance alone. Displacement is a simple example of a large class of physical quantities known collectively as *vector quantities* or *vectors*. A detailed understanding of vectors and how to use them is crucial to many parts of physical science, particularly mechanics and electromagnetism.

   This chapter provides an introduction to the mathematical treatment of vectors. Section 11.2 gives a definition of vector quantities that separates them from simpler scalar quantities – such as mass and distance – that are also defined. This section also describes the graphical representation of vectors and the notation used to distinguish vectors from scalars, both in print and in handwritten work. Section 11.3 introduces some of the basic operations of *vector algebra* such as *scaling* (multiplication by a scalar), *vector addition* and *vector subtraction*. It aims to make clear the meaning of vector equations such as $3a - 2b = 0$. The section ends with a discussion of the way in which a given vector may be split-up (*resolved*) into *component vectors* – a process that is of importance in many practical problems.

   Throughout the chapter the emphasis is on basic ideas and geometric (graphical) methods. The algebraic methods that are more frequently used by those already familiar with vectors are mentioned briefly in a short conclusion (Section 11.4) but their full development is left to other chapters. Since it is those methods that are best suited to tackling three-dimensional problems, most of the questions in this chapter are restricted to two-dimensional situations on idealised planes or flat surfaces. However, even though the discussion is mainly restricted to two dimensions, the significance of vectors is amply demonstrated as is the importance of always using vector notation clearly and correctly.

11.1.2 Fast track questions

Question F1

In terms of a strict interpretation of vector and scalar quantities, what is wrong with each of the following statements?

(a) The velocity of light travelling through a vacuum (usually denoted by c) is approximately 3×10^8 m s^{-1}.

(b) The acceleration due to gravity (usually denoted by g) near the surface of the Earth is approximately 10 m s^{-2}.

(c) The mass of your head pushes down on your neck.

Question F2

a is a displacement of 4.32 km due north, and b is a displacement of 2.40 km due east.

(a) What is the magnitude and the direction of the displacement $3a - 5b$?

(b) Given a specified line, any vector may be expressed as the sum of two (orthogonal) *component vectors*, one parallel to the given line and the other at right angles to that line. What is the magnitude of the (orthogonal) component vector of $3a - 5b$ that points in the direction 20° west of north?

11.1.3 Ready to study?

Study comment

Relatively little background knowledge is required to study this chapter. Terms such as *mass*, *electric charge* and *temperature* are used and it is assumed that you know something about measuring such quantities using SI units, but you do not need to know precise definitions or exact meanings. It is also assumed that you have met *Newton's laws of motion* before, but again you do not need to know those laws in detail in order to study the chapter. (Indeed, this chapter would be good preparation for a thorough study of Newton's laws.) There are some mathematical topics, however, with which you will need to be familiar before starting to study; they are the subject of the following questions.

Question R1

Draw a set of *two-dimensional Cartesian axes*. Label the horizontal axis x and the vertical axis y. Using the axes you have drawn, plot the three points A, B and C, the *coordinates* of which are given in Table 11.1. Draw the *right-angled triangle* that has the points A, B and C at its corners. Write down the lengths of sides AB and BC of that triangle, and use *Pythagoras's theorem* to work out the length of the side CA.

Question R2

Using the triangle given in the answer to Question R1 (Figure 11.15, in the Answers and comments section), and indicating the length of side CA by c and the angle $\hat{CAB}$ by θ, write down expressions for the lengths of sides AB and BC in terms of the length c, the angle θ, and the basic *trigonometric ratios*. What is the value (in degrees) of the angle θ?

Question R3

The *modulus* of a quantity x is written $|x|$ and represents the *absolute value* of x. Evaluate the following expressions and complete the *equations*.

(a) $|3| =$ (b) $|-3| =$

(c) $|(-2)(3.1)| =$ (d) $|(-2.4)/(2)| =$

11.2   Scalars and vectors

11.2.1   Scalars and scalar quantities

Many important physical quantities, such as length, mass and temperature, can be completely specified by a single number together with an appropriate unit of measurement. For instance, it makes perfectly good sense to say that the length of an object is 1.42 m or that the mass of an object is 12.2 kg. Quantities that can be specified in this simple and straightforward way are called *scalar quantities*. Thus:

Scalar quantities are physical quantities that can be completely specified by a single number together with an appropriate unit of measurement.

Different values of a given scalar quantity may be easily added, subtracted, multiplied or divided. In fact, all the elementary operations of arithmetic apply to values of a scalar quantity just as they do to ordinary numbers. So, if a, b and c are three values of a given scalar quantity (three masses, say) then we know that

$$a + b = b + a \tag{1}$$

$$ab = ba \tag{2}$$

$$a + 0 = a \tag{3}$$

Point	x-coordinate	y-coordinate
A	1	1
B	−3	1
C	−3	4

Table 11.1   See Question R1

The brackets show which operations to carry out first.

$$a \times 1 \;=\; a \tag{4}$$

$$(a + b) + c \;=\; a + (b + c) \tag{5}$$

$$a(bc) \;=\; (ab)c \tag{6}$$

$$a(b + c) \;=\; ab + ac \tag{7}$$

Scalar quantities are often referred to simply as **scalars**. There is nothing very surprising about the behaviour of scalars; their properties seem obvious and, indeed, they are obvious. What is much more interesting is that there are many physical quantities that are *not* scalars.

11.2.2 Vectors and vector quantities

Figure 11.1 shows the relative locations of Belfast, Edinburgh and Exeter, three of the major cities in the UK. If you were asked to describe the location of Edinburgh relative to Exeter what would you say? You might start by stating that Edinburgh is 588 km from Exeter, but that would not answer the question. Although you would have defined the *distance* from Exeter to Edinburgh you would still not have fully defined the relative *position*. To provide a full description of the location you have to give the distance *and* the direction.

   The physical quantity that describes the location of one point (Edinburgh, say) with respect to another point (Exeter, say) is called **displacement**. The displacement from Exeter to Edinburgh is 588 km in the direction due north.

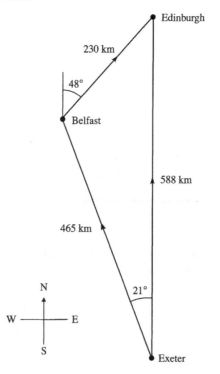

Figure 11.1 An idealised map showing the relative locations of Belfast, Edinburgh and Exeter.

> What is the displacement from Edinburgh to Exeter?
>
> 588 km, due south.   ☐

Displacement, unlike distance, cannot be specified by a number together with an appropriate unit of measurement; an additional ingredient is required – a direction. Of course, the direction can be specified in a number of different ways, but its presence is unavoidable.

Question TI

What is the displacement:

(a)   from Exeter to Belfast?

(b)   from Belfast to Edinburgh?   ☐

It is important to note that displacements are *entirely* specified by *two* items of information:

- A distance (this is called the **magnitude** of the displacement and must be a non-negative scalar quantity). (☞)
- A direction (in a two-dimensional plane this is often specified in terms of compass bearings, but other methods can be used; in three dimensions other methods *must* be used).

Magnitudes are often described as positive quantities, but they may be zero, so we prefer to describe them as *non-negative*.

The two end-points of the displacement do not enter into this specification and are immaterial to the displacement itself. Thus, a displacement of 588 km due north starting from Exeter is the same displacement as one of 588 km due north from Belfast, even though one ends in Edinburgh and the other in the sea. In fact, any two displacements are equal if they have the same magnitude and the same direction.

Question T2

Which of the following displacements is equal to the displacement from Edinburgh to Belfast?

(a)   A displacement of magnitude 230 km, 48° east of north.

(b)   A displacement of magnitude −230 km, 48° west of south.

(c)   A displacement of magnitude 230 km, 42° south of west.   ☐

Displacements are just one example of a large class of physical quantities known as *vector quantities*. The common characteristics of all vector quantities can be summarised as follows:

Vector quantities are physical quantities that can be completely specified by a magnitude and a direction.

Vector quantities are also common in areas of physical science as diverse as fluid dynamics, electromagnetism and quantum physics. In fact they arise in all parts of physics.

The word 'unbalanced' is included here because bodies that are *not* accelerating may be acted upon by several opposing forces that cancel each other out.

Note that the general 'magnitude' referred to here is always a non-negative scalar quantity, just as distance (the magnitude of a displacement) must always be a non-negative scalar quantity. The magnitude of any vector quantity can be thought of as the 'size' or 'length' of that vector.

Apart from displacement, other important vector quantities include *velocity*, *acceleration*, and *force*. You may not be familiar with all of these quantities, so a few words about each should help to define it and to underline its vector nature.

The **velocity** of an object is a vector quantity that describes how fast the object is moving *and* the direction in which it is travelling. The *magnitude* of the velocity is called the **speed** and is a non-negative scalar quantity. The *direction* of the velocity is the direction of motion.

The **acceleration** of an object is a vector quantity that describes the rate of change of the object's velocity. Acceleration is a vector quantity because it involves a *change* in a vector quantity, velocity, and such a change requires both a magnitude and a direction for its complete specification. For example, a car travelling in a fixed direction at ever increasing speed is accelerating because the *magnitude* of its velocity is changing, whereas a car travelling in a circle at a fixed speed is accelerating because the *direction* of its velocity is changing. Any change in the magnitude or the direction of the velocity, or any combination of the two, constitutes an acceleration. Similarly, the rate of change of any other vector quantity must be a vector quantity.

According to *Newton's laws of motion*, when an object accelerates it does so because an (unbalanced) **force** acts upon it (☞). Roughly speaking, force is the 'push' or 'pull' that causes (or tends to cause) acceleration. *Newton's second law* asserts that the acceleration of a body of fixed mass is proportional to the force that acts upon it. We know that the 'effect' of a force depends on its direction as well as its strength, so force is a vector. The 'effect' of a force, according to Newton, is an acceleration, which is also a vector. More specifically, the magnitude of the acceleration is proportional to the magnitude of the force and the direction of the acceleration is the same as the direction of the force.

Question T3

Which of the following are vector quantities? Energy, distance, time, electric charge, rate of change of acceleration, altitude above sea level?   ☐

Just as scalar quantities are often referred to as scalars, so vector quantities are often called **vectors**. It is possible to draw a distinction between vectors and vector quantities on the one hand and scalars and scalar quantities on the other. Vector and scalar quantities are *physical* quantities while vectors and scalars are the mathematical entities that are used to represent those physical quantities. Although it is important to distinguish between real things and their mathematical representations we shall not pursue that distinction here. In what follows the terms vector and vector quantity will be used interchangeably, as will scalar and scalar quantity.

Question T4

How would you counter the following assertions (both of which are wrong)?

(a) The velocity of a particular object (such as my car) cannot be a vector since it is not completely specified by a magnitude and a direction – its complete specification also requires the identification of the moving object (my car).

(b) Displacement cannot be a vector since its complete specification requires the definition of the various compass bearings (or something similar) and not merely the statement of a magnitude and a direction. ☐

11.2.3 Representing scalars and vectors

Everybody knows how to represent a scalar quantity such as mass or length when writing equations or drawing diagrams; just choose an appropriate letter or symbol to represent the quantity concerned (m or l, say) and write it down wherever the scalar quantity is to be represented. It is taken for granted that the chosen symbol indicates the appropriate units of measurement as well as the number of those units, so it makes good sense to write statements such as:

If length $l = 6$ m, then $l^2 = 36$ m^2

Unfortunately, dealing with vectors is a good deal more difficult since directions are involved as well as magnitudes. When drawing diagrams the problem is not too bad since, as in Figure 11.1, a vector can be represented by a line labelled with the appropriate magnitude and pointed in the appropriate direction. The direction of the line is usually indicated by an arrowhead at the end of the line or by a pointer somewhere along the line. For this reason, such lines are usually referred to as *arrows* or **directed line segments**.

Figure 11.2 shows five points O, P, Q, R and S, located in a two-dimensional plane. The positions of the five points can be specified by their *Cartesian coordinates* (x, y) with respect to the axes x and y shown in the figure. (Note that all measurements are expressed in metres.) Also included in the figure is a directed line segment from P to R. This represents the displacement from P to R.

One obvious way of representing the displacement from P to R symbolically is to write it as $\overrightarrow{PR}$. This is a clear if somewhat cumbersome notation that may easily be applied to other displacements.

Question T5

Using Figure 11.2:

(a) Find the magnitude of $\overrightarrow{PR}$.

(b) Describe the direction of $\overrightarrow{PR}$.

(c) Find the magnitude of $\overrightarrow{OQ}$. ☐

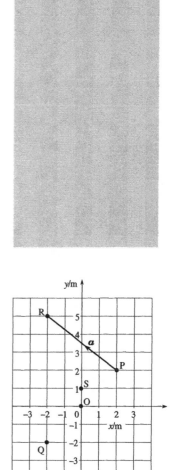

Figure 11.2 A representation of the locations of the five points O, P, Q, R and S in a two-dimensional plane.

Another way of representing vectors in printed material is to symbolise them by bold-face type, thus the bold a in Figure 11.2 is yet another way of representing the displacement $\overrightarrow{PR}$. Indeed, it makes sense to write

$$a = \overrightarrow{PR}$$

This new symbol (the bold a) does not have any obvious link with the points P and R, which may be regarded as an advantage or a disadvantage. However, one very definite bonus is that emboldening is applicable to all kinds of vector quantities (velocities, accelerations, forces, etc.) and not just displacements. The bold-typeface convention is the most widely used method of representing vectors and will be used throughout this book and its companion volume.

Distinguishing vectors by bold-face type is fine in print, but it presents a problem for those using pens or pencils. How can *you* show that a is a vector when *you* write it down? Fortunately there is an easy answer to this problem. When authors are preparing material to be printed they show that an item should be set in bold-face type by putting a wavy underline beneath it. Thus, what is written as $\underset{\sim}{a}$ will be printed as a. You are encouraged to adopt this convention for distinguishing vectors so that you will never make the mistake of writing down a nonsensical equation such as

$$\underset{\sim}{a} = 5 \text{ m}$$

Why is this equation nonsensical?

$\underset{\sim}{a}$ is a vector, it has both magnitude and direction. 5 m is simply a scalar, it has no 'direction'. It makes no sense to equate a quantity that has direction with one that does not. □

Question T6

In Figure 11.2 the vectors b, c, d and e are defined as follows:

$$b = \overrightarrow{PS}, \ c = \overrightarrow{OS}, \ d = \overrightarrow{QO}, \ e = \overrightarrow{QR}$$

(a) Using a pencil, add directed line segments representing each of these vectors to Figure 11.2.

(b) Label each of the directed line segments with the appropriate vector symbol. □

Another advantage of the bold-typeface convention for vectors is that it provides a natural way of representing the magnitude of a vector. Given a vector a, its magnitude is conventionally denoted $|a|$. So, it makes sense to write

$$|a| = 5 \text{ m} \quad \text{(In a manuscript: } |\underset{\sim}{a}| = 5 \text{ m)}$$

even though it was meaningless to write $a = 5$ m.

Sometimes, when there is no risk of confusion, the magnitude of a vector is simply represented by the symbol used for the vector itself, but without the emboldening. Thus, you might see

$$a = |a| = 5 \text{ m, or simply } a = 5 \text{ m}$$

This is obviously much less bother to write, but it is easy to forget that a, being a magnitude, must be non-negative, whereas you are unlikely to forget that something as complicated as $|a|$ *cannot* be negative.

Question T7

Using Figure 11.17 from the Answer to T6, p. 626, (or your copy of Figure 11.2 having completed Question T6) and working to three significant figures, evaluate the following quantities:

(a) $|\vec{RS}|$ (b) $|b|$ and $|c|$ (c) $|b| + |c|$ and $|b||c|$

(d) $|a| + |b| + |c| + |d| + |e|$ (e) $2|a| + \dfrac{|b||c|}{|d|}$ ☐

The main result of this subsection may be summarised as follows:

By convention, bold-face symbols such as a are used to represent vector quantities. The magnitude of such a vector is represented by $|a|$ and is a non-negative scalar quantity. In diagrams, vectors are represented by directed line segments. In handwritten work, vectors are distinguished by a wavy underline ($\underset{\sim}{a}$) and the magnitude of a vector is denoted $|\underset{\sim}{a}|$.

11.3 Introducing vector algebra

Scalar quantities can be added, subtracted, multiplied or divided. So, it is quite possible to form equations and carry out all the normal processes of algebra with scalar quantities. (Finding unknown scalar quantities was probably the first use you ever made of algebra, even if the word 'scalar' wasn't used at the time!) Working with vectors is more difficult because it involves directions as well as magnitudes. Nonetheless, it is possible to make sense of algebraic expressions such as $a + b$ or $3a - 2b$ and to formulate and solve vector equations. This section introduces the basic concepts of **vector algebra** − a subject that is more fully developed in other chapters.

11.3.1 Scaling vectors

The simplest operation of vector algebra is the *scaling* of a vector, that is the multiplication of a vector by a scalar. For example, if v is a velocity of 15 m s^{-1} due north then $2v$ is a velocity of 30 m s^{-1} in the same direction or, to give another example, if a is a displacement of 50 m due east then $-3a$ is a displacement of 150 m due west. Note that in this last example the inclusion of the minus sign has had the effect of *reversing* the direction. Some more examples of scaling are shown in Figure 11.3.

It should be noted that $-1a$ is usually written $-a$ and that $1.5a$ could equally well be written $(1.5)a$ or even $a(1.5)$, so the order of the multiplication is unimportant. Nonetheless, it is more conventional to use the form $1.5a$ since this is more suggestive of 'taking 1.5 lots of a'.

In general, if a is a vector and α is a scalar (which may be positive or negative) then αa, the result of **scaling** a by α, is a vector of magnitude $|\alpha||a|$ that points in the same direction as a if $\alpha > 0$ and points in the opposite direction to a if $\alpha < 0$.

When dealing with vectors, some authors use the term **parallel** to mean 'in the same direction' and the term **antiparallel** to mean 'in the opposite direction'. Thus, the vectors a and $1.5a$ may be described as *parallel*, while a and $(-1.5)a$ are *antiparallel*. Unfortunately the use of antiparallel, though widespread, is not universal. So, whenever you see the term 'parallel' you should take it to mean 'parallel *or* antiparallel' unless the more specialised interpretation is clearly indicated.

Question T8
If a body of mass $m = 6$ kg has velocity v of magnitude 5 m s^{-1} in the direction north-west, what is the magnitude and direction of mv? (In Newtonian mechanics this vector quantity is known as the **momentum** of the body.)　□

Question T9
If a is a vector and α is a non-zero scalar, classify each of the following statements as always true, always false or sometimes true and sometimes false. Explain your answers.

(a)　$|\alpha a|$ cannot be less than zero.

(b)　$\alpha^2 a$ is in the same direction as a.

(c)　$-\alpha a$ is antiparallel to a.

(d)　Provided $|a|$ is not zero,

$$\hat{a} = \frac{a}{|a|}$$

is a vector of unit magnitude in the same direction as a.

(e)　$\left|\dfrac{a}{\alpha}\right| = \dfrac{a}{|\alpha|}$

(f)　$|\alpha a| = |\alpha||a|$　□

Remember, if α is a scalar its modulus, $|\alpha|$, is a non-negative quantity, equal to α itself if $\alpha \geqslant 0$ and $-\alpha$ if $\alpha < 0$. Do not confuse the modulus $|\alpha|$ with the magnitude $|a|$.

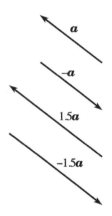

Figure 11.3　A vector a and three of its scaled counterparts $-1a$, $1.5a$ and $-1.5a$.

11.3.2 Adding and subtracting vectors

Two vector quantities of the same type (two displacements say, or two velocities) may be added together to produce another vector quantity of the same type. This operation is called **vector addition**. It is straightforward, though quite different from the everyday operation of (scalar) addition with which you are familiar.

Before examining the general rules for vector addition let's look at an example to see how the process works in practice. Figure 11.4 shows the points O, P, Q, R and S that we considered earlier – we shall use these points to provide the example we need. Imagine yourself to be located at Q, the lowest of the five points, and suppose that you undergo a displacement $\overrightarrow{QO}$. Where would you find yourself? Obviously, at O. Now suppose you undergo another displacement, $\overrightarrow{OS}$. What would your new location be after this second displacement? You would find yourself at point S. So, the overall result of the two successive displacements $\overrightarrow{QO}$ and $\overrightarrow{OS}$ would be to move you from Q to S. But, of course, such an effect could also have been produced by the single displacement $\overrightarrow{QS}$. Thus, it makes sense to write

$$\overrightarrow{QO} + \overrightarrow{OS} = \overrightarrow{QS} \tag{8}$$

The left-hand side of Equation 8 contains a new kind of quantity – *the sum of two vectors*. Thus, Equation 8 is an example of vector addition and the vector $\overrightarrow{OS}$ on the right-hand side is said to be the **(vector) sum** or **resultant** of $\overrightarrow{QO}$ and $\overrightarrow{OS}$.

Question T10

Complete the following (vector) sums.

(a) $\overrightarrow{OS} + \overrightarrow{SP} =$ (b) $\overrightarrow{PS} + \overrightarrow{SR} =$

(c) $(\overrightarrow{PS} + \overrightarrow{SR}) + \overrightarrow{RQ} =$ (d) $\overrightarrow{PS} + (\overrightarrow{SR} + \overrightarrow{RQ}) =$ ☐

So far, all the examples of vector addition that we have considered have involved vectors that are 'nose-to-tail' like $\overrightarrow{OS}$ and $\overrightarrow{SP}$ but, remember, a vector is completely specified by its magnitude and direction; the end-points are immaterial. Points such as O, P, Q, R and S merely provide a convenient way of specifying vectors; the vectors themselves could just as easily be represented by bold-face letters, such as the u and v shown in Figure 11.5 together with their sum w. Moreover, any vector that has the same magnitude and direction as v is equal to v, irrespective of its endpoints, and could be used in place of v in the equation

$$u + v = w \tag{9}$$

Figure 11.6 shows just such a vector, t. Even though t is shown in a different location from v, the fact that $t = v$ means that it is correct to write:

$$u + t = w \tag{10}$$

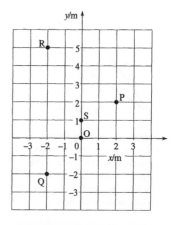

Figure 11.4 The points O, P, Q, R, S and their locations in the (x, y) plane.

Remember, brackets show which operations should be performed first.

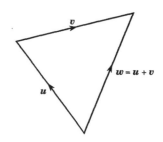

Figure 11.5 Two vectors, u and v, together with their sum $w = u + v$.

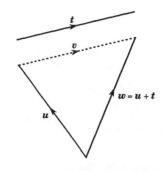

Figure 11.6 The vectors t and v are equal, so $w = u + t$.

So, what is the *general* rule for adding two vectors? The answer is illustrated graphically in Figure 11.7 and may be summarised in the following *triangle rule for vector addition.*

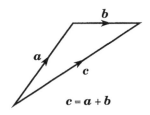

Figure 11.7 The triangle rule for adding vectors. Note the directions of the arrows; the diagram would be incorrect if any one, or any two, of the arrows were reversed.

The **triangle rule** for adding vectors:

Let vectors a and b be represented by appropriate arrows (or directed line segments). If the arrow representing b is drawn from the head of the arrow representing a, then an arrow from the tail of a to the head of b represents the vector sum $a + b$, marked c in Figure 11.7.

Question T11

Figure 11.8 shows three pairs of vectors; A, B; C, D and E, F. Sketch and label some simple diagrams showing how the triangle rule can be used to find the following vector sums: (a) $A + B$, (b) $C + D$, (c) $E + F$. □

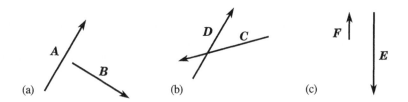

Figure 11.8 See Question T11.

An alternative but equivalent method of adding vectors graphically is provided by the *parallelogram rule*. This has no real advantages over the triangle rule, but it is preferred by some authors. It is illustrated in Figure 11.9 and may be stated as follows.

The **parallelogram rule** for adding vectors:

Let vectors a and b be represented by appropriate arrows (or directed line segments). If the arrows representing a and b are drawn from a common point O so that they form two sides of a parallelogram, when the parallelogram is completed an arrow from O along the diagonal of the parallelogram represents the vector sum $a + b$.

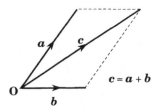

Figure 11.9 The parallelogram rule for adding vectors.

Despite a lengthy discussion of vector addition, nothing has yet been said about vector subtraction. The time has come to remedy that deficiency.

Given two vectors of the same type, such as the vectors a and b shown in Figure 11.10a, you already know how to add them together to form their sum $a + b$ (Figure 11.10b). You also know how to scale the vector b by the number -1 to produce the vector $-b$ that has the same magnitude as b but

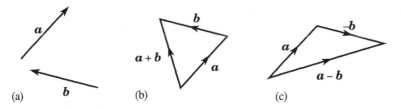

Figure 11.10 (a) Two vectors, (b) their sum, and (c) their difference.

points in the opposite direction. You should therefore be able to convince yourself that Figure 11.10c correctly represents the following summation

$a + (-b)$

This quantity is more usually written as

$a - b$

and is called the *vector difference* of a and b.

The **vector difference** $a - b$ of the vectors a and b is the vector sum of a and $-b$.

Question T12

In Question T11 you were asked to sketch diagrams showing the sums of various pairs of vectors (shown in Figure 11.8). Now sketch similar diagrams for the differences of the same pairs: (a) $A - B$, (b) $C - D$, (c) $E - F$. □

How does the vector $A - B$ in Question T12 differ from the vector $B - A$?

$A - B = -(B - A)$, so the two vectors have equal magnitude but point in exactly opposite directions. □

One final point needs to be discussed before leaving the subject of vector addition and subtraction – the *zero vector*. It has already been stressed that a vector (which has direction and magnitude) cannot be equal to a scalar (which does not have any direction). However, if we consider the result of a vector subtraction such as $a - a$ it's obvious that the result has zero magnitude and does not point in any specific direction. So, how should we symbolise this quantity and how should we refer to it? For the sake of consistency we certainly want to call it a vector so it makes sense to dub it the **zero vector**. As far as its symbolic representation is concerned, different authors take different attitudes. Many simply represent it by 0 as you might expect, but others prefer to use a bold-face zero, **0**, to emphasise its vector nature. In this book we will adopt this somewhat more formal convention, but don't be surprised if you see the other elsewhere.

> The above discussion of the zero vector has not provided a concise definition of $\mathbf{0}$. Write down an equation that *defines* the zero vector.
>
> ---
>
> $\boldsymbol{a} + \mathbf{0} = \boldsymbol{a}$ for any vector $\boldsymbol{a}$.   □

Now that you know about the zero vector you should be able to carry out algebraic manipulations with vectors. Try the following exercise.

> Rearrange the following vector equation to express $\boldsymbol{c}$ in terms of $\boldsymbol{a}$ and $\boldsymbol{b}$:
>
> $$3\boldsymbol{a} - 4\boldsymbol{b} + 2\boldsymbol{c} = \mathbf{0}$$
>
> ---
>
> Subtracting $(3\boldsymbol{a} - 4\boldsymbol{b})$ from each side gives
>
> $$3\boldsymbol{a} - 4\boldsymbol{b} + 2\boldsymbol{c} - (3\boldsymbol{a} - 4\boldsymbol{b}) = \mathbf{0} - (3\boldsymbol{a} - 4\boldsymbol{b})$$
>
> It follows that, $2\boldsymbol{c} = 4\boldsymbol{b} - 3\boldsymbol{a}$
>
> Thus, $\boldsymbol{c} = 2\boldsymbol{b} - \frac{3}{2}\boldsymbol{a}$.   □

<aside>
When you are more familiar with vector algebra you should be able to do some of these steps in your head.
</aside>

Question T13

An angler sitting on a riverbank is watching a duck paddling across the river. The water in the part of the river in which the duck is paddling is moving at velocity $\boldsymbol{v}$ relative to the angler. The duck is moving at velocity $\boldsymbol{u}$ relative to the water.

(a)   What is the velocity of the duck according to the angler?

(b)   What is the velocity of the angler according to the duck?   □

<aside>
In this question you are asked to consider the same situation from different reference points.
</aside>

11.3.3   Resolving vectors

Imagine a ball released from rest on a perfectly smooth inclined plane, as shown in Figure 11.11a. What will happen to the ball immediately after its release? Obviously, the ball will start to move down the plane, accelerating as it does so. Anyone familiar with Newton's laws of motion would say that the acceleration of the ball must be caused by a force pointing down the plane. But what is the origin of the force causing that acceleration? The only 'downward' force that acts on the ball is its **weight**, $\boldsymbol{W}$ − the force that arises from the action of gravity on the ball's mass − and that force acts vertically downwards, *not* parallel to the plane. So, where does the accelerating force come from?

Happily, vector addition provides a simple answer. The weight $\boldsymbol{W}$ of the ball can be regarded as the sum of two other forces as shown in Figure 11.11b, and we can write $\boldsymbol{W} = \boldsymbol{F}_1 + \boldsymbol{F}_2$. The force $\boldsymbol{F}_1$ that is parallel to the plane causes the acceleration, while the force $\boldsymbol{F}_2$ that is normal (i.e. at right

<aside>
The physical 'reality' of $\boldsymbol{F}_1$ and $\boldsymbol{F}_2$ is of no importance here; all that matters is that the vector sum of $\boldsymbol{F}_1$ and $\boldsymbol{F}_2$ is equal to $\boldsymbol{W}$. Note that although Figure 11.11b shows $\boldsymbol{W}$, $\boldsymbol{F}_1$ and $\boldsymbol{F}_2$, the effective force is represented by either $\boldsymbol{W}$ or $\boldsymbol{F}_1 + \boldsymbol{F}_2$.
</aside>

angles) to the plane stops the ball from leaving the plane and accounts for the difference between W and F_1.

This process of splitting a given vector into constituent parts at right angles to each other is called **(orthogonal) resolution**; we speak of *resolving* the vector into its **(orthogonal) component vectors** along the chosen directions. The process of orthogonal resolution is not restricted to the directions shown in Figure 11.11; a vector in a two-dimensional plane can be resolved into orthogonal component vectors along *any* two mutually perpendicular directions in that plane. (In three dimensions it is always possible to choose *three* mutually perpendicular directions, and any given vector in three dimensions can be correspondingly resolved into *three* orthogonal component vectors.) Problems in which a given vector has to be resolved into (orthogonal) component vectors are very common, so it is worth knowing how to perform such resolutions.

The sort of problem you might be asked to solve is illustrated in Figure 11.12. A vector a is given, and a line AB, inclined at an angle θ to a, is specified. The problem is to resolve a into two component vectors, one parallel to AB, the other normal to AB. To solve the problem, just use the parallelogram rule (in the case of orthogonal component vectors it's really a 'rectangle rule'). Construct a rectangle like the one shown in Figure 11.12, with a as its diagonal and one side parallel to AB. Call the component vectors parallel and normal to AB, respectively, a_p and a_n. Applying basic trigonometry to the rectangle, you should then be able to see that the magnitudes of the two orthogonal component vectors are:

$$|a_p| = |a| \cos \theta \tag{11}$$

$$\text{and, } |a_n| = |a| \sin \theta \tag{12}$$

With this information you should be able to find orthogonal component vectors for yourself, though you may need to think carefully about their directions.

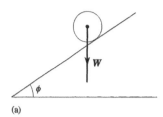

(a)

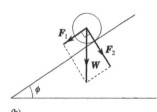

(b)

Figure 11.11 (a) A ball of weight W, released from rest on an inclined plane. (b) The component vectors of W, parallel and normal to the plane.

Note that the angle θ has been chosen in such a way that it is in the range $0° \leqslant \theta \leqslant 90°$.

Question T14

If the weight of the ball in Figure 11.11 is of magnitude 5 N and the angle of inclination of the plane (ϕ) is 30°, what are the magnitudes and directions of the orthogonal component vectors F_1 and F_2? (*Hint*: Pay attention to the definition of ϕ.) ☐

Question T15

An aeroplane is flying on a northwesterly course at 500 km h^{-1}. What are the component vectors of the plane's velocity parallel and normal to a line running north–south? ☐

In principle, the process of resolution is not restricted to situations in which the chosen directions are perpendicular and the component vectors are at right angles to each other. In two dimensions a vector may be resolved into component vectors along *any* two directions that are not parallel, and in three dimensions a vector may be resolved into component vectors along any three

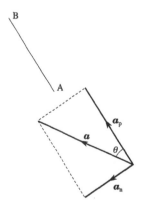

Figure 11.12 A given vector a and a line AB. The vector is to be resolved into component vectors parallel (a_p) and normal (a_n) to the line AB.

directions that do not all lie in the same plane. The process of carrying out such resolutions is similar to that outlined above, though the mathematics is a little more complicated. Fortunately, such non-orthogonal resolutions are relatively rare in elementary work. It was in order to emphasise this restriction to mutually perpendicular directions that we made such widespread use of the word 'orthogonal' in this subsection. However, since orthogonal resolution is so common we do not follow this practice elsewhere. Generally speaking, if you are given a vector and asked to find its component vector along some line you should assume that it is the *orthogonal* component vector along that line which is required unless you are given a clear indication to the contrary.

Important note

Pay close attention to the term *component vector*. As you pursue the study of vectors you will soon encounter *scalar* quantities referred to as the **components** of a vector. Like component vectors, these (scalar) components of a given vector can be used to specify that vector. Nonetheless, (scalar) components are different from (though related to) the component vectors we have just been discussing. Take care not to confuse the (scalar) components of a vector with its component vectors. In this book the two quantities will always be clearly distinguished, but other authors may not be so careful.

11.4 **Working with vectors**

The methods that have been used in this chapter have been largely graphical. Such methods provide a good basis for the introduction of fundamental concepts but they are not well suited to the everyday business of working with vectors. Methods of a more algebraic nature are really needed for that purpose. Such methods exist and are introduced in the next chapter.

Apart from simplifying calculations, the algebraic methods also have the advantage of making it almost as easy to treat three-dimensional problems involving directions in space as it is to treat two-dimensional problems confined to a plane. The use of vectors in describing and analysing three-dimensional situations makes them one of the most valuable mathematical tools at the disposal of the physical scientist.

11.5 **Closing items**

11.5.1 *Chapter summary*

1. *Scalar quantities* (or *scalars*) can be completely specified by a single number together with an appropriate unit of measurement.

2. *Vector quantities* (or *vectors*) can be completely specified by a magnitude and a direction. In diagrams, vectors are usually represented by arrows or *directed line segments*. In print, vectors are distinguished by bold-face type, and in a manuscript, vectors are indicated by a wavy underline.

3. The *magnitude* of a vector is a non-negative scalar that represents the 'length' or 'size' of that vector. The magnitude of a is shown as $|a|$ in print or $|\underline{a}|$ in a manuscript.

4. Any vector a may be multiplied by a scalar α to produce a *scaled* vector αa which points in the same direction as a if $\alpha > 0$ and in the opposite direction if $\alpha < 0$. The magnitude of αa is $|\alpha a| = |\alpha||a|$.

5. Vectors may be added graphically using either the *triangle rule* or the *parallelogram rule* (Figures 11.7 and 11.9).

6. The *zero vector*, $\mathbf{0}$, has the property that $a + \mathbf{0} = a$ for any vector a.

7. A vector may be *resolved* into *component vectors* along appropriately chosen lines. Given a vector a, its *orthogonal component vectors* parallel and normal to a line inclined at an angle θ ($0° \leqslant \theta \leqslant 90°$) to a are of magnitude

$$|a_{\mathrm{p}}| = |a|\cos\theta$$

and,

$$|a_{\mathrm{n}}| = |a|\sin\theta$$

8. Displacement, velocity, acceleration, force, weight and momentum are all examples of vector quantities.

11.5.2 Achievements

Having completed this chapter you should be able to:

A1. Define the terms that are emboldened in the text of the chapter.

A2. Identify unfamiliar quantities as scalars or vectors given the definitions of those quantities and recognise displacement, velocity, acceleration, momentum, force and weight as vector quantities.

A3. Carry out algebraic manipulations with scalar quantities.

A4. Draw diagrams to represent specified vectors in two dimensions using arrows or directed line segments, and determine the magnitude and direction of any vector represented in that way.

A5. Recognise, use and interpret the notations (a, $\underline{a}$, $|a|$, $|\underline{a}|$) used to represent vectors and their magnitudes in print and in a manuscript.

A6. Identify violations of vector notation such as equating vectors and scalars or equating vector magnitudes with negative quantities.

A7. Understand and represent graphically the operations of scaling a vector, adding and subtracting vectors and resolving a vector into component vectors in various directions.

A8. Evaluate the magnitudes of the orthogonal component vectors of a given vector parallel and normal to a given line.

A9. Formulate and rearrange simple vector equations including those which involve the zero vector **0**.

11.5.3   End of chapter questions

Question E1   The magnetic field at a point in space (at any particular time) is completely specified by two items of information: (i) a direction, (ii) a non-negative quantity called the strength of the field that can be expressed as some multiple of an SI unit called the tesla (T). Justify the claim that the magnetic field at a point is a vector quantity.

Question E2   If a number of electrically charged particles are at rest relative to one another in some isolated region of space, and if one of those particles with charge q experiences a force F due to the attraction or repulsion of the others, then we say that the particle is subject to an *electric field* given by F/q. Is the electric field at the location of the particle a vector quantity? Explain your answer.

Question E3   The vectors a and b point north and east, respectively. $|a| = 2$ cm and $|b| = 1$ cm. Draw rough sketches to show the following vectors:

(a) $-\frac{1}{2}a$ (b) $a - b$ (c) $2a + 4b$

Question E4   Figure 11.13 shows five vectors a, b, c, d and e. Express each of the vectors c, d and e in terms of the vectors a and b and hence determine which of the following equations involving those vectors is correct.

(a) $c = 2a + 2b$

(b) $-a + 2b + 2c + e = 0$

(c) $-2a + \frac{1}{2}c + \frac{1}{2}d = 0$

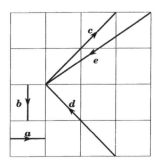

Figure 11.13   See Question E4.

Question E5   An aircraft has a speed relative to still air of 150 m s^{-1}. The pilot wishes to fly due north when the wind velocity is 20 m s^{-1} towards the north-east. Draw a sketch showing the compass bearing on which the pilot should fly. Determine the compass bearing to the nearest degree and calculate the speed of the aircraft relative to the ground, correct to one decimal place. (*Hint*: You might find it useful to recall Question T13.)

Question E6   If a, b and c are non-zero displacement vectors and α and β are non-zero numbers, classify the following statements as; always true, sometimes true and sometimes false, or always false. Explain your answers.

(a) $|\alpha a| = |\alpha||a|$

(b) $|a + b| < |a| + |b|$

(c) $\alpha a = |\alpha a|$

(d) $\alpha\,(\boldsymbol{a} + \boldsymbol{b}) = \alpha\,\boldsymbol{a} + \alpha\,\boldsymbol{b}$

(e) $(\alpha\,\beta)(\boldsymbol{a} + \boldsymbol{b}) = \alpha\,(\beta\,\boldsymbol{a} + \beta\,\boldsymbol{b})$

(f) $(\alpha + \beta)(\boldsymbol{a} + \boldsymbol{b}) = \alpha\,\boldsymbol{a} + \beta\,\boldsymbol{b}$

Question E7 $\boldsymbol{a}$ is a displacement of 30 m, due north. $\boldsymbol{b}$ is a velocity of 20 m s^{-1}, 25° west of north.

(a) What are the orthogonal component vectors of $\boldsymbol{a}$ parallel and normal to $\boldsymbol{b}$?

(b) What are the orthogonal component vectors of $\boldsymbol{b}$ parallel and normal to $\boldsymbol{a}$?

(c) Why are you unable to work out the orthogonal component vectors of $\boldsymbol{a} + \boldsymbol{b}$, parallel and normal to $\boldsymbol{a} - \boldsymbol{b}$?

Chapter 12   Working with vectors

12.1   Opening items

12.1.1   Chapter introduction

When modelling the physical world mathematically we need to use quantities that can be measured. Such physical quantities mainly fall into two classes, *scalars* and *vectors*. The mathematics of scalars such as mass and temperature is simply the familiar mathematics of real numbers and functions, but the mathematics of vector quantities such as force and velocity, which have both a *magnitude* and a *direction*, is quite different, and is the subject of this chapter. Vectors were introduced in Chapter 11 as geometric entities – more like arrows than numbers. Our main aim in this chapter is to show you how the study of vectors may be transformed from geometry to algebra.

Section 12.2 briefly reviews the topics introduced in the last chapter; the definition of scalar and vector quantities, the graphical (or geometric) representation of vectors, and the notation used in referring to vector quantities. The section also reviews the basic operations of *scaling*, *vector addition* and *vector subtraction* in terms of the geometric representation. The section ends with a review of the *resolution* of vector quantities into *component vectors*, using the geometric representation.

Section 12.3 introduces the use of *unit vectors* to denote directions and, in particular, the use of i, j, and k to specify the directions of the x-, y-, and z-axes in a right-handed Cartesian coordinate system. Section 12.3 continues by showing how these Cartesian unit vectors may be used to represent any vector in three dimensions, and then how vectors represented in *Cartesian form* may be manipulated in some of the basic operations of vector algebra, such as scaling and vector addition.

Section 12.4 introduces an alternative algebraic representation of vector quantities in terms of *ordered triples* of numbers. Sections 12.3 and 12.4 each conclude with a specific example which illustrates how, in practice, vector operations may be carried out using the Cartesian form, or the ordered triple representations, respectively.

12.1.2   Fast track questions

Question FI

Three forces F_1, F_2 and F_3 acting on a particle are such as to keep the particle in equilibrium. F_1 and F_2 are given as $F_1 = (-i + 2j + 4k)$ newtons and $F_2 = (-i - 5k)$ newtons.

(a)   Find the vector F_3.

(b)   Suppose the magnitude of F_1 is doubled, and the magnitude of F_3 is tripled (without changing their directions), while F_2 remains unchanged. Find the new resultant force acting on the particle, the magnitude of this resultant force, and the unit vector acting in the direction of this resultant force.

Question F2

In the absence of an air current, the velocity of a particle is given by $v_p = (2, 2, -1)$ m s^{-1}. Find the resultant velocity of the particle, and the time taken for the particle to travel a distance of 5 m along its resultant path, when an air current of velocity $v_c = (-1, 1, 3)$ m s^{-1} is present.

12.1.3   Ready to study?

Study comment

Although the basic concepts of *scalar* and *vector* quantities are reviewed briefly in this chapter, it is assumed that you have some knowledge of these already. For example, you should be familiar with the definitions and methods of representation of *scalar* and *vector quantities*, and terms such as *zero vector*. You should understand the ideas of the *scaling of a vector quantity* by a scalar, the *addition* and *subtraction of vectors*, and the resolution of vectors into *component vectors*. Also, you should be familiar with the terms: *acceleration*, *displacement*, *electric charge*, *mass*, *momentum*, *temperature* and *velocity*. In addition, it is assumed that you have met *Newton's laws of motion*, and that you appreciate the conditions required for a particle to be in *equilibrium*. You should be familiar with basic mathematical concepts including the ideas of *Cartesian coordinate systems*, *Pythagoras's theorem* and the use of *trigonometric functions*. You should understand also the ways in which directions may be given with respect to points of the compass. The following *Ready to study questions* will allow you to establish whether you need to review some of the topics before embarking on this chapter.

Question RI

A right-angled triangle ABC has sides AB and BC of length 12 cm and 5 cm, respectively. Angle $A\hat{B}C$ is 90°. Determine the angles $B\hat{A}C$ and $A\hat{C}B$ and the length of the side AC.

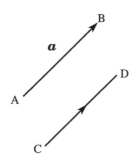

Figure 12.1 Vector **a** or $\overrightarrow{AB}$ is represented by an arrow, and $\overrightarrow{CD}$ by a directed line segment. Notice that while the vectors $\overrightarrow{AB}$ and $\overrightarrow{BA}$ have the same magnitude, they do not have the same direction, and they are therefore distinct vectors.

The dimensions (or units) of the quantities on each side of an equation involving scalars must, of course, be the same. Thus two lengths can be added to give a total length, but the addition of mass to length would have no physical meaning.

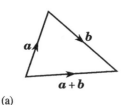

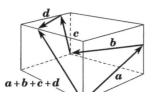

(a)

(b)

Figure 12.2 (a) The sum of two vectors obtained from the triangle rule. (b) The sum of four vectors (in three dimensions).

Question R2

A ladder 5 m long stands on level ground with its highest point against a wall at a point 4.5 m above the ground. Determine the angle of inclination of the ladder to the ground, and the distance of the foot of the ladder from the bottom of the wall.

Question R3

Points A and B are given by their Cartesian coordinates in two dimensions as $(1, 2)$ and $(4, 3)$, respectively. Determine the lengths OA, OB and AB, and the angle AÔB, where O is the origin of the coordinate system.

12.2   A brief review of scalars and vectors

12.2.1   What are scalars and vectors?

A **vector** is a mathematical object that has both *magnitude* and *direction*. Vectors may be represented diagramatically by *arrows* or *directed line segments*, as shown in Figure 12.1. These graphical vectors are sometimes referred to as **geometric vectors**.

The vector from A to B in Figure 12.1 can be denoted by $\overrightarrow{AB}$ and, the vector from C to D can be denoted $\overrightarrow{CD}$. Alternatively, and more commonly, the vector $\overrightarrow{AB}$ is denoted by **a**, where the bold typeface indicates that **a** is a vector with direction as well as magnitude.

It should be noted that geometric vectors are completely specified by their magnitude and direction, so $\overrightarrow{AB} = \overrightarrow{CD}$, even though they have been drawn at different points on the page.

The sum of two vectors may be determined by the **triangle rule** (see Figure 12.2a), and the **vector sum** **a** + **b** is known as the **resultant vector**. The result of adding any number of vectors, such as **a**, **b**, **c**, and **d** in Figure 12.2b, is obtained by simply joining them 'nose to tail'.

Scalar quantities

Many physical quantities can be specified completely by a single number together with an appropriate unit of measurement; for example length, mass, time and temperature. Such quantities are called **scalar quantities** or simply **scalars**, and we commonly denote them by italic symbols such as L, m, t and T. (☜)

Vector quantities

There are many physical quantities, for example *displacement*, *velocity*, *acceleration* and *force*, which cannot be completely defined by their magnitudes (in the appropriate units); each of them is only really specified if we are also told the direction in which they act. The essence of such quantities is that they have both magnitude *and* direction, where the **magnitude** is a nonnegative scalar quantity which gives the 'size' of the quantity.

> **Vector quantities** are physical quantities that have both magnitude *and* direction (and which can be adequately represented by geometric vectors).

Notation

In this chapter we will follow the standard convention of using bold type to indicate vectors, but in handwritten work it is usual to indicate a vector, *a*, by a wavy underline, a̰, (☞) while a geometric vector or displacement vector from A to B may be written $\overrightarrow{AB}$.

In this book we usually denote the magnitude of a vector *a* by $|a|$, while in your written work you should use $|\underset{\sim}{a}|$. (It is important that you remember to include the underline to denote a vector, and the vertical lines for a magnitude, in your written work.) (☞)

You will also encounter an alternative notation, used by many authors, in which the magnitude of the vector *a* is denoted by an unemboldened *a*.

a̰ is often used to indicate bold type to a printer. You may see alternative notation used elsewhere, e.g. a̰, $\underline{AB}$, $\underline{AB}$ or $\overrightarrow{AB}$.

You may understand what you mean, but the reader needs to understand it too!

Which of the following are vector quantities:

(a) your weight; (b) the pressure at a certain depth in the sea; (c) a charging rhinoceros?

(a) Your weight is a force, and is therefore a vector quantity. (Your mass, which is what you try to reduce when you diet, is a scalar.)

(b) The pressure in a liquid doesn't have a direction, and is therefore a scalar.

(c) A charging rhinoceros certainly has magnitude and direction, but it is not a vector; the resultant of two rhinos, one heading north and another east, is not $\sqrt{2}$ rhinos heading north-east. (☞) □

Is it generally true that $|a + b| = |a| + |b|$?

No, it is not generally true, as we can see from Figure 12.2a. The length $|a + b|$ of one side of the triangle is not generally equal to the sum of the lengths of the other two sides. □

The serious point to be made here is that vector quantities (which can be adequately represented by geometric vectors) must add according to the triangle rule.

12.2.2 Scaling a vector, unit vectors and the zero vector

In order to **scale** a vector *a* by a factor 2, say, we simply double its length, and denote the result by 2*a*, which is compatible with the triangle rule of addition because (as we would expect) we then have $a + a = 2a$.

The scaling factor may be positive or negative so that, for example, if s is a displacement of 10 km north, then $3s$ is a displacement of 30 km north; $-2s$ is a displacement of 20 km south and, dividing by 2, we have $\frac{1}{2}s$, or $s/2$, a displacement of 5 km north.

Unit vectors

If s is a displacement of 10 km north, then dividing s by 10 km produces a vector of magnitude 1 in the northerly direction.

In general, if we divide an arbitrary vector a by its magnitude we obtain a (dimensionless) vector in the same direction as a and of magnitude 1 (a dimensionless number). Such a vector is known as a **unit vector** in the direction of the vector a, and is denoted by $\hat{a}$, so that

$$\hat{a} = \frac{a}{|a|} \tag{1}$$

If $\hat{n}$ is a unit vector pointing north, and $\hat{w}$ is a unit vector pointing west, what is the magnitude of $\hat{n} + \hat{w}$?

Adding according to the triangle rule, the magnitude of $\hat{n} + \hat{w}$ is $\sqrt{1^2 + 1^2} = \sqrt{2}$, from Pythagoras's theorem. (Notice that it must be the positive square root, since magnitudes are always positive; and there are no dimensions.) □

Question T1

If F represents a force directed vertically downwards, find a unit vector directed vertically upwards. □

The zero vector

We are intent on establishing an algebra of vectors, so clearly one of our requirements is a **zero vector**, 0, which is defined so that

$$a + (-1)a = 0 \tag{2}$$

for any vector a. The zero vector has magnitude zero, and its direction is undefined (since it is irrelevant). You may have noticed that we have written zero as a vector, since the sum of two vectors must of course be a vector, and you should use a wavy underline with every occurrence of the zero vector in your written work.

12.2.3 Addition and subtraction of vectors

The result of adding two vector quantities, say a and b, is determined by the triangle rule of Figure 12.2a, or equivalently by the **parallelogram rule**

With the addition and subtraction of vectors, it is important that the vectors are of the same physical type, e.g. both forces.

illustrated in Figure 12.3a and is of the same physical type as the original vectors. (The choice of rule is merely a matter of taste.)

We follow the practice of ordinary algebra and usually replace $(-1)a$ by $-a$. The vector $-a$ has the same magnitude as the vector a, but it acts in the opposite direction (it is *antiparallel* to a); for example, in Figure 12.3b $\overrightarrow{JN} = -\overrightarrow{NJ}$. One further point, which we take for granted in ordinary algebra, is that the order in which we add vectors is immaterial, so

$$a + b = b + a \tag{3}$$

and

$$a + (b + c) = (a + b) + c \tag{4}$$

Vector subtraction is achieved by scaling a vector by -1 and then using vector addition as before. For example, in Figure 12.3b $\overrightarrow{OS} = b$ and therefore

$$\overrightarrow{OS} = -b, \text{ then } \overrightarrow{SP} = \overrightarrow{OP} + \overrightarrow{SO} = a + (-b) = a - b$$

With reference to Figure 12.3:

(a) Express the vector $\overrightarrow{JW}$ in terms of a and b.

(b) What is the final position of an object which is placed at R and then displaced by the vector $3a - 2b$?

(c) Write down two vectors equal to $2a + 2b$.

(a) $\overrightarrow{JW} = \overrightarrow{JN} + \overrightarrow{NR} + \overrightarrow{RV} + \overrightarrow{VW} = b + b + b + a = 3b + a.$

(b) $3a - 2b = \overrightarrow{RM}$ and so the final position of the object is M.

(c) $\overrightarrow{NX} = \overrightarrow{KU} = 2a + 2b.$

Remember that the position in which a geometric vector is drawn is not relevant, and the vectors $\overrightarrow{NX}$ and $\overrightarrow{KU}$ in Figure 12.3 are equal, in spite of the fact that they are defined by distinct line segments (and the vectors $\overrightarrow{JT}$ and $\overrightarrow{OY}$ would do equally well). Notice also $2a + 2b = 2(a + b)$, which is a particular case of a more general rule, which states that for any scalar k

$$ka + kb = k(a + b) \tag{5}$$

Question T2

With reference to Figure 12.3:

(a) Express the vectors $\overrightarrow{KU}$, $\overrightarrow{WL}$ and $\overrightarrow{YK}$ in terms of a and b.

(b) Write down two vectors equal to $2a - 3b$. □

A mathematician would recognise Equations 3 and 4 as the *commutative* and *associative* properties of *vector addition.*

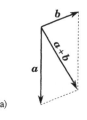
(a)

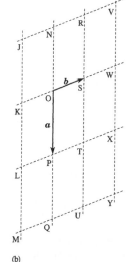
(b)

Figure 12.3 (a) The parallelogram rule of addition. (b) An array of points defining many different vectors.

A mathematician would recognise this as the *distributive rule of scalar multiplication* over vector addition.

12.2.4 Resolution of vectors, and component vectors

Often when dealing with a vector quantity we need to know how much the vector contributes in a given direction. For example, consider the two-dimensional situation in which a boat moves with velocity V across a river at an angle θ to the bank, as shown in Figure 12.4a. To find how long the boat takes to cross the river we need to find the effective velocity normal to (i.e. perpendicular to) the river bank. This can be obtained by treating the vector V as the vector sum of two orthogonal (i.e. mutually perpendicular) **component vectors** normal to the bank and parallel to it. (☞)

Alternatively described as the *vector components* of V by some authors.

If $\hat{v}_n$ and $\hat{v}_p$ are unit vectors normal to, and parallel to the bank, respectively, then V can be written as the sum of two component vectors so that

$$V = |V| \sin \theta \, \hat{v}_n + |V| \cos \theta \, \hat{v}_p$$

The component vectors of V are in this case (see Figure 12.4b): $|V| \sin \theta \, \hat{v}_n$ in a direction normal to the bank, and of magnitude $|V| \sin \theta$, and, $|V| \cos \theta \, \hat{v}_p$ in a direction parallel to the bank, and of magnitude $|V| \cos \theta$.

If the width of the river is d, we can now see that the time taken for the boat to cross is $d/(|V| \sin \theta)$.

This process of **resolution** (i.e. splitting) into *component vectors* can be thought of as the reverse of vector addition; instead of adding two (or more) vectors to produce a single resultant vector, we are replacing a single vector by the sum of two orthogonal *component vectors*. Note that when these component vectors are added together we obtain our original vector.

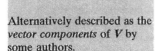
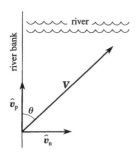

(a)

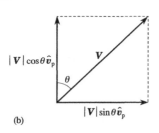

(b)

Figure 12.4 (a) Velocity V at an angle θ to the river bank, with unit vectors $\hat{v}_n$ and $\hat{v}_p$. (b) Velocity V resolved into component velocities $|V| \sin \theta \, \hat{v}_n$ and $|V| \cos \theta \, \hat{v}_p$ normal to, and parallel to, the river bank.

The string of a simple pendulum exerts a force F of magnitude 15 N on the pendulum bob, at an angle of 15° to the (upward) vertical. Resolve this force into horizontal and vertical component vectors.

If $\hat{v}$ and $\hat{h}$ are unit vectors in the (upward) vertical direction and horizontal direction, respectively, then the force in the horizontal direction is

$$|F| \sin \theta \, \hat{h} = (15 \sin 15°) \, \hat{h} = (3.88 \text{ N}) \, \hat{h}$$

and the force in the vertical direction is

$$|F| \cos \theta \, \hat{v} = (15 \cos 15°) \, \hat{v} = (14.5 \text{ N}) \, \hat{v} \quad \square$$

Question T3

A block of wood lies on a rough plane which is inclined at an angle of 25° to the horizontal. It is prevented from sliding down the plane by a static frictional force of magnitude 20 N acting up the plane. Find the horizontal and vertical component vectors of that frictional force. $\square$

12.3 **Vectors in a Cartesian coordinate system**

In the previous section you saw how it was possible to represent the velocity of a boat in terms of the unit vectors $\hat{v}_n$ and $\hat{v}_p$ normal to, and parallel to, the bank of a river. This same idea can be extended to three dimensions, and provides a very convenient means of specifying three-dimensional vectors.

12.3.1 *Cartesian unit vectors*

The use of unit vectors is of particular importance in specifying the positive directions of the axes in a Cartesian coordinate system. In two dimensions the **Cartesian unit vectors** *i* and *j* are used to specify the directions of the *x*- and *y*-axes as shown in Figure 12.5.

In three dimensions a third Cartesian unit vector *k* is added to specify the direction of the *z*-axis. However, there are two possible directions, opposite to one another, in which the *z*-axis can be directed so that it is normal to both the *x*- and *y*-axes. To remove any ambiguity about the choice of direction of the *z*-axis a *right-handed* Cartesian coordinate system is almost invariably used. There are various ways of describing this system, but only one simple method is included here. If the thumb and first two fingers of your *right* hand are arranged approximately mutually perpendicular as in Figure 12.6, then, if the first and second finger point along the *x*- and *y*-axes, respectively, the thumb points along the *z*-axis in a right-handed system.

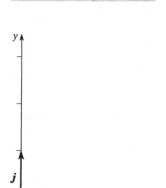

Figure 12.5 Unit vectors *i* and *j* are directed along the *x*- and *y*-axes.

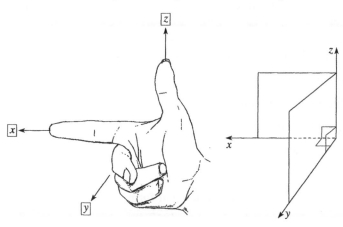

Figure 12.6 Directions of the axes in a right-handed Cartesian coordinate system.

Which of the following is a right-handed coordinate system:

(a) *i* points west, *j* points north and *k* points vertically upward,

(b) *i* is vertically downward, *j* points south and *k* points west?

Case (b) is right-handed, case (a) is not. □ (☞)

If you have difficulty confirming this answer, try drawing some simple diagrams of cases (a) and (b), and compare them with Figure 12.6.

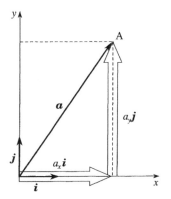

Figure 12.7 The component vectors of vector *a* in terms of *i* and *j*.

Some authors refer loosely to just the 'components' of a vector, meaning either the *Cartesian scalar components* or the *Cartesian component vectors*. The context usually makes the meaning clear.

Now let us consider how these Cartesian unit vectors may be used to specify the Cartesian component vectors of any vector *a* in the (x, y) plane, i.e. an arbitrary vector in two dimensions. The vector *a* can be resolved into two orthogonal component vectors parallel to the x- and y-axes, respectively, as shown in Figure 12.7. If a_x and a_y are the coordinates of the point A, then the component vectors will have magnitudes a_x and a_y. Hence the component vectors will be $a_x i$ and $a_y j$. From vector addition, *a* can be expressed as:

$$a = a_x i + a_y j \tag{6}$$

The important point to notice about Equation 6 is that, once we have specified the directions of *i* and *j* the vector *a* is completely determined (both in magnitude and direction) by the pair of scalar quantities a_x and a_y.

The values a_x and a_y are known as the **Cartesian scalar components** of *a* (or simply the components of *a*); $a_x i$ and $a_y j$ are the **Cartesian component vectors** of *a* (or simply the component vectors of *a*).

We obtain the full benefit from this method of representing vectors when the idea is extended to three dimensions. In Figure 12.8 we show how a vector *a* in three dimensions can be resolved into three orthogonal component vectors parallel to the x-, y- and z-axes. If a_x, a_y and a_z are the coordinates of the point A, then

$$a = a_x i + a_y j + a_z k \tag{7}$$

In this case the Cartesian component vectors of *a* are

$$a_x i, \ a_y j \ \text{and} \ a_z k$$

while the Cartesian scalar components of *a* are

$$a_x, \ a_y \ \text{and} \ a_z$$

and when we write

$$a = a_x i + a_y j + a_z k$$

we say that *a* is expressed in **Cartesian form**.

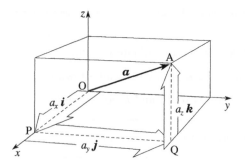

Figure 12.8 The Cartesian component vectors of a vector *a* in terms of *i*, *j* and *k*.

The suffix notation used in the above definitions is quite convenient, and it is natural to express any given vector, r say, in the same form. Thus $r_x i$ is the natural choice of notation for the vector component in the x-direction of the vector r, while $r_y j$ and $r_z k$ denote its y- and z-component vectors.

Most of us have great difficulty visualising objects, and their relative distances from each other, in three dimensions; vector methods largely overcome these problems.

Some authors use the term 'x-component' for both the vector component in the x-direction and the scalar component in the x-direction. In this book it means the scalar component.

A portable rectangular cabin is 3 m wide, 5 m long and 2.5 m high, and it is placed on a horizontal foundation with its long axis pointing north/south. An origin is chosen in the bottom south-west corner of the cabin, with i pointing east, j pointing north, and k pointing vertically upward.

What vector displacement r is required to move an object from the origin to a point at the centre of the ceiling?

The vector $a = (3i + 5j)$ m is the displacement from the origin to the opposite corner of the floor of the cabin. It follows that $\frac{1}{2}a = (1.5i + 2.5j)$ m is the displacement from the origin to the centre of the floor. The centre of the ceiling is 2.5 m above this point, and so its displacement from the origin is given by

$$r = (1.5i + 2.5j + 2.5k) \text{ m}. \quad \square$$

Once we have found the vector displacement r from the origin to the centre of the ceiling, then it is possible to find displacements *from this new point*. For example, the vector displacement of the opposite corner of the floor of the cabin from the origin is $a = 3i + 5j$, and therefore the displacement of the opposite corner of the cabin from the centre of the ceiling is (☞)

$$-r + a = -(1.5i + 2.5j + 2.5k)\,\text{m} + (3i + 5j)\,\text{m} = (1.5i + 2.5j - 2.5k)\,\text{m}$$

We go from the centre of the ceiling to the origin, i.e. $-r$, then from the origin to the far corner of the floor, i.e. a.

Question T4
Is the suggested coordinate system for the cabin described above right-handed? What vector displacement s will move an object from the centre of the ceiling to the mid-point of the northern wall? □

12.3.2 The magnitude and combination of vectors in Cartesian form

We can regard vectors in two dimensions as merely a special case of vectors in three dimensions; their z-component just happens to be zero. So hereafter our results apply equally well to vectors in two or three dimensions.

It is convenient to be able to determine the magnitude of a vector in terms of its Cartesian components, and this may be easily done if we apply

Pythagoras's theorem to vector $\boldsymbol{a}$ in Figure 12.8. First we consider the (x, y) plane only. Triangle OPQ is a right-angled triangle so,

$$OQ^2 = OP^2 + PQ^2$$

But $OP = a_x$ and $PQ = a_y$, so:

$$OQ^2 = a_x^2 + a_y^2$$

Now we consider the right-angled triangle OQA:

$$OA^2 = OQ^2 + QA^2$$

But $\quad OQ^2 = a_x^2 + a_y^2 \quad$ and $\quad QA^2 = a_z^2$

and so $\quad OA^2 = a_x^2 + a_y^2 + a_z^2$

But $\quad OA = |\boldsymbol{a}|$

therefore:

$$|\boldsymbol{a}| = (a_x^2 + a_y^2 + a_z^2)^{1/2} \tag{8}$$

Given that $\boldsymbol{a} = 2\boldsymbol{i} + 3\boldsymbol{j} + 4\boldsymbol{k}$ calculate $|\boldsymbol{a}|$.

$$|\boldsymbol{a}| = (2^2 + 3^2 + 4^2)^{1/2} = \sqrt{29}. \quad \square$$

Question T5

In Question T4 you were asked to determine a vector displacement $\boldsymbol{s}$. Now determine the magnitude of that displacement. $\quad \square$

As we saw in Subsection 12.2.2, the scaling of a vector is simply multiplication of the vector by a scalar quantity, and this operation is very straightforward when a vector is expressed in Cartesian form. For example let α be a scalar quantity and let $\boldsymbol{a} = a_x\boldsymbol{i} + a_y\boldsymbol{j} + a_z\boldsymbol{k}$, then

$$\alpha\,\boldsymbol{a} = \alpha\,a_x\,\boldsymbol{i} + \alpha\,a_y\,\boldsymbol{j} + \alpha\,a_z\,\boldsymbol{k} \tag{9}$$

The sum of two vectors in Cartesian form is equally easy.

Let $\boldsymbol{a} = a_x\boldsymbol{i} + a_y\boldsymbol{j} + a_z\boldsymbol{k}$ and $\boldsymbol{b} = b_x\boldsymbol{i} + b_y\boldsymbol{j} + b_z\boldsymbol{k}$, then

$$\boldsymbol{a} + \boldsymbol{b} = (a_x + b_x)\boldsymbol{i} + (a_y + b_y)\boldsymbol{j} + (a_z + b_z)\boldsymbol{k} \tag{10}$$

and similarly for subtraction

$$\boldsymbol{a} - \boldsymbol{b} = (a_x - b_x)\boldsymbol{i} + (a_y - b_y)\boldsymbol{j} + (a_z - b_z)\boldsymbol{k} \tag{11}$$

At 3 a.m. a hedgehog of mass 1.5 kg starts crossing the M1 from west to east at a speed of 2.0 km h^{-1}. What is the momentum of the hedgehog?

Taking i to be a unit vector pointing east, the momentum is given by $h = (1.5 \text{ kg}) \times (2.0 \text{ km h}^{-1})i = 3.0i \text{ kg km h}^{-1}$. ☐

At 3.05 a.m. a truck of mass 15 000 kg and speed 50 km h^{-1} passes the same spot travelling north. What is the momentum of the truck?

Taking j to be a unit vector pointing north, the momentum of the truck is $t = (15\,000 \text{ kg}) \times (50 \text{ km h}^{-1})\,j = 75 \times 10^4 \,j \text{ kg km h}^{-1}$. ☐

> Momentum is defined as the product of the mass of an object, a scalar, with the velocity of the object, a vector (or, more accurately, the velocity of the centre of mass of the object).

Question T6

What is the magnitude of the combined momentum of the truck and the hedgehog at 3.06 a.m. (and what was the name of the hedgehog)? ☐

Having considered the operations of scaling, addition and subtraction separately, we are now in a position to combine these processes when using the Cartesian component approach.

Suppose, for example, that $a = 2i + 3j + 4k$ and $b = 3i - j + 4k$ then

$$4a - 3b = 4(2i + 3j + 4k) - 3(3i - j + 4k) = -i + 15j + 4k$$

and more generally, for scalars α and β,

$$\alpha a + \beta b = (\alpha a_x + \beta b_x)i + (\alpha a_y + \beta b_y)j + (\alpha a_z + \beta b_z)k \qquad (12)$$

> Notice the greater convenience of the Cartesian component approach in comparison with the geometric vector approach in carrying out these operations on vectors.

Is there any physical restriction on α, β, a and b in order that Equation 12 makes sense?

Yes, the dimensions of $\alpha\,a$ and $\beta\,b$ must be the same. ☐

12.3.3 An application of vectors in Cartesian form

To complete Section 12.3 let us consider how the concepts covered in this section may be applied in a specific example.

A particle travels at velocity v_1 for 3 s and then velocity v_2 for 2 s, where v_1 and v_2 are given by: $v_1 = (2i - 3j + 6k) \text{ m s}^{-1}$ and $v_2 = (4i + 12j - 3k) \text{ m s}^{-1}$. Determine:

Basic vector algebra

(a)   the magnitudes of the velocities v_1 and v_2;

(b)   the displacements s_1 and s_2 over the first and second intervals;

(c)   the distances associated with these displacements;

(d)   the total displacement, s_t;

(e)   the magnitude of the total displacement; and

(f)   the unit vector $\hat{s}_t$ in the direction of s_t.

(a)   To determine the magnitudes of v_1 and v_2 we apply Equation 8 which we deduced from Pythagoras's theorem. For v_1, the components in m s^{-1} are 2, -3 and 6, and therefore:

$$|v_1| = [2^2 + (-3)^2 + 6^2]^{1/2}\ \text{m s}^{-1} = 49^{1/2}\ \text{m s}^{-1} = 7\ \text{m s}^{-1}$$

Similarly:

$$|v_2| = [4^2 + 12^2 + (-3)^2]^{1/2}\ \text{m s}^{-1} = 169^{1/2}\ \text{m s}^{-1} = 13\ \text{m s}^{-1}$$

(b)   The displacement s corresponding to a constant velocity v for a time t is given by $s = tv$, so we can find s_1 and s_2 by scaling the vectors v_1 and v_2 by the appropriate times. Hence:

$$s_1 = (3\ \text{s})v_1 = (6i - 9j + 18k)\ \text{m}$$

and $s_2 = (2\ \text{s})v_2 = (8i + 24j - 6k)\ \text{m}$

(c)   The required distances are simply the magnitudes of s_1 and s_2. The direct approach to the determination of $|s_1|$ and $|s_2|$ would appear to involve the application of Equation 8 to the vectors s_1 and s_2 found in part (b). However, it is easier to make use of the answers obtained for $|v_1|$ and $|v_2|$ in part (a), since $|s_1| = (3\ \text{s})\,|v_1|$ and $|s_2| = (2\ \text{s})|v_2|$. Therefore

$$|s_1| = 3\ \text{s} \times 7\ \text{m s}^{-1} = 21\ \text{m}$$

and $|s_2| = 2\ \text{s} \times 13\ \text{m s}^{-1} = 26\ \text{m}$

(d)   The total displacement is simply the vector sum $s_t = s_1 + s_2$, where s_1 and s_2, are the answers obtained for part (b). Following the Cartesian component vector method for addition, we obtain:

$$s_1 + s_2 = s_t = [(6 + 8)i + (-9 + 24)j + (18 - 6)k]\ \text{m}$$

and therefore $s_t = [14i + 15j + 12k]\ \text{m}$

(e)   The value of $|s_t|$ can be obtained using Equation 8:

$$|s_t| = [14^2 + 15^2 + 12^2]^{1/2}\ \text{m} = \sqrt{565}\ \text{m} \approx 23.8\ \text{m}$$

(f)   The required unit vector is obtained using Equation 1:

$$\hat{s}_t = \frac{s_t}{|s_t|} = \frac{14i + 15j + 12k}{\sqrt{565}}$$

therefore $\hat{s}_t = 0.589i + 0.631j + 0.505k$ □

You should use Equation 8 to check that the magnitude of this vector is indeed unity.

Question T7

Particle A, of mass 2 kg, travelling with velocity $v_A = (i - 2j + 4k)$ m s^{-1} collides with particle B, of mass 3 kg, travelling with velocity $v_B = (3i + j - 2k)$ m s^{-1}. The particles stick together on collision and then move as one combined body with momentum the same as the total momentum of A and B before the collision. Determine:

(a) the momenta of A and B before the collision;

(b) the total (resultant) momentum of A and B before the collision;

(c) the magnitudes of the momenta of A and B before the collision;

(d) the velocity of the combined body after the collision. □

'momenta' is the plural of momentum.

12.4 Vectors as ordered triples

12.4.1 Ordered triples and ordered pairs

A vector, $a = 3i - j + 4k$ say, in Cartesian form, is often abbreviated to three numbers in brackets, $(3, -1, 4)$, in which the order of the numbers is of crucial importance, for otherwise we would not know which number referred to which component. Such a collection of numbers is called an **ordered triple**.

In general we write

$$a = a_x i + a_y j + a_z k = (a_x, a_y, a_z) \qquad (13)$$

An ordered triple, such as $(3, -1, 4)$, can mean the coordinates of a point in three dimensions or the vector $3i - j + 4k$, but the context will make it clear which is intended. (In fact this dual meaning is an advantage, as you will see in Subsection 4.4.)

Express the vector $2i - j + 3k$ as an ordered triple.

$2i - j + 3k = (2, -1, 3)$ □

In the case of two-dimensional vectors, we refer to **ordered pairs** rather than ordered triples, for example, $v = 2i + 5j = (2, 5)$. Therefore $(2, 5)$ is the ordered pair which represents the vector v.

12.4.2 Manipulation of ordered triples

The extension of the calculations of the last section, for vectors in Cartesian form, to vectors represented as ordered triples is very easy, but we will state the results 'for the record'.

These equations are not identical to those of the previous section, but they are so similar in meaning that they do not warrant separate numbers.

If $\boldsymbol{a} = (a_x, a_y, a_z)$ and $\boldsymbol{b} = (b_x, b_y, b_z)$, then

$$|\boldsymbol{a}| = (a_x^2 + a_y^2 + a_z^2)^{1/2} \tag{8}$$

$$\alpha \boldsymbol{a} = \alpha a_x \boldsymbol{i} + \alpha a_y \boldsymbol{j} + \alpha a_z \boldsymbol{k} \tag{9}$$

$$\boldsymbol{a} + \boldsymbol{b} = (a_x + b_x)\boldsymbol{i} + (a_y + b_y)\boldsymbol{j} + (a_z + b_z)\boldsymbol{k} \tag{10}$$

$$\alpha \boldsymbol{a} + \beta \boldsymbol{b} = (\alpha a_x + \beta b_x, \alpha a_y + \beta b_y, \alpha a_z + \beta b_z) \tag{12}$$

Question T8
Find the magnitude of the vector represented by the ordered triple $(5, -2, -1)$. ☐

Express $2(2\boldsymbol{i} - \boldsymbol{j} + 3\boldsymbol{k})$ as an ordered triple and find the ordered triple which represents the sum of $(\boldsymbol{i} + 2\boldsymbol{j} - \boldsymbol{k})$ and $(3\boldsymbol{i} - \boldsymbol{j} + 4\boldsymbol{k})$.

$2(2\boldsymbol{i} - \boldsymbol{j} + 3\boldsymbol{k}) = 2(2, -1, 3) = (4, -2, 6)$

$(\boldsymbol{i} + 2\boldsymbol{j} - \boldsymbol{k}) + (3\boldsymbol{i} - \boldsymbol{j} + 4\boldsymbol{k}) = (1, 2, -1) + (3, -1, 4) = (4, 1, 3)$ ☐

Question T9
Find the ordered triple that represents the resultant of the vectors $(2, -1, -2)$ and $(3, -2, 7)$, and find also the ordered triple which represents the unit vector in the direction opposite to that of the resultant. ☐

12.4.3   An application of ordered triples

The following exercises are intended to illustrate how vectors may be manipulated using the ordered triple notation.

Forces $\boldsymbol{F}_1 = (2, -1, 4)$ N and $\boldsymbol{F}_2 = (1, 3, 1)$ N act simultaneously on a particle. Determine:

(a)   the magnitudes of $\boldsymbol{F}_1$ and $\boldsymbol{F}_2$;

(b)   the resultant of $\boldsymbol{F}_1$ and $\boldsymbol{F}_2$;

(c)   the magnitude of this resultant;

(d)   the vector required to cancel the combined effect of $\boldsymbol{F}_1$ and $\boldsymbol{F}_2$;

(e)   the vectors $2\boldsymbol{F}_1$, $5\boldsymbol{F}_2$ and $2\boldsymbol{F}_1 + 5\boldsymbol{F}_2$.

(a) $|F_1| = [2^2 + (-1)^2 + 4^2]^{1/2}\,\text{N} = [4 + 1 + 16]^{1/2}\,\text{N}$

  $= 21^{1/2}\,\text{N} = 4.58\,\text{N}$

  $|F_2| = [1^2 + 3^2 + 1^2]^{1/2}\,\text{N} = [1 + 9 + 1]^{1/2}\,\text{N} = 11^{1/2}\,\text{N} = 3.32\,\text{N}$

(b) $F_1 + F_2 = (2 + 1, -1 + 3, 4 + 1)\,\text{N} = (3, 2, 5)\,\text{N}.$

(c) $|F_1 + F_2| = [3^2 + 2^2 + 5^2]^{1/2}\,\text{N} = [9 + 4 + 25]^{1/2}\,\text{N}$

  $= 38^{1/2}\,\text{N} = 6.16\,\text{N}.$

(d)   If $F_1 + F_2 = (3, 2, 5)\,\text{N}$, then the force vector required to cancel the effect of $F_1 + F_2$ is $-(F_1 + F_2) = (-3, -2, -5)\,\text{N}.$

(e) $2F_1 = 2(2, -1, 4)\,\text{N} = (4, -2, 8)\,\text{N}.$

  $5F_2 = 5(1, 3, 1)\,\text{N} = (5, 15, 5)\,\text{N}$

so that $2F_1 + 5F_2 = (4 + 5, -2 + 15, 8 + 5)\,\text{N} = (9, 13, 13)\,\text{N}.$ □

Question T10

A particle travels with velocity $v_1 = (2, 1, -2)\,\text{m s}^{-1}$ for 5 s, then with velocity $v_2 = (-1, 2, 3)\,\text{m s}^{-1}$ for 2 s, and finally with velocity $v_3 = (-2, -1, 2)\,\text{m s}^{-1}$ for 3 s. Find the total displacement of the particle over the 10 s, the magnitude of this total displacement, and the unit vector in the direction of the total displacement.   □

12.4.4   Position vectors

There are two main applications of vectors in physical science. The first concerns the mathematical modelling (or representation, if you prefer) of physical quantities such as force, displacement and velocity. The second is equally important, and concerns the location of objects in three dimensions, and is essentially a form of geometry. Generally, we do not associate a vector with a particular point in space (although later you may encounter applications where it is desirable to do just that), and vectors such as $\overrightarrow{JN}$ and $\overrightarrow{UY}$ in Figure 12.3 are defined to be the same vector. On the other hand, the vector a in Figure 12.8 specifies the position of the point A provided that we know that the end of the vector is fixed at the origin. Vectors that are used in this way to determine the positions of points are often known as **position vectors**.

Two points A and B are specified by the position vectors

  $\overrightarrow{OA} = (1, -3, 5)\,\text{cm}$ and $\overrightarrow{OB} = (-2, 2, 4)\,\text{cm}$

Find the distance between the points A and B, and the position vector of the mid-point M of AB.

Is a position vector r identical to a displacement vector r?

The displacement vector from A to B is defined by

$$\overrightarrow{AB} = \overrightarrow{AO} + \overrightarrow{OB} = [(-2, 2, 4) - (1, -3, 5)]\text{ cm} = (-3, 5, -1)\text{ cm}$$

so that the distance from A to B is

$$\overrightarrow{AB} = [(9 + 25 + 1)^{1/2}]\text{ cm} \approx 5.92\text{ cm}.$$

The mid-point of AB has coordinates $\frac{1}{2}[(-2, 2, 4) + (1, -3, 5)]$ cm so that the position vector $\overrightarrow{OM} = \frac{1}{2}[(-2, 2, 4) + (1, -3, 5)]$ cm $= (-0.5, -0.5, 4.5)$ cm.

The position vector **r** and the displacement vector **r** are identical, they are the same vector; but the interpretation of them is different. The fact that we are told that **r** is a displacement vector tells us only the magnitude and direction of the displacement. We can use that displacement to specify a particular point only if we first specify a point from which to make the displacement. On the other hand, if we are told that **r** is a position vector we know that the position of the point A is determined if the vector is placed with its tail at the origin. In a sense, position vectors are simply a special class of displacement vectors since position vectors define displacements from the origin. ☐

Notice that in the above exercise we were quite happy to use exactly the same notation for the coordinates of M and for the position vector $\overrightarrow{OM}$. The context makes it absolutely clear which is intended, and, in any case, there is very little difference between saying that a point P is determined by the coordinates (1, 2, 3) and that P is determined by the position vector (1, 2, 3).

Question T11

A particle is moving with velocity $v = (1, 1, 2)\text{ m s}^{-1}$ and at time $t = 0$ it is at the point P with position vector $\overrightarrow{OP} = (2, 3, -4)$ m. What is the position vector $\overrightarrow{OQ}$ of the particle at time $t = 5$ s? ☐

12.5   Closing items

12.5.1   Chapter summary

1. *Scalar quantities* can be specified completely by a single number together with an appropriate unit of measurement.

2. *Vector quantities* can be specified by a magnitude and a direction. Geometric vectors are represented by arrows or *directed line segments* and vector quantities are often represented pictorially by geometric vectors. In print, vectors are denoted by bold typeface and, in handwritten material, by a wavy underline.

3. The *magnitude* of a vector is a non-negative scalar that represents the 'length' or 'size' of that vector. The magnitude of a is denoted by $|a|$ (or sometimes by a) in print, and by $|\underset{\sim}{a}|$ in handwritten material.

4. Any vector a may be multiplied by a scalar α to produce a scaled vector $\alpha\,a$ which points in the same direction as a if $\alpha > 0$ and in the opposite direction if $\alpha < 0$. The magnitude of $\alpha\,a$ is $|\alpha\,a| = |\alpha|\,|a|$.

5. If any non-zero vector a is divided by $|a|$, a vector of unit magnitude is obtained which points in the same direction as a. Such a vector is called a *unit vector* and is denoted by $\hat{a}$.

6. Vectors may be added geometrically using either the *triangle* or *parallelogram rule* (see Figures 12.2 and 12.3).

7. A vector may be *resolved* into *component vectors* along appropriately chosen directions. Given a vector a, its orthogonal *component vectors* parallel and normal to a direction inclined at an angle θ to a are of magnitude

    $$|a_{\mathrm{p}}| = |a|\cos\theta \quad \text{and} \quad |a_{\mathrm{n}}| = |a|\sin\theta$$

8. The *Cartesian unit vectors* in the directions of the Cartesian axes x, y and z are denoted by i, j and k.

9. A vector a can be expressed in Cartesian form as

    $$a = a_x\,i + a_y\,j + a_z\,k \qquad (7)$$

    The scalars a_x, a_y and a_z are called the Cartesian scalar components of a, whereas the vectors $a_x\,i$, $a_y\,j$ and $a_z\,k$ are called the Cartesian component vectors of a.

    > Both are often described loosely as the 'components' of a.

10. The *magnitude* of the vector a is given by

    $$|a| = (a_x^2 + a_y^2 + a_z^2)^{1/2} \qquad (8)$$

11. The operations of *scaling* and *vector addition* take the following Cartesian algebraic forms (for any scalar α and vectors a and b):

    $$\alpha\,a = \alpha\,a_x\,i + \alpha\,a_y\,j + \alpha\,a_z\,k$$

    $$\text{and} \quad a + b = (a_x + b_x)i + (a_y + b_y)j + (a_z + b_z)k \qquad (10)$$

12. A vector represented by $a = a_x\,i + a_y\,j + a_z\,k$ can also be represented by the abbreviated notation of an *ordered triple*:

    $$a = (a_x, a_y, a_z)$$

13. The operations of scaling and vector addition take the following abbreviated forms (for any scalar α and vectors a and b):

    $$\alpha\,a = (\alpha\,a_x, \alpha\,a_y, \alpha\,a_z)$$

    $$\text{and} \quad a + b = (a_x + b_x, a_y + b_y, a_z + b_z)$$

14.   Vectors may be used to determine the position of points relative to a chosen origin, and they are then known as *position vectors*.

12.5.2   Achievements

Having completed this chapter, you should be able to:

A1.   Define the terms that are emboldened in the text of the chapter.

A2.   Identify quantities as being scalars or vectors, given the definitions of the quantities.

A3.   Recognise and use the notations $(a,\ \underline{a},\ |a|,\ |\underline{a}|)$ to represent vectors and their magnitudes.

A4.   Carry out and represent graphically the operations of scaling, addition and subtraction of vectors, and of resolving a vector into orthogonal component vectors.

A5.   Determine a unit vector in the same direction as a given vector.

A6.   Use Cartesian unit vectors and Cartesian scalar components to represent a given vector in Cartesian form.

A7.   Evaluate the magnitude of any vector in terms of its Cartesian components.

A8.   Scale, add and subtract vectors in Cartesian form.

A9.   Use Cartesian scalar components to represent a given vector as an ordered triple (or an ordered pair in two dimensions).

A10.   Scale, add and subtract vectors using the ordered triple notation for vectors.

A11.   Use a position vector to specify the location of a point.

12.5.3   End of chapter questions

Question E1   The electric field at any point in space can be defined as the electrical force experienced by a positive charge at that point, divided by the magnitude of the charge. Given that charge is a scalar quantity, decide whether electric field is a scalar or a vector quantity, and justify your decision.

Question E2   On Treasure Island, Captain Flint decides to bury his treasure at the point at which he arrives after making a displacement of 20 m west and then a displacement of 10 m north, starting from a distinctive rock which he uses as his reference point. Illustrate these displacements graphically, and find the magnitude and direction of the resultant displacement. Find also the component of the resultant displacement in the north-west direction.

Question E3 Vectors **a** and **b** are given by

$$a = -i + 5j - 2k \text{ and } b = 3i - 2j - 4k$$

Find $a + b$, $|a + b|$ and the unit vector in the direction of $a + b$.

Question E4 Given the forces: $F_1 = (i - 2j + 3k)$ N, and $F_2 = (3i + j - 4k)$ N, find the force given by $3F_1 + 2F_2$, the magnitude of this force and its Cartesian vector component along the z-axis.

Question E5 Given the vectors **a** and **b** in Cartesian form as:

$$a = -i + 4j - 2k \text{ and } b = 2i - j + 2k$$

express the vectors **a**, **b** and $a + b$ in terms of ordered triples. What physical restriction must apply to vectors **a** and **b** for the resultant to have any sensible meaning?

Question E6 Vectors **a** and **b** are given by: $a = (2, 1, -1)$ and $b = (1, -3, 2)$. Find, in ordered triple notation, the vectors $2a + 5b$, $3a - 2b$, and the vector **c** such that $a + b + c = (0, 0, 0)$.

Question E7 Given the forces $F_1 = (2, 0, 2)$ N, $F_2 = (1, -2, 3)$ N and $F_3 = (-3, 1, 2)$ N, determine:

(a) the ordered triple representing $F_1 + F_2 + F_3$;

(b) the magnitude of $F_1 + F_2 + F_3$;

(c) the ordered triple representing the unit vector in the direction of $F_1 + F_2 + F_3$;

(d) the component vector in the z-direction of the force which would completely counteract the force $F_1 + F_2 + F_3$;

(e) the scalar component of $F_1 + F_2 - F_3$ in the y-direction.

Question E8 A pyramid has a square base of length 100 m and its height is 20 m. Cartesian coordinates are chosen with one corner of the base as the origin, with **i** and **j** in the direction of the adjacent edges of the base, and with **k** vertically upward. Find the position vector of the vertex of the pyramid.

Part 4

Basic differentiation

Introducing differentiation

Chapter

13

13.1 Opening items

13.1.1 Chapter introduction

Many of the concepts and laws that physical scientists have to study involve the *rate of change* of one quantity with respect to another. Motion, which involves ideas such as *position*, *velocity* and *acceleration* provides many important examples. The velocity of an object is defined by the *rate of change* of its position with respect to time. The acceleration of an object is similarly defined by the *rate of change* of velocity with respect to time. In fact, rates of change are of such general importance that they have spawned an important branch of mathematics – *differential calculus*. This subject, created by Isaac Newton (1642–1727) and Gottfried Wilhelm Leibniz (1646–1716), allows the everyday notion of 'rate of change' to be made mathematically precise and has many applications throughout science and technology.

This chapter defines the important concept of the *derivative* and explains its role in evaluating rates of change. In Section 13.2 we adopt a graphical approach to rates of change and identify them with the gradients of tangents to graphs. This is done in the context of *linear motion*, introducing both *velocity* and *acceleration* as the gradients of appropriately constructed graphs. Although the discussion in Section 13.2 is restricted to simple cases, this graphical approach is cumbersome and it is quite clear that a simple algebraic technique for evaluating gradients would be of great value. The mathematical process known as *differentiation* provides just such a technique, and it is here that the *derivative* enters the discussion. In Subsection 13.3.1 the link between functions and graphs (essentially the notion that a curve can be described by an equation) is used to introduce the idea of a derivative, while in Subsections 13.3.2 and 13.3.3 differentiation is introduced and used to find some simple derivatives. A formal definition of the derivative is given in Subsection 13.3.4 and applied to velocity and acceleration in Subsection 13.3.5. Linear motion continues to provide the context and motivation throughout this discussion, but Subsection 13.3.6 goes on to show how derivatives can also be used in a wide variety of other circumstances.

Throughout this chapter the emphasis is on basic ideas and understanding rather than on complicated calculations. The main aim is that you should understand the mathematical approach to rates of change, be able to interpret derivatives such as

$$\frac{dx}{dt}, \frac{dv}{dt} \text{ and } \frac{dy}{dx}$$

graphically, and evaluate them algebraically in straightforward situations. The full development of the techniques and applications of differentiation is left to other chapters.

13.1.2 Fast track questions

Question F1

Figure 13.1 is the position–time graph for a car moving along a straight road. Describe the car's journey in everyday language.

During which parts of the journey is the acceleration (a) negative, (b) zero and (c) positive?

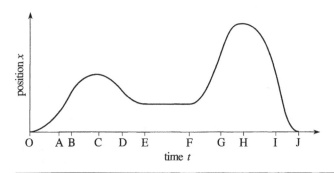

Figure 13.1 A graphical description of a car's journey.

Question F2

The position $x(t)$ of a body at time t is given by $x(t) = -pt + qt^2$ where p and q are constants.

(a) Describe the nature and significance of dx/dt, write down its formal definition and use that definition to find an expression for dx/dt in terms of p, q and t.

(b) If $p = 2 \text{ m s}^{-1}$ and $q = 3 \text{ m s}^{-2}$, at what time is the velocity v_x zero? What is the position at that time? What is the acceleration a_x at that time?

13.1.3 Ready to study?

Study comment

In order to study this chapter you will need to be familiar with the following terms: *absolute value* (or *modulus*), *Cartesian coordinates*, *dependent variable*, *function*, *gradient*, *graph*, *independent variable*, *linear function* and *quadratic function*. In addition, you will need to be familiar with SI units, and you must be able to determine the gradient of a straight-line graph. It is also assumed that you can carry out basic algebraic manipulations. The following *Ready to study questions* will allow you to establish whether you need to review some of the topics before embarking on this chapter.

Question R1

On a set of two-dimensional Cartesian coordinate axes plot the points A = (1, 1), B = (3, 3), C = (2, 6). Calculate the gradients of the lines AB, BC, AC.

Question R2

Draw the graph of the linear function $x(t) = 1.2 + 3t$. What is its gradient?

Question R3

Draw the graph of the quadratic function $x(t) = 5 + t + 3t^2$, for $t \geqslant 0$.

Question R4

For the function $x(t) = -4 + 6t + 2t^2$, evaluate:

(a) $x(0)$,   (b) $x(2)$,   (c) $|x(-1)|$,   (d) $x(2.5) - x(2)$,

(e) $x(2.1) - x(2)$.

(f)   If $\Delta t = 0.50$, evaluate $\dfrac{x(2 + \Delta t) - x(2)}{0.50}$.

(g)   If $\Delta t = 0.10$ and $t = 2$, evaluate $\dfrac{x(t + \Delta t) - x(t)}{\Delta t}$

Question R5

Given that the variables x and t cover the same range of values, which (if any) of the following functions are equivalent?

(a) $s(t) = 1 + 2t + t^2$ (b) $y(x) = (1 + x)^2$

(c) $v(t) = (1 - t)^2 + 4t$.

13.2 Rates of change: graphs and gradients

13.2.1 Position−time graphs and constant velocity

Modern science is based on the idea that the world can only be understood by first making careful observations and measurements. Graphs provide an important way to represent such measurements. Although in the real world many things move along a curved path, it is obviously simpler to start with **linear motion**, i.e. motion along a straight line. This turns out to be less restrictive than it might initially appear, since any motion not involving spinning objects can be expressed as a combination of movements along two or three straight lines.

Figure 13.2 shows a car and a pedestrian moving along a straight line. The line has been designated as the *x-axis* of some appropriately chosen system of *Cartesian coordinates* which has its *origin* at the fixed point indicated in the figure. At any time *t*, the **position** of the car or the pedestrian is completely specified by its *position coordinate x* which determines its **displacement** from the origin at that time. Note that *x* is measured in metres and may be positive or negative according to which side of the origin a point is located. At the particular instant pictured in Figure 13.2 the car is located at the point $x = 30$ m and the pedestrian is at $x = -20$ m.

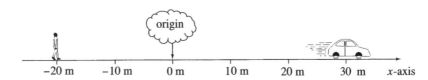

Figure 13.2 An example of one-dimensional motion.

The position of a point automatically determines another quantity – its **distance** from the origin. This is the magnitude of its displacement from the origin. Of course, distances have to be *positive* quantities (it makes no sense to speak of a *distance* of -20 m) so we define the distance from the origin to any point on the line as $|x|$, the *modulus* or *absolute value* of its position coordinate. You will recall that absolute values are always positive, so $|-20$ m$| = 20$ m.

Table 13.1 shows the position of the car at 5-second intervals, starting from the time when it passed through the origin. The corresponding graph, called the **position–time graph** of the motion, is shown in Figure 13.3. As you can see, after about 301s the graph is a straight line (i.e. linear) indicating that the car's position coordinate is increasing by equal amounts in equal intervals of time. In other words, the *rate of change* of the car's

> In three dimensions, three position coordinates x, y and z are required to specify the position of a point relative to a fixed origin. These three quantities define the *position vector r* of the point and are referred to as the x, y and z-components of the vector in that context. The distance from the origin to the point is then given by $|r|$, the *magnitude* of the position vector. When dealing with linear motion in one dimension a single component, x say, suffices to determine the position vector or the vector giving its displacement from the origin. In what follows we shall refer to x as 'the position' even though it is, in reality, only one component of the position vector.

Table 13.1 The position of a car at various times

Time t/s	Position coordinate x/m	Time t/s	Position coordinate x/m
0	0	35	68.0
5	1.7	40	84.0
10	6.8	45	99.0
15	15.0	50	115.0
20	26.0	55	130.0
25	39.0	60	146.0
30	53.0		

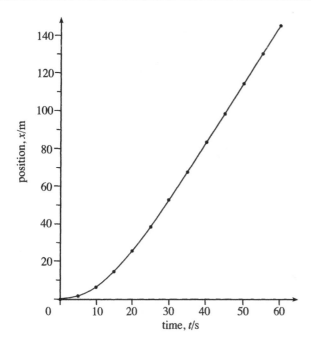

Figure 13.3 Position−time graph from Table 13.1.

position with respect to time is constant after about 30 s. Now, for an object moving along the x-axis, the rate of change of its position coordinate x with respect to time defines its **velocity**, v_x, which tells us how fast it moves and in which direction. (If the object moves in such a way that x increases with time then v_x is positive; if it moves in the opposite direction so that x decreases with time then v_x is negative.) So we can interpret the linear part of position−time graph as indicating that the car has a **constant** (or **uniform**) **velocity** after about 30 s.

Velocity, like position, is a *vector* quantity which requires three *scalar components v_x, v_y*, and v_z for its full specification in three dimensions. The speed of the object is given by $v = |v|$, the *magnitude* of the velocity.

Using the data in Table 13.1 evaluate the car's velocity v_x in the time intervals from 40 s to 50 s and from 50 s to 60 s.

Note the convention that time is plotted horizontally and position vertically.

From 40 s to 50 s, the position coordinate changes from 84 m to 115 m, so the velocity is

$v_x = (115 \text{ m} - 84 \text{ m})/(50 \text{ s} - 40 \text{ s}) = 3.1 \text{ m s}^{-1}$.

From 50 s to 60 s, the velocity is

$v_x = (146 \text{ m} - 115 \text{ m})/(60 \text{ s} - 50 \text{ s}) = 3.1 \text{ m s}^{-1}$

Note that in this case the velocity is independent of the time interval over which it is evaluated, i.e. it is moving at a uniform velocity. Also note that the position coordinate is increasing with time so the velocity is positive. □

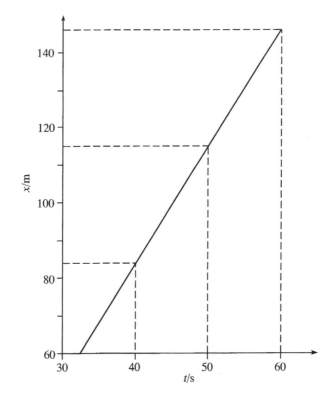

Figure 13.4 Enlargement of the linear portion of Figure 13.3.

Figure 13.4 shows an enlargement of the linear part of Figure 13.3. How would you interpret the calculations performed in the last question in terms of Figure 13.4?

Each of the calculations of velocity involved dividing the 'rise' of the graph (the change in x) by the corresponding 'run' (the corresponding change in t), i.e. finding the *gradient* of the straight line using the general formula

$$\text{gradient} = \frac{\text{rise}}{\text{run}} = \frac{x_2 - x_1}{t_2 - t_1} \quad \square$$

We can summarise this discussion as follows:

For linear motion with uniform velocity, the position–time graph is linear, and the *gradient* of the position–time graph represents the *rate of change* of position with respect to time, i.e. the velocity.

Question TI

Construct a table similar to Table 13.1, and roughly sketch the corresponding position–time graph, when position coordinates are measured:

(a)   from a new origin 20 m to the right of the old origin, with positions to the right defined as positive;

(b)   from the new origin defined in part (a), but this time with positions to the *left* defined as positive.

How do your graphs relate to that of Figure 13.3?   ☐

An important point to notice about motion along the x-axis with uniform velocity v_x is that it implies that the moving object is always travelling in the same *direction*. Sometimes we may want to know how rapidly an object is moving but we may not care about its direction of motion. Under such circumstances the quantity likely to interest us is the **speed** of the object, given by

$$v = |v_x| = \text{magnitude of velocity}$$

Since $|v_x|$ can never be negative it is clear that the speed v can never tell us anything about the direction of motion. Objects moving in opposite directions with velocities of 20 m s^{-1} and -20 m s^{-1} will both have a speed of 20 m s^{-1}.

Question T2

Figure 13.5 shows the position–time graphs for four different bodies each moving with a different constant velocity along the x-axis. If you assume the position and time scales are the same in each case:

(a)   Arrange the bodies in order of increasing speed.

(b)   Which body has the greatest velocity?

(c)   Which body has the least velocity?   ☐

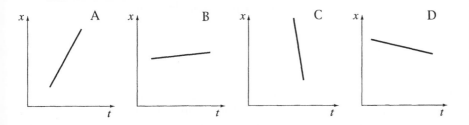

Figure 13.5 See Question T2.

13.2.2   Position–time graphs and instantaneous velocity

You saw in the last subsection that the gradient of a linear position–time graph can represent a constant velocity. This idea of using gradients is also the key to dealing with velocities that are not constant. Figure 13.6 is an enlargement of Figure 13.3 in the interval from 0 to 20 s. In this case the graph is curved which indicates that the velocity is changing. As you will see it is still true that the velocity at any particular time is given by the gradient of the graph at that time, but what exactly do we mean by the gradient at a particular time when the graph is curved? For example, in Figure 13.6, what is the gradient of the curve at $t = 5$ s? The definition of the gradient at a point on a curved graph occupied Newton and Leibniz, the founders of the calculus, more than a little and we will now consider that question in the context of position−time graphs.

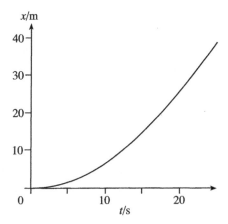

Figure 13.6 Enlargement of the curved portion of Figure 13.3.

How should we set about finding the velocity of the car at $t = 5$ s? Let's try something similar to our approach in the last subsection. In the 15 s interval from 5 s to 20 s the position coordinate changes by 24.3 m, from 1.7 m to 26.0 m. If we divide the change in position coordinate by the time interval we get the **average velocity** $\langle v_x \rangle$ over that time interval.

So, over the interval from 5 s to 20 s,

$$\langle v_x \rangle = 24.3 \text{ m}/15 \text{ s} = 1.6 \text{ m s}^{-1}$$

We can now measure the average velocity over a shorter interval starting at 5 s. For example, over the interval from 5 s to 15 s, the change in position coordinate is 13.3 m.

So, over the interval from 5 s to 15 s,

$$\langle v_x \rangle = 13.3 \text{ m}/10 \text{ s} = 1.3 \text{ m s}^{-1}$$

The values are plotted in Figure 13.7. In principle we could go on like this, using shorter and shorter intervals, but in practice it becomes impossible to continue when the interval becomes so short that the change in position coordinate cannot be reliably determined from the graph. Fortunately, we do not really need to do that as you are about to see.

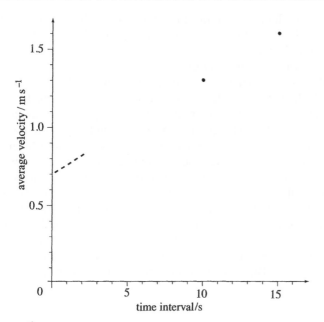

Figure 13.7 The average velocities of the car plotted against the duration of the time intervals over which they have been computed.

Figure 13.7 The average velocities of the car plotted against the duration of the time intervals over which they have been computed.

Question T3

Using Figure 13.6, determine as accurately as you can the average velocities over the following intervals: (a) 5 s to 10 s, (b) 5 s to 9 s, (c) 5 s to 8 s. Then plot the three additional points on the graph in Figure 13.7. ☐

Now imagine using intervals starting at 5 s that are much shorter than those in Question T3 to evaluate some more average velocities.

Where do you think these points would lie on the graph of Figure 13.7?

They would lie on the dashed line shown, which cuts the axis at about 0.70 m s^{-1}. ☐

The velocity of 0.70 m s^{-1} is called the **limit** of the average velocity, i.e. the value that is reached as the time interval from 5 s is made smaller and smaller. This limit gives us a definition of the **instantaneous velocity** at any particular moment:

instantaneous velocity at time t = the limit of the average velocity over an interval around t as that interval is made smaller and smaller

Note that by means of this limiting procedure we are able to make sense of the phrase 'instantaneous velocity' even though the moving object can't really travel any distance at all in an 'instant'. In general, 'the velocity' of an object is taken to refer to its instantaneous velocity at a particular time. Likewise, 'the speed' is generally taken to refer to the **instantaneous speed**, i.e. the magnitude of the (instantaneous) velocity.

It is useful to view what we have done in terms of the gradients of **chords**, i.e. straight lines that cut the curve at two points. In Figure 13.8, the upper straight line passes through the points on the graph corresponding to the times $t = 5$ s and $t = 20$ s. The gradient of this straight line is the *average velocity* over that time interval. Other average velocities over shorter intervals are given by the *gradients* of the other lines in Figure 13.8, each of which cuts the curve at the beginning and endpoint of the relevant interval. As the intervals become smaller and smaller the lines crowd together and become nearly indistinguishable from one another. This corresponds to the average velocity becoming closer and closer to its limit. Figure 13.9 shows the straight line that just touches the graph at 5 s. This line is the 'limit' of the lines drawn in Figure 13.8. It is the **tangent** at $t = 5$ s, and its gradient gives the *instantaneous velocity* at $t = 5$ s. (Tangents were introduced in Chapter 9.)

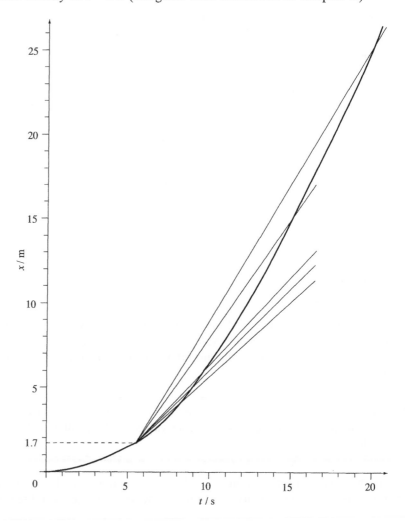

Figure 13.8 The average velocities plotted on Figure 13.7 are the gradients of the lines passing through the point corresponding to time $t = 5$ s.

Measure the gradient of the tangent at 5 s in Figure 13.9.

The answer is approximately 0.70 m s⁻¹, the instantaneous velocity deduced from Figure 13.7. □

Although limited in accuracy because of the difficulty of drawing tangents we now have an alternative definition of the instantaneous velocity. At any time t:

instantaneous velocity = gradient of the tangent to the position–time graph

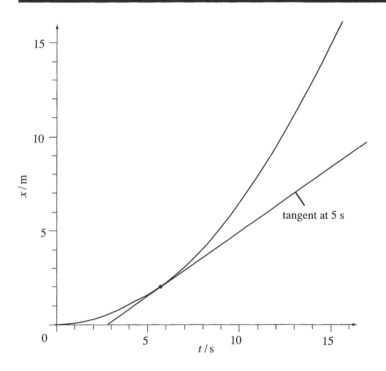

Figure 13.9 The instantaneous velocity of the car at $t = 5$ s is the gradient of the tangent to the graph at $t = 5$ s.

Eventually, this approach will allow us to develop an algebraic technique which is much easier to apply than the graphical method, but before doing that we will consider how the ideas we have already developed allow us to deal with acceleration.

13.2.3 Velocity–time graphs and instantaneous acceleration

Table 13.2 shows the measured velocity v_x at six instants for a ball falling freely from rest. Just as Table 13.1 was used to plot a position–time graph,

Acceleration, like velocity and position, is a *vector* quantity that requires three *scalar components* a_x, a_y and a_z for its full specification in three dimensions. In linear motion a_x alone suffices to specify the acceleration.

Table 13.2 Instantaneous velocity of a falling ball during the first second after being released from rest

Time from release t/s	Velocity of falling ball v_x/m s^{-1}
0.0	0.00
0.2	1.97
0.4	3.93
0.6	6.10
0.8	7.81
1.0	9.80

so we can use Table 13.2 to obtain the **velocity–time graph** shown in Figure 13.10. In this case, the graph is a straight line, i.e. for any given interval of time the velocity changes by a constant amount. The rate of change of velocity with respect to time is called **acceleration**, and it can be found from the gradient of the velocity–time graph. Figure 13.10 shows that for our freely-falling ball the acceleration in the first second is constant or uniform.

For linear motion with **constant** (or **uniform**) **acceleration**, the velocity–time graph is linear, and its gradient represents the rate of change of velocity with respect to time, i.e. the acceleration.

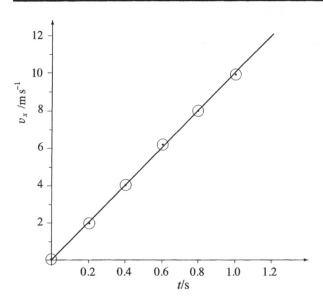

Figure 13.10 The velocity–time graph of a falling ball for the first second after release.

What is the gradient of the linear graph in Figure 13.10? (Take care to include the appropriate SI units.)

The gradient is approximately 10 m s^{-2}. (The units are $(\text{m s}^{-1})/\text{s} = \text{m s}^{-2}$.) □

Acceleration, like velocity, has direction as well as magnitude. By giving the velocities in Table 13.2 a positive sign, we are saying that the downwards direction is positive and, since the velocities are increasing with time, the acceleration is also positive. If the velocities were decreasing in magnitude, but still positive, the acceleration would be negative.

It's worth noting that a negative acceleration does not always slow things down, as the following question shows.

Suppose an object is moving in the negative x-direction and it is subject to a negative acceleration. Does the magnitude of its velocity (i.e. its speed) increase or decrease?

> The initial velocity is negative and negative acceleration will make this more negative − but the magnitude of the velocity (i.e. speed) will increase. □

The term **deceleration** is often used to describe the slowing down of a body. It is sometimes stated that deceleration is negative acceleration but, as the question above shows, that is not necessarily the case − a negative acceleration only causes a reduction in speed if the motion is in the direction defined as positive.

Accurate measurement of the acceleration of freely-falling objects in the absence of air resistance gives a value of $9.8\,\text{m}\,\text{s}^{-2}$ (to two significant figures). However, our ball is not falling in a vacuum so *air resistance* tends to slow its fall, as Figure 13.11 shows. After a few seconds the graph curves, rising less and less steeply. Hence the velocity is increasing less rapidly and the acceleration is no longer constant. (This is because the air resistance becomes greater as the ball's velocity increases.) Faced with this changing acceleration we can take the same approach to **instantaneous acceleration** as we took to instantaneous velocity: draw a tangent to the curve at the relevant time and measure its gradient. In Figure 13.11 a tangent has been drawn at the point on the curve where the time is 3.75 s. The tangent has a gradient of $3.9\,\text{m}\,\text{s}^{-2}$, so this is the instantaneous acceleration of the ball at 3.75 s. So, for linear motion, at any time t:

> instantaneous acceleration = gradient of the tangent to the velocity−time graph

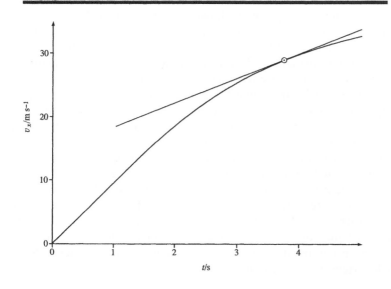

Figure 13.11 The velocity–time graph of a falling ball. The gradient of the tangent line at 3.75 s gives the instantaneous acceleration of the ball at 3.75 s.

Question T4

Figure 13.12 shows the velocity–time graph of a car on a short journey along a straight road. Measure the instantaneous acceleration at 10 s, 40 s and 55 s. Give a one-sentence description of the journey using everyday language. ☐

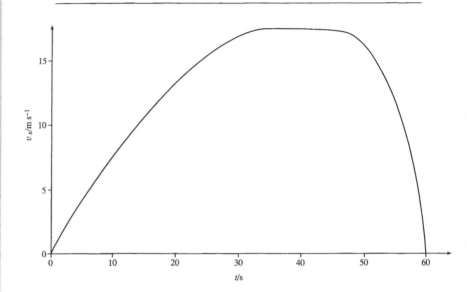

Figure 13.12 See Question T4.

13.2.4 Gradients and rates of change in general

The crucial relationship in our discussion has been between the gradient of a tangent at a point on a curve and a rate of change. The gradient of the tangent at a point on the position–time curve gives the rate of change of position with respect to time, i.e. the instantaneous velocity, at the time corresponding to where the tangent is drawn. The gradient of the tangent at a point on the velocity–time curve gives the rate of change of velocity with respect to time, i.e. the instantaneous acceleration, again at the time corresponding to where the tangent is drawn. This approach works in many other situations. For example, if we have a graph showing how the temperature of an object changes with time then the gradient of the tangent at a point on that graph gives the instantaneous rate of change of temperature with respect to time at that specific time. The gradients of the tangents at different points on the curve give the rates of change of temperature at different times. Although the expression 'rate of change' is used, the quantity plotted horizontally need not be time. If temperature is plotted against altitude then drawing a tangent to the curve and finding its gradient will give the rate of change of temperature with respect to altitude at the altitude corresponding to the point where the tangent is drawn. Another example is the plot of the magnitude of the force exerted by a spring against its extension. Again the gradient at any point gives the rate of change of force magnitude with respect to extension at that point.

Question T5

Figure 13.13 shows a graph of the volume V of a sample of gas against its pressure P (V is measured in litres (l) and P in atmospheres (atm)) (☞). Estimate the rate of change of volume with respect to pressure at $P = 1$ atm. What happens to the rate of change of volume with respect to pressure as the pressure increases? What happens to the rate of change of pressure with respect to volume as the volume increases? ☐

The atm (atmosphere) is a unit of pressure equivalent to 1.013×10^5 pascal.

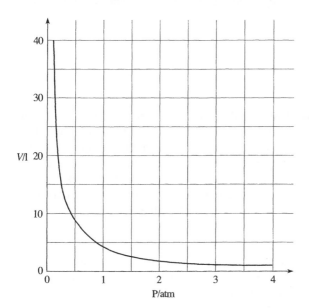

Figure 13.13 See Question T5.

13.3 Rates of change: functions and derivatives

Section 13.2 should have convinced you that determining the gradients of tangents can provide useful information in many situations. However, performing this task graphically is very tedious and not always precise. Fortunately there is a non-graphical technique for calculating gradients that can be applied if we know the equation that describes the shape of a graph. This technique is called **differentiation**. It is part of that branch of mathematics known as **calculus** and was developed initially to study motion. In this section, we will briefly review the concept of a *function* and then introduce differentiation.

13.3.1 Functions, graphs and derivatives

Crudely speaking, a *function* is a rule (usually written as an algebraic equation) that relates two *sets* (usually sets of numbers or values). For example, the rule $x = t^2$ relates any real number t to a non-negative real number x. We show that x is determined by t by writing

Note that x and t are now being used to represent real numbers rather than the physical quantities position and time.

Table 13.3 A table of values for the function $x(t) = t^2$

t	$x(t)$
0.0	0.00
0.5	0.25
1.0	1.00
1.5	2.25
2.0	4.00
2.5	6.25
3.0	9.00

The terms used here to describe functions are discussed more fully in the chapter on *functions and graphs*.

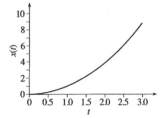

Figure 13.14 The graph of the function $x(t) = t^2$.

This equation is an assertion. You are not expected to be able to show that it is so at this stage, but you will be able to do so later.

$x'(t)$ should be read as 'x prime of t'.

$x(t) = t^2$ for any real number t

and we say that x is a *function* of t. For any value of t we can use the given rule to work out a corresponding value of x; if $t = 2$ then $x = 4$, if $t = 3$ then $x = 9$ and so on. (Note that when we write $x(t)$ in this context we do *not* mean $x \times t$.) When dealing with functions of this kind we call x and t *variables* and we say that x is a *dependent variable*, the value of which depends upon the value of the *independent variable t*. Table 13.3 lists some representative values of $x(t)$ for various values of t while Figure 13.14 shows the *graph* of the function obtained by plotting $x(t)$ along the vertical axis and t along the horizontal axis.

Question T6

Plot the graph of the function $x(t) = 4t + 7$. ☐

A note on mathematical terminology The set of allowed values for the independent variable of a function is called the *domain* of the function. Another set called the *codomain* contains all of the allowed values of the dependent variable (and possibly some other values besides). The formal definition of a function requires that the rule which defines the function should be applicable to *each* value within its domain and that corresponding to each such value there should be a *single* value in the codomain.

It is easy to discuss linear motion in terms of functions. For instance, suppose an object is moving along the x-axis of a coordinate system with constant velocity u_x. If we let t represent the time that has elapsed since the object passed through the origin, we can say that the position x of the object is a *function* of the time t and is given by

$$x(t) = u_x t$$

Similarly, the position of an object that starts from rest at the origin at time $t = 0$ and moves with constant acceleration a_x is given by the function

$$x(t) = \tfrac{1}{2} a_x t^2 \quad (\text{☞})$$

Note that the functions are different in these two cases because the physical circumstances (constant velocity in the first case, constant acceleration in the second) are different. In general we might expect that any particular kind of linear motion will correspond to a particular form for the function $x(t)$.

Now, the process of *differentiation* has the following purpose:

Given a function $x(t)$, the process of differentiation can provide a related function $x'(t)$ such that the value of $x'(t)$ at any particular value of t is equal to the gradient of the tangent to the graph of x against t at that particular value of t.

The new function $x'(t)$ produced by the process of differentiation is called the **derived function** or **derivative** of $x(t)$ and is usually indicated by the symbol $\dfrac{dx}{dt}$ or, more formally, $\dfrac{dx}{dt}(t)$ to remind us that:

1. it is a *function* of t;
2. its value at any particular value of t is equal to the *gradient* of the graph of x against t at that value of t.

Note that $\dfrac{dx}{dt}$ is *not* a ratio of two quantities dx and dt, it is a single symbol that represents a particular function. This is why we sometimes write this as $\dfrac{dx}{dt}(t)$.

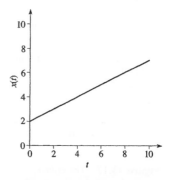

Figure 13.15 Graphical representation of the function $x(t) = 0.5t + 2$.

13.3.2 A linear function and its derivative

This subsection uses the algebraic view of a function to calculate derivatives and hence the gradients of graphs of functions. Soon you will be able to dispense with the graphs altogether and calculate derivatives directly.

The most general form of a *linear function* is $x(t) = mt + c$, where m and c are constants. The graph of such a function is a straight line of gradient m that intersects the vertical axis at c. Figure 13.15 shows the graph of $x(t) = 0.5t + 2$. The gradient of the graph at the point corresponding to $t = 2$ is 0.5, and it is clear that the gradient has the same value at any other point on the graph. In other words the derivative of $x(t) = 0.5t + 2$ is 0.5 at $t = 2$ and at every other value of t. So, in this case, $x'(t) = dx/dt = 0.5$ for all values of t.

> What is the gradient at the point corresponding to $t = 2$ on the graph of the function $x(t) = mt + c$? What is the gradient at *all* points on the graph? What is the derivative (derived function) of $x(t)$?
>
> ---
>
> The gradient is m at $t = 2$ and for all other values of t.
>
> The derivative of $x(t)$ is $dx/dt = m$ for all values of t. □

In Figure 13.15, x and t represented numerical quantities, but if they had represented the position x (in metres) of an object at time t (in seconds), then the gradient of the graph would have represented the velocity v_x (in m s^{-1}) of the object. Thus the derivative $\dfrac{dx}{dt}(t)$ describes the velocity at time t. We can summarise this result in Figure 13.16 and as follows:

The derivative of the linear function $x(t) = mt + c$ is the constant function which takes the value m everywhere. This corresponds to the result that if the position coordinate, x, increases linearly with time, t, then the velocity v_x is constant.

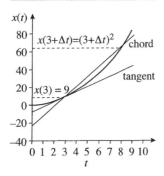

Figure 13.17 The chord joining the points corresponding to $t = 3$ and $t = 3 + \Delta t$ on the curve $x(t) = t^2$.

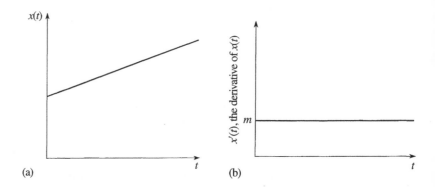

(a)   (b)

Figure 13.16 Graphs of (a) $x(t) = mt + c$ and (b) its derivative, which is the constant function $x'(t) = m$.

Question T7

The velocity, v_x, of an object moving along a straight line is given by the function $v_x(t) = u_x + a_x t$ where u_x and a_x are constants. Find the derivative of $v_x(t)$ at $t = 2$ s. What is the derivative at an arbitrary value of t? In the light of your result interpret the meaning of the constants a_x and u_x. ☐

13.3.3   A quadratic function and its derivative

Now that you have seen how to differentiate linear functions let us consider *quadratic functions*. Recall Figure 13.8 in which we used a sequence of *chords* 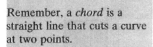 to approximate the tangent to a curve and thus found the gradient of the curve at a point. Let us apply this technique to the differentiation of a simple quadratic function.

Suppose the function we wish to differentiate is $x(t) = t^2$, and suppose we want to find the gradient when $t = 3$. The corresponding value of x is $x(3) = 3^2$. Imagine drawing a chord that cuts the graph of $x(t)$ at $t = 3$ and cuts it again at another value which we denote $t = 3 + \Delta t$ (see Figure 13.17). Since $x(t) = t^2$ the value of x corresponding to $t = 3 + \Delta t$ is $x = (3 + \Delta t)^2$. The gradient of the chord is therefore

$$\text{gradient of chord} = \frac{\text{rise}}{\text{run}} = \frac{(3 + \Delta t)^2 - 3^2}{(3 + \Delta t) - 3}$$

$$= \frac{9 + 6\Delta t + (\Delta t)^2 - 9}{\Delta t} = \frac{6\Delta t + (\Delta t)^2}{\Delta t} = 6 + \Delta t$$

Let us use this expression to calculate the gradients of a sequence of different chords that provide closer and closer approximations to the tangent at $t = 3$. The chord which intersects the curve at $t = 3$ and $t = 4$ has $\Delta t = 1$ and therefore a gradient of $6 + \Delta t = 6 + 1 = 7$. The gradients of the other chords with smaller Δt approach the gradient of the tangent as Δt approaches zero. But

Remember, a *chord* is a straight line that cuts a curve at two points.

Δt represents a small increment in t. It denotes a single number — it does *not* mean a number Δ multiplied by a number t.

Note that the square of Δt is written $(\Delta t)^2$ to reinforce the idea that Δt is a single number.

the gradient of each chord is given by $6 + \Delta t$, which approaches 6 as Δt approaches zero. So 6 is the value for the gradient of the tangent at $t = 3$. You can demonstrate this for yourself in Question T8.

Question T8

Complete Table 13.4, where the third column is the gradient of the chord joining the points on the curve corresponding to $t = 3$ and $t = (3 + \Delta t)$. □

Table 13.4 See Question T8

$t + \Delta t$	Δt	Gradient
3.5	0.5	6.5
3.4	...	...
...	0.2	...
...	...	6.01

We can now generalise this process to obtain the gradient at *any* value of t. Pick a value t and a value close to it denoted by $t + \Delta t$. The corresponding values of x are

$$x(t) = t^2$$

and $x(t + \Delta t) = (t + \Delta t)^2$

The gradient of the chord that cuts the graph at t and $t + \Delta t$ is given by

$$\text{gradient of chord} = \frac{x(t + \Delta t) - x(t)}{(t + \Delta t) - t} = \frac{(t + \Delta t)^2 - t^2}{(t + \Delta t) - t}$$

$$= \frac{t^2 + 2t(\Delta t) + (\Delta t)^2 - t^2}{\Delta t} = 2t + \Delta t$$

Now let the chord approach the tangent at t, i.e. let Δt approach zero. This gives $2t$ as the gradient at time t. When $t = 3$ the gradient given by this expression is 6 ($= 2 \times 3$) as we saw before. But we now have a general expression for the gradient at any value of t and hence an expression for the derived function

$$x'(t) = \frac{dx}{dt}(t) .$$

So, the derivative of a quadratic function $x(t) = t^2$ is

$$\frac{dx}{dt}(t) = 2t .$$

Note that although $x(t) = t^2$ is a quadratic function it is *not* the most general quadratic function — we deal with that shortly.

Note that the derivative $dx/dt = 2t$ is zero when $t = 0$ and is negative when t is less than zero. This is exactly what we expect from the gradient of the graph of $x(t) = t^2$ — sketch the graph yourself if you are not convinced.

13.3.4   Formal definition of a derivative

In the linear and quadratic examples we have investigated so far, we found that at any value of t the derivative dx/dt of a function $x(t)$ is approximated by the gradient $\Delta x/\Delta t$ of a chord. (In this expression $\Delta x = x(t + \Delta t) - x(t)$

and this represents the change in x in the interval from t to $t + \Delta t$.) We also found that the gradient $\Delta x/\Delta t$ provides an increasingly good approximation to dx/dt as Δt approaches zero. Indeed, in the limit as Δt goes to zero it will become exactly equal to the derivative of $x(t)$ at t.

In general, given *any* function $x(t)$ we define the derivative of $x(t)$ by

$$\frac{dx}{dt} = \lim_{\Delta t \to 0} \left(\frac{\Delta x}{\Delta t} \right) = \lim_{\Delta t \to 0} \left[\frac{x(t + \Delta t) - x(t)}{\Delta t} \right] \tag{1}$$

Note that at any particular value of t, $\Delta x/\Delta t$ is a ratio of two quantities (Δx and Δt), but dx/dt is *not* a ratio of dx and dt. Rather you should think of dx/dt at any particular value of t as the value of a function − the derived function $x'(t)$ − at that value of t.

The definition of the derivative in Equation 1 is subject to the additional proviso 'if a unique limit exists'. The proviso is necessary because there are some functions that cannot be differentiated at every point in their domain. (☜) Sometimes $\dfrac{dx}{dt}(t)$ is written rather than dx/dt to stress that the derivative is a function.

A mathematical note   A precise definition of the limit of an expression as Δt approaches zero is surprisingly complicated. The limit is an expression that does not involve Δt but which can be approached as closely as we wish by making Δt small enough. The complication comes in giving precise meaning to the phrases 'as closely as we wish' and 'small enough'. In the cases we will meet, the limit is obvious. For example, $\lim_{\Delta t \to 0} (6 + \Delta t) = 6$, but there are cases where the limit is much less clear.

Although we are now thinking purely in terms of functions, rather than graphs, the process of finding a derivative looks much the same written down as it did before. For example,

if $x(t) = 4 - 2t + 3t^2$

then $x(t + \Delta t) = 4 - 2(t + \Delta t) + 3(t + \Delta t)^2$

thus $\Delta x = x(t + \Delta t) - x(t)$

$= 4 - 2t - 2(\Delta t) + 3t^2 + 6t(\Delta t) + 3(\Delta t)^2 - (4 - 2t + 3t^2)$

$= -2(\Delta t) + 6t(\Delta t) + 3(\Delta t)^2$

Hence, $\Delta x/\Delta t = -2 + 6t + 3\Delta t$

So, $\dfrac{dx}{dt} = \lim_{\Delta t \to 0} \left(\dfrac{\Delta x}{\Delta t} \right) = -2 + 6t$

Question T9

Find dx/dt for each of the following functions:

(a) $x(t) = 14 - t$ (b) $x(t) = 1 + t + t^2$

(c) $x(t) = (1 - t)^2$ □

The graphs of these non differentiable functions often have sharp kinks in them and are generally rather 'unphysical'. Nonetheless, it is important to appreciate that such functions exist.

13.3.5 Velocity and acceleration as derivatives

In this subsection we will gather together our results concerning derivatives and apply them to position, velocity and acceleration.

Suppose that $x(t)$, $v_x(t)$, $a_x(t)$ are, respectively, the position, velocity and acceleration of an object at time t as it moves along the x-axis of a coordinate system. Since velocity is the rate of change of position with respect to time we can write

$$v_x(t) = \frac{dx}{dt} = \text{rate of change of position with respect to time} \quad (2)$$

Remember, in three dimensions the quantities, x, v_x and a_x are each components of a vector quantity.

Similarly, since acceleration is the rate of change of velocity with respect to time

$$a_x(t) = \frac{dv_x}{dt} = \text{rate of change of velocity with respect to time} \quad (3)$$

Thus by two successive differentiations we can go from an expression giving the position to one giving the acceleration. The following questions illustrate this, and also yield some useful general results about differentiation.

Question T10

At time t, the position x of an object moving along a straight line is given by $x(t) = p + qt + rt^2$ where p, q and r are constants. Show that the velocity at time t is $v_x(t) = q + 2rt$. ☐

In order to find the acceleration of this object we must differentiate again.

If $v_x(t) = q + 2r\,t$ find $\dfrac{dv_x}{dt}$.

$$\frac{dv_x}{dt} = \lim_{\Delta t \to 0} \left[\frac{v_x(t + \Delta t) - v_x(t)}{\Delta t} \right]$$

$$= \lim_{\Delta t \to 0} \left[\frac{q + 2r(t + \Delta t) - q - 2rt}{\Delta t} \right] = \lim_{\Delta t \to 0} \left(\frac{2r\Delta t}{\Delta t} \right) = 2r \quad ☐$$

If $x(t)$ is the quadratic function

$$x(t) = p + qt + rt^2 \quad (4)$$

where p, q and r are constants, then

Of course, we could have obtained the same result by noticing that $v_x(t)$ is a *linear function* and $2r$ is its gradient.

$$dx/dt = q + 2rt \tag{5}$$

and if $v_x(t)$ is the linear function

$$v_x(t) = q + 2rt \tag{6}$$

then

$$dv_x/dt = a_x(t) = 2r \tag{7}$$

What are the physical interpretations of the constants p, q and r?

If we look at the expressions for x and v_x when $t = 0$, we can see that p must be the initial position $x(0)$ and q must be the initial velocity $v_x(0)$. The expression for dv_x/dt shows that r is half the (constant) acceleration. ☐

Since r is a constant we have found that a body whose position is a quadratic function of time must be moving with constant acceleration. In fact the converse also holds, and the position is a quadratic function if and only if the acceleration is a constant.

Question T11

If $p = -2 \, \text{m}$, $q = -3 \, \text{m s}^{-1}$, and $r = 1 \, \text{m s}^{-2}$ in Equations 4–7, what is t when the velocity is zero? ☐

Finally, we note that when discussing linear motion it is a common practice to describe the location of the moving object in terms of its *displacement* $s_x(t)$ from its initial position $x(0)$. (☞) Thus, rather than basing the discussion on the position $x(t)$, the emphasis is on the *difference* in positions:

$$s_x(t) = x(t) - x(0)$$

This has the advantage that when $t = 0$ we are certain to find $s_x(0) = 0$.

Since $x(0)$ is a *fixed* point it is bound to be true that $\dfrac{ds_x}{dt} = \dfrac{dx}{dt}$, so at any time t the velocity of the moving object will be

$$v_x(t) = \frac{ds_x}{dt}$$

Question T12

The motion of an object travelling with constant acceleration a_x along the x-axis of a coordinate system is described by the equations

$$s_x(t) = u_x t + \tfrac{1}{2}a_x t^2 \tag{8}$$

$$v_x(t) = u_x + a_x t \tag{9}$$

where u_x is the velocity at time $t = 0$. Show that Equation 9 is a consequence of Equation 8. ☐

Like position, displacement is a *vector* quantity that requires three *scalar components* s_x, s_y and s_z for its specification in three dimensions.

Sometimes displacements are measured from a *moving* reference point. In such cases $\dfrac{ds_x}{dt}$ is the velocity of the object *relative* to the *moving* reference point and will differ from $\dfrac{dx}{dt}$ which represents the velocity relative to the *fixed* origin.

13.3.6 Derivatives and rates of change in general

We have seen that both velocity and acceleration are rates of change with respect to time. However 'rate of change' is a concept with wider application and does not have to be restricted to change with respect to time. For example, Figure 13.18 shows a spring aligned along the x-axis of a system of coordinates in such a way that when the spring is extended so that its free end is at x (where $x \geqslant 0$) the force F_x (directed along the negative x-direction) exerted by the spring at its free end is $F_x(x) = -kx$ where k is a constant. (In practice such an expression for $F_x(x)$ might hold true over a limited range of values for x.) If you wanted to know how difficult it would be to increase the extension significantly you might well ask 'what is the rate of change of the F_x with respect to x?' In other words, 'what is the derivative of $F_x(x)$ with respect to x?'

dF_x/dx is calculated in the same way as dx/dt — only the letters have changed, not the underlying ideas. Since $F_x(x) = -kx$ is a *linear* function of gradient k it follows that, $dF_x/dx = -k$. Note that in this case x is the *independent variable* and F_x is the *dependent variable*.

Although we have been using x to represent a position coordinate throughout much of this chapter there is no reason why it should always have that meaning. In fact, it is common practice to use the letter x to represent a general independent variable, whilst a general dependent variable is commonly represented by y. (It is in this spirit that we refer to the horizontal and vertical axes of a graph as the x and y axes, respectively.) If we adopt this general notation and regard y as a function of x then, irrespective of the particular meaning we attach to x and y, we can define the rate of change of y with respect to x by

$$\frac{dy}{dx} = \lim_{\Delta x \to 0}\left(\frac{\Delta y}{\Delta x}\right) = \lim_{\Delta x \to 0}\left[\frac{y(x + \Delta x) - y(x)}{\Delta x}\right] \tag{10}$$

Remember that if a unique limit exists for all values of x in the domain of $y(x)$ this procedure defines a function — the *derived function* $y'(x)$ — which has the same domain as the original function.

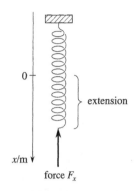

Figure 13.18 A spring aligned along the x-axis of a coordinate system.

In this general notation, notice that x is taken as the independent variable with y as the dependent variable. This is a change from our earlier usage of $x(t)$, where x was the dependent variable and t the independent variable.

Remember that dy/dx is a *function* of x.

Question T13

Suppose the variation of air temperature, T, with altitude, h, is given by $T(h) = ah + b$ where a and b are constants. Find dT/dh, the rate of change of temperature with respect to altitude. What are the physical interpretations of the constants a and b, and what would be their SI units? (T is measured in kelvin K or degrees Celsius °C.) □

Question T14

The kinetic energy, E_{kin}, of a body of mass m moving with speed v is $E_{kin}(v) = \frac{1}{2}mv^2$. Write down or work out an expression for dE_{kin}/dv and explain its physical meaning. □

13.4 **Conclusion**

13.4.1 *The importance of differentiation*

Much of physical science is concerned with the dynamical processes involved with change and development. Differential calculus is the mathematics of change. It is hardly surprising therefore that many physical scientists would say that the development of calculus has been the single most important development in mathematics since the time of the ancient Greeks. Nor is it surprising that mastering the techniques of calculus is an important step in the development of every physical scientist.

In this chapter we have been mainly concerned with definitions and we have limited our considerations to the differentiation of linear and quadratic functions. In practice physical scientists rarely have to bother about limits and such mathematical niceties as domains and codomains; they rely on knowing the derivatives of some simple functions and a few basic rules that enable them to differentiate more complicated functions. Armed with this knowledge they are able to use differentiation to tackle an astonishing range of problems that would otherwise be quite intractable. Those wider skills and their applications are the subject of other chapters.

13.5 **Closing items**

13.5.1 *Chapter summary*

1. In *linear motion* the *position* of a moving object can be specified by a single position coordinate x at any time t.

2. The *(instantaneous) velocity* v_x of an object in linear motion is the rate of change of its position coordinate with respect to time.

3. In linear motion, a plot of x against t is called a *position−time graph*. The gradient of the tangent to such a graph at any particular value of t determines the (instantaneous) velocity of the moving object at that time. The velocity is constant if and only if the position−time graph is a straight line.

4. The *(instantaneous) acceleration* a_x of an object in linear motion is the rate of change of its velocity with respect to time.

5. In linear motion, a plot of v_x against t is called a *velocity−time graph*. The gradient of the tangent to such a graph at any particular value of t determines the (instantaneous) acceleration of the moving object at that time. The acceleration is constant if and only if the velocity−time graph is a straight line.

6. If a single value of a dependent variable x can be associated with each value of an independent variable t in some specified domain, then we say that x is a *function* of t, and we speak of the function $x(t)$ (pronounced 'x of t').

7. Given a function $x(t)$, the rate of change of x with respect to t at any particular value of t is given by

$$\frac{dx}{dt} = \lim_{\Delta t \to 0} \left(\frac{\Delta x}{\Delta t} \right) = \lim_{\Delta t \to 0} \left[\frac{x(t + \Delta t) - x(t)}{\Delta t} \right] \tag{1}$$

If a unique limit exists for all values of t in some domain, then this formula defines a function called the *derivative* or *derived function*, written $x'(t)$ or $\dfrac{dx}{dt}(t)$.

8. The value of the derivative dx/dt of a function $x(t)$ at any given value of t is equal to the gradient of the tangent to the graph of x against t at that value of t.

9. In linear motion, the position coordinate of a moving object may be regarded as a function of time and written $x(t)$. For such an object

$$v_x(t) = dx/dt = \text{rate of change of position coordinate with respect to time}$$

$$a_x(t) = dv_x/dt = \text{rate of change of velocity with respect to time}$$

10. Generally, if y is a function of x, the derivative dy/dx describes the rate of change of y with respect to x.

If $y(x)$ is the linear function $y(x) = ax + b$, then $dy/dx = a$

If $y(x)$ is the quadratic function $y(x) = ax^2 + bx + c$, then

$$dy/dx = 2ax + b$$

13.5.2 Achievements

Having completed this chapter, you should be able to:

A1. Define the terms that are emboldened in the text of the chapter.

A2. Draw position–time and velocity–time graphs.

A3. Understand the relationship between rates of change and gradients of tangents and how they can be approximated by gradients of chords.

A4. Interpret the gradient of a tangent to a position–time or a velocity–time graph as velocity or acceleration, respectively.

A5. Recognise a linear position–time graph as describing constant velocity and a linear velocity–time graph as describing constant acceleration.

A6. Know the definition of a derivative as a limit and give a graphical interpretation of this definition.

A7. Use the definition of a derivative as a limit to calculate the derivatives of linear and quadratic functions.

A8. Write down an expression for the derivative of a given linear or quadratic function.

A9. Express velocity and acceleration as derivatives for the case of linear motion and solve simple linear motion problems.

A10. Interpret rates of change in terms of the derivative of one quantity with respect to another.

13.5.3 End of chapter questions

Question E1 Sketch roughly the position–time graph of a stone released from rest at a height of 10 m above the ground. Take downwards as the positive x-direction, and note that (neglecting air resistance) such a stone has a constant acceleration $a_x = 9.8$ m s^{-2}. (Take the origin as the initial position of the stone.)

Question E2 Figure 13.19 shows the position–time graph of five different bodies each moving with different constant velocity. If you assume the position and time scales are the same in each case:

(a) Arrange the bodies in order of increasing speed.

(b) Which body has the greatest velocity?

(c) Which body has the least velocity?

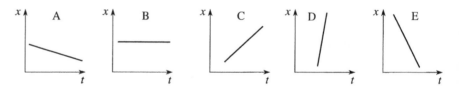

Figure 13.19 See Question E2.

Question E3 Table 13.5 shows position–time measurements of a ball falling from rest under gravity.

(a) Plot the position–time graph.

(b) Using the graph determine the instantaneous velocity at 0.2 s.

(c) Repeat the procedure at 0.6 s.

Table 13.5 See Question E3

Time/s	Position/m
0.0	0.00
0.1	0.05
0.2	0.20
0.3	0.40
0.4	0.80
0.5	1.20
0.6	1.80
0.7	2.40
0.8	3.20

Question E4 Define the term velocity and explain two methods of obtaining instantaneous velocity from a position–time graph.

Question E5 Write down the definition of the derivative as a limit and use it to find dx/dt when:

(a) $x(t) = 3t + 6$ (b) $x(t) = 6t^2 + 4t$ (✎)

Question E6 Use the formulae for the derivative of linear and quadratic functions to write down the derivatives with respect to t of: (☞)

(a) $x(t) = -4 + 7t - 3t^2$

(b) $x(t) = (5t + 7)(2t - 6)$

(c) $x(t) = u_x t - \frac{1}{2}gt^2$

where u_x and g are constants.

For the expressions in Questions E5, E6(a) and E6(b) t is to be taken as a dimensionless parameter, not as time.

Question E7 If $x(t) = p - qt + rt^2$ find the acceleration at time $t = 2$ s if $p = 13$ m, $q = 11$ m s^{-1}, and $r = 7$ m s^{-2}.

Question E8 The area, A, of a circle radius r is πr^2. (a) What is meant by dA/dr? Find an expression for dA/dr. (b) Write down an expression for the difference, D, between the area of a square of side $2r$ and a circle of radius r. (c) Find an expression for dD/dr and interpret it.

Chapter 14 Differentiating simple functions

14.1 Opening items

14.1.1 Chapter introduction

Being able to *differentiate* accurately and confidently is a necessary skill for any physicist. Differentiation is such a powerful tool with so many applications that it is worth some effort to ensure that you have mastered the techniques explained in this chapter.

The main objective of this chapter is to answer the following question:

'If we know how to differentiate two functions $f(x)$ and $g(x)$, how do we differentiate their sum, difference, product and quotient?'

To answer this question we discuss first the idea of a *function* and how functions can be combined to produce new ones. Section 14.2 briefly reviews the mathematical basis of *differentiation*, giving first a graphical interpretation and then a formal definition.

Section 14.3 answers our question and is the core of the chapter. It provides a list of derivatives of various standard functions (such as power functions (x^n) and trigonometric functions) and introduces the various rules that will enable you to differentiate a wide range of combinations of these standard functions. In particular it will explain how to differentiate *sums*, *constant multiples*, *products* and *quotients* of the standard functions. Sometimes the method is straightforward, as with the derivative of a *sum*, but it may also be quite complicated, as with the derivative of a *quotient*.

We also discuss the *logarithmic* and *exponential* functions and, while $\log_e x$ and e^x are the most important of such functions, it is sometimes necessary to be able to differentiate $\log_a x$ and a^x where a is some positive number other than e. In Section 14.4 we consider why e^x is so important and then add a^x and $\log_a x$ to the list of functions that can be differentiated.

14.1.2 Fast track questions

Question F1

Differentiate each of the following functions (it is not necessary to simplify your answers).

(a) $f(x) = x^2 \sin x + \log_e x$

(b) $f(x) = (x^3 - 6x)(2x^2 + 5x - 1)$ (use the *product rule*)

(c) $f(x) = \dfrac{1}{\sin x}$

(d) $f(x) = \dfrac{x^2 + 4x - 1}{x^2 + 9}$

(e) $f(x) = \dfrac{x \tan x}{3x^2 - 2x + 1}$

(f) $f(x) = 2^x + x^2 + e^2$

(g) $f(x) = (e^x + e^{-x})^2$

Question F2

If at time t the displacement, $s_x(t)$, of a particle along the x-axis from some particular reference point is given by

$$s_x(t) = e^{-\alpha t}[A \cos(\omega t) + B \sin(\omega t)]$$

where A, B, α and ω are constants, find expressions for $v_x(t)$, the velocity at time t, and $a_x(t)$, the acceleration at time t, and show that

$$a_x(t) + 2\alpha v_x(t) + (\omega^2 + \alpha^2)s_x(t) = 0$$

14.1.3 Ready to study?

Study comment

To begin the study of this chapter you will need to be familiar with the following terms: *base* (of a *logarithm*), *exponential function*, *inequality* (in particular *greater than* (>) and *less than* (<)), *integer*, *inverse function*, *logarithmic function*, *modulus* (or *absolute value*, $|x|$), *product*, *quotient*, *radian*, *real number*, *reciprocal*, *set*, *square root*, *sum*, *trigonometric function* and *trigonometric identities*. (The *trigonometric identities* that you require are repeated in this chapter, as is the definition of the *modulus* function.) In addition you will need to have some familiarity with the concept of a *function*, and with related terms such as *argument*, *codomain*, *domain* and *variable* (both *dependent* and *independent*), but the precise meaning of these terms is briefly reviewed in the chapter. Similar comments apply to physics concepts used in this chapter, such as *acceleration*, *displacement*, *force*, *speed*, *velocity* and *Newton's second law*. The following *Ready to study questions* will allow you to establish whether you need to review some of the topics, or to improve your general algebraic skills, before embarking on this chapter.

Question R1

Write the following sums, products and quotients of logarithms and exponentials as compactly as you can:

(a)　$\log_{10} x + \log_{10} y$　　　　　　　(b)　$(\log_2 4 \times \log_2 4)^3$

(c)　$\log_e x - \log_e y$　　　　　　　　　(d)　e^x/e^y

(e)　$(e^x/e^y)^2$　　　　　　　　　　　　(f)　$\log_e x + \log_e x + \log_e x$

The notations $\sin x$ and $\sin (x)$ are identical in meaning. On the whole we will follow customary practice and omit the brackets in this chapter unless the argument of the function consists of something more than a single letter. Similar comments apply to other functions, such as $\tan (x)$ and $\log_e (x)$.

Question R2

If $f(x) = \cos x$, $g(x) = \sin x$, and $h(x) = \tan x$, rewrite the following expressions in terms of trigonometric functions and simplify them where possible:

(a)　$[f(x) + g(x)]^2$

(b)　$f(x)h(x) + g(x)/h(x)$

(c)　$[f(x) - g(x)]/[1 + h(x)]$

Question R3

Given that $f(x) = \dfrac{x^2}{4} - 1$, find the following values, $f(1)$, $f(0)$, $f(-2)$, $|f(0)|$, $|f(0) - f(-2)|$ and sketch the graph of $y = f(x)$

Question R4

Given that a^x and $\log_a x$ are inverse functions, simplify the following:

On some calculators $\log_e$ corresponds to the ln key.

(a)　$\log_e[e^{2 \log_{10}(x)}]$　　(b)　$\exp(\log_e(10^x))$　(c)　$\log_2(2^2)$.

14.2　Functions and derivatives – a brief review

14.2.1　Functions and variables

As stressed in the last chapter, a *function f* is a rule that assigns a single value $f(x)$ in a set called the *codomain* to each value x in a set called the *domain*.

Functions are very often defined by formulae, for example $f(x) = x^2$, and in such cases we assume, unless we are told otherwise, that the domain is the largest set of real values for which the formula makes sense. In the case of $f(x) = x^2$ the domain of the function is the set of all real numbers.

The function $g(x) = \dfrac{1}{1 - x}$ is not defined when $x = 1$, since 1/0 has no meaning, and so we take the set of all real numbers x with $x \neq 1$ as its domain.

The function $f(x) = 1 + x + x^2$　　　　　　　　　　　　　　(1a)

is another example of a function that is defined for all values of x. One may think of the function as a sort of 'machine' with x as the input and $1 + x + x^2$ as the output. The input x is known as the *independent variable* and, if we write $y = f(x)$, the output y is known as the *dependent variable*.

It is important to note that the same function f could equally well be defined using some other symbol, such as t, to represent the independent variable:

$$f(t) = 1 + t + t^2 \qquad (1b)$$

This freedom to relabel the independent variable is often of great use, though it is vital that such changes are made consistently throughout an equation.

We may *evaluate* the function in Equation 1, whether we call it $f(x)$ or $f(t)$, for any value of the independent variable; for example, if we choose to use x to denote the independent variable, and set $x = 1$, we have

$$f(1) = 1 + 1 + 1^2 = 3$$

Similarly, if $x = \pi$ $f(\pi) = 1 + \pi + \pi^2$

and, if $x = 2a$ $f(2a) = 1 + (2a) + (2a)^2 = 1 + 2a + 4a^2$

When we write expressions such as $f(\pi)$ or $f(2a)$, whatever appears within the brackets is called the *argument* of the function. The value of $f(x)$ is determined by the *value* of its argument, irrespective of what we call the argument.

Question T1

If $f(x) = x^2 + 1$ and $g(x) = 2x$,

(a) write down expressions for $f(\sqrt{x})$ and $g\left(\dfrac{x}{2}\right)$.

(b) For which values of the independent variable x is $f(x) = g(x)$?

(c) For which value of the independent variable x is the following true?

$$\frac{f(x + 0.1) - f(x)}{0.1} = 0.2 \quad \square$$

14.2.2   Rates of change, gradients and derivatives

The graph of the function $g(t) = t^2$ is shown in Figure 14.1. It is clear from the graph that as the value of t increases from $t = 0$, the value of $g(t)$ also increases from 0. In particular, note that as t increases from 0 to 1 the value of $g(t)$ changes from 0 to 1, and that as t increases from 1 to 2 the value of $g(t)$ changes from 1 to 4. Thus, the change in the value of $g(t)$ that corresponds to a change of one unit in the value of t depends on the initial value of t. For this particular function, if the initial value of t is large, then the change in $g(t)$ will also be large; for example, the change in $g(t)$ from $t = 100$ to $t = 101$ is

$$g(101) - g(100) = 101^2 - 100^2 = 201$$

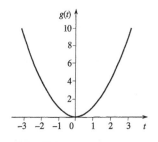

Figure 14.1   The graph of $g(t) = t^2$.

Clearly, as t increases the *rate* at which $g(t)$ is changing is itself changing, becoming greater and greater. How are we to measure the rate at which a function changes? If you imagine yourself walking up a hill in the shape of the graph shown in Figure 14.1, starting at $t = 0$ and travelling to the right, your path would become steeper and steeper. It would become increasingly difficult to make progress, because, with each step you took, the slope would become greater and your walk would become more and more of a climb. The *slope* of the graph is the key; once we can describe the slope precisely for any value of t, we will be able to measure the *rate of change* of the function $g(t)$ − or of any other function.

The graph of an arbitrary function $f(x)$ is shown in Figure 14.2a. Roughly speaking, the slope of this graph at the point P, where $x = a$ and $f(x) = f(a)$, is the same as the slope of the dashed straight line PQ where Q is a nearby point corresponding to $x = a + h$ and $f(x) = f(a + h)$. Of course, the two slopes are not exactly the same; the point P is fixed and its location determines the slope of $f(x)$ at P, whereas the slope of the line PQ will depend on the location of Q as well as that of P. Nonetheless, the two slopes are similar, so we can roughly describe the slope of the curve in terms of the **gradient** of the line PQ which is defined by

$$\text{gradient of the line PQ} = \frac{\text{rise}}{\text{run}} = \frac{f(a + h) - f(a)}{(a + h) - a}$$

$$= \frac{f(a + h) - f(a)}{h} \quad (☜) \quad (2)$$

It is quite common to use Δx in place of h to indicate a small change in the variable x. (Δ is the upper case Greek letter delta.) Moreover, if we let $y = f(x)$ then we can use Δy to represent $f(a + h) - f(a)$, and we can write gradient of the line

$$PQ = \frac{\Delta y}{\Delta x}$$

This is what was done in the last chapter.

Now, if we let the point Q get closer and closer to P then this **chord** becomes more and more like the **tangent** in Figure 14.2b that just touches the graph at P. It is the gradient of this tangent that really represents the slope of the curve at P. This is clear from the figure and from the fact that the gradient of the tangent (unlike that of the chord) is determined by the location of P alone. Fortunately, the gradient of the tangent is fairly easy to work out, since all we have to do is to consider what happens to the gradient of the chord PQ as Q gets closer and closer to P. As Q approaches P, h gets smaller and the gradient of the chord PQ approaches the gradient of the tangent at P ever more closely. Expressed more formally, the gradient of the tangent at P is given by Equation 2 in the **limit** as h tends to zero. Thus

$$\text{gradient of tangent at P} = \lim_{h \to 0} \left[\frac{f(a + h) - f(a)}{h} \right] \quad (3)$$

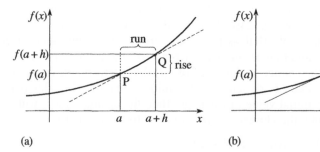

(a)   (b)

Figure 14.2 The graph of an arbitrary function.

We call this gradient the *derivative* of $f(x)$ at $x = a$ and denote it by $f'(a)$ so that

$$f'(a) = \lim_{h \to 0} \left[\frac{f(a + h) - f(a)}{h} \right] \qquad (4)$$

The derivative at $x = a$ defined in this way also represents the **rate of change** of $f(x)$ with respect to x when $x = a$. This last point is even easier to appreciate if we introduce a dependent variable y such that $y = f(x)$, for we can then represent the vertical *rise* in Figure 14.2a by $\Delta y = f(a + h) - f(a)$, and the horizontal *run* by $\Delta x = h$. With these definitions it follows from Equation 4 that the derivative at $x = a$ is given by

$$f'(a) = \lim_{\Delta x \to 0} \left(\frac{\Delta y}{\Delta x} \right)$$

Even greater emphasis can be given to the idea that the derivative is a rate of change with respect to x by replacing $f'(a)$ by the alternative symbol $\frac{dy}{dx}(a)$.

$$\text{Thus,} \quad \frac{dy}{dx}(a) = \lim_{\Delta x \to 0} \left(\frac{\Delta y}{\Delta x} \right) = \lim_{h \to 0} \left[\frac{f(a + h) - f(a)}{h} \right] \qquad (5)$$

$\frac{dy}{dx}$ is read as 'dee y by dee x'.

The definition of the derivative at $x = a$ represented by Equations 4 and 5 can be applied at any point on the graph provided that Δy and Δx can be defined at that point and that their quotient $\Delta y/\Delta x$ has a unique limit as $\Delta x \to 0$ at that point. Thus, provided the derivative of $f(x)$ exists at every point within some domain (i.e. some set of x values) it is possible to define a new function on that domain that associates any given value of x with the gradient of $f(x)$ at that point. This new function is the **derived function** or **derivative** of $f(x)$ and is written $f'(x)$ or $\frac{df}{dx}(x)$. If $y = f(x)$, the derived function may also be written $\frac{dy}{dx}(x)$ or just $\frac{dy}{dx}$. (☞) Thus, provided unique limits exist,

Although this terminology may blur the important distinction between a function and the value of the function at a particular point, in what follows we will often refer to derived functions as derivatives and represent them by $\frac{dy}{dx}$.

$$\frac{dy}{dx} = \lim_{h \to 0} \left[\frac{f(x + h) - f(x)}{h} \right] \qquad (6)$$

When using this formula it is important to remember that dy/dx is *not* a quotient of two quantities dy and dx even though it may look like one. At a given value of x, the derivative dy/dx represents the gradient of the graph of $y = f(x)$ at that value of x.

Apart from a slight change of notation, this is the main conclusion we arrived at in the last chapter. Using it we can easily show (as we effectively did in Subsection 13.3.4) that the derivative of $f(x) = x^2$ is $f'(x) = 2x$, and that the derivative of $f(x) = 1/x$ is

$$f'(x) = -\frac{1}{x^2} \qquad\qquad (7)$$

The function $f(x) = 1/x$ has the set of all non-zero real numbers as its domain, and its derivative $f'(x) = -1/x^2$ has the same domain.

Question T2

(a) If $f(x) = 1/(\omega x)$ where ω is a constant, use the definition of the derivative to find $f'(x)$.

(b) If $g(t) = 1/(\omega t)$ use the answer to part (a) to write $g'(t)$ and $g'(2t)$.   □

$f(x)$

x

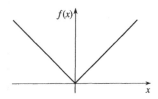

Figure 14.3   The graph of $f(x) = |x|$.

Differentiation is a powerful technique, but don't be misled into thinking that if a function exists then its derivative must also exist. The function $f(x) = |x|$ ('the *modulus* of x') illustrates the situation where the derivative has a smaller domain than the function (see Figure 14.3). The modulus of x is defined by

$$|x| = x \quad \text{if } x > 0$$

and $|x| = -x$ if $x < 0$

For positive x the graph of $f(x) = |x|$ is a straight line with gradient 1, so

$$f'(x) = 1 \text{ if } x > 0$$

while for negative x the graph is a straight line with gradient -1, so

$$f'(x) = -1 \text{ if } x < 0$$

However, when $x = 0$ no unique limit exists since we get different values for the limit depending on how we calculate it. In particular, if we approach $x = 0$ from the positive side, where $x > 0$ and h is positive

$$\lim_{h\to0+}\left[\frac{f(0+h)-f(0)}{h}\right] = \lim_{h\to0+}\left(\frac{|h|-|0|}{h}\right) = \lim_{h\to0+}\left(\frac{|h|}{h}\right) = 1$$

whereas, if we approach $x = 0$ from the negative side, where $x < 0$ and h is negative

$$\lim_{h\to0-}\left[\frac{f(0+h)-f(0)}{h}\right] = \lim_{h\to0-}\left(\frac{|h|-|0|}{h}\right) = \lim_{h\to0-}\left(\frac{|h|}{h}\right) = -1$$

So, if h approaches zero through positive values the limit is 1, but if h approaches zero through negative values the limit is -1. A *unique* limit exists only if we obtain the *same* result no matter how h approaches 0. Since *different* answers have been obtained in this case there is no unique limit and hence no derivative at $x = 0$. Although the original function is defined for all real values of x the derivative is only defined for non-zero values of x.

On the whole, the domains of functions are of more concern to mathematicians than to physical scientists. Nonetheless, it is important that you should be aware of their significance, and prepared to investigate them if necessary.

Question T3

(a) Use your calculator to convince yourself that $\dfrac{\exp(h) - 1}{h}$ is very nearly equal to 1 when h is very small.

(b) Let $f(x) = \exp(x)$. What does part (a) tell you about $f'(0)$?

(c) Use the fact that $\lim\limits_{h \to 0}\left[\dfrac{\exp(h) - 1}{h}\right] = 1$ and the definition of the derivative to show that $f'(x) = \exp(x)$. □

exp (x) is often written in the alternative form e^x where e $= 2.718\,281\,828\,46$ to 11 decimal places. When written in this form it is easy to see that exp $(0) = 1$ and that exp $(x + h) = $ exp (x) exp (h).

14.2.3 Notation for derivatives

Various notations are in common use for the derivative of a given function. Two such notations have already been introduced in this chapter and you will probably meet a third elsewhere. Here we provide a brief summary.

Function notation
This is the notation we have mainly used so far, in which a function is represented by f or $f(x)$ or some similar symbol and its derivative (the derived function) is represented by f' or $f'(x)$. This is a neat and compact notation but care is needed in handwritten work to make sure that the all-important prime ($'$) is clearly visible.

Leibniz notation
This is the dy/dx notation that is especially popular among physicists. We will make much use of it in what follows.

If we let $y = f(x)$, then in Leibniz notation the derivative is written $\dfrac{dy}{dx}(x)$ or $\dfrac{df}{dx}(x)$. Sometimes these are abbreviated to $\dfrac{dy}{dx}$ (in which case it is assumed that we are regarding the variable y as a function $y(x)$ of x), or $\dfrac{df}{dx}$ (in which case it is assumed that we are discussing a function $f(x)$).

An advantage of Leibniz notation is that it is also possible to write the derivative as

$$\frac{d}{dx}[f(x)]$$

so that for a specific function, $f(x) = x^2$ say, we could write

$$\frac{d}{dx}(x^2) = 2x$$

Gottfried Wilhelm Leibniz (1646–1716) was a philosopher and mathematician and one of the co-inventors of *calculus*.

This notation developed from the use of Δx and Δy to represent small increments in the variables x and y, respectively. Leibniz intended dy/dx to represent the ratio $\Delta y/\Delta x$ when both increments were 'infinitesimally small'. (The concept of a limit was not developed until much later.)

Isaac Newton (1642–1727), co-inventor of *calculus* and founder of Newtonian mechanics, is widely regarded as the genius of his age.

Newtonian notation

A different notation that is especially common in mechanics, but which will not be used in this chapter, uses a dot to denote derivatives. Thus, if $x(t)$ represents the x coordinate of a moving particle at time t, then $\dot{x}(t)$ represents the rate of change of x with respect to time, i.e. the x component of the particle's velocity.

Given that $y = f(x) = 1/x$, and recalling Equation 7, determine the following:

(a) $\dfrac{dy}{dx}(2)$ (b) $\dfrac{df}{dx}(2a)$ (c) $\dfrac{d}{dt}\left(\dfrac{1}{t}\right)$ (d) $f'(2 + 3x)$

We know from Equation 7 that if $f(x) = 1/x$, then $f'(x) = -\dfrac{1}{x^2}$, so that

(a) $\dfrac{dy}{dx}(2) = -\dfrac{1}{2^2} = -\dfrac{1}{4}$ (b) $\dfrac{df}{dx}(2a) = -\dfrac{1}{(2a)^2} = -\dfrac{1}{4a^2}$

(c) $\dfrac{d}{dt}\left(\dfrac{1}{t}\right) = -\dfrac{1}{t^2}$ (d) $f'(2 + 3x) = -\dfrac{1}{(2 + 3x)^2}$ □

14.3   Derivatives of simple functions

Now that the definitions are out of the way we can get down to the real business of *doing* calculus. The procedure by which derivatives are determined is called **differentiation**. In practice everybody who uses differentiation regularly knows the derivatives of various standard functions (x, $\sin x$, $\exp(x)$, etc.) and knows some simple rules for finding the derivatives of various combinations (sums, differences, products, quotients and reciprocals) of those standard derivatives. The formal definition of a derivative is rarely used in practice. This section introduces the standard derivatives and the basic rules for combining them.

14.3.1   Derivatives of basic functions

Tables 14.1a and b list a number of functions with which you should already be familiar along with their derivatives. Each of these derivatives can be deduced from the definition given in the last section, though the proof is not always easy. If you are going to use calculus frequently you will need to know these derivatives or at least know where you can look them up quickly. When using the tables it is important to remember the following points:

- n and k are constants.
- The functions in Table 14.1b are special cases of those in Table 14.1a, corresponding to $k = 1$.

- In each of the trigonometric functions x must be an angle in radians or a dimensionless real variable.

Table 14.1 Some standard derivatives

(a)

$f(x)$	$f'(x)$
k (constant)	0
kx^n	nkx^{n-1}
$\sin kx$	$k\cos kx$
$\cos kx$	$-k\sin kx$
$\tan kx$	$k\sec^2 kx$
$\operatorname{cosec} kx$	$-k\operatorname{cosec} kx\cot kx$
$\sec kx$	$k\sec kx\tan kx$
$\cot kx$	$-k\operatorname{cosec}^2 kx$
$\exp(kx)$	$k\exp(kx)$
$\log_e(kx)$	$1/x$

(b) When $k = 1$

$f(x)$	$f'(x)$
1	0
x^n	nx^{n-1}
$\sin x$	$\cos x$
$\cos x$	$-\sin x$
$\tan x$	$\sec^2 x$
$\operatorname{cosec} x$	$-\operatorname{cosec} x\cot x$
$\sec x$	$\sec x\tan x$
$\cot x$	$-\operatorname{cosec}^2 x$
$\exp(x)$	$\exp(x)$
$\log_e(x)$	$1/x$

Question T4

Use the definition of the derivative to show that if $f(x) = x^n$ and n is a positive integer, then $f'(x) = nx^{n-1}$. (☞) ☐

Answer T4 uses the binomial *theorem. If you are not familiar with this just read the answer, do not attempt the question.*

The best way to get to know the standard derivatives is to use them frequently. Here are a few questions to start the process of familiarisation.

Find the derivatives (using Table 14.1) of the following functions:

(a) $f(x) = x^3$ (b) $g(t) = 1/t^3$ (c) $h(x) = h(x) = \dfrac{1}{\sqrt[5]{x^2}}$

In each case we express the function in the form x^n so that we may apply the rule discussed in Question T4.

(a) For $f(x) = x^3$ we have $n = 3$, so $f'(x) = 3x^{3-1} = 3x^2$.

(b) If we rewrite $g(t) = 1/t^3$ as $g(t) = t^{-3}$, we have $n = -3$, so

$$g'(t) = (-3)t^{-3-1} = -3t^{-4} = -3/t^4.$$

(c) If we rewrite

$$h(x) = \frac{1}{\sqrt[5]{x^2}} \text{ as } h(x) = \frac{1}{x^{2/5}} = x^{-2/5}, \text{ we have } n = -\frac{2}{5} \text{ so}$$

$$h'(x) = \left(-\frac{2}{5}\right)x^{(-2/5)-1} = -\frac{2}{5}x^{-7/5}$$

$$= -\frac{2}{5}\frac{1}{x^{7/5}} = -\frac{2}{5}\frac{1}{\sqrt[5]{x^7}} \quad \square$$

Remember that angles must be measured in *radians* if Table 14.1 is to apply.

Which function in Table 14.1 is its own derivative?

$$y = \exp(x) \text{ has } \frac{dy}{dx} = \exp(x).$$

(It is this property of the exponential function which makes it so important.) □

Question T5

Find where the gradients of the tangents to the graph of $y = \sin x$ have the following values: (a) 0, (b) −1, (c) +1. (☜) □

Question T6

(a) Use your calculator (making sure that it is in radian mode) to investigate the limits

$$\lim_{h \to 0} \left[\frac{\cos(h) - 1}{h} \right] \text{ and } \lim_{h \to 0} \left[\frac{\sin h}{h} \right]$$

Using successively smaller values of h (e.g. $h = \pm 0.1, \pm 0.01$, etc.) try to estimate the limit in each case.

(b) Use the limits obtained in part (a) to find $\frac{d}{dx}(\sin x)$ from the general definition of the derived function.

(*Hint*: Use the trigonometric identity $\sin(A + B) = \sin A \cos B + \cos A \sin B$.) □

Question T7

Which functions in Table 14.1b have derivatives which:

(a) are always positive;

(b) are always negative;

(c) can be positive or negative? □

14.3.2   Derivative of a sum of functions

While it is possible, in theory, to find the derivative of any given function from the definition, this would in fact be an arduous process. In practice, those using calculus employ a set of simple rules which can be applied to combinations of functions to find the derivatives of a wide variety of functions with relative ease. The first of these simple rules is called the **sum rule**, which states:

The derivative of a sum of functions is the sum of the derivatives of the individual functions.

This is an important rule and should be committed to memory.

If $f(x)$ and $g(x)$ are two functions, this can be written as:

sum rule $\quad \dfrac{d}{dx}[f(x) + g(x)] = \dfrac{d}{dx}[f(x)] + \dfrac{d}{dx}[g(x)]$ $\qquad$ (8)

Alternatively we may write this rule in the form

$$[f(x) + g(x)]' = f'(x) + g'(x) \text{ or just } (f + g)' = f' + g'$$

This rule enables us to differentiate functions such as $y = x^{1/2} + \log_e x$. To do so we first note that it is the sum of two standard functions $f(x) = x^{1/2}$ and $g(x) = \log_e x$; we then differentiate each of these functions and finally add the derivatives to obtain the required answer.

So, if $\qquad f(x) = x^{1/2}$ and $g(x) = \log_e x$

then, from Table 14.1

$$\frac{df}{dx} = \frac{1}{2}x^{-1/2} \quad \text{and} \quad \frac{dg}{dx} = \frac{1}{x}$$

so, $\quad \dfrac{dy}{dx} = \dfrac{df}{dx} + \dfrac{dg}{dx} = \dfrac{1}{2}x^{-1/2} + \dfrac{1}{x}$

and if $\quad y = \sqrt{x} + \log_e x \quad$ then $\quad \dfrac{dy}{dx} = \dfrac{1}{2\sqrt{x}} + \dfrac{1}{x}$

It is usually a good idea to *think* before applying the rule, as in the following question.

Find the derivative of $h(x) = \dfrac{\cos x}{\sin x} + \dfrac{\sin x}{\cos x}$.

The function $h(x)$ can be written as $h(x) = \cot x + \tan x$ and is therefore a sum of two of the standard functions that appear in Table 14.1.

Let $f(x) = \cot x$ to obtain $f'(x) = -\operatorname{cosec}^2 x$.

Let $g(x) = \tan x$ to obtain $g'(x) = \sec^2 x$.

Since $h(x) = \cot x + \tan x$ we have $h'(x) = -\operatorname{cosec}^2 x + \sec^2 x$. $\quad \square$

The rule 'derivative of a sum equals the sum of the derivatives' applies equally well if more than two functions are added together. This allows us to obtain the derivatives of functions such as those in the following question.

Question T8

Use the sum rule to find dy/dx in each of the following cases:

(a) $y = 2 + \sqrt[3]{x} + e^x$

(b) $y = (1 + \sqrt{x})^2$

(c) $y = \log_e (xe^x)$

(d) $y = \cot x \sin x + \tan x \cos x$

(*Hint*: Write each function as a sum of functions appearing in Table 14.1.)   ☐

Question T9

If you were to use the definition of the derived function (Equation 6) to show that the 'derivative of the sum is the sum of the derivatives', what assumption must you make about the limit of a sum?   ☐

14.3.3   Derivative of a constant multiple of a function

Table 14.1 gives the derivative of $\sin x$, for example, but what about $2 \sin x$? In other words, what is the derivative of a constant times a function? The answer is given by the **constant multiple rule,** which states:

The derivative of a constant times a function is equal to the constant times the derivative of the function.

Using k for the constant and $f(x)$ for the function, this result can be written as:

$$\text{constant multiple rule} \quad \frac{d}{dx}[k\, f(x)] = k\,\frac{d}{dx}[f(x)] \tag{9}$$

i.e. $[kf(x)]' = kf'(x)$ or just $(kf)' = kf'$

Therefore, for the above example,

$$\frac{d}{dx}(2\sin x) = 2\frac{d}{dx}(\sin x) = 2\cos x$$

similarly, $\dfrac{d}{dx}(\pi e^x) = \pi\dfrac{d}{dx}(e^x) = \pi \times e^x = \pi e^x$

Differentiate $h(x) = \left(x + \dfrac{1}{x}\right)^3$

If we expand the function, we find

$$\left(x + \frac{1}{x}\right)^3 = x^3 + 3x^2 \times \frac{1}{x} + 3x \times \frac{1}{x^2} + \frac{1}{x^3}$$

$$= x^3 + 3x + \frac{3}{x} + \frac{1}{x^3}$$

$$= x^3 + 3x + 3x^{-1} + x^{-3}$$

So $\quad \dfrac{d}{dx}[h(x)] = \dfrac{d}{dx}(x^3 + 3x + 3x^{-1} + x^{-3})$

$$= 3x^2 + 3 \times \frac{d}{dx}(x) + 3\frac{d}{dx}(x^{-1}) + (-3)x^{-4} \quad (\text{☞})$$

$$= 3x^2 + 3 \times 1 + 3 \times (-1)x^{-2} - 3x^{-4}$$

$$= 3x^2 + 3 - \frac{3}{x^2} - \frac{3}{x^4} \qquad \square$$

The derivative of kx^n was included in Table 14.1a, but as you can see, you could have worked it out from the derivative of x^n and the constant multiple rule.

Use the constant multiple rule to show that $[f(x) - g(x)]' = f'(x) - g'(x)$.

We may certainly write, $f(x) - g(x) = f(x) + (-1)g(x)$,

so, using the sum rule we may write

$$\frac{d}{dx}[f(x) - g(x)] = \frac{d}{dx}[f(x)] + \frac{d}{dx}[(-1)g(x)]$$

and using the constant multiple rule we can write

$$\frac{d}{dx}[f(x) - g(x)] = \frac{d}{dx}[f(x)] - \frac{d}{dx}[g(x)]$$

Thus, $[f(x) - g(x)]' = f'(x) - g'(x) \quad \square$

This last result shows that *the derivative of the difference of two functions is the difference of the derivatives*, i.e.

$$\frac{d}{dx}[f(x) - g(x)] = \frac{d}{dx}[f(x)] - \frac{d}{dx}[g(x)] \qquad (10)$$

Question T10

Find the derivatives of the following functions:

(a) $\left(\sqrt{x} - \dfrac{1}{\sqrt{x}}\right)^4$

(b) $\log_e \sqrt[3]{x}$

(c) $\left(\sin\dfrac{x}{2} + \cos\dfrac{x}{2}\right)^2$

(d) $(\pi - x)^2$

(e) $\sin(x + 2)$

(f) $\left[\tan\left(x + \dfrac{\pi}{4}\right)\right](1 - \tan x)$ ☐

Hint: $\tan(A + B) =$
$$= \left(\frac{\tan A + \tan B}{1 - \tan A \tan B}\right)$$

Question T11

If k is a constant and $f(x)$ a function, write down the definition of $f'(x)$ and $[kf(x)]'$. What assumption must you make about limits to justify the conclusion $[kf(x)]' = kf'(x)$?   ☐

14.3.4   Derivative of a product of functions

This subsection introduces yet another way of combining functions and obtaining the derivative of the result. This is the *product* of two functions, $f(x)g(x)$. Our aim is to express the derivative of such a product in terms of the derivatives of the functions $f(x)$ and $g(x)$ that are multiplied together to form the product, but the answer is not as obvious as it was for the sum or difference of functions. The 'obvious' answer that the derivative of a product is the product of the derivatives is *wrong*.

You cannot simply multiply derivatives.

To convince yourself of this consider $f(x) = x^2$ and $g(x) = x^3$. Then $f(x)g(x) = x^2 x^3 = x^5$. We can differentiate each of these functions using Table 14.1:

thus, $f(x) = x^2$ implies $f'(x) = 2x$

              $g(x) = x^3$ implies $g'(x) = 3x^2$

and $f(x)g(x) = x^5$ implies $[f(x)g(x)]' = 5x^4$

Clearly, $[f(x)g(x)]' \ne f'(x) \times g'(x)$

Let $f(x) = 2x$ and $g(x) = \dfrac{1}{4x}$

Find $f'(x)$, $g'(x)$, $[f(x)g(x)]'$ and show that $[f(x)g(x)]' \ne f'(x) \times g'(x)$.

First we have $f'(x) = 2$,

and $g(x) = \dfrac{1}{4x} = \dfrac{1}{4}x^{-1}$ so $g'(x) = -\dfrac{1}{4}x^{-2}$

Therefore $f'(x) \times g'(x) = 2 \times \left(-\dfrac{x^{-2}}{4} \right) = -\dfrac{1}{2} x^{-2}$

However, $f(x) g(x) = 2x \times \dfrac{1}{4x} = \dfrac{1}{2}$

and the derivative of this is, $[f(x)g(x)]' = 0$

So in this case $[f(x)g(x)]' \ne f'(x) \times g'(x)$. $\square$

The correct determination of $[f(x)g(x)]'$ involves not only $f'(x)$ and $g'(x)$ *but also* $f(x)$ and $g(x)$. It is given by the **product rule** and is easier to express in symbols than in words:

$$\text{product rule} \quad \dfrac{d}{dx}[f(x)\, g(x)] = g(x)\dfrac{df}{dx} + f(x)\dfrac{dg}{dx} \qquad (11)$$

i.e. $[f(x)g(x)]' = f'(x)g(x) + f(x)g'(x)$ or just $(fg)' = f'g + fg'$

The expression of this rule in terms of words is probably not very helpful:

The derivative of f times g equals the derivative of f times g plus f times the derivative of g.

You may find Figure 14.4 a more useful memory aid.

In the case of $f(x) = x^2$ and $g(x) = x^3$ (discussed earlier) where

$$f'(x) = 2x \text{ and } g'(x) = 3x^2$$

the application of the product rule to $f(x)g(x)$ gives

$$[f(x)g(x)]' = f'(x)g(x) + f(x)g'(x)$$
$$= 2x \times x^3 + x^2 \times 3x^2$$
$$= 2x^4 + 3x^4 = 5x^4$$

and $5x^4$ is what we expect for the derivative of $f(x)g(x) = x^2x^3 = x^5$.

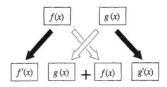

Figure 14.4 Differentiating a product of functions.

If $f(x) = 2x$ and $g(x) = \dfrac{1}{4x}$ apply the product rule to find the derivative of $f(x)g(x)$.

$$f(x) = 2x \quad \text{and} \quad g(x) = \dfrac{1}{4x}$$

so $f'(x) = 2$ and $g'(x) = -\dfrac{1}{4x^2}$

thus $[f(x)g(x)]' = f'(x)g(x) + f(x)g'(x)$

$$= 2 \times \frac{1}{4x} + 2x\left(-\frac{1}{4x^2}\right) = \frac{1}{2x} - \frac{1}{2x} = 0$$

which is correct since $f(x)\,g(x) = 2x\dfrac{1}{4x} = \dfrac{1}{2}$ is a constant and has derivative zero. $\square$

Illustrations of the product rule

With the aid of the product rule we can now differentiate many more functions. The important step is to express the function that is to be differentiated in terms of the sum, difference or product of functions with known derivatives. Here are some examples.

Example 1

Differentiate $2\sin x \cos x$.

Solution $2\sin x \cos x$ may be written as a product of the functions
$f(x) = 2\sin x$ and $g(x) = \cos x$

for which $f'(x) = 2\cos x$ and $g'(x) = -\sin x$

Using the product rule we can therefore write

$$\frac{d}{dx}(2\sin x \cos x) = (2\cos x)(\cos x) + (2\sin x)(-\sin x) \quad (\text{✍})$$

$$= 2(\cos^2 x - \sin^2 x) \quad \square$$

Note that we have deliberately switched to Leibniz notation at this point.

Example 2

Differentiate $xe^x + \sec^2 x$.

Solution $xe^x + \sec^2 x$ is the sum of two functions xe^x and $\sec^2 x$, each of which is a product. So we differentiate xe^x first and then $\sec^2 x$ (using the product rule for each) and then add the answers using the sum rule.

First, xe^x may be written as the product of the functions

$$f(x) = x \text{ and } g(x) = e^x$$

for which $f'(x) = 1$ and $g'(x) = e^x$

then, using the product rule, we can write

$$\frac{d}{dx}(x\,e^x) = e^x \times \frac{d}{dx}(x) + x \times \frac{d}{dx}(e^x)$$

$$= e^x \times 1 + x \times e^x = (1 + x)e^x$$

Second, $\sec^2 x = \sec x \times \sec x$, therefore

$$f(x) = \sec x \quad \text{and} \quad g(x) = \sec x$$

giving us $f'(x) = \sec x \tan x$ and $g'(x) = \sec x \tan x$

Then, again using the product rule, we find

$$\frac{d}{dx}(\sec^2 x) = (\sec x \tan x) \times (\sec x) + (\sec x) \times (\sec x \tan x)$$

$$= 2 \sec^2 x \tan x$$

Thus, combining these results, and using the sum rule, we find

$$\frac{d}{dx}(x e^x + \sec^2 x) = \frac{d}{dx}(x e^x) + \frac{d}{dx}(\sec^2 x)$$

$$= (1 + x)e^x + 2 \sec^2 x \tan x \quad \square$$

The *kinetic energy* of a particle of mass m moving with a speed $v(t)$ that varies with time t, is $m[v(t)]^2/2$. Use the product rule to show that the rate of change of $m[v(t)]^2/2$ with time is $mv(t)\dfrac{dv}{dt}(t)$.

Writing $[v(t)]^2 = v(t) \times v(t)$, and using the product rule, we obtain

$$\frac{d}{dt}[v(t)]^2 = v'(t)\, v(t) + v(t)\, v'(t) = 2v(t)\, v'(t)$$

Using the constant multiple rule, it follows that

$$\frac{d}{dt}\left\{\frac{1}{2}m[v(t)]^2\right\} = \frac{1}{2}m\frac{d}{dt}\{[v(t)]^2\}$$

$$= \frac{1}{2}m\,[2v(t)v'(t)] = mv(t)\frac{dv}{dt}(t) \quad \square$$

Question T12

Find the derivative of each of the following functions:

(a) $x^2 \log_e x$ (b) $(\sin x - 2 \cos x)^2$

(c) $(1 - \sqrt{x})(1 + \sqrt[3]{x})$ (d) e^{2x} (e) $\sec x \cot x$

(f) $x^3 F(x)$, where $F(x)$ is an arbitrary function $\square$

Question T13

The instantaneous motion of a particle moving along a straight line (call it the x-axis) can be described, at time t, in terms of its *displacement* $s_x(t)$ from a fixed reference position, its *velocity* $v_x(t)$ and its *acceleration* $a_x(t)$. (☞) These quantities are defined in such a way that

$$a_x(t) = \frac{dv_x}{dt}(t) \quad \text{and} \quad v_x(t) = \frac{ds_x}{dt}(t)$$

Strictly speaking, displacement is a *vector* quantity that requires three *components* for its full specification in three dimensions. However, when dealing with one-dimensional motion it is conventional to refer to the single varying component (s_x in this case) as the displacement. Similar comments apply to velocity and acceleration.

You may find Question T13 quite challenging. It isn't difficult if you tackle it the right way, but the algebra will get rather fierce if you don't. Look at the answer if the going gets tough.

If the displacement is given as a function of time by

$$s_x(t) = [\cos(\omega t) - 2\sin(\omega t)]\exp(\alpha t)$$

where ω and α are constants

show that $a_x(t) = 2\alpha v_x(t) - (\omega^2 + \alpha^2)s_x(t)$ (☜) ☐

Question T14

This question extends the product rule to three functions. If $f(x)$, $g(x)$, $h(x)$ are any three functions, show that

$$\frac{d}{dx}[f(x)\,g(x)\,h(x)] = f'(x)\,g(x)\,h(x) + f(x)\,g'(x)\,h(x) +$$

$$+ f(x)\,g(x)\,h'(x) \quad ☐$$

Question T15

Use the result developed in Question T14 to find

$$\frac{d}{dx}(e^x \log_e x \sin x) \quad ☐$$

14.3.5 Derivative of a quotient of functions

It is frequently the case that we need to find the derivative of the quotient $\dfrac{f(x)}{g(x)}$ of two functions $f(x)$ and $g(x)$. (☜) The derivative of a quotient, like that of a product, depends not only on the values of $f'(x)$ and $g'(x)$, but also on $f(x)$ and $g(x)$. The **quotient rule** states:

We must of course ensure that $g(x) \neq 0$ if the quotient is to be well defined.

This is an important rule and should be committed to memory.

The quotient rule can be derived from the basic definition of the derivative. We will, however, state the rule here without deriving it.

$$\text{quotient rule} \quad \frac{d}{dx}\left[\frac{f(x)}{g(x)}\right] = \frac{g(x)\dfrac{df}{dx} - f(x)\dfrac{dg}{dx}}{[g(x)]^2} \tag{12}$$

i.e. $\left[\dfrac{f(x)}{g(x)}\right]' = \dfrac{f'(x)g(x) - f(x)g'(x)}{[g(x)]^2}$ or just $\left(\dfrac{f}{g}\right)' = \dfrac{f'g - fg'}{g^2}$

You may find Figure 14.5 a useful memory aid.

To apply this result to the function $y = \dfrac{\sin x}{x}$

take $f(x) = \sin x$ and $g(x) = x$

so that $f'(x) = \cos x$ and $g'(x) = 1$

Then $\dfrac{dy}{dx} = \dfrac{(\cos x)x - (\sin x) \times 1}{x^2} = \dfrac{x \cos x - \sin x}{x^2}$

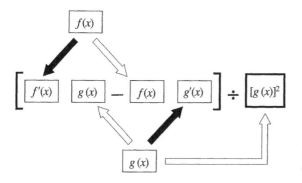

Figure 14.5 The quotient rule.

The same result can be obtained by writing

$$y = \dfrac{\sin x}{x} = \sin x \times x^{-1}$$

then using the product rule

$$\dfrac{dy}{dx} = \cos x \times x^{-1} +$$

$$\sin x \times (-x^{-2})$$

$$= \dfrac{x \cos x - \sin x}{x^2}$$

If you cannot remember the quotient rule it is always possible to use the product rule in this way.

Write $y = x^3$ as x^5/x^2 and use the quotient rule to find dy/dx.

Set $f(x) = x^5$ and $g(x) = x^2$

then $f'(x) = 5x^4$ and $g'(x) = 2x$

and $\dfrac{dy}{dx} = \dfrac{f'(x)g(x) - f(x)g'(x)}{[g(x)]^2}$

$$= \dfrac{(5x^4)(x^2) - (x^5)(2x)}{(x^2)^2} = \dfrac{5x^6 - 2x^6}{x^4} = 3x^2$$

(and from Table 14.1 we know that the derivative of x^3 is $3x^2$, so we have verified that the quotient rule gives the correct answer in this particular case.) □

Use the quotient rule to find $\dfrac{d}{dx}\left[\dfrac{1}{F(x)}\right]$ in terms of $F'(x)$ and $F(x)$.

Putting $f(x) = 1$ and $g(x) = F(x)$ in the quotient formula, we obtain

the **reciprocal rule** $\dfrac{d}{dx}\left[\dfrac{1}{F(x)}\right] = -\dfrac{F'(x)}{[F(x)]^2}$ (13) □

You may find this useful, but there is no point in remembering it if you know the quotient rule since it is just a special case of that more general rule.

Question T16

Find dy/dx in each of the following cases:

(a) $y = \dfrac{x}{e^x}$

(b) $y = \dfrac{1 - x^2}{\cos x}$

(c) $y = \dfrac{x^5 - x^4 + x^3 - x^2}{(x + 1)^2}$

(d) $y = \dfrac{x \log_e x}{\tan x}$ □

Question T17

If $f(x) = (x - 1)^2$ and $g(x) = xe^x$, find the following:

(a) $\dfrac{d}{dx}[f(x) - g(x)]$

(b) $\dfrac{d}{dx}[f(x)\,g(x)]$

(c) $\dfrac{d}{dx}\left[\dfrac{f(x)}{g(x)}\right]$

(d) $\dfrac{d}{dx}\left[\dfrac{g(x)}{f(x)}\right]$ □

In part (d) of Question T17, try using the reciprocal rule and the result from part (c).

The aim of the next question is to derive the quotient rule from the product rule. It begins by showing how to obtain the reciprocal rule *without* using the quotient rule.

Question T18

(a)   Let $f(x)$ be an arbitrary function and let $h(x) = 1/f(x)$ so that

$$f(x)h(x) = 1$$

Differentiate both sides of this equation and hence obtain a formula for $\dfrac{d}{dx}\left[\dfrac{1}{f(x)}\right]$ (similar to that given in Equation 13).

(b)   Let $f(x)$ and $g(x)$ be any two functions, and write

$$\frac{f(x)}{g(x)} = f(x) \times \frac{1}{g(x)},$$

then use the answer to part (a) and the product rule to derive the quotient rule.   □

Question T19

Use $\dfrac{d}{dx}(\sin x) = \cos x$, $\dfrac{d}{dx}(\cos x) = -\sin x$, and the quotient rule to differentiate (a) $\tan x$, and (b) $\cot x$.   □

Question T20

(a) Use the reciprocal rule to find $\dfrac{d}{dx}(e^{-x})$.

(b) The function $f(x) = \dfrac{1}{2}(e^x + e^{-x})$ is sometimes known as the *hyperbolic cosine* of x (and is denoted by $\cosh x$).

Find $\dfrac{df}{dx}$ and show that $[f(x)]^2 - \left(\dfrac{df}{dx}\right)^2 = 1$. □

Question T21

Let $F(x) = G\dfrac{Mm}{(x-a)^2}$ where G, M, m and a are constants.

Find $F'(x)$. □

Question T22

Using $\dfrac{d}{dx}(\sin x) = \cos x$ and $\dfrac{d}{dx}(\cos x) = -\sin x$, along with the reciprocal rule, find the derivatives of (a) $\sec x$ and (b) $\csc x$.

□

14.4 More about logarithmic and exponential functions

14.4.1 Why exp(x) is considered to be special

Of all the functions listed in Table 14.1, $f(x) = \exp(x) = e^x$ is the only one which is its own derivative. Far from being a technical curiosity this fact turns out to be of great significance and ultimately explains why the exponential function plays such an important role in physics. (☞) It also has an interesting interpretation in terms of the graph of $y = e^x$, shown in Figure 14.6 (see p. 395). Since $f'(x) = e^x$ the gradient of the tangent to the graph at the point A which has coordinates $x = a$ and $y = e^a$ is $f'(a) = e^a$ which is the same as the height of A above C.

So the rate of change of $f(x)$ when $x = a$ is simply $f(a)$; that is, the rate of change of the exponential function at any point is equal to the value of the function itself at that point. In Figure 14.6 the point B has coordinates $(a - 1, 0)$. Since A is the point (a, e^a), the line AB has a gradient of

$$\frac{CA}{BC} = \frac{e^a}{1} = e^a$$

and is therefore the tangent to the graph at A. So, to draw a tangent to $y = e^x$ at (a, e^a), simply join (a, e^a) to $(a - 1, 0)$ and the resulting line is bound to have a gradient of e^a.

Exponential functions describe the decay of radioactive samples, the growth of unrestricted populations, the dying oscillations of an appropriately damped pendulum and many other physically important phenomena.

Is $f(x) = e^x$ the only function which is its own derivative? (Think about the rules of differentiation that you already know.)

No. The derivative of a constant multiple of a function is the constant multiple times the derivative, so we have

$$\frac{d}{dx}(Ae^x) = Ae^x \quad \text{for any constant } A$$

Thus, any constant multiple of the exponential function, such as Ae^x, is its own derivative. However, apart from functions of this kind there are no other functions that are their own derivatives. (✎) □

The function $f(x) = 0$ is equal to its own derivative (0) at every point, but this is a special case of Ae^x corresponding to $A = 0$.

Using the product rule find the derivatives of (a) e^{2x}, (b) e^{3x}, (c) e^{4x}. Deduce the formula for the derivative of e^{nx} where n is any positive integer.

(a) Since $e^{2x} = e^x \times e^x$ the product rule gives

$$\frac{d}{dx}(e^{2x}) = 2e^{2x}$$

(b) $e^{3x} = e^x \times e^{2x}$, so $\dfrac{d}{dx}(e^{3x}) = e^x \times 2e^{2x} + e^x e^{2x} = 3e^{3x}$

(c) $e^{4x} = e^{2x} \times e^{2x} = e^{2x} \times 2e^{2x} + 2e^{2x}\, e^{2x} = 4e^{4x}$

Therefore, generally, if n is any positive integer $\dfrac{d}{dx}(e^{nx}) = ne^{nx}$. □

A similar result holds true even when n is not a positive integer. This was given in Table 14.1, but it deserves to be emphasised again here.

$$\frac{d}{dx}(e^{kx}) = ke^{kx} \text{ for any constant } k \tag{14}$$

Question T23

If $y = \dfrac{Ce^x}{1 + Ce^x}$ for some constant C, show that $\dfrac{dy}{dx} = y(1 - y)$. □

Question T24

Differentiate $f(x) = (e^x + e^{-x})(e^x - e^{-x})$. □

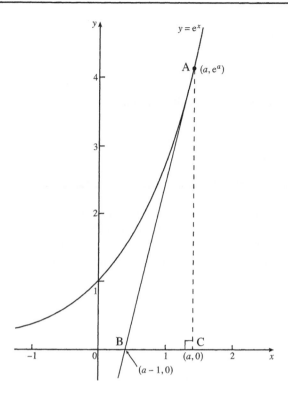

Figure 14.6 The graph of $y = e^x$.

14.4.2 Derivative of a^x

The previous subsection considered the derivative of e^x. However, functions such as 2^x and π^x sometimes arise in physical applications, and so it is necessary to know how to differentiate any function of the form a^x, where the positive number a being raised to the power x is not necessarily e. As an example we will consider the differentiation of 2^x, but the method is of general applicability. First note that exp is the *inverse function* of $\log_e$, which means that exp reverses the effect of $\log_e$ so that

$$e^{\log_e 2} = \exp(\log_e 2) = 2$$

since

$$2 = \exp(\log_e 2)$$

$$2^x = [\exp(\log_e 2)]^x = \exp(x \log_e 2)$$

So,

$$\frac{d}{dx}(2^x) = \frac{d}{dx}[\exp(x \log_e 2)]$$

Now, from Equation 14 (or Table 14.1) we know that

$$\frac{d}{dx}(e^{kx}) = k e^{kx} \quad \text{for any constant } k$$

and, if we choose k to be $\log_e 2$, this gives us

$$\frac{d}{dx}(2^x) = \frac{d}{dx}[\exp(x\log_e 2)]$$

$$= (\log_e 2)[\exp(x\log_e 2)] = (\log_e 2)2^x$$

A similar argument applies with any positive number, a say, in place of 2. So we have the result:

$$\frac{d}{dx}(a^x) = (\log_e a)\,a^x \quad (a>0) \tag{15}$$

Rather than trying to remember this result you would probably be wiser to try to remember the method that was used to obtain it. Notice the way that the properties of exponentials and logarithms have been used to express a function of the form a^x in terms of e^{kx}, and how the simple properties of e^{kx} have then been exploited.

Find $\dfrac{d}{dx}(\pi^x)$.

Here $a = \pi$ so the derivative is $(\log_e \pi)\pi^x$.   □

Apply the formula for differentiating a^x (Equation 14) when $a = e$.

The result is $(\log_e a)a^x$ and if $a = e$ then, since $\log_e e = 1$, we have $\dfrac{d}{dx}(e^x) = e^x$ as before.   □

Question T25

Differentiate each of the following functions:

(a) $f(x) = 3^x$ (b) $f(x) = 3^{x-1}$

(c) $f(x) = (2+3)^x$ (d) $f(x) = 2^x\,e^x$ □

14.4.3   Derivative of $\log_a x$

As stated in Table 14.1:

$$\frac{d}{dx}(\log_e x) = \frac{1}{x}$$

From this it is easy to deduce the derivative of $\log_e kx$, where k is a constant

$$\log_e (kx) = \log_e k + \log_e x$$

so $\quad \dfrac{d}{dx} (\log_e (kx)) = \dfrac{d}{dx} (\log_e k) + \dfrac{d}{dx} (\log_e x) = \dfrac{1}{x}$

Note that k plays no part in the derivative because k is constant, so

$$\dfrac{d}{dx} (\log_e k) = 0.$$

If we had been compelled to differentiate $\log_{10} x$ or $\log_2 x$ the problem would have been a little more difficult, but still not intractable. The general method for differentiating logs to an arbitrary base, such as $\log_a x$, is remarkably similar to that for dealing with a^x: it involves expressing $\log_a x$ in terms of $\log_e x$ which can be differentiated easily. As an example, let us consider $\log_{10} x$. To express it in terms of $\log_e x$ we write

$$y = \log_{10} x$$

then $\quad x = 10^y$

and, taking logarithms to the base e of both sides of this equation, we obtain

$$\log_e x = \log_e (10^y) = y \log_e 10$$

and recalling that $y = \log_{10} x$ this can be rearranged to give

$$y = \log_{10} x = \dfrac{\log_e x}{\log_e 10}$$

In more general terms

$$\log_c b = \dfrac{\log_a b}{\log_a c}$$

We may now easily differentiate $\log_{10} x$ because $1/\log_e 10$ is a constant, so

$$\dfrac{d}{dx} (\log_{10} x) = \dfrac{d}{dx} \left(\dfrac{\log_e x}{\log_e 10} \right)$$

$$= \left(\dfrac{1}{\log_e 10} \right) \dfrac{d}{dx} (\log_e x)$$

$$= \left(\dfrac{1}{\log_e 10} \right) \times \dfrac{1}{x} = \dfrac{1}{x \log_e 10}$$

A similar argument applies for any positive base a giving the general result:

$$\dfrac{d}{dx} (\log_a x) = \dfrac{1}{x \log_e a} \qquad (16)$$

Once again, you should try to remember the general method rather than just the result.

If $a = e$, check that the above formula gives our previous result for the derivative of $\log_e x$.

$$\dfrac{d}{dx} (\log_a x) = \dfrac{1}{x \log_e a}$$

If $a = e$ then $\log_e e = 1$, and the result is $1/x$ as expected. $\quad \square$

Differentiate $\log_\pi x$.

From Equation 16:

$$\frac{d}{dx}(\log_\pi x) = \frac{1}{x \log_e \pi} \quad \square$$

Question T26
Differentiate each of the following functions:

(a) $f(x) = \log_2 x$

(b) $f(x) = \log_{10} x^2$

(c) $f(x) = \dfrac{\log_2 x}{\log_{10} x}$ [*Hint*: $\log_a b = (\log_a c)(\log_c b)$]

(d) $f(x) = a^x \log_a x$, where a is a positive constant $\square$

14.4.4   Resisted motion under gravity: an example

Imagine a parachutist falling to Earth with velocity, $v_x(t)$ downwards, at time t. For convenience we will choose the positive x-axis to be the downward vertical, with the origin at the point where the parachute opens. The parachutist is subject to two forces, gravity acting downwards and the resistance of the air on his parachute acting upwards. Suppose that the resistive force is proportional to $[v_x(t)]^2$ with a constant of proportionality k (which has to be determined experimentally). It follows from *Newton's second law* that the equation governing the motion is

$$m\frac{d}{dt}[v_x(t)] = mg - k[v_x(t)]^2 \tag{17}$$

where m is the mass of the parachutist and g is the magnitude of the acceleration due to gravity. The forces on the right-hand side have opposite sign because they act in opposite directions.

Now, suppose you want to check that the following expression for $v_x(t)$ satisfies Equation 17

$$v_x(t) = \sqrt{\frac{mg}{k}\left[\frac{1 + B\exp\left(-2t\sqrt{gk/m}\right)}{1 - B\exp\left(-2t\sqrt{gk/m}\right)}\right]} \tag{18}$$

where B is a constant (determined by the speed of the parachutist at time $t = 0$). It appears that we must first differentiate $v_x(t)$, which is a rather complicated multiple of the quotient of two functions.

We could certainly proceed directly, and just differentiate the function as it stands, but a little thought should tell you that the algebra is going to get very nasty; so it is worth trying to simplify the expression. To simplify the notation let $A = \sqrt{mg/k}$ and $\alpha = 2\sqrt{gk/m}$, and let us write v_x rather than $v_x(t)$. So we have a simplified version of Equation 18

$$v_x = A \left[\frac{1 + B \exp(-\alpha t)}{1 - B \exp(-\alpha t)} \right] \tag{19}$$

Now, using the quotient rule to differentiate this equation with respect to t, we find

$$\frac{dv_x}{dt} = A \left\{ \frac{\left[\dfrac{d}{dt}(1 + Be^{-\alpha t})\right](1 - Be^{-\alpha t}) - \left[\dfrac{d}{dt}(1 - Be^{-\alpha t})\right](1 + Be^{-\alpha t})}{(1 - Be^{-\alpha t})^2} \right\}$$

$$\frac{dv_x}{dt} = A \left[\frac{(-B\alpha e^{-\alpha t})(1 - Be^{-\alpha t}) - (B\alpha e^{-\alpha t})(1 + Be^{-\alpha t})}{(1 - Be^{-\alpha t})^2} \right]$$

$$\frac{dv_x}{dt} = A \left[\frac{(-B\alpha e^{-\alpha t} + B^2\alpha e^{-2\alpha t}) - (B\alpha e^{-\alpha t} + B^2\alpha e^{-2\alpha t})}{(1 - Be^{-\alpha t})^2} \right]$$

$$\frac{dv_x}{dt} = A \left[\frac{-2B\alpha e^{-\alpha t}}{(1 - Be^{-\alpha t})^2} \right]$$

Substituting this expression for $dv_x(t)/dt$ into the left-hand side of Equation 17 and using the fact that $A\alpha = 2g$ we obtain

$$mA \left[\frac{-2B\alpha e^{-\alpha t}}{(1 - Be^{-\alpha t})^2} \right] = \frac{-4mgBe^{-\alpha t}}{(1 - Be^{-\alpha t})^2}$$

and substituting the same expression for $dv_x(t)/dt$ into the right-hand side of Equation 17 and using the fact that $kA^2 = mg$ we obtain

$$mg - mg \left(\frac{1 + Be^{-\alpha t}}{1 - Be^{-\alpha t}} \right)^2 = mg \left[\frac{(1 - Be^{-\alpha t})^2 - (1 + Be^{-\alpha t})^2}{(1 - Be^{-\alpha t})^2} \right]$$

$$= \frac{-4mgBe^{-\alpha t}}{(1 - Be^{-\alpha t})^2}$$

Thus, the two sides of Equation 17 are indeed equal if v_x has the form given in Equation 18.

This really completes the differentiation and subsequent manipulation, but having introduced an expression (Equation 18) for the downward velocity under gravity in the presence of a resistive force it is worth noting at least one of its mathematical properties.

What is the behaviour of $v_x(t)$ as t becomes large? In other words, what is $\lim_{t \to \infty} [v_x(t)]$?

As t becomes large $Be^{-\alpha t}$ becomes small, and the velocity approaches the value

$$\sqrt{\frac{mg}{k}} \frac{1+0}{1-0} = \sqrt{\frac{mg}{k}}$$

This final constant value of the velocity is called the **terminal velocity**. Its existence is a striking feature of resisted motion.   ☐

What is the terminal velocity if $m = 80$ kg, $k = 30$ kg m^{-1} and $g = 9.81$ m s^{-2}?

Terminal velocity =

$$\sqrt{\frac{mg}{k}} = \sqrt{\frac{(80 \text{ kg}) \times (9.81 \text{ m s}^{-2})}{(30 \text{ kg m}^{-1})}} \approx 5.1 \text{ m s}^{-1}. \quad ☐$$

Question T27

Show that $y(x) = \dfrac{\log_e 2 - x}{\log_e 2 + x}$ satisfies the equation

$$2y'(x) \log_e 2 + [1 + y(x)]^2 \quad ☐$$

14.5   Closing items

14.5.1   Chapter summary

1. A function, f say, is a rule that assigns a single value of the dependent variable, y say, (in the codomain) to each value of the independent variable x in the domain of the function, such that $y = f(x)$.

2. The *rate of change*, or *derivative* of a function $f(x)$ at $x = a$, can be interpreted as the *gradient* of the graph of the function at the point $x = a$. Some standard derivatives are given in Table 14.1a (repeated here for reference).

3. The derivative is defined more formally in terms of a *limit*, and may be represented in a variety of ways. If $y = f(x)$

$$f'(x) = \frac{dy}{dx} = \lim_{\Delta x \to 0} \left(\frac{\Delta y}{\Delta x} \right) = \lim_{h \to 0} \left[\frac{f(x+h) - f(x)}{h} \right]$$

4. The *sum*, *constant multiple*, *product* and *quotient* rules for differentiating combinations of functions are as follows:

sum rule $\qquad\qquad (f+g)' = f' + g'$

constant multiple rule $\quad (kf)' = kf'$

product rule $\qquad\quad (fg)' = f'g + fg'$

quotient rule $\qquad\quad \left(\dfrac{f}{g}\right)' = \dfrac{f'g - fg'}{g^2}$

5. The derivatives of the standard functions are given in Table 14.1a, repeated here.

6. The derivatives of a^x and $\log_a x$ are given by

$$\frac{d}{dx}(a^x) = (\log_e a)\, a^x \qquad\qquad\qquad (15)$$

and

$$\frac{d}{dx}(\log_a x) = \frac{1}{x \log_e a} \qquad\qquad\qquad (16)$$

7. You should be aware that logarithms to different bases are related by

$$\log_a b = (\log_a c)(\log_c b)$$

14.5.2 Achievements

Having completed the chapter, you should be able to:

A1. Define the terms that are emboldened in the text of the chapter.

A2. Define the derivative of a function (at a point) as a limit and give the definition a graphical interpretation.

A3. Use the definition of the derivative as a limit to calculate the derivative of simple functions.

A4. Know the derivatives of a range of standard functions.

A5. Use addition, subtraction, multiplication and division to express a given function in terms of 'simpler' ones in appropriate cases.

A6. Differentiate a constant multiple of a function.

A7. Apply the rule for differentiating the sum (and difference) of functions.

A8. Apply the product rule of differentiation.

A9. Apply the quotient rule of differentiation.

A10. Use the properties of logarithms and exponentials to determine the derivatives of $\log_a x$ and a^x.

Table 14.1(a) Some standard derivatives

$f(x)$	$f'(x)$
k (constant)	0
kx^n	nkx^{n-1}
$\sin kx$	$k \cos kx$
$\cos kx$	$-k \sin kx$
$\tan kx$	$k \sec^2 kx$
$\operatorname{cosec} kx$	$-k \operatorname{cosec} kx \cot kx$
$\sec kx$	$k \sec kx \tan kx$
$\cot kx$	$-k \operatorname{cosec}^2 kx$
$\exp(kx)$	$k \exp(kx)$
$\log_e(kx)$	$1/x$

14.5.3 End of chapter questions

Question E1   Write down the derivative of $f(x) = \log_e x$. As x increases from zero how does the gradient of the tangent to the graph change? What is the behaviour of $\log_e x$ as x approaches infinity?

Question E2   Using the definition of the derivative as a limit, differentiate $f(x) = 1/x^2$.

Question E3   Find dy/dx for each of the following functions:

(a) $y = 2x(x - 1)(x + 2)$

(b) $y = \sin(x + \pi/4)$

(*Hint*: $\sin(A + B) = \sin A \cos B + \cos A \sin B$)

(c) $y = \log_e x^2$

Question E4   Find $f'(x)$ for each of the following functions:

(a) $f(x) = x \cos x + x^2 \sin x$

(b) $f(x) = \left(x - \dfrac{1}{x}\right)^2 e^x$

(c) $f(x) = \dfrac{1 + x + x^2}{1 - x^2}$

(d) $f(x) = \dfrac{\sqrt{x} \sin x}{\log_e x}$

(e) $f(x) = \dfrac{\exp(x + \log_e x)}{\sqrt{x}}$

(f) $f(x) = a^x x^a$

Think about this question. The solution is easy if you handle it in the right way.

Question E5   If n is a positive integer, the series

$$1 + x + x^2 + x^3 + \ldots + x^n$$

has the sum $\dfrac{x^{n+1} - 1}{x - 1}$.

Show that the sum of the series

$$1 + 2x + 3x^2 + \ldots + nx^{n-1}$$

is $\dfrac{nx^{n+1} - (n + 1)x^n + 1}{(x - 1)^2}$

Question E6   If f and g are functions show that

$$\left(\frac{1}{f(x)\,g(x)}\right)' = \left(\frac{1}{f(x)}\right)'\frac{1}{g(x)} + \frac{1}{f(x)}\left(\frac{1}{g(x)}\right)'$$

Chapter 15 Differentiating composite functions

15.1 Opening items

15.1.1 Chapter introduction

Applications of differentiation are widespread throughout physical science. For example, given the displacement from the origin of a moving object as a function of time, the corresponding velocity can be found by differentiating that function. Or, given a simple electrical circuit consisting of a resistor and an inductor connected in series with a charged capacitor and a switch (i.e. an *LCR* circuit), the potential differences across the inductor and resistor are, respectively, $L\dfrac{dI}{dt}(t)$ and $RI(t)$ where $I(t)$ is the current in the circuit at time t after the switch is closed and L and R are constants. It is also the case that in such a circuit $I(t) = \dfrac{dq}{dt}(t)$ where $q(t)$ is the charge on the capacitor at time t. Because the need to differentiate arises so often in physical science it is essential that you should be able to differentiate readily a wide variety of functions. The techniques of basic differentiation should enable you to differentiate certain standard functions such as powers, trigonometric functions, exponentials and logarithms, or certain combinations of those standard functions such as sums, constant multiples, products and quotients. This chapter goes further in that it introduces techniques that, when combined with those of basic differentiation, will allow you to differentiate almost all the functions of a single variable that you are likely to meet.

It is often the case that a physical quantity can be represented by a function of a variable that is itself a function of another variable. For example, the displacement from equilibrium of a one-dimensional simple harmonic oscillator may be represented by $x = A \cos(\omega t + \phi)$, but $(\omega t + \phi)$ is itself a function of t, so we might write $\theta(t) = (\omega t + \phi)$, in which case we can write $x = A \cos(\theta(t))$. Such an expression makes it clear that x is a *function of a function* of t, and that to find the velocity of such an oscillator we need to be able to differentiate a function of a function of t. Section 15.2 of this chapter describes functions of a function in more detail and introduces (in Subsection 15.2.2) the *chain rule* for their differentiation. The chain rule is one of the most useful techniques of calculus; among other things it enables us to differentiate *implicit functions*, *inverse functions* and *parametric functions*, all of which are discussed in Subsections 15.2.3, 15.2.4 and 15.2.5.

If we want to find the acceleration of an object given its position then we must proceed via its velocity; the acceleration is the rate of change of the velocity, which is itself the rate of change of the position. Hence the process requires two successive differentiations. We say that the acceleration is the *second derivative* of the position of the object. Similarly, the potential difference across the inductor in the series *LCR* circuit mentioned earlier is proportional to dI/dt but $I = dq/dt$, so the potential difference is proportional to the second derivative of the charge. Clearly, *second* (and *higher*) *derivatives* are also of interest to physical scientists, so they form the second major theme of this chapter, discussed in Section 15.3.

15.1.2   Fast track questions

Question F1

Find the first four derivatives of the following functions:

(a) $F(x) = 3 \sin x + 4 \cos x$

(b) $y = \log_e x \quad x > 0$

> The alternative notations ln (x) and e^x are often used for $\log_e x$ and exp (x), respectively.

Question F2

If $x = a(\theta - \sin \theta)$ and $y = a(1 - \cos \theta)$, where a is a constant, find

$$\frac{dy}{dx} \quad \text{and} \quad \frac{d^2y}{dx^2}.$$

Question F3

Find the first and second derivatives of the following functions:

(a) $F(t) = \exp[\alpha t + \sin (\omega t)]$ with respect to t, where α and ω are constants

(b) $F(x) = 5 \log_e (\sin^2 x + 1)$ with respect to x

(c) $F(x) = \arcsin (5x)$ with respect to x

> Some authors use the alternative notation $\sin^{-1} (x)$ for arcsin (x).

15.1.3   Ready to study?

Study comment

In order to study this chapter you should be familiar with the following terms: *derivative*, *differentiate*, *domain*, *function* and *inverse function*. You will need to be familiar with the following specific types of function: *exponential functions*, *logarithmic functions*, *power functions*, *trigonometric functions* (*cosine, sine, tangent, cosec, sec, cot*) and *inverse trigonometric functions* (*arccos, arcsin, arctan*). You should also know the meaning of the term *radian* and be aware of the need to express angles in radians when using them in the *arguments* of trigonometric functions. You should be able to differentiate *powers, exponential functions, logarithmic functions* and *trigonometric functions*, and to use the *sum rule, constant multiple rule, product rule* and *quotient rule of differentiation*. The following *Ready to study questions* will let you establish whether you need to review some of these topics before working through this chapter.

Question R1

Which of the following formulae can be used to define a function of x? What restrictions are there on the admissible values of x?

(a) $f(x) = x^5$

(b) $y = \arcsin x$ (i.e. $\sin y = x$)

(c) $y^2 = 4x^2$

Question R2

For each of the following, express x as a function of y:

(a) $y = 9x^3$ (b) $y = \tan x$ (c) $y = e^{2x}$

(d) $y = \log_e (3x)$

Question R3

Define the derivative of a function $f(x)$ at $x = a$.

Question R4

For each of the functions $f(x)$ below find the derivative $f'(x)$:

(a) $8x^6 + 6x^3 - 5x^2 - 2$ (b) $3 \sin x + 4 \cos x$

Question R5

Find dy/dx for each of the following:

(a) $y = 6e^x$ (b) $y = x^2 - 5 \log_e x$

Question R6

Differentiate the following functions of x:

(a) $x^2 e^x$ (b) $x^3 \sin x$ (c) $e^x \cos x$

(d) $\dfrac{\log_e x}{3x^2 + x}, x > 0$ (e) $\dfrac{3 \cos x}{2 + \sin x}$ (f) $\dfrac{x}{2 + \cos x}$

15.2 Derivatives of composite functions

15.2.1 A function of a function

In this subsection, we review the concept and notation of a *function of a function* (first introduced in Chapter 3) by considering the example of a *simple harmonic oscillator*, such as a (frictionless) simple pendulum or a weight on an (ideal) spring. At time t the displacement x of such a simple harmonic oscillator from its equilibrium position ($x = 0$) is given by

$$x = A \cos (\omega t + \phi) \tag{1}$$

where A is the *amplitude* of the oscillation, ω the *angular frequency* and ϕ the *phase constant*. The expression inside the brackets, $\omega t + \phi$, is called the *phase*; it is a function of time, and its value determines the stage the oscillator has reached in its cycle of motion. The oscillator returns to the same position each time the phase increases by 2π. (☞) Now, suppose that $A = 5.00$ cm, $\omega = 0.60$ s^{-1} and $\phi = \pi$, and that you are asked to calculate the displacement of the oscillator at $t = 10$ s. To answer this question you would first need to evaluate the phase when $t = 10$ s. This is given by

$$\omega t + \phi = 0.60 \text{ s}^{-1} \times 10 \text{ s} + \pi = 9.144 \quad \text{(to two decimal places)}$$

and then substitute this value into Equation 1 to find

$$x = 5.00 \text{ cm} \times \cos (9.14) = -4.94 \text{ cm}$$

In this example, x is a function of the expression ($\omega t + \phi$), which is itself a function of t. For this reason x is said to be a **function of a function** of t, or a **composite function** of t. In Subsection 15.2.2 we will see how to differentiate such expressions, but for the moment we concentrate on reviewing the concept of and notation for a function of a function.

The use of boxes to represent the action of functions is helpful here. Figure 15.1 shows, in this form, the three functions $f(t) = 5t + 1$, $f(t) = t^6$, $f(t) = \sin t$. In each case the function converts an input value (the *argument* of the function) into an output value of the function. It is an inherent property of functions that a *single* output value corresponds to each valid input.

For the function of a function we use two boxes in series. Figure 15.2 shows the action of the two composite functions $\sin (5t + 1)$ and $\cos (\omega t + \phi)$. In each case the output from the first box is used as the input for the second box.

The phase here is dimensionless; however, the *phase* can be interpreted as an angle in which case it is measured in *radians*, and an increase by 2π radians represents the completion of another cycle.

To evaluate cos (9.14) correctly your calculator must be in the *radian mode*.

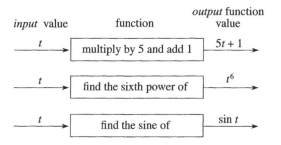

Figure 15.1 Box representation of a function.

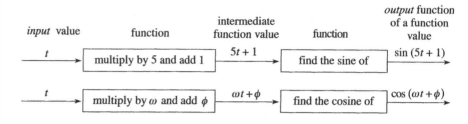

Figure 15.2 Composite
functions.

In the left margin:

The notations sin x and
sin (x) are identical in
meaning. On the whole we
will follow customary practice
and omit the brackets in this
chapter unless the argument
of the function consists of
something more than a single
letter. Similar comments
apply to other functions, such
as tan (x) and $\log_e (x)$.

Figure 15.2 Composite functions.

In general, we can consider a composite function $f(g(t))$ as depicted in
Figure 15.3. $f(g(t))$ is read as 'f of g of t'.

The composite function can be made more user-friendly by writing
$u = g(t)$ so that the output is $f(u)$. Hence, if we consider the composite function $\sin(5t + 1)$, we can write $g(t) = 5t + 1$ so that $f(g(t)) = \sin(g(t))$ and
we can denote $g(t)$ by u so that $f(u) = \sin u$ where, in this case, $u = 5t + 1$.

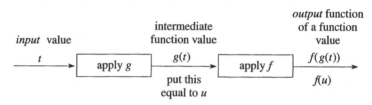

Figure 15.3 General composite function.

Question T1

Draw a box representation of the following functions:

(a) $\cos(3t - 4)$ (b) $(t^4 + 2)^3$ □

There are two observations that you may find useful when trying to determine whether or not a function is a composite. First, decide if the function
definition can be read naturally by using the word 'of'. For example, in

$$h(x) = \log_e (x^2) \quad (x \neq 0)$$

the right-hand side can be read as 'log to base e of (x squared)', which helps
to identify $h(x)$ as a composite $f(g(x))$, where

$$f(x) = \log_e x$$

and $g(x) = x^2$

Second, think about how you would calculate function values on a calculator.
For $h(x) = \log_e (x^2)$, you would proceed as follows:

first enter the value of x and square it,

then find the natural logarithm $\log_e$ of the result.

A function of a function always involves applying *first* one process, *then*
another to the result of the first. Table 15.1 shows some examples of
composite functions.

In the left margin:

The restriction $x \neq 0$ ensures
that $\log_e (x^2)$ i.e. 'f of g of x'
is defined.

Sometimes the notation
$(f \circ g)(x)$ is used for $f(g(x))$.

Table 15.1   Some composite functions

Composite function	Read as ...	First ...	then ...
$\tan (3x)$	'tan *of* 3x'	find $3x$	take tan (of the result)
$\log_e (1/x)$	'log$_e$ *of* 1/x'	find $1/x$	take log$_e$ (of the result)
$e^{\sin x}$	'exp *of* sin x'	find $\sin x$	take exp (of the result)

The following is a very important question. Make sure that you understand it before continuing.

Question T2

For each of the following functions, decide whether $F(x)$ is a composite function. If $F(x)$ is composite, express F in the form $f(g(x))$, by writing down explicit expressions for $u = g(x)$ and $f(u)$.

(a) $F(x) = e^{2x}$

(b) $F(x) = e^x \sin x$

(c) $F(x) = \log_e (x^3)$

(d) $F(x) = \sin (x^2 + 2x + 1)$

(e) $F(x) = \sin^2 x$ (Remember that $\sin^2 x$ means $(\sin x)^2$)

(f) $F(x) = \sin^3 x$

(g) $F(x) = (\cos x)^{0.5}$ ☐

Two points are worthy of comment to save you from possible errors.

In general $f(g(x)) \neq g(f(x))$. The order in which the functions are applied is important.

The $\neq$ symbol is read as 'is not equal to'.

This is easily illustrated by the following example. When f is the function 'add 3' and g is the function 'square' then $g(f(x)) = (x + 3)^2$ but $f(g(x)) = x^2 + 3$. The only value of x for which $(x + 3)^2 = x^2 + 3$ is $x = -1$, as you can verify, so it is generally true that $f(g(x)) \neq g(f(x))$. There are *some* pairs of functions for which $f(g(x)) = g(f(x))$; for example if g is 'cube' and f is 'square' then $f(g(x)) = (x^3)^2 = x^6$ and $g(f(x)) = (x^2)^3 = x^6$, also. These are exceptions, however.

Does $f(g(x)) = g(f(x))$ for the functions $g(x) = 2x + 1$, $f(x) = 3x + 2$?

$f(g(x)) = f(u) = 3u + 2 = 3(2x + 1) + 2 = 6x + 3 + 2 = 6x + 5$

$g(f(x)) = g(u) = 2u + 1 = 2(3x + 2) + 1 = 6x + 4 + 1 = 6x + 5$

Hence $f(g(x)) = g(f(x))$ in this case.   ☐

The function
$g(f(x)) = \sin(\log_e(x))$
presents no such problems.

When dealing with a composite function $f(g(x))$, care must be taken to ensure that the function is only applied to admissible values of x, i.e. to values of x that are within the *domain* of the composite function.

Consider the following three examples:

1. $g(x) = x^2 + 1, f(x) = \log_e(x)$

   Here $f(g(x)) = \log_e(x^2 + 1)$.

   The logarithmic function $\log_e(x)$ is defined only if $x > 0$, that is, only if the input to the function is positive. However, in this case $x^2 + 1 > 0$ for all x, so the composite function is well defined for all (real) values of x.

2. $g(x) = x^2 - 1, f(x) = x^{1/2}$

   Here $f(g(x)) = (x^2 - 1)^{1/2}$.

   Any input value x will lead to a value of $g(x)$, but the input to the function f must be non-negative. Hence values of x that satisfy $x \geqslant 1$ or $x \leqslant -1$ are admissible, but values such that $-1 < x < 1$ are excluded from the domain of $f(g(x))$.

3. $g(x) = \sin(x), f(x) = \log_e(x)$

   Here $f(g(x)) = \log_e(\sin(x))$.

   The values of x for which $\sin(x) \leqslant 0$ are not admissible since they provide an unacceptable input for the $\log_e$ function.
   The box diagram presents a useful check. In Figure 15.4 we show the function $\log_e(\sin(x))$.

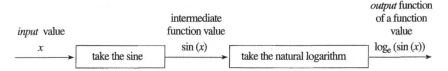

Figure 15.4 The function $\log_e(\sin(x))$.

What are the entry requirements for each box in Figure 15.4?

The input to the first box can be any real number but the input to the second box must be positive. Therefore, only those values of x which lead to a positive value of $\sin x$ are acceptable and will allow a successful evaluation of $\log_e(\sin(x))$. □

Question T3

If $f(x) = (x + 1)^{1/2}$ and $g(x) = \log_e x$ what are the restrictions on x in order that we may define $f(g(x))$ and $g(f(x))$? ☐

15.2.2   The chain rule

It is often necessary to differentiate a function of a function. This can sometimes be achieved by using basic differentiation techniques. Here, we will consider two special cases where this is possible, and then we will generalise the results to provide a powerful general rule (the *chain rule*) that can be used to differentiate any function of a function.

Using the product rule

Consider the function $\sin^2 x$.

This can be regarded as a function of a function, for if we write $g(x) = \sin(x)$ and $f(x) = x^2$

then $\qquad f(g(x)) = (\sin x)^2 = \sin^2 x$

However, it can also be regarded as a product since $\sin^2 x = \sin x \times \sin x$, so it can be differentiated by means of the product rule to give

$$\frac{d}{dx}(\sin^2 x) = \frac{d}{dx}(\sin x \times \sin x) = (\cos x \times \sin x) + (\sin x \times \cos x)$$

i.e. $\quad \dfrac{d}{dx}(\sin^2 x) = 2 \sin x \cos x$

Now, the derivative of a function does not depend on how we choose to 'regard' the function. So, in this particular case, we have been able to use the product rule to find the derivative of a function of a function.

Similarly, the function $\cos^2 x$ can be considered as a product, so

$$\frac{d}{dx}(\cos^2 x) = \frac{d}{dx}(\cos x \times \cos x)$$

$$= (-\sin x \times \cos x) + [\cos x \times (-\sin x)] = 2 \cos x \times (-\sin x)$$

$$= -2 \cos x \times \sin x$$

and $\quad \dfrac{d}{dx}(x^3 + 4)^2 = \dfrac{d}{dx}[(x^3 + 4) \times (x^3 + 4)]$

$$= 3x^2 (x^3 + 4) + (x^3 + 4) \times 3x^2$$

$$= 2(x^3 + 4) \times 3x^2$$

In each of these three cases the pattern is the same; we start with an expression of the form

$$q(x) = (\text{function of } x)^2$$

and we find that

$$\frac{dq}{dx} = 2(\text{function of } x) \times (\text{derivative of function})$$

So, if $\quad q(x) = [p(x)]^2$

then $\quad q'(x) = 2p(x) \times p'(x)$

Question T4

Find the derivatives of the following composite functions:

(a) $\quad q(x) = (2x - 3)^2$

(b) $\quad q(x) = (3x^4 + 1)^2$

(c) $\quad q(x) = (x + \sin x)^2$ □

Using the properties of the logarithmic function

The function $\log_e (x^3)$ can be regarded as a function of a function, but thanks to the special properties of logarithmic functions it can also be written as $3 \log_e x$, and it then follows from the constant multiple rule that

$$\frac{d}{dx} [\log_e (x^3)] = 3 \times \frac{d}{dx} [\log_e (x)] = 3 \times \frac{1}{x} = \frac{3}{x}$$

So, once again, we have been able to use a technique of basic differentiation to find the derivative of a function of a function. In this case it would have been tempting to have taken the result that the derivative of $\log_e (x)$ is $1/x$ and to have concluded that the derivative of $\log_e x^3$ is $1/x^3$. *But this is clearly wrong.* In fact, the correct answer,

$$\frac{3}{x} = \frac{1}{x^3} \times 3x^2$$

and the 'correction factor' of $3x^2$ is the derivative of the function x^3.

What is the derivative of $\log_e (x^5)$?

$\log_e (x^5) = 5 \log_e x$ and the derivative is $5/x$. (Once again it is worth noticing that this answer may be written as

$$\frac{1}{x^5} \times (5x^4)$$

where the final factor $(5x^4)$ is the derivative of the 'inner' function x^5.) □

We can summarise this result as follows:

if $\qquad q(x) = \log_e [p(x)]$

then $\qquad q'(x) = \dfrac{1}{p(x)} \times p'(x)$

The general case — the chain rule

The two boxed results given above are special cases of a general rule for differentiating functions of functions. That general rule is known as the *chain rule.*

The **chain rule**

If $\quad q(x) = f(g(x))$

then $q'(x) = f'(g(x)) \times g'(x)$ $\hspace{4cm}$ (2a)

Alternatively, if we write

$\qquad u = g(x)$ and $q = f(u)$

then we can express the chain rule in the form

$\qquad \dfrac{dq}{dx} = \dfrac{dq}{du} \times \dfrac{du}{dx}$ $\hspace{4cm}$ (2b)

The chain rule expressed in either of these equivalent forms is also known as the **function of a function rule**. You will also see the rule written in various 'mixed' notations such as

$$\frac{dq}{dx} = f'(u) \times \frac{du}{dx}.$$

When we use the chain rule in the form of Equation 2b, it is almost as if we are multiplying together two fractions, dq/du and du/dx, and cancelling the du terms to leave dq/dx. We most definitely are *not* doing this, but it must be admitted that the idea makes the rule easier to remember!

All of the results obtained earlier in this subsection using the product rule, or the special properties of logarithmic functions, can also be obtained from the chain rule. For example, if we look again at the function $q(x) = \sin^2 x$ but we now identify $g(x) = u = \sin x$ and $q = u^2$, the chain rule tells us that

$$q'(x) = \frac{dq}{du} \times \frac{du}{dx} = 2u \times \cos x = 2 \sin x \cos x$$

Similarly, in the second example, if $q(x) = (x^3 + 4)^2$ and we identify $u = x^3 + 4$ and $q = u^2$

then $q'(x) = \dfrac{dq}{du} \times \dfrac{du}{dx} = 2u \times (3x^2) = 2(x^3 + 4) \times 3x^2 = 6x^2(x^3 + 4)$

Also, if $q(x) = \log_e(x^3)$ and we identify $u = x^3$ and $q = \log_e u$

then $q'(x) = \dfrac{dq}{du} \times \dfrac{du}{dx} = \dfrac{1}{u} \times (3x^2) = \dfrac{3x^2}{x^3} = \dfrac{3}{x}$

Moreover, the chain rule can be applied to many other composite functions that could not be differentiated using more elementary techniques. We can summarise the steps involved in applying the chain rule as follows:

To differentiate a function of a function $q(x) = f(g(x))$ using the chain rule:

1.   *Identify* $u = g(x)$ and $q = f(u)$

2.   *Differentiate* to find $\dfrac{du}{dx}$ and $\dfrac{dq}{du}$.

3.   *Multiply* the last two expressions to find $\dfrac{dq}{dx}$.

4.   *Rewrite* u in terms of x and simplify.

Pay particular attention to the last step in this procedure. Forgetting to re-express the final answer in terms of the original variables is a very common mistake, even among those who are proficient at differentiation.

Here are some worked examples to show how this procedure works in practice.

Example 1
Find the derivative of $e^{\sin x}$.

> *Solution*   Let $u = \sin x$ and $q = e^u$
>
> then $\dfrac{du}{dx} = \cos x$, $\dfrac{dq}{du} = e^u$ so $\dfrac{dq}{dx} = \dfrac{dq}{du} \times \dfrac{du}{dx} = e^u \times \cos x$
>
> thus $\dfrac{dq}{dx} = e^{\sin x} \times \cos x = \cos x \, e^{\sin x}$ □

Example 2
Find the derivative of $\sin(x^2 + 1)$.

> *Solution*   Let $u = x^2 + 1$ and $q = \sin u$
>
> then $\dfrac{du}{dx} = 2x$, $\dfrac{dq}{du} = \cos u$ so $\dfrac{dq}{dx} = \dfrac{dq}{du} \times \dfrac{du}{dx} = \cos u \times 2x$
>
> thus $\dfrac{dq}{dx} = \cos(x^2 + 1) \times 2x = 2x \cos(x^2 + 1)$ □

Question T5

Using the chain rule, find the derivatives of the following functions:

(a) e^{2x} (b) $\exp(x^2)$ (c) $(x^2 + 1)^5$ (d) $\sin(6x)$ (e) $\cos(3x)$

(f) $\log_e(e^x)$ (g) $\log_e(1 + x^2)$ (h) $\cos(\omega t + \phi)$ □

In practice you may have to deal with much nastier functions, and they may include functions of functions of functions or worse! This is where the chain rule really justifies its name since we can go on applying it as though adding links to a chain. For example, if $q = f(g(h(x)))$ then we can let $v = h(x)$, $u = g(v)$ and $q = f(u)$ and use the chain rule to obtain

$$\frac{dq}{dx} = \frac{dq}{du} \times \frac{du}{dv} \times \frac{dv}{dx} = f'(u) \times g'(v) \times h'(x)$$

In practice, when faced with this kind of problem, it is often easier to just repeatedly apply the usual two-step form of the chain rule. Part (c) of the next question will give you the chance to try this for yourself.

Differentiate the following composite functions, which all arise in physical problems, with respect to the indicated variable:

(a) The *Gaussian function*, $y = \dfrac{1}{\sigma\sqrt{2\pi}} \exp[-(x - m)^2/(2\sigma^2)]$, with respect to x, where σ, m and π are constants.

(b) The *Boltzmann factor*, $y = e^{-E/kT}$, with respect to T, where E and k are constants.

(c) The *Planck function*,

$$y = \frac{2hc^2}{\lambda^5} \frac{1}{(e^{hc/\lambda kT} - 1)},$$

with respect to λ, where h, c, k and T are all constants.

(a) $\dfrac{dy}{dx} = \dfrac{1}{\sigma\sqrt{2\pi}} \exp\left[\dfrac{-(x - m)^2}{2\sigma^2}\right] \times \dfrac{d}{dx}\left[\dfrac{-(x - m)^2}{2\sigma^2}\right]$

$= \dfrac{1}{\sigma\sqrt{2\pi}} \exp\left[\dfrac{-(x - m)^2}{2\sigma^2}\right] \times \left[\dfrac{-2(x - m)}{2\sigma^2}\right]$

$= \dfrac{-(x - m)}{\sigma^3\sqrt{2\pi}} \exp\left[\dfrac{-(x - m)^2}{2\sigma^2}\right]$

(b) $\dfrac{dy}{dT} = e^{-E/kT} \times \dfrac{d}{dT}\left(\dfrac{-E}{kT}\right) = \dfrac{E}{kT^2} e^{-E/kT}$

(c) Using the product rule, we obtain

$$\frac{dy}{d\lambda} = \frac{d}{d\lambda}\left(\frac{2hc^2}{\lambda^5}\right) \times \frac{1}{(e^{hc/\lambda kT} - 1)} + \frac{2hc^2}{\lambda^5} \times \frac{d}{d\lambda}\left[\frac{1}{(e^{hc/\lambda kT} - 1)}\right]$$

$$= \frac{-10hc^2}{\lambda^6} \times \frac{1}{(e^{hc/\lambda kT} - 1)} + \frac{2hc^2}{\lambda^5} \times \frac{d}{d\lambda}[(e^{hc/\lambda kT} - 1)^{-1}]$$

Now, applying the chain rule to the final term, we find

$$\frac{dy}{d\lambda} = \frac{-10hc^2}{\lambda^6} \times \frac{1}{(e^{hc/\lambda kT} - 1)}$$

$$+ \frac{2hc^2}{\lambda^5} \times \left[-(e^{hc/\lambda kT} - 1)^{-2}\right] \times \frac{d}{d\lambda}(e^{hc/\lambda kT} - 1)$$

Applying the chain rule again, to the new final term

$$\frac{dy}{d\lambda} = \frac{-10hc^2}{\lambda^6} \times \frac{1}{(e^{hc/\lambda kT} - 1)}$$

$$+ \frac{2hc^2}{\lambda^5} \times \left[\frac{-1}{(e^{hc/\lambda kT} - 1)^2}\right] \times e^{hc/\lambda kT} \frac{d}{d\lambda}\left(\frac{hc}{\lambda kT}\right)$$

$$= \frac{-10hc^2}{\lambda^6} \times \frac{1}{(e^{hc/\lambda kT} - 1)}$$

$$+ \frac{2hc^2}{\lambda^5} \times \left[\frac{-1}{(e^{hc/\lambda kT} - 1)^2}\right] \times e^{hc/\lambda kT} \left(\frac{-hc}{\lambda^2 kT}\right)$$

$$= \frac{2hc^2}{\lambda^6(e^{hc/\lambda kT} - 1)}\left[\frac{hc}{\lambda kT}\frac{e^{hc/\lambda kT}}{(e^{hc/\lambda kT} - 1)} - 5\right] \qquad \square$$

15.2.3 Differentiating implicit functions

Functions are often specified *explicitly* in terms of equations or formulae such as $f(x) = ax^2$ or $\theta(t) = \omega t + \phi$, where the dependent variable is isolated on the left-hand side as the *subject* of the equation. However, it is also possible to specify a function *implicitly* by means of a formula that relates the dependent and independent variables, but does not isolate the dependent variable as the subject. For instance, the formula

$$\sqrt{\frac{y}{a}} - x = 0 \tag{3}$$

implicitly defines y as a function of x, even though it does not do so explicitly. In this particular case a little algebraic manipulation soon leads to the explicit relationship $y = ax^2$, but there are also cases where an implicitly defined function has no explicit representation at all. (☞) In general, a dependent variable y will be defined implicitly as a function of an independent variable x if x and y are related by an equation of the form

$$F(x, y) = 0$$

where $F(x, y)$ is a function of both x and y. A function $y(x)$ defined in this way is called an **implicit function**.

The chain rule makes it possible to differentiate implicit functions by means of a technique known as **implicit differentiation**. As a simple example, suppose that y is defined as an implicit function of x by the equation

$$y^3 = x^5$$

Differentiating both sides of this equation (with respect to x) gives us

$$\frac{d}{dx}(y^3) = 5x^4$$

But, if we put $u = y^3$ and use the chain rule $\dfrac{du}{dx} = \dfrac{du}{dy} \times \dfrac{dy}{dx}$ we find

$$\frac{d}{dx}(y^3) = 3y^2 \frac{dy}{dx}$$

thus $\quad 3y^2 \dfrac{dy}{dx} = 5x^4 \quad$ and hence $\quad \dfrac{dy}{dx} = \dfrac{5x^4}{3y^2}$

We can now use the original relationship between x and y to express the right-hand side entirely in terms of x. In this case, $y^2 = x^{10/3}$

so $\quad \dfrac{dy}{dx} = \dfrac{5}{3} x^{2/3}$

In physical science, implicit differentiation often arises in proofs and derivations, and frequently involves functions that have no exact explicit representation.

Question T6
> Use the technique of implicit differentiation to find dy/dx if $y^5 = \sin x$. Leave your answer as an expression in both x and y. □

15.2.4 Differentiating inverse functions

This subsection shows how the chain rule can be used to differentiate *inverse functions* such as the inverse trigonometric function $\arcsin x$. However, before introducing the general method for differentiating such functions we consider a simple problem that illustrates a useful ancillary rule.

For example, the angular position of a planet moving around the Sun is determined from a quantity called the *eccentric anomaly E* that is defined implicitly as a function of time t by *Kepler's equation*

$$E - e \sin E = 2\pi(t - T)/a^{3/2}$$

where e, T and a are constants. Although it is possible to find approximate expressions for E, there is no exact explicit representation.

Remember that if $y = x^a$, then $dy/dx = ax^{(a-1)}$.

If $y = x^{1/2}$, write down dy/dx. Then write down an expression for x in terms of y, and hence find dx/dy as a function of y. Finally, express dx/dy in terms of x.

Using the rule for differentiating powers,

$$\frac{dy}{dx} = \frac{1}{2}x^{-1/2} = \frac{1}{2\sqrt{x}}$$

$x = y^2$, so $\dfrac{dx}{dy} = 2y$. In terms of x, $\dfrac{dx}{dy} = 2\sqrt{x}$. □

Notice that the expression for $\dfrac{dy}{dx}$ in the last question is the reciprocal of that

for $\dfrac{dx}{dy}$. This is illustrative of a very useful general rule, the **inversion rule**:

Check that this rule works by trying some other functions that involve powers of x. A general proof is given in the chapter devoted to *solving first-order differential equations* in the companion volume.

The *inversion rule*: $\quad \dfrac{dx}{dy} = 1 \Big/ \left(\dfrac{dy}{dx}\right) \qquad\qquad (4)$

This looks as if we are finding the reciprocal of a fraction dy/dx. That is definitely *not* what we are doing, but it does make the rule easy to remember.

In a few cases the inversion rule is sufficiently powerful on its own to enable us to differentiate some inverse functions. For instance, using the inversion rule we can differentiate $\log_e x$, the inverse function of e^x. If $y = \log_e x$, then $x = e^y$ and $dx/dy = e^y$. Now, using this result together with the inversion rule,

$$\frac{d(\log_e x)}{dx} = \frac{dy}{dx} = 1 \Big/ \left(\frac{dx}{dy}\right) = \frac{1}{e^y} = \frac{1}{x}$$

The inverse trigonometric function arcsin x is defined for $-1 \leqslant x \leqslant 1$ by the requirement that if $y = $ arcsin x, then y is that value in the range $-\pi/2$ to $\pi/2$ such that $\sin y = x$.

Often, though, the inversion rule is not enough and we have to use the chain rule to differentiate inverse functions. As an example, consider the problem of finding the derivative dy/dx when

$$y = \arcsin x$$

The graph of this function is shown in Figure 15.5. Taking the sine of both sides gives us

$$\sin y = \sin(\arcsin x) = x$$

Now differentiate both the right and left-hand sides of this equation with respect to x. The right-hand side is simply 1, but to differentiate the left-hand side we have to apply the chain rule.

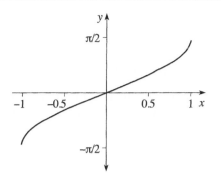

Figure 15.5 The graph of $y = \arcsin x$.

If we let $z = \sin y$,

$$\frac{dz}{dx} = \frac{dz}{dy} \times \frac{dy}{dx} = \cos y \times \frac{dy}{dx}$$

Hence, the result of differentiating both sides of the equation $\sin y = x$ is

$$\cos y \times \frac{dy}{dx} = 1$$

so that $\dfrac{dy}{dx} = \dfrac{1}{\cos y}$

All that remains is to express the right-hand side in terms of x. To do this note that $\cos^2 y = 1 - \sin^2 y = 1 - x^2$, so $\cos y = \pm \sqrt{1 - x^2}$.

Now, an examination of Figure 15.5 shows that the graph of $y = \arcsin x$ has a *positive* gradient throughout its domain of definition, so only *positive* values of $\cos y$ are admissible and hence we must use the positive square root when writing the final answer.

Thus $\dfrac{dy}{dx} = \dfrac{1}{\sqrt{1 - x^2}}$

Table 15.2 records this result together with the derivatives of some other inverse functions.

Finally let us look at a more complicated example that uses both the chain rule and the inversion rule.

Example 3

Figure 15.6 shows the trajectory of a falling object released at point O from an aircraft moving horizontally with a constant speed u. The tangent at any point on the trajectory is inclined at an angle θ to the horizontal as shown, and this angle increases throughout the fall. What is the rate of change of θ with time?

Solution Assuming that air resistance may be neglected, and using the coordinate system shown in Figure 15.6, in which x increases to the right and y increases downwards, the components of velocity in the horizontal and vertical direction will be

Table 15.2 Derivatives of some inverse functions

$f(x)$	$f'(x)$
arcsin x	$\dfrac{1}{\sqrt{1-x^2}}$
arccos x	$\dfrac{-1}{\sqrt{1-x^2}}$
arctan x	$\dfrac{1}{1+x^2}$
arcsec x	$\dfrac{1}{x\sqrt{x^2-1}}$
arccosec x	$\dfrac{-1}{x\sqrt{x^2-1}}$
arccot x	$\dfrac{-1}{1+x^2}$

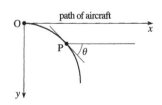

Figure 15.6 Motion of an object released from an aircraft.

$$\frac{dx}{dt} = u \quad \text{and} \quad \frac{dy}{dt} = gt$$

where g is the magnitude of the acceleration due to gravity.

Now, you should be able to see from Figure 15.6 that $\tan\theta = dy/dx$, but using the chain rule and the inversion rule we find

$$\frac{dy}{dx} = \frac{dy}{dt} \times \frac{dt}{dx} = \frac{dy}{dt} \bigg/ \frac{dx}{dt} = \frac{gt}{u}$$

Thus,

$$\tan\theta = \frac{gt}{u}$$

and from this we obtain

$$\theta = \arctan\left(\frac{gt}{u}\right)$$

So, to find the rate of change of θ with t, i.e. $\dfrac{d\theta}{dt}$, we need to differentiate the inverse function $\arctan\left(\dfrac{gt}{u}\right)$.

To do this we first take the tangent of both sides:

$$\tan\theta = \tan\left[\arctan\left(\frac{gt}{u}\right)\right] = \frac{gt}{u}$$

and then differentiate with respect to t. The right-hand side gives g/u, so applying the chain rule to the left-hand side we obtain

$$\sec^2\theta \times \frac{d\theta}{dt} = \frac{g}{u}$$

so

$$\frac{d\theta}{dt} = \frac{g}{u\sec^2\theta}$$

Now we only need to express $\sec^2\theta$ in terms of t, and we can do this using a standard trigonometric identity and the expression for $\tan\theta$ we obtained earlier

Trigonometric identities are fully discussed in Chapter 6.

$$\sec^2\theta = 1 + \tan^2\theta = 1 + \frac{g^2t^2}{u^2}$$

Thus

$$\frac{d\theta}{dt} = \frac{g}{u} \times \frac{1}{\left(1 + \dfrac{g^2t^2}{u^2}\right)} = \frac{gu}{u^2 + g^2t^2} \qquad \square$$

A word of caution

If you differentiate the function $\theta = \arctan(t)$ then, as Table 15.2 indicates, you should find that

$$\frac{d\theta}{dt} = \frac{1}{1 + t^2}.$$

It would be quite wrong to conclude from this that the process of differentiating

$$\theta = \arctan\left(\frac{gt}{u}\right)$$

can be accomplished by simply replacing t on the right-hand side by gt/u. You can obtain the right answer by replacing t by gt/u on both sides, but the safest way is to go through the full method as we have just done. □

Question T7

Find the derivative of each of the following functions:

(a) $\arcsin(x^2)$

(b) $\arccos(x^2 + 4)$ (☞)

(c) $\arctan(3 - x)$ □

You may find it helpful to know that the gradient of $\arccos x$ is negative throughout its domain of definition. You will also need to think carefully about part (b).

15.2.5   Differentiating parametric functions

It is sometimes useful to express two variables x and y each in terms of a third variable called a **parameter**. For example, in *coordinate geometry* the equation that represents a circle of radius a, centred on the origin, is usually written in the form

Coordinate geometry is discussed in Chapters 9 and 10.

$$x^2 + y^2 = a^2$$

However, thanks to the trigonometric identity $\cos^2 \theta + \sin^2 \theta = 1$, the same circle can also be represented in terms of the parameter θ by the pair of equations

$x = a \cos \theta$	(5a)
$y = a \sin \theta$	(5b)

where θ may take any value in the range 0 to 2π.

Eliminating θ from Equations 5a and 5b (by squaring and adding the two equations) immediately yields the usual equation of the circle. Equations 5a and 5b are called the **parametric equations** of the circle. Functions defined in terms of parametric equations are called **parametric functions**. Whenever we describe the motion of a moving object by specifying its position coordinates x, y and z as functions of time, we are really writing down the parametric equations of the object's pathway, with time t as the parameter.

Returning to the case of the circle $x^2 + y^2 = a^2$, if we want to find dy/dx, the gradient of the tangent at a point on the circle, we can use the parametric equations and employ a variation of the chain rule, namely

$$\frac{dy}{dx} = \frac{dy}{d\theta} \times \frac{d\theta}{dx} = \frac{dy}{d\theta} \bigg/ \frac{dx}{d\theta} \qquad\qquad (6)$$

Now, $\quad \dfrac{dy}{d\theta} = a\cos\theta \quad$ and $\quad \dfrac{dx}{d\theta} = -a\sin\theta$

so we obtain $\quad \dfrac{dy}{dx} = \dfrac{a\cos\theta}{-a\sin\theta} = -\cot\theta$

This is an example of **parametric differentiation**. (✎)

Example 4

As illustrated in Figure 15.7, a projectile is fired into the air at an angle of $60°$ to the horizontal with a speed of $u = 6\,\mathrm{m\,s^{-1}}$, so its initial horizontal velocity is $u_x = u\cos(60°) = 3\,\mathrm{m\,s^{-1}}$, and its initial vertical velocity is $u_y = u\sin(60°) = 3\sqrt{3}\,\mathrm{m\,s^{-1}}$. What is the flight time for the projectile to reach the top of its trajectory (i.e. how long does it take to reach the point at which the tangent to the trajectory is horizontal)?

Solution   If we neglect air resistance, and take upwards as the positive y-direction, the equations describing its trajectory are (✎)

$$x = u_x t \quad \text{and} \quad y = u_y t - \tfrac{1}{2} g t^2$$

where x and y are, respectively, the horizontal and vertical displacements from the point of projection at time t, and g is the magnitude of the acceleration due to gravity. The horizontal and vertical components of the velocity are given by

$$\frac{dx}{dt} = u_x \quad \text{and} \quad \frac{dy}{dt} = u_y - gt$$

At the top of the trajectory the tangent to the trajectory is horizontal, therefore

$$\frac{dy}{dx} = 0$$

but, using the chain rule and the inversion rule

$$\frac{dy}{dx} = \frac{dy}{dt} \times \frac{dt}{dx} = \frac{dy}{dt} \bigg/ \frac{dx}{dt} = \frac{u_y - gt}{u_x}$$

so, at the top of the trajectory

$$\frac{u_y - gt}{u_x} = 0, \quad \text{i.e.} \quad t = \frac{u_y}{g} \qquad \square$$

An equivalent result

$$\frac{dy}{dx} = -\frac{x}{y}$$

can be obtained by applying the technique of *implicit differentiation* to the usual equation of a circle.

If you are unfamiliar with the physics of projectile motion, just take these first two equations on trust.

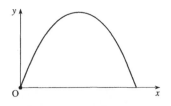

Figure 15.7   Trajectory of a projectile.

Question T8

Find the derivative dy/dx for each of the following pairs of parametric equations:

(a) $x = a \cos \theta$, $y = b \sin \theta$

(b) $x = at^2$, $y = 2at$

(c) $x = ct$, $y = c/t$ ☐

15.3   Second and higher derivatives

15.3.1   The derivative of a derivative

In the same way that we can differentiate a function $f(x)$ to obtain its derivative (or derived function) $f'(x)$, so we can differentiate the function $f'(x)$ to obtain its derivative which is denoted $f''(x)$. This is called the **second derivative** of $f(x)$.

Hence if $f(x) = 4x^3 + 7x^2 + 6x - 8$ it follows that

$$f'(x) = 12x^2 + 14x + 6$$

and $f''(x) = 24x + 14$

$f''(x)$ should be read as 'f double prime of x'.

If we write $y = f(x)$ and denote its derivative (sometimes called its *first derivative* in this context) by dy/dx, then an alternative notation for the second derivative is $\dfrac{d^2y}{dx^2}$. For example, if $y = 4 \sin(2x) + e^{3x}$, then the first derivative is

$\dfrac{d^2y}{dx^2}$ is read as 'dee two y by dee x squared'.

$$\frac{dy}{dx} = 8 \cos(2x) + 3e^{3x}$$

and the second derivative is

$$\frac{d}{dx}\left(\frac{dy}{dx}\right) = \frac{d^2y}{dx^2} = 8 \times [-2 \sin(2x)] + 3 \times (3e^{3x})$$

i.e. $\dfrac{d^2y}{dx^2} = -16 \sin(2x) + 9e^{3x}$

Notice that the notation $\dfrac{d^2y}{dx^2}$ is a rather natural one since it suggests that the operation of differentiation has been repeated twice. Indeed, the symbol d/dx is sometimes referred to as a **differential operator** since it conveys the instruction to differentiate whatever appears immediately to its right. Also notice the precise position of the superscript 2s in the second derivative symbol; it is the differential operator that is 'squared', not the independent variable y.

Some important physical quantities are often thought of as second derivatives. For instance, in *kinematics* − the study of motion − *velocity* is defined as the rate of change of *position*, so

$$v_x = \frac{dx}{dt}$$

for example, but *acceleration* is defined as the rate of change of velocity, so

$$a_x = \frac{dv_x}{dt}$$

and consequently we can regard acceleration as a second derivative, with

$$a_x = \frac{d^2x}{dt^2}$$

Indeed, derivatives with respect to time arise so often in physical science that you will often see them denoted by a special notation that uses a dot to indicate differentiation; thus $v_x = \dot{x}(t)$ and $a_x = \ddot{x}(t)$.

Question T9

Find $f''(x)$ for each of the following functions:

(a) $f(x) = x^6 - 2x^2 + 7$

(b) $f(x) = ax + b,$ where a and b are constants

(c) $f(x) = a \sin x + b \cos x$

(d) $f(x) = e^{3x}$

(e) $f(x) = \log_e (x^2 + 1)$

(f)   Rewrite your answers to parts (a) to (e) using $\frac{d^2y}{dx^2}$ notation.   □

Question T10

(a)   If the x-coordinate of a moving object is given as a function of time by $x(t) = pt + qt^2$, find expressions for the components of velocity and acceleration in the x-direction.

(b)   Find $\ddot{x}(t)$ when $x(t) = A \sin (\omega t)$ and when $x(t) = B \cos (\omega t)$.   □

15.3.2   Higher derivatives

Assuming that the necessary derivatives exist, we can differentiate a function $y = f(x)$ as many times as we want. The third derivative is the derivative of the second derivative, the fourth derivative is the derivative of the third derivative and so on. All these are known collectively as **higher derivatives**, and we define them by

$$\frac{d^3y}{dx^3} = \frac{d}{dx}\left(\frac{d^2y}{dx^2}\right), \quad \frac{d^4y}{dx^4} = \frac{d}{dx}\left(\frac{d^3y}{dx^3}\right),$$

and so on.

In general, the nth derivative is defined by

$$\frac{d^ny}{dx^n} = \frac{d}{dx}\left(\frac{d^{n-1}y}{dx^{n-1}}\right) \qquad (7)$$

To be consistent with this notation we should write

$$\frac{dy}{dx} \quad \text{as} \quad \frac{d^1y}{dx^1}$$

but this is never done in practice.

In the alternative functional notation, derivatives higher than the third are usually denoted by lower case Roman numerals: $f'(x)$, $f''(x)$, $f'''(x)$, $f^{(iv)}(x)$ ..., but the nth derivative is usually denoted $f^{(n)}(x)$.

Example 5

Find the first three derivatives of $y = \dfrac{1}{x^3}$.

Solution

$$\frac{dy}{dx} = \frac{-3}{x^4}, \quad \frac{d^2y}{dx^2} = \frac{+12}{x^5}, \quad \frac{d^3y}{dx^3} = \frac{-60}{x^6} \quad \square$$

Find the first three derivatives of x^n, where $n \geq 3$. What happens if $n = 2$?

$$\frac{dy}{dx} = nx^{n-1}, \quad \frac{d^2y}{dx^2} = n(n-1)x^{n-2}, \quad \frac{d^3y}{dx^3} = n(n-1)(n-2)x^{n-3}$$

If $n = 2$ the derivatives are $2x$, 2, and 0. $\quad \square$

Example 6

Find the first four derivatives of $f(x) = 3x^4 - 8x^3 + 5x^2 - 6$.

Solution $\quad f'(x) = 12x^3 - 24x^2 + 10x$

$f''(x) = 36x^2 - 48x + 10$

$f'''(x) = 72x - 48$

$f^{(iv)}(x) = 72 \quad (\text{☞}) \quad \square$

The nth derivative is sometimes called the nth *order* derivative.

Occasionally you may see $f(x)$ written as $f^{(0)}(x)$, usually to obtain a neat form for a general formula. For example, in the series

$$\sum_{n=0}^{\infty} f^{(n)}(0)\frac{x^n}{n!}$$

when $n = 0$.

Note that $f^{(n)}(x) = 0$ for $n \geq 5$.

Example 7

Find the nth derivative of $f(x) = e^{3x}$.

> *Solution* The first, second and third derivatives are
>
> $$f'(x) = 3e^{3x}, f''(x) = 3 \times 3e^{3x} = 3^2 e^{3x}, \text{ and}$$
>
> $$f'''(x) = 3^2 \times 3e^{3x} = 3^3 e^{3x}.$$
>
> It is clear from these that, in general, the nth derivative is given by
>
> $$f^{(n)}(x) = 3^n e^{3x} \quad \square$$

Question T11

> Find the first four derivatives and the nth derivative of the following functions:
>
> (a) $y = f(x) = 4e^{5x}$
>
> (b) $y = 3x^3$
>
> (c) $y = A \cos(2x) + B \sin(2x)$, where A and B are constants.
>
> (d) $y = \log_e x$, where $x \neq 0$ $\square$

15.3.3   Second derivatives of implicit and parametric functions

The techniques we have accumulated so far to obtain $f'(x)$ from $f(x)$ can, in most cases, be applied to $f'(x)$ to obtain $f''(x)$ with no major difficulties. There are, however, two points which need great care. First is the fact that

$$\frac{dx}{dy} = 1 \bigg/ \left(\frac{dy}{dx}\right) \text{ is an exception}$$

$$\frac{d^2 x}{dy^2} \text{ and } 1 \bigg/ \left(\frac{d^2 y}{dx^2}\right) \text{ are } not \text{ generally equal.}$$

Consider for example the relationship

$$y = \frac{1}{x}$$

Differentiating this expression twice we obtain

$$\frac{dy}{dx} = -\frac{1}{x^2} \text{ and } \frac{d^2 y}{dx^2} = +\frac{2}{x^3}$$

Now, if $x = \dfrac{1}{y}$, $\dfrac{dx}{dy} = -\dfrac{1}{y^2}$ and $\dfrac{d^2x}{dy^2} = +\dfrac{2}{y^3}$

So, while it is true that $\dfrac{dx}{dy} = \dfrac{-1}{y^2} = -x^2 = 1 \Big/ \left(\dfrac{dy}{dx}\right)$

it is also the case that $\dfrac{d^2x}{dy^2} \neq 1 \Big/ \left(\dfrac{d^2y}{dx^2}\right)$ (☞)

Note that we have used the symbol ≠ to show that the two terms are unequal.

since $\dfrac{d^2x}{dy^2} = \dfrac{2}{y^3} = 2x^3$,

whereas $1 \Big/ \left(\dfrac{d^2y}{dx^2}\right) = 1 \Big/ \left(\dfrac{2}{x^3}\right) = \dfrac{x^3}{2}$

We also need to take great care over parametric differentiation. We have seen that for the pair of equations

$x = a \cos\theta$ and $y = a \sin\theta$

$$\dfrac{dy}{dx} = \dfrac{dy}{d\theta} \times \dfrac{d\theta}{dx} = \dfrac{dy}{d\theta}\Big/\dfrac{dx}{d\theta} = \dfrac{a\cos\theta}{-a\sin\theta} = -\cot\theta$$

But to obtain $\dfrac{d^2y}{dx^2}$ we need to determine

$$\dfrac{d^2y}{dx^2} = \dfrac{d}{dx}(-\cot\theta).$$

This can be done with the aid of the chain rule, either by writing $z = -\cot\theta$ and using

$$\dfrac{dz}{dx} = \dfrac{dz}{d\theta} \times \dfrac{d\theta}{dx}$$

or more directly, by writing

$$\dfrac{d}{dx}(-\cot\theta) = \dfrac{d}{d\theta}(-\cot\theta) \times \dfrac{d\theta}{dx}$$

In either case we use the inversion rule to replace $d\theta/dx$ by the reciprocal of $dx/d\theta$ and hence obtain

$$\dfrac{d^2y}{dx^2} = \dfrac{d}{d\theta}(-\cot\theta)\Big/\dfrac{dx}{d\theta} = \dfrac{+\operatorname{cosec}^2\theta}{-a\sin\theta} = -\dfrac{1}{a}\operatorname{cosec}^3\theta$$

Example 8

If $x = at^2$ and $y = 2at$, determine $\dfrac{d^2y}{dx^2}$.

Solution $\dfrac{dx}{dt} = 2at$ and $\dfrac{dy}{dt} = 2a$

so $\dfrac{dy}{dx} = \dfrac{dy}{dt} \times \dfrac{dt}{dx} = \dfrac{dy}{dt} \Big/ \dfrac{dx}{dt} = \dfrac{2a}{2at} = \dfrac{1}{t}$

hence $\dfrac{d^2y}{dx^2} = \dfrac{d}{dx}\left(\dfrac{dy}{dx}\right) = \dfrac{d}{dx}\left(\dfrac{1}{t}\right)$

using the chain rule, we find

$$\dfrac{d^2y}{dx^2} = \dfrac{d}{dt}\left(\dfrac{1}{t}\right) \times \dfrac{dt}{dx} = -\dfrac{1}{t^2} \times \dfrac{dt}{dx}$$

and using the inversion rule, we find

$$\dfrac{d^2y}{dx^2} = \dfrac{-1}{t^2} \Big/ \dfrac{dx}{dt} = \dfrac{-1}{t^2}\dfrac{1}{2at} = \dfrac{-1}{2at^3}$$

It is worth observing that implicit differentiation provides an alternative way of dealing with this example. To adopt this approach start by noting that $y^2 = 4a^2t^2 = 4a \times at^2 = 4ax$.

So implicit differentiation of this with respect to x gives us

$$2y\dfrac{dy}{dx} = 4a$$

A second implicit differentiation, in which $y\dfrac{dy}{dx}$ is treated as a product, gives

$$2y\dfrac{d^2y}{dx^2} + \dfrac{2dy}{dx} \times \dfrac{dy}{dx} = 0$$

so $2 \times 2at \times \dfrac{d^2y}{dx^2} + 2\dfrac{1}{t} \times \dfrac{1}{t} = 0$

i.e. $4at\dfrac{d^2y}{dx^2} + \dfrac{2}{t^2} = 0$

hence, as before $\dfrac{d^2y}{dx^2} = \dfrac{-2}{t^2} \times \dfrac{1}{4at} = \dfrac{-1}{2at^3}$ □

Question T12

Find $\dfrac{d^2y}{dx^2}$ for each of the following pairs of parametric equations:

(a) $x = \cos\theta$, $y = \sin\theta$

(b) $x = a\cos\theta$, $y = b\sin\theta$

(c) $x = ct$, $y = c/t$ □

15.4 A further example

We will end by tackling a set of related electrical problems – all of which involve a circuit containing an inductor and capacitor (an *LC* circuit). Working through each of the steps in these problems for yourself will help you to review what you have learned in this chapter.

15.4.1 An example from electrical circuits

In a *series LC circuit* the charge $q(t)$ on the capacitor at time $t > 0$ satisfies the equation

$$L\frac{d^2q}{dt^2} + \frac{1}{C}q = 0 \tag{8}$$

You do not need to be familiar with the physics of electrical circuits to follow the essential steps of this discussion.

where L and C are constants, known as the inductance and capacitance, respectively.

The first problem we consider is that of showing that

$$q(t) = A\cos\left(\frac{t}{\sqrt{LC}}\right) + B\sin\left(\frac{t}{\sqrt{LC}}\right) \tag{9}$$

satisfies Equation 8 for arbitrary values of the constants A and B.

To do this we need to deduce an expression for d^2q/dt^2 from Equation 9 and substitute it into Equation 8. The first step is to find dq/dt from Equation 9.

Now, $q(t)$ is the sum of two terms, each of which is a function of a function. Taking just the second term and calling it y we have

$$y = B\sin\left(\frac{1}{\sqrt{LC}}\right) \text{ and if we put } u = \frac{t}{\sqrt{LC}}$$

we find $y = B\sin u$. From this, we see

$$\frac{du}{dt} = \frac{1}{\sqrt{LC}} \quad \text{and} \quad \frac{dy}{du} = B\cos u ,$$

and if we then use the chain rule, we find

$$\frac{dy}{dt} = \frac{dy}{du} \times \frac{du}{dt} = B \cos u \times \frac{1}{\sqrt{LC}} = \frac{B}{\sqrt{LC}} \cos u$$

$$= \frac{B}{\sqrt{LC}} \cos\left(\frac{t}{\sqrt{LC}}\right)$$

If we now repeat the process for the first term in Equation 9, what is the derivative of the term?

$$y = A \cos\left(\frac{t}{\sqrt{LC}}\right)$$

If we put $y = A \cos u$, where $u = \dfrac{t}{\sqrt{LC}}$, it follows from the chain rule that

$$\frac{dy}{dt} = \frac{dy}{du} \times \frac{du}{dt} = -A \sin u \times \frac{1}{\sqrt{LC}}$$

$$= -\frac{A}{\sqrt{LC}} \sin\left(\frac{t}{\sqrt{LC}}\right) \quad \square$$

From the above discussion, what is dq/dt from Equation 9?

$$\frac{dq}{dt} = \frac{-A}{\sqrt{LC}} \sin\left(\frac{t}{\sqrt{LC}}\right) + \frac{B}{\sqrt{LC}} \cos\left(\frac{t}{\sqrt{LC}}\right) \quad \square$$

So we now know dq/dt. We now need to find $\dfrac{d^2q}{dt^2}$

What is the derivative of $y = \dfrac{-A}{\sqrt{LC}} \sin\left(\dfrac{t}{\sqrt{LC}}\right)$?

Putting $y = \dfrac{-A}{\sqrt{LC}} \sin u$, where $u = \dfrac{t}{\sqrt{LC}}$, we find using the chain rule,

$$\frac{dy}{dt} = \frac{dy}{du} \times \frac{du}{dt} = \frac{-A}{\sqrt{LC}} \cos u \times \frac{1}{\sqrt{LC}} = \frac{-A}{LC} \cos\left(\frac{t}{\sqrt{LC}}\right) \quad \square$$

What is the derivative of $y = \dfrac{B}{\sqrt{LC}} \cos\left(\dfrac{t}{\sqrt{LC}}\right)$?

Using the same method we find, $\dfrac{dy}{dt} = \dfrac{-B}{LC} \sin\left(\dfrac{t}{\sqrt{LC}}\right)$ $\square$

What is $\dfrac{d^2q}{dt^2}$?

$$\dfrac{d^2q}{dt^2} = \dfrac{-A}{LC} \cos\left(\dfrac{t}{\sqrt{LC}}\right) - \dfrac{B}{LC} \sin\left(\dfrac{t}{\sqrt{LC}}\right) \quad \square$$

Therefore, if we substitute this result (and Equation 9) into Equation 8, we find

$$L\dfrac{d^2q}{dt^2} + \dfrac{1}{C}q = L\left[\dfrac{-A}{LC}\cos\left(\dfrac{t}{\sqrt{LC}}\right) - \dfrac{B}{LC}\sin\left(\dfrac{t}{\sqrt{LC}}\right)\right]$$

$$+ \dfrac{1}{C}\left[A\cos\left(\dfrac{t}{\sqrt{LC}}\right) + B\sin\left(\dfrac{t}{\sqrt{LC}}\right)\right]$$

$$= \dfrac{-A}{C}\cos\left(\dfrac{t}{\sqrt{LC}}\right) - \dfrac{B}{C}\sin\left(\dfrac{t}{\sqrt{LC}}\right) + \dfrac{A}{C}\cos\left(\dfrac{t}{\sqrt{LC}}\right) + \dfrac{B}{C}\sin\left(\dfrac{t}{\sqrt{LC}}\right) = 0$$

Thus the given expression (Equation 9) does satisfy the equation

$$L\dfrac{d^2q}{dt^2} + \dfrac{1}{C}q = 0 \text{ , irrespective of the values of } A \text{ and } B.$$

Now we can consider a related problem. Given that Equation 9 provides a solution to Equation 8 for *any* values of A and B, what are the specific choices of A and B needed to ensure that at time $t = 0$ the charge on the capacitor has the value q_0 and the rate of flow of charge in the circuit is zero? In other words, what choices of A and B ensure that

$$q(0) = q_0 \quad \text{and} \quad \dfrac{dq}{dt}(0) = 0$$

To satisfy the second of these conditions we require

$$-\dfrac{A}{\sqrt{LC}} \sin(0) + \dfrac{B}{\sqrt{LC}} \cos(0) = 0$$

i.e. $\dfrac{B}{\sqrt{LC}} = 0$

We must therefore choose $B = 0$, so

$$q = A \cos\left(\frac{t}{\sqrt{LC}}\right)$$

To satisfy the additional condition that $q(0) = q_0$ we require

$$A \cos (0) = q_0$$

i.e. $A = q_0$

so the solution satisfying both the given initial conditions is

$$q(t) = q_0 \cos\left(\frac{t}{\sqrt{LC}}\right)$$

15.5 Closing items

15.5.1 Chapter summary

1. A *function of a function* produces an output by applying *first* one process, *then* another to an input.

2. In general $f(g(x)) \ne g(f(x))$. The order in which the functions are applied is important.

3. When dealing with a *composite function* $f(g(x))$, care must be taken to ensure that the function is only applied to admissible values of x, i.e. to values of x that are within the *domain* of the composite function.

4. A function of a function can be differentiated using the *chain rule*, which states that if $u = g(x)$ and $q = f(u)$, then

$$\frac{dq}{dx} = \frac{dq}{du} \times \frac{du}{dx} = f'(u) \times g'(x) \tag{2}$$

5. An *implicit function* is a function that is defined by means of a relation such as $F(x, y) = 0$. The chain rule makes it possible to differentiate implicit functions by the technique of *implicit differentiation*.

e.g. if $y^3 = x^5$ then $3y^2 \dfrac{dy}{dx} = 5x^4$, so $\dfrac{dy}{dx} = \dfrac{5x^4}{3y^2} = \dfrac{5}{3} x^{2/3}$

6. According to the *inversion rule*

$$\frac{dx}{dy} = 1 \bigg/ \left(\frac{dy}{dx}\right) \tag{4}$$

7. *Inverse functions* can be differentiated by applying the inversion rule and the chain rule.

8. If x and y are both defined in terms of a *parameter* θ, then the derivative $\frac{dy}{dx}$ may be determined by the technique of *parametric differentiation*

$$\frac{dy}{dx} = \frac{dy}{d\theta} \times \frac{d\theta}{dx} = \frac{dy}{d\theta} \bigg/ \frac{dx}{d\theta} \tag{6}$$

9. If $y = f(x)$, then the *nth derivative* of y is defined by

$$\frac{d^n y}{dx^n} = \frac{d}{dx}\left(\frac{d^{n-1}y}{dx^{n-1}}\right) \tag{7}$$

The first four derivatives of $f(x)$ may also be represented by $f'(x), f''(x),$ $f'''(x), f^{(iv)}(x)$, and the nth derivative by $f^{(n)}(x)$.

15.5.2   Achievements

Having completed this chapter, you should be able to:

A1. Define the terms that are emboldened in the text of the chapter.

A2. Identify a function of a function.

A3. Use the chain rule to find the derivative of a function of a function.

A4. Find the derivative of an inverse function.

A5. Find the derivative of a function defined parametrically or implicitly.

A6. Find the second derivative of a function of a function.

A7. Find the second derivative of an inverse function.

A8. Find the second derivative of a function defined parametrically or implicitly.

A9. Find the nth derivative of a function.

A10. Recognise and use the different notations for higher derivatives.

15.5.3   End of chapter questions

Question E1   Which of the following is a function of a function? If it is necessary to restrict the values of x, then say so.

 (a) $F(x) = \sin(3x)$

 (b) $F(x) = \log_e(1 - x^2)$

 (c) $y^3 = x$

Question E2   Find the derivative of:

 (i)   each of the functions in Question E1,

 (ii) $f(g(x))$ and $g(f(x))$, where $u = g(x) = x^3 + 2$ and $f(x) = 5x + 4$.

Question E3   Find the derivative dy/dx of $y = \arccos(3x - 2)$.

Question E4   Find the derivative dy/dx when

$$x = 2 + 3\sqrt{2}t \quad \text{and} \quad y = 1 + 3\sqrt{2}t - 4.9t^2$$

Questions E5   Find the second derivatives of the functions in Questions E1, E3 and E4. If it is necessary to restrict the values of x, then say so.

Question E6   Find the first five derivatives and the nth derivative of the following functions:

 (a) $f(x) = x^6 - 2x^4 - x^3 + 6x$

 (b) $f(x) = xe^x$

Stationary points and graph sketching

Chapter

16

16.1 Opening items

16.1.1 Chapter introduction

The temperature in any week rises and falls each day. Given appropriate equipment you could easily record the temperature at various times during a typical week and plot a graph to show the variation. If you did so, you could then answer a question such as 'what was the lowest temperature on Wednesday?' But you could also answer more general questions about the week's readings such as 'What was the lowest temperature recorded at any time?' Both these questions seek minimum values of the temperature; the question about Wednesday is seeking a *relative* or *local minimum*, the question about the whole week is seeking an *absolute* or *global minimum*.

This chapter is mainly concerned with the problem of identifying and locating maxima and minima using the techniques of *differentiation*. It presumes that you have been given physical information in the form of a function, $y = f(x)$ say, and that you have to determine the salient features of that function, particularly the location of its maxima and minima. Problems of this kind are common in physical science. For example, you might well be asked to find any of the following:

- The angle at which a projectile should be launched to ensure that it travels as high as possible for a given launch speed.
- The separation between two atoms that will minimise the energy arising from their mutual interaction.
- The lowest temperature on a metal rod that is heated at both ends.

Subsection 16.2.1 poses a practical problem of this sort; that of providing maximum illumination at a given point on a dining table. The mathematical description of the illumination, using graphs and derivatives, is considered in Subsection 16.2.2, which also reviews some of the basic concepts needed later. Subsections 16.2.3 and 16.2.4 explain what we mean by the maxima and minima (both global and local) of a function and explain their relationship to the *stationary points* at which the function's first derivative is zero. Subsection 16.2.5 describes the *first derivative test*, which is often the simplest way to identify and locate local maxima and minima. Another test, which may be even easier to use in some situations – the *second derivative test* – is considered in Subsection 16.3.2. Other aspects of the second derivative of a function, particularly its general graphical significance and the nature of *points of inflection* at which it changes sign, are considered in Subsection 16.3.3. The chapter ends with a discussion of graph sketching in Section 16.4 which shows how a knowledge of the stationary points and other major features of a function can help you to draw its graph.

16.1.2   Fast track questions

Question FI

Under certain circumstances, when sound travels from one medium to another, the fraction of the incident energy that is transmitted across the interface is given by

$$E(r) = \frac{4r}{(r+1)^2}$$

where r is the ratio of the *acoustic resistances* of the two media. Find any stationary points of $E(r)$, classify them as local maxima, local minima or points of inflection, and sketch the graph of $E(r)$ for $r \geqslant 0$.

Question F2

When air resistance is ignored the horizontal range of a projectile is given by

$$R = \frac{u^2}{g} \sin(2\theta)$$

where u is the initial speed of projection, g is the magnitude of the acceleration due to gravity and θ is the angle (in radians) of projection (with $0 \leqslant \theta \leqslant \pi/2$). Without differentiating the equation, find the value of θ which, for given u, gives the maximum range of the projectile. Confirm that your answer corresponds to a local maximum by investigating the derivatives $dR/d\theta$ and $d^2R/d\theta^2$.

Question F3

Sketch the graph for the function

$$f(x) = x^2 + \frac{1}{x^2} \quad \text{for} \quad -3 \leqslant x \leqslant 3$$

16.1.3   Ready to study?

Study comment

In order to study this chapter you should be familiar with the following terms: *cosine*, *derivative*, *differentiate*, *domain*, *exponential function*, *function*, *gradient* (of a graph), *infinity symbol* (∞), *logarithmic function*, *inequality symbols* ($>$, $<$, $\geqslant$, $\leqslant$), *product rule*, *quotient rule* (*of differentiation*), *second derivative*, *sine*, *tangent* and *variable*. In addition you will need to be able to *factorise* a simple *polynomial*, solve a *quadratic equation* and differentiate a range of functions using the techniques of *basic differentiation* and *functions of functions* (i.e. *composite functions*) using the *chain rule*. The following *Ready to study questions* will help you to establish whether you need to review some of these topics before embarking on the chapter.

Question RI

Differentiate the following functions:

(a) $f(x) = x^4 + 3x^2 - 7x + 6$ (b) $f(x) = \sin(2x)$

(c) $f(x) = \dfrac{3x}{x^2 + 4}$ (d) $f(x) = \dfrac{2x - 5}{(x^2 + a^2)^{1/2}}$

(e) $f(x) = x^2 e^x$

Question R2

Find the second derivative of each of the functions (a) to (c) in Question R1.

Question R3

Solve the equation $x^4 - x^3 = 0$.

16.2   Stationary points and the first derivative

16.2.1   The dinner party problem

Suppose that you and your friends are about to sit down to dinner at a table illuminated by a single lamp suspended at an adjustable height over the centre of the table (see Figure 16.1a). One of your friends notes that the plates are poorly lit and asks you to adjust the lamp so that more light falls on them. What would you do? If you lower the lamp you will decrease the distance r between the lamp and the plates which might improve the situation, but you will also reduce the angle ϕ between the light and the plate; so the plates will present a smaller effective area to the light (Figure 16.1b), and that will make matters worse. Overall, it is not obvious what the effect of lowering the lamp will be. The case is no clearer if you raise the lamp. You will increase r and ϕ and these changes will also produce opposing effects, so again it is

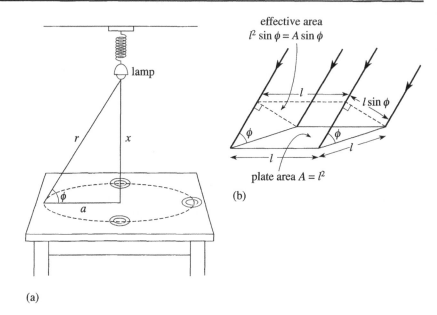

(b)

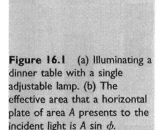

(a)

Figure 16.1   (a) Illuminating a dinner table with a single adjustable lamp. (b) The effective area that a horizontal plate of area A presents to the incident light is $A \sin \phi$.

unclear whether this will increase the light falling on the plates or not. As a physical scientist, faced with this dilemma you would probably do what anybody else would do and move the lamp up and down until you had the best result. But as a student of physical science you should also appreciate that this simple problem presents an interesting challenge. The experimental fact that there *is* an optimum position for the lamp shows that there must be a unique solution to the corresponding theoretical problem. The question is, how do you find that solution?

The best way to start looking for this solution is to make sure that the question is clear. Simplified to its bare essentials, the problem can be stated as follows:

> Given a horizontal circle of radius a and a lamp that emits energy uniformly in all directions at rate P, at what height x above the centre of the circle must the lamp be located to ensure that the rate at which energy falls on a small horizontal disc of area A, at the perimeter of the circle, is a maximum?

The rate at which energy falls on the disc is given by

$$L = \frac{PA \sin \phi}{4\pi r^2}$$

Note   You are not expected to understand the origin of this formula, but $A \sin \phi$ represents the projected area of the disc perpendicular to the incident light, and $(P/4\pi r^2)$ represents the power per unit area in the same plane of projection at a distance of r from the lamp.

The next thing to note is that the only variables on the right-hand side are ϕ and r, and they can both be expressed in terms of x (the height above the centre of the circle) and various constants. From Figure 16.1 we can see

$$\sin \phi = \frac{x}{r}$$

and $\quad r = \sqrt{x^2 + a^2}$

So, if we substitute these two equations into the expression for L we find in turn

$$L = \frac{PAx}{4\pi r^3}$$

and $\quad L(x) = \dfrac{PAx}{4\pi(x^2 + a^2)^{3/2}}$ $\qquad\qquad$ (1)

where we have replaced L by $L(x)$ to emphasise that the value of L depends on the single variable x, i.e. L is a function of x.

The original physical problem has now been reduced to the following mathematical problem:

> Given the values of P, A and a, find the value of x at which $L(x)$ is a maximum.

Solving problems of this general kind is the central theme of this chapter. In particular we will concentrate on using the techniques of *differentiation* to investigate functions such as $L(x)$, and to determine the locations of their maxima, minima and any other points of special interest. The solution of the 'dinner party problem' that has been posed above will be considered from various perspectives at different points in this section.

The watt is the SI unit of power, $1\ \text{W} = 1\ \text{J s}^{-1}$.

16.2.2 Graphs, gradients and derivatives

One way of solving the dinner party problem is to plot the graph $L(x)$ and then determine the optimum value of x. Such a graph is shown in Figure 16.2 for the case $P = 100$ W, $A = 1 \times 10^{-4}$ m^2 and $a = 1$ m. As you can see, $L(x)$ is a maximum when x is about 0.7 m. This method works, but it is slow, and it can be very inaccurate unless the graph is plotted with care. Furthermore, it requires specific values for the various constants. It would be much better if we could find a non-graphical technique for determining the location of the maximum — especially if that technique were to give an algebraic answer that remained true whatever the values of P, A and a.

The key to developing such a non-graphical (i.e. algebraic) technique lies in recognising that the *gradient* (or *slope*) of a graph provides valuable insight into the behaviour of the graph and of the function that it represents.

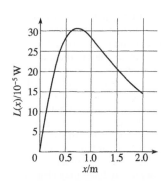

Figure 16.2 The graph of $L(x)$ for particular choices of P, A and a.

As Figure 16.3 shows, the graph of a function may contain regions where the value of the function increases as x increases or regions where it decreases as x increases. In those regions where its value increases, $f(x)$ is said to be an **increasing function** and in those regions where it decreases it is said to be a **decreasing function**.

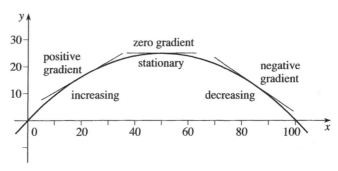

Figure 16.3 The gradient of $y = f(x)$ at various values of x.

In general, regions where a function is increasing or decreasing are associated with different signs for the gradient of that graph. The gradient of a graph is *positive* where the function is increasing and *negative* where the function is decreasing. It should also be clear that at the point where the function reaches its maximum (or minimum) the tangent is horizontal and the gradient is zero. As you will see shortly, it is this observation that provides the basis for an algebraic method of locating maxima and minima.

As discussed in Chapters 14 and 15, given a function $y = f(x)$ it is usually possible to find a related function called the *derived function* or *derivative* that may be denoted $f'(x)$ or $df(x)/dx$, or simply dy/dx. (✒) Your ability to determine derivatives was tested in Subsection 16.1.3, but what matters here is your ability to *interpret* derivatives. In graphical terms the derivative of a function evaluated at a point is nothing other than the gradient of that function at that point. Thus, given an appropriate **interval**, i.e. a range of values such as $0 < x < 0.7$ or $0.8 \leqslant x \leqslant 2$ we can say that

---

$f(x)$ is *increasing* over an interval if $df/dx > 0$ for all points in that interval (✒)

---

and

$f(x)$ is *decreasing* over an interval if $df/dx < 0$ for all points in that interval.

The special case $df/dx = 0$ leads to another definition:

$f(x)$ is **stationary** at any point where $df/dx = 0$, and such points are called **stationary points**.

Just as there are some functions that do not have a gradient at every point, so there are some functions that do not have a derivative at every point. These are considered later.

When referring to an interval mathematicians generally have in mind parts of the number line along which all real numbers are sequentially arranged along the horizontal axis of a graph. For this reason they naturally refer to numbers as being to the left or to the right of other numbers according to their respective values. When an interval is defined they also speak of the right-hand and left-hand end of that interval.

How would you describe the behaviour of the function $y = x^2$ (a small section of this graph is shown in Figure 16.4) in each of the following cases:

(a) $0 < x < 3$ (b) $-3 < x < 0$ (c) when $x = 0$

$dy/dx = 2x$, so:

(a) Over $0 < x < 3$, dy/dx is positive everywhere so $y = x^2$ is increasing.

(b) Over $-3 < x < 3$, dy/dx is negative everywhere so $y = x^2$ is decreasing.

(c) When $x = 0$, dy/dx is zero, so that $y = x^2$ has a stationary points at $x = 0$. (We will have more to say about these points shortly.) □

Question T1

By considering their derivatives, determine the intervals over which each of the following functions is increasing, and the intervals over which each is decreasing. What are the stationary points in each case?

(a) $f(x) = 6x^2 - 3x$

(b) $g(x) = x^2 + 5x + 6$

(c) $h(x) = 6 + x - 3x^2$

(d) $k(x) = x^3 - 6x^2 + 9x + 1$

(e) $l(x) = 1 + 4x - x^3/3$

(f) $m(x) = x^3$ □

16.2.3 Global maxima and minima

The greatest value of a function in a given interval is known as the **global maximum** over that interval, and the smallest value of the function in the interval is known as the **global minimum**. (The equivalent terms **absolute maximum** and **absolute minimum** are also widely used.)

Often we wish to determine the global maximum or global minimum of a given function, and the values of the independent variable at which they occur. The simplest case is when the function is either increasing or decreasing throughout the interval, for then the global maximum or minimum must occur at one end of the interval or the other.

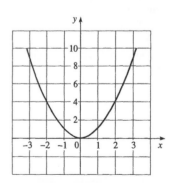

Figure 16.4 The graph of $y = x^2$ over the interval $-3 \leqslant x \leqslant 3$.

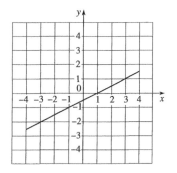

Figure 16.5 The graph of $y = (x - 1)/2$.

Try substituting some specific values of x into

$$\frac{dy}{dx} = 1 - 0.02x$$

if you cannot see that this is the case.

Consider the function $y = (x- 1)/2$ over the interval $-4 \leqslant x \leqslant 4$. Where does it take its greatest and least values?

Differentiating the function, we obtain $dy/dx = \frac{1}{2}$. Since this is positive irrespective of the value of x, the function must be increasing throughout the interval. The *global minimum* of y is -2.5, and occurs when $x = -4$; the *global maximum* is 1.5, and occurs when $x = 4$. (See Figure 16.5.) □

In general, an increasing function has its global minimum at the left-hand end of the interval and its global maximum at the right-hand end (and vice versa for a decreasing function).

However, more care is needed when dealing with a function that both increases and decreases in a given interval. The function $y = x - 0.01x^2$ over the interval $0 \leqslant x \leqslant 100$ is a case in point. (This was the function plotted in Figure 16.3.) In this case it is clear that the global maximum does not occur at an end of the interval. In fact, from the graph, we can see that the function appears to be increasing for $0 \leqslant x < 50$ and decreasing for $50 < x \leqslant 100$, and the global maximum (which turns out to be $y = 25$) occurs at $x = 50$. Of course we were fortunate enough in this case to have been given the graph, but we could have managed without it by employing the following strategy.

First we would have differentiated the function to obtain

$$\frac{dy}{dx} = 1 - 0.02x$$

then we could have seen that

$$\frac{dy}{dx} > 0 \quad \text{if } 0 \leqslant x < 50 \qquad \text{so the function is increasing}$$

$$\frac{dy}{dx} = 0 \quad \text{if } x = 50 \qquad \text{so the function is stationary} \quad \left. \right\} (\text{☜}) \ (2)$$

$$\frac{dy}{dx} < 0 \quad \text{if } 50 < x \leqslant 100 \quad \text{so the function is decreasing}$$

Clearly, the stationary point at which dy/dx changes sign corresponds to the global maximum in this case.

Find the global minimum for $y = x^2$ over the interval $-3 \leqslant x \leqslant 3$.

$dy/dx = 2x$. So y is decreasing if $-3 \leqslant x < 0$, stationary at $x = 0$, and increasing if $0 < x \leqslant 3$. Clearly the global minimum is at $x = 0$. (The graph of this function was plotted in Figure 16.4.) □

It should be clear from the above that finding stationary points can help in locating global maxima and minima but it is important to realise that stationary points are not necessarily maxima or minima. This is demonstrated by the following question.

What are the stationary points of the function $y = x^3$ over the interval $-3 \leqslant x \leqslant 3$? What are the global maximum and minimum values on this interval and where are they located?

First we differentiate the function to obtain $\dfrac{dy}{dx} = 3x^2$. It follows that $\dfrac{dy}{dx} = 0$ when $x = 0$ so this is the only stationary point. (See Figure 16.6.) However, the function is increasing everywhere else in the interval, so the global maximum is $y = 27$ and occurs at $x = 3$, while the global minimum is $y = -27$ and occurs at $x = -3$. □

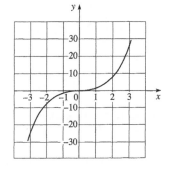

Figure 16.6 The graph of $y = x^3$.

So, although stationary points are important in the process of finding global maxima and minima, they do not automatically determine those maxima and minima. In fact, as long as the derivative exists, a general procedure that always works is as follows:

A strategy for finding the global maximum or minimum over an interval:

1. Given a function $y = f(x)$ over an interval, first differentiate to obtain the derivative $dy/dx = f'(x)$; then determine the stationary points by finding the values of x for which $f'(x) = 0$.

2. Calculate the value of $f(x)$ at the stationary points that lie in the interval, and then calculate the values of $f(x)$ at the ends of the interval.

3. The greatest of these calculated values is the global maximum, the least is the global minimum. (☞)

You can use this strategy to solve the next question. Notice that there is no need to consider the graph of the function.

Note that a particular global maximum or minimum may occur more than once on a given interval. For instance $y = 2x^2$ over $-1 \leqslant x \leqslant 1$, has global maxima of 2 at $x = -1$ *and* at $x = 1$. The constant function $y = a$ attains both its maximum and minimum values at every point in any interval since its maximum and minimum are both equal

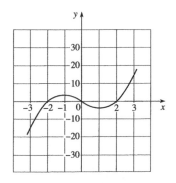

Figure 16.7 The graph of $f(x) = x^3 - 4x$.

Find the global maximum and the global minimum of the following function:

$$f(x) = x^3 - 4x \tag{3}$$

over the interval $-2 \leqslant x \leqslant 3$.

Differentiating the function we obtain $f'(x) = 3x^2 - 4$, so the stationary points occur when $3x^2 - 4 = 0$ which gives $x = \pm 2/\sqrt{3} \approx \pm 1.155$. Notice that both stationary points lie in the given interval.

The values of the function at the stationary points and at the ends of the intervals are: $f(-2) = 0$, $f(-2/\sqrt{3}) \approx -3.079$, $f(3) = 15$.

Hence, the global maximum is 15 (occurring at $x = 3$) and the global minimum is (approximately) -3.079 (occurring at $x = 2/\sqrt{3}$). $\square$

We didn't need the graph of the function in order to reach the above conclusions, but it is instructive to see what is happening in graphical terms (see Figure 16.7). As we proceed along the x-axis from -2, the function increases, until we reach $x = -2/\sqrt{3}$, it then decreases between $x = -2/\sqrt{3}$ and $x = 2/\sqrt{3}$, and then increases thereafter.

In the above example both stationary points were in the given interval, but it would not have caused a problem if they had not been; in fact it would have made the process easier, as the following question illustrates.

Find the global maximum and the global minimum values of the following function:

$$f(x) = x^3 - 4x \quad \text{over the interval } 0 \leqslant x \leqslant 4.$$

(This is the same function as in the last question, but the interval is different.)

Differentiating the function we find $f'(x) = 3x^2 - 4$, so the stationary points are at $x = \pm 2/\sqrt{3}$ (as determined above).

Only one of these points, $x = 2/\sqrt{3}$, lies in the given interval, so we need only calculate the values $f(0) = 0$, $f(2/\sqrt{3}) \approx -3.079$, $f(4) = 48$.

The global maximum value is 48 (occurring at $x = 4$), and the global minimum value is (approximately) -3.079 (occurring at $x = 2/\sqrt{3}$). $\square$

Question T2
Find the global maximum and the global minimum values of the function $f(x) = x^4 - 4x^3$ in the interval $-1 \leqslant x \leqslant 4$. $\square$

Question T3

Find the global maximum and global minimum values of each of the following functions in the intervals given (you should not need to plot the graphs):

(a) $f(x) = 4x + 3$, (i) in $-1 \leqslant x \leqslant 4$ and (ii) in $-3 \leqslant x \leqslant 2$.

(b) $g(x) = x^2 + 4$, (i) in $-2 \leqslant x \leqslant 1$, (ii) in $2 \leqslant x \leqslant 5$ and (iii) in $-3 \leqslant x \leqslant 2$.

(c) $h(x) = 4 - (x - 1)^2$, (i) in $0 \leqslant x \leqslant 3$ and (ii) in $2 \leqslant x \leqslant 3$.

The strategy that we have developed in this section will work very well for a specific function, such as $f(x) = 4x + 3$ in a specific interval such as $-1 \leqslant x \leqslant 4$, because we can simply compare the sizes of numerical values. However, it is less satisfactory (though still correct) for a function involving parameters, such as that which arose in the dinner party problem

$$L(x) = \frac{PAx}{4\pi(x^2 + a^2)^{3/2}} \tag{1}$$

Try comparing the sizes of this function at the ends of the interval $0.5 \leqslant x \leqslant 3$ if you doubt that this is so. We will need to develop some more sophisticated tools in order to deal with such cases.

16.2.4 Local maxima and minima

In the last subsection we saw that in the interval $-2 \leqslant x \leqslant 3$ the function $f(x) = x^3 - 4x$ (illustrated in Figure 16.7) has a global maximum at $x = 3$. However, if we restrict our attention to the region around the stationary point at $x = -2/\sqrt{3}$, then we find that the graph is higher at that value of x than at nearby values of x. At such a point we say that the function has a *local* maximum. Similarly, at the other stationary point, $x = 2/\sqrt{3}$, the graph is lower at that value of x than at nearby values, and at such we say that the function has a *local* minimum. More formally we say that:

The function $f(x)$ has a **local maximum** at $x = a$ if for all x near to a we have $f(x) < f(a)$, i.e. $f(a)$ is locally the largest value of $f(x)$.

The function $f(x)$ has a **local minimum** at $x = a$ if for all x near to a we have $f(x) > f(a)$, i.e. $f(a)$ is locally the smallest value of $f(x)$.

Local maxima and *minima* are sometimes referred to as **relative maxima** and **relative minima**, and are collectively called **local extrema**. They mark the points at which the function changes from increasing to decreasing (or vice versa). Points of this kind, at which the derivative changes sign, are known as **turning points**, so that *local extrema* and *turning points* are different names for essentially the same thing. (*Local extrema* is a plural. A single local maximum or minimum is referred to as a *local extremum*.)

Some exceptional cases where turning points are not stationary points will be considered later in Subsection 16.2.6.

For **smooth functions**, whose graphs exhibit no sharp kinks or sudden discontinuous jumps, local extrema are always *stationary points* at which $df/dx = 0$.

However, the converse is not true. It is *not* the case that all stationary points are local extrema — even in the case of smooth functions. The function $y = x^3$ that we considered in the last section again provides a simple example. It is stationary at $x = 0$, since $dy/dx = 3x^2$ which vanishes (i.e. is equal to zero) at $x = 0$. Nonetheless, as the graph of $y = x^3$ shows (see Figure 16.6), there is neither a local maximum nor a local minimum at $x = 0$.

So, given a function $y = f(x)$, how do we find its local maxima and minima without plotting a graph? There are actually three main methods. In all three methods the first step is the same; differentiate the function, set the derivative equal to zero, and solve the resulting equation, $df/dx = 0$, to find the stationary points. Once this has been done the next step is to investigate (or 'test') the stationary points to see which, if any, are local maxima and which are local minima. This is where the three methods differ. One method uses the behaviour of the first derivative as a test; this is described in the next subsection. Another uses the behaviour of the second derivative; this is described in Section 16.3. The third method is less useful but simpler in some cases and is described below. The most appropriate method to use in any given situation will depend on the details of the function being investigated.

The most obvious way of testing a stationary point $x = a$ to see if it is a local maximum or a local minimum is to evaluate the function at that point and then to compare this value with the value of the function at nearby points either side of $x = a$. (☜) If the function has smaller values either side of $x = a$ then $x = a$ is a local maximum. If the values are greater either side of $x = a$ then it is a local minimum. If neither of these conditions is true then the stationary point is neither a local maximum nor a local minimum.

The problem with this method is that of knowing what constitutes 'nearby points' without plotting the graph. Using points that are too far from the stationary point can easily lead to misleading answers.

Use the above method to find and classify the stationary points of the following functions:

(a) $f(x) = 2x^3 + 1$ (b) $f(x) = x^4 - 2$ (c) $f(x) = -3x^2 + 6$

(a) $\dfrac{df}{dx} = 6x^2$, so the only stationary point is $x = 0$.

At the stationary point, $f(0) = 1$. Immediately to the left, at $x = -\delta$ where δ is a small positive quantity $f(-\delta) = -2\delta^3 + 1$ so $f(-\delta) < f(0)$; immediately to the right at $x = +\delta$, we have $f(+\delta) = 2\delta^3 + 1$, so $f(+\delta) > f(0)$. Thus the stationary point is neither a maximum nor a minimum.

(b) $\dfrac{df}{dx} = 4x^3$, so the only stationary point is $x = 0$.

At the stationary point, $f(0) = -2$. Immediately to the left, $f(-\delta) = \delta^4 - 2$, so $f(-\delta) > f(0)$. Immediately to the right, $f(+\delta) = \delta^4 - 2$, so $f(+\delta) > f(0)$. Thus, in this case the stationary point is a local minimum.

(c) $\dfrac{df}{dx} = -6x$, so the only stationary point is $x = 0$.

At the stationary point, $f(0) = 6$. Immediately to the left, $f(-\delta) = -3\delta^2 + 6$, so $f(-\delta) < f(0)$. Immediately to the right, $f(-\delta) = -3\delta^2 + 6$, so $f(-\delta) < f(0)$. Thus, in this case the stationary point is a local maximum. $\square$

The method we have just used has worked well for the simple cases we have chosen, but it involves quite a lot of work (three function evaluations in each case) and could well have yielded results that were difficult to interpret. The method described in the next subsection is usually quicker and easier.

Returning to the dinner party problem, for the case $P = 100$ W, $A = 1 \times 10^{-4}$ m^2 and $a = 1$ m, determine the stationary point of $L(x)$ in the interval $0 \leqslant x \leqslant 2$ m and confirm that it is a local maximum.

$$L(x) = \frac{PAx}{4\pi(x^2 + a^2)^{3/2}} = \frac{PA}{4\pi} x (x^2 + a^2)^{-3/2}$$

So, using the *product rule* of differentiation we find

$$\frac{dL}{dx} = \frac{PA}{4\pi} (x^2 + a^2)^{-3/2} + \frac{PA}{4\pi} x \frac{d}{dx}\left[(x^2 + a^2)^{-3/2}\right]$$

using the *chain rule* (i.e. the function of a function rule) to differentiate the final term

$$\frac{dL}{dx} = \frac{PA}{4\pi} (x^2 + a^2)^{-3/2} - \left(\frac{3}{2}\right)\frac{PA}{4\pi} x (x^2 + a^2)^{-5/2}(2x)$$

i.e. $\quad \dfrac{dL}{dx} = \dfrac{PA}{4\pi}\left[\dfrac{(x^2 + a^2) - 3x^2}{(x^2 + a^2)^{5/2}}\right] = \dfrac{PA(a^2 - 2x^2)}{4\pi(x^2 + a^2)^{5/2}}$

If we set $\dfrac{dL}{dx} = 0$, we see that the only stationary points are those for which $(a^2 - 2x^2) = 0$, i.e. $x = \pm a/\sqrt{2}$ and only one of these ($x = a/\sqrt{2}$) is in the given interval. Since $a = 1$ m, the stationary point is at $x = 0.707$ m.

In this case it is easier to proceed numerically rather than algebraically, so we will note that at the stationary point

$$L(0.707 \text{ m}) = \frac{100 \text{ W} \times 1 \times 10^{-4} \text{ m}^2 \times 0.707 \text{ m}}{4\pi \times (1.50 \text{ m}^2)^{3/2}} = 3.06 \times 10^{-4} \text{ W}$$

It is the need to select 'nearby' numerical values in this way that makes this method less reliable than the alternatives (unless we carry out some additional investigations).

To the left, at $x = 0.600$ m say, (☜)

$$L(0.600 \text{ m}) = \frac{100 \text{ W} \times 1 \times 10^{-4} \text{ m}^2 \times 0.600 \text{ m}}{4\pi \times (1.36 \text{ m}^2)^{3/2}} = 3.01 \times 10^{-4} \text{ W}$$

To the right, at $x = 0.800$ m say,

$$L(0.800 \text{ m}) = \frac{100 \text{ W} \times 1 \times 10^{-4} \text{ m}^2 \times 0.800 \text{ m}}{4\pi \times (1.64 \text{ m}^2)^{3/2}} = 3.03 \times 10^{-4} \text{ W}$$

Thus the value of the function is lower either side of the stationary point, so $x = 1/\sqrt{2}$ m is a local maximum. (This confirms the approximate result deducted from Figure 16.2 in Subsection 16.2.2.) □

16.2.5 The first derivative test

In order to find the stationary points of a function you usually have to differentiate the function and set the resulting first derivative equal to zero. The **first derivative test** makes further use of the first derivative by using it to determine the nature of the stationary point. In particular it uses the fact, already evident in Figures 16.3, 16.4 and 16.7, that if the first derivative changes sign at a stationary point then that point must be either a local maximum or a local minimum (i.e. a local extremum).

The possible behaviour of the first derivative $f'(x)$ in the neighbourhood of a stationary point is analysed more systematically in Table 16.1

Table 16.1   The behaviour of the first derivative near a stationary point at $x = a$

Sign of $f'(x)$ to left of a	*Sign of $f'(x)$ to right of a*	Graphical behaviour	Conclusion
+	−	⌒	The stationary point is a turning point and corresponds to a local maximum
−	+	⌣	The stationary point is a turning point and corresponds to a local minimum
+	+	⌡	The stationary point is not a turning point and so the point is not a local extremum
−	−	⌐	The stationary point is not a turning point and so the point is not a local extremum

The small graphs given in Table 16.1 represent the local graphical behaviour of the function. We will have more to say about the last two cases shortly. For the moment we note the following test.

The first derivative test for local extrema

If $f'(a) = 0$ and $f'(x)$ changes sign from positive to negative at $x = a$, then there is a local maximum at a.

If $f'(a) = 0$ and $f'(x)$ changes sign from negative to positive at $x = a$, then there is a local minimum at a.

If $f'(a) = 0$ but $f'(x)$ does not change sign at $x = a$, then further investigation is required. (See Subsection 16.3.2.) (☞)

Such investigations may indicate a *point of inflection* (with horizontal tangent), as described in Subsection 16.3.3.

Find the locations of the *local extrema* of the function $f(x) = x^3 - 3x$.

We see that $f'(x) = 3x^2 - 3$ so that the stationary points occur when $3x^2 - 3 = 0$, i.e. at $x = -1$ and $x = 1$. Factorising the function, we have

$$f'(x) = 3x^2 - 3 = 3(x - 1)(x + 1)$$

and we see that the derivative changes sign at $x = -1$ and $x = 1$ so that the local extrema occur at these points. Alternatively we can obtain the same result from a table of values of $f'(x)$ (see Table 16.2). Notice that we do not need a graph of the function (although it is in fact similar to that of Figure 16.7). □

Table 16.2 A table of values of $f'(x) = 3x^2 - 3 = 3(x - 1)(x + 1)$

x	$f'(x)$	Sign of the derivative
−1.1	0.63	positive
−1.0	0.00	sign change
−0.9	−0.57	negative
0.0	−3.00	negative
0.9	−0.57	negative
1.0	0.00	sign change
1.1	0.63	positive

Question T4

(a) Find the stationary points of the function

$$f(x) = x^3 - 3x^2$$

and decide whether each corresponds to a local maximum, a local minimum or neither.

(b) Find the stationary points of the function

$$g(x) = x^3/3 + 2x^2 + 3x + 1$$

and classify each as a local maximum, a local minimum or neither.

(c) Show that at $x = -1$ there is a local extremum of the function

$$h(x) = x^4 - 4x^3 + 16x$$

and decide whether it is a local maximum or local minimum. Are there any other local extrema? [*Hint*: $(x^3 - 3x^2 + 4) = (x + 1)(x^2 - 4x + 4)$] □

16.2.6   Exceptional cases

Functions that are not differentiable

Occasionally you may come across functions that cannot be differentiated at various points; in fact in some situations we go out of our way to construct such functions. The 'square wave' and 'saw-tooth wave' that are so useful in the control of oscilloscopes are cases in point (see Figure 16.8). The techniques that have been developed in this chapter require only a minor modification to deal with such cases, as the following example illustrates.

Consider the function $f(x) = |x|$, which is defined as follows:

$$f(x) = \begin{cases} x, & \text{for } x > 0 \\ 0, & \text{for } x = 0 \\ -x, & \text{for } x < 0 \end{cases}$$

The derivative $f'(x)$ does not exist at $x = 0$. For $x > 0$, $f'(x) = 1$ (positive) while for $x < 0$, $f'(x) = -1$ (negative) so there is no unique value that can be associated with $f'(x)$ at $x = 0$. However, the graph of the function, Figure 16.9, shows that it has a local minimum at $x = 0$.

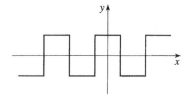

Figure 16.8   A square wave.

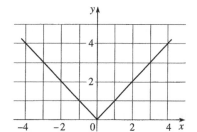

Figure 16.9   The graph of the function $f(x) = |x|$.

To test (without drawing the graph) the point at which the derivative fails to exist, in order to see if it is a local maximum or local minimum, we simply examine the sign of the gradient on either side of the point as before. This strategy will succeed provided that the graph of the function does not have a 'jump' or 'break' at the point in question. A function whose graph has no breaks is described as **continuous**.

Similarly we can adapt this method to determine the global maxima and minima. We just include the values of the function at the points where the derivative fails to exist in our list of values to be tested.

Finding the maxima and minima without using calculus

It is not always necessary to use calculus to find the global maximum and minimum values of a function; in fact in some instances it is far easier not to.

For example, suppose that we place no restriction on the values of x, then the greatest and least values of the function $y = \sin x$ are 1 and -1 (see

Figure 16.10   The graph of sin x.

Figure 16.10) and so no calculation is required in order to discover that the greatest and least values of the function $y = 1 + \sin(2x)$ are 2 and 0 and that they occur alternately at values of x that are odd integer multiples of $\pi/4$.

As another illustration, suppose that a is some positive constant, and that we wish to find the global maximum of the function

$$f(x) = \frac{1}{(x^2 + a^2)^2}$$

and again we place no restriction on the values of x.

We could certainly use the methods that we have developed in this chapter, but it would involve a little work since we would need to differentiate the function. In fact it is much easier to proceed as follows.

The function attains its greatest value when $x^2 + a^2$ is least, i.e. when $x = 0$. Thus the global maximum value of $f(x)$ is $f(0) = 1/a^4$.

What would have happened if we had been asked for the global minimum value?

Clearly we want $x^2 + a^2$ to be as large as possible but it has no greatest value since we can make it as large as we please by choosing x sufficiently large. It follows that $f(x)$ may be made as close to zero as we please. Although the value of $f(x)$ decreases towards zero as $x \to \pm\infty$, it never achieves the value of zero and so, strictly, the function does not achieve a global minimum value. The graph of $y = f(x)$ is shown in Figure 16.1.1.

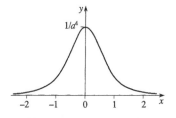

Figure 16.11 The graph of

$$y = \frac{1}{(x^2 + a^2)^2}.$$

The arrow symbol ($\to$) can be read as 'approaches' or 'tends to'.

Question T5

Without differentiating the function, use the fact that $x^2 - 4x + 5 = (x - 2)^2 + 1$, to find the global maximum value of the function

$$f(x) = \frac{1}{(x^2 - 4x + 5)^2}$$

in the interval $-\infty \leqslant x \leqslant \infty$. □

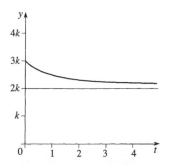

Figure 16.12 The graph of

$$y = k\left(\frac{2t + 3}{t + 1}\right)$$

Functions that do not have a global maximum or minimum

Not all functions have a global maximum or global minimum in a given region, as the above example shows. As an easier example, if we are allowed to choose any real value for x, the function $f(x) = x$ can be as large positive (or negative) as we please and so has no global maximum or minimum value.

Perhaps not quite so obvious is the case of $f(x) = 1/x$ in the interval $-1 \leqslant x \leqslant 1$. Here the function is not defined at $x = 0$, but we can obtain positive or negative values as large as we please by choosing x sufficiently close to 0, and so the function does not achieve a global maximum or minimum.

Some cases are more subtle yet. Suppose, for example, that you are told that the strength of a certain signal is given by $k\left(\dfrac{2t + 3}{t + 1}\right)$, where t is some kind of positive numerical variable and k is a positive constant, and that you are asked to find the global minimum value of the signal strength. The graph of the function is shown in Figure 16.12. For very large values of t, the function value is very nearly equal to $2k$, so you might be tempted to give this as your answer. Nonetheless, in strict mathematical terms the function does not have a global minimum and does not achieve the value $2k$ for any value of t.

16.2.7   Back to the dinner party

Finally in this section, we return to the dinner party problem from Subsection 16.2.1. The task was to find the value of $x \geqslant 0$ for which the function

$$L(x) = \frac{PAx}{4\pi(x^2 + a^2)^{3/2}}$$

attains its greatest value, where P, A and a are positive constants.

We saw earlier that differentiating $L(x)$ gave us

$$\frac{dL}{dx} = I\frac{(a^2 - 2x^2)}{(x^2 + a^2)^{5/2}}$$

where we have written $I = PA/4\pi$, and that the stationary points, given by $dL/dx = 0$, satisfy $x^2 = a^2/2$ so the only meaningful solution is $x = a/\sqrt{2}$. When $0 < x < (a/\sqrt{2})$ we have $dL/dx > 0$ and when $x > (a/\sqrt{2})$ we have $dL/dx < 0$, so from the first derivative test we see that $x = (a/\sqrt{2})$ is a local maximum. In fact we can see more than this, for the function is increasing throughout $0 < x < (a/\sqrt{2})$ and decreasing for all $x > (a/\sqrt{2})$. It follows that $x = (a/\sqrt{2})$ must actually be a global maximum for all positive values of x.

16.3   Stationary points and the second derivative

16.3.1   Graphical significance of the second derivative

The second derivative of $f(x)$ is the *rate of change* of the gradient of $f(x)$. Thus, a positive second derivative implies that the function is becoming steeper 'upwards'.

Figure 16.13 shows the graphs of two functions $y = h(x)$ and $y = j(x)$. One might loosely describe the graph of $h(x)$ to be 'bowl-shaped' upwards, while the graph of $j(x)$ is 'bowl-shaped' downwards. The technical term of this property of a graph is *concavity*, and it is closely related to the behaviour of the *second derivative* of the function, $\frac{d^2y}{dx^2}$. The first graph is **concave upwards**, while the second is **concave downwards**. It is possible for the graph of a function to be concave upwards on part of its domain and concave downwards elsewhere, for example the graph shown in Figure 16.7.

You will see shortly that the concept of concavity leads us to another method of classifying local extrema and also to the important notion of a *point of inflection*.

The concavity of a graph is easily determined from its second derivative as follows:

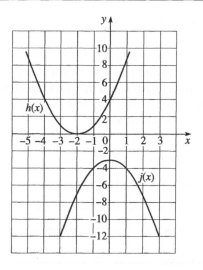

Some find it helpful to remember that $f''(x) > 0$ gives a 'happy' curve

i.e. concave upwards, while $f''(x) < 0$ gives a 'sad' curve

i.e. concave downwards.

Figure 16.13 Graphs of functions that are concave upwards (the top graph), and concave downwards (the bottom graph).

Let a function $f(x)$ be such that $f''(x)$ is defined for all x in an interval then:

If $f''(x) > 0$ throughout the interval then the graph of $f(x)$ is *concave upwards*;

If $f''(x) < 0$ throughout the interval then the graph of $f(x)$ is *concave downwards*.

What is the concavity of the graph of $f(x) = x^3 + x^2$?

Differentiating the function we have $f'(x) = 3x^2 + 2x$ so that $f''(x) = 6x + 2$. It follows that $f''(x)$ is positive if $x > -1/3$, so the graph is concave upwards, while $f''(x)$ is negative if $x < -1/3$ and the graph is concave downwards. $\square$

Question T6

For each of the following functions use the second derivative to determine the intervals over which the function is concave upwards and the intervals over which it is concave downwards:

(a) $f(x) = -x^2 + 7x$

(b) $f(x) = -(x + 2)^2 + 8$

(c) $f(x) = -x^3 + 6x^2 x - 1$

(d) $f(x) = (x + 5)^3$

(e) $f(x) = x(x - 4)^3$

(f) $f(x) = xe^x$ $\square$

16.3.2   The second derivative test

At a local minimum the graph of a function is concave upwards while at a local maximum the graph is concave downwards, and this provides us with a useful method known as the **second derivative test** for classifying local extrema. This method is probably easier to use than the first derivative test, *provided that the second derivative is easy to calculate.*

The second derivative test for local extrema

If $f'(a) = 0$ and $f''(a) < 0$, then there is a local maximum at a.

If $f'(a) = 0$ and $f''(a) > 0$, then there is a local minimum at a.

If $f'(a) = 0$ and $f''(a) = 0$ then *no decision can be made* without further investigation. (✍)

We have emphasised this point because it is a common source of errors.

It requires effort to stretch or compress an elastic spring and the *potential energy* stored in the deformed spring is a measure of this effort. The units of potential energy are joules (J), so the constant k has units $\mathrm{J\,m^{-2}}$.

The *potential energy* stored in an ideal elastic spring of natural length L is given by

$$V(x) = \frac{K(x - L)^2}{2}$$

when it is stretched (or compressed) to a length x, where K is a positive constant. For some real materials, such as a rectangular block of rubber of natural length L, the following model is more accurate

$$U(x) = \frac{K}{2}\left(x^2 + \frac{2L^3}{x} - 3L^2\right) \tag{4}$$

(where x is again the stretched or compressed length). Find and classify any local extrema of the function $U(x)$.

Differentiating $U(x)$ gives us

$$U'(x) = K\left(x - \frac{L^3}{x^2}\right).$$

Thus, there is a stationary point when $x = L$. Now we differentiate again to obtain

$$U''(x) = K\left(1 + \frac{2L^3}{x^3}\right)$$

It is always a good idea to check your results with the physical meaning if you can. In this case we conclude, from the shape of the graphs near $x = L$, that deforming the block of rubber by either compressing or stretching it will increase the potential energy, and this is in accord with our physical intuition. Notice that both the ideal model and the realistic model predict this result. However, according to the ideal model it requires a finite amount of energy to compress the rubber to zero length, while the realistic model suggests that this would require an infinite amount of energy.

and we see that $U''(L) = 3K$ which is greater than zero. It follows from the second derivative test that the function has a local minimum at $x = L$. The graphs of $U(x)$ and $V(x)$ are shown in Figure 16.14. The negative values of x have no physical significance.   □

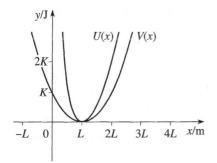

Figure 16.14 The graphs of $y = U(x)$, the inner curve, and $y = V(x)$, the outer curve.

16.3.3 Points of inflection

A point on a graph where the concavity changes from upwards to downwards or vice versa is called a **point of inflection**. At a point of inflection *the second derivative changes sign*.

Notice particularly that we require the second derivative to *change sign*; it is not sufficient for the second derivative to be zero at the point.

Does the function $f(x) = x^3 - 4x$ have a point of inflection? If so, where is it?

If $f(x) = x^3 - 4x$ then $f''(x) = 6x$ and the second derivative changes sign at $x = 0$, so that this point is a point of inflection. The function is concave upwards if $x > 0$ and concave downwards if $x < 0$ (see Figure 16.7). $\square$

It is important to note that just because a section of a graph is concave upwards (or downwards) it does not mean it will have a minimum (or maximum).

Does the function $f(x) = x^4$ have a point of inflection? If so, where is it?

For the function $f(x) = x^4$ we have $f''(x) = 12x^2$ which is positive for $x < 0$ and also positive for $x > 0$. In this case it is certainly true that $f''(0) = 0$, but the second derivative does not change sign at $x = 0$, and so there is no point of inflection at $x = 0$. (In fact this function is positive everywhere except at $x = 0$ where it is zero, and so has a local minimum at $x = 0$. The graph is similar in shape to Figure 16.4 but slightly 'blunter' at the origin.) $\square$

At a point of inflection, say $x = a$, the second derivative $f''(x)$ changes sign and so $f''(a) = 0$. However the condition $f''(a) = 0$ is not *sufficient* to ensure that the second derivative changes sign and that a point of inflection exists at $x = a$.

Note that a point of inflection need not be a stationary point, although it may be. Those points of inflection that are also stationary points (i.e. those at which $f''(x)$ changes sign and $f'(x) = 0$) are called *points of inflection with horizontal tangent*.

The function $f(x) = x^3$ has a first derivative $f'(x) = 3x^2$ and a second derivative $f''(x) = 6x$, so that the first derivative is zero at $x = 0$ and the second derivative changes sign there. In this case we have a point of inflection that happens to occur at a stationary point. In such a case, the fact that the stationary point is *not* a turning point would be sufficient to tell us that the point is in fact a point of inflection. (See the last two cases in Table 16.1).

The function $f(x) = x^3 - 3x$ has, as we have seen, a point of inflection at $x = 0$ but $f'(x) = 3x^2 - 3$ is *not* zero at $x = 0$.

Does the graph of the potential energy function

$$U(x) = \frac{K}{2}\left(x^2 + \frac{2L^3}{x} - 3L^2\right)$$

(4)

have any points of inflection? (K is a positive constant.)

$$U'(x) = K\left(x - \frac{L^3}{x^2}\right) \quad \text{and} \quad U''(x) = K\left(1 + \frac{2L^3}{x^3}\right)$$

so that the second derivative is zero if

$$\left(1 + \frac{2L^3}{x^3}\right) = 0,$$

i.e. if $x^3 = -2L^3$ which gives us

$$x = -\sqrt[3]{2}\, L = -2^{1/3}L \approx -1.2599\, L\,.$$

We must now check the sign of $U''(x)$ immediately to either side of this value. Since K is a positive constant we have

$$U''(-1.2594L) \approx -0.0012K < 0$$

while $\quad U''(-1.2604L) \approx +0.0011K > 0$

and so the second derivative changes sign and there is a point of inflection at $x = -2^{1/3}L$. This mathematical result has no physical meaning for the block of natural rubber since only positive x values (distorted length) are physically acceptable. $\square$

Question T7

Find the points of inflection of the following functions

(a) $f(x) = -x^3 + x^2$

(b) $f(x) = \dfrac{x^4}{4} - \dfrac{2x^3}{3} + \dfrac{x^2}{2} + 2x$

(c) $f(x) = 2x^3 - 9x^2 + 12x - 4$

(d) $f(x) = x^5 - \dfrac{5}{3}x^3 + 4$

(e) $f(x) = \sin x$

(f) $f(x) = x^5 - \dfrac{5x^4}{3}$ ☐

16.3.4 Which test should you use?

Given a function $y = f(x)$, which test should you use to determine whether its stationary points are local maxima, local minima or points of inflection? In practice the choice of method is often a matter of taste, but it is sometimes easier to use the first derivative test when the second derivative involves a large calculation.

Examine the function $f(x) = 3x^5 - 5x^3$ for local extrema and points of inflection.

In this case the second derivative is easy to calculate, so differentiating the function, we find

$$f'(x) = 15x^4 - 15x^2 = 15x^2(x^2 - 1)$$

and $f''(x) = 60x^3 - 30x = 30x(2x^2 - 1)$

Stationary points are located where $f'(x) = 0$, i.e. where $15x^2(x^2 - 1) = 0$, so that $x = 0$, $x = 1$ or $x = -1$.

At $x = -1$, $f''(x) = -30$ which is < 0 hence we have a local maximum.

At $x = 1$, $f''(x) = 30$ which is > 0 so that we have a local minimum.

At $x = 0$, $f''(x) = 0$ so that we cannot come to an immediate conclusion. However $f''(x)$ changes sign at $x = 0$ so this point must be a point of inflection. (Another indication of this is the presence of a local maximum on one side of $x = 0$, and a local minimum on the other).

The second derivative $30x(2x^2 - 1)$ also changes sign at $x = \pm 1/\sqrt{2}$ so that these are two further points of inflection though neither of them is a stationary point. Figure 16.15 shows a sketch of the graph of $f(x)$. ☐

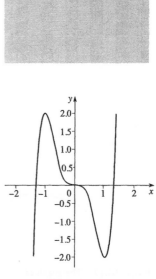

Figure 16.15 Graph of $f(x) = 3x^5 - 5x^3$.

Find and classify the extrema of $f(x) = \sqrt{x^2 + 1}$.

Differentiating the function we obtain

$$f'(x) = \frac{x}{\sqrt{x^2 + 1}}$$

In this case it is certainly possible to calculate the second derivative, but the algebra is not pleasant and it is easier to use the first derivative test. First we see that the only stationary point occurs at $x = 0$, then for $x < 0$ the derivative is negative while for $x > 0$ the derivative is positive. It follows that there is a local minimum at $x = 0$, and this is confirmed by the graph of the function (see Figure 16.16). $\square$

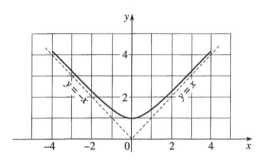

Figure 16.16 The graph of $f(x) = \sqrt{x^2 + 1}$.

The '6−12 function' is sometimes used to represent the potential energy that arises from the interaction of two atoms separated by a distance r

$$U(r) = \frac{-A}{r^6} + \frac{B}{r^{12}}$$

where A and B are positive constants. Left to themselves, the atoms will be at equilibrium when their separation r corresponds to a local minimum of $U(r)$. Find and classify the local extrema of $U(r)$.

Differentiating the function we have

$$U'(r) = \frac{6A}{r^7} - \frac{12B}{r^{13}} = \frac{6}{r^{13}} (Ar^6 - 2B).$$

The derivative is zero if

$$r = \left(\frac{2B}{A}\right)^{1/6}.$$

The derivative is positive if r is greater, and negative if r is less, than this value; and it follows that there is a local minimum at this point. (The graph is shown in Figure 16.17.) $\square$

Figure 16.17 The graph of $U(r) = \frac{-A}{r^6} + \frac{B}{r^{12}}$.

Question T8

For each of the functions in Question T1 (Subsection 16.2.2), locate the local extrema, and classify them as local maxima or local minima. □

16.4 Graph sketching

You will have noticed that we have made great use of graphs to illustrate the algebraic results obtained in this chapter, but the reverse process can be quite productive: we can often use knowledge of local extrema and points of inflection, obtained algebraically, to help us to construct the graph of a function.

16.4.1 The value of graph sketching

Given a function, for example $f(x) = 3x^5 - 5x^3$, we can plot its graph by taking a selection of values for x, calculating the corresponding values of $y = 3x^5 - 5x^3$ and drawing a smooth curve through the points (x, y) so generated. With the advent of advanced scientific calculators and graph plotting computer programs, very often the simplest method of producing a graph is to plot it on a machine. However this is not always the case, and it is quite possible that using a machine without some prior knowledge of the graph will be totally unproductive. You may like to try plotting the graph of the function

$$f(x) = \frac{1}{(10^6 x - 1)(x - 10^6)}$$

on a computer, or calculator, if you doubt that this is the case. Another problem that often arises is that of determining the behaviour of a function that involves several algebraic parameters for all values of the parameters rather than for some particular numerical choices.

The purpose of **graph sketching**, as opposed to graph *plotting*, is to determine the important features of the graph with the minimum of effort. Of course, if more detailed information is required it will be necessary to carry out further investigation.

16.4.2 A scheme for graph sketching

In this subsection we present a number of techniques that will produce the required detail for the majority of functions that you are likely to encounter.

I. Symmetry about the y-axis

A function $f(x)$ is said to be **even** if $f(-x) = f(x)$. Examples of even functions are $f(x) = 5x^2$, $f(x) = x^{2n}$ (for any integer n) and $f(x) = \cos x$. The graph of an even function is symmetrical about the y-axis; for an example see the graph of $y = x^2$ in Figure 16.4.

A function $f(x)$ is said to be **odd** if $f(-x) = -f(x)$. Examples of odd functions are $f(x) = 2x$, $f(x) = x^3$, $f(x) = x^{2n+1}$ (for any integer n), $f(x) = \sin x$ and $f(x) = \tan x$.

If $f(x)$ is odd and $f(0)$ is defined then $f(-0) = -f(0)$, i.e. $f(0) = -f(0)$ which implies that $f(0) = 0$ and therefore the graph of any odd function must pass through the origin. If the portion of the graph for which $x \geqslant 0$ is rotated $180°$ clockwise about the origin, it coincides with that portion for which $x \leqslant 0$; for an example see Figure 16.6. Notice that a function may be neither odd nor even. In fact, odd and even functions are rather special cases.

Classify the following functions as odd, even or neither:

(a) $f(x) = 3x^2 + 2$ (b) $g(x) = 2x^3 - x$ (c) $h(x) = 3x^2 + x$

(a) $f(-x) = 3(-x)^2 + 2 = 3x^2 + 2 = f(x)$, therefore it is even.

(b) $g(-x) = 2(-x)^3 - (-x) = -2x^3 + x = -g(x)$, therefore it is odd.

(c) $h(-x) = 3(-x)^2 + (-x) = 3x^2 - x \neq h(x)$ or $-h(x)$, therefore it is neither odd nor even.   □

We notice some special general rules here. All *even powers* of x are *even functions*; all *odd powers* of x are *odd functions*. Functions which combine odd and even powers of x may or may not have special symmetry. Also the product of two odd functions or two even functions is an even function; the product of an odd and even function is odd.

Question T9

Classify the following functions as odd, even or neither:

(a) $x + \sin x$ (b) $x^2 - 2 \cos x$ (c) $x^2 + \sin x$

(d) $x \sin x$ (e) $x \cos x$ □

---

2.   Forbidden regions

Consider the curve whose equation is

$$y = (x^2 - 1)^{1/2}$$

A *real* number is one which is not *complex*, and does not involve $i = \sqrt{-1}$. *Complex numbers* are discussed in the companion volume.

In order to obtain a *real* value for y we required that $x^2 \geqslant 1$. This means that either $x \geqslant 1$ or $x \leqslant -1$ and there is no part of the curve in the interval $-1 < x < 1$.

3.   Intercepts with the axes

Any graph that meets the y-axis does so when $x = 0$, hence to find such intercepts we simply put $x = 0$ in the equation. For example, given $y = x^3 + 4$ we put $x = 0$ to obtain $y = 4$.

Similarly, any curve that meets the x-axis does so where $y = 0$. For the curve whose equation is $y = x^2(x - 1)$, putting $y = 0$ yields $x = 0$ or $x = 1$.

Such solutions are generally called **roots** and in this case there may be said to be a *double root* at $x = 0$ and a *single root* at $x = -1$. Similarly $x^3(x + 1)^2 - (x - 3) = 0$ has a *triple root* at $x = 0$, a *double root* at $x = -1$, and a *single root* at $x = 3$. The single root at $x = 1$ represents a simple crossing point on the axis *and the function changes sign there*, but at the double root $x = 0$ the graph touches the x-axis (in a similar fashion to the graph $y = x^2$, Figure 16.4) and the function does not change sign. At a triple root the graph would touch the x-axis and change sign (as in the graph of $y = x^3$, Figure 16.6). Such roots, corresponding to a factor raised to a power, are often known as **multiple roots**.

4. Isolated points where the function is not defined

The function $h(x) = 1/x$ is defined for all values of x except $x = 0$. In such cases it is often informative to investigate the values of the function immediately above and below the isolated point. Reference to the graph of $h(x)$ in Figure 16.18 shows that for values of x approaching zero from the positive side, $h(x)$ is large and positive; whereas for values of x approaching zero from the negative side, $h(x)$ is large and negative. A line to which a curve approaches in the *limit* (as the curve is *extrapolated* to infinity) but never touches, is called an **asymptote**. In Figure 16.18 the line $x = 0$ be described as a **vertical asymptote** and the line $y = 0$ as a **horizontal asymptote**.

Figure 16.18 The graph of $y = 1/x$.

5. Behaviour for large values of $|x|$

It is sometimes possible to get an idea of a function's approximate behaviour by ignoring terms in the function that are small in magnitude compared to others. For example, the function

$$f(x) = \sqrt{x^2 + 1} = \sqrt{x^2}\sqrt{\left(1 + \frac{1}{x^2}\right)} = |x|\sqrt{\left(1 + \frac{1}{x^2}\right)} \quad (\text{☞})$$

Remember that here we are using the convention that square roots are positive so that $\sqrt{x^2} = |x|$.

can be approximated by $g(x) = |x|$ for values of $|x| \gg 1$ because $1/x^2$ is then much smaller in magnitude than 1. We may write $f(x) \approx |x|$ as $x \to \pm\infty$ which means that the function is approximately equal to $|x|$ when x is large (positive or negative). Figure 16.16 showed the graphs of $y = \sqrt{x^2 + 1}$ and $y = |x|$. In such a case the function $f(x) = \sqrt{x^2 + 1}$ is said to approach the function $g(x) = |x|$ *asymptotically* as $x \to \pm\infty$.

6. Behaviour for small values of $|x|$

When x is small and positive, or small and negative, it is sometimes possible to approximate the function by ignoring terms that are small in magnitude. As an example $f(x) = x^3 - 9x$ is approximated by the function $j(x) = -9x$ when $|x| \ll 1$ because x^3 is much smaller in magnitude than $-9x$. This tells us at once that the tangent to the graph at $x = 0$ is the line $y = -9x$.

While it is true that the second derivative changes sign as x increases through the value 1, this point is not a candidate for a point of inflection since the function is undefined there. However, we can conclude that the graph changes from concave downwards to concave upwards at this point.

7. Use of the derivative(s)

Applying the previous techniques we can examine the function

$$f(x) = \frac{x^2}{x-1}.$$

We first notice that the function is undefined for $x = 1$; then we see that it has a double root at $x = 0$ (so the graph touches the x-axis at $x = 0$). For values of x a little less than 1 the function is very large and negative, while for values of x a little more than 1 the function is very large and positive.

For values of $|x| \gg 1$ we see that

$$f(x) = \frac{x^2}{x-1} \approx x$$

(because $x - 1$ is 'roughly the same size as x when $|x| \gg 1$'), and so $y = x$ is an asymptote.

Notice that all of the above information was obtained without the use of calculus, but if we need more detail then we can differentiate to obtain (after some manipulation)

$$f'(x) = \frac{x(x-2)}{(x-1)^2}.$$

So we can see from this that there are stationary points at $x = 0$ (confirming what we already suspected) and at $x = 2$.

We must now choose to use either the first or second derivative tests. In this case it is not too difficult to show that

$$f''(x) = \frac{2}{(x-1)^3}$$

so we use the second derivative test. We have $f''(0) = -2 < 0$, so that there is a local maximum at $x = 0$, and $f''(2) = 2 > 0$. There are no points of inflection since the second derivative is never zero. (☜) Combining all the information we obtain the sketch graph shown in Figure 16.19.

8. Concavity

In some circumstances you may find that the concavity provides useful extra information about the curve; however, this is not usually the case.

16.4.3   Examples

In this subsection we sketch the graphs of three functions. In any one example we need some, but not necessarily all, of the techniques outlined in the previous subsection.

Remember that you may not need to obtain all the information that we derive in these examples. The general principle is that you should obtain enough information to produce a sketch that is sufficient to your needs.

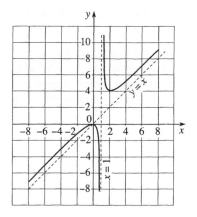

Figure 16.19 A sketch graph of $f(x) = x^2/(x - 1)$.

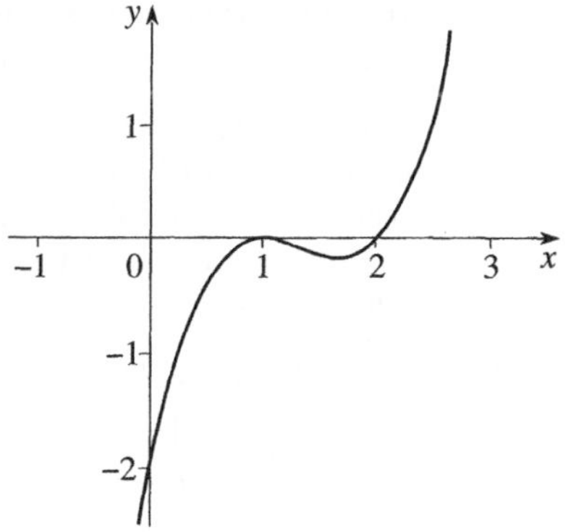

Figure 16.20 A sketched graph of the function $f(x) = (x - 1)^2(x - 2)$.

Example 1

Sketch the graph of $f(x) = (x - 1)^2(x - 2)$.

Solution

(a) We notice that $y = f(x) = 0$ when $x = 1$ (a double root) or $x = 2$ so we have found two intercepts on the x-axis. Also $f(0) = -2$ gives us the intercept on the y-axis.

(b) If $x \gg 1$ then $(x - 1)^2 \approx x^2$ and $x - 2 \approx x$ so that $f(x) \approx x^3$, and so we see that the graph approaches $y = x^3$ when $|x| \gg 1$.

(c) Differentiating the function we have $f'(x) = 3x^2 - 8x + 5 = (3x - 5)(x - 1)$ and so there are stationary points at $x = 1$ (where $y = 0$) and at $x = 5/3$ (where $y = -4/27$).

(d) Differentiating again we have $f''(x) = 6x - 8$, and since $f''(1) = -2 < 0$ there is a local maximum at $x = 1$, while $f''(5/3) = 2 > 0$ so there is a local minimum at $x = 5/3$. The graph is therefore concave upwards if $6x - 8$ is positive, i.e. $x > 4/3$, and concave downwards if $6x - 8$ is negative, i.e. $x < 4/3$.

We can now complete the sketch (see Figure 16.20). □

Example 2

Sketch the graph of $g(x) = \dfrac{x}{1 + x^2}$.

Solution

(a) We first notice that $g(0) = 0$ and this is the only intercept on the x-axis and on the y-axis.

If $x \gg 1$ then $1 + x^2 \approx x^2$ so that $g(x) \approx 1/x$ and the graph approaches $y = 1/x$.

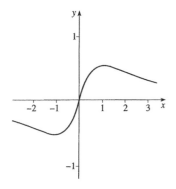

Figure 16.21 A sketched graph of $g(x) = x/(1 + x^2)$.

Try some large positive and negative values for x/e^x on your calculator.

If $x \ll 1$ then x^2 is small compared to 1 and x, so that $g(x) \approx x$, and therefore the tangent at $x = 0$ is $y = x$.

(b) Since $g(-x) = -g(x)$ the function is odd (despite mixing odd and even powers of x); moreover $g(x) > 0$ when $x > 0$, and $g(x) < 0$ when $x < 0$. It follows that no part of the graph appears in the region where $y > 0$ when $x < 0$, nor in the region where $y < 0$ when $x > 0$.

(c) Differentiating we see that

$$g'(x) = \frac{1 - x^2}{(1 + x^2)^2}$$

and so there are stationary points (when $1 - x^2 = 0$) at $x = 1$ (where $y = 1/2$) and at $x = -1$ (where $y = -1/2$).

(d) In this case we may choose to use the first derivative test, and we see that $g'(x) < 0$ if $x > 1$ or $x < -1$ and $g'(x) > 0$ if $-1 < x < 1$. The function is therefore decreasing for $x < -1$, increasing for $-1 < x < 1$ and decreasing for $x > 1$. It follows that there is a local minimum at $x = -1$ and a local maximum at $x = 1$.

(e) At $x = 0$ we have $g'(x) = 1$, confirming that the tangent is $y = x$ at the origin.

(f) If necessary we could find the second derivative

$$g''(x) = \frac{2x(x^2 - 3)}{(x^2 + 1)^3}$$

and the fact that $g''(-1) = 0.5 > 0$ while $g''(1) = -0.5$ to confirm that $x = -1$ is a local minimum and $x = 1$ is a local maximum, from the second derivative test. The second derivative changes sign at $x = 0$ and at $x = \pm\sqrt{3}$ so that these are points of inflection.

All of this information can now be used to produce a sketch of the graph as in Figure 16.21. ☐

Example 3
Sketch the graph of $h(x) = x/e^x$.

Solution
(a) We notice that $h(x) = 0$ at $x = 0$ (and this is the only intercept on the x-axis and on the y-axis).

(b) For $x \gg 1$ the value of e^x is considerably larger than x so that $h(x) \approx 0$. On the other hand, for $x \ll 1$, e^x is very small and positive so that $h(x)$ is then large and negative. (✆)

(c) Differentiating we see that

$$h'(x) = \frac{1 - x}{e^x}$$

so that there is a stationary point at $x = 1$.

(d) Differentiating again we see that

$$h''(x) = \frac{x-2}{e^x}$$

and, since $h''(1) = -1/e < 0$, it follows that there is a local maximum at $x = 1$. The second derivative changes sign at $x = 2$ so that this is a point of inflection. See Figure 16.22. □

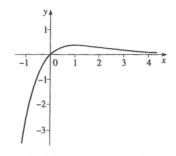

Figure 16.22 A sketched graph of $h(x) = \dfrac{x}{e^x}$.

Question T10

Sketch the graphs of the following functions:

(a) $f(x) = x - \dfrac{1}{x}$

(b) $f(x) = x^4 - 4x^3 + 10$

(c) $f(x) = x^4 - x^2$

16.5   Closing items

16.5.1   Chapter summary

1.   In order to discover the *global maximum* (or *minimum*) of a given function in an *interval* $a \leqslant x \leqslant b$:

(a) First ensure that the function is defined at every point of the interval. If this is not the case there may be no global maximum or minimum.

(b) Evaluate the function at any points where the derivative is undefined.

(c) Evaluate the function at the end points of the interval.

(d) Evaluate the function at the *stationary points* (i.e. points where the derivative of the function is zero).

The global maximum is the greatest, and the global minimum the least, of these values.

2.   To find the *local maxima* and *minima* of a given function, $f(x)$ say:

(a) Find the stationary points (the points at which $f' = 0$).

(b) If it is easy to find the second derivative, evaluate $f''(x)$ at the stationary points. Then:
    (i)   A positive second derivative indicates a local minimum
    (ii)   A negative second derivative indicates a local maximum
    (iii) If the second derivative is zero at the stationary point one must investigate further:

If $f''(x)$ changes sign at the stationary point then the point is a *point of inflection* (with horizontal tangent).

If $f''(x)$ does not change sign at the stationary point then either: $f''(x) > 0$ near the point, which means that the graph is *concave upwards*, and there is a local minimum; or, $f''(x) < 0$ near the point, which means that the graph is *concave downwards*, and there is a local maximum.

(c) If it is not easy to calculate $f''(x)$, examine the behaviour of $f'(x)$ near the stationary point.

   (i) If $f'(x)$ changes sign at the stationary point then this point is a *turning point*, and must therefore be a local maximum or a local minimum. If $f'(x)$ changes from negative to positive there is a local minimum. If $f'(x)$ changes from positive to negative there is a local maximum.

   (ii) If $f'(x)$ does not change sign at the stationary point then this point is a point of inflection.

(d) If the first and second derivative tests are too difficult to apply, simply investigate the value of $f(x)$ at values of x that are close to the stationary point.

3. In order to *sketch the graph* of a function, the following points may be considered:

(a) The symmetry of the function. Is it *even* $(f(-x) = f(x))$ or *odd* $(f(-x) = -f(x))$?

(b) Intervals or points where the function is undefined.

(c) Points at which the graph crosses the axes. Do the intersections on the x-axis correspond to *multiple roots*?

(d) Behaviour of the graph near points at which the function is undefined. Are there any *asymptotes*?

(e) Behaviour of the graph when $|x| \gg 1$. Does the graph approach some simple curve?

(f) Behaviour of the graph where $|x| \ll 1$. What is the tangent at the origin?

(g) Points at which the function is stationary. Are these turning points or points of inflection (with a horizontal tangent)? Where is the function *increasing* and where is it *decreasing*?

(h) Where is the graph *concave upwards* and where is it *concave downwards*?

16.5.2 Achievements

Having completed this chapter, you should be able to:

A1. Define the terms that are emboldened in the text of the chapter.

A2. Define increasing and decreasing functions.

A3. Define global and local extrema.

A4. Locate the stationary points of a function.

A5. Define a point of inflection.

A6. Identify the nature of the stationary points of a function using both first and second derivatives.

A7. Distinguish between odd and even functions, and recognise the graphs of such functions.

A8. Sketch the graph of a function from its formula showing its main features.

16.5.3   End of chapter questions

Question E1   Find the stationary points on the graph of $y = x^2 + x$ and on the graph of $y = x^2 - x$.

Explain why the function $f(x) = x^2 + |x|$ has no stationary points, and find its global maximum value in the interval $-1 \leqslant x \leqslant 1$.

Question E2   Find any local maxima, local minima and points of inflection of the function

$$f(x) = 6x^5 - 10x^3$$

Question E3   A closed container is to be constructed in the shape of a circular cylinder of volume 16π (metre)^3. What dimensions should it have if the total surface area is to be a minimum? (☞)

Questions E4   Sketch the graph of the function

$$f(x) = \frac{x^2 + 3}{x + 1}$$

The volume of a circular cylinder of radius r and height h is $\pi r^2 h$. The area of the curved surface of such a cylinder is $2\pi rh$.

Part 5

Basic integration

Introducing integration

Chapter

17

17.1 **Opening items**

17.1.1 *Chapter introduction*

The notion of a *limit of a sum* arises in many applications of mathematics to physical science. For example, suppose we want to find the total mass of a cylindrical column of air, the density of which decreases with height in a known way. One approach is to divide the air column into a number of horizontal discs – like a stack of coins – each sufficiently thin that we may regard it as having constant density. We can then work out the mass of each disc by multiplying its volume by its (constant) density, and add all those disc masses together to find an approximate value for the mass of the entire column. Our final answer will only be an approximation to the correct value because our assumption that each disc has constant density is only approximately true. Nonetheless, we may obtain a good estimate in this way and we may improve its accuracy by increasing the number of discs in the sum. Indeed, if we consider what happens as the thickness of the discs approaches zero (and the number of discs approaches infinity) then we see that the limit of the sum of disc masses is the exact value of the air column's total mass. This is typical of the way in which the limit of a sum may arise in a physical problem. Limits of sums of this kind are referred to by scientists and mathematicians as *definite integrals*. The *formulation* of physical problems in terms of definite integrals is a major theme of this chapter.

Another major theme is the *evaluation* of definite integrals. Sometimes the only way of evaluating the limit of a sum is to perform the sum for a very large number of terms and then try to work out what would happen 'in the limit' as the number of terms approached infinity. This 'head-on' approach called *numerical integration* can provide accurate answers, but it can also be time consuming and very hard work. Fortunately there are many situations in which the evaluation of definite integrals can be performed more easily thanks to an important mathematical result known as the *fundamental theorem of calculus*. According to the fundamental theorem there is a remarkable link between the evaluation of the limit of a sum and the process of differentiation. In fact, many definite integrals may be evaluated by a procedure in which the key step amounts to little more than the reversal of the usual process of differentiation. This process of *inverse differentiation,* or *indefinite integration* as it is more commonly called, is a vital part of the mathematical tool kit of every physical scientist and engineer.

In Section 17.2 we introduce the basic concepts of definite and indefinite integration. We start by recalling the definition of a derivative and then go on to define the *inverse derivative* that emerges from the process of inverse differentiation. Using the physical concepts of position, velocity and acceleration we show how inverse derivatives may be used to evaluate quantities such as the distance travelled in a given time by an object moving with varying velocity. This discussion leads us to formulate the concept of a definite integral defined as the limit of a sum. We then state (and prove) the fundamental theorem of calculus which provides the formal link between the evaluation of limits of sums and inverse differentiation. Finally, we round off the section by explaining why it makes sense to refer to inverse derivatives as indefinite integrals.

Section 17.3 is concerned with the formulation of problems in terms of definite integrals and their evaluation using indefinite integrals. Section 17.4 attempts to put the topic into its scientific context and provides some exercises for you to do.

17.1.2   Fast track questions

Question F1

Evaluate the following indefinite integrals:

(a) $\int (1 + 2x + 3x^2)\, dx$ (b) $\int [\sin(2x) + \cos(3x)]\, dx$

(c) $\int \left(e^t + \dfrac{1}{t^2}\right) dt$ (d) $\int dw$

> Note that x, t and w are dimensionless variables in Questions F1 and F2. They have no physical significance in those questions.

Question F2

Evaluate the following definite integrals:

(a) $\displaystyle\int_{-1/4}^{1/4} \cos(2\pi x)\, dx$ (b) $\displaystyle\int_0^3 (2t - 1)^2\, dt$

(c) $\displaystyle\int_1^2 \dfrac{(1 + e^t)^2}{e^t}\, dt$ (d) $\displaystyle\int_4^9 \sqrt{x}\left(x - \dfrac{1}{x}\right) dx$

Question F3

An object starts from rest at time $t = 0$, and moves along the x-axis with acceleration $a_x(t)$ given, at time t, by

$$a_x(t) = a - bt \quad \text{where } a = 5 \text{ m s}^{-2} \text{ and } b = 3 \text{ m s}^{-3}$$

State whether each of the following statements is true or false and explain why.

(a) The object comes to rest when $t = (5/3)$ s.
(b) The displacement of the object from its initial position is zero when $t = 5$ s.
(c) The magnitude of the area bounded by the velocity–time graph and the t-axis, from $t = 0$ to $t = 1$ s, represents the distance the object travels in the first second.
(d) The magnitude of the area bounded by the graph of $y = a_x(t)$ and the t-axis between $t = 0$ s and $t = 5/3$ s represents the maximum speed of the object in the first ten seconds.

17.1.3 Ready to study?

Study comment

In order to study this chapter you will need to be familiar with the following terms: *Cartesian coordinate system*, *derivative* (or *derived function*), *function*, *graph*, *limit* and *magnitude* (of a vector, as in *magnitude of a force*). You should be able to *differentiate* a range of functions, and be able to *expand*, *simplify* and *evaluate* expressions, including those that involve *exponential* and *trigonometric functions*.

You will need to know the following derivatives:

$$\frac{d}{dt}(t^n) = nt^{n-1} \qquad\qquad \frac{d}{dt}[\sin(at)] = a\cos at$$

$$\frac{d}{dt}[\cos(at)] = -a\sin(at) \qquad\qquad \frac{d}{dt}(e^{at}) = ae^{at}$$

(where n and a are constants); and to remember that a *function* $f(x)$ is often given by a formula such as $y = f(x)$, and is a rule that assigns a unique value of y to each value of x (see the note below). The following *Ready to study questions* will allow you to establish whether you need to review some of the topics before embarking on this chapter.

A special note on functions

The function

$$f(x) = 1 + x + x^2$$

is an example of a function that is defined for all values of x. One may think of the function as a sort of machine with x as the input and $1 + x + x^2$ as the output. The input x is known as the *independent variable* and, if we write $y = f(x)$, the output y is known as the *dependent variable*, since the function $f(x)$ determines the way in which y depends on x.

The same function f could equally well be defined using some other symbol, such as t, to represent the independent variable:

$$f(t) = 1 + t + t^2$$

This freedom to relabel the independent variable is often of great use, though it is vital that such changes are made consistently throughout an equation.

We may *evaluate* this function, whether we call it $f(x)$ or $f(t)$, for any value of the independent variable; for example, if we choose to use x to denote the independent variable, and set $x = 1$, we have

$$f(1) = 1 + 1 + 1^2 = 3$$

Similarly, if $x = \pi$ $\quad f(\pi) = 1 + \pi + \pi^2$

and, if $x = 2a$ $\quad f(2a) = 1 + 2a + (2a)^2 = 1 + 2a + 4a^2$

When we write expressions such as $f(\pi)$ or $f(2a)$ whatever appears within the brackets is called the *argument* of the function. The value of $f(x)$ is determined by the *value* of its argument, irrespective of what we call the argument.

Question R1

For each of the following functions $f(t)$ find the derived function or derivative, $f'(t) = \dfrac{df}{dt}$:

(a) $\quad f(t) = t^{2/3}$ (b) $\quad f(t) = \sin\left(\dfrac{\pi t}{2}\right)$ (c) $\quad f(t) = 5e^{at}$

Question R2

Expand the following expressions:

(a) $\dfrac{(1 + e^x)^2}{e^x}$ (b) $\sqrt{x}\left(x - \dfrac{1}{x}\right)$

Question R3

If $F(x) = x\left[\sin\left(\dfrac{\pi x}{2}\right) - 1\right]$, evaluate the following:

(a) $\quad F(-1) - F(-2)$ (b) $\quad F(1) - F(-1)$ (c) $\quad F(1) - F(-2)$

17.2 **The concept of integration**

17.2.1 *Inverse derivatives: reversing differentiation*

From a mathematical point of view, *differentiation* is a rather straightforward procedure. Given a function $f(x)$, its *derivative* is another function that may be denoted by $f'(x)$ or $\dfrac{df}{dx}$ and which is defined by

$$\frac{df}{dx} = \lim_{h \to 0}\left(\frac{f(x + h) - f(x)}{h}\right) \tag{1}$$

for all values of x at which a unique limit exists. This simple rule makes it possible to find derivatives for an enormous range of functions.

The importance of differentiation in the study of physical science would be hard to overestimate. Not only is it used in the analysis and solution of a great many problems, it even plays a fundamental role in the definition of many basic concepts. Nowhere is this more obvious than in **kinematics**, the study of motion. For example, consider an object moving in a straight line along the x-axis of a *Cartesian coordinate system*. Such an object is said to be undergoing **linear motion**, and its **position** at any time t is determined by the single position coordinate x. We can represent the position by $x(t)$ to emphasise that it is a *function* of t, since its value changes with time as the object moves along the line. Now, for any moving object the two important physical quantities that help to characterise the motion are **velocity** and **acceleration**, and these may both be defined as derivatives. In the case of linear motion along the x-axis we have:

velocity $\qquad v_x = \dfrac{dx}{dt}$

acceleration $\qquad a_x = \dfrac{dv_x}{dt}$

Thus, given the position of a moving object as a function of time, i.e. given the explicit form of the function $x(t)$, it is usually fairly easy to determine the corresponding velocity $v_x(t)$ and acceleration $a_x(t)$, and hence obtain some insight into the nature of the motion.

The position of a certain object undergoing linear motion is specified by $x(t) = at^3 + bt + c$, where a, b and c are constants. What are the functions $v_x(t)$ and $a_x(t)$ that describe the velocity and acceleration of the object? By considering the values of the functions $x(t)$ and $v_x(t)$ when $t = 0$, explain the physical significance of constants b and c.

$$v_x(t) = \frac{dx}{dt} = 3at^2 + b \quad \text{and} \quad a_x(t) = \frac{dv_x}{dt} = 6at$$

When $t = 0$, $x(0) = c$, so c represents the *initial position* of the moving object.

Similarly, $v_x(0) = b$, so b represents the *initial velocity* of the moving object. $\square$

It is comforting to know that it is relatively easy to determine the velocity and acceleration that correspond to a given form for $x(t)$, but unfortunately this is not a problem that often confronts a physical scientist. More common is the problem that arises if we turn the previous discussion on its head and ask how we might determine the position of an object if we are told its

---

In practice, derivatives are usually found by combining well-known 'standard derivatives' in a variety of ways, but both the standard derivatives and the rules for combining them are based on Equation 1.

Strictly speaking, *position, velocity* and *acceleration* are *vector quantities* in that each require three *scalar components* for their complete specification in three dimensions. However, in the case of linear motion each of these quantities is effectively specified by a single scalar component. Thus, in the context of linear motion we will refer to x, v_x and a_x as position, velocity and acceleration, respectively, even though each is, in reality, only a single component of the corresponding vector.

For reasons that will become clear later, *inverse differentiation* is usually referred to as *indefinite integration*.

velocity, or perhaps its acceleration, as a function of time? The latter problem is particularly common since, according to **Newton's second law** of motion, the acceleration of a body of fixed **mass** is proportional to the total **force** acting on that body, and we often start the analysis of a problem knowing only the forces that are involved. Clearly, in order to deal with such problems and determine $x(t)$ from $v_x(t)$ or $a_x(t)$ we need to reverse the process of differentiation. This reverse process is known as **inverse differentiation** and is generally much harder than ordinary differentiation. (✎)

As an example of inverse differentiation, and its pitfalls, let us try inverting the problem we considered above. Specifically, let us suppose we are told that the velocity of a particular object moving in one dimension is given by $v_x(t) = 3at^2 + b$ and that we want to find its position as a function of time. How should we do this?

In view of the earlier discussion, your first thought might be the right one that inverse differentiation would give $x(t) = at^3 + bt + c$. However, you will not usually be in possession of such 'privileged' information. Normally the best that you could do would be to try to think of the most general function with a derivative equal to the given form of $v_x(t)$. Looking at the form of $v_x(t)$ it is pretty clear that its inverse derivative must include the expression $at^3 + bt$, but there is no evidence whatsoever that a constant c should be added to this. However, since the derivative of *any* constant is zero, you might well say that it is quite *possible* that there is an additional constant since we may add *any* constant to $at^3 + bt$ and the resulting expression will still have the property that its derivative is identical to $v_x(t)$. Thus, if we only know that

$$v_x(t) = \frac{dx}{dt} = 3at^2 + b \tag{2}$$

then the most we can say about $x(t)$ is that it is of the general form

$$x(t) = at^3 + bt + \text{arbitrary constant} \tag{3}$$

The point here is that a knowledge of the velocity alone is not sufficient to completely determine the position of the object as a function of time because we do not know its initial position on the x-axis (i.e. the value of the arbitrary constant). Of course, if we are given some extra information, such as the location of the object at $t = 0$, then we may be able to determine the value of the arbitrary constant, but, in the absence of such additional information, Equation 3 is the most complete answer we can find.

As you can see, in the case of inverse differentiation there is no simple rule to apply, no reliable formula to use; all you can do is to 'inspect' the given function and to use your knowledge of differentiation to 'suggest' the form of the inverse derivative. You can check your answer by differentiating it to make sure you recover the function you started with, but you must always remember to add an arbitrary constant to your answer since the derivative of any constant is zero.

We can sum up the process of inverse differentiation in the following way.

Given a function $f(x)$, its **inverse derivative** is any function $F(x)$ such that

$$\frac{dF}{dx} = f(x)$$

An *inverse derivative* of a given function is also known as a **primitive** of that function, though for reasons that will become clear later, physical scientists usually refer to them as *indefinite integrals*. Note that a function may (and generally does) have infinitely many inverse derivatives, corresponding to the infinitely many possible choices for the value of the arbitrary constant. It therefore makes sense to speak of *an* inverse derivative rather than *the* inverse derivative of a given function.

Write down three different inverse derivatives of $f(x) = 2x$.

Three suitable functions would be

$$F(x) = x^2 - 2, \quad F(x) = x^2 \quad \text{and} \quad F(x) = x^2 + 1 \qquad (\text{☞})$$

since in each case $\dfrac{dF}{dx} = f(x)$. □

Any other function of the form $F(x) = x^2 + C$, where C is any constant would be equally acceptable. C is then called an *integration constant* or a *constant of integration*.

In view of the many possible answers to the question 'what is an inverse derivative of $f(x) = 2x$?' it is customary to present the answer in the 'general' form $F(x) = x^2 + C$, where C is an arbitrary constant. This leads to the following observation:

If $F(x)$ and $F_2(x)$ are both inverse derivatives of the same function $f(x)$, then there exists a constant K such that

$$F(x) = F_2(x) + K$$

The constant K here represents the difference between the integration constants C_1 and C_2 associated with the inverse derivatives $F(x)$ and $F_2(x)$, i.e. $K = C_1 - C_2$.

If you feel that you have grasped the principle of inverse differentiation, try the following question. If you're not so confident, treat the first two parts as a worked example and then try to answer the remaining part.

Write down an inverse derivative $F(x)$ for each of the following functions:

(a) $f(x) = x^2$ (b) $f(x) = 3x^2 - 2x$ (c) $f(x) = ax/2$,

where a is a constant

In each of the following C represents an arbitrary constant:

(a) $F(x) = \dfrac{x^3}{3} + C$ since $\dfrac{d}{dx}\left(\dfrac{x^3}{3} + C\right) = \dfrac{3x^2}{3} = x^2$

(b) $F(x) = x^3 - x^2 + C$ since $\dfrac{d}{dx}(x^3 - x^2 + C) = 3x^2 - 2x$

(c) $F(x) = \dfrac{ax^2}{4} + C$ since $\dfrac{d}{dx}\left(\dfrac{ax^2}{4} + C\right) = \dfrac{2ax}{4} = \dfrac{ax}{2}$ □

Write down an inverse derivative $F(t)$ for each of the following functions:

(a) $f(t) = t^{-3}$ (b) $f(t) = -3t^{-1.5} + 5t^{1.5}$

In each of the following C represents an arbitrary constant:

(a) $F(t) = \dfrac{-t^{-2}}{2} + C$

since $\dfrac{d}{dt}\left(\dfrac{-t^{-2}}{2} + C\right) = \dfrac{-(-2)\,t^{-3}}{2} = t^{-3}$

(b) $F(t) = 6t^{-0.5} + 2t^{2.5} + C$

since $\dfrac{d}{dx}(6t^{-0.5} + 2t^{2.5} + C) = -3t^{-1.5} + 5t^{1.5}$ □

At time t, an object undergoing linear motion has acceleration $a_x(t) = 4At^3$, where $A = 1$ m s^{-5}. Given that the object is at rest at $t = 0$, find the velocity of the object at $t = 3$ s.

We know that $a_x(t) = \dfrac{dv_x}{dt}$ and, using inverse differentiation, we can see that $v_x(t) = At^4 + C$, where C is an arbitrary constant, since differentiating $At^4 + C$ produces $4At^3$. But we also know that $v_x = 0$ when $t = 0$, and since $v_x(0) = C$, it follows that $C = 0$ in this case. Hence $v_x(t) = At^4$ and consequently $v_x(3\text{ s}) = (1\text{ m s}^{-5})\ (3\text{ s})^4 = 81$ m s^{-1}. □ (☜)

Note that in this case we have used additional information about the problem, the condition $v_x(0) = 0$, to determine the relevant value of C. This is often necessary in physical problems.

Motion with uniform acceleration under gravity

To stress the importance of inverse differentiation (i.e. indefinite integration) let us use it to derive some well-known equations — those that describe an object moving vertically near the surface of the Earth, with constant acceleration due to gravity. In this case, if we let the x-axis point vertically downwards the object will experience a constant acceleration given by

$$a_x(t) = g \qquad (4)$$

where g is the magnitude of the acceleration due to gravity, which has a value of about 9.81 m s^{-2}.

At $t = 0$, when the object has position coordinate $x(0)$, its initial velocity is $v_x(0) = u_x$, where u_x is a constant. Applying inverse differentiation to Equation 4 we see that $v_x(t) = gt + C$ for some constant C. It follows that $v_x(0) = C$, but we know that $v_x(0) = u_x$, so $C = u_x$, and therefore

$$v_x(t) = u_x + gt \qquad (5)$$

We have now found the velocity, but we can apply the same argument again to find how the position varies with time. This time we require a function of t that, when differentiated, produces $u_x + gt$. It is not difficult to see that the appropriate function is

$$x(t) = u_x t + \frac{gt^2}{2} + D$$

for some constant D. Now since the initial position of the object is $x(0)$ we see that $D = x(0)$ and hence

$$x(t) = u_x t + \frac{gt^2}{2} + x(0) \qquad (6)$$

In fact, when dealing with problems of this kind we usually want to know how far the object is from its initial position and in which direction (i.e. upwards or downwards). This information is given by the **displacement** of the object from its initial position, which in this case is defined by

$$s_x(t) = x(t) - x(0) \qquad (7)$$

Using this definition to eliminate $x(t)$ and $x(0)$ from Equation 6 we have

$$s_x(t) = u_x(t) + \frac{gt^2}{2} \qquad (8)$$

and this equation gives us the displacement of the object at any time t. The **distance** $s(t)$ of the object from its initial position, at time t, is a positive quantity given by the magnitude of its displacement, so $s(t) = |s_x(t)|$, and it too can be found from Equation 8.

For the sake of completeness we note that Equation 5 may be rearranged to give $t = (v_x - u_x)/g$ and that upon substituting this into Equation 8 and rearranging we obtain

$$v_x^2 + u_x^2 + 2gs_x \qquad (9)$$

Note the distinction between *displacement* and *distance*: in one dimension displacement may be positive or negative according to direction, but distance must always be positive.

If x is a positive quantity then $|x| = x$ and $|-x| = x$.

Equations 5, 8 and 9 are a particular case of the *uniform acceleration equations*. The simplicity of the above derivation of these fundamentally important kinematic relationships is ample proof of the power of inverse differentiation.

Motion with non-uniform acceleration

It is important to appreciate that the above method can be applied to any motion in which the acceleration is a known function of time − the acceleration does not have to be uniform (i.e. constant). We have already seen some examples of this kind, but to emphasise the point, and to provide a further illustration, we now consider **simple harmonic motion**, a form of linear motion in which the acceleration of an object is given by

$$a_x(t) = -A\omega^2 \sin(\omega t) \tag{10}$$

ω is the Greek letter omega.

where A and ω are positive constants, known respectively as the **amplitude** and **angular frequency** of the motion. Suitable SI units for these quantities would be m for the amplitude and s^{-1} for the angular frequency, so the combination $A\omega^2$ can be expressed in units of $m\ s^{-2}$, the same units as acceleration.

Now, the general properties of the sine function are such that the term $\sin(\omega t)$ in Equation 10 varies repeatedly between $+1$ and -1 as t increases, so both the magnitude and the direction of the acceleration fluctuate with time. As you might expect, these regular changes of sign result in back and forth (oscillatory) motion; a fact we shall now demonstrate using inverse differentiation.

The first step is to find the velocity $v_x(t)$, an inverse derivative of $a_x(t)$. In this case it must involve a function whose derivative is $-A\omega^2 \sin(\omega t)$. The only obvious choices involve $\cos(\omega t)$, the derivative of which is $-\omega \sin(\omega t)$. Bearing this in mind, and remembering the need to include an arbitrary additive constant C in the answer, it is not too difficult to see that the required inverse derivative has the general form

You can confirm that this is correct by showing that its derivative is $-A\omega^2 \sin(\omega t)$.

$$v_x(t) = A\omega \cos(\omega t) + C \qquad (\text{☞}) \tag{11}$$

In order to find the position as a function of time, $x(t)$, we must find the appropriate inverse derivative of $v_x(t)$. This is one of those cases where 'we' means 'you'.

Question T1

Write down the inverse derivative of $v_x(t)$ in this case, and then use the conditions $x = 0$ when $t = 0$, and $v_x = 0$ when $t = \pi/(2\omega)$ to show that

$$x(t) = A \sin(\omega t) \tag{12} \quad \square$$

As t increases and the function $\sin(\omega t)$ varies smoothly between $+1$ and -1, the value of $x(t)$ varies smoothly between $+A$ and $-A$. Moreover due to the periodic (repeating) nature of the sine function, the motion is repeated every time t increases by $2\pi/\omega$. Thus Equation 12 does indeed describe oscillations, as promised, and the quantity $T = 2\pi/\omega$ is known as the **period** of the motion. From this you can see that the conditions $x = 0$ when $t = 0$, and

$v_x = 0$ when $t = \pi/(2\omega)$ given in Question T1 are equivalent to saying that the oscillating object passes through the origin at $t = 0$, and is momentarily at rest a quarter of a period later (at $t = T/4 = \pi/(2\omega)$) when its displacement from the origin attains its maximum value, A. (☞)

Question T2

Write down inverse derivatives of x^2, x^3 and x^4. Write down a rule for determining an inverse derivative of x^p, where the power p may have any real value except -1. ☐

Question T3

Write down an inverse derivative of (a) e^{-x}, and (b) $x + e^x$. ☐

17.2.2   Inverse derivatives and the area under a graph

We now extend our study of inverse derivatives by exposing an important relationship between an inverse derivative of a function and the area under the graph of that function. The relationship is a general one, but we will introduce it in the physical context of kinematics.

Consider an object moving along a straight line with velocity $v_x(t)$. Provided we know the explicit form of the function $v_x(t)$ it is a fairly straightforward matter to plot the graph of v_x against t. Such a graph is called the **velocity–time graph** of the motion.

Figure 17.1a shows a particularly simple velocity–time graph, that of an object released from rest at time $t = 0$ and falling vertically downwards with constant acceleration under the influence of gravity. As we saw in the last subsection, if we take vertically downwards to be the positive x-direction then the velocity of such an object, at time t, is

$$v_x = gt \tag{13}$$

where the constant g is the magnitude of the acceleration due to gravity. This simple relationship is reflected in the velocity–time graph where the value of v_x increases in proportion to t, i.e. it is a straight-line graph.

If we choose two particular times, t_1 and t_2 say, where $t_2 > t_1$, then we can see from Equation 13 (or from Figure 17.1b) that the corresponding values of the velocity are $v_x(t_1) = gt_1$ and $v_x(t_2) = gt_2$. It follows that the average velocity of the falling object over the period from t_1 to t_2 is

$$\tfrac{1}{2}[v_x(t_1) + v_2(t_2)] = \tfrac{1}{2}(gt_1 + gt_2)$$

and that the distance travelled during that time is

$$\tfrac{1}{2}(gt_1 + gt_2)(t_2 - t_1)$$

Now, this distance represents the change in the position coordinate of the object over the interval $t_2 - t_1$ so we can write

By combining Equations 10 and 12 it is possible to characterise simple harmonic motion by the requirement that $a_x = -\omega^2 x$. This is the usual starting point for the analysis of simple harmonic motion.

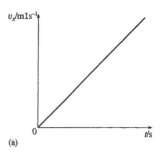

(a)

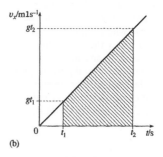

(b)

Figure 17.1   (a) A simple velocity–time graph. (b) The area under the graph between $t = t_1$ and $t = t_2$.

$$x(t_2) - x(t_1) = \tfrac{1}{2}(gt_1 + gt_2)(t_2 - t_1) \tag{14}$$

However, the right-hand side of this equation can be interpreted graphically in terms of Figure 17.1b as the shaded area under the velocity–time graph between $t = t_1$ and $t = t_2$. (☜) Note that the 'area' referred to here must be expressed in terms of the scale units that appear on the graph axes, *not* in terms of the actual area of paper in mm^2 or any other unit. In view of this interpretation we can say that

> The shaded area in Figure 17.1b is a trapezium, the area of which is generally given by
>
> $\tfrac{1}{2}(\text{height}_1 + \text{height}_2) \times \text{base}$

$$\begin{aligned} x(t_2) - x(t_1) &= \text{the area under the graph of } v_x(t) \\ &\text{between } t_1 \text{ and } t_2 \end{aligned} \tag{15}$$

If you take g to be 10 m s^{-2}, what is the hatched area in Figure 17.1b if $t_1 = 5$ s and $t_2 = 10$ s? What is the corresponding change in position $x(t_2) - x(t_1)$?

The hatched area is given by

$$\tfrac{1}{2}(50 \text{ m s}^{-1} + 100 \text{ m s}^{-1}) \times (10 \text{ s} - 5 \text{ s}) = 375 \text{ m}$$

Consequently, $x(t_2) - x(t_1) = 375$ m   □

If we recall that $x(t)$ is an *inverse derivative* of $v_x(t)$, we can express Equation 15 in another way:

The area under the graph of $v_x(t)$ between $t = t_1$ and $t = t_2$ is equal to the corresponding change in its inverse derivative $x(t_2) - x(t_1)$.

Although we have only deduced this relationship for one particular form of $v_x(t)$ it is actually a general relationship. In fact, it is a very general relationship since it not only applies to any form of $v_x(t)$ but to any function at all, provided that function has an inverse derivative and provided we take due care over signs (see later). Subject to these provisos we can say quite generally that:

If $F(x)$ is any one of the inverse derivatives (i.e. indefinite integrals) of $f(x)$, then the **area under the graph** of $f(x)$, between a and b where $b > a$, is given by $F(b) - F(a)$.

> Note that whichever inverse derivative of $f(x)$ we choose to use when evaluating the area under the graph, the associated constant of integration C will play no part in the final answer since it will cancel when we calculate the *difference* $F(b) - F(a)$.

In case you are worried that this result, even if true for one inverse derivative of $f(x)$, might not be true for *all* the inverse derivatives of $f(x)$, just take note of the following argument.

If $F_1(x)$ and $F_2(x)$ are both inverse derivatives of $f(x)$ then we know that $F_1(x) = F_2(x) + K$ for some constant K. It therefore follows that

$$F_1(b) - F_1(a) = [F_2(b) + K] - [F_2(a) + K] = F_2(b) - F_2(a)$$

So if the boxed result is true for one inverse derivative of $f(x)$ it is true for *any* inverse derivative of $f(x)$.

A note on notation

As you can see, expressions such as $F(b) - F(a)$ and $x(t_2) - x(t_1)$ are very common in these discussions of inverse derivatives and the area under a graph. In order to simplify the process of writing such expressions we will henceforth indicate them by means of square brackets, as follows

$$[F(x)]_a^b = F(b) - F(a)$$

$$\text{and} \quad [x(t)]_{t_1}^{t_2} = x(t_2) - x(t_1)$$

The boxed statement given above is very important, but so far it is little more than an assertion. Before applying it to any physical or mathematical problems we should really give you a good reason for believing that it is true for *any* $f(x)$.

Figure 17.2 represents the graph of some general function $f(x)$; the variable x here need not represent position, it could be any quantity. If we let a and b represent particular values of x then the shaded region in Figure 17.2 represents the area under the graph between $x = a$ and some general value x that is less than or equal to b. This area (measured in the appropriate scale units) will be denoted by $A(x)$.

If we now consider what happens to $A(x)$ when x increases by some small amount Δx, we can see that $A(x)$ will also increase by an amount ΔA. We can approximate this increase in area by the area of the small rectangle shown in Figure 17.2, so

$$\Delta A \approx f(x)\Delta x$$

and therefore $\quad \dfrac{\Delta A}{\Delta x} \approx f(x)$

Now, this approximation will become increasingly accurate as Δx becomes smaller, so in the limit as Δx tends to zero, we can use the definition of the derivative (discussed in Chapter 13) to say

$$\frac{dA}{dx} = f(x)$$

In other words, the area function $A(x)$ is an *inverse derivative* of the function $f(x)$ whose graph has been drawn. But we have already seen that if the boxed statement is true for one inverse derivative of $f(x)$ it is true for *any* inverse derivative of $f(x)$. Since it is certainly true that the area under the graph between $x = a$ and $x = b$ is equal to $A(b) - A(a)$, it follows that it is also equal to $F(b) - F(a)$ where $F(x)$ is *any* inverse derivative of $f(x)$, as claimed.

Before leaving the general principles and looking at some applications, there is one more point that must be made about the area under a graph and that concerns its sign. So far, both the graphs we have considered have had the property of being positive at all points. However, as you are well aware, it is perfectly possible for a velocity $v_x(t)$, or a general function $f(x)$, to be negative over all or part of its domain of definition. Would our results about inverse derivatives and areas under graphs still be true if all or part of the graph had been below the horizontal axis, as in Figure 17.3?

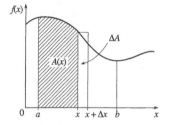

Figure 17.2 The graph of a general function $f(x)$.

The symbol $\approx$ means 'approximately equal to'.

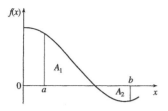

Figure 17.3 The area under a graph in any region where the graph is below the horizontal axis, such as A_2, will be negative. According to the 'signed area' convention we use, the total area under the graph in Figure 17.3 is $A_1 + A_2$ where A_2 is a negative quantity. Other authors sometimes adopt a different convention in which the area of each region is defined to be positive and the area under the graph is then given by $A_1 - A_2$.

The answer is yes, though in such cases the 'area' below the horizontal axis (such as A_2 in Figure 17.3) must be treated as a negative quantity. This is a sufficiently important point that it deserves a box of its own.

> When calculating the area under a graph, any regions that are below the horizontal axis should be regarded as having negative areas.

The above results can be used in two ways. If we are given, or can easily find, the area under the graph of a function, then we can determine the corresponding change in the inverse derivative. On the other hand, if we know or can easily find the inverse derivative of a function, then we can determine the area under the graph of that function. Both these applications are illustrated in the following questions which you should now attempt.

Figure 17.4 shows the velocity–time graph of a moving object. Determine the displacement of the object from its initial position after 7 seconds. How far does the object travel during the final second of its journey? (You will need to know that the area of a triangle of base length a and height h is $\frac{1}{2}\,ha$.)

The areas under the graph, A_1, A_2 and A_3, shown in Figure 17.4 are

$$A_1 = (10 \text{ m s}^{-1}) \times (4 \text{ s}) = 40 \text{ m}$$

$$A_2 = \tfrac{1}{2}\,(10 \text{ m s}^{-1}) \times (2 \text{ s}) = 10 \text{ m}$$

$$A_3 = \tfrac{1}{2}\,(-5 \text{ m s}^{-1}) \times (1 \text{ s}) = -2.5 \text{ m}$$

The displacement s_x from the initial position after 7 s is therefore

$$s_x(7 \text{ s}) = x(7 \text{ s}) - x(0) = (40 + 10 - 2.5) \text{ m} - 0 \text{ m} = 47.5 \text{ m}$$

During the final second of its journey s_x changes by -2.5 m. The distance travelled during that final second is therefore 2.5 m. ☐

Remember, the *distance* between two points is equal to the *magnitude* of the *displacement* from one point to the other.

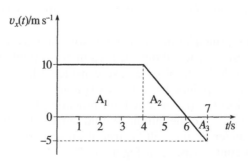

Figure 17.4
A velocity–time graph in which $v_x(t)$ changes sign.

Find the area under the graph of $f(x) = 3x^2$ between $x = 1$ and $x = 3$ (see Figure 17.5).

Using inverse differentiation, we see that $F(x) = x^3 + C$, where C is a constant, is an inverse derivative of $f(x) = 3x^2$ since

$$\frac{dF}{dx} = f(x)$$

It follows that the required area under the graph is

$$[F(x)]_1^3 = [x^3 + C]_1^3 = (27 + C) - (1 + C) = 26 \quad \square$$

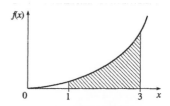

Figure 17.5 The graph of $f(x) = 3x^2$.

Note that in answering this last question the value of C makes no difference to the final result. Since $F(x)$ could be *any* inverse derivative of $f(x)$ it makes sense to choose the simplest and take $C = 0$ in such cases.

Find the magnitude of the area under the graph of $y = 9 - x^2$ between $x = 4$ and $x = 7$.

The graph of $y = 9 - x^2$ is shown in Figure 17.6. In this case we note that an inverse derivative of $y = 9 - x^2$ is

$$Y(x) = 9x - \frac{x^3}{3} + C \quad \text{where } C \text{ is a constant}$$

Choosing $C = 0$, it follows that the area under the curve between $x = 4$ and $x = 7$ is

$$[Y(x)]_4^7 = \left[9x - \frac{x^3}{3}\right]_4^7 = \left(63 - \frac{343}{3}\right) - \left(36 - \frac{64}{3}\right) = -66$$

However, the question asks for the *magnitude* of the area under the curve, so the answer to the question is $+66$. $\quad \square$

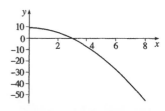

Figure 17.6 The graph of $y = 9 - x^2$.

Question T4

(a) An object moves along a fixed x-axis at time t with velocity $v_x(t) = v_0 - bt^2$ where $v_0 = 2 \text{ m s}^{-1}$ and $b = 1 \text{ m s}^{-3}$. Determine the displacement of the object from its initial position (i.e. its position at $t = 0$) after two seconds, and explain how that displacement is related to a graph of v_x against t.

(b) By noting that the velocity changes sign at $t = \sqrt{2}\,\text{s}$, find the total distance travelled by the object in the first two seconds and explain how this relates to a graph of v_x against t. $\quad \square$

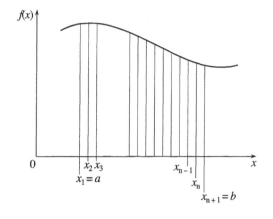

Figure 17.7 The area under the graph of an arbitrary function may be approximated by a sum of rectangles. ($x_1 = a$, $x_{n+1} = b$)

17.2.3 Definite integrals: the limit of a sum

We have seen how inverse differentiation can be used to find the area under the graph of a function; however, there is an alternative method of evaluating areas which does not require any knowledge of inverse derivatives. This alternative technique is based on the idea that the area under a graph can be estimated by breaking it up into a large number of very thin rectangles with easily calculable areas that can be added together. This idea is illustrated in Figure 17.7.

In Figure 17.7 the area under the graph of $f(x)$ from $x_1 = a$ to $x_{n+1} = b$ has been divided into n vertical strips, each of which can be approximated by rectangles. The first rectangle is of height $f(x_1)$ and width $x_2 - x_1$, and the last is of height $f(x_n)$ and width $x_{n+1} - x_n$.

If we let $\Delta x_i = x_{i+1} - x_i$ where i may be any whole number in the range $1 \le i \le n$, then we can say that the area of the ith rectangle is $f(x_i)\Delta x_i$, and that the total area under the graph between a and b is approximately

$$f(x_1)\Delta x_1 + f(x_2)\Delta x_2 + f(x_3)\Delta x_3 + \ldots + f(x_n)\Delta x_n = \sum_{i=1}^{n} f(x_i)\,\Delta x_i$$

where we have used the **summation symbol** (Σ) as a shorthand way of indicating the sum. (In this case the summation symbol should be read as 'the sum from i equals 1 to n' and is an instruction to add together terms similar to those immediately following the symbol for the relevant values of the **summation variable** i.) The approximation to the area under the graph represented by this sum will become increasingly accurate as the number of rectangles increases *and* the width of the rectangles decreases. We can ensure that both requirements are met by letting Δx denote the width of the widest rectangle, and then taking the limit of the above sum as Δx tends to zero. Thus we have

$$\text{the area under the graph of } f(x) \text{ from } a \text{ to } b = \lim_{\Delta x \to 0}\left(\sum_{i=1}^{n} f(x_i)\,\Delta x_i \right)$$

The symbol $\le$ should be read as 'is less than or equal to'.

Now this limit of a sum is very important, but also very cumbersome, so it is given a special name. It is called a **definite integral**, and it is denoted by a special symbol, a sort of distorted 'S' (for summation). Thus,

$$\int_a^b f(x)\, dx = \lim_{\Delta x \to 0} \left(\sum_{i=1}^n f(x_i)\, \Delta x_i \right) \qquad (16)$$

A note on terminology
Definite integrals are of such great importance that each part of the symbol has its own name. The distorted S is called an **integral sign**, the values a and b are known as the lower and upper **limits of integration**, respectively, the function that is being integrated ($f(x)$ in this case) is called the **integrand** and the dx which indicates that the integration is to be performed 'with respect to the variable x' is called the **element of integration** or **integration element**. The use of the terms *integral* and *integration* in this context makes good sense, since they imply bringing things together to make a whole and that is exactly what a summation does.

The process of evaluating a sum and taking its limit can be very tedious, so you will be pleased to learn that there are easier ways of evaluating integrals that will be explained in the next subsection (though you might already guess what they are). However, the idea of a limit of a sum is one that arises quite naturally in many scientific problems and that is why definite integrals are an essential part of the physical scientist's tool kit. In the remainder of this subsection we will show you how a finite sum can be used to estimate the value of a definite integral, and then invite you to express various physical problems in terms of definite integrals, without asking you to evaluate them.

Example 1
(a) Given that $f(x) = x^2 + 1$, write down the values of $f(x)$ at the following values of x: $x_1 = 1.00$, $x_2 = 1.50$, $x_3 = 2.00$, $x_4 = 2.50$, $x_5 = 3.00$, $x_6 = 3.50$ and $x_7 = 4.00$.

Table 17.1   Values of $f(x) = x^2 + 1$

i	x_i	$f(x_i)$	Δx_i	$f(x_i)\Delta x_i$
1	1.00	2.00	0.50	1.00
2	1.50	3.25	0.50	1.63
3	2.00	5.00	0.50	2.50
4	2.50	7.25	0.50	3.63
5	3.00	10.00	0.50	5.00
6	3.50	13.25	0.50	6.63
7	4.00	17.00		

$$\sum_{i=1}^{6} f(x_i)\, \Delta x_i = 20.39$$

Notice that, while it is not essential that the points dividing the interval are equally spaced, it is very convenient if they are.

(b) By letting $\Delta x_i = x_{i+1} - x_i$, write down the value of Δx_i for $i = 1$ to 6.

(c) Use your answers to (a) and (b) to work out the corresponding values of $f(x_i)\Delta x_i$ and hence evaluate $\sum_{i=1}^{6} f(x_i)\,\Delta x_i$.

(d) Explain the relationship between your answer to (c) and the definite integral $\int_{1}^{4} (x^2 + 1)\,dx$.

Solution The answers to parts (a), (b.) and (c) are given in Table 17.1.

(d) The sum that has been evaluated is a crude approximation (✎) to the given definite integral. By dividing the interval from $x = 1.0$ to $x = 4.0$ into finer parts (i.e. by using more values of x that are more narrowly separated), the quality of the approximation can be improved. □

In this case a *very* crude approximation since the actual value of the integral is 24.

The method used to estimate the value of the definite integral in Example 1 is a crude example of a process known as **numerical integration**. We will not pursue that topic in this chapter except to note that there are far more sophisticated methods of using a finite sum to estimate the value of a definite integral, many of which can be very precise but require a great deal of repetitive labour and are therefore usually implemented on a computer.

Now try the following questions.

(a) A car moves along a straight road with constant speed V for a time $T = t_2 - t_1$. How far does it travel in that time?

(b) At time $t = t_2$ the car starts to slow down, and it eventually comes to rest at time $t = t_3$. Throughout this period its speed is given by

$$v(t) = at - bt^2$$

where a and b are constants. By considering how far the car travels during a short time Δt_i, deduce the form of the definite integral that represents the total distance travelled between t_2 and t_3.

(a) Distance travelled $= VT$

(b) If the interval t_2 to t_3 is subdivided into n short intervals of duration Δt_i then the distance travelled in any one of those intervals will be approximately

$$v(t_i)\,\Delta t_i$$

and the total distance travelled will be approximately

$$\sum_{i=1}^{n} v(t_i)\,\Delta t_i$$

The accurate value of the total distance travelled will be given by the limit of the above sum as the number of intervals increases and the duration of the longest approaches zero, i.e. as $\Delta t \to 0$, but that is just a definite integral, so the total distance travelled between t_2 and t_3 is just

$$\int_{t_2}^{t_3} v(t)\, dt = \int_{t_2}^{t_3} (at - bt^2)\, dt \quad \square$$

Figure 17.8 shows a column of atmospheric air of cross-sectional area A and total height H. At a distance h above the ground, the density of the air in the column is given by

$$\rho(h) = \rho_0 e^{-h/\lambda} \quad \text{where } \rho_0 \text{ and } \lambda \text{ are constants.}$$

Use the fact that any small part of the column extending from h to $h + \Delta h$ has volume $A\Delta h$ and a mass approximately given by

$$\rho(h) A\, \Delta h$$

to deduce an expression (involving a definite integral) for the total mass of the column.

If the height of the column, from $h = 0$ to $h = H$, is divided into n subintervals of height Δh_i we can approximate the mass of the column by

$$\sum_{i=1}^{n} \rho(h_i)\, A\, \Delta h_i$$

In the limit as the width of the largest slab tends to zero, i.e. as $\Delta h \to 0$, this sum becomes the definite integral

$$\int_0^H \rho(h)\, A\, dh = \int_0^H \rho_0\, e^{-h/\lambda}\, A\, dh \quad \square$$

Figure 17.8 A column of air of height H, cross-sectional area A and variable density $\rho(h)$.

An elastic string of natural length L can be maintained at a stretched length $L + x$ by applying a force

$$F_x(x) = \frac{\lambda x}{L} \quad \text{where } \lambda \text{ is a constant.}$$

In stretching the string the force is said to do *work*, and the work done in increasing the extension x by a small amount Δx is approximately

$$F_x(x)\, \Delta x$$

The extension x increases from 0 to L, and we imagine this interval to be divided into a large number of subintervals by points $x_1, x_2, \ldots, x_{n+1}$ where $x_1 = 0$ and $x_{n+1} = L$. The sum

$$\sum_{i=1}^{n} F_x(x_i) \, \Delta x_i$$

is thus a good approximation to the work done if the Δx_i are small. With practice it is unnecessary to write down this intermediate step, and we go straight to the integral as in this example.

If this approximation becomes more accurate as Δx becomes smaller, write down a definite integral for the work done in stretching the string from its natural length to twice its natural length.

The work done will be

$$\int_0^L F_x(x) \, dx = \int_0^L \frac{\lambda x}{L} \, dx$$

(Note that although the length of the string is increased from L to $2L$, the extension x only increases from 0 to L, so these are the limits of integration, with respect to x, in this case.) □

So much for the formulation of definite integrals by considering the limits of sums. Let us now turn to their evaluation.

17.2.4   The fundamental theorem of calculus

In the last subsection we were mainly concerned with the definition of a definite integral as the limit of a sum. But we started the last subsection by showing that the limit of a sum could be used to describe the area under a graph. Thus, a definite integral can be interpreted in terms of an area under a graph. However, we already know (from Subsection 17.2.2) that the area under the graph of a function $f(x)$ between $x = a$ and $x = b$ is equal to the corresponding change in the value of its inverse derivative $F(b) - F(a)$, where $\frac{dF}{dx} = f(x)$. It follows that we can write:

If $F(x)$ is any inverse derivative of a given function $f(x)$, so that

$$\frac{dF}{dx} = f(x), \text{ then}$$

$$\int_a^b f(x) \, dx = [F(x)]_a^b = F(b) - F(a) \tag{17}$$

As you perform more and more integrals this result will become very familiar. Nonetheless it is worth pausing to note how remarkable it is that the limit of a sum represented by the symbol on the left can be evaluated with the aid of (inverse) differentiation as shown by the difference on the right. The link between summation and differentiation is far from obvious to most physical scientists.

This remarkable result, relating the limit of a sum to a difference in inverse derivatives, is known as the **fundamental theorem of calculus**. It provides the key to evaluating limits of sums and provides another reason for our interest in inverse derivatives.

Example 2
Use the fundamental theorem of calculus to find the value of

$$\int_1^3 t^4 \, dt.$$

Solution   Since $\dfrac{d}{dt}\left(\dfrac{t^5}{5}\right) = t^4$

$$\int_1^3 t^4 \, dt = \left[\frac{t^5}{5}\right]_1^3 = \left(\frac{3^5}{5}\right) - \left(\frac{1^5}{5}\right) = \frac{243}{5} - \frac{1}{5} = 48.4 \qquad (\text{☞})$$

Note that the fundamental theorem applies to *any* inverse derivative of the integrand, so we have deliberately chosen to use the simplest in which the arbitrary constant C is zero.   □

Notice that

$$\int_1^3 t^4 \, dt$$

is *defined* to be a limit of a sum — we can *evaluate* it using inverse differentiation.

Use Equation 17 to evaluate the following three integrals that were obtained at the end of Subsection 17.2.3.

(a) $\displaystyle\int_{t_2}^{t_3} (at - bt^2) \, dt$ (b) $\displaystyle\int_0^H \rho_0 \, e^{-h/\lambda} A \, dh$ (c) $\displaystyle\int_0^L \frac{\lambda x}{L} \, dx$

(a)   In this case the integrand is $at - bt^2$.

An inverse derivative of this function is $\dfrac{at^2}{2} - \dfrac{bt^3}{3} + C$ (☞)

You can confirm this by differentiating.

Choosing $C = 0$, it follows from the fundamental theorem of calculus that

$$\int_{t_2}^{t_3} (at - bt^2) \, dt = \left[\frac{at^2}{2} - \frac{bt^3}{3}\right]_{t_2}^{t_3}$$

$$= \left[\frac{at_3^2}{2} - \frac{bt_3^3}{3}\right] - \left[\frac{at_2^2}{2} - \frac{bt_2^3}{3}\right]$$

i.e. $\displaystyle\int_{t_2}^{t_3} (at - bt^2) \, dt = \frac{a}{2}(t_3^2 - y_2^2) - \frac{b}{3}(t_3^3 - t_2^3)$

(b)   In this case the integrand is $A\rho_0 e^{-h/\lambda}$.

An inverse derivative of this function is $-A\rho_0 \lambda e^{-h/\lambda} + C$.

Choosing $C = 0$, it follows from the fundamental theorem of calculus that

$$\int_0^H \rho_0\, e^{-h/\lambda}\, A\; dh \;=\; [-A\rho_0\lambda\, e^{-h/\lambda}]_0^H$$

$$= (-A\rho_0\lambda\, e^{-H/\lambda}) - (-A\rho_0\lambda)$$

i.e. $\displaystyle\int_0^H \rho_0\, e^{-h/\lambda}\, A\; dh \;=\; A\rho_0\lambda(1 - e^{-H/\lambda})$

(c)   In this case the integrand is $\dfrac{\lambda x}{L}$.

An inverse derivative of this function is $\dfrac{\lambda x^2}{2L} + C$.

Choosing $C = 0$, it follows from the fundamental theorem of calculus that

$$\int_0^L \frac{\lambda x}{L}\, dx = \left[\frac{\lambda x^2}{2L}\right]_0^L = \frac{\lambda L^2}{2L} - 0 = \frac{\lambda L}{2} \quad \square$$

There are two important points to note about the answers to this question.

1.   In every case the arbitrary constant C that was included in the inverse derivative played no part in the final answer since the answer only concerned a *difference* of the form $[F(x)]_a^b$. That is why you were able to put $C = 0$ in each case without any fear of error.

2.   In each case the final answer only involved constants, not variables. Thus, even though we may not know the numerical values of all these constants, the result of each definite integral is itself a constant.

Question T5

Evaluate the following definite integrals:

(a) $\displaystyle\int_1^2 x^{1/2}\, dx$ (b) $\displaystyle\int_{-2}^0 4x^2\, dx$ (c) $\displaystyle\int_{-\pi/2}^{\pi/2} \sin x\, dx$ $\square$

Study comment

In view of its importance it is appropriate to provide some further justification for the fundamental theorem of calculus. However such justification will not actually help you to evaluate integrals. We therefore present this justification in the form of an aside that you may regard as optional reading.

Aside — justifying the fundamental theorem of calculus

An informal justification for the fundamental theorem of calculus, without reference to the area under a graph, can be obtained as follows:

Since $f(x) = \dfrac{dF}{dx} = \lim\limits_{\Delta x \to 0}\left[\dfrac{F(x + \Delta x) - F(x)}{\Delta x}\right]$

for each value of x we have

$$f(x)\Delta x \approx F(x + \Delta x) - F(x)$$

Now, the definite integral from a to b is obtained as the limit of a sum of terms of the form $f(x_i) \Delta x_i$ where x_i takes values

$$x_1, x_2, \ldots x_n \text{ with } a = x_1 < x_2 < x_3 < \ldots < x_n < x_{n+1} = b$$

and $\Delta x_i = x_{i+1}$ so that

$$\int_a^b f(x) \, dx = \sum_{i=1}^n f(x_i) \, \Delta x_i = \sum_{i=1}^n [F(x_{i+1}) - F(x_i)]$$

$$= [F(x_{n+1}) - F(x_n)] + [F(x_n) - F(x_{n-1})] + \ldots$$

$$+ [F(x_3) - F(x_2)] + [F(x_2) - F(x_1)]$$

But this last expression is equal to $F(x_{n+1}) - F(x_1)$, since all the intermediate terms, such as $F(x_n)$ and $F(x_2)$ cancel.

However, $x_1 = a$ and $x_{n+1} = b$, so $F(x_{n+1}) - F(x_1) = F(b) - F(a)$.

So, in the limit we have $\int_a^b f(x) \, dx = F(b) - F(a)$

thus justifying the fundamental theorem of calculus.

17.2.5 Indefinite integrals

Indefinite integrals are nothing new, in fact we have already remarked that 'indefinite integral' is an alternative term for *inverse derivative*; only the notation used is new.

If $F(x)$ is an inverse derivative of $f(x)$, so that $F'(x) = f(x)$, then we write

$\int f(x) \, dx = F(x)$, and we call $\int f(x) \, dx$ an **indefinite integral** of $f(x)$.

This is the notation and terminology that physical scientists generally use when dealing with inverse derivatives. All of our earlier results involving inverse derivatives can be expressed in terms of *indefinite integrals* and henceforth that is what we will do.

Thus, using C to represent an arbitrary constant, we can now write:

$$\int 3x^2 \, dx = x^3 + C \qquad \text{since } \frac{d}{dx}(x^3 + C) = 3x^2$$

$$\int e^{5x} \, dx = \frac{e^{5x}}{5} + C \qquad \text{since } \frac{d}{dx}\left(\frac{e^{5x}}{5} + C\right) = e^{5x}$$

This notation arises from the fundamental theorem of calculus, from which it is evident that definite integrals are closely related to inverse derivatives.

$$\int \sin\left(\frac{x}{2}\right) dx = -2 \cos\left(\frac{x}{2}\right) + C \text{ since } \frac{d}{dx}\left[-2\cos\left(\frac{x}{2}\right) + C\right] = \sin\left(\frac{x}{2}\right)$$

In this context it clearly makes good sense to call C the **constant of integration**.

Notice that the indefinite integral of a function is another function. This should be contrasted with the fact that the definite integral of a function between given limits of integration is a constant.

Question T6

Find the following indefinite integrals:

(a) $\int t^3\, dt$ (b) $\int \cos x\, dx$ (c) $\int e^{2x}\, dx$

(*Hint*: Use your knowledge of differentiation to guess the answer and then check the result.) □

Question T7

At time t the acceleration of an object moving along a straight line may be represented by $a_x(t) = a_0(e^{kt} - e^{-kt})$, where a_0 and k are constants.

(a) Evaluate $\int a_x(t)\, dt$ and explain its physical interpretation.

(b) If $k = 1\,\text{s}^{-1}$, and if the position of the object, $x(t)$ at time t, satisfies $x(0) = 0$ and $x(1\,\text{s}) = 1\,\text{m}$, find the constant of integration in $\int a_x(t)\, dt$, in terms of a_0. □

17.2.6   Summary of Section 17.2

We now have two types of integral:

(a) The *definite integral*: $\int_a^b f(x)\, dx$, a specific value, defined in terms of the limit of a sum and interpreted as the area under the graph of $f(x)$ between $x = a$ and $x = b$ (with regions below the x-axis regarded as having negative area.)

(b) The *indefinite integral*: $\int f(x)\, dx$, a function $F(x)$, defined by the requirement that $\dfrac{dF}{dx} = f(x)$ and interpreted as the result of reverse differentiation. A given function $f(x)$ generally has infinitely many indefinite integrals any two of which will differ by a constant.

These two types of integral are related by the *fundamental theorem of calculus* which says that if $F(x)$ is any indefinite integral of $f(x)$, then

$$\int_a^b f(x)\, dx = [F(x)]_a^b = F(b) - F(a)$$

17.3   The applications of integration

17.3.1   Some uses of integration

Integration, the analysis and evaluation of definite and indefinite integrals, is of great importance in mathematics, science and technology. We have already seen three ways in which integration can arise in physical problems:

1.   as a way of reversing the effect of differentiation;
2.   as a way of determining the limit of a sum; and
3.   as a way of determining the area under a graph (a special case of (2)).

In this subsection we look at more examples of integration to illustrate its power. It is not necessary for you to fully understand all of the physical concepts that are used in these examples in order to appreciate the mathematical points that are being made.

The mass of an object of variable density

Suppose we need to determine the mass of a solid metal bar, of length L, whose density $\rho(x)$ at a distance x from one end is given by $\rho(x) = k(x + a)^2$ for some particular constants k and a. We suppose that the bar has a uniform square cross section of side b.

The mass of the bar between x and $x + \Delta x$ is approximately $b^2\rho(x)\Delta x$ and the total mass is approximately obtained by adding together the masses of all these small pieces. The exact mass is given by the definite integral

$$\int_0^L b^2\rho(x)\, dx = \int_0^L kb^2 (x + a)^2\, dx$$

Since $\dfrac{d}{dx}\left(kb^2\,\dfrac{(x + a)^3}{3}\right) = kb^2(x + a)^2$

it follows that

$$\int_0^L b^2\rho(x)\, dx = \left[kb^2\,\frac{(x + a)^3}{3}\right]_0^L = kb^2\,\frac{(L + a)^3}{3} - kb^2\,\frac{a^3}{3}$$

$$= \frac{kb^2L}{3}\,(L^2 + 3La + 3a^2)$$

Question T8

A bar of length $2L$ has a circular cross section of radius R. Its density, $\rho(x)$, when x is measured from the mid-point of the bar, is $\rho(x) = Ae^{2Bx}$. Find an expression for the mass of the bar. (*Hint*: The bar is not symmetrical.)   □

The density is defined as the mass per unit volume.

This force acts perpendicular to the dam wall.

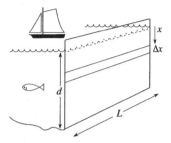

Figure 17.9 A rectangular dam.

Summing a force and calculating a total moment

Consider a simplified mathematical model of a dam in which we assume that it is rectangular in shape, of width L and depth d and placed with its plane vertical, as in Figure 17.9.

The *pressure* at a depth x below the surface of the water is $P(x) = kx$ for some constant k, and so the magnitude of the force acting on a rectangular element of the dam wall of width L and height Δx (and consequently area $\Delta A = L\Delta x$) at a depth x is approximately

$$\Delta F \approx P(x)\Delta A = Lkx\Delta x$$

Since this approximation becomes more accurate as Δx becomes smaller, the total force on the dam is of magnitude

$$\int_0^d Lkx\, dx = \left[\frac{Lkx^2}{2}\right]_0^d = \frac{Lkd^2}{2}$$

The designers of such a dam have many other factors to take into account in addition to the total force that it must resist. For example, the pressure of the water also has a twisting effect on the structure. This twisting effect is known as the *torque* of the force, and for the element of area discussed above this *torque* about the line where the dam emerges from the water has a magnitude given by

$$x \times Lkx\Delta x = Lkx^2\Delta x$$

The total torque about the waterline is given by

$$\int_0^d Lkx^2\, dx = \left[\frac{Lkx^3}{3}\right]_0^d = \frac{Lkd^3}{3}$$

Finding an average

Suppose that a car completes a journey in such a way that it covers a total distance S in a time T, but at a speed $v(t)$ that varies throughout the journey. (Remember the speed of an object is defined as the *magnitude* of its velocity, so although the speed may vary it may never be negative.) Since the car covers a distance S in a time T, it is clear that its average speed is given by

$$\langle v \rangle = S/T \tag{18}$$

The angular brackets $\langle\rangle$ mean 'the average of' the term enclosed.

However, since the distance covered during any small time interval Δt will be $v(t)\Delta t$, it is also clear that the total distance covered during the journey will be

$$S = \int_{t_1}^{t_1+T} v(t)\, dt \tag{19}$$

where t_1 denotes the time at which the journey started. It follows from Equations 18 and 19 that

$$\langle v \rangle = \frac{1}{T}\int_{t_1}^{t_1+T} v(t)\, dt \tag{20}$$

Thus, the integral provides a way of *defining* the average of a quantity that varies continuously. This definition has many applications.

Question T9

A bar of length $2L$ has a temperature $T(x)$ that varies along its length, when x is measured from the mid-point of the bar. If $T(x) = A \cos(Bx)$, find an expression for the average temperature of the bar. What is the value of this average temperature if $L = 2.0$ m, $A = 30$ K and $B = 0.4$ m^{-1}? ☐

17.3.2 The stability of a satellite

To end this section we consider one final example, concerning the stability of an orbiting satellite, that brings together some of the ideas that have been introduced earlier.

The motion of a satellite in orbit about the Earth is generally very complicated, so we will consider a highly idealised case of such motion. First we will suppose that the satellite can be modelled by a uniform thin rod, of length $2L$ and uniform mass per unit length ρ, which means that its total mass is $2L\rho$ and that on the Earth's surface it would balance about its mid-point (its *centre of mass*). We will also suppose that the satellite is placed in a circular *geostationary orbit* above the equator, in which it takes 24 hours to complete an orbit and thus appears fixed in the sky from any point on the Earth's surface. The distance from the centre of the Earth to the satellite will be denoted by H. Finally, we will assume that when placed in its orbit the satellite is made to rotate about an axis through its centre at a rate of exactly 360° in 24 hours in such a way that it remains parallel to the Earth's surface throughout its orbit (as indicated in Figure 17.10a). (☞)

Once the satellite is placed in orbit (and given its initial rotation) all the stabilisers are switched off, and it is subject only to the effects of the Earth's gravity. The question we wish to address is this: 'Will the satellite

If you imagine the arrow in Figure 17.10a rotating through 360° and carrying the satellite with it, then you can see that the satellite must also rotate through 360°.

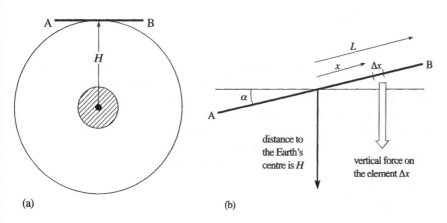

(a)   (b)

Figure 17.10   A satellite (AB) in geostationary orbit.

remain with its orientation parallel to the Earth's surface (as in Figure 17.10a), or will it begin to tumble?'

To answer the question we need to know a little about the force due to gravity. If two particles of mass M and m are a distance r apart, then each is attracted towards the other with a force of magnitude GmM/r^2 (where G is known as *Newton's gravitational constant*). A large spherical object, such as the Earth, exerts the same gravitational force as a particle of equal mass placed at its centre, and, since the satellite is small compared to the size of its orbit, we can assume that all the forces act vertically downwards, i.e. parallel to the line through the centre of the rod (satellite) and the centre of the Earth.

We will now suppose that the satellite is disturbed very slightly, so that it is at an angle α to the horizontal (as in Figure 17.10b), and we will attempt to find the torque, or turning effect, of the forces due to gravity. If we find that the torque tends to restore the satellite to its horizontal position we can conclude that the satellite is stable, but if we find that the torque tends to increase the angle α we must conclude that the satellite is unstable since a small disturbance may send it into a spin from which it will not recover.

If we consider a small element of the satellite of length Δx with position coordinate x (measured along the satellite, from its centre), then the mass of that element will be

(density per unit length) $\times$ (length of element) $= \rho \Delta x$

and the distance of the element from the centre of the Earth will be $H + x \sin \alpha$.

Now the downward vertical force acting on the element will have a magnitude given approximately by

$$\frac{\text{(gravitational constant)} \times \text{(mass of Earth)} \times \text{(mass of element)}}{\text{(distance of element from centre of Earth)}^2}$$

So, if we represent this magnitude by ΔF and let M represent the mass of the Earth we can write

$$\Delta F \approx \frac{GM\rho \, \Delta x}{(H + x \sin \alpha)^2}$$

The torque, or turning effect, of the vertical force on the element, about a line that passes through the centre of the satellite and is perpendicular to the plane of Figure 17.10, is $x\Delta F$. In this case ΔF (being a magnitude) must be positive, but x (and consequently $x\Delta F$) may be positive or negative. A positive torque will tend to reduce α and restore stability, a negative torque will tend to increase α and lead to instability. To find the total torque on the satellite we need to consider the definite integral

$$\int_{-L}^{L} \frac{GM\rho x}{(H + x \sin \alpha)^2} \, dx$$

This integral can certainly be evaluated using an appropriate indefinite integral, but the techniques required to deduce that indefinite integral are beyond the scope of this chapter. However, we can use the interpretation of a definite integral as an area under a graph to obtain the result we need. If we put

$$f(x) = \frac{GM\rho x}{(H + x \sin \alpha)^2}$$

then, for suitably chosen values of the constants, we may plot the graph of $f(x)$, as in Figure 17.11.

The required integral is the area under the graph of $y = f(x)$ between $-L$ and L. From the graph it is easy to see that the negative contribution, from the interval $-L \le x \le 0$, is greater in magnitude than the positive contribution, from the interval $0 \le x \le L$. This means that the net result must be *negative* and therefore the total moment acts so as to increase the angle α.

So our conclusion is that any small disturbance of the satellite will lead to an even greater disturbance: in other words, it is *unstable*.

We have only dealt with the simplest of cases. In practice one would need to consider more complicated structures, and the effects of other factors such as expansion due to solar heating, but nevertheless the example illustrates the importance of integration and the value of the methods that we have introduced in this chapter in solving physical problems.

17.4 Conclusion

17.4.1 The importance of integration

The concept of a *definite integral* (the *limit of a sum*) is fundamental to much of applicable mathematics, and is essential to a proper study of physical science.

In this chapter we have seen how a variety of physical problems can be formulated in terms of definite integrals and we have also seen some of the ways in which such integrals can be evaluated. Finite sums and areas under graphs can both provide estimates of the values of definite integrals, but normally the most convenient method is to use the fundamental theorem of calculus to relate the definite integral to a difference in values of an appropriate indefinite integral, if such a function can be found. Almost all the definite integrals considered in this chapter have corresponded to indefinite integrals that have been easy to express in terms of elementary functions (such as e^x, $\sin x$, x^n, and so on). However, this is not always so. As your studies continue it is quite certain that you will encounter integrals, both definite and indefinite, of a more challenging kind. Some of the methods that can be used to deal with these integrals are explored in later chapters.

The following questions are intended to improve your basic skill in evaluating integrals, and to remind you of some of the important ideas introduced in this chapter.

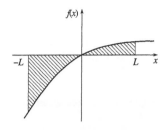

Figure 17.11 Areas under the graph representing the positive and negative torques acting on the satellite.

You might have anticipated this result since in Figure 17.10b the section of the satellite to the left of centre is brought nearer to the Earth and so will be pulled further in this direction – not compensated sufficiently by the other half of the satellite, on which the force is weakened, as it is further from the Earth.

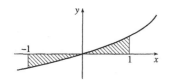

Figure 17.12 See Question T12.

Question T10

Evaluate the following indefinite integrals:

(a) $\int (1 + x + x^2) \, dx$

(b) $\int (\sin x + \cos x) \, dx$

(c) $\int \left(e^{-x} + \frac{1}{x^3} \right) dx$

(d) $\int 2 \, dw$ □

Question T11

Evaluate the following definite integrals:

(a) $\int_{-1/2}^{1/2} \sin (\pi x) \, dx$

(b) $\int_0^{3t} (3t + 2)^2 \, dt$

(c) $\int_0^1 \frac{(1 + e^x)}{e^x} \, dx$

(d) $\int_1^4 \sqrt{x} \, (x + 1) \, dx$ □

Question T12

Find the sum of the magnitudes of the two shaded areas in Figure 17.12, i.e. find the magnitude of the area enclosed by the graph of the function $y = e^x - 1$, the x-axis and the lines $x = -1$ and $x = 1$. Note that in this case you are not being asked for 'the area under the graph' which may be negative, but for the *magnitude* of the area *enclosed*, which must be positive. □

Question T13

The acceleration, velocity and position coordinate of an object moving along a straight line are represented by $a_x(t)$, $v_x(t)$ and $x(t)$, respectively. What physical meaning, if any, can you associate with the following

$$\int_{t_1}^{t_2} a_x(t) \, dt \qquad \int_{t_1}^{t_2} v_x(t) \, dt \qquad \int_{t_1}^{t_2} x(t) \, dt$$ □

17.5 Closing items

17.5.1 Chapter summary

1. If $f(x)$ is a given function and $F(x)$ is any function such that $\dfrac{dF}{dx} = f(x)$

   then we write $F(x) = \int f(x)\, dx$ and we call $F(x)$ an *indefinite integral* (or *inverse derivative* or *primitive*) of $f(x)$.

2. If $F_1(x)$ and $F_2(x)$ are both indefinite integrals of the same function $f(x)$, then there exists a constant K such that

   $$F_1(x) = F_2(x) + K$$

3. For motion in a straight line, if the *position*, *velocity* and *acceleration* at time t are respectively $x(t)$, $v_x(t)$ and $a_x(t)$ then

   $$v_x(t) = \frac{dx}{dt} \quad \text{and} \quad a_x(t) = \frac{dv_x}{dt}$$

   so, $\quad x(t) = \int v_x(t)\, dt \quad \text{and} \quad v_x(t) = \int a_x(t)\, dt$

4. The *definite integral* of a function $f(x)$ from $x = a$ to $x = b$ is defined by the limit of a sum and may be written as

   $$\int_a^b f(x)\, dx = \lim_{\Delta x \to 0} \left(\sum_{i=1}^{n} f(x_i)\, \Delta x_i \right) \tag{16}$$

   where Δx is the width of the largest subinterval, and where $x_1 = a$ and $x_{n+1} = b$.

5. According to the *fundamental theorem of calculus*, if $F(x)$ is any indefinite integral of a given function $f(x)$, so that $\dfrac{dF}{dx} = f(x)$, then

   $$\int_a^b f(x)\, dx = [F(x)]_a^b = F(b) - F(a)$$

   This often provides a convenient way of evaluating definite integrals.

6. The definite integral $\int_a^b f(x)\, dx$, with $b > a$, may be interpreted as *the area under the graph* of $f(x)$ between $x = a$ and $x = b$. It follows from the fundamental theorem of calculus that the area under the graph of $f(x)$ between $x = a$ and $x = b$ is equal to the corresponding change in the indefinite integral, $F(b) - F(a)$. When calculating the area under a graph, any regions that are below the horizontal axis should be regarded as having negative areas.

This statement is true for the functions treated in this chapter, but the definition of the definite integral as the limit of a sum allows it to be applied in cases where simple graphical interpretation is not possible.

17.5.2   Achievements

Having completed this chapter, you should be able to:

A1. Define the terms that are emboldened in the text of the chapter.

A2. Define an indefinite integral of a given function in terms of a process that reverses the effect of differentiation, and use that definition to express physical quantities (such as position and velocity, in the case of linear motion) as indefinite integrals of other physical quantities (such as velocity and acceleration).

A3. Determine (by inspection) indefinite integrals of a range of simple functions in various mathematical and physical contexts, recognising that the process of indefinite integration generally introduces an arbitrary constant.

A4. Define the definite integral of a given function between given limits in terms of the limit of a sum, and use that definition to formulate various physical problems in terms of definite integrals.

A5. Recognise that the definition of a definite integral as the limit of a sum can be used to justify the use of a finite sum to estimate the value of a given definite integral.

A6. Interpret the definite integral of a given function between given limits in terms of the area under the graph of that function between those limits (with due regard to signs), and use this identification to determine areas and/or evaluate definite integrals in sufficiently simple cases.

A7. State the fundamental theorem of calculus that relates definite and indefinite integrals, and use it to evaluate definite integrals in sufficiently simple cases.

17.5.3   End of chapter questions

Question E1   The acceleration of an object moving along the x-axis is given at time t by

$$a_x(t) = a_0 \left(e^{t/k} + e^{-t/k} \right)$$

where a_0 and k are constants. If $a_0 = 2.00$ m s^{-2} and $k = 1.00$ s, obtain expressions for the velocity $v_x(t)$ and the position $x(t)$ given that $v_x(0) = x(0) = 0$.

What is the area under the graph of $a_x(t)$ against t between $t = 1$ s and $t = 2$ s, and what does this area represent?

exp(a) is an alternative way of writing e^a.

Question E2   The indefinite integral $\int \exp(x^2)\, dx$ cannot be expressed in terms of elementary functions, so it is not possible to evaluate the definite integral $\int_0^1 \exp(x^2)\, dx$ in a straightforward way using the fundamental theorem of calculus. Describe two other ways in which you might estimate the value of this definite integral. (You are not required to perform these evaluations.)

Question E3 Evaluate the following indefinite integrals:

(a) $\int x^{1/2} \, dx$ (b) $\int (x + 3)(x + 2) \, dx$

(c) $\int 6 \cos (3x) \, dx$ (d) $\int (e^x + e^{-x})(e^x - e^{-x}) \, dx$

(*Hint*: By expanding where appropriate, you should be able to determine by inspection (i.e. guess) a function which, when differentiated, produces the given integrand in each case.)

Question E4 Use the fundamental theorem of calculus to evaluate each of the integrals in Question E3 between 0 and 1, i.e. with lower limit 0 and upper limit 1.

Question E5 Find the sum of the magnitudes of the shaded areas in Figure 17.13, i.e. find the area enclosed by the graph of the function $y = x^3 + 1$, the x-axis and the lines $x = -2$ and $x = 2$. (Note that in this case you are not being asked for 'the area under the graph,' which would be the difference of these magnitudes.)

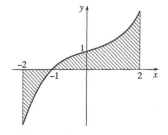

Figure 17.13 See Question E5.

Chapter 18 Integrating simple functions

18.1 Opening items

18.1.1 Chapter introduction

Integration enters into almost every area of physical science, and it does so in two quite different ways. On the one hand, *indefinite integration* allows us, up to a point, to reverse the effect of differentiation and is therefore of importance wherever differentiation arises, i.e. almost everywhere. On the other hand, the process of *definite integration* allows us to extend the idea of summation to the addition of continuous distributions. For example, a column of air will have a density that increases continuously from its top to its bottom, yet definite integration allows us to add together the mass of each layer in the column and, in an appropriate limit, evaluate the total mass of the column. Both aspects of integration – reversing differentiation and finding limits of sums – are important in their own right, but they take on increased significance when they are brought together by the *fundamental theorem of calculus*, since it allows us to use indefinite integrals in the evaluation of many definite integrals.

This chapter contains three main sections. The first (Section 18.2) reviews the most important concepts introduced in the last chapter, including indefinite and definite integration, and the relation of the latter to the (signed) area under a graph. It also provides several more examples of integration, many in the form of questions.

Section 18.3 is mainly concerned with the determination of indefinite integrals. It contains two tables of standard indefinite integrals, (the more basic in Subsection 18.3.1 and the more advanced in Subsection 18.3.3) and sandwiched between them are various rules for integrating combinations of functions whose individual indefinite integrals are already known. It almost goes without saying that this section, with its emphasis on simple functions, is only scratching the surface of a very large topic; other techniques of integration are dealt with in the next chapter, and in the companion volume.

Section 18.4 concerns definite integrals. It lists the general mathematical properties of definite integrals and looks at the special simplifications that occur in the physically important cases where the functions being integrated are *odd*, *even* or *periodic*. The section concludes with a discussion of *improper integrals* that may involve integrating over an infinite range of values, or integrating a function which itself becomes infinite at some point in the range of integration.

18.1.2   Fast track questions

Question F1

A function $f(x)$ is positive in an interval $a \leqslant x \leqslant b$. Explain how the area under the graph of $f(x)$ between $x = a$ and $x = b$ can be represented as a definite integral. What are the magnitudes of the various areas enclosed by the graph of $y = x^3$, the x-axis and the lines $x = -2$ and $x = 1$? (Note that this graph crosses the horizontal axis at $x = 0$.)

Question F2

Find the following indefinite integrals:

(a) $\int (4x^2 + 7x - 5) \, dx$ (b) $\int e^{-6x} \, dx$

(c) $\int [3 \cos(4x) - 5 \sin(4x)] \, dx$ (d) $\int 3 \log_e (2x) \, dx$

Question F3

Evaluate the following definite integrals (to four decimal places):

(a) $\displaystyle\int_{-2}^{3} (3 - 2x - x^2) \, dx$

(b) $\displaystyle\int_{\pi/4}^{\pi/3} [6 \cos (3x) - 10 \cos (2x)] \, dx$

(c) $\displaystyle\int_{1}^{2} 4 \log_e x \, dx$

18.1.3   Ready to study?

Study comment

To study this chapter you will need to be familiar with the following terms: *constant, derivative, differentiation, function, graph, inequality, limit, magnitude* (of a *vector*, in the sense of the strength of a *force*), *modulus* (as in $|-3| = 3$) and *summation* (including the *summation symbol*, Σ). In addition you will need to be familiar with the properties of the *elementary functions, powers, roots, reciprocals, exponentials, logarithms* and *trigonometric functions* (including the *reciprocal* and *inverse trigonometric functions* such as $sec(x)$ and $arcsec(x)$). It is assumed that you are reasonably proficient at differentiation and that you know how to differentiate *sums* and *products* of the elementary functions as well as being able to use the *chain rule* to differentiate *functions of functions. Implicit differentiation* is used in Subsection 18.3.3 but lack of familiarity with that technique should not prevent you from studying the chapter. It is also assumed that you are already familiar with the terminology of *functions* and that you can perform simple *indefinite integrals* and evaluate simple *definite integrals*; however, each of these topics is briefly reviewed in Section 18.2 in an effort to make this chapter as self-contained as possible. The following *Ready to study questions* will allow you to establish whether you need to review some of the topics before embarking on this chapter.

Question R1

For each of the following functions find the derivative $f'(x) = df/dx$:

(a) $f(x) = 8x^6 + 6x^3 - 5x^2 - 2$

(b) $f(x) = 3\sin(x) + 4\cos(x)$

(c) $f(x) = 5e^{2x} - 2e^{-2x}$

(d) $f(x) = 2\log_e x + 3\log_e(2x)$

Question R2

Find dy/dx for each of the following equations:

(a) $y = 5\cos(x) + 2x^2 - 7x$ (b) $y = 6e^{-3x}$

(c) $y = x^2 - 5\log_e x$

Question R3

Evaluate the following

(a) $\lim_{x \to 1}(2 - x)$ (i.e. the limit as x tends to 1 of $(2 - x)$)

(b) $\sum_{i=1}^{3} 2i$ (i.e. the sum from $i = 1$ to $i = 3$ of $2i$)

(c) $\arcsin(-1)$ (i.e. the inverse sine of -1)

18.2 Simple integrals – principles and practice

This section repeats the main results of the last chapter and applies them in various contexts. It also gives you the chance to practice your integration skills.

18.2.1 Functions and graphs

Mathematically, a **function** f is a rule that assigns a single value $f(x)$ in a set called the *codomain* to each value x in a set called the *domain*.

Functions are usually defined by formulae. An example is

$$f(x) = 1 + x + x^2 \qquad \text{(1a)}$$

the domain of which consists of all values of x.

The same function f could equally well be defined using some other symbol, such as t, to represent the *independent variable*:

$$f(t) = 1 + t + t^2 \qquad \text{(1b)}$$

This freedom to relabel the *independent variable* is often of great use, though it is vital that such changes are made consistently throughout an equation.

The **graph** of a function $f(x)$ is a plot of the points $(x, f(x))$. Often, we choose to let $y = f(x)$, in which case the graph will consist of all the points (x, y) that satisfy the relation $y = f(x)$. For example, Figure 18.1a depicts part of the graph of $f(x) = x^3 + 4$; where, for the purposes of drawing the graph, we have written the relationship as $y = x^3 + 4$. Part (b) of the same figure shows the points that satisfy the relation

$$x^2 + y^2 = 4$$

This second curve is *not* the graph of a function, since there are values of x that correspond to more than one value of y.

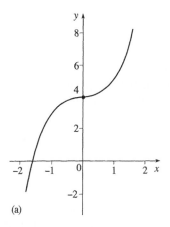

(a)

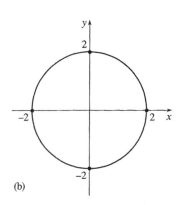

(b)

Figure 18.1 (a) Graph of $y = x^3 + 4$. (b) The circle $x^2 + y^2 = 4$. (Part (b) is an example of a relationship that does not fit the strict definition of a function.)

Question TI

Which of the following expressions define y as a function of x?

(a) $y = (x + 2)^2$

(b) $y = \sqrt{x}$ (where $\sqrt{x}$ is the *positive* square root of x)

(c) $y^2 = x$ (d) $y = 2$ (e) $y^2 = 2$ □

18.2.2 Indefinite integrals: reversing differentiation

The process of **indefinite integration** (sometimes referred to as **inverse differentiation** or **antidifferentiation**) reverses the effect of differentiation, at least up to a point.

If $f(x)$ is a given function and $F(x)$ is any function such that

$$\frac{dF}{dx} = f(x) \quad \text{then we write} \quad F(x) = \int f(x)\, dx$$

and we call $F(x)$ an **indefinite integral** of $f(x)$.

Indefinite integrals are also known as *inverse derivatives*, *antiderivatives* or *primitives*.

The reversal of differentiation is not quite complete because any given function will have an infinite number of indefinite integrals, any two of which may differ by a constant. For example, if the height x (in metres) of an object above ground level at time t (in seconds) is given by the function

$$x(t) = (400\,\text{m}) - \tfrac{1}{2}\,gt^2 \tag{2}$$

where $g = 9.81\,\text{m s}^{-2}$, then the velocity v_x of the object is given by the derivative

$$v_x(t) = \frac{dx}{dt} = -gt \tag{3}$$

The negative sign in Equation 3 indicates that the velocity is directed downwards, since we chose x to represent the height, which increases in the upwards direction.

However, using indefinite integration to reverse this process is not possible, i.e. given $v_x = -gt$ we cannot deduce that Equation 2 is true. On the basis of Equation 3 alone, any one of the following indefinite integrals of the velocity *might* have been the correct expression for the position as a function of time, even though, as it happens, none of them actually agrees with Equation 2.

$$x(t) = (3\,\text{m}) - \tfrac{1}{2}\,gt^2 \tag{4}$$

$$x(t) = -\tfrac{1}{2}\,gt^2 \tag{5}$$

$$x(t) = (-5\,\text{m}) - \tfrac{1}{2}\,gt^2 \tag{6}$$

The point is that all these indefinite integrals differ by a constant, and the most we can really say about $x(t)$ on the basis of Equation 3 is that it is of the form

$$x(t) = C - \tfrac{1}{2}\,gt^2 \tag{7}$$

where C is an arbitrary constant.

Similarly, we can say that

$$\int 4y^2\, dy = \frac{4}{3}y^3 + C \quad \text{since} \quad \frac{d}{dy}\left(\frac{4}{3}y^3 + C\right) = 4y^2 \tag{8}$$

but we cannot determine the arbitrary constant of integration unless we are given additional relevant information.

A general consequence of this is that if $F(x)$ is an indefinite integral of $f(x)$ then $F(x) + C$, where C is any constant, is also an indefinite integral of $f(x)$.

Question T2

Show that if $F(x)$ is an indefinite integral of $f(x)$ then $aF(x)$ is an indefinite integral of the function $af(x)$ where a is a constant. Further, if $G(x)$ is an indefinite integral of another function $g(x)$ show that $F(x) + G(x)$ is an indefinite integral of the function $f(x) + g(x)$. Write down an indefinite integral of the function $af(x) + bg(x)$ where b is a constant. □

Question T3

Find the following indefinite integrals:

(a) $\int 3\,dx$ (b) $\int x^5\,dx$ (c) $\int 5x^{1/4}\,dx$ □

Question T4

Verify the following results by differentiation:

(a) $\displaystyle\int \frac{1}{\sqrt{2x+2}}\,dx = \sqrt{2x+2} + C$

(b) $\displaystyle\int (x^2 - 2x)^8\,(x-1)\,dx = \frac{1}{18}\,(x^2 - 2x)^9 + C$

(c) $\displaystyle\int \frac{\cos x}{(2 + \sin x)^2}\,dx = -\frac{1}{2 + \sin x} + C$ □

18.2.3   Definite integrals: the limit of a sum

Apart from *indefinite* integration, which reverses differentiation, there is another kind of integration, *definite* integration, that is closely related to summation.

Given a function $f(x)$ and two values $x = a$, and $x = b$, the **definite integral** of $f(x)$ with respect to x, from a to b is defined by

$$\int_a^b f(x)\,dx = \lim_{\Delta x \to 0}\left(\sum_{i=1}^{n} f(x_i)\,\Delta x_i\right)$$

where $x_1 < x_2 < x_3 < \ldots x_{n-1} < x_n < x_{n+1}$, and $\Delta x_i = x_{i+1} - x_i$, with $x_1 = a$ and $x_{n+1} = b$, and Δx is the largest of the Δx_i.

a and *b* are referred to respectively as the **lower** and **upper limits of integration**.

To see how this second kind of integration arises in a physical context, consider the problem of calculating the *work* done by an expanding gas. For the sake of simplicity, let us suppose that the gas is confined in a cylindrical vessel (Figure 18.2a) of cross-sectional area A, by a piston which is free to move in the x-direction. When the position coordinate of the piston is x (measured to the right from the left-hand end of the cylinder), the volume of the gas will be $V = xA$ and we may denote the pressure in the gas by $P(V)$.

Figure 18.2 (a) Expansion of a gas by a small amount $\Delta V = A\Delta x$. (b) A more substantial expansion from an initial value V_a to a final volume V_b.

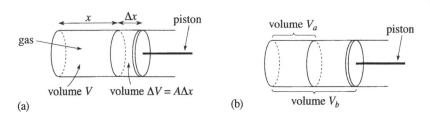

(a)

(b)

For a fixed mass of gas at a fixed temperature, this pressure will decrease as V increases. Whatever its value, the pressure will exert a force in the x-direction on the piston and for a given value of V that force will be

$$F_x(V) = A \times P(V)$$

If the gas causes the piston to move to the right through a small distance Δx then the gas will have done a small amount of *work* ΔW. If the force F_x remained constant throughout this small expansion we could say

$$\Delta W = F_x(V) \times \Delta x$$

However, F_x will *not* remain constant throughout the expansion; it will decrease as x increases and the pressure falls. Nonetheless, if we make Δx very small then the pressure will change very little during the expansion and we can use the approximation

$$\Delta W \approx F_x(V) \times \Delta x \approx A \times P(V) \times \Delta x$$

As a result of this expansion the volume of the gas will have increased by a small amount, ΔV, which will be equal to $A\Delta x$, so we may rewrite the last result as

$$\Delta W \approx P(V)\Delta V \tag{9}$$

Now, if we want to calculate the work done by the gas when it increases its volume from an initial value V_a to a final volume V_b (Figure 18.2b), we can do so, at least approximately, by dividing the expansion into many small steps and using Equation 9 to find the work done in each step. Adding together the work done in each small step leads to the following *approximate* expression for the total work done in the expansion:

$$P(V_1)\Delta V_1 + P(V_2)\Delta V_2 + P(V_3)\Delta V_3 + \ldots + P(V_n)\Delta V_n$$
$$= \sum_{i=1}^{n} P(V_i)\,\Delta V_i \tag{10}$$

If we take the limit of the above sum as ΔV, the width of the widest step, tends to zero, we obtain the following exact result

$$W = \lim_{\Delta V \to 0}\left[\sum_{i=1}^{n} P(V_i)\,\Delta V_i\right] \tag{11}$$

This limit of a sum may be written as the definite integral of $P(V)$ with respect to V, from V_a to V_b, so we have

$$W = \int_{V_a}^{V_b} P(V)\,dV \tag{12}$$

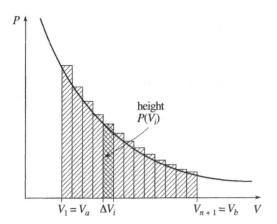

Figure 18.3 A visualisation of Equation 10. The curve shows the pressure P as a function of the volume V. The work done in any small part of the expansion, from V_i to V_{i+1} is approximately represented by the area of the corresponding rectangular strip, $P(V_i)\Delta V_i$. The total amount of work done in expanding from V_a to V_b is approximately represented by the sum of the areas of the rectangles. Note that as the number of rectangles increases and they become narrower, their total area approaches the area under the graph between V_a and V_b; this point is explored in the next subsection.

The task of evaluating the limit of a sum may seem a daunting one. However, there are many cases in which the evaluation is really rather straightforward, thanks to the existence of a deep link between definite integrals and indefinite integrals. This link is embodied in the **fundamental theorem of calculus**:

If $F(x)$ is any indefinite integral of $f(x)$, so that $\int f(x)\,dx = F(x)$, then

$$\int_a^b f(x)\,dx = F(b) - F(a) \qquad (13)$$

Example 1

Evaluate the definite integral $\displaystyle\int_1^4 3x^2\,dx$.

Solution   In this case the integrand of the definite integral is $3x^2$, and an indefinite integral of this is

$$F(x) = \int 3x^2\,dx = x^3 + C$$

Consequently,

$$\int_1^4 3x^2\,dx = F(4) - F(1)$$

i.e. $\displaystyle\int_1^4 3x^2\,dx = (4^3 + C) - (1^3 + C)$

so $\displaystyle\int_1^4 3x^2\,dx = 63$ $\square$

In this example, the fact that x does not appear in the final answer means that the x which appeared in the definite integral could have been replaced by any other symbol without affecting the answer. Hence, we can be sure that

$$\int_1^4 3x^2 \, dx = \int_1^4 3X^2 \, dX = \int_1^4 3\xi^2 \, d\xi$$

They are all alternative ways of writing the number 63. A variable that can be replaced in this way without altering the value of an expression is called a **dummy variable**. In a definite integral the integration variable is *always* a dummy variable.

Question T5

Evaluate the following definite integrals:

(a) $\displaystyle\int_{-1}^{2} 3 \, dx$ (b) $\displaystyle\int_{0}^{2} x^5 \, dx$ (c) $\displaystyle\int_{16}^{81} 5x^{1/4} \, dx$ □

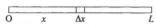

Figure 18.4 See Question T6.

Γ is the upper case Greek letter gamma.

Question T6

Figure 18.4 shows a thin horizontal beam of length L and uniform mass per unit length λ (i.e. the mass of the whole beam is λL). Such a beam can be thought of as an assembly of small elements of length Δx, each of which has a *weight* of magnitude $\lambda g \Delta x$ (where g is the magnitude of the acceleration due to gravity), and exerts a *torque* Γ of magnitude $x\lambda g \Delta x$ about an axis through O that is perpendicular to the plane of Figure 18.4. Write down an expression, involving the limit of a sum, for the magnitude of the total torque about the axis through O due to the weight of the entire beam. Rewrite this expression as a definite integral and evaluate it.   □

18.2.4   Definite integrals: the area under a graph

Given a function $f(x)$ and two values of x, such as a and b with $a < b$ (see Figure 18.5), it is fairly easy to see that if $f(x)$ is never negative between a and b then

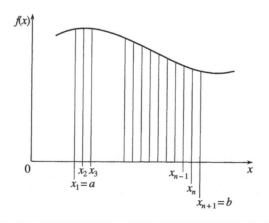

Figure 18.5 The area under the graph of an arbitrary function may be approximated by a sum of rectangles.

$$\int_a^b f(x)\, dx = \text{the area under the graph of } f(x)$$
$$\text{between } x = a \text{ and } x = b$$

provided we measure the area in the scale units used on the graph's axes and take it to mean the area below the curve but above the horizontal axis between the given limits.

The situation is a little more complicated if $a > b$, or if $f(x)$ is negative between those limits since we then have to take care over signs. However, our convention is clear:

The **area under the graph** of a function $f(x)$ between $x = a$ and $x = b$ is equal to the corresponding definite integral,

$$\int_a^b f(x)\, dx = F(b) - F(a) .$$

It is a consequence of this definition that if, as in Figure 18.6, $a < b$, but part of the graph is below the horizontal axis, then the area under the graph in that region will be a *negative* quantity.

Find the area under the graph of $y = \cos x$ (Figure 18.7) in each of the following cases.

(a)   between $x = 0$ and $x = \pi/2$;

(b)   between $x = \pi/2$ and $x = \pi$;

(c)   between $x = 0$ and $x = \pi$;

(d)   between $x = \pi/2$ and $x = 0$.

(a)   In this case the area under the graph is

$$\int_0^{\pi/2} \cos(x)\, dx = [\sin(x)]_0^{\pi/2} = \sin(\pi/2) - \sin(0) = 1$$

(b)   In this case the area under the graph is

$$\int_{\pi/2}^{\pi} \cos(x)\, dx = [\sin(x)]_{\pi/2}^{\pi} = \sin(\pi) - \sin(\pi/2) = -1$$

(c)   In this case the area under the graph is

$$\int_0^{\pi} \cos(x)\, dx = [\sin(x)]_0^{\pi} = 0$$

(d)   In this case the area under the graph is

$$\int_{\pi/2}^0 \cos(x)\, dx = [\sin(x)]_{\pi/2}^0 = \sin(0) - \sin(\pi/2) = -1 \quad \square$$

According to the convention used in this book, the total area under the graph in Figure 18.6 is $A_1 + A_2$ where A_2 is a negative quantity. Other authors sometimes adopt a different convention in which the area of each region is defined to be positive and the area under the graph is then given by $A_1 - A_2$.

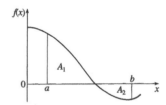

Figure 18.6   The area under a graph in any region where the graph is below the horizontal axis, such as A_2, will be negative.

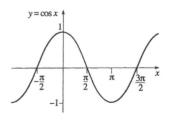

Figure 18.7   The graph of the function $\cos x$.

Question T7
Find the sum of the magnitudes of the areas enclosed by the graph of $y = bx^5$ and the x-axis, between $x = -1$ m and $x = 1$ m, given that $b = 2$ m^{-4}. □

18.3 Integrals of simple functions

Integration, the process of analysing and evaluating definite and indefinite integrals, can be very challenging. In practice everybody who uses integration regularly remembers the indefinite integrals of a variety of basic functions (x^n, $\sin x$, $\exp (x)$, etc.) and knows where to look up the indefinite integrals of some slightly more complicated functions. They also know various rules and techniques for finding the indefinite integrals of combinations of those basic functions. This section introduces the standard integrals and the simplest rules for combining them. Some additional techniques are introduced in the next chapter. Scientists are increasingly making use of algebraic computing packages to carry out the task of integration, but it is still necessary to appreciate the basic results of the subject in order to understand the warnings and limitations that often attend computer evaluations.

18.3.1 *Some standard integrals*

Tables 18.1a and 18.1b list a number of functions with which you should already be familiar along with their indefinite integrals. These may be classed as **standard integrals**. Each of the table entries can be checked by differentiating the indefinite integral to show that it is equal to the integrand. If you are going to use calculus frequently you will need to know these integrals or at least know where you can look them up quickly. When using the tables it is important to remember the following points:

- n and k are (given) constants and C is an arbitrary constant.
- The functions in Table 18.1b are special cases of those in Table 18.1a, corresponding to $k = 1$.
- In each of the trigonometric functions, x must be an angle in radians or a dimensionless real variable.

If you are reasonably proficient at differentiation many of the entries in Table 18.1 should come as no surprise. Indeed, we have used some of them earlier in the chapter on this assumption. The integrals of the reciprocal trigonometric functions (cosec, sec and cot) may seem a bit strange, but they are rarely encountered in most fields of physics so we will not dwell on them here. Far more common, and definitely worth dwelling on, is the integral of $1/x$. As you can see it has been listed as $\log_e |x| + C$; the use of the *modulus*, $|x|$, should remind you that the argument of the logarithmic function must be positive, but it also indicates that the result holds true even when x itself is negative.

Table 18.1   Some standard indefinite integrals

(a)		(b)	
$f(x)$	$\int f(x)\,dx$	$f(x)$	$\int f(x)\,dx$
k (a constant)	$kx + C$	1	$x + C$
$kx^n,\ \ n \neq -1$	$k\dfrac{x^{n+1}}{n+1} + C$	$x^n,\ \ n \neq -1$	$\dfrac{x^{n+1}}{n+1} + C$
$k\dfrac{1}{x},\ \ x \neq 0$	$k\log_e |x| + C$	$\dfrac{1}{x},\ \ x \neq 0$	$\log_e |x| + C$
$\sin(kx)$	$-\dfrac{1}{k}\cos(kx) + C$	$\sin(x)$	$-\cos(x) + C$
$\cos(kx)$	$\dfrac{1}{k}\sin(kx) + C$	$\cos(x)$	$\sin(x) + C$
$\tan(kx)$	$\dfrac{1}{k}\log_e |\sec(kx)| + C$	$\tan(x)$	$\log_e |\sec(x)| + C$
$\operatorname{cosec}(kx)$	$\dfrac{1}{k}\log_e \left|\tan\left(\dfrac{kx}{2}\right)\right| + C$	$\operatorname{cosec}(x)$	$\log_e \left|\tan\left(\dfrac{x}{2}\right)\right| + C$
$\sec(kx)$	$\dfrac{1}{k}\log_e |\sec(kx) + \tan(kx)| + C$	$\sec(x)$	$\log_e |\sec(x) + \tan(x)| + C$
$\cot(kx)$	$\dfrac{1}{k}\log_e |\sin(kx)| + C$	$\cot(x)$	$\log_e |\sin(x)| + C$
$\sec^2(kx)$	$\dfrac{1}{k}\tan(kx) + C$	$\sec^2(x)$	$\tan(x) + C$
e^{kx}	$\dfrac{1}{k}e^{kx} + C$	e^x	$e^x + C$
$\log_e(kx)$	$x\log_e(kx) - x + C$	$\log_e(x)$	$x\log_e(x) - x + C$

Using Table 18.1 find the following:

(a) $\displaystyle\int dx$ (b) $\displaystyle\int x^4\,dx$ (c) $\displaystyle\int_0^2 x^4\,dx$ (d) $\displaystyle\int_{-2}^2 x^4\,dx$

(e) $\displaystyle\int t^4\,dt$ (f) $\displaystyle\int_{-2}^2 t^4\,dt$ (g) $\displaystyle\int_1^2 \dfrac{dx}{x}$

(a) $\displaystyle\int dx = \int 1dx = x + C$ from Table 18.1b.

(b)   With $n = 4$ the result for x^n in Table 18.1b gives

$$\int x^4\,dx = \frac{x^5}{5} + C$$

(c) $\displaystyle\int_0^2 x^4\,dx = \left[\frac{x^5}{5}\right]_0^2 = \frac{2^5}{5} - 0 = \frac{32}{5}$

Table 18.1b is obtained from Table 18.1a by setting $k = 1$.

If you are going to integrate frequently, you should commit to memory at least the first five and the last two entries in Table 18.1b.

You should note that you may often see expressions such as

$$\int_1^2 \frac{dx}{x} = \Big[\log_e x \Big]_1^2$$

i.e. the modulus bars on the x are omitted since we know x is positive here.

(d) $\displaystyle\int_{-2}^2 x^4 \, dx = \left[\frac{x^5}{5} \right]_{-2}^2 = \frac{32}{5} - \left(\frac{-32}{5} \right) = \frac{64}{5}$

Now t and x are just *dummy variables* so that the results for parts (e) and (f) are identical to those of parts (b) and (d), respectively:

(e) $\displaystyle \frac{t^5}{5} + C$

(f) $\displaystyle \frac{32}{5} - \left(\frac{-32}{5} \right) = \frac{64}{5}$

(g) $\displaystyle\int_1^2 \frac{dx}{x} = \left[\log_e x \right]_1^2 = \log_e 2 - \log_e 1 = \log_e 2 = 0.69$ ☐

By differentiating the given indefinite integrals, confirm the correctness of the table entries for (a) e^{kx} and (b) $\log_e (kx)$.

(a) $\displaystyle \frac{d}{dx} \left(\frac{1}{k} e^{kx} + C \right) = \frac{k}{k} e^{kx} = e^{kx}$

(b) Using the product rule of differentiation

$$\frac{d}{dx} \left[x \log_e (kx) - x + C \right]$$

$$= \log_e (kx) \frac{d}{dx} (x) + x \frac{d}{dx} \left[\log_e (kx) \right] - 1$$

and remembering that

$$\frac{d}{dx} \left[\log_e (kx) \right] = \frac{d}{dx} \left(\log_e k + \log_e x \right) = \frac{d}{dx} (\log_e x) = \frac{1}{x}$$

we find

$$\frac{d}{dx} \left[x \log_e (kx) - x + C \right] = \log_e (kx) + x \left(\frac{1}{x} \right) - 1$$

$$= \log_e (kx) \qquad\qquad ☐$$

Question T8

Using Table 18.1 find:

(a) $\int 8x^{-1}\,dx$

(b) $\int_1^3 8x^{-1}\,dx$

(c) $\int \sec^2(4x)\,dx$

(d) $\int_0^{\pi/16} \sec^2(4x)\,dx$ □

The following question illustrates how such calculations may arise in physics.

In Subsection 18.2.3 we discussed the expansion of a gas in a cylinder and we showed that W, the work done by the gas as it expands, can be expressed in the form

$$W = \int_{V_a}^{V_b} P(V)\,dV$$

The exact form of the function $P(V)$ depends on the nature of the gas, but for a fixed quantity of *ideal gas*, expanding at a fixed temperature

$$P(V) = \frac{A}{V} \quad \text{for some constant } A$$

and hence $\qquad W = \int_{V_a}^{V_b} \frac{A}{V}\,dV$

For a fixed quantity of *ideal gas* at fixed temperature, $PV = nRT$, where n, R and T are constant, so $A = nRT$.

Evaluate W for such a gas, in terms of V_a and V_b.

$$W = \int_{V_a}^{V_b} \frac{A}{V}\,dV = A\left[\log_e |V|\right]_{V_a}^{V_b} \quad (\text{☞})$$

i.e. $W = A[\log_e(V_b) - \log_e(V_a)] = A\log_e(V_b/V_a)$ (☞) □

V_a and V_b are both positive so that

$$|V_a| = V_a$$

and

$$|V_b| = V_b$$

Note that in the final answer for W, the argument of the logarithmic function is dimensionless. This must always be the case in any kind of physical calculation. After all, it makes sense to write $\log_e(6)$ but what sense could be made of $\log_e(6\,\text{m})$ or $\log_e(6\,\text{kg})$? It is worth noting in this context that the process of integration may sometimes lead you to expressions of the kind $\log_e(x) + C$, where x is a quantity that has dimensions. When this happens you can recover a dimensionless argument by writing the constant of integration in the form $-\log_e D$, for an appropriate constant D, and then using the identity

$$\log_e x - \log_e D = \log_e(x/D)$$

The argument is then dimensionless, providing x and D have the same dimensions.

When we arrive at such an answer it is always wise to check that it is sensible. For example, consider what happens when V_b is small, large, a little bigger than V_a, or a little less than V_a. Does the answer make sense?

18.3.2 Combining integrals

Very seldom do we have functions to integrate which are exactly those listed in Table 18.1. The useful general results which follow allow us much more flexibility.

Integrating the sum of two functions

Suppose that we wish to obtain an indefinite integral of the function

$$f(x) = 4x^3 + 2x$$

The function x^4 has a derivative $4x^3$, and the function x^2 has a derivative $2x$, so that the function $x^4 + x^2$ has a derivative $4x^3 + 2x$. Hence

$$\int (4x^3 + 2x)\, dx = x^4 + x^2 + C$$

Note the appearance of only one constant of integration. Also note that the integral of the sum is a sum of the indefinite integrals of the terms in the sum. This is a particular case of a general rule:

The **sum rule** $\int [g(x) + h(x)]\, dx = \int g(x)\, dx + \int h(x)\, dx$ (14)

Find the following integrals:

(a) $\int (\sin x + \cos x)\, dx$ (b) $\displaystyle\int_0^{\pi/2} (\sin x + \cos x)\, dx$

Using Table 18.1 we obtain:

(a) $\int (\sin x + \cos x)\, dx = -\cos x + \sin x + C$

(b) $\displaystyle\int_0^{\pi/2} [\sin x + \cos x]\, dx = [-\cos x + \sin x]_0^{\pi/2}$

$$= \left(-\cos\frac{\pi}{2} + \sin\frac{\pi}{2}\right) - (-\cos 0 + \sin 0)$$

$$= (-0 + 1) - (-1 + 0)$$

$$= 1 + 1 = 2 \quad \square$$

Notice once again that we can choose the arbitrary constant C to be 0 when evaluating definite integrals.

Integrating a constant multiple of a function

Another general rule for integrating combinations, albeit of a rather trivial sort, is given by the following constant multiple rule:

The **constant multiple rule** $\int kf(x)\, dx = k\int f(x)\, dx$ (15)

This result has already been used in several places without comment.

Integrating a linear combination of functions

A **linear combination** of two functions $f(x)$ and $g(x)$ is a function of the form $kf(x) + lg(x)$ where k and l are constants. We can find a general rule for integrating such combinations by combining the last two results (given in Equations 14 and 15)

$$\int [kf(x) + lg(x)] \, dx = k \int f(x) \, dx + l \int g(x) \, dx \qquad (16)$$

Using Table 18.1 and Equation 16 evaluate the following:

(a) $\int [6 \cos(x) - 2 \sin(x)] \, dx$ (b) $\displaystyle\int_{\pi/4}^{\pi/2} [6 \cos(x) - 2 \sin(x)] \, dx$

(a) $\int [6 \cos(x) - 2 \sin(x)] \, dx = 6 \sin(x) + 2 \cos(x) + C$

(b) $\displaystyle\int_{\pi/4}^{\pi/2} [6 \cos(x) - 2 \sin(x)] \, dx = [6 \sin(x) + 2 \cos(x)]_{\pi/4}^{\pi/2}$

$$= \left(6 \sin \frac{\pi}{2} + 2 \cos \frac{\pi}{2}\right) - \left(6 \sin \frac{\pi}{4} + 2 \cos \frac{\pi}{4}\right)$$

$$= (6 + 0) - \left(\frac{6}{\sqrt{2}} + \frac{2}{\sqrt{2}}\right) = 6 - \frac{8}{\sqrt{2}} = 0.343 \quad \square$$

Question T9

Find the following integrals:

(a) $\int \sqrt{x}\,(x^2 - 3) \, dx$ (*Hint*: Expand the brackets.)

(b) $\displaystyle\int \left(\frac{5}{t^{2/3}} + \frac{2}{t^{1/3}}\right) dt$ (c) $\displaystyle\int_1^3 \frac{(x+2)^3}{\sqrt{x}} \, dx$

(d) $\displaystyle\int \frac{t^2 - 6t + 1}{t^4} \, dt$ (e) $\int [-3 \cos(x) + 2 \sec^2(x)] \, dx$

(f) $\displaystyle\int_0^{\pi/4} \tan^2(\theta) \, d\theta$ (*Hint*: $\sec^2(\theta) = \tan^2(\theta) + 1$) $\square$

Question T10

Evaluate the integral

$$\int_1^2 [\sin(\pi x) + \pi \log_e(3x)] \, dx \quad \square$$

18.3.3 Further standard integrals

The integrals listed in Table 18.1 may all be classed as standard integrals, that is to say they are well known results that may be used in their own right or in the evaluation of other, more complicated, integrals. In this subsection we extend the list of standard integrals by adding the more complicated cases covered in Table 18.2. More extensive lists of standard integrals can be found in various reference works. Some of these contain hundreds of pages of integrals!

A note about inverse trigonometric functions and hyperbolic functions

The *inverse trigonometric functions* arcsin (x), arccos (x) and arctan (x) are the inverses of the functions sin (x), cos (x) and tan (x), respectively, so that, for example arcsin (sin (θ)) = θ, provided $-\pi/2 \leqslant \theta \leqslant \pi/2$. The alternative notations: $\sin^{-1}(x)$, $\cos^{-1}(x)$ and $\tan^{-1}(x)$ and asin (x), acos (x) and atan (x) are also in common use.

The *hyperbolic functions* are defined by

$$\sinh(x) = \frac{e^x - e^{-x}}{2} \qquad \text{(pronounced 'shine } x\text{')}$$

$$\cosh(x) = \frac{e^x + e^{-x}}{2} \qquad \text{(pronounced 'cosh } x\text{')}$$

$$\tanh(x) = \frac{e^x - e^{-x}}{e^x + e^{-x}} = \frac{\sinh(x)}{\cosh(x)} \qquad \text{(pronounced 'than } x\text{', 'th' as in 'beneath')}$$

Their inverses arcsinh (x), arccosh (x) and arctanh (x) are often written as $\sinh^{-1}(x)$, $\cosh^{-1}(x)$ and $\tanh^{-1}(x)$. For further information see Chapter 7. ☐

It should be noted that several of the integrals given in Table 18.2 are subject to restrictions. For instance, arcsin (ax/b) is only defined for $-|b/a| \leqslant x \leqslant |b/a|$, and those expressions enclosed by square roots must be positive for all relevant values of x. We will not spell all these restrictions out in full, though we will have more to say on this topic in Subsection 18.4.3. However, it is important to keep your wits about you whenever you use tabulated results of this kind or the equivalent results that might be supplied by a computer program.

Using Table 18.2, evaluate $\displaystyle\int_{-1}^{1} \frac{1}{1+x^2} \, dx$.

From Table 18.2, with $a = b = 1$, we find

$$\int_{-1}^{1} \frac{1}{1+x^2} \, dx = \left[\arctan x \right]_{-1}^{1} = \arctan(1) - \arctan(-1)$$

$$= \frac{\pi}{4} - \left(\frac{-\pi}{4}\right) = \frac{\pi}{2} \quad \square$$

Table 18.2 Some standard indefinite integrals

This formidable list is included for your convenience. You are not expected to memorise the results!

$f(x)$	$\int f(x)\,dx$
$\dfrac{1}{ax+b}$	$\dfrac{1}{a}\log_e \lvert ax+b\rvert + C$
$\dfrac{1}{b^2+a^2x^2}\quad (a>0)$	$\dfrac{1}{ab}\arctan\left(\dfrac{ax}{b}\right)+C$
$\dfrac{1}{a^2x^2-b^2}\quad (a>0)$	$\dfrac{1}{2ab}\log_e\left\lvert\dfrac{ax-b}{ax+b}\right\rvert+C$
$\dfrac{1}{\sqrt{b^2+a^2x^2}}\quad (a>0)$	$\dfrac{1}{a}\operatorname{arcsinh}\left(\dfrac{ax}{b}\right)+C=\dfrac{1}{a}\log_e\left\lvert ax+\sqrt{a^2x^2+b^2}\right\rvert+C$
$\dfrac{1}{\sqrt{b^2-a^2x^2}}\quad (a>0)$	$\dfrac{1}{a}\arcsin\left(\dfrac{ax}{b}\right)+C$
$\dfrac{1}{\sqrt{a^2x^2-b^2}}\quad (a>0)$	$\dfrac{1}{a}\operatorname{arccosh}\left(\dfrac{ax}{b}\right)+C=\dfrac{1}{a}\log_e\left\lvert ax+\sqrt{a^2x^2-b^2}\right\rvert+C$
$\dfrac{x}{\sqrt{b^2+a^2x^2}}$	$\left(\dfrac{b^2}{a^2}+x^2\right)\dfrac{1}{\sqrt{b^2+a^2x^2}}+C=\dfrac{1}{a^2}\sqrt{b^2+a^2x^2}+C$
$\dfrac{x}{\sqrt{b^2-a^2x^2}}$	$\left(-\dfrac{b^2}{a^2}+x^2\right)\dfrac{1}{\sqrt{b^2-a^2x^2}}+C=\dfrac{1}{a^2}\sqrt{b^2-a^2x^2}+C$
$\dfrac{x}{\sqrt{a^2x^2-b^2}}$	$\left(-\dfrac{b^2}{a^2}+x^2\right)\dfrac{1}{\sqrt{a^2x^2-b^2}}+C=\dfrac{1}{a^2}\sqrt{a^2x^2-b^2}+C$
$\sqrt{b^2+a^2x^2}$	$\dfrac{1}{a}\left[\dfrac{1}{2}ax\sqrt{a^2x^2-b^2}-\dfrac{1}{2}b^2\log_e\left\lvert ax+\sqrt{a^2x^2-b^2}\right\rvert\right]+C$
$\sqrt{b^2-a^2x^2}\quad (a>0)$	$\dfrac{1}{a}\left[\dfrac{1}{2}ax\sqrt{b^2-a^2x^2}+\dfrac{1}{2}b^2\arcsin\left(\dfrac{ax}{b}\right)\right]+C$
$\sqrt{a^2x^2-b^2}$	$\dfrac{1}{a}\left[\dfrac{1}{2}ax\sqrt{a^2x^2-b^2}-\dfrac{1}{2}b^2\log_e\left\lvert ax+\sqrt{a^2x^2-b^2}\right\rvert\right]+C$
$\exp(ax)\sin(bx)$	$\dfrac{\exp(ax)}{a^2+b^2}\left[a\sin(bx)-b\cos(bx)\right]+C$
$\exp(ax)\cos(bx)$	$\dfrac{\exp(ax)}{a^2+b^2}\left[a\cos(bx)-b\sin(bx)\right]+C$

Study comment
Many of the results in Table 18.2 can be derived by considering inverse trigonometric functions, though the derivations can be tricky and often involve the technique of *implicit differentiation*. We give an example of one of these derivations below (in Example 2) and then ask you to perform another for yourself. However, if you are unfamiliar with implicit differentiation you may wish to omit both and go directly to Question T11.

Example 2
Confirm by differentiation that (as implied by Table 18.2) the indefinite integral of $\dfrac{1}{\sqrt{1 - x^2}}$ is arcsin $(x) + C$.

Solution Let $y = \arcsin(x)$ then $\sin(y) = x$ and, upon differentiating both sides of this equation with respect to x, we obtain

$$\cos(y)\,\frac{dy}{dx} = 1 \qquad\qquad (\text{☞})$$

Here we have used the technique of *implicit differentiation*, which is based on the *chain rule* for differentiating a *function of a function*. Essentially, we are saying that if $u = \cos(y)$, and y is a function of x, then

$$\frac{du}{dx} = \frac{du}{dy} \times \frac{dy}{dx}$$

But $x = \sin(y)$, so we can use the identity $\sin^2(y) + \cos^2(y) = 1$ to write

$$\cos(y) = \pm\sqrt{1 - \sin^2(y)} = \pm\sqrt{1 - x^2}$$

The sign ambiguity $(\pm)$ has arisen because $\sqrt{x}$ represents a *positive* square root in this chapter.

We also know that $\dfrac{dy}{dx} = \dfrac{1}{\cos(y)}$

so, $\dfrac{dy}{dx} = \dfrac{\pm 1}{\sqrt{1 - x^2}}$ (17)

However, the arcsin function (see Figure 18.8) is conventionally defined in such a way that its gradient (i.e. its first derivative) is positive everywhere, so we must select the positive sign in Equation 17.

Thus $\dfrac{d}{dx}[\arcsin(x)] = \dfrac{1}{\sqrt{1 - x^2}}$

It follows that

$$\int \frac{1}{\sqrt{1 - x^2}}\,dx = \arcsin(x) + C$$

as implied by Table 18.2. □

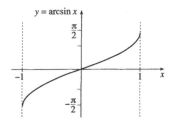

$y = \arcsin x$

Figure 18.8 The graph of arcsin (x).

Show that (as implied by Table 18.2) $\int \dfrac{1}{1+x^2}\,dx = \arctan(x) + C$

Let $y = \arctan(x)$ then $\tan(y) = x$ and, differentiating both sides of the equation with respect to x, we have

$$\sec^2(y)\,\frac{dy}{dx} = 1$$

But $x = \tan(y)$, so we can use the identity $\sec^2(y) = 1 + \tan^2(y)$ to write

$$\sec^2(y) = 1 + \tan^2(y) = 1 + x^2$$

thus $\dfrac{d}{dx}\arctan(x) = \dfrac{1}{1+x^2}$ (☞)

so $\displaystyle\int \dfrac{1}{1+x^2}\,dx = \arctan(x) + C$ □

Note that there is no sign ambiguity in this case because we have not taken a square root.

Question T11
Find the following integrals:

(a) $\displaystyle\int_0^2 e^{3-2x}\,dx$

(b) $\displaystyle\int \left(\frac{3x+2}{x+4}\right)\,dx$

(*Hint*: Write $3x + 2 = 3(x + 4) - 10$)

(c) $\displaystyle\int_2^4 \frac{1}{5x^2 + 20}\,dx$

(d) $\displaystyle\int_0^{1/3} \frac{3}{(8 - 9x^2)^{1/2}}\,dx$ □

18.4   **More about definite integrals**

18.4.1   *Properties of definite integrals*

The definite integral has the following mathematical properties, each of which makes good sense if you interpret it in terms of the (signed) area under a graph.

Properties of definite integrals:

1. $\displaystyle\int_a^a f(x)\,dx = 0$ (18)

2. $\displaystyle\int_b^a f(x)\,dx = -\int_a^b f(x)\,dx$ (19)

3. $$\int_a^b f(x)\,dx = \int_a^c f(x)\,dx + \int_c^b f(x)\,dx \quad \text{where } a \leqslant c \leqslant b \qquad (20)$$

4. $$\int_a^b [kf(x) + lg(x)]\,dx = k\int_a^b f(x)\,dx + l\int_a^b g(x)\,dx \qquad (21)$$

5. If $f(x) \geqslant 0$ for all x in the interval $a \leqslant x \leqslant b$ then

$$\int_a^b f(x)\,dx \geqslant 0 \qquad (22)$$

6. If $m \leqslant f(x) \leqslant M$ for all x in the interval $a \leqslant x \leqslant b$ then

$$m(b - a) \leqslant \int_a^b f(x)\,dx \leqslant M(b - a) \qquad (23)$$

Properties 1 to 3 are derived below using the fundamental theorem of calculus. Property 4 is the extension to definite integrals of the rules for combining indefinite integrals that were introduced in Subsection 18.3.2. It too can be derived using the fundamental theorem of calculus. Properties 5 and 6 follow from the definition of a definite integral as the limit of a sum. They are not derived in this chapter, but they may be treated as tutorial exercises.

1. If $b = a$ then $\displaystyle\int_a^a f(x)\,dx = F(a) - F(a) = 0$

2. $$\int_b^a f(x)\,dx = F(a) - F(b) = -(F(b) - (F(a)))$$

$$= -\int_a^b f(x)\,dx$$

3. $$\int_a^c f(x)\,dx + \int_c^b f(x)\,dx = F(c) - F(a) + F(b) - F(c)$$

$$= F(b) - F(a) = \int_a^b f(x)\,dx$$

Property 6 can be useful in cases where it is impossible to calculate the integral exactly. In such cases it is often desirable to obtain an estimate for the integral. The following example illustrates the method:

By considering the maximum and minimum values of $\sqrt{x^2+1}$ in the interval $1 \leqslant x \leqslant 3$, show that

$$2\sqrt{2} \leqslant \int_1^3 \sqrt{x^2+1}\, dx \leqslant 2\sqrt{10}\,.$$

Figure 18.9 shows the graph of $y = \sqrt{x^2+1}$, and clearly the least value of $\sqrt{x^2+1}$ in the given interval is $\sqrt{2}$ (when $x = 1$) and the greatest value is $\sqrt{10}$ (when $x = 3$). Thus, from Property 6, we have

$$2\sqrt{2} \leqslant \int_1^3 \sqrt{x^2+1}\, dx \leqslant 2\sqrt{10} \quad \square$$

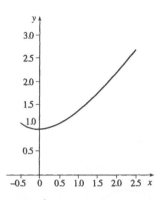

Figure 18.9 The graph of $y = \sqrt{x^2+1}$.

Question T12

If $f(x) \geqslant g(x)$ for all x where $a \leqslant x \leqslant b$ show that

$$\int_a^b f(x)\, dx \geqslant \int_a^b g(x)\, dx$$

(*Hint*: Consider $f(x) - g(x)$.) $\quad \square$

Question T13

A bullet of mass m is fired horizontally into a box full of sand. As the bullet travels through the sand its motion is opposed by a resistive force of magnitude

$$F(x) = \sqrt[3]{(a^3 - b^3 x^3)}\, m$$

where a and b are positive constants, and x is the distance the bullet has travelled through the sand. In a particular test, it is found that the bullet travels a distance $L = a/b$ through the sand before being brought to rest.

Using an appropriate limit of a sum, write down an expression for the work done in bringing the bullet to rest. Rewrite your answer as a definite integral, but do not attempt to evaluate it. Given that the work done in bringing the bullet to rest is equal to minus the initial kinetic energy of the bullet when it enters the sand, write down an upper estimate of that initial kinetic energy, in terms of a, b, and m. (*Hint*: The work done when a constant force F_x moves its point of application through a displacement Δx is generally $\Delta W = F_x \Delta x$. In this case the force is variable, and only its magnitude is given, but the fact that it brings the bullet to rest shows that it acts in the opposite direction to the displacement and therefore does a *negative* amount of work.) $\quad \square$

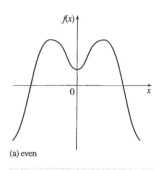

(a) even

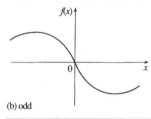

(b) odd

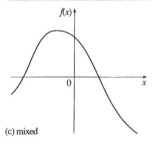

(c) mixed

Figure 18.10 (a) An even function. (b) An odd function. (c) A function of mixed symmetry.

This condition ensures that if $f(x)$ is meaningful then so is $f(-x)$.

18.4.2   Odd, even and periodic integrands

Odd and even functions

When looking at the graphs of functions it is easy to see that some functions are symmetric about a vertical line through the origin and others are not. Three examples of this are shown in Figure 18.10. Part (a) of the figure shows a function $f(x)$ that is obviously symmetric about the origin – the part of the graph drawn to the left of the origin is the mirror image of the part to the right of the origin. Mathematically, such a function is called an **even function** and is characterised by the property

$$f(-x) = f(x) \quad \text{for all } x \text{ in the domain of } f(x) \qquad \text{(even function)}$$

Even functions are sometimes referred to as **symmetric functions.**

Figure 18.10b shows a different function. It certainly isn't an even function yet there is clearly some sort of relationship between the parts of the function shown to the left of the origin and those to the right. Functions of this sort are called **odd functions** and are characterised by the property

$$f(-x) = -f(x) \text{ for all } x \text{ in the domain of } f(x) \qquad \text{(odd function)}$$

Odd functions are sometimes referred to as **antisymmetric functions**.

Figure 18.10c shows a function $f(x)$ which is neither odd nor even. There is no simple relationship between $f(-x)$ and $f(x)$ for this function, though it is interesting (and occasionally useful) to note that any function with a symmetric domain, (☞) including this one, can be written as the sum of an even function and an odd function. To see why this is so consider the following identity which holds true for any function $f(x)$

$$f(x) = \tfrac{1}{2}[f(x) + f(-x)] + \tfrac{1}{2}[f(x) - f(-x)] \qquad (24)$$

The first term on the right-hand side of this identity, $[f(x) + f(-x)]/2$, is an even function as you can see if you replace x by $-x$ everywhere within it; and, by the same argument, the second term, $[f(x) - f(-x)]/2$, is an odd function. Thus, an arbitrary function can indeed be written as a sum of even and odd parts. Functions which are neither purely even nor purely odd are sometimes referred to as functions of **mixed symmetry**.

Classify each of the following functions as even, odd or of mixed symmetry:

(a) x^3 (b) $2x + 4x^4 + 1$ (c) $\sin(x)$ (d) $2\cos(3x)$

(a)   If $f(x) = x^3$, then $f(-x) = (-x)^3 = -x^3 = -f(x)$. So x^3 is an odd function of x.

(b) Similarly, if $f(x) = 2x + 4x^4 + 1$, then

$$f(-x) = 2(-x) + 4(-x)^4 + 1 = -2x + 4x^4 + 1$$

which is neither $f(x)$ nor $-f(x)$. So $2x + 4x^4 + 1$ is of mixed symmetry.

(c) $\sin(-x) = -\sin(x)$, so $\sin(x)$ is an odd function.

(d) $2\cos(-3x) = 2\cos(3x)$, so $2\cos(3x)$ is an even function. ☐

The examples we have just considered have been fairly straightforward. In more complicated cases involving the products of two or more functions it is often useful to recall the following rules:

● even function × even function = even function

● odd function × odd function = even function

● odd function × even function = odd function

These are similar to the rules for products of signs.

Integrals of even and odd functions
If you think of a definite integral in terms of the area under a graph, then it should be clear from Figure 18.10 that

If $f(x)$ is an *odd* function, then

$$\int_{-a}^{a} f(x)\,dx = 0 \qquad (25)$$

If $f(x)$ is an *even* function, then

$$\int_{0}^{a} f(x)\,dx = 2\int_{-a}^{a} f(x)\,dx \qquad (26)$$

Note that in both these integrals the **range of integration** (from $-a$ to a) is symmetric about the origin. Don't expect these results to work in more general situations.

These two results can often be of help when evaluating definite integrals, especially the first result (Equation 25) which sometimes removes the need to do any hard work at all.

Evaluate $\displaystyle\int_{-\pi}^{\pi} 4x^2 \sin^3(2x)\,dx$

In this case the range of integration is symmetric about the origin and the integrand is an *odd function*. Consequently, we can say that the integral is zero without any further work. ☐

$\sin x$ is odd, $\sin^2 x$ is even, $\sin^3 x$ is odd and x^2 is even. So, together $x^2 \sin^3 x$ is odd.

Odd and even functions arise in many fields of physical science so you should always ask yourself if you can use symmetry to simplify the process of definite integration. This is especially true in quantum physics where you will often be asked to integrate functions of known symmetry. Another property of many of the functions that have to be integrated in a course on quantum physics is *periodicity*.

Periodic functions and their integrals

A function $f(x)$ is said to be **periodic** if there exists a constant $k\,(>0)$ such that

$$f(x) = f(x + nk) \quad \text{for all } x \text{ and for all integers}$$
$$\text{(i.e. whole numbers) } n$$

The smallest value of k for which this condition holds true is called the **period** of the function.

The most obvious examples of periodic functions are the trigonometric functions $\sin(x)$ and $\cos(x)$, both of which have period 2π, since $\sin(x) = \sin(x + 2n\pi)$ and $\cos(x) = \cos(x + 2n\pi)$. The graphs of all periodic functions, like the graphs of $\sin(x)$ and $\cos(x)$ in Figure 18.11, consist of regular repetitions of a single basic unit that covers one period. As a consequence of their repetitive nature we can state the following:

If $f(x)$ is a periodic function of period k, then

$$\int_a^{a+k} f(x)\, dx = \int_0^k f(x)\, dx \tag{27}$$

and, if n is any integer, then $\displaystyle\int_0^{nk} f(x)\, dx = n\int_0^k f(x)\, dx$ (28)

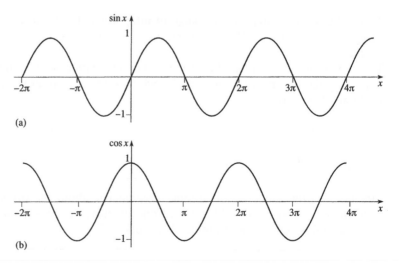

Figure 18.11 The graphs of the periodic functions (a) sin (x), and (b) cos (x).

The essential point being made in Equations 27 and 28 is that when dealing with periodic functions we learn just as much by integrating over one full period as we do by integrating over several full periods. Moreover, if we are integrating over a full period, it doesn't matter where that period begins, a general point a is just as good as the origin $x = 0$.

Question T14

Which of the following are equal to zero? Explain your answers. (This is an easy question, it should not require any long calculations but you will need to think.)

(a) $\displaystyle\int_{-5}^{5} 3x \, dx$ (b) $\displaystyle\int_{-\pi}^{\pi} \sin^2 (x/2) \, dx$

(c) $\displaystyle\int_{3\pi}^{7\pi} [3 \sin(3x) + \sin^3 (x)] \, dx$ $\square$

18.4.3   Improper and divergent integrals

In the study of physical science it is not uncommon to encounter definite integrals that involve infinity. In this subsection we consider two of the ways in which such integrals arise: integrals in which one or both of the limits tends towards infinity, and integrals in which the integrand itself tends towards infinity at some point in the range of integration. These are somewhat specialised topics that your tutor might advise you to omit at this stage.

Integration over an infinite range

Infinity (∞) is not a real number so, strictly speaking, we cannot use it as one of the limits of integration that we introduced earlier when discussing the limit of a sum. However, you will often see integrals with an infinite upper or lower limit. Such usage is justified by the following definitions

$$\int_{-\infty}^{b} f(x) \, dx = \lim_{a \to -\infty} \left(\int_{a}^{b} f(x) \, dx \right) \tag{29}$$

$$\int_{a}^{\infty} f(x) \, dx = \lim_{b \to \infty} \left(\int_{a}^{b} f(x) \, dx \right) \tag{30}$$

The integrals on the left are called **improper integrals**. Although written as definite integrals they are not properly defined by the usual process of taking the limit of a sum; rather one or more additional limits must be taken (as indicated on the right). If the limits on the right exist and are finite the corresponding improper integral is said to be **convergent**. If the relevant limit does not exist, or if it tends to $\pm\infty$, then the improper integral is said to be **divergent**.

It is easy to find divergent integrals; for example consider the following:

$$\int_{0}^{\infty} \cos (x) \, dx = \lim_{b \to \infty} \left(\int_{0}^{b} \cos (x) \, dx \right) = \lim_{b \to \infty} \left(\left[\sin (x) \right]_{0}^{b} \right) = \lim_{b \to \infty} \left[\sin (b) \right]$$

In this case the integral diverges because as b tends to infinity the value of $\sin(b)$ oscillates between 1 and -1. There is no unique limiting value, so the limit does not exist.

On the other hand, it is equally easy to find integrals that converge. For instance, if we ignore all forces other than the Earth's gravitational pull, the minimum launch speed v that will just allow an object of fixed mass m to escape from the surface of the Earth to a point infinitely far away (i.e. the escape speed from the Earth) is given by

$$\frac{1}{2}mv^2 = \int_R^\infty \frac{GmM}{r^2}\, dr = \left[-\frac{GmM}{r}\right]_R^\infty = (0) - \left(-\frac{GmM}{R}\right) = \frac{GmM}{R}$$

so $v = \sqrt{\dfrac{2GM}{R}}$

where G is the gravitational constant, M is the mass of the Earth and R is the radius of the Earth. In this case the improper integral converges and, as is customary in such cases, we have omitted the formal step of writing down the limits that are required to define properly that integral. In effect we have treated infinity as though it were a very large real number.

Improper integrals must always be treated with care and generally require case by case consideration. However, one general rule which it is useful to keep in mind is the following:

If $I = \displaystyle\int_a^\infty f(x)\, dx$

where $f(x)$ is continuous over the range of integration and if, for large values of x,

$$f(x) = \frac{g(x)}{x^n} \text{ where } g(x) \text{ is finite and non-zero, and } n > 1,$$

then the integral I is *convergent*.

Integration with an infinite integrand

The expression $\dfrac{1}{\sqrt{1 - x^2}}$ is well defined only if $-1 < x < 1$, yet it is not unusual to see integrals such as

$$I = \int_0^1 \frac{1}{\sqrt{1 - x^2}}\, dx \quad \text{or} \quad I = \int_{-1}^0 \frac{1}{\sqrt{1 - x^2}}\, dx \quad \text{or even} \quad I = \int_{-1}^1 \frac{1}{\sqrt{1 - x^2}}\, dx$$

What can such integrals mean when the integrand isn't even defined at all points in the range of integration? In fact, they are all improper integrals which can be interpreted as limits of properly defined definite integrals. For example,

$$I = \int_0^1 \frac{1}{\sqrt{1 - x^2}}\, dx = \lim_{\varepsilon \to 0}\left(\int_0^{1-\varepsilon} \frac{1}{\sqrt{1 - x^2}}\, dx\right)$$

where $\lim\limits_{\varepsilon \to 0}$ indicates that we are considering the limit as the positive quantity ε tends to zero, i.e. as b, the upper limit, tends to 1 from below. With this understanding, the integrand is one of the standard integrands listed earlier in Table 18.2, so we can use the corresponding indefinite integral given in that table to write

$$I = \int_0^1 \frac{1}{\sqrt{1 - x^2}}\, dx = \lim_{\varepsilon \to 0}\left(\left[\arcsin x\right]_0^{1-\varepsilon}\right)$$

$$= \lim_{\varepsilon \to 0}\,[\arcsin (1 - \varepsilon)] - (\arcsin 0) = \arcsin (1) = \frac{\pi}{2}$$

In this case the improper integral is convergent and gives a sensible answer that can, as usual, be interpreted as (the limit of) an area under a graph, even though the graph itself cannot be drawn at $x = 1$.

Without using the formal notation of limits, evaluate the following improper integrals:

(a) $I = \int_{-1}^0 \frac{1}{\sqrt{1 - x^2}}\, dx$ (b) $I = \int_{-1}^1 \frac{1}{\sqrt{1 - x^2}}\, dx$

(a) $I = \int_{-1}^0 \frac{1}{\sqrt{1 - x^2}}\, dx = [\arcsin x]^0_{-1}$

$$= [\arcsin (0)] - [\arcsin (-1)] = \frac{\pi}{2}$$

(b) $I = \int_{-1}^1 \frac{1}{\sqrt{1 - x^2}}\, dx = [\arcsin x]^1_{-1}$

$$= [\arcsin (1)] - [\arcsin (-1)] = \pi \quad (☞) \quad \square$$

Although we have just considered a convergent integral it is easy to find divergent cases. For example $\int_0^1 \frac{1}{x}\, dx$ is divergent because

$$\int_0^1 \frac{1}{x}\, dx = \lim_{\varepsilon \to 0}\left(\left[\log_e |x|\right]_\varepsilon^1\right) = \lim_{\varepsilon \to 0}\,(0 - \log_e \varepsilon)$$

and $\log \varepsilon$ has no finite limit as ε tends to zero. Similarly the integral $\int_0^2 \frac{1}{(1 - x^2)}\, dx$ is divergent because of the behaviour of the integrand as x tends to 1. As a general rule:

Part (b) of this answer could also have been obtained by using part (a) and the fact that the integrand is an even function.

You should note that you may often see expressions such as

$$\int_1^2 \frac{dx}{x} = \left[\log_e x\right]_1^2$$

i.e. the modulus bars on the x are omitted since we know the x is positive.

If $I = \int_a^\infty f(x)\,dx$ where $f(x)$ can be written as $f(x) = \dfrac{g(x)}{(x-p)^n}$ with $g(x)$ finite over the range of integration and non-zero at $x = p$, then there is said to be a **singularity** (or an *infinity*) of **order** n at $x = p$ and the integral I is convergent if the order of that singularity is less than 1 (i.e. $n < 1$).

One final warning about integrals that involve a singularity: remember that even if they are written to look like ordinary definite integrals they are still improper and should be considered (at least in the back of your mind) as limits of proper integrals. Forget this and you may make the following kind of mistake

$$\int_{-1}^1 \frac{1}{x^2}\,dx = \left[-\frac{1}{x}\right]_{-1}^1 = (-1) - (1) = -2 \qquad\qquad \text{(WRONG)}$$

It looks plausible, but there is a singularity in the middle of the range of integration. By considering the separate limits as x tends to zero from above and below it is easy to see that the integral is actually divergent.

Question T15

Rewrite the improper integral $\int_{-1}^1 \frac{1}{x}\,dx$ as the sum of the limits of two properly defined definite integrals. Will it be convergent? If so, what is its value?   □

18.5   Closing items

18.5.1   Chapter summary

1. A *function* f is a rule that assigns a single value $f(x)$ to each value of x in a set called the *domain*.

2. If $f(x)$ is a given function and $F(x)$ is any function such that
    $$\frac{dF}{dx} = f(x) \quad \text{then we write} \quad F(x) = \int f(x)\,dx$$
    and we call $F(x)$ an *indefinite integral* of $f(x)$.

3. If $F_1(x)$ and $F_2(x)$ are both indefinite integrals of the same function $f(x)$, then there exists a constant K such that $F_1(x) = F_2(x) + K$.

4. The *definite integral* of a function $f(x)$ from $x = a$ to $x = b$ is defined by the limit of a sum and may be written as
    $$\int_a^b f(x)\,dx = \lim_{\Delta x \to 0} \left(\sum_{i=1}^n f(x_i)\,\Delta x_i\right)$$

where $\Delta x_i = x_{i+1} - x_i$ with $x_1 = a$ and $x_{n+1} = b$, and Δx is the largest of the Δx_i.

5. According to the *fundamental theorem of calculus*, if $F(x)$ is any indefinite integral of a given function $f(x)$, such that $\int f(x)\,dx = F(x)$, then

$$\int_a^b f(x)\,dx = F(b) - F(a) \tag{13}$$

6. The (signed) *area under the graph* of a function $f(x)$ between $x = a$ and $x = b$ is equal to the corresponding definite integral, $\int_a^b f(x)\,dx = F(b) - F(a)$. When calculating the area under a graph between a and b, with $a < b$, any regions that are below the horizontal axis should be regarded as having negative areas.

7. Several (indefinite) *standard integrals* are listed in Tables 18.1 and 18.2.

8. Integrals (both definite and indefinite) may be combined using the following rules:

the *sum rule* $\quad \int [g(x) + h(x)]\,dx = \int g(x)\,dx + \int h(x)\,dx \quad$ (14)

the *constant multiple rule* $\quad \int k f(x)\,dx = k \int f(x)\,dx \quad$ (15)

9. Definite integrals have the following six properties:

1. $\displaystyle \int_a^a f(x)\,dx = 0 \tag{18}$

2. $\displaystyle \int_b^a f(x)\,dx = -\int_a^b f(x)\,dx \tag{19}$

3. $\displaystyle \int_a^b f(x)\,dx = \int_a^c f(x)\,dx + \int_c^b f(x)\,dx \quad \text{where } a \leqslant c \leqslant b \tag{20}$

4. $\displaystyle \int_a^b [k f(x) + l g(x)]\,dx = k \int_a^b f(x)\,dx + l \int_a^b g(x)\,dx \tag{21}$

   where k and l are constants

5. If $f(x) \geqslant 0$ for all x in the interval $a \leqslant x \leqslant b$ then

$$\int_a^b f(x)\,dx \geqslant 0 \tag{22}$$

6. If $m \leqslant f(x) \leqslant M$ for all x in the interval $a \leqslant x \leqslant b$ then

$$m(b - a) \leqslant \int_a^b f(x)\,dx \leqslant M(b - a) \tag{23}$$

10. If $f(x)$ is an *odd function* so that $f(-x) = -f(x)$, then

$$\int_{-a}^{a} f(x)\,dx = 0 \tag{25}$$

If $f(x)$ is an *even function* so that $f(-x) = f(x)$, then

$$\int_{-a}^{a} f(x)\,dx = 2 \int_{0}^{a} f(x)\,dx \tag{26}$$

If $f(x)$ is a *periodic function* of *period k*, so that $f(x) = f(x + nk)$, where n is any integer, then

$$\int_{a}^{a+k} f(x)\,dx = \int_{0}^{k} f(x)\,dx \tag{27}$$

and $$\int_{0}^{nk} f(x)\,dx = n \int_{0}^{k} f(x)\,dx \tag{28}$$

11. *Improper integrals* may have infinite upper or lower limits and/or *singularities* in their integrands. Such integrals should be interpreted as appropriate limits of (proper) definite integrals, and may be *convergent* or *divergent*. They should be treated with care.

18.5.2 Achievements

Having completed this chapter, you should be able to:

A1. Define the terms that are emboldened in the text of the chapter.

A2. Define an indefinite integral of a given function in terms of a process that reverses the effect of differentiation, explain the significance of the constant of integration introduced by that process, and determine a variety of indefinite integrals.

A3. Define the definite integral of a given function between given limits in terms of the limit of a sum, state the fundamental theorem of calculus that relates definite and indefinite integrals, and use it to evaluate a variety of definite integrals.

A4. Interpret the definite integral of a given function between given limits in terms of the area under the graph of that function between those limits (with due regard to signs).

A5. Use tables of standard integrals to determine definite and indefinite integrals of the tabulated functions and of linear combinations of those functions. Also, use the techniques of differentiation to justify the entries in such tables.

A6. Recognise, write down and use the general mathematical properties of definite integrals.

A7. Recognise, write down and use the general mathematical properties of definite integrals with integrands that are even, odd or periodic.

A8. Recognise improper integrals, determine whether or not they converge in simple cases, and evaluate those that do.

18.5.3   End of chapter questions

Question E1 (a) What is the meaning of each of the symbols in the expression $\int_a^b f(x)\,dx$?

(b) How may we use an indefinite integral of $f(x)$ to evaluate the integral?

(c) Give a geometrical interpretation of the integral if $f(x) > 0$ for all x such that $a \leqslant x \leqslant b$.

Question E2 Find the following integrals:

(a) $\displaystyle \int \left(x^2 + \frac{1}{x^2} \right) dx$
   (b) $\displaystyle \int_2^1 \frac{1}{x}\,dx$

(c) $\displaystyle \int_{\pi/6}^{\pi/2} 8 \cos\left(\frac{x}{2}\right) dx$
   (d) $\displaystyle \int_1^2 \tfrac{1}{2} \log_e\left(\tfrac{1}{2} x\right) dx$

Question E3 Find the area under the graph *and* the sum of the magnitudes of the enclosed areas between the graph $y = f(x)$, the x-axis and the lines $x = a$ and $x = b$ for each of the following:

(a) $f(x) = e^{-x}$ $a = 0,\ b = 4$

(b) $f(x) = x - x^2$ $a = 0,\ b = 2$

(c) $f(x) = $ $a = -2,\ b = -1.$

Question E4 Using Tables 18.1 and 18.2 where appropriate, determine the following integrals:

(a) $\displaystyle \int \left(\frac{4y}{\sqrt{6 - 2y^2}} \right) dy$

(b) $\displaystyle \int \left(\frac{12t^2 - 21}{4t^2 - 9} \right) dt$ (*Hint*: Rewrite the numerator.)

(c) $\displaystyle \int \left(\frac{2x + 3}{4x^2 - 9} \right) dx$ (*Hint*: Rewrite the denominator.)

Question E5　Without performing any integrals, determine which of the following statements are true and explain your reasoning:

(a)　$\displaystyle\int_1^2 \log_e x \, dx = \int_3^2 \log_e x \, dx + \int_1^3 \log_e x \, dx$

(b)　$\displaystyle\int_a^b |f(x)| \, dx > \int_a^b f(x) \, dx$

(c)　$\displaystyle\int_{-\pi/2}^{\pi/2} (x^3 + \sin^2 x) \, dx = 0$

Integrating by parts and by substitution

Chapter 19

19.1 Opening items

19.1.1 Chapter introduction

Obtaining the indefinite integral of a function $f(x)$ amounts to finding a function $F(x)$ whose derivative $\dfrac{dF}{dx}$ is equal to $f(x)$. For example

$$\int \cos x \, dx = \sin x + C$$

precisely because

$$\frac{d}{dx}(\sin x + C) = \cos x$$

Every time we manage to differentiate a specific function $F(x)$ to obtain the answer $\dfrac{dF}{dx} = f(x)$ we automatically discover the solution to a corresponding problem in integration, namely 'what is the indefinite integral $\int f(x) \, dx$?' Integrals of simple functions, i.e. those that can be easily recognised as the derivatives of other functions, can quickly be found in this way. For example, to find $\int \cos(2x) \, dx$, you might guess that the result was something like $\sin(2x) + C$. On differentiating $\sin(2x) + C$, you would find that the derivative was in fact $2\cos(2x)$. So your initial guess has to be adjusted by a factor of $\frac{1}{2}$, and the answer is

$$\int \cos(2x)dx = \tfrac{1}{2}\sin(2x) + C$$

Whereas there are simple rules that enable us to differentiate almost any function we meet, it is a sad fact of life that there are no such rules for integration. This is not to say that there aren't methods for finding integrals: there are plenty; it is just that these methods do not always work and, what is more, there are many simple looking integrals for which no method could possibly work. For example, it is not possible to express the integral $\int e^{x^2} \, dx$ in terms of the elementary functions, sine, cosine, exp and so on.

The methods discussed in this chapter will enable you to find many useful integrals, but all integration methods ultimately amount to recognising the integrand $f(x)$ as the derivative of some function. Indeed, many simple integrals, such as $\int \cos(2x) \, dx$, can be found by making an intelligent guess at the function $F(x)$, then differentiating your guess to see if it gives $f(x)$, and if not, adjusting it appropriately (this may involve, for example, multiplication by a constant). Of course there are many functions $f(x)$ for which it would be very difficult to guess the form of the indefinite integral straight away. In this chapter, we discuss two standard methods of integration which can sometimes be used to manipulate an integral so as to bring it into a form where intelligent guesswork (or use of a table of standard integrals) will give you the final answer.

The first of these methods, *integration by parts*, is based on the *product rule* for differentiation, and is often useful when the function $f(x)$ to be integrated is the product of two simpler functions. Thus, for example, it enables you to integrate functions like xe^x.

The second method, *integration by substitution*, involves rewriting the integral $\int f(x)\, dx$ in terms of a new variable $y = g(x)$, and is based on the *chain rule* for differentiation. It will often enable you to turn a really unpleasant-looking integral – such as $\int x\,(1 - x^2)^{5/2}\, dx$ – into one which can be evaluated more easily.

19.1.2   Fast track questions

Question F1
Find the indefinite integrals

(a) $\int x^2\, e^{-x}\, dx$

(b) $\int e^{-x} \sin (3x)\, dx$

Question F2
Find the indefinite integral $\int \sin x\, (1 + \cos x)^4\, dx$.

Question F3
Evaluate the integral $\int_0^1 \dfrac{x}{\sqrt{1 + x}}\, dx$ to three decimal places.

19.1.3   Ready to study?

Study comment
In order to study this chapter, you will need to be familiar with the following terms: *chain rule, constant of integration, definite integral, derivative, function of a function, fundamental theorem of calculus, indefinite integral, integrand, inverse derivative, inverse trigonometric functions, product rule,* and *trigonometric identities*. In addition, you will need to have a good knowledge of differentiation (including the use of the *product* and *chain* rules). You should be able to recognise an *indefinite integral* as an *inverse derivative*, and you need to be familiar with the integrals of simple functions such as x^n, $\sin x$,

$\cos x$, e^x. You should also be able to find integrals of functions like $3x$ and $\cos (2x)$ using intelligent guesswork, and you should be familiar with the procedure of evaluating *definite integrals*. The following *Ready to study questions* will allow you to establish whether you need to review some of these topics before embarking on this chapter.

The alternative notations: exp (x) for e^x, ln (x) for $\log_e x$ and $\sin^{-1}$ for arcsin are in common use. You may also see asin for arcsin on some calculators.

Question R1

Use the product rule to find $\dfrac{dy}{dx}$ in each of the following cases:

(a) $y = x^6 \log_e x$ (b) $y = e^x \cos (2x)$

Question R2

Use the chain rule to find $f'(x)$ when

(a) $f(x) = \cos (x^4)$ (b) $f(x) = \log_e (1 + x^2)$

Question R3

Differentiate the inverse trigonometric function arcsin $(2x)$.

Use your answer to find the indefinite integral of $\dfrac{1}{\sqrt{1 - 4x^2}}$.

Question R4

Find the indefinite integrals of

(a) $x^{-4/3}$ (b) $\sin (2x) + 3 \cos x$ (c) $\exp (x/4)$

Question R5

Evaluate the definite integral $\displaystyle\int_1^2 x^2 \, dx$.

Question R6

Without using a calculator, find the possible values of

(a) $\cos(x)$ if $\sin(x) = \dfrac{1}{3}$ (b) $\tan(x)$ if $\sec(x) = \sqrt{5}$

Question R7

If $f(x) = \dfrac{(x + 1)^2}{x + 3}$ and $y = x + 3$, express $f(x)$ in terms of y.

19.2   **Integration by parts**

19.2.1   The method deduced from the differentiation of a product

In this subsection, we shall look at the product rule for differentiation, and derive from it the important integration technique known as **integration by parts**. We begin with an example.

Example 1
Find the integral $\int x \cos x \, dx$.

> *Solution*   As you will see shortly, we can find this integral by looking at the derivative of the product $x \sin x$, the reason being that the integrand $x \cos x$ appears in this derivative. To be precise, on using the product rule to differentiate $x \sin x$, we find that
>
> $$\frac{d}{dx}(x \sin x) = \sin x + x \cos x$$
>
> An alternative way of expressing this result is to say that $x \sin x$ is the *indefinite integral* of $(\sin x + x \cos x)$, so that we can write
>
> $$\int (\sin x + x \cos x) \, dx = x \sin x + C$$
>
> where C is an arbitrary constant. (☞) Using the rule for the sum of two integrals, it follows that
>
> $$\int \sin x \, dx + \int x \cos x \, dx = x \sin x + C$$
>
> and a little rearrangement gives us
>
> $$\int x \cos x \, dx = x \sin x - \int \sin x \, dx + C$$
>
> Using the fact that the indefinite integral of $(-\sin x)$ is $\cos x + C$, we obtain the result
>
> $$\int x \cos x \, dx = x \sin x + \cos x + C \qquad\qquad (☞) \;\square$$

Thus, by looking at the derivative of the product $x \sin x$, we have been able to find $\int x \cos x \, dx$, the indefinite integral of another product.

We will show you shortly how to generalise this method, but first you should try the following question.

Question T1
> Find the derivative of the product xe^x,
> and use it to find $\int xe^x \, dx$. $\square$

Throughout this chapter, the symbol C denotes an arbitrary constant of integration.

Note that there is no need to include two arbitrary constants of integration in our final result.

We have seen that looking at *derivatives* of products has enabled us to find some *integrals* that also involve products; however, you may feel that the method is rather 'hit or miss'. The question is:

What product should we differentiate in order to find the integral of a given product?

In order to answer this question, we must first investigate the derivative of two *arbitrary* functions $F(x)$ and $g(x)$. From the product rule for differentiation, we have

$$\frac{d}{dx}(F(x)\,g(x)) = \frac{dF(x)}{dx}\,g(x) + F(x)\,\frac{dg(x)}{dx} \tag{1}$$

which means that $F(x)g(x)$ is an inverse derivative, or indefinite integral, of the function

$$\frac{dF(x)}{dx}\,g(x) + F(x)\,\frac{dg(x)}{dx}$$

on the right-hand side of Equation 1. So we may write

$$\int\left[\frac{dF(x)}{dx}\,g(x) + F(x)\,\frac{dg(x)}{dx}\right]dx = F(x)\,g(x) + C$$

i.e. $\displaystyle\int\frac{dF}{dx}\,g(x)\,dx + \int F(x)\,\frac{dg}{dx}\,dx = F(x)\,g(x) + C$

and, rearranging this equation (and absorbing the arbitrary constant C into the integrals), it follows that

$$\int\frac{dF}{dx}\,g(x)\,dx = F(x)\,g(x) - \int F(x)\,\frac{dg}{dx}\,dx$$

We can omit the constant C here, since both indefinite integrals already implicitly contain an arbitrary constant.

When applying this equation, we are usually given the left-hand side, i.e. we are given $\dfrac{dF}{dx}$ and $g(x)$, and for this reason it is normal to represent the function $\dfrac{dF}{dx}$ by $f(x)$, so that the equation becomes

$$\int f(x)\,g(x)\,dx = F(x)\,g(x) - \int F(x)\,\frac{dg}{dx}\,dx \tag{2}$$

where $\dfrac{dF}{dx} = f(x)$

Equation 2 is used so frequently that it is worth committing to memory. Some people achieve this by remembering the following words: 'the integral of the product of two functions is equal to the primitive of the first times the second,

The *primitive* of a function is synonymous with its *indefinite integral* (see Chapter 17).

In this section, when referring to $f(x)$, $g(x)$ and $F(x)$, we shall always do so in the context of Equation 2.

Here we chose $F(x) = \sin x$, but we might have taken $F(x) = \sin x + C$. If we had done so we would have noticed that the constant C introduced at this stage makes no difference to the final answer. In applying Equation 2, we take $F(x)$ to be the simplest possible indefinite integral of $F(x)$, and this usually means choosing the constant to be zero. The arbitrary constant does of course reappear at the final stage.

minus the integral of the primitive of the first times the derivative of the second.' If this is of no help to you, the best advice is to find your own method.

Use Equation 2, with $f(x) = \cos x$ and $g(x) = x$, to find $\int x \cos x \, dx$.

We know that $g(x) = x$. Also we have $\dfrac{dF}{dx} = f(x) = \cos x$ so that we may choose $F(x) = \sin x$. Thus Equation 2 gives

$$\int x \cos x \, dx = \int \underbrace{(\cos x)}_{f(x)} \times \overset{g(x)}{x} \times dx = \underbrace{(\sin x)}_{F(x)} \times \overset{g(x)}{x} - \int \underbrace{(\sin x)}_{F(x)} \times \overset{g'(x)}{1} \, dx$$

$$= x \sin x + \cos x + C$$

as we saw in Example 1. □

As you can see, Equation 2 converts the problem of finding $\int f(x) g(x) \, dx$ into that of finding $\int F(x) \dfrac{dg}{dx} \, dx$. In order for this equation to be useful in practice, there are clearly two necessary requirements:

1. We need to be able to find $F(x)$, an indefinite integral of $f(x)$.

2. The integral on the right-hand side of Equation 2, i.e. $\int F(x) \dfrac{dg}{dx} \, dx$, should be easier to find than our original integral, $\int f(x) g(x) \, dx$.

The technique of using Equation 2 to find $\int f(x) g(x) \, dx$ is called the method of *integration by parts*. Some questions and examples should help to make this method clearer, and will also illustrate the importance of the two requirements just mentioned.

Example 2
Use Equation 2 to find the indefinite integral $\int x \sin x \, dx$.

Solution   We start by choosing functions $f(x)$ and $g(x)$ so that our integral $\int x \sin x \, dx$ is the integral $\int f(x) \, g(x) \, dx$ appearing on the left-hand side of Equation 2. There are two possibilities: we can either choose

$$f(x) = \sin x \quad \text{and} \quad g(x) = x$$

or $f(x) = x \quad \text{and} \quad g(x) = \sin x.$

Since we know how to integrate both x and $\sin x$, the requirement that $f(x)$ should be a function that is easy to integrate does not help us in our choice. Let us, therefore, decide to take $f(x) = \sin x$ and $g(x) = x$, and see what happens. With this choice of $f(x)$ and $g(x)$, we have

$$F(x) = -\cos x \text{ and } \frac{dg}{dx} = 1.$$

Substituting our expressions for $f(x)$, $g(x)$, $F(x)$ and $\frac{dg}{dx}$ into Equation 2, we obtain

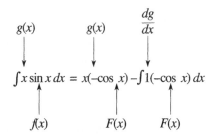

so that $$\int x \sin x \, dx = -x \cos x + \int \cos x \, dx$$

and since we know that the integral of $\cos x$ is $\sin x + C$, the final result is

$$\int x \sin x \, dx = -x \cos x + \sin x + C \quad \square \tag{3}$$

Let us see what would have happened in this example if we had chosen $f(x) = x$ and $g(x) = \sin x$. In that case, $F(x)$, the integral of x, is $x^2/2$ while dg/dx, the derivative of $\sin x$, is $\cos x$. Substituting these expressions into Equation 2 we obtain

$$\underset{\uparrow f(x)}{\int} \overset{\overset{g(x)}{\downarrow}}{x} \overset{\overset{g(x)}{\downarrow}}{\sin} x \, dx = \underset{\uparrow F(x)}{\frac{x^2}{2}} (\sin x) - \int \underset{\uparrow F(x)}{\frac{x^2}{2}} (\cos x) \, dx \tag{4}$$

You can always check your answer by differentiating the right-hand side. We shall not do so in the examples we give here, but *you* should.

As you can see, the integral appearing on the right-hand side of Equation 4 is no easier to find than the one we started with (in fact, it is harder!). As we mentioned earlier, when we apply Equation 2, the choice of $f(x)$ and $g(x)$, in any particular situation, should normally be such that the integral $\int F(x) \dfrac{dg}{dx}\, dx$ is *easier* to find than the original integral, $\int f(x)\, g(x)\, dx$. If, in using

integration by parts, you find that you are left with a more complicated integral than the one you started with, this probably means that you have chosen $f(x)$ and $g(x)$ the wrong way round. Do not despair − simply change them over and start again.

What would be the right choice of $f(x)$ and $g(x)$ to find the indefinite integral $\int x\, e^{-x}\, dx$?

If we choose $f(x) = e^{-x}$, $g(x) = x$, then Equation 2 gives

$$\int x e^{-x}\, dx = x e^{-x} + \int e^{-x}\, dx$$

leaving us with an easy integral to find. But if we choose $f(x) = x$, $g(x) = e^{-x}$, then on applying Equation 2 we find

$$\int x e^{-x}\, dx = \tfrac{1}{2} x^2 e^{-x} + \int \tfrac{1}{2} x^2 e^{-x}\, dx$$

which leaves us with a harder integral to find. Thus we should choose $f(x) = e^{-x}$ and $g(x) = x$. ◻

Example 3
Use integration by parts to find $\int x \log_e(x)\, dx$.

Solution   Here we could either choose

$$f(x) = x \qquad \text{and} \qquad g(x) = \log_e x$$
$$\text{or} \quad f(x) = \log_e x \quad \text{and} \qquad g(x) = x$$

Remember that we need to be able to integrate $f(x)$ in order to obtain $F(x)$. It is much easier to integrate x than it is to integrate $\log_e x$; so we choose $f(x) = x$ and $g(x) = \log_e x$.

Then $\qquad F(x) = \dfrac{1}{2} x^2 \quad$ and $\quad \dfrac{dg}{dx} = \dfrac{1}{x}$

Substituting our expressions for $f(x)$, $g(x)$, $F(x)$ and $\dfrac{dg}{dx}$ into Equation 2, we obtain

$$\int x \log_e x \, dx = \frac{1}{2} x^2 \log_e x - \int \left(\frac{x^2}{2}\right)\left(\frac{1}{x}\right) dx$$

$$= \frac{1}{2} x^2 \log_e x - \int \frac{1}{2} x \, dx$$

Since the integral of $\frac{1}{2} x$ is

$$\frac{1}{2}\left(\frac{1}{2} x^2\right) + C = \frac{1}{4} x^2 + C$$

we obtain as our final answer

$$\int x \log_e x \, dx = \frac{1}{2} x^2 \log_e x - \frac{1}{4} x^2 + C \quad \square$$

Question T2

Use Equation 2 to find the indefinite integral $\int x^2 \log_e x \, dx$. ☐

Definite integrals

So far, we have used integration by parts to find indefinite integrals, but the method can just as easily be used to evaluate definite integrals. According to Equation 1,

$$\frac{d}{dx}[F(x)\, g(x)] = \frac{dF}{dx} g(x) + F(x) \frac{dg}{dx} \tag{1}$$

so, the *fundamental theorem of calculus* implies that

$$\int_a^b \left[\frac{dF}{dx} g(x) + F(x) \frac{dg}{dx}\right] dx = [F(x)\, g(x)]_a^b$$

This can be rewritten as

$$\int_a^b \frac{dF}{dx} g(x) \, dx + \int_a^b F(x) \frac{dg}{dx} \, dx = [F(x)\, g(x)]_a^b$$

Rearranging this equation, and introducing again the notation $f(x)$ for $\frac{dF}{dx}$, we arrive at the following formula:

$$\int_a^b f(x)\, g(x) \, dx = [F(x)\, g(x)]_a^b - \int_a^b F(x) \frac{dg}{dx} \, dx$$

where $\frac{dF}{dx} = f(x)$

Example 4

Evaluate the integral $\int_1^2 x e^{2x} \, dx$, correct to four decimal places.

Solution We use Equation 5, taking $f(x) = e^{2x}$ *and* $g(x) = x$,

then $\frac{dg}{dx} = 1$ and $F(x) = \frac{1}{2} e^{2x}$.

Substituting into Equation 5 we obtain

$$\int_1^2 x e^{2x}\, dx = \left[\frac{1}{2} x e^{2x}\right]_1^2 - \frac{1}{2}\int_1^2 e^{2x}\, dx$$

$$= (54.598150 - 3.694528) - \frac{1}{2}(27.299075 - 3.694528)$$

$$= 39.1013 \quad \square$$

Question T3

Evaluate the integral $\displaystyle\int_0^1 x e^{3x}\, dx$, correct to four decimal places. $\quad\square$

19.2.2   Integrating by parts more than once

If you look back over the integrals appearing in Subsection 19.2.1, both in the text and in the *Text questions*, you will see that many of them are of the form $\int x f(x)\, dx$ where $f(x)$ is a function that is straightforward to integrate, such as an exponential, a sine or cosine. These are, perhaps, the easiest integrals to find using integration by parts, and it is not hard to see why. Referring to Equation 2, we see that if $g(x) = x$, then $dg/dx = 1$ and so an integral of this form becomes

$$\int x f(x)\, dx = x F(x) - \int F(x)\, dx \tag{6}$$

Provided that we can not only integrate $f(x)$ to find $F(x)$, but also integrate $F(x)$ itself, then the right-hand side of Equation 6 is easy to find. This will certainly be the case if $f(x)$ is an exponential function, a sine or cosine − for such functions can be integrated readily, to give functions of the same form which can also be integrated.

What would happen if instead of $\int x f(x)\, dx$, we had to find an integral of the form $\int x^2 f(x)\, dx$?

Example 5

Find $\int x^2 \cos x\, dx$.

Solution   We will try taking $f(x) = \cos x$ and $g(x) = x^2$. (Remember that if this should prove unhelpful, we can always make the opposite choice, and see whether that works better.) With this choice, we have $F(x) = \sin x$ and $dg/dx = 2x$.

Substituting these expressions into Equation 2, we find

$$\int x^2 \cos x\, dx = x^2 \sin x - \int 2 x \sin x\, dx$$

$$= x^2 \sin x - 2\int x \sin x\, dx \tag{7}$$

At first sight, the integral appearing on the right-hand side (i.e. $\int x \sin x\, dx$) may not look much better than the one we started with, $\int x^2 \cos x\, dx$, but you may recognise it as an integral that you know how to do by *integration by parts*! In fact, we have already found it, in Example 2. Equation 3 tells us that

$$\int x \sin x \, dx = -x \cos x + \sin x + C \qquad (3)$$

and substituting this into Equation 7, we find

$$\int x^2 \cos x \, dx = x^2 \sin x - 2(-x \cos x + \sin x + C)$$
$$= x^2 \sin x + 2x \cos x - 2 \sin x + C \quad \Box \ (\mathbb{I}\mathbb{F})$$

Any integral of the form $\int x^n f(x) \, dx$ containing the product of some positive integer power n of x and a function $f(x)$ which is an exponential, or a sine or cosine (and so can easily be integrated as many times as is necessary), can be found by repeated application of the method of integration by parts. Note that — as our notation suggests — it is important to take x^n to be $g(x)$, for it is this that ensures that every time we integrate by parts, we are left with a similar integral, but one in which the power of x has been reduced by 1.

Question T4

Find the integral $\int x^2 e^{2x} \, dx$. $\Box$

19.2.3 Integrals of various functions, including $\log_e x$ and $e^x \sin x$

You now know how to use integration by parts to find any integral of the form $\int x^n f(x) \, dx$, where $f(x)$ is an exponential function, a sine or cosine function. In this subsection, we shall show you some clever tricks that will enable you to use integration by parts to find some other integrals. We will concentrate here on indefinite integrals, and so use Equation 2; it is equally easy to use these methods to evaluate definite integrals, using Equation 5. ($\mathbb{I}\mathbb{F}$)

The first trick simply involves the realisation that the function $f(x)$ in Equation 2 may be taken to be 1. This means that integration by parts, which is a technique that essentially applies to integrals of *products*, can in fact be applied to an integral of *any* function; we simply regard the integrand as the product of the original function and 1. This approach may often lead to integrals worse than the one we start with, but in some cases it is extremely useful, as in the following example.

Example 6

Find the integral $\int \log_e x \, dx$.

Solution Here we choose $f(x) = 1$, $g(x) = \log_e x$. Then we have

$$F(x) = x, \frac{dg}{dx} = \frac{1}{x} .$$

Notice that there is no need to write $-2C$ in the final answer, since C is an *arbitrary constant*. We could, if we like, define a new arbitrary constant C' equal to $-2C$, and then decide to relabel it C, but it is unnecessary to go through these steps explicitly.

Some students find such clever tricks rather daunting since they assume, mistakenly, that they will be required to invent equally clever ideas. The fact is that the first person to come up with these ideas was indeed clever, but the rest of us simply follow along behind and file the tricks away in our minds for future use.

Substituting these expressions for $f(x)$, $g(x)$, $F(x)$ and $dg/dx = g'(x)$ into Equation 2, we obtain

$$\int \underbrace{1}_{f(x)} \times \overbrace{\log_e x}^{g(x)} \, dx = x \times \overbrace{\log_e x}^{g(x)} - \int \underbrace{x}_{F(x)} \overbrace{\left(\frac{1}{x}\right)}^{g'(x)} dx$$

$$= x \log_e x = \int 1 \, dx = x \log_e x - x + C$$

So our final answer is

$$\int \log_e x \, dx = x \log_e x - x + C \quad \square$$

It is not always easy to know in advance whether this trick will work, but it is worth bearing it in mind as part of your repertoire of integration methods. Try applying it to the integral in the next question.

Question T5

Given that

$$\int \frac{x}{1 + x^2} \, dx = \frac{1}{2} \log_e (1 + x^2) + C$$

find the integral $\int \arctan x \, dx$ (☜) $\square$

> You will learn how to find integrals of this sort in Section 19.3. For the moment, you should simply check that
>
> $$\frac{d}{dx}\left[\frac{1}{2} \log_e (1 + x^2)\right] = \frac{x}{1 + x^2}$$

The second clever trick that you will often find useful involves applying Equation 2 twice, this time in such a way as to obtain a simple algebraic equation for the unknown integral, which can then be solved. An example is the best way to present the method. We will let I denote the integral that we wish to evaluate, and for our example we choose

$$I = \int e^x \sin (2x) \, dx$$

If we take

$$f(x) = \sin (2x) \text{ and } g(x) = e^x \tag{8}$$

then $F(x) = -\frac{1}{2} \cos (2x)$ and $\frac{dg}{dx} = e^x$

Equation 2 then gives us

$$I = \int e^x \sin (2x) \, dx = e^x \left[-\frac{1}{2} \cos (2x)\right] - \int \left[-\frac{1}{2} \cos (2x)\right] e^x \, dx$$

i.e. $I = -\frac{1}{2} e^x \cos (2x) + \frac{1}{2} \int e^x \cos (2x) \, dx$ \hfill (9)

At this point, it may seem that integration by parts has failed; the integral remaining on the right-hand side of Equation 9 is no simpler than the one we started with. However, let us see what happens when we apply integration by parts to this integral, $\int e^x \cos(2x) dx$.

We will take $f(x) = \cos(2x)$ and $g(x) = e^x$.   (☞)

Then we have $F(x) = \dfrac{1}{2} \sin(2x)$ and $\dfrac{dg}{dx} = e^x$,

so $\int e^x \cos(2x) dx = e^x \left[\dfrac{1}{2} \sin(2x) \right] - \int \left[\dfrac{1}{2} \sin(2x) \right] e^x dx$

$$= \dfrac{1}{2} e^x \sin(2x) - \dfrac{1}{2} \underbrace{\int e^x \sin(2x) dx}_{\text{This is the integral } I} \qquad (10)$$

Notice that, as in Equation 8, we have chosen $g(x)$ to be the exponential function.

which contains the integral I with which we started! Substituting Equation 10 into Equation 9, we obtain

$$I = -\dfrac{1}{2} e^x \cos(2x) + \dfrac{1}{2} \left[\dfrac{1}{2} e^x \sin(2x) - \dfrac{1}{2} \int e^x \sin(2x) dx \right]$$

so that $I = -\dfrac{1}{2} e^x \cos(2x) + \dfrac{1}{4} e^x \sin(2x) - \dfrac{1}{4} I$ (☞)

We can now collect all the terms in I together on the left-hand side to give

This is an important step, because we now have an equation containing I and *no other integrals*.

$$\dfrac{5}{4} I = -\dfrac{1}{2} e^x \cos(2x) + \dfrac{1}{4} e^x \sin(2x)$$

i.e. $I = \dfrac{4}{5} \left[-\dfrac{1}{2} e^x \cos(2x) + \dfrac{1}{4} e^x \sin(2x) \right]$

We have now found an expression for the indefinite integral I, but we find a more general expression by adding an arbitrary constant C. Thus;

$$\int e^x \sin(2x) dx = -\dfrac{2}{5} e^x \cos(2x) + \dfrac{1}{5} e^x \sin(2x) + C$$

This method, using integration by parts twice and then solving the resulting equation to obtain the required integral, will work for any product of an exponential function with either a sine or cosine function. For such cases, it does not matter whether you choose the exponential function to be $f(x)$ or $g(x)$, as long as you make the *same* choice in both your integrations by parts. If you do not, your resulting equation will simply reduce to $0 = 0$ – true, but not very useful!

Question T6

Find the integral $\int e^{2x} \cos x \, dx$.

(*Hint*: Start by taking $f(x) = \cos x$ and $g(x) = e^{2x}$, then let $J = \int e^{2x} \cos x \, dx$.) □

It is sometimes possible to obtain a simple equation for your desired integral by integrating by parts just once, as in the next example.

Example 7

Find the integral $\int \cos^2 x \, dx$.

Solution   If we let $f(x) = \cos x$ and $g(x) = \cos x$ then $F(x) = \sin x$ and $dg/dx = -\sin x$.

Applying Equation 2 then gives us

$$\int \cos^2 x \, dx = \sin x \cos x + \int \sin^2 x \, dx \tag{11}$$

We can now use the trigonometric identity $\cos^2 x + \sin^2 x = 1$, so that $\sin^2 x = 1 - \cos^2 x$, in Equation 11, so that it becomes

$$\int \cos^2 x \, dx = \sin x \cos x + \int (1 - \cos^2 x) dx$$
$$= \sin x \cos x + \int 1 \, dx - \int \cos^2 x \, dx$$

Putting $\int 1 \, dx = x + C$ and collecting all $\int \cos^2 x \, dx$ terms together on the left-hand side, we find

$$2 \int \cos^2 x \, dx = \sin x \cos x + C$$

or finally, $\int \cos^2 x \, dx = \frac{1}{2} \sin x \cos x + \frac{1}{2}x + C$ □

Question T7

In the integral $\int \cos^4 x \, dx$, take $f(x) = \cos x$, $g(x) = \cos^3 x$, and show that $\int \cos^4 x \, dx = \frac{1}{4} \sin x \cos^3 x + \frac{3}{4} \int \cos^2 x \, dx$.

(*Hint*: You will need to use the identity $\cos^2 x + \sin^2 x = 1$). □

19.3   **Integration by substitution**

19.3.1   *An introduction to the method of substitution*

Evaluating an integral of the form $\int f(x) \, dx$ amounts to recognising $f(x)$ as the derivative of some function $F(x)$. However, if $F(x)$ is a rather complicated function of x, perhaps a function of a function of x, it may not be at all easy to spot that $f(x)$ is its derivative.

Use the chain rule to find the derivative of $\frac{2}{3}(1 + x)^{3/2}$.

This function can be written as $f[y(x)]$, where $f(y) = \frac{2}{3}y^{3/2}$ and $y(x) = 1 + x$.

The chain rule tells us that

$$\frac{d}{dx}\{f[y(x)]\} = \frac{df(y)}{dy}\frac{dy(x)}{dx}$$

and here,

$$\frac{df}{dy} = y^{1/2} \quad \text{and} \quad \frac{dy}{dx} = 1$$

So we find

$$\frac{d}{dx}\left[\frac{2}{3}(1 + x)^{3/2}\right] = (1 + x)^{1/2}. \quad \square$$

Since the derivative of $\frac{2}{3}(1 + x)^{3/2}$ is $(1 + x)^{1/2}$, it follows that

$$\int (1 + x)^{1/2}\, dx = \frac{2}{3}(1 + x)^{3/2} + C \tag{12}$$

This result is probably not one that would have been obvious if you had not first calculated the derivative of $\frac{2}{3}(1 + x)^{3/2}$. In this section, we explain an integration method that can often be used to bring integrals like $\int (1 + x)^{1/2}\, dx$ into a much simpler form, such that you will probably be able to evaluate them by inspection, or by reference to a table of standard integrals.

We will first show you how the method works, by applying it to the integral that we have just found, i.e. $\int (1 + x)^{1/2}\, dx$. The first step is to introduce a new variable y, equal to $1 + x$. The integrand $(1 + x)^{1/2}$ is then simply equal to $y^{1/2}$. We must now decide what to do with the 'dx' inside the integral sign. Since $y = 1 + x$ it follows that $dy/dx = 1$, and if dy/dx could be regarded as the *quotient* of the two quantities dy and dx, we could rearrange this equation to read

$$dy = dx \tag{13}$$

Strictly speaking, dy/dx is not a quotient, and the symbols dy and dx have not been defined to have any meaning on their own. However the procedure that we outline here can certainly be justified (though a proof of its veracity lies outside the scope of this book and properly belongs in a course on mathematical analysis). We use Equation 13 to transform the integral $\int (1 + x)^{1/2}\, dx$ into another integral written just in terms of y, and putting $dx = dy$ and $(1 + x)^{1/2} = y^{1/2}$, we find that

$$\int (1 + x)^{1/2}\, dx = \int y^{1/2}\, dy = \frac{2}{3}y^{3/2} + C \tag{☞}$$

Here we have used the standard integral

$$\int y^n\, dy = \frac{y^{n+1}}{n + 1} + C$$

with $n = \dfrac{1}{2}$

Finally, our answer must be written in terms of x, and so we replace y by $(1 + x)$ on the right-hand side, to obtain

$$\int (1 + x)^{1/2}\, dx = \tfrac{2}{3}\, (1 + x)^{3/2} + C \tag{14}$$

Although you may be worried by the use of Equation 13 in deriving Equation 14, you can see that the result is *correct* − it is exactly the same as Equation 12.

The method used here to find the integral $\int (1 + x)^{1/2}\, dx$ is known as **integration by substitution** because we *substitute* the new variable y into our unknown integral to obtain another integral that we hope will be easier to evaluate. Try the method that we used to derive Equation 14 on the next question.

Question T8

Find the integral $\int (x - 2)^{4/3}\, dx$. Check your answer by differentiation. ☐

Here is a slightly harder example of integration by substitution:

Example 8

Find the integral $\int x^2 \cos (x^3)\, dx$ by means of the substitution $y = x^3$.

Solution   We note first that if $y - x^3$, then $dy/dx = 3x^2$. Again, treating dy and dx as though they were ordinary algebraic quantities, we rewrite this equation as $dy = 3x^2 dx$. We can rearrange this equation further if we want to and, at this point, it is convenient to notice that the 'product' $x^2 dx$ appears in the integral we are trying to find. So we write

$$x^2\, dx = \frac{1}{3}\, dy$$

We can also write $\cos (x^3) = \cos y$ so that, on making the substitution $y = x^3$, we find that

$$\int x^2 \cos (x^3)\, dx = \frac{1}{3} \int \cos y\, dy = \frac{1}{3} \sin y + C$$

Finally, we substitute $y = x^3$ into the right-hand side of this result, to obtain an answer for our integral in terms of x

$$\int x^2 \cos (x^3)\, dx = \frac{1}{3} \sin (x^3) + C$$

You can easily check that this result is correct by differentiating it. ☐

We will now list the steps involved in the method of substitution. As we do so, we shall note how we applied them to the integral in Example 8. This will show you how you can set your working out when you come to do problems of this sort.

Method	**Example:** $\int x^2 \cos(x^3)dx$
Step 1 Decide what substitution $y = g(x)$ to try.	We make the substitution $y = x^3$.
Step 2 Calculate $\dfrac{dg}{dx}$ and write down the equation $dy = \dfrac{dg}{dx}\,dx$. It may be convenient to rearrange this equation.	Here, $\dfrac{dg}{dx} = 3x^2$, so $dy = 3x^2 dx$. We rearrange this equation into the form $x^2\,dx = \dfrac{1}{3}\,dy$
Step 3 Make the substitution, to obtain an integral in terms of y. (Leave no x at all.)	$\displaystyle\int x^2 \cos(x^3)\,dx$ $= \displaystyle\int \underbrace{\cos(x^3)}_{\cos y} \times \underbrace{x^2\,dx}_{\frac{1}{3}\,dy}$ $= \dfrac{1}{3}\displaystyle\int \cos y\,dy$
Step 4 Integrate with respect to y.	$= \dfrac{1}{3}\sin y + C$
Step 5 Substitute $g(x)$ for y in order to obtain a final answer that is a function of x.	$= \dfrac{1}{3}\sin(x^3) + C$

Question T9

Use integration by substitution to find the following integrals:

(a) $\int x(5 + 2x^2)^{16}\,dx$, making the substitution $y = 5 + 2x^2$

(b) $\int \cos^4(x)\sin(x)\,dx$, making the substitution $y = \cos x$ ☐

Much the hardest part of the method of integration by substitution is to know what substitution to make – after all, we begin with an integral that we cannot find, and the point of the method is to arrive in Step 3 at an integral which is easier, and this obviously depends on our choice of $g(x)$. Skill at picking a suitable substitution will come with practice and experience, and in Subsections 19.3.2 and 19.3.3, we will present you with several examples of particular substitutions which will work for certain types of integrals.

As you will see in many subsequent examples, if the integrand contains some function of x – like $1 + x^2$ here – it is often a good idea to try that as $g(x)$.
Notice also that integration by parts is not *necessarily* the best way to tackle the integral of a product.

19.3.2   Integrals of 'a function and its derivative'

In this subsection, we shall introduce you to a particular class of integrals that can always be simplified by a suitable substitution. We will start with an example.

Example 9
Find the integral $\int x \sqrt{1 + x^2}\, dx$.

    Solution

    Step 1   We will try the substitution $y = g(x) = 1 + x^2$.

    Step 2 $\dfrac{dg}{dx} = 2x$, so $dy = 2x\,dx$. As $x\,dx$ appears in the integral, we

                rearrange this equation to read $x\,dx = \tfrac{1}{2}dy$.

    Step 3   We also have $\sqrt{1 + x^2} = y^{1/2}$. Substituting this, and $x\,dx = \tfrac{1}{2}dy$ into the integral, we obtain

$$\int x \sqrt{1 + x^2}\, dx = \frac{1}{2} \int \sqrt{y}\, dy$$

    Step 4   Integrating with respect to y gives us

$$\frac{1}{2} \int \sqrt{y}\, dy = \frac{1}{3} y^{3/2} + C$$

    Step 5   Finally, we substitute $y = 1 + x^2$ into the answer obtained in Step 4, to obtain

$$\int x \sqrt{1 + x^2}\, dx = \frac{1}{3} (1 + x^2)^{3/2} + C. \quad \square$$

Question T10
Find the integral $\int (\cos x) e^{\sin x}\, dx$ by making the substitution $y = \sin x$. $\square$

You may have noticed that the integral in Example 9, $\int x (1 + x^2)^{1/2}\, dx$, and that of Question T10, $\int (\cos x) e^{\sin x}\, dx$, have something in common. In both cases, the integrand is not simply a function of the suggested $g(x)$ – instead it is a product of some function of $g(x)$ and the *derivative* $g'(x)$ $(= dg/dx)$.

For example, we can write the integrand $x(1 + x^2)^{1/2}$ as $2x \times \tfrac{1}{2}(1 + x^2)^{1/2}$, i.e. as a product of the function of a function $\tfrac{1}{2}[g(x)]^{1/2}$ and the derivative $g'(x) = 2x$, so that

$$\int x\,(1 + x^2)^{1/2}\,dx = \int \underbrace{2x}_{g'(x)}\,\underbrace{\tfrac{1}{2}\,(1 + x^2)^{1/2}}_{\tfrac{1}{2}\,[g(x)]^{1/2}}\,dx$$

$$\underbrace{\phantom{\tfrac{1}{2}\,[g(x)]^{1/2}}}_{\text{function of } g(x)}$$

Write $(\cos x)\,e^{\sin x}$ in the form '$g'(x) \times$ [function of $g(x)$]', where $g(x) = \sin x$.

Here $g'(x) = \cos x$, so $(\cos x)\,e^{\sin x} = e^{g(x)} \times g'(x)$ □

Write $e^{2x}(1 + e^{2x})^3$ in the form '$g'(x) \times$ [function of $g(x)$]', where $g(x) = e^{2x}$.

Here $g'(x) = 2e^{2x}$ so that

$$e^{2x}\,(1 + e^{2x})^3 = 2e^{2x} \times \tfrac{1}{2}\,(1 + e^{2x})^3$$
$$= g'(x) \times \tfrac{1}{2}\,(1 + e^{2x})^3$$
$$= g'(x) \times \tfrac{1}{2}\,[1 + g(x)]^3 \quad □$$

Integrals of the form $\int g'(x)\,p[g(x)]\,dx$, where the integrand is the product of some composite function $p[g(x)]$ and the derivative $g'(x)$, are sometimes called 'integrals of a function and its derivative'. Such integrals can always be simplified by making the substitution $y = g(x)$. Let us see why. When we make this substitution in the integral

$$\int p[g(x)]\,g'(x)\,dx$$

the expression $p[g(x)]$ becomes simply $p(y)$, and we can also write

$$dy = g'(x)\,dx$$

Thus the integral becomes

$$\int p(y)\,dy$$

and so, if we know how to integrate the function $p(y)$, the substitution $y = g(x)$ will enable us to find the original integral $\int p[g(x)]\,g'(x)\,dx$.

Indeed, if we introduce the notation $P(y) = \int p(y)\,dy$, and remember that our final answer must be given as a function of x by writing $y = g(x)$ in $P(y)$, we can write down a neat formula for integrals of this sort:

$$\int p[g(x)]\,g'(x)\,dx = P[g(x)] \quad \text{where} \quad P(y) = \int p(y)\,dy \qquad (15)$$

Example 10

Find $\int e^{2x}(1 + e^{2x})^3\, dx$.

> *Solution* We make the substitution $y = g(x) = e^{2x}$ so that $g'(x) = 2e^{2x}$.
>
> The integral then becomes $\int g'(x) \times \frac{1}{2}[1 + g(x)]^3\, dx$
>
> so that in this case the function $p(y) = \frac{1}{2}(1 + y)^3$.
>
> The integral $P(y)$ is therefore given by
>
> $$P(y) = \int p(y)\, dy = \int \tfrac{1}{2}(1 + y)^3\, dy$$
>
> which can be simplified to
>
> $$P(y) = \tfrac{1}{8}(1 + y)^4 + C$$
>
> and this means that
>
> $$P[g(x)] = \tfrac{1}{8}(1 + e^{2x})^4 + C$$
>
> Finally, Equation 15 becomes
>
> $$\int e^{2x}(1 + e^{2x})^3\, dx = \tfrac{1}{8}(1 + e^{2x})^4 + C \quad \square$$

Whenever you encounter an integral that you think requires a substitution, you should first check whether it is of the form given on the left-hand side of Equation 15. If it is, the substitution $y = g(x)$ may well enable you to find it. There is no need to memorise Equation 15, though — the *method* is the important thing. Don't forget that $g'(x)$ may simply be a constant. Thus, for example, the integral $\int (3 + 2x)^{7/2}\, dx$ is of this form — we can write it as $\int [\frac{1}{2}(3 + 2x)^{7/2}] \times 2\, dx$, observing that the derivative of $g(x) = (3 + 2x)$ is simply 2.

Question T11

Find the following integrals:

(a) $\int (2 + \cos x)^7 \sin x\, dx$ (b) $\displaystyle\int \frac{x}{x^2 + 1}\, dx$ (☎) □

This was the integral that you needed to use in the answer to Question T5.

As you gain more practice with integrals of a function and their derivative, you will probably find that you can write down the answer almost immediately, by intelligent guesswork.

19.3.3 More examples of integration by substitution

In this subsection, we will show you some other examples of integration by substitution. Each example will be followed by a similar *Text question*, which should help you to become familiar with the lines of attack likely to work with different types of integral.

Example 11

Find the integral $\displaystyle\int \frac{x}{(1+2x)^2}\,dx$

Solution

Step 1 Here we will try the substitution $y = g(x) = 1 + 2x$

Step 2 Then $\dfrac{dg}{dx} = 2$, so $dy = 2\,dx$, i.e. $dx = \frac{1}{2}\,dy$.

Step 3 The integral can now be written in the form

$$\int \left[\frac{x}{(1+2x)^2} \right] \frac{1}{2}\,dy$$

The integrand, which is in square brackets, must be converted into a function of y, and in order to do this we need to invert the relation $y = 1 + 2x$ to obtain x in terms of y. (☞) This gives $x = \frac{1}{2}(y-1)$. Thus on making the substitution $y = 1 + 2x$, the integral becomes

$$\int \frac{x}{(1+2x)^2}\,dx = \int \left[\frac{1}{y^2} \times \frac{(y-1)}{2} \right] \times \left(\frac{1}{2}\,dy \right)$$

$$= \frac{1}{4} \int \left(\frac{1}{y} - \frac{1}{y^2} \right) dy$$

Notice that the method uses the function $y = 1 + 2x = g(x)$ *and* its inverse. We will come back to this point later.

Step 4 Evaluating the integral with respect to y (on the right) gives

$$\frac{1}{4} \int \left(\frac{1}{y} - \frac{1}{y^2} \right) dy = \frac{1}{4} \left(\log_e y + \frac{1}{y} \right) + C$$

Step 5 Substituting $y = 1 + 2x$ in this result gives the final answer

$$\int \frac{x}{(1+2x)^2}\,dx = \frac{1}{4} \left(\log_e (1+2x) + \frac{1}{1+2x} \right) + C$$

This example illustrates the point that if a particular function of x, in this case $(1 + 2x)$, features prominently in the integrand, it is worth trying that function as $g(x)$. □

Question T12

Find $\int x(1+x)^{5/2}\,dx$ by making the substitution $y = 1 + x$. □

In Example 11 and Question T12, it was necessary at one stage to find x in terms of y. Sometimes it is simpler to *start* with x as a function of y, and write the substitution in the form $x = h(y)$. In that case, we simply substitute $h(y)$ into the integrand, and replace dx by $(dh/dy)\,dy$. The next example is of this type.

Example 12

Find the integral $\displaystyle\int \frac{1}{\sqrt{1-x^2}}\, dx$.

Solution

Step 1   The presence of $\sqrt{1-x^2}$ in the denominator suggests we should make use of the trigonometric relationship $\cos^2 y = 1 - \sin^2 y$.

So we try the substitution $x = h(y) = \sin y$.

Step 2   Then $\dfrac{dh}{dy} = \cos y$, and so $dx = \cos y\, dy$

and the integral can be written in the form

$$\int \left[\frac{1}{\sqrt{1-x^2}} \right] (\cos y\, dy)$$

where again the expression in square brackets must be converted into a function of y.

Here we have used the identity $\cos^2 y = \sin^2 y = 1$.

Step 3   With this substitution, $\sqrt{1-x^2} = \sqrt{1 - \sin^2 y} = \cos y$ so the integral becomes (✎)

$$\int \frac{1}{\sqrt{1-x^2}}\, dx = \int \frac{\cos y}{\cos y}\, dy = \int 1\, dy$$

Step 4   The integral is easy: $\int 1\, dy = y + C$.

Step 5   Finally, we express the result in terms of x, using the fact that if $x = \sin y$, then $y = \arcsin x$, and so

$$\int \frac{1}{\sqrt{1-x^2}}\, dx = \int 1\, dy = y + C = \arcsin x + C \quad \square$$

You may already have known the integral in Example 12 — it appears in every table of standard integrals. The method used to find it in Example 12 opens the way to finding many other integrals of the same type. Consider, for example, the integral

$$\int \frac{1}{\sqrt{9-x^2}}\, dx$$

What substitution might we make here?

The experience that we gained in Example 12 suggests that we want something like $x = a \sin y$, where a is a suitable constant. We <u>must</u> choose a so that the identity $\cos^2 y + \sin^2 y = 1$ can be used to turn $\sqrt{9-x^2}$ into a multiple of $\cos y$. Then this will cancel the $\cos y$ term in $dx = a \cos y\, dy$, leaving us with a very simple integral.

Putting $x = a \sin y$ into $\sqrt{9 - x^2}$, it becomes $\sqrt{9 - a^2 \sin^2 y}$, and if we choose $a = 3$ it will be equal to $\sqrt{9 - 9 \sin^2 y} = 3 \cos y$. So the substitution $x = 3 \sin y$ can be used to simplify the integral.

The following integral is of the same type; so try replacing x by $a \sin y$ for a suitable choice of the constant a.

Question T13

Find the integral $\int \dfrac{1}{\sqrt{1 - 16x^2}}\, dx$. ☐

The substitution $x = \sin y$ can also be used to find integrals where the integrand is some other function of $\sqrt{1-x^2}$, as in the next question.

Question T14

Find the integral $\int \sqrt{1-x^2}\, dx$. (*Hint*: You will find the solution to Example 7 helpful here; you will also need to use the identity $\cos^2 y + \sin^2 y = 1$, both in making the substitution and in expressing the final answer in terms of x). ☐

Our next example is one of another class of integrals that can be found by means of different trigonometric substitution.

Example 13

Find the integral $\int \dfrac{1}{1 + x^2}\, dx$.

Solution

Step 1 Here, we will make the substitution $x = h(y) = \tan y$.

Step 2 Then

$$\frac{dx}{dy} = \frac{dh}{dy} = \sec^2 y, \quad \text{so } dx = \sec^2 y\, dy$$

Step 3 With this substitution, $1 + x^2 = 1 + \tan^2 y$. At this point, we use the trigonometric identity $1 + \tan^2 y = \sec^2 y$. Thus the integrand is simply

$$\frac{1}{\sec^2 y}$$

Using this, and $dx = \sec^2 y\, dy$, we find that the integral becomes

$$\int \frac{1}{1 + x^2}\, dx = \int \frac{\sec^2 y}{\sec^2 y}\, dy = \int 1\, dy$$

Step 4 $\int 1\, dy = y + C$

Step 5 If $x = \tan y$, then $y = \arctan x$, so the final answer is

$$\int \frac{1}{1 + x^2}\, dx = \arctan x + C \quad \square$$

The integral in Example 13 may also be one that you have seen before – but again, the point is that the method can be generalised to other integrals of the same form.

Consider

$$\int \frac{1}{4 + x^2}\, dx \ .$$

Here, we might reason that a substitution of the form $x = a \tan y$ should work; we choose the constant a so that the identity $1 + \tan^2 y = \sec^2 y$ can be used to transform $(4 + x^2)$ into a multiple of $\sec^2 y$. Putting $x = a \tan y$ into $(4 + x^2)$, it becomes $(4 + a^2 \tan^2 y)$; and if we choose $a = 2$, this becomes $4(1 + \tan^2 y) = 4 \sec^2 y$. So $x = 2 \tan y$ is the substitution we want here.

Question T15

Find the integral $\displaystyle\int \frac{1}{5 + 4x^2}\, dx \ . \quad \square$

19.3.4 Definite integrals – transforming the limits

So far in this section, we have dealt exclusively with *indefinite* integrals. Of course, once you have found the appropriate indefinite integral, finding a definite integral is just a matter of putting in the limits of integration. For example, let us return to the integral of Example 8,

$$\int x^2 \cos (x^3)\, dx = \frac{1}{3} \sin (x^3) + C$$

If, say, we want to evaluate the *definite* integral

$$\int_0^2 x^2 \cos (x^3)\, dx$$

we simply put the limits in:

$$\int_0^2 x^2 \cos (x^3)\, dx = \left[\frac{1}{3} \sin (x^3) \right]_0^2 \tag{☞}$$

$$= \frac{1}{3}\left(\sin (2^3) - \sin (0) \right)$$

$$= \frac{1}{3} \sin 8 = 0.3298 \text{ (to 4 decimal places)} \tag{16}$$

Here we have used the *fundamental therorem of calculus*:

$$\int_a^b f(x)\, dx = [F(x)]_a^b$$

Remember to set your calculator to radian mode.

However, there is sometimes slightly less work involved if we find the definite integral directly, in terms of the new variable y. This involve *transforming the limits* of the definite integral. To see how it works, let us go through the process of substituting $y = x^3$ into the above integral

$$\int_0^2 x^2 \cos(x^3)\, dx$$

As before, we write $\cos(x^3) = \cos y$ and $x^2\, dx = \frac{1}{3}dy$. We must now note that if x ranges from 0 to 2, then $y = x^3$ ranges from 0^3 to 2^3, i.e. between 0 and 8. So the limits of integration in the integral with respect to y are 0 and 8. Thus we obtain

$$\int_0^2 x^2 \cos(x^3)\, dx = \frac{1}{3}\int_0^8 \cos y\, dy$$

$$= \frac{1}{3}\left[\sin y\right]_0^8 = \frac{1}{3}(\sin 8 - \sin 0) = \frac{1}{3}\sin 8$$

as in Equation 16.

In Example 9, you saw that the indefinite integral $\int x(1 + x^2)^{1/2}\, dx$ could be found by means of the substitution $y = 1 + x^2$. If we want to evaluate the *definite* integral

$$\int_1^3 x(1 + x^2)^{1/2}\, dx$$

using this substitution, what are the limits in the integral with respect to y?

When $x = 1$, $y = 1 + 1^2 = 2$; and when $x = 3$, $1 + 3^2 = 10$. So the lower limit is 2 and the upper limit is 10. □

So, for definite integrals, when we make the substitution $y = g(x)$, we have two ways of proceeding. We can either follow Steps 1–5 of Subsection 19.3.1, finding the indefinite integral first, and putting in the x limits of integration at the end of the calculation. Alternatively, when carrying out Step 3 and writing the integral in terms of y, we can at the same time transform the x limits of integration (let us call the lower limit a and the upper limit b) into the corresponding y limits of integration, $g(a)$ and $g(b)$. Often this will save a little time because there is then no need to carry out Step 5. Use the second procedure to answer the following question.

We haven't established the validity of the method of integration by substitution but we should mention that it depends upon the fact that the function $g(x)$ has an inverse.

Question T16

Evaluate the integral

$$\int_0^4 x\sqrt{4-x}\,dx$$

by means of the substitution $y = 4 - x$.   □

An application to electricity and magnetism

We end this subsection with an application of integration by substitution to electricity and magnetism. (☎)

Example 14

Figure 19.1 shows a wire AB carrying a current I (denoted i on the figure). Such a current produces a magnetic field around the wire, and each small element of the wire, Δx say, makes a small contribution to this field. The object of this example is to determine the magnitude B of the resulting field at P due to the contributions from all such elements from A to B.

Solution   First we need to know that the current through the element Δx produces a contribution to the field at P perpendicular to the plane of the paper, and of magnitude

$$\frac{I\sin\theta\,\Delta x}{r^2}$$

Thus the total magnitude B of the field at P is

$$\int_{x_A}^{x_B}\frac{I\sin\theta}{r^2}\,dx$$

and, since $h = r\sin\theta$ while I and h are constants, we have

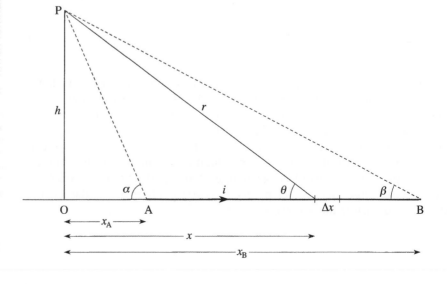

Figure 19.1   An electrical current through a segment of wire AB.

$$B = \int_{x_A}^{x_B} \frac{I \sin \theta}{r^2} \, dx = I \int_{x_A}^{x_B} \frac{h}{r^3} \, dx = I \int_{x_A}^{x_B} \frac{h}{(x^2 + h^2)^{3/2}} \, dx$$

In order to evaluate the integral on the right-hand side, we make the substitution $x = h \tan \phi$ so that $dx = h \sec^2 \phi \, d\phi$, and therefore

$$I \int \frac{h}{(x^2 + h^2)^{3/2}} \, dx = I \int \frac{h^2 \sec^2 \phi}{(h^2 \tan^2 \phi + h^2)^{3/2}} \, d\phi$$

$$= I \int \frac{h^2 \sec^2 \phi}{h^3 \sec^3 \phi} \, d\phi = \frac{I}{h} \int \cos \phi \, d\phi$$

$$= \frac{I}{h} \sin \phi + C \qquad (17)$$

In deriving this equation we have made use of the identity

$$\sec^2 \phi = 1 + \tan^2 \phi.$$

Now we substitute for ϕ in terms of x in Equation 17 using the fact that $\tan \phi = x/h$ implies that

$$\sin \phi = \frac{x}{\sqrt{h^2 + x^2}} \, ,$$

see Figure 19.2.

It follows that

$$I \int \frac{h}{(x^2 + h^2)^{3/2}} \, dx = \frac{I}{h} \sin \phi + C = \frac{I x}{h\sqrt{h^2 + x^2}} + C$$

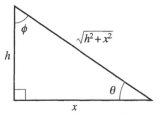

Figure 19.2 Geometrical meaning the substitution $x = h \tan \phi$. Notice that $\sin \phi = \cos \theta$.

We can now evaluate the magnitude of the field at P, and write

$$B = \int_{x_A}^{x_B} \frac{I \sin \theta}{r^2} \, dx = \left[\frac{I x}{h \sqrt{h^2 + x^2}} \right]_{x_A}^{x_B}$$

$$= \frac{I}{h} \left[\frac{x_B}{\sqrt{h^2 + x_B^2}} - \frac{x_A}{\sqrt{h^2 + x_A^2}} \right]$$

Using the angles α and β defined by Figure 19.1, we obtain the neat result

$$B = \frac{I}{h} (\cos \beta - \cos \alpha) \qquad (18)$$

There is one circumstance that is of particular interest, namely when we wish to estimate the field at points close to a very long straight wire; in which case we can assume that the wire is of infinite length (since the contributions from distant sections of the wire are insignificant). This corresponds to choosing $\alpha = \pi$ and $\beta = 0$, and we then have the useful result

$$B \approx \frac{2I}{h} \qquad (19) \quad \square$$

19.4 Which method to try?

In the *Text questions* you have done so far in this chapter, either you have been told what integration method to use, or it has been clear by comparison with what has gone immediately before. It is time for you now to start deciding for yourself what the best line of attack will be and, if you decide to use substitution, to choose for yourself what substitution to try. To help you with this, we shall list below some types of integrals and comment on the best methods of finding them. (As we are concerned here with the *method* of integration, we shall consider only indefinite integrals.)

19.4.1 Some helpful advice

1. Integrals of products of certain functions

Integration by parts will always work if the integrand is a product of any two of: an integer power of x; a sine or cosine; an exponential. There are certain other products for which it works: for example, $x \log_e x$. However, integration by parts is certainly not applicable to all products (try applying it, for example, to $\int x^2 \cos (x^3)\, dx$ and you will soon discover that it doesn't work).

2. Integrals of the form $\int p[g(x)]g'(x)\, dx$

Make the substitution $y = g(x)$. This will transform the integral into $\int p(y)\, dy$.

3. Integrals of the form $\int [1/(\sqrt{A - Bx^2})]dx]$, or $\int \sqrt{A - Bx^2}\, dx$, where A and B are positive constants

Use of substitution of the form $x = a \sin y$, where a is chosen so that $\sqrt{A - Bx^2}$ turns into a multiple of $\cos y$ when you make the substitution. In this case, we can guarantee that this substitution will work

4. Integrals of the form $\int [1/(A + Bx^2)]\, dx$ where A and B are positive constants

Use of a substitution of the form $x = a \tan y$, where a is chosen so that $A + Bx^2$ turns into a multiple of $\sec^2 y$ when you make the substitution. Here too, a substitution of this sort will certainly work.

5. Integrals in which the integrand contains some particular function of x

For example, in the integral $\int e^{2x} (1 + e^{2x})^3\, dx$ we might try $g(x) = e^{2x}$. Of course, there may be several functions of x appearing in the integrand. However, if a particular function of x appears in the denominator, or is raised to some power, that is a good one to try as $g(x)$.

19.4.2 Exercises

In the following question, all the integrals except one can be found using the methods introduced in this chapter; we leave it to you to choose which method to use. The exception is an integral that can be done in a very simple fashion, it doesn't require integration by parts or the method of substitution.

Question T17
Find the indefinite integrals

(a) $\displaystyle\int \frac{(x+3)^2}{(x+5)^4}\,dx$

(b) $\displaystyle\int 3x \cos\left(\tfrac{1}{2}\,x\right) dx$

(c) $\displaystyle\int \frac{1}{\sqrt{2-x^2}}\,dx$

(d) $\displaystyle\int \frac{x}{\sqrt{2-x^2}}\,dx$

(e) $\displaystyle\int \sqrt{x}\,(1+x)\,dx$ □

As you carry your study of mathematics and physical science further, you will probably encounter more substitutions that can be used for certain classes of integral; you may also learn how to combine various types of substitution with clever algebraic manipulations, or crafty use of trigonometric identities, and this will enlarge your integration repertoire even more. However, the most important techniques of integration are the two that have been presented in this chapter: integration by parts and integration by substitution. Consequently, if you have mastered these (as we hope) you have, in a sense, nothing further to learn about integration; all that remains for you to do is to gain greater familiarity with the ingenious tricks that can be used to bring an even greater number of integrals under your control.

19.5   Closing items

19.5.1   Chapter summary

1. A product $f(x)\,g(x)$ may often be integrated using the method of *integration by parts*, which is based on the formula

$$\int f(x)\,g(x)\,dx = F(x)\,g(x) - \int F(x)\,\frac{dg}{dx}\,dx \qquad (2)$$

where $\dfrac{dF}{dx} = f(x)$

2. If this formula is to be useful, we must be able to integrate the function $f(x)$ to obtain $F(x)$; also, we want the integral on the right-hand

side to be easier than the original integral. These considerations must be borne in mind when choosing $f(x)$ and $g(x)$ in any given case.

3. In the case of a definite integral, the formula becomes

$$\int_a^b f(x)\, g(x)\, dx = \left[F(x)\, g(x) \right]_a^b - \int_a^b F(x)\, \frac{dg}{dx}\, dx \tag{5}$$

where $\dfrac{dF}{dx} = f(x)$

4. It is sometimes necessary to apply the technique of integrating by parts more than once. This can either be because the remaining integral to be evaluated also requires the use of integration by parts; or, in the case of integrals like $\int e^x \sin x\, dx$, by integrating by parts twice we can arrive at a simple equation involving the original integral, which can then be solved.

5. A small class of integrals can be found by taking the function $f(x)$ to be 1.

6. *Integration by substitution.* An integral of the form $\int f(x)dx$ may often be transformed into a simpler integral by writing

$$y = g(x) \quad \text{and} \quad dy = \frac{dg}{dx}\, dx$$

and substituting these relations into the integral to obtain an integral in terms of the new variable y which (if $g(x)$ has been intelligently chosen) should be easy to find. Once it is found the answer may then be expressed in terms of x by substituting $y = g(x)$.

7. Integrals of a function and its derivative, of the form $\int p[g(x)]g'(x)\, dx$ may always be simplified by means of the substitution $y = g(x)$.

8. If some function of x appears raised to some power, or as a denominator, in the integrand, that may be a good choice for $g(x)$.

9. There are also certain types of substitution which work for particular classes of integrals; for example, a substitution of the form $x = a \sin y$, where a is a suitable constant, can be used to find integrals of the form

$$\int \frac{1}{\sqrt{A - Bx^2}}\, dx$$

where A and B are positive constants.

10. When a substitution $y = g(x)$ is used to evaluate definite integrals, it is necessary to transform the limits of integration when writing down the integral with respect to y.

19.5.2 Achievements

Having completed this chapter, you should be able to:

A1. Define the terms that are emboldened in the text of the chapter.

A2. Use the method of integration by parts to find indefinite integrals involving products of simple functions, such as $\int x \sin x \, dx$ or $\int x^3 \log_e x \, dx$.

A3. Apply integration by parts more than once to find indefinite integrals of the form $\int x^n f(x) \, dx$, where $f(x)$ is an exponential, or a sine or cosine.

A4. Apply integration by parts in such a way as to obtain a simple equation for the original unknown integral, which can then be solved, and recall that this method is particularly useful when the integrand is a product of an exponential function and a sine or cosine function.

A5. Remember that, in the formula for integration by parts (Equation 2), the function $f(x)$ may be taken to be 1, and use this trick to find certain integrals.

A6. Use integration by parts to evaluate definite integrals.

A7. Use a given substitution to transform a difficult indefinite integral into an easier one, and so find the original integral.

A8. Recognise an integral of the form $\int p[g(x)]g'(x) \, dx$ and use the substitution $y = g(x)$ to find it.

A9. Use the method of substitution to evaluate definite integrals, remembering to transform the limits correctly.

A10. Choose, for a given integral, a suitable substitution that will enable you to evaluate it.

19.5.3 End of chapter questions

Question E1 Find the integral $\int x^2 \sin(kx) \, dx$ where k is a constant.

Question E2 Find the integral $\int e^{3x} \cos(3x) \, dx$. (*Hint*: Start by taking $f(x) = e^{3x}$ and $g(x) = \cos(3x)$.)

Question E3 Evaluate the integral

$$\int_0^1 x \cos(5x) \, dx$$

to four decimal places.

Question E4   Use the substitution $y = x^2 + 2$ to find the integral

$\int (x^2 + 2)^6 x^3 \, dx$

Question E5   Find the following integrals:

(a) $\int \dfrac{2}{3x + 4} \, dx$ (b) $\int x \exp (x^2) \, dx$ (c) $\int e^{2x} \sin (e^{2x}) \, dx$

Question E6   Evaluate the integral

$\displaystyle\int_{1}^{2} x \, (2 + x)^{3/2} \, dx$

to four decimal places.

Question E7   Find the integral $\int \arcsin x \, dx$. (*Hint*: Start by integrating by parts).

Question E8   Find the following integrals:

(a) $\int \sec^2 (3x) \tan (3x) \, dx$ (b) $\int \dfrac{1}{1 + 9x^2} \, dx$ (c) $\int \dfrac{x^5}{1 + 2x^3} \, dx$

Appendix –
Maths handbook

Introduction

The *Maths handbook* is intended as a quick reference to the mathematics introduced in *Basic Mathematics for the Physical Sciences*, and as a source of additional reference material.

The handbook is structured for easy reference; it is divided into sections and subsections, with these numbered, for example, as 1 or 3.1 respectively. Section 0 *Units, symbols and statistics* is intended to put mathematics in a physical context by refering to general items such as SI units and errors of observation. The remaining sections are (loosely) associated with the correspondingly numbered parts and chapters of this book. Thus, Section 1 covers Part 1, Chapters 1 to 7, Section 2 covers Part 2 and so on. Within each section the material is arranged in a mathematically logical sequence with references to relevant chapters provided in square brackets where appropriate. A gallery of graphs, curves and functions appears near the end of the handbook (Section 6), but is referred to from several points within the handbook. The handbook ends with a table of physical and mathematical constants.

In some instances you may find the same note arising in two or more sections reflecting the structure of the chapters. Some material is split between sections, for example different aspects of the equations of lines are discussed in Sections 1 and 2 (and cross-referenced).

0   Units, symbols and statistics

0.1   *SI base units*

SI is an abbreviation for *Système International d'Unités*. The seven basic units are given in Table 1.

Table 1

Physical quantity	Name of unit	Symbol for SI unit	Physical quantity	Name of unit	Symbol for SI unit
Length	metre	m	Temperature	kelvin	K
Mass	kilogram	kg	Luminous intensity	candela	cd
Time	second	s	Amount of substance	mole	mol
Electric current	ampere	A			

Table 2

Physical quantity	Unit	Symbol	Definition
Energy	joule	J	$kg\,m^2\,s^{-2}$
Force	newton	N	$kg\,m\,s^{-2} = J\,m^{-1}$
Power	watt	W	$kg\,m^2\,s^{-3} = J\,s^{-1}$
Electric charge	coulomb	C	As
Electric potential difference	volt	V	$kg\,m^2\,s^{-3}\,A^{-1} = J\,A^{-1}\,s^{-1}$
Electric resistance	ohm	Ω	$kg\,m^2\,s^{-3}\,A^{-2} = V\,A^{-1}$
Electric capacitance	farad	F	$A^2\,s^4\,kg^{-1}\,m^{-2} = A\,s\,V^{-1}$
Magnetic flux	weber	Wb	$kg\,m^2\,s^{-2}\,A^{-1} = V\,s$
Inductance	henry	H	$kg\,m^2\,s^{-2}\,A^{-2} = V\,s\,A^{-1}$
Magnetic field	tesla	T	$kg\,s^{-2}\,A^{-1} = Wb\,m^{-2}$
Frequency	hertz	Hz	s^{-1}
Pressure	pascal	Pa	$kg\,m^{-1}\,s^{-2} = N\,m^{-2} = J\,m^{-3}$

Table 3

Multiple	Prefix	Symbol for prefix	Multiple	Prefix	Symbol for prefix
10^{12}	tera	T	10^{-3}	milli	m
10^9	giga	G	10^{-6}	micro	μ
10^6	mega	M	10^{-9}	nano	n
10^3	kilo	k	10^{-12}	pico	p
10^0			10^{-15}	femto	f

Table 4

Physical quantity	Unit	Symbol	Definition
Mass	atomic mass unit	u	$1.661 \times 10^{-27}\,kg$
Length	light year	ly	$9.462 \times 10^{15}\,m$
Length	angstrom	Å	$10^{-10}\,m$
Length	parsec	pc	$3.086 \times 10^{16}\,m$
Volume	litre	ℓ	$10^{-3}\,m^3$
Time	year	yr	$3.156 \times 10^7\,s$
Temperature	degree Celsius	°C	$1\,K*$
Pressure	atmosphere	atm	$1.013 \times 10^5\,Pa$
Mass	tonne	t	$10^3\,kg$
Energy	electronvolt	eV	$1.602 \times 10^{-19}\,J$
Energy	kilocalorie	kcal	$4187\,J$
Power	horsepower	hp	$7.457 \times 10^2\,W$

* The zeros of the Kelvin and Celsius scales are different: 0 °C is equivalent to 273.15 K.

Table 5

Physical quantity	Unit	Symbol	Definition
Plane angle	radian	rad	$\theta/\mathrm{rad} = s/r$
Plane angle	degree of arc	°	$1° = (\pi/180)$ rad
Plane angle	minute of arc	′	$1' = (1°/60)$
Plane angle	second of arc	″	$1'' = (1°/3600)$

Angular values quoted without units should be assumed to be in rads.

Table 6

Symbol		Name	Symbol		Name	Symbol		Name
lower case	upper case		lower case	upper case		lower case	upper case	
α*	A	alpha	ι	I	iota	ρ	P	rho (roe)
β	B	beta	κ	K	kappa	σ	Σ	sigma
γ	Γ	gamma	λ	Λ	lambda	τ	T	tau (taw)
δ	Δ	delta	μ	M	mu (mew)	υ	Y	upsilon
ε	E	epsilon	ν†	N	nu (new)	ϕ	Φ	phi (fie)
ζ	Z	zeta	ξ	Ξ	xi (csi)	χ	X	chi
η	H	eta	o	O	omicron	ψ	Ψ	psi
θ	Θ	theta	π	Π	pi	ω	Ω	omega

* Be careful to distinguish α (Greek alpha) from $\propto$ ('proportional to') and a (italic a).
† Be careful also to distinguish ν (Greek nu) from v (italic vee).

0.2   SI derived units

A number of derived units are given special names. The more common examples of these are given in Table 2.

0.3   SI multiples

The standard SI multiples and submultiples are given in Table 3.

0.4   Non-SI units

A number of non-SI units are popular with scientists and continue to be prevalent in appropriate contexts. Some of these are listed in Table 4.

0.5   Angular units

The quantities associated with angular measurement have units which revert to being simple numbers when expressed in terms of the basic units. This follows from the definition of the angle θ (measured in radians) subtended at the centre of a circle of radius r, by an arc of length s, as $\theta/\mathrm{rad} = s/r$. Angles may also be expressed in degrees of arc, and/or minutes and seconds of arc. To distinguish the scale being used, the appropriate unit must be appended to the simple number (Table 5).

0.6   The Greek alphabet (and its pronunciation)

See Table 6 above.

0.7 Mathematical symbols

$\propto$	proportional to	$\boldsymbol{v}$ or $\underset{\sim}{v}$	a vector (printed or handwritten)	$\displaystyle\int_a^b f(x)\,dx$	the (definite) integral of $f(x)$ with respect to x, from a to b
$=$	equal to	$\boldsymbol{0}$	the zero vector		
$\neq$	not equal to	$|\boldsymbol{v}|$	magnitude of the vector $\boldsymbol{v}$	$f(x, y)$	f, a function of x and y
$>$	greater than				
$\geqslant$	greater than or equal to	$\hat{\boldsymbol{r}}$	a unit vector in the direction of $\boldsymbol{r}$	$\dfrac{\partial f}{\partial x}$	the first partial derivative of $f(x, y)$ with respect to x
$>>$	much greater than	$\times$	vector (cross) product		
$+ - \times \div$	basic operations; add, subtract, multiply, divide	i	square root of -1, $\sqrt{-1}$	$\hat{\mathrm{p}}_x$	the operator representing p_x
$\pm$	plus or minus	z^*	complex conjugate of z	$\exp(x)$	the exponential function
$(\ )$	parentheses	$\arg(z)$	argument of z	$\log_a(x)$	the logarithmic function to the base a
$\{\ \}$	braces	$\rightarrow$	tends to (as in a limiting process)		
$\{x \in A\colon P\}$	the set of elements of A which have the property P	$f(x)$	f, a function of x	$\log_e(x)$	logarithmic function to the base e (sometimes written as $\ln x$, but not here)
$\mathbb{Z}$	the set of all integers $\ldots,$ $-2, -1, 0, 1, 2, 3, \ldots$	$f'(x)$	f prime, the derivative of $f(x)$		
		$\mathbb{Q}$	the set of all rational numbers	(v_x, v_y, v_z)	ordered triple form of a vector
$\mathbb{R}$	the set of all real numbers	$\mathbb{C}$	the set of all complex numbers	$\overrightarrow{PQ}$	the displacement from P to Q, a directed line segment
e	the base of natural logarithms; e ≈ 2.718	π	pi, the ratio of a circle's circumference to its diameter; $\pi \approx 3.142$	$\boldsymbol{i}, \boldsymbol{j}, \boldsymbol{k}$	Cartesian unit vectors
∞	infinity				
$r!$	r factorial, the number $r \times (r-1) \times (r-2)$ $\ldots \times 3 \times 2 \times 1$	$0!$	zero factorial, defined to be 1	$\cdot$	scalar (dot) product
				$z = x + iy$	Cartesian form of a complex number
$\sum$	the sum of all terms of the immediately following kind	^nC_r	the number of ways of choosing r items from n items irrespective of order $(= n!/(n-r)!\,r!)$	$|z|$	modulus of z
				$r\,e^{i\theta}$	exponential form of a complex number
$\%$	percent				
$\equiv$	identical to	$\displaystyle\sum_{i=1}^N$	the sum of terms of the following kind for $i = 1$ to $i = N$	$\displaystyle\lim_{h\to 0}$	the limit as h tends to zero
$\approx$	approximately equal to			$f(a)$	the function f, evaluated at $x = a$
$<$	less than	(x, y, z)	Cartesian coordinates of a point	$f^{(n)}(x)$	the nth derivative of $f(x)$
$\leqslant$	less than or equal to	$x(t)$	x, a function of t		
$<<$	much less than	$\ddot{x}(t)$	x double dot, the second derivative of x with respect to t	$\dot{x}(t)$	x dot, the first derivative of x with respect to t
$/$	alternative notation for division			$\dfrac{df}{dx}(x)$	the first derivative of f with respect to x (same as $f'(x)$)
$\ldots$	and so on (ellipsis)	$\dfrac{df}{dx}(a)$	the first derivative of f with respect to x evaluated at $x = a$		
$[\ ]$	square brackets			$\dfrac{df(x)}{dx}$	alternative notation to $\dfrac{df}{dx}(x)$
$\in$	an element of (the set $\ldots$)	$\dfrac{d^2f(x)}{dx^2}$	the second derivative of f with respect to x		
$\mathbb{N}$	the set of all natural numbers $1, 2, 3, \ldots$				

$\int f(x)\,dx$	the (indefinite) integral of $f(x)$ with respect to x
$[F(x)]_a^b$	the difference $F(b) - F(a)$
$|x|$	the modulus (or absolute value) of x
$\sqrt[n]{x}$	the nth root of x
e^x	alternative form for the exponential function
$\log_{10}(x)$	logarithmic function to the base 10 (sometimes written as $\log x$, but not here)

Parentheses are often omitted in the following expressions:

$\sin(x)$	the sine function
$\cos(x)$	the cosine function
$\tan(x)$	the tangent function
$\operatorname{cosec}(x)$	the cosecant function
$\sec(x)$	the secant function
$\cot(x)$	the cotangent function
$\arcsin(x)$	the inverse sine function (sometimes written as $\sin^{-1}x$, but not here)
$\arccos(x)$	the inverse cosine function (sometimes written as $\cos^{-1}x$, but not here)
$\arctan(x)$	the inverse tangent function (sometimes written as $\tan^{-1}x$, but not here)
$\operatorname{arccosec}(x)$	the inverse cosecant function (sometimes written as $\operatorname{cosec}^{-1}x$, but not here)
$\operatorname{arcsec}(x)$	the inverse secant function (sometimes written as $\sec^{-1}x$, but not here)
$\operatorname{arccot}(x)$	the inverse cotangent function (sometimes written as $\cot^{-1}x$, but not here)

0.8   Statistics

Arithmetic mean
The mean of n values, $x_1, x_2, x_3, x_4, \ldots, x_{n-2}$, x_{n-1}, x_n, of a quantity x

$$\langle x \rangle =$$

$$\frac{x_1 + x_2 + x_3 + x_4 + \ldots x_{n-2} + x_{n-1} + x_n}{n}$$

$$= \frac{1}{n}\sum_{i=1}^{n} x_i$$

Standard deviation
Given n values, $x_1, x_2, x_3, x_4, x_5, \ldots x_n$, of a quantity x, with mean $\langle x \rangle$ (this is sometimes written as $\bar{x}$, the deviation of the ith value is $d_i = x_i - \langle x \rangle$, and the standard deviation σ_n is defined by

$$\sigma_n = \sqrt{\frac{d_1^2 + d_2^2 + \ldots + d_n^2}{n}} = \left(\frac{1}{n}\sum_{i=1}^{n} d_i^2\right)^{1/2}$$

According to *Bessel's correction*, if the values $x_1, \ldots, x_n$ are regarded as a sample of an infinite population of possible measurements, the best estimate of the standard derivation of that population is

$$\sigma = \left(\frac{n}{n-1}\right)^{1/2} \sigma_n$$

The distinction between σ and σ_n is insignificant for large n.

The method of least squares
The equation of the least squares line through the data points $(x_1, y_1), (x_2, y_2), \ldots, (x_n, y_n)$ is given by $y = ax + b$, where

$$a = \frac{\left(\sum_{i=1}^{n} x_i y_i\right) - n\langle x\rangle\langle y\rangle}{\left(\sum_{i=1}^{n} x_i^2\right) - n\langle x\rangle^2}$$

and

$$b = \langle y \rangle - a\langle x \rangle$$

The point $(\langle x\rangle, \langle y\rangle)$ lies on the least squares line.

Normal (or Gaussian) distribution
A quantity is said to be normally distributed if the probability (i.e. relative likelihood) that a single determination of its value will yield a result in the narrow range Δx, centred on the value x is

$$y\,\Delta x = \frac{1}{\sqrt{2\pi}\,\sigma}\exp\left[\frac{-(x - \langle x\rangle)^2}{2\sigma^2}\right]\Delta x$$

where $\langle x \rangle$ is the mean of the distribution and σ its standard deviation.

Table 7   Straightening curves

Points plotted		Straight line ($Y = mX + C$)	Relation between x and y
$X = x$,	$Y = \log_e y$	$\log_e y = mx + \log_e A$	$y = A\,e^{mx}$
$X = \log_e x$,	$Y = \log_e y$	$\log_e y = m \log_e x + \log_e A$	$y = Ax^m$
$X = f(x)$,	$Y = g(y)$	$g(y) = mf(x) + C$	$y = g^{-1}(mf(x) + C)$
$X = x^n$,	$Y = y$	$y = mX + C$	$y = mx^n + C$

Given n readings of a normally distributed quantity, with (sample) mean $\langle x \rangle$ and (sample) standard deviation σ_n, then about 68% of the readings will lie within $\pm\sigma_n$ of the mean value, 95% within $\pm 2\sigma_n$, and 99.7% within $\pm 3\sigma_n$, provided n is sufficiently large.

An estimate of the *standard error in the mean*, s_m, is given by

$$s_m \approx \frac{\sigma_n}{\sqrt{n-1}}$$

The standard error in a sample mean is

$$s_m = \frac{\sigma}{\sqrt{n}}$$

Errors

Independent errors $\pm e_1$, $\pm e_2$, $\pm e_3, \ldots \pm e_n$ in a measured quantity will give rise to an overall error $\pm E$ of:

$$E = \sqrt{e_1^2 + e_2^2 + e_3^2 + \ldots + e_n^2}$$
$$= (e_1^2 + e_2^2 + e_3^2 + \ldots + e_n^2)^{1/2}$$

For quantities A and B with independent errors ΔA and ΔB:

If $X = A + B$, or $X = A - B$, the error in X is given by:

$$\Delta X = \sqrt{(\Delta A)^2 + (\Delta B)^2}$$

If $X = AB$, or $X = A/B$, the error in X is given by:

$$\frac{\Delta X}{X} = \sqrt{\left(\frac{\Delta A}{A}\right)^2 + \left(\frac{\Delta B}{B}\right)^2}$$

If $X = A^n$, the error in X is given by:

$$\frac{\Delta X}{X} = n\left(\frac{\Delta A}{A}\right)$$

If $X = kA$, where k is a constant, the error in X is given by:

$$\Delta X = k\,\Delta A$$

Straight-line graphs from data

Logarithms and other functions may be used to obtain straight-line fits to data and hence provide insight into the functional relation between the measured quantities x and y.

(In Table 7, m, n, C and A denote constants and g^{-1} is the inverse function to g.)

1   Basic arithmetic and algebra

See also Subsection 2.2 and Figures 1 to 3 on p. 590.

1.1   Introducing algebra and arithmetic [Chapter 1]

Order of priority for basic mathematical operations: anything in brackets, multiplication ($\times$) and division ($\div$), then addition ($+$) and subtraction ($-$)

Powers: a^b denotes a to the power b

Roots: $a^{p/q}$ denotes the qth root of a^p, i.e. $\sqrt[q]{a^p}$

Reciprocals:

$$a^{-b} = \frac{1}{a^b} \quad (a \neq 0)$$

$$a^0 = 1$$

$$\frac{a/b}{c/d} = \frac{ad}{bc}$$

$a^1 = a$

$a^m \times a^n = a^{m+n}$

$\dfrac{a}{b} + \dfrac{c}{d} = \dfrac{ad + cb}{bd} \quad \left(not \; \dfrac{a+c}{b+d} \right)$

$a^m/a^n = a^{m-n}$

$(a^m)^n = a^{mn}$

$(a - b)^2 = a^2 - 2ab + b^2$

$(a + b)^2 = a^2 + b^2 + 2ab$

$(a + b)(a - b) = a^2 - b^2$

$(a + b)(c + d) = ac + ad + bc + bd$

The modulus or absolute value of a number p is denoted by $|\,p\,|$, and if p is a positive number then $|\,p\,| = p$ and $|\,-p\,| = p$.

1.2 Equality, proportionality and inequality [Chapter 1]

Proportionality
If $y \propto x$ then $y = Kx$ where K is a constant of proportionality.

Inequality
If $x > y$ and a is any number,
  then $x + a > y + a$

If $x > y$ and k is a positive number,
  then $kx > ky$

If $x > y$ and k is a negative number,
  then $kx < ky$

1.3 Functions and graphs [Chapter 3]

A function is any combination of two sets (called the domain and the codomain) and a rule that meets the following conditions:

- the rule may be applied to every element of the first set,
- the rule associates a *single* element of the second set with each element of the first set.

If the first set consists of values of an independent variable x, and the second set consists of values of y (called the dependent variable) then we write $y = f(x)$ and say that y is a function of x.

Inverse functions
The inverse function of $f(x)$ is a function $g(y)$ such that if $y = f(x)$, then $x = g(y)$ for every value of x in the domain of $f(x)$. If $y = f(x)$, its inverse function is often denoted by f^{-1}, so that $x = f^{-1}(y)$.

Straight lines
A linear function is any function that may be written in the form

$$f(x) = mx + c$$

where m and c are constants, called the gradient and intercept, respectively. Its graph is a straight line.

The equation of a straight line can be represented in various ways:

- gradient–intercept form

  $$y = mx + c$$

- point–gradient form

  $$y - y_0 = m(x - x_0)$$

- two-point form

  $$\frac{y - y_1}{x - x_1} = \frac{y_2 - y_1}{x_2 - x_1}$$

- intercept form

  $$\frac{x}{a} + \frac{y}{b} = 1$$

gradient, $m = \dfrac{\Delta y}{\Delta x} = \dfrac{\text{rise}}{\text{run}}$

Quadratics
See also Subsection 2.3 and Figures 4 to 8 on pp. 590–1.

A quadratic function is any function that may be written in the form

$$f(x) = ax^2 + bx + c$$

where a, b and c are constants, and $a \neq 0$. Its graph is a parabola.

A quadratic function may be written in various ways:

- completed square form

$$f(x) = a(x - p)^2 + q$$

where $p = \dfrac{-b}{2a}$ and $q = \dfrac{-b^2}{4a} + c$

- factorised form

$$f(x) = a(x - \alpha)(x - \beta)$$

where $\alpha = \dfrac{-b + \sqrt{b^2 - 4ac}}{2a}$

and $\beta = \dfrac{-b - \sqrt{b^2 - 4ac}}{2a}$ $(a \neq 0)$

Polynomials

A polynomial function of degree n is any function of the form

$$f(x) = a_0 + a_1 x + a_2 x^2 + \ldots$$
$$+ a_{n-2} x^{n-2} + a_{n-1} x^{n-1} + a_n x^n$$

where n is an integer, and $a_n \neq 0$. The $n+1$ constants $a_0, a_1, a_2, \ldots a_{n-2}, a_{n-1}$ and a_n are called the coefficients of the polynomial.

1.4   Solving equations [Chapter 4]

These are often written as

$$\frac{-b \pm \sqrt{b^2 - 4ac}}{2a}$$

The quadratic equation $ax^2 + bx + c = 0$ has roots (i.e. solutions)

$$\alpha = \frac{-b + \sqrt{b^2 - 4ac}}{2a}$$

and $\beta = \dfrac{-b - \sqrt{b^2 - 4ac}}{2a}$ $(a \neq 0)$

- if $b^2 > 4ac$ there are two real roots
- if $b^2 = 4ac$ there is one real root
- if $b^2 < 4ac$ there are no real roots

See Figures 6 to 8.

The roots satisfy $\alpha + \beta = -b/a$ and $\alpha\beta = c/a$

The fundamental theorem of algebra states that *every* polynomial equation of degree n has precisely n roots provided complex roots and repeated roots are counted.

(For numerical solutions of equations see the Newton–Raphson formula in the companion volume, *Further Mathematics for the Physical Sciences*, Subsection 9.3.7.)

1.5   Trigonometric functions [Chapter 5]

Radian measure

Remember to set your calculator to radian mode or degree mode as appropriate.

$$\phi = \frac{s}{r} \text{ rad} \quad 2\pi \text{ radians} = 360°$$

Pythagoras's theorem

For any right angled triangle

$$(\text{adjacent})^2 + (\text{opposite})^2 = (\text{hypotenuse})^2$$

Trigonometric ratios

For a right-angled triangle, and for $0° \leqslant \theta \leqslant 90°$:

$$\sin \theta = \frac{\text{opposite}}{\text{hypotenuse}}$$

$$\cos \theta = \frac{\text{adjacent}}{\text{hypotenuse}}$$

$$\tan \theta = \frac{\text{opposite}}{\text{adjacent}}$$

Sine rule

For any triangle with sides of length a, b, c:

$$\frac{a}{\sin \hat{A}} = \frac{b}{\sin \hat{B}} = \frac{c}{\sin \hat{C}}$$

Beware of ambiguity when given two sides and an angle not between them, especially if the angle is opposite the shorter side.

Table 8 Angles and their trigonometric ratios

θ/degrees	θ/radians	sin θ	cos θ	tan θ
0	0	0	1	0
30	$\pi/6$	1/2	$\sqrt{3}/2$	$1/\sqrt{3}$
45	$\pi/4$	$1/\sqrt{2}$	$1/\sqrt{2}$	1
60	$\pi/3$	$\sqrt{3}/2$	1/2	$\sqrt{3}$
90	$\pi/2$	1	0	undefined

Table 9 Positive trigonometric functions

Angular range	Positive trigonometric functions
$0° \leqslant \theta < 90°$	all
$90° \leqslant \theta < 180°$	sin θ
$180° \leqslant \theta < 270°$	tan θ
$270° \leqslant \theta < 360°$	cos θ

sin	all
positive	positive
tan	cos
positive	positive

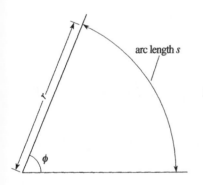

Figure A Radian measure.

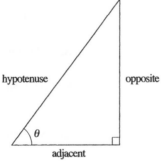

Figure B A right-angled triangle.

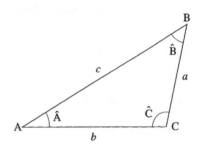

Figure D The sides and angles of an arbitrary triangle.

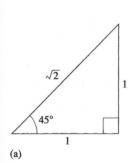

(a)

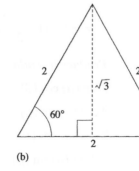

(b)

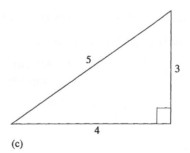

(c)

Figure C Useful triangles.

Cosine rule

See Figure D.
For any triangle with sides of length a, b, c:
$$a^2 = b^2 + c^2 - 2bc \cos \hat{A}$$

Trigonometric functions and identities

See Figures 18 to 20.

The trigonometric functions, $\sin(\theta)$, $\cos(\theta)$ and $\tan(\theta)$, are periodic functions of period 2π (period π for $\tan(\theta)$) that are defined to agree with the corresponding trigonometric ratios in the range $0° \leqslant \theta \leqslant 90°$. They satisfy the following conditions.

Symmetry relations:

$$\sin(-\alpha) = -\sin(\alpha)$$

$$\cos(-\alpha) = \cos(\alpha)$$

$$\tan(-\alpha) = -\tan(\alpha)$$

Reciprocal trigonometric functions:

See Figures 24 to 26.

$$\sec(\alpha) = 1/\cos(\alpha)$$

$$\operatorname{cosec}(\alpha) = 1/\sin(\alpha)$$

$$\cot(\alpha) = 1/\tan(\alpha)$$

Basic identities:

$$\sin^2(\alpha) + \cos^2(\alpha) = 1$$

$$\sec^2(\alpha) = 1 + \tan^2(\alpha)$$

$$\tan(\alpha) = \sin(\alpha)/\cos(\alpha)$$

Addition formulae:

$$\sin(\alpha + \beta) = \sin(\alpha)\cos(\beta) + \cos(\alpha)\sin(\beta)$$

$$\cos(\alpha + \beta) = \cos(\alpha)\cos(\beta) - \sin(\alpha)\sin(\beta)$$

$$\tan(\alpha + \beta) = \frac{\tan(\alpha) + \tan(\beta)}{1 - \tan(\alpha)\tan(\beta)}$$

$$\text{if} \quad \alpha + \beta \neq \frac{(2n+1)\pi}{2}$$

$$\sin(\alpha - \beta)$$

$$= \sin(\alpha)\cos(\beta) - \cos(\alpha)\sin(\beta)$$

$$\cos(\alpha - \beta)$$

$$= \cos(\alpha)\cos(\beta) + \sin(\alpha)\sin(\beta)$$

$$\tan(\alpha - \beta) = \frac{\tan(\alpha) - \tan(\beta)}{1 + \tan(\alpha)\tan(\beta)}$$

$$\text{if} \quad \alpha - \beta \neq \frac{(2n+1)\pi}{2}$$

Double-angle formulae:

$$\sin(2\alpha) = 2\sin(\alpha)\cos(\alpha)$$

$$\cos(2\alpha) = \cos^2(\alpha) - \sin^2(\alpha)$$

$$= 1 - 2\sin^2(\alpha) = 2\cos^2(\alpha) - 1$$

Half-angle formulae:

If $t = \tan(\alpha/2)$ then

$$\sin(\alpha) = \frac{2t}{1 + t^2}$$

$$\cos(\alpha) = \frac{1 - t^2}{1 + t^2}$$

$$\text{and} \quad \tan(\alpha) = \frac{2t}{1 - t^2}$$

Sum formulae:

$$\sin(\alpha) + \sin(\beta) = 2\sin\left(\frac{\alpha + \beta}{2}\right)\cos\left(\frac{\alpha - \beta}{2}\right)$$

$$\sin(\alpha) - \sin(\beta) = 2\cos\left(\frac{\alpha + \beta}{2}\right)\sin\left(\frac{\alpha - \beta}{2}\right)$$

$$\cos(\alpha) + \cos(\beta) = 2\cos\left(\frac{\alpha + \beta}{2}\right)\cos\left(\frac{\alpha - \beta}{2}\right)$$

$$\cos(\alpha) - \cos(\beta)$$

$$= -2\sin\left(\frac{\alpha + \beta}{2}\right)\sin\left(\frac{\alpha - \beta}{2}\right)$$

Product formulae:

$$2\sin(\alpha)\cos(\beta) = \sin(\alpha + \beta) + \sin(\alpha - \beta)$$

$$2\cos(\alpha)\sin(\beta) = \sin(\alpha + \beta) - \sin(\alpha - \beta)$$

$$2\cos(\alpha)\cos(\beta) = \cos(\alpha + \beta) + \cos(\alpha - \beta)$$

$$-2\sin(\alpha)\sin(\beta)$$

$$= \cos(\alpha + \beta) - \cos(\alpha - \beta)$$

Table 10   Trigonometric functions between 0 and 2π

θ/radians	sin (θ)	cos (θ)	tan (θ)
0	0	1	0
$\pi/6$	1/2	$\sqrt{3}/2$	$1/\sqrt{3}$
$\pi/4$	$1/\sqrt{2}$	$1/\sqrt{2}$	1
$\pi/3$	$\sqrt{3}/2$	1/2	$\sqrt{3}$
$\pi/2$	1	0	undefined
$2\pi/3$	$\sqrt{3}/2$	$-1/2$	$-\sqrt{3}$
$3\pi/4$	$1/\sqrt{2}$	$-1/\sqrt{2}$	-1
$5\pi/6$	1/2	$-\sqrt{3}/2$	$-1/\sqrt{3}$
π	0	-1	0
$7\pi/6$	$-1/2$	$-\sqrt{3}/2$	$1/\sqrt{3}$
$5\pi/4$	$-1/\sqrt{2}$	$-1/\sqrt{2}$	1
$4\pi/3$	$-\sqrt{3}/2$	$-1/2$	$\sqrt{3}$
$3\pi/2$	-1	0	undefined
$5\pi/3$	$-\sqrt{3}/2$	1/2	$-\sqrt{3}$
$7\pi/4$	$-1/\sqrt{2}$	$1/\sqrt{2}$	-1
$11\pi/6$	$-1/2$	$\sqrt{3}/2$	$-1/\sqrt{3}$
2π	0	1	0

Small angle approximations

For a small angle θ, measured in radians:

$$\cos \theta \approx 1 - \frac{\theta^2}{2} \quad \text{and} \quad \sin \theta \approx \tan \theta \approx \theta$$

Inverse trigonometric functions

See Figures 21 to 23 and 27 to 29.

The abbreviations asin (θ), acos (θ) and atan (θ) or alternatively $\sin^{-1}$ (θ), $\cos^{-1}$ (θ) and $\tan^{-1}$ (θ), are sometimes used for the inverse trigonometric functions, but not here.

$\arcsin (\sin \theta) = \theta \qquad -\pi/2 \leqslant \theta \leqslant \pi/2$

$\arccos (\cos \theta) = \theta \qquad 0 \leqslant \theta \leqslant \pi$

$\arctan (\tan \theta) = \theta \qquad -\pi/2 < \theta < \pi/2$

$\text{arccosec} (\text{cosec} \, \theta) = \theta \;\; -\pi/2 \leqslant \theta \leqslant \pi/2, \, \theta \neq 0$

$\text{arcsec} (\sec \theta) = \theta \qquad 0 \leqslant \theta \leqslant \pi, \, \theta \neq \pi/2$

$\text{arccot} (\cot \theta) = \theta \qquad 0 < \theta < \pi$

1.6   Exponential and logarithmic functions [Chapter 6]

See Figure 16.

For more on derivatives of exp and log see Section 4.

Exponential functions

$$e = \lim_{n \to \infty} \left(1 + \frac{1}{n} \right)^n$$

$$\approx 2.718\ 281\ 828\ 459\ 05 \ldots$$

(to 14 decimal places)

$\exp (x) = e^x$

$\dfrac{d}{dx} (e^{kx}) = k e^{kx}$

$e^0 = 1$

$e^1 = e$

$e^{-1} = 1/e$

$e^x \times e^y = e^{x+y}$

$e^x/e^y = e^{x-y}$

$(e^x)^y = e^{xy}$

a^x can be written as e^{kx} where $e^k = a$

Logarithmic functions

See Figure 17.

If $x = a^y$ then $y = \log_a x$ $(x > 0$ and $a > 0)$

$\log_a (1) = 0$

$\log_a (a) = 1$

$\log_a (x^y) = y \log_a (x)$

$\log_a (x) + \log_a (y) = \log_a (xy)$

$\log_a (a^x) = x$

$\log_a (x) - \log_a (y) = \log_a (x/y)$

$\log_a (x) = \log_b (x)/\log_b (a)$

Inverse relation of exp and log$_e$

exp and log$_e$ are inverse functions.

If $\quad y = \exp (x) \quad$ then $\quad x = \log_e (y)$

$\quad\quad \exp (\log_e x) = x \quad\quad \log_e (e^x) = x$

1.7 Hyperbolic functions [Chapter 7]

See Figures 30 to 32.

Basic definitions

$\sinh (x) = \dfrac{e^x - e^{-x}}{2}$

$\cosh (x) = \dfrac{e^x + e^{-x}}{2}$

$\tanh (x) = \dfrac{\sinh (x)}{\cosh (x)} = \dfrac{e^x - e^{-x}}{e^x + e^{-x}}$

Relationship to the hyperbola

A hyperbola can be defined by means of the parametric equations

$x = a \cosh (t)$

$y = b \sinh (t)$

Reciprocal hyperbolic functions

$\text{cosech} (x) = \dfrac{1}{\sinh (x)} \quad$ provided $x \neq 0$

$\text{sech} (x) = \dfrac{1}{\cosh (x)}$

$\coth (x) = \dfrac{1}{\tanh (x)} \quad$ provided $x \neq 0$

Inverse hyperbolic functions

See Figures 33 to 35.

$\text{arcsinh} (\sinh (x)) = x$

$\text{arccosh} (\cosh (x)) = x \quad$ for $x \geqslant 0$

$\text{arctanh} (\tanh (x)) = x$

In terms of the logarithmic function

$\text{arcsinh} (x) = \log_e (x + \sqrt{x^2 + 1})$

$\text{arccosh} (x) = \log_e (x + \sqrt{x^2 - 1}) \quad$ for $x \geqslant 1$

$\text{arctanh} (x) = \dfrac{1}{2} \log_e \left(\dfrac{1 + x}{1 - x}\right) \quad$ for $-1 < x < 1$

Identities

Symmetry relations:

$$\begin{aligned} \sinh (-x) &= -\sinh (x), \\ \cosh (-x) &= \cosh (x), \\ \tanh (-x) &= -\tanh (x) \end{aligned}$$

Basic identities:

$\tanh (x) = \dfrac{\sinh (x)}{\cosh (x)}$

$\cosh^2 (x) - \sinh^2 (x) = 1$

$1 - \tanh^2 (x) = \text{sech}^2 (x)$

$\coth (x) = \dfrac{\text{cosech} (x)}{\text{sech} (x)} \quad$ for $x \neq 0$

Addition identities:

$\sinh (x + y) = \sinh (x) \cosh (y) + \cosh (x) \sinh (y)$

$\cosh (x + y) = \cosh (x) \cosh (y) + \sinh (x) \sinh (y)$

$$\tanh (x + y) = \frac{\tanh (x) + \tanh (y)}{1 + \tanh (x) \tanh (y)}$$

Double-argument identities:

$$\sinh (2x) = 2 \sinh (x) \cosh (x)$$

$$\cosh (2x) = \cosh^2 (x) + \sinh^2 (x)$$

$$\cosh (2x) = 1 + 2 \sinh^2 (x)$$

$$\cosh (2x) = 2 \cosh^2 (x) - 1$$

$$\tanh (2x) = \frac{2 \tanh (x)}{1 + \tanh^2 (x)}$$

Half-argument identities:

$$\cosh^2 \left(\frac{x}{2}\right) = \frac{1}{2} [\cosh (x) + 1]$$

$$\sinh^2 \left(\frac{x}{2}\right) = \frac{1}{2} [\cosh (x) - 1]$$

and if $t = \tanh (x/2)$ then

$$\sinh (x) = \frac{2t}{1 - t^2}$$

$$\cosh (x) = \frac{1 + t^2}{1 - t^2}$$

$$\tanh (x) = \frac{2t}{1 + t^2}$$

Sum identities:

$$\sinh (x) + \sinh (y)$$
$$= 2 \sinh \left(\frac{x + y}{2}\right) \cosh \left(\frac{x - y}{2}\right)$$

$$\sinh (x) - \sinh (y)$$
$$= 2 \cosh \left(\frac{x + y}{2}\right) \sinh \left(\frac{x - y}{2}\right)$$

$$\cosh (x) + \cosh (y)$$
$$= 2 \cosh \left(\frac{x + y}{2}\right) \cosh \left(\frac{x - y}{2}\right)$$

$$\cosh (x) - \cosh (y)$$
$$= 2 \sinh \left(\frac{x + y}{2}\right) \sinh \left(\frac{x - y}{2}\right)$$

Product identities:

$$2 \sinh (x) \cosh (y)$$
$$= \sinh (x + y) + \sinh (x - y)$$

$$2 \cosh (x) \cosh (y)$$
$$= \cosh (x + y) + \cosh (x - y)$$

$$2 \sinh (x) \sinh (y)$$
$$= \cosh (x + y) - \cosh (x - y)$$

2 Basic geometry

2.1 Introducing geometry [Chapter 8]

The sum of the interior angles of an n sided polygon is $(2n - 4) \times 90°$, i.e. $(2n - 4)$ right angles.

Conditions for congruent triangles

Lengths of all three sides. (SSS)

Lengths of two sides and the angle between them. (SAS)

Length of one side and the angles at each end of it. (ASA)

A right angle, the hypotenuse and the length of one other side. (RHS)

Areas and volumes
See Table 11.

2.2 Coordinate geometry [Chapter 9]

See Figure 13.

Lines
General equation of a line

$$ax + by + c = 0$$

The equation of a straight line can be represented in various ways:

Table 11   Useful formulae for areas

Figure	Surface area
Rectangle	base $\times$ height
Parallelogram	base $\times$ height
Triangle	$\frac{1}{2} \times$ base $\times$ height
	If $s = \frac{1}{2}(a + b + c)$ (see Figure D) the area can also be written in the form $$\sqrt{s(s-a)(s-b)(s-c)}$$
Trapezium	$\frac{1}{2} \times$ (base$_1$ + base$_2$) $\times$ height
Circle	πr^2
Sector of a circle	$\frac{1}{2}\, \theta\, r^2$
Ellipse	πab
Cylinder (*not* including ends)	$2\pi r \times$ height*
Cone (*not* including base)	$\frac{1}{2} \times$ base perimeter $\times$ slant height
Sphere	$4\pi r^2$

*The circumference of a circle of radius r is $2\pi r$.

Table 12   Useful formulae for volumes

Figure	Volume
Rectangular prism	base area $\times$ height
Cylinder	$\pi r^2 \times$ height
Cone (or pyramid)	$\frac{1}{3} \times$ base area $\times$ height
Sphere	$\frac{4}{3}\pi r^3$

- gradient–intercept form
$$y = mx + c$$

- point–gradient form
$$y - y_0 = m(x - x_0)$$

- two-point form
$$\frac{y - y_1}{x - x_1} = \frac{y_2 - y_1}{x_2 - x_1}$$

- intercept form
$$\frac{x}{a} + \frac{y}{b} = 1$$

$$\text{gradient, } m = \frac{\Delta y}{\Delta x} = \frac{\text{rise}}{\text{run}}$$

Two lines with gradients m_1 and m_2 are perpendicular if $m_1 m_2 = -1$.

The distance PQ between the points $P(x_1, y_1)$ and $Q(x_2, y_2)$ is given by
$$PQ = \sqrt{(x_2 - x_1)^2 + (y_2 - y_1)^2}$$
(See also Subsection 1.3 and Figures 1 to 3.)

Circles

See Figure 14.

Equation of a circle of radius R with centre at the origin
$$x^2 + y^2 = R^2$$

Equation of a circle of radius R with centre at the point (x_0, y_0)
$$(x - x_0)^2 + (y - y_0)^2 = R^2$$

Equation of the tangent to the circle $x^2 + y^2 = R^2$ at the point (x_1, y_1)
$$x_1 x + y_1 y = R^2$$

Polar coordinates

See Figure G in Subsection 3.2.

Polar to Cartesian conversion:
$$x = r \cos \theta \quad \text{and} \quad y = r \sin \theta$$

Cartesian to polar conversion:
$$r^2 = x^2 + y^2, \quad \sin \theta = y/r$$
$$\text{and} \quad \cos \theta = x/r \quad r \neq 0$$

Coordinate geometry in three dimensions

The distance between two points

$$d^2 = (x_2 - x_1)^2 + (y_2 - y_1)^2 + (z_2 - z_1)^2$$

The equation of a plane

$$ax + by + cz = d$$

The equations of a line in three dimensions

$$\frac{x - a}{l} = \frac{y - b}{m} = \frac{z - c}{n}$$

2.3   Conic sections [Chapter 10]

See Table 13.

3   Basic vector algebra

3.1   Introducing scalars and vectors [Chapter 11]

Representing scalars and vectors

A scalar quantity has magnitude alone. It can be specified by a single number (together with the appropriate units of measurement).

Vector quantities are physical quantities that have both magnitude and direction and which can be adequately represented by 'geometric' vectors (i.e. directed line segments).

Table 13   (See also Figures 5, 9, 13 and 15)

Property	Circle	Parabola	Ellipse	Hyperbola
eccentricity	$e = 0$	$e = 1$	$0 \leqslant e < 1$	$e > 1$
focus		$(a, 0)$	$(\pm ae, 0)$	$(\pm ae, 0)$
directrix		$x = -a$	$x = \pm a/e$	$x = \pm a/e$
standard equation	$x^2 + y^2 = a^2$	$y^2 = 4ax$	$\dfrac{x^2}{a^2} + \dfrac{y^2}{b^2} = 1$	$\dfrac{x^2}{a^2} - \dfrac{y^2}{b^2} = 1$
	centre $(0, 0)$ radius a		$b = a\sqrt{1 - e^2}$	$b = a\sqrt{e^2 - 1}$
asymptotes	none	none	none	$y = \pm\dfrac{b}{a}x$
parametric form	$x = a \cos \theta$ $y = a \sin \theta$	$x = at^2$ $y = 2at$	$x = a \cos \theta$ $y = b \sin \theta$	$x = a \cosh \theta$ $y = b \sinh \theta$
tangent at (x_1, y_1)	$x_1 x + y_1 y = a^2$	$y_1 y = 2a(x + x_1)$	$\dfrac{x_1 x}{a^2} + \dfrac{y_1 y}{b^2} = 1$	$\dfrac{x_1 x}{a^2} - \dfrac{y_1 y}{b^2} = 1$
polar form	$r = a$, centre $(0, 0)$, radius a	$\dfrac{L}{r} = 1 + \cos \theta$	$\dfrac{L}{r} = 1 + e \cos \theta$	$\dfrac{L}{r} = 1 + e \cos \theta$
	$r = 2a \cos \theta$, centre $(a, 0)$, radius a		$0 \leqslant e < 1$	$e > 1$
some other forms and special cases	$(x - p)^2 + (x - q)^2 = a^2$ centre (p, q) radius a	$y = Ax^2 + Bx + C$		rectangular hyperbola $x^2 - y^2 = a^2$
	$x^2 + y^2 + 2Gx + 2Fy + C = 0$ centre $(-G, -F)$ radius $\sqrt{G^2 + F^2 - C}$			rectangular hyperbola $y = \dfrac{k}{x}$

In print vectors are usually denoted by bold type, e.g. a. In handwritten work it is usual to use a wavy underline $\underset{\sim}{a}$.

Vectors may be represented algebraically in the following ways:

- ordered triple form

$$v = (v_x, v_y, v_z)$$

- Cartesian unit vector form

$$v = v_x i + v_y j + v_z k$$

- component vector form

$$v = v_x + v_y + v_z$$

The Cartesian unit vectors i, j and k are defined below.

The magnitude of v, is then the positive scalar quantity

$$|v| = (|v_x|^2 + |v_y|^2 + |v_z|^2)^{1/2}$$

$$= (v_x^2 + v_y^2 + v_z^2)^{1/2}$$

Given a vector v, its orthogonal component vectors parallel and normal to a line inclined at an angle θ ($0° \leqslant \theta \leqslant 90°$) to v are of magnitude

$$|v_p| = |v| \cos \theta$$

and $|v_n| = |v| \sin \theta$

3.2 Working with vectors [Chapter 12]

Scaling a vector

For any vector v and scalar α

$$|\alpha v| = |\alpha| |v|$$

- ordered triple form

$$\alpha v = (\alpha v_x, \alpha v_y, \alpha v_z)$$

- Cartesian unit vector form

$$\alpha v = \alpha v_x i + \alpha v_y j + \alpha v_z k$$

- component vector form

$$\alpha v = \alpha v_x + \alpha v_y + \alpha v_z$$

Adding vectors

The triangle rule: Let vectors a and b be represented by appropriate arrows (or directed line segments). If the arrow representing b is drawn from the head of the arrow representing a, then an arrow from the tail of a to the head of b represents the vector sum $a + b$.

- ordered triple form

$$a + b = ((a_x + b_x), (a_y + b_y), (a_z + b_z))$$

- Cartesian unit vector form

$$a + b = (a_x + b_x)i + (a_y + b_y)j + (a_z + b_z)k$$

- component vector form

$$a + b = (a_x + b_x) + (a_y + b_y) + (a_z + b_z)$$

Properties of vector addition and scaling

The zero vector, 0, has the property that $v + 0 = v$

$$a + b = b + a$$

$$a + (b + c) = (a + b) + c$$

$$a + (-1)a = 0$$

$$\alpha (a + b) = \alpha a + \alpha b$$

for any scalar α

Unit vectors

A unit vector in the direction of the vector a is

$$\hat{a} = \frac{a}{|a|}$$

Note that $|\hat{a}| = 1$ (*Not* 1 unit, just 1)

The Cartesian unit vectors in the positive x, y and z directions are denoted i, j and k.

4 Basic differentiation

4.1 Introducing differentiation [Chapter 13]

Definition of the derivative
The derivative of the function $f(x)$ at $x = a$ is denoted by $df/dx(a)$ or $f'(a)$, and is defined by

$$f'(a) = \frac{df}{dx}(a) = \lim_{\Delta x \to 0} \left(\frac{f(a + \Delta x) - f(a)}{\Delta x} \right)$$

provided the limit exists and is unique.

If $y = f(x)$, the derivative dy/dx is the *rate of change* of y with respect to x.

4.2 Differentiating simple functions [Chapters 13 and 14]

Standard derivatives

Note: k must be such that the function arguments are dimensionless.

$d/dx\,[\log_e (kx)]$ is independent of k.

Basic rules of differentiation

The sum rule:

$$\frac{d}{dx}(f(x) + g(x)) = \frac{d}{dx}(f(x)) + \frac{d}{dx}(g(x))$$

i.e. $(f(x) + g(x))' = f'(x) + g'(x)$

The constant multiple rule:

$$\frac{d}{dx}(kf(x)) = k\frac{d}{dx}(f(x))$$

i.e. $(k f)'(x) = k f'(x)$

The product rule:

$$\frac{d}{dx}(f(x)\,g(x)) = \frac{df}{dx}g(x) + f(x)\frac{dg}{dx}$$

i.e. $(f(x)\,g(x))' = f'(x)\,g(x) + f(x)\,g'(x)$

The quotient rule:

$$\frac{d}{dx}\left(\frac{f(x)}{g(x)}\right) = \frac{\dfrac{df}{dx}g(x) - f(x)\dfrac{dg}{dx}}{(g(x))^2}$$

i.e. $\left(\dfrac{f(x)}{g(x)}\right)' = \dfrac{f'(x)\,g(x) - f(x)\,g'(x)}{g^2(x)}$

4.3 Differentiating composite functions [Chapter 15]

Example
If $y = \sin(ax^2)$ we can write

$$y = \sin(u) \text{ where } u = ax^2$$

Then

$$\frac{dy}{dx} = \cos u \times 2ax$$

i.e.

$$\frac{dy}{dx} = 2ax \cos(ax^2)$$

Table 14 Standard derivatives

$f(x)$	$f'(x)$	$f(x)$	$f'(x)$
k (constant)	0	$\sec(kx)$	$k \sec(kx) \tan(kx)$
kx^n	nkx^{n-1}	$\cot(kx)$	$-k \operatorname{cosec}^2(kx)$
$\sin(kx)$	$k \cos(kx)$	$\exp(kx)$	$k \exp(kx)$
$\cos(kx)$	$-k \sin(kx)$	$\log_e(kx)$	$1/x$ (☞)
$\tan(kx)$	$k \sec^2(kx)$	a^x	$(\log_e a)a^x$
$\operatorname{cosec}(kx)$	$-k \operatorname{cosec}(kx) \cot(kx)$	$\log_a x$	$1/(x \log_e a)$

The chain rule: differentiating functions of functions

If y is a function of u, so $y = f(u)$, and if in turn u is a function of x, so $u = g(x)$, then

$$\frac{dy}{dx} = \frac{dy}{du} \times \frac{du}{dx} = f'(u) \times g'(x)$$

Implicit differentiation

An implicit function is a function that is defined by means of a relation such as $F(x, y) = 0$. If the equation $F(x, y) = 0$ can be written $g(x) = f(y)$ then, using the chain rule, we can say

$$\frac{dg}{dx} = \frac{df}{dy}\frac{dy}{dx}$$

If df/dy can be expressed in terms of x, this can be rearranged to give dy/dx in terms of x.

Example

If $y^3 = x^5$ then

$$3y^2 \frac{dy}{dx} = 5x^4$$

so

$$\frac{dy}{dx} = \frac{5x^4}{3y^2} = \frac{5}{3}x^{2/3}$$

The inversion rule

$$\frac{dx}{dy} = 1 \bigg/ \left(\frac{dy}{dx}\right)$$

Parametric differentiation

If x and y are both defined in terms of a parameter θ, then the derivative dy/dx may be determined by

$$\frac{dy}{dx} = \frac{dy}{d\theta} \times \frac{d\theta}{dx} = \frac{dy}{d\theta} \bigg/ \frac{dx}{d\theta}$$

Higher derivatives

If $y = f(x)$, then the nth derivative of y is defined by

$$\frac{d^n y}{dx^n} = \frac{d}{dx}\left(\frac{d^{n-1}y}{dx^{n-1}}\right) \quad n \text{ is an integer, } n > 1$$

The first four derivatives of $f(x)$ may also be represented by $f'(x), f''(x), f'''(x), f^{(iv)}(x)$, and the nth derivative by $f^{(n)}(x)$.

4.4 Stationary points and graph sketching [Chapter 16]

Derivatives and the shape of the graph

$f(x)$ is *increasing* throughout an interval if $df/dx > 0$ for all points in that interval.

$f(x)$ is *decreasing* throughout an interval if $df/dx < 0$ for all points in that interval.

$f(x)$ is *stationary* at any point where $df/dx = 0$, and such points are called *stationary points*.

If $f''(x) > 0$ throughout an interval the graph of $f(x)$ is concave upwards.

If $f''(x) < 0$ throughout an interval the graph of $f(x)$ is concave downwards.

Global maximum (or minimum) in an interval

In order to find the global maximum (or minimum) of a given function in an interval $a \leqslant x \leqslant b$:

(a) First ensure that the function is defined at every point of the interval. If this is not the case there may be no global maximum or minimum.

(b) Evaluate the function at every point where the derivative is undefined.

(c) Evaluate the function at the end points of the interval.

(d) Evaluate the function at the stationary points (i.e. points where the derivative of the function is zero).

The global maximum is the greatest, and the global minimum the least, of these values.

Local maxima and minima

To find the local maxima and minima of a given function, $f(x)$ say:

(a) If you can, find the stationary points (the points at which $f'(x) = 0$).

(b) If it is easy to find the second derivative, evaluate $f''(x)$ at the stationary points. Then:

   (i) A positive second derivative indicates a local minimum.

   (ii) A negative second derivative indicates a local maximum.

   (iii) If the second derivative is zero at the stationary point one must investigate further:

     If $f''(x)$ changes sign at the stationary point then the point is a point of inflection (with horizontal tangent).

     If $f''(x)$ does not change sign at the stationary point then either: $f''(x) > 0$ near the point, which means that the graph is *concave upwards*, and there is a local minimum; or, $f''(x) < 0$ near the point, which means that the graph is concave downwards, and there is a local maximum.

(c) If it is not easy to calculate $f''(x)$, examine the behaviour of $f'(x)$ near the stationary point.

   (i) If $f'(x)$ changes sign at the stationary point then this point is a turning point, and must therefore be a local maximum or a local minimum. If $f'(x)$ changes from negative to positive there is a local minimum. If $f'(x)$ changes from positive to negative there is a local maximum.

   (ii) If $f'(x)$ does not change sign at the stationary point then this point is a point of inflection.

(d) If for a given function the first and second derivative tests are too difficult to apply, useful information can often be obtained by finding the values of $f(x)$ at values of x that are close to the stationary point and by plotting the graph of $f(x)$ on a calculator.

Graph sketching

In order to sketch the graph of a function, the following points may be considered:

(a) The symmetry of the function.

Is it *even* $(f(-x) = f(x))$ or *odd* $(f(-x) = -f(x))$?

(b) Intervals or points where the function is undefined.

(c) Points at which the graph crosses the axes. Do the intersections on the x-axis correspond to multiple roots?

(d) Behaviour of the graph near points at which the function is undefined. Are there any asymptotes?

(e) Behaviour of the graph when $|x| \gg 1$. Does the graph approach some simple curve?

(f) Behaviour of the graph where $|x| \ll 1$. What is the tangent at the origin?

(g) Points at which the function is stationary. Are these turning points (or points of inflection with a horizontal tangent)? Where is the function increasing and where is it decreasing?

(h) Where is the graph concave upwards and where is it concave downwards?

(i) Are there any *points of* inflection (i.e. points at which the second derivative of the function changes sign)?

5   Basic integration

5.1   Introducing integration
[Chapter 17]

Indefinite integrals

If $f(x)$ is a given function and $F(x)$ is a function such that its derivative $F'(x) = f(x)$ then $F(x)$ is known as an indefinite integral of $f(x)$, which we indicate by writing

$$\int f(x)\, dx = F(x)$$

A given function $f(x)$ may have many indefinite integrals, but any two of them may only differ by a constant, so in general:

If $\dfrac{dF}{dx} = f(x)$ then $\int f(x)\, dx = F(x) + C$

where C is a constant of integration and $f(x)$ is called the integrand.

Definite integrals

The definite integral of $f(x)$ from $x = a$ to $x = b$ is   defined   as   a   limit   of a sum

$$\int_a^b f(x)\, dx = \lim_{\Delta x \to 0} \left[\sum_{n=1}^{n} f(x_i)\, \Delta x_i \right]$$

with $\Delta x_i = x_{i+1} - x_i$

where $x_1 = a$, and $x_{n+1} = b$ and Δx is the largest of the intervals Δx_i.

Definite integrals may be evaluated using indefinite integrals thanks to the fundamental theorem of calculus which says that if $F(x)$ is an indefinite integral of $f(x)$ so that $F'(x) = f(x)$, then

$$\int_a^b f(x)\, dx = [F(x)]_a^b = F(b) - F(a)$$

5.2   Integrating simple functions
[Chapter 18]

Properties of definite integrals

1. $\displaystyle\int_a^a f(x)\, dx = 0$

2. $\displaystyle\int_b^a f(x)\, dx = -\int_a^b f(x)\, dx$

3. $\displaystyle\int_a^b f(x)\, dx = \int_a^c f(x)\, dx + \int_c^b f(x)\, dx$

where $a \leqslant c \leqslant b$

4. If k is any constant

then $\displaystyle\int_a^b k\, f(x)\, dx = k \int_a^b f(x)\, dx$

5. If $f(x) \geqslant 0$ for all x in the interval

$a \leqslant x \leqslant b$ then $\displaystyle\int_a^b f(x)\, dx \geqslant 0$

6. If $m \leqslant f(x) \leqslant M$ for all x in the interval $a \leqslant x \leqslant b$ then

$$m(b - a) \leqslant \int_a^b f(x)\, dx \leqslant M(b - a)$$

7. $\displaystyle\int_a^b (f(x) + g(x))\, dx = \int_a^b f(x)\, dx + \int_a^b g(x)\, dx$

8. If $f(x) \geqslant g(x)$ for all values of x in the interval $a \leqslant x \leqslant b$ then

$$\int_a^b f(x)\, dx \geqslant \int_a^b g(x)\, dx$$

Standard integrals

Note   The constant k must be such that the arguments of the various functions are dimensionless. Also, $a > 0$.

Table 15   Some standard integrals (C is a constant)

Function $f(x)$	Indefinite integral $\int f(x)\,dx$	Function $f(x)$	Indefinite integral $\int f(x)\,dx$
k (a constant)	$kx + C$	$\sec(kx)$	$\dfrac{1}{k}\log_e |\sec(kx) + \tan(kx)| + C$
$kx^n,\ n \neq -1$	$k\dfrac{x^{n+1}}{n+1} + C$	$\cot(kx)$	$\dfrac{1}{k}\log_e |\sin(kx)| + C$
$k\dfrac{1}{x},\ x \neq 0$	$k\log_e |x| + C$	$\log_e(kx)$	$x\log_e(kx) - x + C$
$\sin(kx)$	$-\dfrac{1}{k}\cos(kx) + C$	e^{kx}	$\dfrac{1}{k}e^{kx} + C$
$\cos(kx)$	$\dfrac{1}{k}\sin(kx) + C$	$\dfrac{1}{a^2x^2 + b^2}$	$\dfrac{1}{ab}\arctan\left(\dfrac{ax}{b}\right) + C$
$\tan(kx)$	$\dfrac{1}{k}\log_e |\sec(kx)| + C$	$\dfrac{1}{a^2x^2 - b^2}$	$\dfrac{1}{2ab}\log_e\left|\dfrac{ax - b}{ax + b}\right| + C$
$\sec^2(kx)$	$\dfrac{1}{k}\tan(kx) + C$	$\dfrac{1}{\sqrt{b^2 + a^2x^2}}$	$\dfrac{1}{a}\operatorname{arcsinh}\left(\dfrac{ax}{b}\right) + C$
$\operatorname{cosec}(kx)$	$\dfrac{1}{k}\log_e\left|\tan\left(\dfrac{kx}{2}\right)\right| + C$	$\dfrac{1}{\sqrt{b^2 - a^2x^2}}$	$\dfrac{1}{a}\arcsin\left(\dfrac{ax}{b}\right) + C$

5.3   Integrating by parts and by substitution [Chapter 19]

Integration by parts

$$\int f(x)\, g(x)\, dx = F(x)\, g(x) - \int F(x)\, \frac{dg}{dx}\, dx\,,$$

Integration by substitution

An integral of the form $\int f(x)\, dx$ may often be transformed into a simpler integral by writing $x = g(y)$ and $dx = g'(y)\, dy$. Substituting these relationsinto the original integral gives a new integral in terms of the new variable y

$$\int f(x)\, dx = \int f(g(y))\, g'(y)\, dy$$

When applying this technique to a definite integral, transform the limits of integration along with the integration variable.

If an integral is recognised as being of the form $\int f(g(y))\, g'(y)\, dy$ it may be simpler to use the substitution $x = g(y)$.

Example:

$$\int \frac{dx}{\sqrt{1 - x^2}}$$

If $x = \sin y$ then $dx = \cos y\, dy$ and the integral may be written

$$\int \frac{1}{\sqrt{1 - \sin^2 y}}\, \cos y\, dy = \int dy = y + C$$

$$= \arcsin x + C$$

6   Graphs, curves and functions

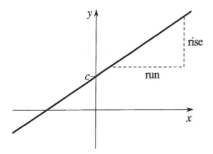

Figure 1 $y = mx + c$ where $m = $ rise/run.

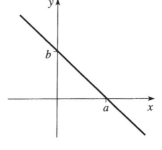

Figure 2 $(x/a) + (y/b) = 1$.

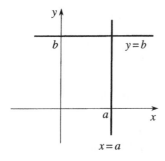

Figure 3   Horizontal line $y = b$ and vertical
line $x = a$.

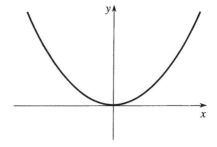

Figure 4 $y = x^2$.

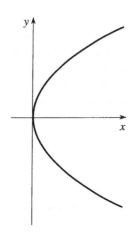

Figure 5 $y^2 = 4ax$.

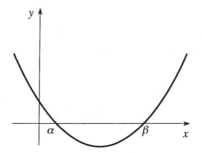

Figure 6 $y = (x - \alpha)(x - \beta)$.

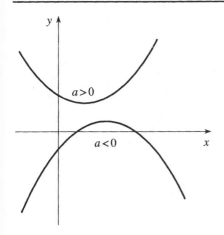

Figure 7 $y = ax^2 + bx + c$ with $a > 0$ (above) and $a < 0$ (below).

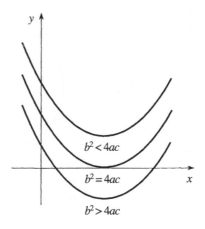

Figure 8 $y = ax^2 + bx + c$.

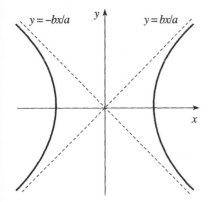

Figure 9 $(x^2/a^2) - (y^2/b^2) = 1$.

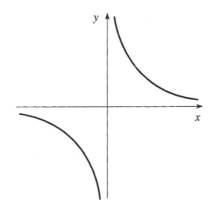

Figure 10 $y = 1/x$.

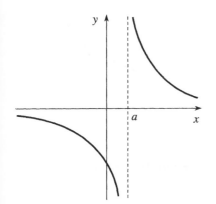

Figure 11 $y = 1/(x - a)$.

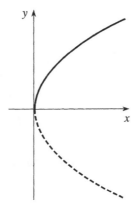

Figure 12 $y = \sqrt{x}$ and $y = -\sqrt{x}$ (dashed).

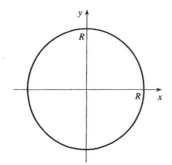

Figure 13 $x^2 + y^2 = R^2$.

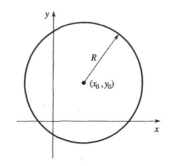

Figure 14
$(x - x_0)^2 + (y - y_0)^2 = R^2$.

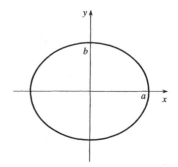

Figure 15 $(x^2/a^2) + (y^2/b^2) = 1$.

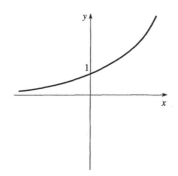

Figure 16 $y = a^x$.

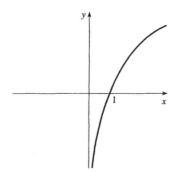

Figure 17 $y = \log_a x$.

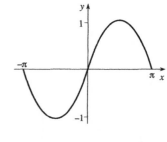

Figure 18 $y = \sin x$.

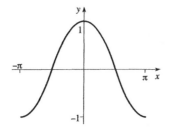

Figure 19 $y = \cos x$.

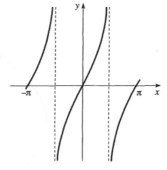

Figure 20 $y = \tan x$.

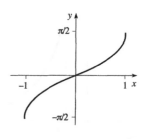

Figure 21 $y = \arcsin x$.

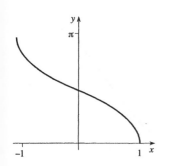

Figure 22 $y = \arccos x$.

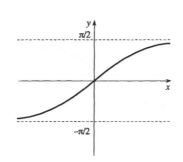

Figure 23 $y = \arctan x$.

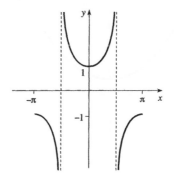

Figure 24 $y = \sec x$.

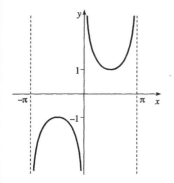

Figure 25 $y = \operatorname{cosec} x$.

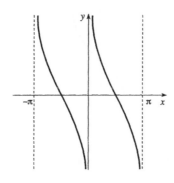

Figure 26 $y = \cot x$.

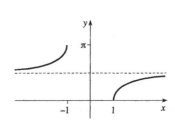

Figure 27 $y = \operatorname{arcsec} x$.

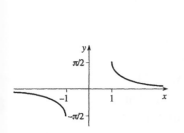

Figure 28 $y = \operatorname{arccosec} x$.

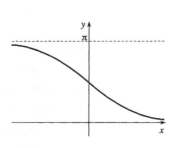

Figure 29 $y = \operatorname{arccot} x$.

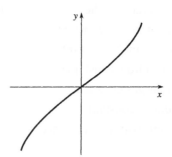

Figure 30 $y = \sinh x$.

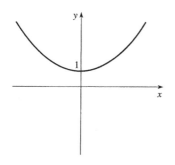

Figure 31 $y = \cosh x$.

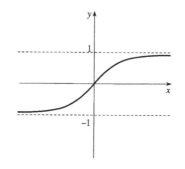

Figure 32 $y = \tanh x$.

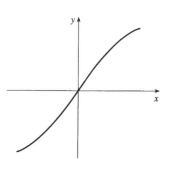

Figure 33 $y = \text{arcsinh } x$.

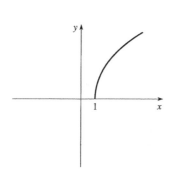

Figure 34 $y = \text{arccosh } x$.

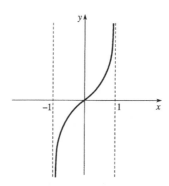

Figure 35 $y = \text{arctanh } x$.

Table of constants

Useful constants

Magnitude of the acceleration due to gravity	$g = 9.81 \text{ m s}^{-2}$
Gravitational constant	$G = 6.673 \times 10^{-11} \text{ N m}^2 \text{ kg}^{-2}$
Boltzmann's constant	$k = 1.381 \times 10^{-23} \text{ J K}^{-1}$
Avogadro's constant	$N_A = 6.022 \times 10^{23} \text{ mol}^{-1}$
Molar gas constant	$R = 8.314 \text{ J K}^{-1} \text{ mol}^{-1}$
Stefan's constant	$\sigma = 5.670 \times 10^{-8} \text{ W m}^{-2} \text{ K}^{-4}$
Permittivity of free space	$\varepsilon_0 = 8.854 \times 10^{-12} \text{ C}^2 \text{ N}^{-1} \text{ m}^{-2}$
	$1/4\pi\varepsilon_0 = 8.988 \times 10^9 \text{ N m}^2 \text{ C}^{-2}$
Permeability of free space	$\mu_0 = 4\pi \times 10^{-7} \text{ T m A}^{-1}$
Speed of light in vacuum	$c = 2.998 \times 10^8 \text{ m s}^{-1}$
Planck's constant	$h = 6.626 \times 10^{-34} \text{ J s}$
	$\hbar = h/2\pi = 1.055 \times 10^{-34} \text{ J s}$
Rydberg constant	$R = 1.097 \times 10^7 \text{ m}^{-1}$
Bohr radius	$a_0 = 5.292 \times 10^{-11} \text{ m}$

Charge of proton	$e = 1.602 \times 10^{-19}$ C
Charge of electron	$-e = -1.602 \times 10^{-19}$ C
Electron rest mass	$m_e = 9.109 \times 10^{-31}$ kg
Electron charge to mass ratio	$-e/m_e = -1.759 \times 10^{11}$ C kg^{-1}
Proton rest mass	$m_p = 1.673 \times 10^{-27}$ kg
Neutron rest mass	$m_n = 1.675 \times 10^{-27}$ kg
Mass of the Earth	5.98×10^{24} kg
Mass of the Moon	7.35×10^{22} kg
Average radius of Earth orbit	1.50×10^{11} m
Average radius of Moon orbit	3.84×10^{8} m

SI units and conversions

[The base units are: m; kg; s; A; K; mol; cd]

Quantity	Unit	Conversion	Useful conversions
Speed	m s^{-1}		1 degree $\approx$ 0.0174 5 radian
Acceleration	m s^{-2}		1 radian $\approx$ 57.30 degrees
Angular speed	rad s^{-1}		Absolute zero:
Angular acceleration	rad s^{-2}		0 K $\approx$ −273.15 °C
Linear momentum	kg m s^{-1}		1 electronvolt
Angular momentum	kg m^2 s^{-1}		(eV) $= 1.602 \times 10^{-19}$ J
Force	newton (N)	1 N = 1 kg m s^{-2}	
Energy	joule (J)	J = 1 N m = 1 kg m^2 s^{-2}	
Torque	N m		
Power	watt (W)	1 W = J s^{-1}	
Pressure	pascal (Pa)	1 Pa = 1 N m^{-2}	
Frequency	hertz (Hz)	1 Hz = 1 s^{-1}	
Charge	coulomb (C)	1 C = 1 A s	
Potential difference	volt (V)	1 V = J C^{-1}	
Resistance	ohm (Ω)	1 Ω = 1 V A^{-1}	
Capacitance	farad (F)	1 F = 1 A s V^{-1}	
Inductance	henry (H)	1 H = 1 V s A^{-1}	
Conductivity	siemens (S)	1 S = 1 Ω^{-1} m^{-1}	
Electric field	N C^{-1}	1 N C^{-1} = 1 V m^{-1}	
Magnetic field	tesla (T)	1 T = N s m^{-1} C^{-1} = 1 kg s^{-2} A^{-1}	
Magnetic flux	weber (Wb)	1 Wb = 1 T m^2	
Angular frequency	hertz (Hz)	1 Hz = 1 s^{-1}	
Angular wavenumber	m^{-1}		
Activity	becquerel (Bq)	1 Bq = 1 s^{-1}	
Absorbed dose	gray (Gy)	1 Gy = 1 J kg^{-1}	
Dose equivalent	sievert (Sv)	1 Sv = 1 J kg^{-1}	

Answers and comments

Chapter 1

Fast track answers

F1 (a) $z = [4-(1+2)^2]^2 = (4-9)^2 = (-5)^2 = 25$

(b) $z = 4x^2y^2 - 4xy(x+y)^2 + (x+y)^4$

$= 4x^2y^2 - 4xy(x^2 + 2xy + y^2) +$

$\qquad (x^2 + 2xy + y^2)(x^2 + 2xy + y^2)$

$= 4x^2y^2 - 4x^3y - 8x^2y^2 - 4xy^3 + x^4 + 4x^3y$

$\qquad + 6x^2y^2 + 4xy^3 + y^4$

$= x^4 + 2x^2y^2 + y^4$

F2 (a) $2mu - \dfrac{2}{u}d = \dfrac{2mu^2 - 2d}{u} = \dfrac{2(mu^2 - d)}{u}$

(b) $\dfrac{2l^2}{3h} + \dfrac{h}{2} = \dfrac{4l^2 + 3h^2}{6h}$

(c) $\dfrac{3l(h+f)^2}{2a^2} + \dfrac{3a(h^2-f^2)}{-2l^2}$

$= \dfrac{3l(h+f)^2}{2a^2} \times \dfrac{(-2l^2)}{3a(h^2-f^2)}$

$= \dfrac{-6l^3(h+f)^2}{6a^3(h+f)(h-f)}$

$= \dfrac{-l^3(h+f)}{a^3(h-f)} = -\left(\dfrac{l}{a}\right)^3 \dfrac{(h+f)}{(h-f)}$

F3 The value of v that will just permit escape is given by $m_1 v^2/2 = Gm_1 m_2/r$. Cancelling a common factor of m_1 on both sides and multiplying by 2 gives $v^2 = 2Gm_2/r$. Taking the positive square root gives $v = \sqrt{2Gm_2/r}$. The inequality $v > \sqrt{2Gm_2/r}$ gives the condition for escape.

Text answers

T1 (a) $3 + 6/2 - 4 = 3 + 3 - 4 = 2$

(b) $4.2 \times 3.0 - 1.4 = 12.6 - 1.4 = 11.2$

(c) $ab + c/2 = 2 \times 1/2 + (-4)/2 = 1 - 2 = -1$

T2 (a) $(6 + 3/(9 - 7))/2 = (6 + 3/2)/2$

$= (6 + 1.5)/2 = 7.5/2 = 3.75$

(b) $2[(3a - 4)b + 2a(3b + 1)]$

$= 2[(3 - 4)(-1) + 2(-3 + 1)]$

$= 2[(-1)(-1) + 2(-2)] = 2(1 - 4) = -6.$

(c) $2\{[2 - 2(2 + 3)/(2 - 1)] + [1 + (2 + 3)/2]/2\}$

$= 2[(2 - 10) + (1 + 2.5)/2] = 2(-8 + 1.75)$

$= -12.5$

T3 (a) $2 + 4/3^2 = 2 + 4/9 = 2 + 0.44 = 2.44$

(b) $2 + (4 - 3a)^2 = 2 + (4 - 3)^2 = 2 + 1 = 3$

(c) $(2 + (2 + x)^2)^2 = (2 + (2 + 2)^2)^2$

$= (2 + 4^2)^2 = (2 + 16)^2$

$= 18^2 = 324$

T4 (a) $3x(4 + 3y) + 2x = 12x + 9xy + 2x$

$= 14x + 9xy$

(b) $(x + 2y)^2 = (x + 2y)(x + 2y)$

$= x^2 + 2xy + 2yx + 4y^2$

$= x^2 + 4xy + 4y^2$

(c) $(x + y)(x - y) = x^2 - xy + yx - y^2 = x^2 - y^2$

(d) $(3x + 6y)/2 = 3x/2 + 6y/2 = 3x/2 + 3y$

(e) $(p + q)/q = p/q + q/q = p/q + 1$

(f) $(2xy)^2/2 = 4x^2y^2/2 = 2x^2y^2$

T5 (a) $3x^2y + 6xy = 3xy(x + 2)$

(b) $8x^3 + 4x^2y + 2xy^2 + 2x$

$= 2x(4x^2 + 2xy + y^2 + 1)$

T6 (a) $7/42 = 7/(6 \times 7) = 1/6$

(b) $6x/(2x^2) = (3 \times 2x)/(2x \times x) = 3/x$

(c) $8(x + 3)/(2x) = 4(x + 3)/x$

T7 (a) The denominator is $(3x^2y + 6xy) = 3xy(x + 2)$, so the numerator and denominator have a common factor $(x + 2)$ and the fraction becomes $(x + 3)/(3xy)$.

(b) The numerator is $(3xy + 3y^2) = 3y(x + y)$, and the denominator is $(4x^2y + 4x^3) = 4x^2(y + x)$, so cancelling a common factor of $(x + y)$ in the fraction gives $3y/4x^2$.

T8 (a) $\dfrac{2}{3} + \dfrac{3}{5} = \dfrac{10}{15} + \dfrac{9}{15} = \dfrac{19}{15}$

(b) $\dfrac{x}{3} + \dfrac{x}{2} = \dfrac{2x}{6} + \dfrac{3x}{6} = \dfrac{5x}{6}$

(c) $\dfrac{1}{u} + \dfrac{1}{v} = \dfrac{v}{uv} + \dfrac{u}{uv} = \dfrac{v + u}{uv}$

(d) $\dfrac{1 + a}{3 + a}$ cannot be simplified

(e) $\dfrac{2(1 + x)}{(1 + x)^2} = \dfrac{2}{(1 + x)}$

(f) $\dfrac{5/7}{2/3} = \dfrac{5}{7} \times \dfrac{3}{2} = \dfrac{15}{14}$

(g) $\dfrac{3x}{x/3} = 3x \times \dfrac{3}{x} = \dfrac{9x}{x} = 9$

(h) $\dfrac{3}{a} + \dfrac{7}{2a} = \dfrac{6}{2a} + \dfrac{7}{2a} = \dfrac{13}{2a}$

(i) $(-7x^2) \div (-3x) = \dfrac{7x^2}{3x} = \dfrac{7x}{3}$

T9 (a) $2^2 \times 2^3 = 2^5 = 32$

(b) $2^3 \times 2^4 \times 2^5 = 2^{12} = 4\,096$

(c) $(-2)^4 \times (-2)^2 = (-2)^6 = 64$

(d) $(2^3)^2 = 2^6 = 64$

(e) $[(-0.2)^2]^3 = (-0.2)^6 = 64 \times 10^{-6} = 6.4 \times 10^{-5}$

(f) $[(-1)^{17}]^{23} = (-1)^{17 \times 23} = (-1)^{391} = -1$

T10 (a) $2^{-4} = 1/2^4 = 1/16 = 0.062\,5$

(b) $10^{-3} = 1/10^3 = 0.001$

(c) $\left(\dfrac{1}{2}\right)^{-3} = 2^3 = 8$

(d) $(0.2)^{-2} = 1/(0.2)^2 = 1/0.04 = 25$

T11 (a) $5^2 \times 5^{-5} = 5^{2 - 5} = 5^{-3}$

(b) $(5^2)^{-3} = 5^{-6}$

(c) $2^4/2^6 = 2^4 \times 2^{-6} = 2^{4 - 6} = 2^{-2}$

(d) $(3^{-1})^{-1} = 3^1$ (i.e. 3)

(e) $\left(\dfrac{1}{2}\right)^{-3}$ is already in the required form, with

$a = \dfrac{1}{2}$. However, it may be rewritten as $(2^{-1})^{-3} = 2^3$ if desired. Also, 8^1 is correct.

(f) $\left(\dfrac{2^5}{2^6}\right)^{-1} \times 2 \times 2^3 = \left(\dfrac{2^6}{2^5}\right) \times 2 \times 2^3$

$= \dfrac{2^{10}}{2^5} = 2^{10-5} = 2^5$

T12 (a) $9^{1/2} = 3$ (-3 is excluded by our convention)

(b) $16^{1/2} = 4$ (-4 is excluded by our convention)

(c) $16^{1/4} = 2$ (-2 is excluded by our convention)

(d) $27^{1/3} = 3$

(e) $\sqrt{64} = 8$ (-8 is excluded by our convention)

(f) $\sqrt[4]{49} = \sqrt{7} = 2.65$ (-2.65 is excluded by our convention)

T13 (a) $(-8)^{1/3} = -2$, there is no positive root.

(b) $\left(-\dfrac{1}{16}\right)^{1/2}$ cannot be found since it is the square root of a negative number.

(c) $(-2)^{1/2}$ cannot be found. It too is the square root of a negative number.

T14 (a) $7^{4/9} = 2.37$

(b) $(2.7)^3/(2.7)^2 = (2.7)^{3-2} = (2.7)^1 = 2.7$

(c) $\sqrt{2} + \sqrt[3]{2} = 2^{1/2} \times (2^{1/3}) = 2^{1/2+1/3} = 2^{5/6} = 1.78$

(d) $\dfrac{\sqrt{3}}{\sqrt[5]{3}} = 3^{1/2} \times (3^{1/5})^{-1} = 3^{1/2 - 1/5}$
$= 3^{3/10} = 1.39$

T15 (a) $1/\sqrt{r} = 1/r^{1/2} = r^{-1/2}$

(b) $r^3\sqrt{r} = r^3 \times r^{1/2} = r^{7/2}$

(c) $r^5/\sqrt{r} = r^5/r^{1/2} = r^5 \times r^{-1/2} = r^{9/2}$

T16 (a) $4x^2y^2z^4 = (2xyz^2)^2$

(b) $\dfrac{h^2n^2}{8m} (\sqrt[3]{L^3})^{-2} = \dfrac{h^2n^2L^{-2}}{8m} = \dfrac{1}{8m}\left(\dfrac{hn}{L}\right)^2$

(c) $\left(\dfrac{8\pi^2 Z\sqrt{l^5}}{3^3\sqrt{Z}}\right)^{1/2} = \left(\dfrac{8\pi^2 Z^{2/3}l^{5/2}}{3}\right)^{1/2}$

$= \dfrac{2\sqrt{2}\,\pi Z^{2/6}l^{5/4}}{\sqrt{3}} = \left(\dfrac{8}{3}\right)^{1/2}\pi Z^{1/3}l^{5/4}$

T17 If a common factor of $\dfrac{\theta}{b^2 - a^2}$ is extracted from each term on the right-hand side

$\phi = \left(b^2Vr - \dfrac{b^2Va^2}{r} - a^2Ur + \dfrac{a^2Ub^2}{r}\right)\dfrac{\theta}{b^2 - a^2}$

If a common factor of b^2V is extracted from the first two terms and a common factor of $-a^2U$ from the second two terms

$\phi = \left[b^2V\left(r - \dfrac{a^2}{r}\right) - a^2U\left(r - \dfrac{b^2}{r}\right)\right]\dfrac{\theta}{b^2 - a^2}$

T18 $F = 2\pi r\left[\dfrac{2}{3}r^2g(\rho - \sigma) - 3\eta v\right]$

So, when $F = 0$ and $v = v_t$ we have

$0 = \dfrac{2}{3}r^2 g(\rho - \sigma) - 2\eta v_t$

If $3\eta v_t$ is added to both sides

$3\eta v_t = \dfrac{2}{3}r^2g(\rho - \sigma) - 3\eta v_t + 3\eta v_t$

The last two terms on the right-hand side cancel, so if both sides are divided by 3η

$v_t = \dfrac{2r^2}{9\eta} g(\rho - \sigma)$

T19 (a) $\dfrac{F_b}{F_g} = \dfrac{(4/3)\pi r^3 \sigma g}{(4/3)\pi r^3 \rho g} = \dfrac{\sigma}{\rho}$

$\dfrac{F_v}{F_g} = \dfrac{6\pi\eta rv}{(4/3)\pi r^3 \rho g} = \dfrac{3\eta v}{2r^2\rho g/3} = \dfrac{9\eta v}{2r^2\rho g}$

$\dfrac{F_b}{F_v} = \dfrac{(4/3)\pi r^3 \sigma g}{6\pi\eta rv} = \dfrac{2r^2\sigma g}{9\eta v}$

(b) If $\dfrac{F_v}{F_b} = 10$ when $v = v_t$

$10 = \dfrac{F_v}{F_b} = \dfrac{1}{(F_b/F_v)} = \dfrac{9\eta v_t}{2r^2\sigma g}$

If we multiply both sides by $\dfrac{r^2}{10}$,

$r^2 = \dfrac{9\eta v_t}{20\sigma g}$

Since r^2 is positive we can take square roots of both sides, and since we know that r itself (a radius) is also positive we can neglect negative roots. Thus,

$r = \sqrt{\dfrac{9\eta v_t}{20\sigma g}}$

(c) If $F_g = F_v$ (i.e. $\dfrac{F_v}{F_g} = 1$)

when $v = v_t$ $\quad 1 = \dfrac{9\eta v_t}{2r^2\rho g}$

If both sides are multiplied by r^2,

$r^2 = \dfrac{9\eta v_t}{2\rho g}$

If positive square roots are taken as before

$r = \sqrt{\dfrac{9\eta v_t}{2\rho g}}$

T20 (a) If $y = 2ax + b^2$

then $\quad 2ax = y - b^2$

so $\quad x = \dfrac{y - b^2}{2a}$ provided $a \neq 0$

(b) If $(y - b) + (x - a) = 5xh^2 + 3$

then $\quad (x - a) - 5xh^2 = 3 - (y - b)$

so $\quad x - 5xh^2 = 3 - (y - b) + a$

i.e. $\quad x(1 - 5h^2) = 3 - y + b + a$

If both sides are divided by $(1 - 5h^2)$

$x = \dfrac{3 - y + a + b}{(1 - 5h^2)}$ provided $(1 - 5h^2) \neq 0$

(The condition $(1 - 5h^2) \neq 0$ is required to make sure we have not unwittingly divided by zero.)

(c) If $\dfrac{1}{x - a} + \dfrac{1}{y - b} = 3t^2$

then $\dfrac{1}{x - a} = 3t^2 - \dfrac{1}{y - b}$

so $\dfrac{1}{x - a} = \dfrac{3t^2(y - b) - 1}{y - b}$

If both sides are multiplied by

$$\dfrac{(x - a)(y - b)}{3t^2(y - b) - 1}$$

$$\dfrac{y - b}{3t^2(y - b) - 1} = x - a$$

provided $3t^2(y - b) - 1 \neq 0$

thus $x = \dfrac{y - b}{3t^2(y - b) - 1} + a$

provided $3t^2(y - b) - 1 \neq 0$

(d) If $t + a = \sqrt{\dfrac{3b}{(x - y)}}$

then $(t + a)^2 = \dfrac{3b}{x - y}$

If both sides are multiplied by $\dfrac{x - y}{(t + a)^2}$

$x - y = \dfrac{3b}{(t + a)^2}$ provided $(t + a) \neq 0$

thus $x = \dfrac{3b}{(t + a)^2} + y$ provided $(t + a) \neq 0$

T21 If $F = \dfrac{Gm_1m_2}{r^2}$ then $r^2 = \dfrac{Gm_1m_2}{F}$

All of the quantities involved are positive, so we can take positive square roots of both sides to find

$$r = \sqrt{\dfrac{Gm_1m_2}{F}}$$

To check this out with some simple numbers, let $m_1 = 2$, $m_2 = 3$, $r = 4$ and $G = 5$. (This is nothing like the correct value of the constant G, but that does not matter; we are not carrying out a physical calculation, we are just checking algebraic manipulation.) With the assumed values for G, m_1, m_2 and r we find

$$F = \dfrac{5 \times 2 \times 3}{16} = \dfrac{30}{16}$$

If these same values are used in the rearranged equation

$$r = \sqrt{\dfrac{5 \times 2 \times 3}{30/16}} = \sqrt{16} = 4$$

which is consistent.

T22 (a) $V = IR$ where R is a constant

(b) $R = \dfrac{kL}{r^2}$ where k is a constant

(c) $E + \phi = hf$ where h is a constant

(d) $F_{grav} = \dfrac{Gm_1m_2}{r^2}$ where G is a constant

(Of course you may have used different symbols for the constants.)

T23 (a), (b), (c), (d), (e) and (g) are true, (f) and (h) are false. Only the convention that $\sqrt{2}$ is positive avoids ambiguity in the case of (g).

T24 (a) If $2x + 4 > 6$ then $2x > 6 - 4$

i.e. $2x > 2$ so $x > 1$

(b) If $2(a - x) \leqslant 7$

then $a - x \leqslant \dfrac{7}{2}$, i.e. $-x \leqslant \dfrac{7}{2} - a$

If both sides are multiplied by -1 and the inequality is reversed

$$x \geqslant a - \dfrac{7}{2}$$

(c) If $2(a - x) \leqslant 3x + b$ then $2a \leqslant 3x + b + 2x$

i.e. $2a - b \leqslant 5x$ so $\dfrac{2a - b}{5} \leqslant x$

which may be rewritten $x \geqslant \dfrac{2a - b}{5}$

T25 If $R = \dfrac{\rho l}{A}$, then $R > R_0$ implies that $\dfrac{\rho l}{A} > R_0$

(a) It follows that if ρ and A are given $l > \dfrac{R_0 A}{\rho}$

(b) It also follows that if l and ρ are given

$$A < \dfrac{\rho l}{R_0}$$

Chapter 2

Fast track answers

FI (a) $293.45 = 2.9345 \times 10^2$, (b) $1\,380 = 1.380 \times 10^3$, (c) $-2\,804 = -2.804 \times 10^3$, (d) $0.00567 = 5.67 \times 10^{-3}$.

F2 (a) 1.23×10^6, 727 and 0.0432 all have three significant figures. (In some contexts 727 may be intended to represent an *exactly* known quantity.)

(b) 8.5×10^6 and 1.148×10^7 both have the same order of magnitude (i.e. the *nearest* 'power of ten' to each is 10^7); none of the others have matching orders of magnitude.

F3 In the SI system, the unit of frequency f is hertz (Hz), which is equivalent to s^{-1}, X is dimensionless and therefore has no units, and the unit of speed c is $m\,s^{-1}$. Appropriate units for R are therefore

$$\frac{s^{-1}}{m\,s^{-1}} = \frac{1}{m} = m^{-1}$$

Any quantity expressed in these units must have the dimensions of length^{-1}.

Ready to study answers

RI (a) $x^3 \times x^4 = x^7$,

(b) $x^5 \times x^{-2} = x^3$,

(c) $1/x^2 = x^{-2}$,

(d) $1/x^{-6} = x^6$,

(e) $x^8/x^3 = x^5$,

(f) $(x^4)^{-6} = x^{-24}$,

(g) $\sqrt{x^3}\,\sqrt{1/x^5} = \sqrt{x^3/x^5} = x^{-1}$. Remember that a positive quantity generally has two square roots, one positive, the other negative. However, by convention, the root symbol indicates the positive square root.

R2 Dividing both sides of the equation by $6\pi r v$ gives $F/(6\pi r v) = \eta$.

Text answers

TI The complete table is shown in Table 2.5.

T2 (a) The number of zeros after the 1, the number of tens multiplied together and the power of ten are all the same.

(b) The (negative) power of ten gives the position of the 1 after the decimal point, for example $10^{-2} = 0.01$ and the 1 is in the second place after the point. (Note that in part (b) the number of zeros after the decimal point is *not* the same as the (negative) power of 10.)

T3 (a) $3.2 \times 10^6 = 3.2 \times 1\,000\,000 = 3\,200\,000$ (You will see later that such numbers may be misleading, hence the need for scientific notation.)

(b) $8.76 \times 10^{-4} = 8.76 \times 0.0001 = 0.000876$.

T4 (a) $98\,765 = 9.8765 \times 10\,000$

$$= 9.8765 \times 10^4.$$

(b) $0.00432 = 4.32 \times 0.001$

$$= 4.32 \times 10^{-3}.$$

Table 2.5 See Answer TI.

Number written in full		Number written using a product of 10		Number written as a power of 10
100 000		$10 \times 10 \times 10 \times 10 \times 10$		10^5
10 000		$10 \times 10 \times 10 \times 10$		10^4
1 000		$10 \times 10 \times 10$		10^3
100		10×10		10^2
10		10		10^1
1		1		10^0
0.1	1/10	1/10	$1/10^1$	10^{-1}
0.01	1/100	$1/(10 \times 10)$	$1/10^2$	10^{-2}
0.001	1/1000	$1/(10 \times 10 \times 10)$	$1/10^3$	10^{-3}

T5 (a) $1.40 \times 10^4 \times 5.50 \times 10^{13} \times 6.20 \times 10^{-15}$

$= 1.40 \times 5.50 \times 6.20 \times 10^2.$

(b) $2 \times 10^6 \times 3 \times 10^5/(8 \times 10^7)$

$= (2 \times 3 \times 1/8) \times 10^4.$

T6 (a) $1.40 \times 10^4 \times 5.50 \times 10^{13} \times 6.20 \times 10^{-15}$

$= 4.774 \times 10^3.$

In evaluating this you should have collected together terms before punching any keys on your calculator.

(b) $2 \times 10^6 \times 3 \times 10^5/(8 \times 10^7) = 7.5 \times 10^3.$

(You will see later when *significant figures* are discussed that these answers may not be reliable.)

T7 (a) $|2.3| = 2.3$

(b) $|-3.1| = 3.1$

(c) $|(4 - 7)/2| = |-3/2| = 3/2$

(d) $|(4 - 5)(7 - 9)| = |(-1)(-2)| = |2| = 2$

(e) $|2| / |4| = 1/2$

(f) $|-6| |3| = 18.$

T8 (a)   65.43 has four significant figures, two decimal places;

(b) 0.003 56 has three significant figures, five decimal places;

(c) 2 278 has four significant figures, no decimal place;

(d) 3.04×10^{-5} has three significant figures, seven decimal places. It should be noted that this last answer needs careful interpretation since the decimal part of the number (3.04) is only specified to two decimal places.

T9 (a)   Only two figures are significant since one of the dimensions (0.23 m) is only known to two significant figures. The volume is therefore 4.1×10^{-3} m^3.

(b) Since the volume has only two significant figures, the density will have only two significant figures.

T10 (a)   Mass of the Sun $\sim 10^{30}$ kg;

(b) radius of the Sun $\sim 10^9$ m;

(c) mass of a proton $\sim 10^{-27}$ kg;

(d) mass of an electron $\sim 10^{-30}$ kg.

T11 $z \approx 7 \times 9 \times 10^3 \times 10^2/(2 \times 4 \times 10^4 \times 10^5)$

$\approx 7 \times 10^{-4}$, so $z \sim 10^{-3}.$

T12 (a)   True, the set of integers is a subset of the set of real numbers.

(b) True, since $\sqrt{8} = 2\sqrt{2}$ and $\sqrt{2}$ is irrational.

(c) False, all natural numbers are positive.

(d) The use of three significant figures indicates that 2.00 is a real number in the range 1.995 000 ... to 2.004 999 ..., so the statement is false. However, in practice you will meet many cases where the intended meaning is unclear, so you should be prepared to use all available information to arrive at the most reasonable interpretation in any given situation.

T13 (a)   2 is an integer and π is an irrational number.

(b) The value of 2 is exact, so does not affect the precision of the answer. The value of π can be expressed to at least as many figures as l and g so will not adversely affect the precision of the answer.

T14 If the equation is rearranged

$$\varepsilon_0 = q_1 q_2/(4\pi F_{el} r^2) = \text{(numerical factor)}$$
$$\times \text{ C}^2/\text{N m}^2$$

The SI units of ε_0 are therefore C^2 N^{-1} m^{-2}.

T15 (a)   [density] = [mass/volume]

$$= [\text{mass/length}^3]$$
$$= \text{M L}^{-3}$$

(b) $[G] = [F_{grav} r^2/(m_1 m_2)]$

$$= [\text{force} \times \text{length}^2/\text{mass}^2]$$
$$= \text{M L T}^{-2} \times \text{L}^2 \text{M}^{-2} = \text{M}^{-1} \text{L}^3 \text{T}^{-2}$$

(The dimensions of force were worked out earlier in the text.)

(c) [speed] = [length/time] = L T^{-1}.

T16 The dimensions of energy can be determined by examining the definition of the SI unit of energy (the joule) in Table 2.4; [energy] = M L^2 T^{-2}. Therefore the implausible expressions are (c) because $[mv/4] = \text{M L T}^{-1}$, (d) because $[mh/2] = \text{M L}$, and (g) because the second term has dimensions $[F^2 h/(mv^2)] = \text{M}^2 \text{L}^2 \text{T}^{-4} \text{ L}/(\text{M L}^2 \text{T}^{-2}) = \text{M L T}^{-2}.$

Chapter 3

Fast track answers

FI The graph of the function $f(x) = x^3 - 4x$ is given in Figure 3.9.

F2 $H(x) = x^2 - 4x + 6 = x^2 - 4x + 4 + 2$
$= (x - 2)^2 + 2$.

This last expression is the completed square form. The vertex is situated at the point (2, 2). The graph does not intersect the x-axis, because the discriminant $b^2 - 4ac$ is negative. $H(x)$ does not have an inverse function because its graph is a parabola, so different values of x do not necessarily correspond to different values of $H(x)$.

F3 As x approaches 1 from above (i.e. decreasing from even greater values of x), $g(x)$ becomes a larger and larger positive number. As x approaches 1 from below (i.e. from lower values of x), $g(x)$ becomes a larger and larger negative number. Thus x = 1 is an asymptote. When x is a large number, positive or negative, $g(x) \approx x^2/x = x$. Thus the line $y = x$ is also an asymptote. These are the only asymptotes, as you can confirm by sketching the graph.

Ready to study answers

RI All the numbers are *real numbers*. 2 and 7 are *integers* (7.0 is ambiguous, but it would not be generally regarded as an integer).

R2 (a), (b), (c) and (d) are all true but (e) is false. Note that the *modulus* (or *absolute value*) of any quantity, $|x|$, is never negative. For the purposes of these comparisons any negative number is regarded as being less than any positive number, and negative large numbers are regarded as being less than negative small numbers.

R3 (a) $5^2 = 25$

(b) $(-4)^3 = -64$

(c) $\sqrt{49} = 7$

(d) $\sqrt[3]{-2000} = -12.60$ (to two *decimal places*)

(e) $\sqrt{0.09} = 0.3$ (to one *significant figure*)

(f) $|5| = 5$

(g) $|-3.2| = 3.2$.

R4 (a) $a^2 \times a = a^{2+1} = a^3$,

(b) $a^3 \times a^2 = a^{3+2} = a^5$,

(c) $b^3/b^2 = b^{3-2} = b$,

(d) $c^4 \times c^2/c^6 = c^{4+2-6} = c^0 = 1$.

R5 $a^2 + 3a - 4a + 7 = a^2 - a + 7 = a(a - 1) + 7$.

R6 (a) $(u + 3)(u - 3) = u^2 + 3u - 3u - 9$
$= u^2 - 9$.

(b) $(p + 2)(p - 4) + (p + 1)^2$
$= p^2 - 4p + 2p - 8 + p^2 + 2p + 1 = 2p^2 - 7$.

Text answers

TI Yes. There is a single definite value of your height for every value of the date. The fact that it may be the same value for many different dates doesn't matter.

T2 Yes. There are two possible values, 0 and 1. For some dates (in the winter) the value will be 0 (i.e. the sun doesn't rise), for others (during the rest of the year) the value will be 1.

T3 No. Simply being told the time of day does not determine the date. (Being told the time of some particular daily event might allow you to determine the date, but the time alone is not enough; the same times recur every day.)

T4 D is (-2, 1), E is (-2, -2), F is (-1, -1), G is (1, -2) and O is (0, 0).

T5 See Figure 3.15. There are real discontinuities where the rate changes, so you should not try to join the lines up.

T6 See Figure 3.16. Draw the curve 'with a flowing hand'. (It does not pass through the origin.) The points supplied by Table 3.1 have been plotted as a guide to the curve.

T7 $y - y_0 = m(x - x_0)$,
therefore $y - y_0 = mx - mx_0$

So, $y = mx - mx_0 + y_0$, hence the gradient is m and the intercept $y_0 - mx_0$.

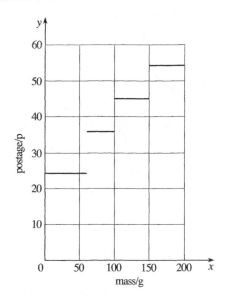

postage/p

mass/g

Figure 3.15 See Answer T5.

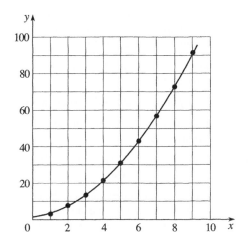

Figure 3.16 See Answer T6.

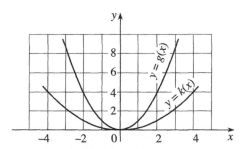

Figure 3.17 See Answer T10.

T8 First of all note that, from Equation 6, the right-hand side of Equation 8 is just equal to m. So we can write Equation 8 as $(y - y_1)/(x - x_1) = m$.

If this is then rearranged, $y - y_1 = m(x - x_1)$, which is in the same form as Equation 7.

Therefore the answer follows from Answer T7, the intercept is $y_1 - mx_1$.

T9 When $x = 0$, $y/b = 1$, so $y = b$. When $y = 0$, $x/a = 1$, so $x = a$. Hence a and b are the intercepts where the straight line cuts the x- and y-axes, respectively.

T10 The graphs of $k(x)$ and $g(x)$ are sketched in Figure 3.17.

T11 (a) As you know that the vertex of $a(x - p)^2$ is at $(p, 0)$ you should be able to see that the required quadratic function has the form

$$f(x) = a(x - p)^2 + q$$

since the additional q will have the effect of moving each point on the parabola, including the vertex, q units upwards.

(b) The quadratic functions (and vertices) are:

(i) $f(x) = (x - 1)^2 + 3$ vertex at $(1, 3)$

(ii) $f(x) = (x + 2)^2 + 1$ vertex at $(-2, 1)$

(iii) $f(x) = (x - 2)^2 - 1$ vertex at $(2, -1)$

They are sketched in Figure 3.18. Note that each has the same 'width' since a is the same in each case.

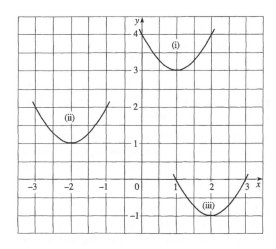

Figure 3.18 See Answer T11.

T12   (a) $f(x) = 3x^2 - 9x + 11$ does not intersect the x-axis since its discriminant is negative. In completed square form (see Equation 14)

$$f(x) = 3\left(x - \frac{3}{2}\right)^2 + \frac{17}{4}$$

So its vertex (a minimum) is at (3/2, 17/4).

(b) $f(t) = -t^2 - 2t - 6$ does not intersect the t-axis since its discriminant is negative. In completed square form

$$f(t) = -(t + 1)^2 - 5$$

so its vertex (a maximum) is at $(-1, -5)$.

(c) $f(R) = -3R^2 + 15R - 18$ intersects the R-axis twice since its discriminant is greater than zero. In completed square form

$$f(R) = -3\left(R - \frac{5}{2}\right)^2 + \frac{3}{4}$$

so its vertex (a maximum) is at (5/2, 3/4).

T13   (a) $y = \dfrac{-5 \pm \sqrt{25 + 4}}{2} = -5.193$ and 0.193

(b) The *discriminant* is less than zero, so there are no points of intersection.

(c) $Z = \dfrac{-0.5 \pm \sqrt{0.25 + 3}}{6} = 0.217$ and -0.384

(d) $x = \dfrac{5 \pm \sqrt{25 - 24}}{2} = 3$ and 2

In simple cases such as (d) it is not always necessary to use the formula to find the points of intersection. With practice you should be able to see 'by inspection' the factorised form of the function: $x^2 - 5x + 6 = (x - 2)(x - 3)$.

T14   (a) See Figure 3.19. The asymptotes are the lines $y = 1$ and $x = -1$, i.e. as $x \to \infty$, $y \to 1$ and when $x = -1$ the denominator $= 0$.

(b) $g(x)$ also has a vertical asymptote, at $x = -1/3$, but there is no horizontal asymptote. Instead, as $x \to \pm\infty$, $g(x) \approx 2x^2/(3x) = 2x/3$. So, the line $y = 2x/3$ is also an asymptote.

T15   The graph of Figure 3.8 shows that each value of y corresponds to a unique value of x. Thus there is no difficulty about defining the inverse function of Equation 20. In fact it's just $G(x) = \sqrt[3]{x}$.

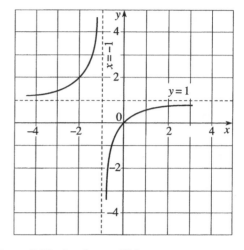

Figure 3.19   See Answer T14.

Figure 3.9 shows a different situation. Each value of y does not correspond to a unique value of x, so it is not possible to define an inverse function.

T16   No. A particular value of $f(x)$ may correspond to more than one value of x, so it is not possible to find an inverse function.

T17 $g(f(x)) = g(x + 2) = 1/(x + 2)$ where x is any real number except -2.

$f(g(x)) = f(1/x) = (1/x) + 2$ where $x \neq 0$.

Be careful to carry out the operations step by step and in the right order.

T18 $g(h(f(y))) = g(h(y + 2)) = g((y + 2)^2)$
$= 1/(y + 2)^2$ where $y \neq -2$.

T19 $G(F(x)) = x$ since $G(x)$ 'undoes' the effect of $F(x)$ for every value of x in the domain of $F(x)$. More formally, if we let $y = F(x)$ and $x = G(y)$,

$H(x) = G(F(x)) = G(y) = x$.

Chapter 4

Fast track answers

F1 $x = -3$.

F2 $x = 2$, $y = 1$.

F3 $x = \frac{1}{2}(-1 \pm \sqrt{5})$. The sum of the roots is -1, and their product is also -1. (If you do not understand the previous sentence you should take this as an indication that you should read the chapter.)

Ready to study answers

R1 (a) $a^2 - 4b^2$

(b) $x^2 - x - 12$

(c) $p^2 - 6p + 9$.

R2 $12x$.

R3 $f(2y) = 3(2y) + 2 = 6y + 2$;

$f(y - 1) = 3(y - 1) + 2 = 3y - 1$.

R4 $+x^2$ and $-x^2$.

R5 The graphs of the two functions are given in Figure 4.6.

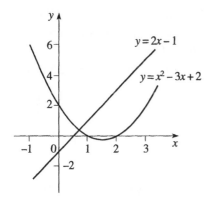

Figure 4.6 See Answer R5.

R6 $\dfrac{1}{1 - x} - \dfrac{1}{1 + x} = \dfrac{(1 + x) - (1 - x)}{(1 - x)(1 + x)} = \dfrac{2x}{1 - x^2}$

Text answers

T1 $2(x + 1) = x - 5$; this is equivalent to $2x + 2 = x - 5$, therefore $2x - x = -5 - 2$ and $x = -7$.

To check this solution, substitute the value for x in the equation: left-hand side $= 2(-7 + 1) = -12$; right-hand side $= -7 - 5 = -12$.

T2 $z - 2 + 6z + 3 - 5z + 7 = 0$; if the terms are collected, $z + 6z - 5z - 2 + 3 + 7 = 0$; $2z + 8 = 0$; so $z = -4$.

To check this solution, substitute the value for x in the equation: $-6 + 3(-7) - (-20 - 7) = -6 - 21 + 27 = 0$.

T3 In terms of a and b the solution for p is $p = -\frac{1}{2}(a + 2b)$; now if we substitute for a and b, we obtain $p = -\frac{1}{2}(1 + 8) = -9/2$.

T4 Notice that we do not need to solve the equation again. We simply substitute the new values for a and b in the equation found in Answer T3, therefore, $p = \frac{1}{2}(4 - 4) = 0$.

T5 (a) $\dfrac{1}{x - 1} = 6 = 6$, so $1 = 6(x - 1)$ or $6x = 7$

and therefore $x = 7/6$.

(b) There is no cancellation here; multiplying both sides by $(x + 1)$ gives $(x + 1)(x^2 - 1) = 2$ which expands to give $x^3 + x^2 - x - 1 = 2$, so this equation cannot be reduced to linear form.

(c) If we expand the bracket, $x^2 + 2 = x^2 - 3x$, so $2 = -3x$ or $x = -2/3$.

(d) If we expand the left-hand side we get

$$\frac{x^2 - 4}{x^2 - 4} = x$$

which gives (after cancellation) $x = 1$.

T6 If we put

$$y = \frac{x^2 + 1}{x^2 - 1}$$

then $2(y + 1) = y + 5$; $2y + 2 = y + 5$; so $y = 3$.

If we then substitute for y, we have

$$\frac{x^2 + 1}{x^2 - 1} = 3; \quad x^2 + 1 = 3(x^2 - 1); \quad 2x^2 = 4; \quad x^2 = 2;$$

so $x = \pm\sqrt{2}$. If we put $x = +\sqrt{2}$ in the original equation to check this, we find the left-hand side

$$= 2\left(\frac{2 + 1}{2 - 1} + 1\right) = 8$$

and the right-hand side

$$= \frac{2 + 1}{2 - 1} + 5 = 8$$

This can then be repeated for $x = -\sqrt{2}$.

T7 (a) Multiply the first equation by 2 and the second by 5:

$$10x + 4y = 0$$

$$10x + 25y = 105$$

Next subtract the first equation from the second:

$$21y = 105$$

so $y = 5$; substituting this value in either equation gives $x = -2$; you can then use the other one to check the answer.

(b) Multiply the first equation by 2:

$$8x + 2y = 20$$

$$3x - 2y = 13$$

this time add the two equations:

$$11x = 33$$

so $x = 3$; substituting this value in either equation gives $y = -2$. Did you remember to check the solution?

T8 We have $a = 1$, $b = 2$, $c = 2$ and $d = -3$, so that $D = 1 \times (-3) - (2 \times 2) = -7$. Also $p = 8$ and $q = -5$ so that (from Equation 12)

$$x = \frac{[(-3) \times 8] - [2 \times (-5)]}{-7} = 2$$

and (from Equation 13)

$$y = \frac{-5 - (2 \times 8)}{-7} = 3$$

As you can see, the amount of work using this general method is *not* significantly less than the previous methods. Hence our reluctance to use it.

T9 The intersection for various values of k is shown in Figure 4.7.

$$k = 2;\ x \approx -0.6,\ y \approx 1.3$$

$$k = 4;\ x \approx 0.3,\ y \approx 1.8$$

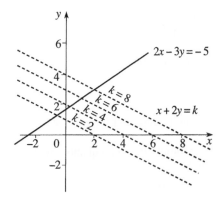

Figure 4.7 See Answer T9.

$$k = 6;\ x \approx 1.1,\ y \approx 2.4$$

$$k = 8;\ x \approx 2.0,\ y \approx 3.0$$

T10 (a) $(x - 1)(x - 2) = 0$, so $x = 1$ or $x = 2$.

(b) $(x - 1)(x + 2) = 0$, so $x = 1$ or $x = -2$.

(c) $(x + 3)(x + 4) = 0$, so $x = -3$ or $x = -4$.

(d) $(x - 4)(x + 3) = 0$, so $x = 4$ or $x = -3$.

(e) $(x - 5)(x + 3) = 0$, so $x = 5$ or $x = -3$.

(f) $(x + 2)(x - 6) = 0$, so $x = -2$ or $x = 6$.

T11 (a) $(2x - 1)(x + 5)$, so $x = 1/2$ or $x = -5$.

(b) $(3x + 1)(x - 4) = 0$, so $x = -1/3$ or $x = 4$.

(c) $(3x + 2)(x - 1) = 0$, so $x = -2/3$ or $x = 1$.

(d) $(4x + 1)(x - 1) = 0$, so $x = -1/4$ or $x = 1$.

(e) $(2x - 1)(2x - 1) = 0$, so $x = 1/2$
(repeated root).

(f) $(3x - 2)(2x + 3) = 0$, so $x = 2/3$ or $x = -3/2$.

T12 (a) $x^2 - x - 2 = x^2 - x + \dfrac{1}{4} - \dfrac{1}{4} - 2$

$$= \left(x - \frac{1}{2}\right)^2 - \frac{9}{4}$$

hence $x - \dfrac{1}{2} = \pm\sqrt{9/4}$; $x = \dfrac{1}{2} \pm \dfrac{3}{2} = 2$ or -1

(b) $x^2 - 2x - 8 = x^2 - 2x + 1 - 1 - 8$

$$= (x - 1)^2 - 9$$

hence $x - 1 = \pm\sqrt{9}$; $x = 1 \pm 3 = 4$ or -2.

(c) $2x^2 + x - 10 = 2\left(x^2 + \dfrac{x}{2} - 5\right)$

$$= 2\left(x^2 + \frac{x}{2} + \frac{1}{16} - \frac{1}{16} - 5\right)$$

$$= 2\left[\left(x + \frac{1}{4}\right)^2 - \frac{81}{16}\right]$$

hence $\left(x + \dfrac{1}{4}\right)^2 = \dfrac{81}{16}$; $x + \dfrac{1}{4} = \pm\dfrac{9}{4}$;

$$x = +2 \text{ or } -2.5.$$

(d) $3x^2 - 18x + 24 = 3(x^2 - 6x + 8)$

$$= 3(x^2 - 6x + 9 - 9 + 8)$$

$$= 3[(x - 3)^2 - 1].$$

Hence $(x - 3)^2 - 1 = 0$; $x - 3 = \pm 1$; $x = 4$ or 2.

T13 The answers are those given in Answer T12.

T14 (a) $x = \frac{1}{2}(6 \pm \sqrt{36 - 4}) = 3 \pm \sqrt{8}$

Sum of roots $= (3 + \sqrt{8}) + (3 - \sqrt{8}) = 6 = -b/a$;

product of roots $= (3 - \sqrt{8})(3 + \sqrt{8}) = 9 - 8$

$$= 1 = c/a.$$

(b) $x = \frac{1}{6}(9 \pm \sqrt{81 - 4 \times 3 \times 2}) = \frac{1}{6}(9 \pm \sqrt{57})$

Sum of roots $= 3 = -b/a$; product of roots $= (1/36)$
$(81 - 57) = 2/3 = c/a$.

In both of these cases we use $a^2 - b^2 = (a - b)$
$(a + b)$ to find the product.

T15 (a) The discriminant

$$= (-20)^2 - 4 \times 4 \times 25 = 0;$$

there are two equal roots.

(b) The discriminant $= 1 - 4 \times 6 = -23$;

there are no real roots.

(c) The discriminant $= 1 - 4 \times (-6) = 25$;

there are two real roots.

T16 The proof is exactly the same as for $\frac{1}{2}(1 - \sqrt{-3})$, except that the sign in front of the square root is changed wherever the root appears.

T17 (a) $x^2 = -25$;

$$x = \sqrt{-25} = \sqrt{25 \times (-1)}$$

$$= \sqrt{25} \times \sqrt{-1} = \pm 5i$$

(b) From Equation 23,

$$x = \frac{1}{2}(-1 \pm \sqrt{1 - 4 \times 6}) = \frac{1}{2}(-1 \pm i\sqrt{23})$$

T18 One of your first trials should show that $x = -1$ is a solution. Then taking out the factor $(x + 1)$ gives

$$x^3 + x^2 - 4x - 4 = (x + 1)(x^2 - 4)$$

$$= (x + 1)(x - 2)(x + 2)$$

and the roots are therefore -1, 2 and -2.

T19 The root is $x = 1.860\,81$.

T20 $x_1 = 0.450\,00$, $x_2 = 0.454\,44$, $x_3 = 0.453\,08$,

$x_4 = 0.453\,50$, $x_5 = 0.453\,37$, $x_6 = 0.453\,41$,

$x_7 = 0.453\,39$, $x_8 = 0.453\,40$, $x_9 = 0.453\,40$.

T21 The sequence is 1, 1.2, 1.248, 1.249\,9968, 1.250\,000\,00. It is characteristic of this method that the number of correct figures in the answer doubles at each stage.

Chapter 5

Fast track answers

F1 Pythagoras's theorem states that the square of the hypotenuse in a right-angled triangle is equal to the sum of the squares of the other two sides.

F2 With the notation of Figure 5.8 (Subsection 5.2.2), we have:

$$\sin \theta = \frac{\text{opposite}}{\text{hypotenuse}}$$

$$\cos \theta = \frac{\text{adjacent}}{\text{hypotenuse}}$$

$$\tan \theta = \frac{\text{opposite}}{\text{adjacent}}$$

From Pythagoras's theorem, the hypotenuse of the 45° triangle in Figure 5.10 (Subsection 5.2) has length $\sqrt{2}$, and therefore:

$$\sin (45°) = \frac{\text{opposite}}{\text{hypotenuse}} = \frac{1}{\sqrt{2}}$$

An angle of $\pi/4$ radians is equivalent to 45° and therefore we have:

$$\tan \left(\frac{\pi}{4}\right) = 1$$

F3 The sine rule $\dfrac{a}{\sin \hat{A}} = \dfrac{b}{\sin \hat{B}} = \dfrac{c}{\sin \hat{C}}$

The cosine rule $a^2 = b^2 + c^2 - 2bc \cos \hat{A}$

where the notation is given in Figure 5.28 (Subsection 5.4.1).

Suppose $b = c = 2$ m and $a = 3$ m. Then using the cosine rule we get:

$$3^2 = 2^2 + 2^2 - 2 \times 2 \times 2 \cos \hat{A}$$

and so:

$$\cos \hat{A} = -1/8$$

There are an infinite number of solutions to this equation for $\hat{A}$, but only one is relevant to our triangle problem:

$$\hat{A} = \arccos(18) = 97.2°$$

Since sides b and c of the triangle are equal, we have:

$$\hat{B} = \hat{C} = \frac{180° - 97.2°}{2} = 41.4°$$

F4 The graphs for these functions are given in Figures 5.21–5.23.

F5 The graphs for these functions are given in Figures 5.26a and 26c.

Ready to study answers

R1 (a) $1/x = h/a$

(b) $x/y = (a/h)/(b/h) = a/b$,

(c) $x^2 + y^2 = a^2/h^2 + b^2/h^2 = (a^2 + b^2)/h^2 = 1$.

R2 Multiplying the first of the given equations by 2, the pair become

$$4x + 10y = 12 \text{ and } 3x - 10y = 9$$

Adding the corresponding sides of the two equations in order to *eliminate* y, we find

$$7x = 21, \text{ so } x = 3$$

Substituting $x = 3$ into either of the original equations shows that $y = 0$.

R3 A function is a rule that assigns a single value from a set called the *codomain* to each value from a set called the *domain*. Thus, saying that the position of an object is a function of time implies that at each instant of time the object has one and only one position.

Text answers

T1 An angle of π radians is equivalent to 180°, so 1 radian is equivalent to $180/\pi \approx 57.3°$. This means that the conversion factor from radians to degrees is (to three significant figures) 57.3 degrees radian^{-1}. Similarly, an angle of 1° is equivalent to

$(\pi/180)$ rad $= 0.0175$ rad and so the conversion factor from degrees to radians is approximately 0.0175 radians degree^{-1}.

T2 First find the number of times 2π rad divides into 201π rad:

$$\frac{201\pi}{2\pi} = 100.5$$

So, in order to convert to the given range we need to add $101 \times 2\pi$ rad to the given angle. Thus, -201π rad reduces to -201π rad $+ 101 \times 2\pi$ rad $= \pi$ rad.

T3 In both cases we calculate $\sqrt{b^2 + c^2}$ where b and c are the two shorter sides, obtaining:

$$\sqrt{5^2 + 12^2} = \sqrt{25 + 144} = \sqrt{169} = 13$$

and

$$\sqrt{8^2 + 15^2} = \sqrt{64 + 225} = \sqrt{289} = 17$$

So both triangles are right-angled, by the converse of Pythagoras's theorem.

T4 If the hypotenuse of a right-angled triangle is labelled, a, and the other two sides are b and c, then:

$$a^2 = b^2 + c^2$$

where a, b and c are positive. Since $c^2 > 0$ we have:

$$a^2 > b^2 \text{ and consequently } a > b.$$

Also, $b^2 > 0$, and so we obtain:

$$a^2 > c^2 \text{ and consequently } a > c.$$

T5 By Pythagoras's theorem, the perpendicular has length, $\sqrt{2^2 + 1} = \sqrt{3}$. Therefore the completed table is as given in Table 5.4.

Table 5.4 See Answer T5.

θ/degrees	θ/radians	$\sin\theta$	$\cos\theta$	$\tan\theta$
0	0	0	1	0
30	$\pi/6$	$1/2$	$\sqrt{3}/2$	$1/\sqrt{3}$
45	$\pi/4$	$1/\sqrt{2}$	$1/\sqrt{2}$	1
60	$\pi/3$	$\sqrt{3}/2$	$1/2$	$\sqrt{3}$
90	$\pi/2$	1	0	undefined

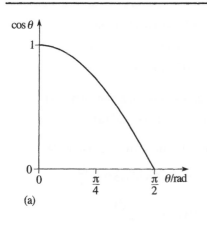

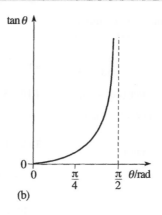

Figure 5.32 See Answer T6.

T6 Graphs of (a) cos θ and (b) tan θ for $0 < \theta < \pi/2$ are given in Figure 5.32.

T7 The angle $A\hat{B}C = 90° - \theta$, so the angle $B\hat{A}C$ must be θ. Since the side AB is the hypotenuse of the right-angled triangle ABC, it follows that

$$\sin \theta = \lambda/d$$

and $\lambda = 2 \times 10^{-6}\,\text{m} \times \sin (17.46°) = 6 \times 10^{-7}\,\text{m}$

T8 From Figure 5.11:

$$\text{cosec} (30°) = 2,\ \sec (30°) = 2/\sqrt{3},\ \cot (30°)$$
$$= \sqrt{3}$$
$$\text{cosec} (60°) = 2/\sqrt{3},\ \sec (60°) = 2,\ \cot (60°)$$
$$= 1/\sqrt{3}$$

T9 $\text{cosec} (23°) = 1/\sin (23°) = 2.559$;

$\sec (56°) = 1/\cos (56°) = 1.788$;

$\cot (\pi/6) = 1/\tan (\pi/6) = 1.732$;

$\cot (1.5) = 1/\tan (1.5) = 0.071$.

T10 Graphs of (a) sec θ (b) cot θ for $0 \leqslant \theta < \pi/2$ are given in Figure 5.33.

T11 $\phi = 0.5° = (0.5 \times \pi/180)\ \text{rad} = (0.5 \times 0.0175)\ \text{rad} = 8.73 \times 10^{-3}\ \text{rad}$ (see Answer T1 for the origin of the conversion factor).

Since ϕ is a small angle, $\phi/\text{rad} \approx s/d$ so $s \approx d \times \phi/\text{rad} = d \times 8.73 \times 10^{-3}$.

T12 The sin (θ) graph repeats itself so that sin $(2\pi + \theta) = \sin (\theta)$. It is antisymmetric, i.e. sin $(\theta) = -\sin (-\theta)$ and continuous, and any value of θ gives a unique value of sin (θ).

The cos (θ) graph repeats itself so that cos $(2\pi + \theta) = \cos (\theta)$. It is symmetric, i.e. cos $(\theta) = \cos (-\theta)$ and continuous, and any value of θ gives a unique value of cos (θ).

T13 The function tan (θ) is antisymmetric, that is tan $(\theta) = -\tan (-\theta)$; it is periodic with period π; it is *not* sinusoidal. The graph of tan $(\theta + \phi)$ has the same shape as that of tan (θ), but is shifted to the left by ϕ.

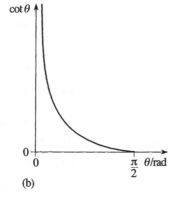

Figure 5.33 See Answer T10.

T14 (a) With a calculator in degree mode, arcsin (0.65) = 40.5°, and in radian mode, arcsin (0.65) = 0.708 rad.

(b) Using Table 5.3 and/or Figures 5.10 and 5.11:

tan (π/4) = 1, so arctan (1) = π/4 (i.e. 45°);

sin (π/3) = $\sqrt{3}$/2 and so arcsin ($\sqrt{3}$/2) = π/3 (i.e. 60°);

cos (π/4) = 1/$\sqrt{2}$, therefore arccos (1/$\sqrt{2}$) = π/4 (i.e. 45°).

T15 (a) Using Snell's law we find:

$$\frac{\mu_g}{\mu_w} = \frac{\sin \theta_r}{\sin \theta_i} = \frac{\sin (60°)}{\sin (50.34°)} = 1.123\,934$$

= 1.12 (to two decimal places)

(b) Again using Snell's law, we have:

$$\sin \theta_i = \frac{\sin (90°)}{1.124\,934}$$

and so

$$\theta_c = \arcsin \left(\frac{1}{1.124\,934} \right) = 62.7°$$

(to one decimal place)

Notice that if you use the approximate value μ_g/μ_w = 1.12 in (b) you obtain sin θ_i = 1/1.12 and θ_c = 63.2°. To avoid such *rounding errors*, you should always work to a higher degree of precision (i.e. keep more figures) at intermediate stages than will be justified in your final answer, and round to a sensible number of figures only at the last step. Also notice that, although there are an infinite number of solutions to the equation for θ_c, only the one given is physically admissible.

T16 Using the cosine rule (Equation 28):

$$x^2 = (40\text{ km})^2 + (30\text{ km})^2$$
$$- 2 \times 40\text{km} \times 30\text{km} \times \cos (180° - 120°)$$
$$= [2500\text{ km}^2 - 2400 \cos (60°)]\text{ km}^2$$
$$= 1300\text{ km}^2$$

and hence: $x = \sqrt{1300}$ km = 36.1 km

T17 Using the cosine rule (Equation 28):

$$x^2 = (4\text{ m})^2 + (5\text{ m})^2 - 2 \times 4\text{ m} \times 5\text{ m}$$
$$\times \cos (120°)$$

= 41 m^2 − 40 cos (120°) m^2

= 41 m^2 + (40/2) m^2 = 61 m^2

and so: $x = \sqrt{61}$ m = 7.81 m

T18 Setting $\alpha = \beta$ in Equation 30 we get:

$$\cos (2\alpha) = \cos^2 (\alpha) - \sin^2 (\alpha)$$

T19 Let $\alpha = 15°$ then, from Equation 34:

$$\cos (30°) = 1 - 2 \sin^2 (15°) = \frac{\sqrt{3}}{2}$$

so that $2 \sin^2 (15°) = \dfrac{2 - \sqrt{3}}{2}$

and therefore $\sin (15°) = \dfrac{\sqrt{2 - \sqrt{3}}}{2}$

T20 Using Equation 44 with $\alpha = \omega t + \phi_1$ and $\beta = \omega t + \phi_2$, we find

$$x(t) = 2A \cos \left(\frac{\omega t + \phi_1 - \omega t - \phi_2}{2} \right) \times$$
$$\cos \left(\frac{\omega t + \phi_1 + \omega t + \phi_2}{2} \right)$$

so $x(t) = 2A \cos \left(\dfrac{\phi_1 - \phi_2}{2} \right) \cos \left(\omega t + \dfrac{\phi_1 + \phi_2}{2} \right)$

Chapter 6

Fast track answers

F1 See Figure 6.4. At $x = 2$ the gradient of the tangent is 4. As this value has been derived from a graph, your value for the gradient may differ slightly from this.

F2 $\lim_{x \to \infty}$ is an instruction to evaluate the expression that follows as x tends to infinity. In this case 1/x tends to 0 as x tends to infinity.

F3 e = 2.718 ... (to three decimal places).

F4 The gradient is k.

F5 (a) $\log_e [(e^x)^y] = \log_e (e^{xy}) = xy$.

(b) This equation cannot be simplified.

(c) exp [($\log_e (x) + 2 \log_e (y)$)]
$$= \exp [(\log_e (x)] \times \exp [(\log_e (y^2)] = xy^2.$$

(d) $\exp[2 \log_e(x)] = \exp[\log_e(x^2)] = x^2$.

(e) $a^{\log_a(x)} = x$.

F6 $\log_{10}(P) = -a \log_{10}(f) + \log_{10}(k)$, thus, a graph of $\log_{10}(P)$ against $\log_{10}(f)$ is a straight line of gradient $-a$ and intercept $\log_{10}(k)$.

Ready to study answers

R1 (a) a^m, (b) $5^0 = 1$.

R2 The expression reduces to $a^{20} \times a^6 = a^{26}$. So $x = 26$.

R3 (a) $16^{-1/4} = (2^4)^{-1/4} = 2^{-1} = 1/2$.

(b) $16^{3/4} = (2^4)^{3/4} = 2^3 = 8$.

(c) $4^{5/2} = (2^2)^{5/2} = 2^5 = 32$.

(d) $27^{-2/3} = (3^3)^{-2/3} = 3^{-2} = 1/9$.

(e) $1/(3^{-2}) = 3^2 = 9$.

R4 The *inverse function* $G(x)$ 'undoes' the effect of $F(x)$, so $G[F(x)] = x$, for any value of x in the *domain* of $F(x)$. If $F(x) = x^3$, then $G(x) = x^{1/3}$.

R5 The *graph* of $y = x^2$ is given in Figure 6.4. The approximate *solutions* are $x = 1.65$ and $x = -1.65$.

R6 The expressions in (a), (c), and (d) will give *straight-line* graphs. The gradients of these graphs are as follows: (a) m, (c) -1, (d) p.

R7 The 'rise' is 8 ($= 13 - 5$) and the 'run' is 2 ($= 3 - 1$) so the *gradient* ($=$ rise/run) is 4.

Text answers

T1 Number of disintegrations per second is given by $\lambda \times$ number of polonium nuclei $= 0.0133 \text{ s}^{-1} \times 6.0 \times 10^{18} = 8.0 \times 10^{16} \text{ s}^{-1}$. (In fact, this only represents the *average* number of decays per second, since nuclear decay is a statistical process in which it is not possible to predict the exact number expected in any particular second.)

T2 $k = -1/(RC)$. The rate of discharge is equal to the current, $I = Q/(RC)$, and Q is decreasing with time so k must be negative.

T3 For the tangent at $t = 10 \text{ s}$, the 'rise' ≈ -20 and 'run' ≈ 28, so the gradient $\approx -20/28 \approx -0.71$, so the

flow rate $\Delta V/\Delta t \approx -0.71$ litre s^{-1}. The line passing through $t = 0$ s, $V = 30$ litre, and $t = 15$ s, $V = 0$ litre, is an approximate tangent to the curve at $t = 5$ s, so $\Delta V/\Delta t \approx -2$ litre s^{-1}. (The exact value you find will depend on how you draw the tangent.)

T4 The accurate values for the currents determined for the tangents are 0.5 A, 0.25 A and 0.125 A, i.e. $\Delta Q/\Delta t \approx 0.25 \text{ s}^{-1} \times Q$. (Your values should approximate these.) Therefore the decay is exponential. (Of course, rather imprecise measurements made at a few points on a graph do not really verify that the graph is exponential, but at least they haven't given any evidence to the contrary, or rather, they shouldn't have done so!)

T5 One way would be to show that the gradient is not proportional to the value of V. For instance, you could measure the gradient at $x = 1$ (where $V = 8$) and at $x = 2$ (where $V = 2$) and show that the gradients at those two points, -16 and -2, respectively, are not in the ratio $8 : 2$.

T6 $e^2 = 7.389$, $e^3 = 20.09$, $e^{1.43} = 4.179$, $e^{-1} = 0.3679$ (all to four significant figures), $e^0 = 1$ ($a^0 = 1$ for all values of a).

T7 $k = -1/RC$ (see Answer T2) and so $Q(t) = Q_0 e^{-t/RC}$. Note that this could also be written as $Q(t) = Q_0 \exp(-t/RC)$

T8 $N_0 = 20 \times 10^{10}$ ($= 2 \times 10^{11}$). When $N = 10 \times 10^{10}$, the gradient $\Delta N/\Delta t = -5 \times 10^4 \text{ s}^{-1}$, so

$$k = (\Delta N/\Delta t)/N = -5 \times 10^{-7} \text{ s}^{-1}.$$

T9 The completed table is given in Table 6.6.

Table 6.6 See Answer T9.

m	a
2	2.2500
5	2.4883
10	2.5937
10^2	2.7048
10^3	2.7169
10^4	2.7181
10^5	2.7183

T10 (a) $100 = 10^2$, so $\log_{10}(100) = 2$.

(b) $1000 = 10^3$, so $\log_{10}(1000) = 3$.

(c) $0.1 = 10^{-1}$, so $\log_{10}(0.1) = -1$.

(d) $0.001 = 10^{-3}$, so $\log_{10}(0.001) = -3$.

(e) $10^{1/2} = 10^{0.5}$, so $\log_{10}(10^{1/2}) = 0.5$.

(f) $10 = 10^1$, so $\log_{10}(10) = 1$.

(g) $1 = 10^0$, so $\log_{10}(1) = 0$.

(h) $\log_{10}(10^{1.52}) = 1.52$.

T11 (a) If $10^x = 6.8$, then $x = \log_{10}(6.8) = 0.8325$.

(b) $10^x = 537$, $x = \log_{10}(537) = 2.730$.

(c) $10^x = 0.34$, $x = \log_{10}(0.34) = -0.4685$.

(All to four significant figures.)

T12 (a) $\log_{10}(4.725) = 0.6744$.

(b) $\log_{10}(47.25) = 1.6744$.

(c) $\log_{10}(472.5) = 2.6744$.

(d) A pattern has emerged – namely, $\log_{10}(4.725 \times 10^n) = n + \log_{10}(4.725)$. Therefore $\log_{10}(4725) = 3.6744$ and $\log_{10}(4.725 \times 10^7) = 7.6744$. (This pattern is discussed further in Subsection 6.3.3.)

T13 $\log_e(1) = 0$, since $e^0 = 1$ ($a^0 = 1$ for *any* value of a). $\log_e(e) = 1$, since $e^1 = e$.

T14 Figure 20 shows the graph of the function $\log_e(x)$ drawn on the same axes as $\log_{10}(x)$. The graph of $\log_e(10)$ is similar in shape to Figure 6.16b. Like $\log_{10}(x)$, it cuts the x-axis at $x = 1$. The rest of the curve can be estimated by drawing a curve through $\log_e(10) \approx 2.303$, $\log_e(e) = 1$ ($e \approx 2.718$) and $\log_e(4.5) \approx 1.5$ (see the seeded question above

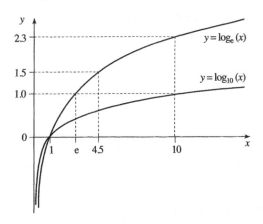

Figure 6.20 See Answer T14.

Question T13). (The exact relationship between $\log_{10}(x)$ and $\log_e(x)$ is discussed in Subsection 6.3.3.)

T15 (a) $e^x = 4.8$, $x = \log_e(4.8) = 1.569$.

(b) $e^x = 10$, $x = \log_e(10) = 2.303$.

(c) $e^x = 0.56$, $x = \log_e(0.56) = -0.5798$. (All to four significant figures.)

T16 $N = N_0 \exp(kt)$. Dividing both sides by N_0, and using the fact that $N/N_0 = 2$, gives us $2 = \exp(kt)$ so $\log_e(2) = kt$ and hence $t = \log_e(2)/k = 0.693/(0.02\ \text{h}^{-1}) = 34.7\ \text{h}$. This applies both when $N_0 = 1000$ and $N = 2000$, and when $N_0 = 2000$ and $N = 4000$ – the time for the population to double is independent of the initial number N_0.

T17 $N = N_0 \exp(-\lambda t)$. Dividing both sides by N_0 gives us $N/N_0 = \exp(-\lambda t)$ where $N/N_0 = 1/2$ so that $1/2 = \exp(-\lambda t)$ and therefore (taking logs of both sides) we find $\log_e(1/2) = -\lambda t$ so that

$$t = [\log_e(1/2)]/(-\lambda)$$
$$= 0.693/(0.0133\ \text{s}^{-1}) = 52.11\ \text{s}.$$

T18 $e^k = 3$, so $k = \log_e(3) \approx 1.099$ (to four significant figures). $y = \exp(1.099x)$. $3^2 = 9$, and $\exp(1.099 \times 2) \approx 9.007 \approx 9$. (The agreement becomes even better if k is taken to more significant figures.)

T19 $\log_{10}(6) = x \log_{10}(2)$, so $x = \log_{10}(6)/\log_{10}(2) \approx 2.585$.

T20 From Equation 27, with $b = 10$, $a = 2$ and $x = 5$: $\log_2(5) \times \log_{10}(2) = \log_{10}(5)$ and so

$$\log_2(5) = \log_{10}(5)/\log_{10}(2) \approx 2.322.$$

T21 Figure 6.18b shows exponential decay, since the graph of $\log_{10}(y)$ plotted against t is a straight line. The gradient of this graph is $-2/5 = -0.4$. Thus the exponential law is $y = y_0 e^{-0.4t}$.

T22 Taking $\log_{10}$ of both sides gives us,

$$2\log_{10}(T) = \log_{10}(k) + 3\log_{10}(R)$$

so $\log_{10}(T) = 3\log_{10}(R)/2 + \log_{10}(k)/2$

The gradient would be $3/2$, irrespective of the base of the logs, since Equations $22-25$ apply to logs to any base.

T23 If $T = k\sqrt{m}$, then $\log_{10}(T) = \log_{10}(k) + \log_{10}(\sqrt{m}) = \log_{10}(k) + [\log_{10}(m)]/2$, so a graph of $\log_{10}(T)$ against $\log_{10}(m)$ would have gradient $1/2$. (Note that logs to any base could be used and the gradient would still be the same.)

Chapter 7

Fast track answers

FI $\sinh (x) = \dfrac{e^x - e^{-x}}{2}, \quad \cosh (x) = \dfrac{e^x + e^{-x}}{2}$

and $\tanh (x) = \dfrac{e^x - e^{-x}}{e^x + e^{-x}}$

F2 Let $y = \text{arccoth} (x)$ then

$$x = \coth (y) = \frac{\cosh (y)}{\sinh (y)} = \frac{e^y + e^{-y}}{e^y - e^{-y}}$$

which can be rearranged as follows:

$$xe^y - xe^{-y} = e^y + e^{-y}$$

$$e^y(x - 1) = e^{-y}(x + 1)$$

$$e^{2y} = \left(\frac{x + 1}{x - 1}\right)$$

Taking the logarithm of both sides gives us

$$2y = \log_e \left(\frac{x + 1}{x - 1}\right)$$

and therefore

$$\text{arccoth} (x) = y = \frac{1}{2} \log_e \left(\frac{x + 1}{x - 1}\right)$$

The condition $|x| > 1$ is necessary so that the argument of the logarithmic function is positive (and this is consistent with Figure 7.14).

Ready to study answers

RI (a) $e^2 = 7.3890$, $e^{-2} = 0.1353$. (This is just a calculator exercise to make sure you know how to evaluate exponential functions.)

(b) Using the general formula for the solution of a quadratic equation we have

$$x = \frac{a \pm \sqrt{a^2 - 4}}{2}$$

and since $a \geq 2$ we see that both roots will be real. The solutions of the equation $e^{2w} - ae^w + 1 = 0$ are

$$w = \log_e (a \pm \sqrt{a^2 - 4}) - \log_e 2$$

R2 (a) $\arcsin (1/3) \approx 0.3398$

and $\dfrac{1}{\sin (1/3)} \approx 3.06$

Notice that these results are *not* the same. Whereas $\arcsin (x)$ is the inverse of the sine function,

$$\frac{1}{\sin (x)}$$

is the reciprocal of the sine function.

(b) $\sin [\sin (1/3)] \approx 0.321$ and $\sin^2 (1/3) \approx 0.107$

Notice again that these results are not the same. This is because the expression

$$\sin [\sin (x)]$$

means 'take the sine of x and then take the sine of the result'.

By contrast

$$\sin^2 (x)$$

means $\sin (x) \times \sin (x)$, i.e. the square of $\sin (x)$.

R3 (a) The inverse function of a given function is, roughly speaking, the function that 'undoes' the effect of the given function. More precisely, if f is the given function and g is its inverse, then $g(f(x)) = x$ for all x in the domain of f. In order for a function to have an inverse it is necessary that each value of $f(x)$ should correspond to a different value of x.

Text answers

TI (a) $\sinh (x) = \dfrac{e^{-x} - e^x}{2}$

$= -\dfrac{e^x - e^{-x}}{2} = -\sinh (x)$

therefore it is an *odd* function

and $\cosh (-x) = \dfrac{e^{-x} + e^x}{2} = \cosh (x)$

therefore it is an *even* function.

(b) Adding the expressions for $\sinh (x)$ and $\cosh (x)$

$$\sinh (x) + \cosh (x) =$$

$$\frac{e^x - e^{-x}}{2} + \frac{e^x + e^{-x}}{2} = e^x$$

(c) $\cosh^2(x) - \sinh^2(x) =$

$$\left(\frac{e^x + e^{-x}}{2}\right)^2 - \left(\frac{e^x - e^{-x}}{2}\right)^2$$

$$= \frac{1}{4}(e^{2x} + 2 + e^{-2x}) - \frac{1}{4}(e^{2x} - 2 + e^{-2x}) = 1$$

T2 $v = \sqrt{A\lambda \tanh\left(\frac{6.3d}{\lambda}\right)}$

$$= \sqrt{(1.8 \text{ m s}^{-2}) \times 18.9 \text{ m} \times \tanh\left(\frac{6.3 \times 6 \text{ m}}{18.9 \text{ m}}\right)}$$

$$\approx 5.73 \text{ m s}^{-1}$$

T3 Dividing the identity $\cosh^2(x) - \sinh^2(x) = 1$, given in Question T1(c), by $\sinh^2(x)$ gives

$$\frac{\cosh^2(x)}{\sinh^2(x)} - 1 = \frac{1}{\sinh^2(x)}$$

and therefore

$$\coth^2(x) - 1 = \operatorname{cosech}^2(x),$$

i.e. $\coth^2(x) - \operatorname{cosech}^2(x) = 1$.

T4 (a)   Let $x = \operatorname{cosech}(y)$ then $\sinh(y) = 1/x$

and therefore $y = \operatorname{arcsinh}(1/x)$.

However, from the definition of the arccosech function, we have

$$\operatorname{arccosech}(x) = y = \operatorname{arcsinh}(1/x)$$

(b) Let $x = \operatorname{sech}(y)$ then $\cosh(y) = 1/x$ so that $y = \operatorname{arccosh}(1/x)$.

However, from the definition of the arcsech function, we have

$$\operatorname{arcsech}(x) = y = \operatorname{arccosh}(1/x)$$

(c) Let $x = \coth(y)$ then $\tanh(y) = 1/x$ so that $y = \operatorname{arctanh}(1/x)$.

However, from the definition of the arccoth function, we have

$$\operatorname{arccoth}(x) = y = \operatorname{arctanh}(1x)$$

T5 Let $y = \operatorname{arctanh}(x)$

then $x = \tanh(y) = \dfrac{e^y - e^{-y}}{e^y + e^{-y}}$

and therefore $e^y(1 - x) = e^{-y}(1 + x)$

which gives $e^{2y} = \left(\dfrac{1 + x}{1 - x}\right)$

The condition

$$\left(\frac{1 + x}{1 - x}\right) = e^{2y} > 0$$

ensures that $-1 < x < 1$, and taking logarithms of both sides gives

$$2y = \log_e\left(\frac{1 + x}{1 - x}\right)$$

which gives the required result.

T6 Using the identity

$$\cosh^2(x) - \sinh^2(x) = 1$$

we can write the equation as

$$(\sinh^2(x) + 1) + \sinh(x) = 3$$

which can be rearranged to give

$$\sinh^2(x) + \sinh(x) - 2 = 0$$

and then factorised to read

$$(\sinh(x) - 1)(\sinh(x) + 2) = 0$$

This equation has the solutions

$$x = \operatorname{arcsinh}(1) \text{ and } x = \operatorname{arcsinh}(-2)$$

Using $\operatorname{arcsinh}(x) = \log_e(x + \sqrt{x^2 + 1})$

the two solutions can be written as

$$x = \log_e(1 + \sqrt{2}) \quad \text{and} \quad x = \log_e(\sqrt{5} - 2)$$

T7 $2\sinh(x)\cosh(x) =$

$$2 \times \left(\frac{e^x - e^{-x}}{2}\right) \times \left(\frac{e^x + e^{-x}}{2}\right)$$

$$= \frac{e^{2x} - e^{-2x}}{2} = \sinh(2x)$$

T8 (a) If we set $y = x$ in the identity

$$\cosh(x + y) = \cosh(x)\cosh(y)$$
$$+ \sinh(x)\sinh(y)$$

we find $\cosh(2x) = \cosh^2(x) + \sinh^2(x)$

But we also have the identity

$$\cosh^2(x) - \sinh^2(x) = 1$$

and so, substituting for $\sinh(x)$ we obtain

$$\cosh(2x) = 2\cosh^2(x) - 1$$

(b) Setting $y = x$ in the identity

$$\tanh(x + y) = \frac{\tanh(x) + \tanh(y)}{1 + \tanh(x)\tanh(y)}$$

leads directly to

$$\tanh(2x) = \frac{2\tanh(x)}{1 + \tanh^2(x)}$$

$$= 2\sinh(x)\cosh(x)\exp(\sinh^2(x))$$

$$= \sinh(2x)\exp(\sinh^2(x))$$

Chapter 8

Fast track answers

F1 $D\hat{H}F = H\hat{F}G = 40°$
(these are alternate angles)

$B\hat{H}D = A\hat{B}H = 75°$
(these are alternate angles)

$B\hat{H}F = B\hat{H}D + D\hat{H}F = 115°$,
so $P\hat{H}F = \frac{1}{2} \times 115° = 57.5°$.

Since $P\hat{H}D + D\hat{H}F = P\hat{H}F$, it follows that
$P\hat{H}D = 57.5° - 40° = 17.5°$.

F2 $B\hat{D}C + B\hat{D}F + F\hat{D}E = 180°$ (angles on a
straight line)

$\therefore B\hat{D}C + F\hat{D}E = 90°$. (The symbol $\therefore$ is often used
for 'therefore'.)

But $F\hat{D}E + D\hat{F}E = 90°$ and $\therefore D\hat{F}E = B\hat{D}C$.

Hence $C\hat{B}D = F\hat{D}E$ and it follows that the triangles BCD and DEF are similar.

We have

$$\frac{BC}{CD} = \frac{FD}{DB}$$

so that $BC = (15 \times 3/5)$ m $= 9$ m.

Also $$\frac{EF}{CD} = \frac{FD}{DB}$$

so that $$EF = \frac{FD}{DB}CD = \frac{FD}{DB}(CE - DE)$$

$$= \frac{5}{15}(15 - 3) \text{ m} = 4 \text{ m}$$

F3 Area of triangle $ABC = \frac{1}{2}BC \times AD = \frac{1}{2}AC \times BE$, but $AC = BC$ and therefore $AD = BE$.
It follows that triangles ADC and BEC are
congruent (see (d) in the four minimal specifications for unique triangles in Subsection 8.3.1).

F4 The cross-sectional area of the pipe wall is
the difference between the areas of the circles of
diameter $50 \text{ mm} = 0.05 \text{ m}$ and diameter 40 mm
$= 0.04 \text{ m}$,

i.e. $$\pi\left(\frac{0.05 \text{ m}}{2}\right)^2 - \pi\left(\frac{0.04 \text{ m}}{2}\right)^2$$

$$= (6.25 - 4.0) \times 10^{-4}\, \pi \text{ m}^2$$

The volume is $10 \text{ m} \times (6.25 - 4.0) \times 10^{-4} \pi \text{ m}^2 = 2.25 \times 10^{-3}\, \pi \text{ m}^3 \approx 7.069 \times 10^{-3} \text{ m}^3$.

(Notice how it is often easier to convert all the
lengths into the appropriate units before calculating
the volume.)

Ready to study answers

R1 (a) Substituting the given values into the
equation gives us

$$\frac{x}{6} = \frac{15}{10}$$

If we then rearrange this we obtain

$$x = \frac{6 \times 15}{10} = \frac{90}{10} = 9$$

(b) If we multiply the given equation by yz, we
find

$$\frac{x}{y} \times yz = \frac{y}{z} \times yz, \quad \text{i.e. } xz = y^2$$

If $x = 6$ and $z = 24$ then $y^2 = 6 \times 24 = 144$, from
which we calculate $y = \pm\sqrt{144} = \pm12$.

(c) $x = 5^{1/3}$ and therefore $x^2 = 5^{2/3} \approx 2.924$.

R2 The lines PQ and AB meet at right angles
(90°) at the mid-point M of AB (so that $AM = MB = 3.5$ cm). The angle between PA and PB is
60° while the angle between PA and QA is 120°.

Text answers

T1 The argument is identical to that for the triangle except that there are now four vertices, and therefore four anti-clockwise turns. Letting A, B, C and D denote the four interior angles, it follows that

$$(180° - A) + (180° - B) + (180° - C)$$
$$+ (180° - D) = 360°$$

and therefore $A + B + C + D = 360°$.

T2 Triangles ADB and ADC are congruent since they share a common side (AD), AB = AC and the angles BÂD and CÂD are equal. Hence the angles AB̂C and AĈB are equal. Further, since angles AD̂B and AD̂C add up to 180° and are equal, their common value is 90°. Finally, the sides BD and CD must be of equal length.

T3 Refer to Figure 8.23 in which ABC is an equilateral triangle. Since AB = AC the triangle is isosceles (with BC as base) and therefore AB̂C = AĈB. Also AB = BC and so the triangle is isosceles (with AC as base) so that AĈB = BÂC. Hence the three angles AB̂C, AĈB and BÂC are of equal size. Since they are the interior angles of a triangle their sum is 180°, and so each is equal to 60°.

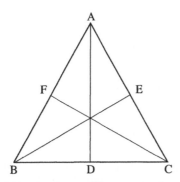

Figure 8.23 See Answer T3.

T4 In Figure 8.9b the triangles ACM and BCM are both isosceles so that CÂM = AĈM = x say, while MĈB = MB̂C = y say. However, the angles in the triangle ABC sum to 180°, i.e. $2x + 2y = 180°$ and therefore AĈB = $x + y = 90°$.

T5 Refer to Figure 8.11. In 3 s the driver travelling at 30 m s⁻¹ would have travelled about 90 m

hence the lettering on the sign should be capable of being read *first* at a distance of 190 m so AC = 190 m. The driver can read a number plate at 25 m so that AB = 25 m. Assuming that what matters is the angle that the eye is scanning (i.e. we assume that focusing is not a problem), we need to ensure that to the driver CD appears to be the same height as EB (a number plate). Since angles EÂB and DÂC are equal, and EB and DC are both vertical, triangles ABE and ACD are similar.

Hence $\dfrac{DC}{EB} = \dfrac{AC}{AB}$

therefore

$$DC = EB \times \frac{AC}{AB} = \left(8 \times 10^{-2} \times \frac{190}{25}\right) m$$

$$= 61 \text{ cm.}$$

T6 The height of the tree is

$$\left(\frac{3}{1} \times 2\right) m = 6 \text{ m.}$$

T7 In Figure 8.24, the tree is represented by AB, its shadow by BE, the pole by CD and its shadow by DE. Triangles ABE and CDE are similar (because AB and CD are both vertical and therefore parallel) and

$$\frac{AB}{CD} = \frac{BE}{DE}$$

so that the height of the tree is given by

$$AB = CD \times \frac{BE}{DE} = \left(2 \times \frac{4.5}{1.5}\right) m = 6 \text{ m .}$$

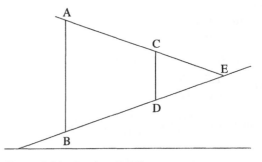

Figure 8.24 See Answer T7.

T8 The line from A to D is a diameter of the circle on the left in Figure 8.15. The portion of the belt which touches the pulley on the left is therefore half of the circumference.

The same is true for that part of the belt which touches the pulley on the right. The length from A to B is a radius + 5 m + a radius. The same is true for DC. The total length of the belt is therefore

$$\left[2 \times 2 \times \frac{2\pi}{2} + 2 \times (2 + 5 + 2)\right] \text{m}$$

$$= [2\pi \times 2 + 2 \times 9] \text{ m}$$

$$= (4\pi + 18) \text{ m} \approx 30.6 \text{ m}$$

T9 (a) $\theta = \dfrac{\text{arc DE}}{r} = \dfrac{\pi}{2} = 90°$

(b) If we take DC as the base of triangle DEC, the perpendicular height is EC, and it follows that the area of the triangle is $\frac{1}{2} \times (4 \text{ m}) \times (4 \text{ m}) = 8 \text{ m}^2$.

(c) Since DEC is a quadrant its area is $\frac{1}{4}\pi r^2 = \frac{1}{4}\pi(4 \text{ m})^2 = 4\pi \text{ m}^2$.

(d) The area of the circle is $16\pi \text{ m}^2$ so that the required area is $16\pi - (4\pi - 8) = (12\pi - 8) \text{ m}^2$.

(e) The area of the segment = area of sector DEC − area of triangle DEC which is $(4\pi - 8) \text{ m}^2 \approx 4.566 \text{ m}^2$.

T10 The cross-sectional area is the area of the annulus between the circles of radius 1 m and 2 m. The area of the larger circle is $\pi(2 \text{ m})^2 = 4\pi \text{ m}^2$ while the area of the smaller circle is $\pi (1 \text{ m})^2 = \pi \text{ m}^2$, thus the area of the annulus is $3\pi \text{ m}^2$. The volume of the prism is therefore $5 \times 3\pi \text{ m}^3 = 15\pi \text{ m}^3$.

T11 Let the radius of the smaller sphere be R and that of the larger sphere be S.

Then $\dfrac{4}{3}\pi S^3 = 2 \times \dfrac{4}{3}\pi R^3$

therefore $S^3 = 2R^3$

and it follows that

$$S = (2R^3)^{1/3} = 2^{1/3} R$$

The ratio of surface areas is

$$\frac{4\pi S^2}{4\pi R^2} = \frac{S^2}{R^2} = (2^{1/3})^2 = 2^{2/3} \text{ or } 4^{1/3} = 1.587$$

Chapter 9

Fast track answers

F1 (a) $y = -2x + 3$

(b) $y - 5 = 3(x - (-1))$, i.e. $y = 3x + 8$

(c) $\dfrac{y - (-3)}{1 - (-3)} = \dfrac{x - (-2)}{-3 - (-2)}$

i.e. $\dfrac{y + 3}{4} = \dfrac{x + 2}{-1}$

i.e. $y = -4x - 11$

(d) $\dfrac{x}{-1} + \dfrac{y}{2} = 1$, i.e. $y = 2x + 2$

F2 (a) $x^2 + y^2 = 25$

(b) $(x - (-1))^2 + (y - (-1))^2 = 2^2$

i.e. $(x + 1)^2 + (y + 1)^2 = 4$

(c) A line through $(4, 4)$ is of the form

$$y - 4 = m (x - 4)$$

i.e. $y = mx + (4 - 4m)$

and it meets the circle $x^2 + y^2 = 25$ where

$$x^2 + [mx + (4 - 4m)]^2 = 25$$

i.e. $(1 + m^2) x^2 + 2m (4 - 4m) x$
$$+ 16 (1 - m)^2 - 25 = 0$$

This quadratic equation has equal roots if

$$4m^2(4 - 4m)^2 = 4[16 (1 - m)^2 - 25] (1 + m^2)$$

Expanding gives

$$9m^2 + 32m + 9 = 0$$

Hence $m = -\dfrac{16 \pm 5\sqrt{7}}{9}$

The equations of the tangents to the circle through the point $(4, 4)$ are

$$y - 4 = \left(-\frac{16 + 5\sqrt{7}}{9}\right)(x - 4)$$

and $y - 4 = \left(-\dfrac{16 - 5\sqrt{7}}{9}\right)(x - 4)$

(d) The tangent to the first circle at the point (x_1, y_1) on the circle is $x_1x + y_1y = 25$.

At $(-3, 4)$ this becomes $-3x + 4y = 25$.

F3 (a) From Equation 16, the distance is found to be

$$\sqrt{[1 - (-1)]^2 + (2 - 3)^2 + [3 - (-2)]^2}$$

$$= \sqrt{2^2 + (-1)^2 + 5^2} = \sqrt{30}$$

(b) Using Equation 20 we can write down the equations

$$\frac{x - 1}{l} = \frac{y - 2}{m} = \frac{z - 3}{n}$$

But the point $(-1, 3, -2)$ also lies on the line, therefore

$$\frac{-1 - 1}{l} = \frac{3 - 2}{m} = \frac{-2 - 3}{n}$$

i.e. $\dfrac{-2}{l} = \dfrac{1}{m} = \dfrac{-5}{n}$

Therefore $l = -2$; $m = 1$; $n = -5$.

The following equations therefore determine the line

$$\frac{x - 1}{-2} = \frac{y - 2}{1} = \frac{z - 3}{-5}$$

(c) The point $(1, 2, 3)$ lies in the plane $x - 3y - z = -8$ (because $1 - 6 - 3 = -8$) as does the point $(-1, 3, -2)$ (because $-1 - 9 + 2 = -8$).

Since both points lie on the plane so does the line joining them.

Ready to study answers

R1 (a) True. If we expand the equation we find

$$(x - 1)(y - 2) - 3(x - 5)(y - 4) - 12$$

$$= -2xy + 10x + 14y - 70 = -2(x - 7)(y - 5)$$

(b) False. (c) False. (d) False.

R2 This is only true if the unknown side is the hypotenuse. If the side of length 12 m is the hypotenuse then the remaining side will be $\sqrt{119}$ m.

R3 A circle is a curve, lying in a plane, every point of which is equidistant from a given point. (Not the apocryphal schoolboy definition 'a straight round line, with a hole in the middle'.)

R4 A tangent is the line that meets the circle at one point only. In effect, it just grazes the edge of the circle, rather than crossing it.

Text answers

T1 D is $(-2, 1)$, E is $(-2, -2)$, F is $(-1, -1)$, G is $(1, -2)$ and O is $(0, 0)$: this is the origin.

T2 (a) Equation can be rearranged to

$y = -\dfrac{1}{3}x + 2$ so that $m = -\dfrac{1}{3}$ and $c = 2$

(b) $y = 2x - 2.5$ so that $m = 2$ and $c = -2.5$

(c) $y = -\dfrac{2}{3}x + \dfrac{5}{3}$ so that $m = -\dfrac{2}{3}$ and $c = \dfrac{5}{3}$

(d) $y = 3x - 3$ so that $m = 3$ and $c = -3$

(e) $y = x + 5$ so that $m = 1$ and $c = 5$

(f) $y = -\dfrac{1}{2}x - \dfrac{7}{2}$ so that $m = \dfrac{1}{2}$ and $c = -\dfrac{7}{2}$

T3 (a) The equation is the same form as Equation 2, so the gradient is clearly 5.

(b) The line does not pass through $(-1, 2)$ (but it does pass through $(1, -2)$).

(c) The equation in gradient−intercept form is $y = 5x - 7$.

(d) The y-intercept is -7.

T4 When $x = 1$, $y = -5$, and when $x = -1$, $y = 1$. Hence the form corresponding to Equation 4 is

$$\frac{y + 5}{x - 1} = \frac{1 + 5}{-1 - 1} = -\frac{6}{2} = -3$$

The intercepts are $(0, -2)$ and $(-2/3, 0)$ so that the form corresponding to Equation 5 is

$$\frac{x}{-2/3} + \frac{y}{-2} = 1$$

Adding $3x + 2$ to both sides of the given equation gives us $3x + y + 2 = 0$, which is in the form of Equation 6.

T5 (a) Substitute $y = 3x - 1$ into $y = -2x + 2$ to obtain

$$3x - 1 = -2x + 2,$$

i.e. $x = \dfrac{3}{5}$ then $y = \dfrac{9}{5} - 1 = \dfrac{4}{5}$

Intersection point is $\left(\dfrac{3}{5}, \dfrac{4}{5}\right)$

(b) The second equation can be rearranged to give $y = 3x + 4$. Substituting this into the first equation gives:

$$x + 2(3x + 4) = -1$$

i.e. $x = -\dfrac{9}{7}$ then $y = -\dfrac{27}{7} + 4 = \dfrac{1}{7}$

Intersection point is $\left(-\dfrac{9}{7}, \dfrac{1}{7}\right)$

(c) $2x + 5y = 10$ (i)

　　$-x + 4y = 2$ (ii)

Add twice (ii) to (i)

$$13y = 14 \quad \text{so that} \quad y = \dfrac{14}{13}$$

and from Equation (i)

$$2x + \dfrac{70}{13} = \dfrac{130}{13} \quad \text{and hence} \quad x = \dfrac{30}{13}$$

Intersection point is $\left(\dfrac{30}{13}, \dfrac{14}{13}\right)$

(d) The second equation is a rearrangement of the first so that any point on the line represents a solution. Substitution would yield $0 = 0$ which, whilst true, is unhelpful.

T6 (a) Numbering the equations as follows

　　$2x + y = 3$ (i)

　　$x + 0.501y = 6$ (ii)

$2 \times$ (ii) $-$ (i) gives us $0.002y = 9$

Hence $y = 4500$, and then from Equation (i) we have

$$x = -\dfrac{4497}{2} = -2248.5$$

(b) From the second pair of equations

　　$2x + y = 3$ (i)

　　$x + 0.499y = 6$ (ii)

$2 \times$ (ii) $-$ (i) gives us $-0.002y = 9$

Hence $y = -4500$ and then from Equation (i) we have

$$x = \dfrac{4503}{2} = 2251.5$$

Both pairs of equations are ill-conditioned.

T7 Let $A = (2, 3)$, $B = (-1, 2)$, $C = (3, -1)$, $D = (-2, -2)$.

A sketch would certainly help us to decide which pairs of points to check, but if we just list all the possibilities, we have

$$AB = \sqrt{(2 + 1)^2 + (3 - 2)^2} = \sqrt{10}$$

$$AC = \sqrt{(2 - 3)^2 + (3 + 1)^2} = \sqrt{17}$$

$$AD = \sqrt{(2 + 2)^2 + (3 + 2)^2} = \sqrt{41}$$

$$BC = \sqrt{(-1 - 3)^2 + (2 + 1)^2} = \sqrt{25}$$

$$BD = \sqrt{(-1 + 2)^2 + (2 + 2)^2} = \sqrt{17}$$

$$CD = \sqrt{(3 + 2)^2 + (-1 + 2)^2} = \sqrt{26}$$

(a) Therefore A and B are closest, (b) A and D are farthest apart.

T8 The equations of the circles are

$$(x - 1)^2 + (y - 2)^2 = 25 \qquad \text{(i)}$$

and $(x + 3)^2 + (y + 2)^2 = 9$ 　　(ii)

Expanding these expressions we obtain

$$x^2 - 2x + y^2 - 4y - 20 = 0 \qquad \text{(iii)}$$

and $x^2 + 6x + y^2 + 4y + 4 = 0$ 　　(iv)

We can eliminate both x^2 and y^2 by subtracting (iii) from (iv) to give

$$8y + 8x + 24 = 0$$

which simplifies to $y + x + 3 = 0$ (v)

which is the equation of a line. Remember that all of these equations are true provided that we are referring to the point, or points (x, y) where the circles meet. This means that such points must also lie on the line (v).

Substituting for y from Equation (v) into Equation (i) we obtain

$$(x - 1)^2 + (-x - 3 - 2)^2 = 25$$

which reduces to the quadratic equation

$$2x^2 + 8x + 1 = 0$$

Hence

$$x = \frac{-8 \pm \sqrt{64 - 8}}{4} = \frac{-8 \pm \sqrt{56}}{4}$$

$$= -2 \pm \frac{1}{2}\sqrt{14}$$

with corresponding values

$$y = -1 \mp \frac{1}{2}\sqrt{14}$$

The intersection points are

$$\left(-2 + \frac{1}{2}\sqrt{14}, \; -1 - \frac{1}{2}\sqrt{14}\right)$$

and

$$\left(-2 - \frac{1}{2}\sqrt{14}, \; -1 + \frac{1}{2}\sqrt{14}\right)$$

T9 (a) We can show directly that the given line is a tangent by substituting the equation of the line into that of the circle, and we obtain

$$x^2 + (33.8 - 2.4x)^2 = 169$$

i.e. $(1 + 5.76)x^2 - 2 \times 2 \times 2.4 \times 33.8x$

$$+ (33.8)^2 - 169 = 0$$

This has equal roots if

$$(2 \times 2.4 \times 33.8)^2 = 4 \times 6.76 \times [(33.8)^2 - 169]$$

and you can verify that this is so by evaluating both sides.

Alternatively we can rearrange the equation of the line first into the form $2.4x + y = 33.8$, then multiply both sides by 169/33.8 (to write it in the same form as Equation 13) to obtain $12x + 5y = 169$. Since $(12)^2 + (5)^2 = 169$ it follows that the point $(12, 5)$ is on the circle and hence, from Equation 13, the line is indeed a tangent.

(b) The point of contact is $(12, 5)$.

(c) The point $(-12, -5)$ lies at the opposite end of a diameter, and, from Equation 13, the tangent at this point has equation $-12x - 5y = 169$, which is clearly parallel to the original line (since it has the same gradient).

T10 (a) Since

$$x^2 + y^2 = a^2 \cos^2(\omega t) + a^2 \sin^2(\omega t) = a^2$$

it follows that P lies on a circle of radius a.

(b) The point returns to its original position in one day.

1 day = $(24 \times 60 \times 60)$ s = 86 400 s, and therefore $(86\,400\,\text{s})\omega = 2\pi$ radians, so that $\omega = (2\pi/86\,400)$ radians s^{-1}.

(c) The point P is moving directly towards, or away from, the satellite if the tangent to the circle $x^2 + y^2 = a^2$ at $P(a \cos(\omega t), \; a \sin(\omega t))$ passes through the point $S(R \cos(\Omega t), \; R \sin(\Omega t))$. The equation of the tangent at P is

$$x_1 x + y_1 y = a^2$$

with $x_1 = a \cos(\omega t)$ and $y_1 = a \sin(\omega t)$, so that the equation becomes:

$$ax \cos(\omega t) + ay \sin(\omega t) = a^2$$

which simplifies to $x \cos(\omega t) + y \sin(\omega t) = a$

The point $S(R \cos(\Omega t), \; R \sin(\Omega t))$ lies on this line if

$$R \cos(\Omega t) \cos(\omega t) + R \sin(\Omega t) \sin(\omega t) = a$$

so that $\cos(\Omega t) \cos(\omega t) + \sin(\Omega t) \sin(\omega t) = a/R$

i.e. $\cos(\Omega - \omega)t = a/R$

Finally, the point P is moving directly towards (or away from) S if

$$t = \frac{1}{\Omega - \omega} \arccos\left(\frac{a}{R}\right)$$

Notice that for solutions to exist we require $\Omega \neq \omega$, so that the satellite is moving relative to the Earth, and also $a \leqslant R$. (A subterranean satellite is of little practical value.)

T11 Let A = $(1, 2, 3)$, B = $(-1, -2, -1)$, C = $(2, 2, -2)$, D = $(3, 0, 1)$.

$$AB = \sqrt{2^2 + 4^2 + 4^2} = \sqrt{36}$$

$$AC = \sqrt{(-1)^2 + 0^2 + 5^2} = \sqrt{26}$$

$$AD = \sqrt{(-2)^2 + 2^2 + 2^2} = \sqrt{12}$$

$$BC = \sqrt{(-3)^2 + (-4)^2 + 1^2} = \sqrt{26}$$

$$BD = \sqrt{(-4)^2 + (-2)^2 + (-2)^2} = \sqrt{24}$$

$$CD = \sqrt{(-1)^2 + 2^2 + (-3)^2} = \sqrt{14}$$

(a) A and D are closest, and (b) A and B are farthest apart.

T12 The required equation must be of the form $3x - 4y + z = d$ for some suitably chosen value of d. The point $(1, 2, 3)$ lies on the plane, so that the values $x = 1$, $y = 2$ and $z = 3$ must satisfy the equation. It follows that $(3 \times 1) - (4 \times 2) + 3 = d$ and therefore $d = -2$, and the required equation is $3x - 4y + z = -2$.

T13 The value $s = 0$ gives $x = 1, y = -2, z = 2$ so the point P $(1, -2, 2)$ lies on the line. Let the required point be Q $(1 + 2s, -2 + s, 2 - 3s)$, and notice that this point *must* lie on the line. Then, from Equation 16, we have $PQ^2 = (2s)^2 + s^2 + (3s)^2 = 14$ so that $s^2 = 1$ and $s = \pm 1$. It follows that the possible choices for Q are $(3, -1, -1)$ and $(-1, -3, 5)$.

$$s = \frac{x - 1}{2} = \frac{y + 2}{1} = \frac{z - 2}{-3}$$

The direction ratios are 2, 1, and -3.

Chapter 10

Fast track answers

F1 (a) Rewrite the equation as $(x - 1)^2 + (y - 3)^2 = 4$; the radius is 2 and the coordinates of the centre are $(1, 3)$.

(b) Substituting $x/2$ into the equation of the circle, we obtain $x^2 + (x^2/4) - 8x + 3x - 15 = 0$, i.e. $5x^2 - 20x - 60 = 0$, i.e. $x^2 - 4x - 12 = 0$ so that $(x - 6)(x + 2) = 0$.

When $x = 6$, $y = 3$ and when $x = -2$, $y = -1$ so that the coordinates of P and Q are $(6, 3)$ and $(-2, -1)$.

The equation of the required circle can be written as

$$x^2 + y^2 + 2Gx + 2Fy + C = 0$$

Since $(1, 1)$ lies on the circle then $1 + 1 + 2G + 2F + C = 0$

i.e. $2 + 2G + 2F + C = 0$ (i)

Since $(6, 3)$ lies on the circle then $36 + 9 + 12G + 6F + C = 0$

i.e. $45 + 12G + 6F + C = 0$ (ii)

Since $(-2, -1)$ lies on the circle then $4 + 1 - 4G - 2F + C = 0$

i.e. $5 - 4G - 2F + C = 0$ (iii)

Equations (i), (ii) and (iii) can be solved simultaneously to find G, F and C. Probably the simplest method is

(ii) − (i) gives us $43 + 10G + 4F = 0$

and (ii) − (iii) gives us $40 + 16G + 8F = 0$

Then multiply the first of these equations by 2 and subtract to give $G = -23/2$, from which we find $F = 18$ and $C = -15$.

The equation of the circle is $x^2 + y^2 - 23x + 36y - 15 = 0$.

(c) The line $y = -\frac{3}{4}x + c$ is parallel to the given line. It meets the circle $x^2 + y^2 = 25$ where $x^2 + (-\frac{3}{4}x + c)^2 = 25$

i.e. $\dfrac{25}{16}x^2 - \dfrac{3c}{2}x + x^2 - 25 = 0$

This quadratic equation has equal roots if

$$\left(\frac{-3c}{2}\right)^2 = 4 \times \left(\frac{25}{16}\right) \times (c^2 - 25)$$

i.e. $\dfrac{9c^2}{4} = \dfrac{25(c^2 - 25)}{4}$

so that $16c^2 - 25 \times 25 = 0$, or $c^2 = (\frac{25}{4})^2$, hence $c = \pm 25/4$ and the equations of the tangent lines can be written $3x + 4y = 25$ and $3x + 4y = -25$.

F2 (a) Comparing the equation of this given parabola to that of the standard form $y^2 = 4ax$ we see that $a = 4$. For the point (x_1, y_1) the *tangent* to the parabola at this point is $y_1y = 2a(x + x_1)$. At the point $(16, 16)$ this becomes $16y = 8(x + 16)$ or $2y = x + 16$. The *normal* has gradient -2 and passes through $(16, 16)$; its equation is $y - 16 = -2(x - 16)$ or $2x + y = 48$.

Repeating the process for the point $(1, -4)$ we obtain

for the *tangent* $2x + y + 2 = 0$

and for the *normal* $2y = x - 9$

The tangents meet at T $= (-4, 6)$ and the normals at R $= (21, 6)$.

Since T and R have the same y-coordinate the line joining them must be parallel to the x-axis.

(b) The parametric representation of a point on a parabola is $(at^2, 2at)$. The two points on the parabola at the ends of a chord can be represented as $(at_1^2, 2at_1)$ and $(at_2^2, 2at_2)$. The equation of the line joining them is

$$\frac{y - 2at_1}{2at_2 - 2at_1} = \frac{x - at_1^2}{at_2^2 - at_1^2}$$

which reduces to $2x - (t_1 + t_2)y + 2at_1t_2 = 0$.

If this chord passes through the focus $(a, 0)$ we have $2a + 2at_1t_2 = 0$ so that $t_1t_2 = -1$.

The gradient of the tangent at the point $(at^2, 2at)$ is $1/t$, so that the product of the gradients at the ends of the chord through the focus is $1/t_1 \times 1/t_2 = 1/(t_1t_2) = -1$. It follows that the tangents at the end points of a chord through the focus are perpendicular.

F3 (a) The tangent to the given ellipse at (x_1, y_1) is

$$\frac{x_1 x}{a^2} + \frac{y_1 y}{b^2} = 1$$

where $a = 4$ and $b = 3$.

For the point $(\frac{16}{5}, \frac{9}{5})$ this equation becomes

$$\frac{16}{5}\frac{x}{16} + \frac{9}{5}\frac{y}{9} = 1 \text{ or } x + y = 5.$$

This line meets the x-axis $(y = 0)$ where $x = 5$, and meets the y-axis $(x = 0)$ where $y = 5$. Hence the intercepts are equal.

(b) From Equation 32 the tangent to the hyperbola $xy = c^2$ at the point $(ct, c/t)$ is $t^2y + x - 2tc = 0$. Putting $c = 2\sqrt{3}$ and $t = \sqrt{3}/2$, we have $x = ct = 2\sqrt{3} \times \sqrt{3}/2 = 3$ and $y = c/t = 2\sqrt{3} \times 2/\sqrt{3} = 4$ so that the tangent line at $(3, 4)$ is

$$\left(\frac{\sqrt{3}}{2}\right)^2 y + x - 2\left(\frac{\sqrt{3}}{2}\right)(2\sqrt{3}) = 0$$

which simplifies to give $(3/4)y + x - 6 = 0$ or $3y + 4x = 24$.

Ready to study answers

R1 $PQ = \sqrt{(3 - 0)^2 + (1 - (-2))^2} = \sqrt{9 + 9} = 3\sqrt{2}$

Gradient $= (1 - (-2))/(3 - 0) = 1$.

R2 $y - 1 = 3(x - (-2))$ or $y = 3x + 7$. The perpendicular line has gradient $-\frac{1}{3}$ and equation

$$y - 1 = -\tfrac{1}{3}(x - (-2)) \quad \text{or} \quad x + 3y - 1 = 0$$

R3 $\dfrac{y - (-1)}{x - 2} = \dfrac{4 - (-1)}{-1 - 2}$ or $\dfrac{y + 1}{x - 2} = \dfrac{5}{-3}$

so that $3y + 3 + 5x - 10 = 0$ and therefore $3y + 5x - 7 = 0$. This intersects $y = 3x$ where $3(3x) + 5x - 7 = 0$ i.e. $x = 1/2$ and $y = 3/2$ so that the point of intersection is $(1/2, 3/2)$.

R4 (a) The quadratic equation $Ax^2 + Bx + C = 0$ has a repeated root at $x = -B/(2A)$ if $B^2 = 4AC$.

Recall that

$$x = \frac{-B \pm \sqrt{B^2 - 4AC}}{2A}$$

(b) The quadratic equation $x^2 + (mx + c)^2 = a^2$ can be rearranged into the form $(1 + m^2)x^2 + 2mcx + c^2 - a^2 = 0$, in which case $A = (1 + m^2)$, $B = 2mc$ and $C = c^2 - a^2$ and the equation has a repeated root if $(2mc)^2 = 4(1 + m^2)(c^2 - a^2)$ which rearranges to $m^2 c^2 = m^2 c^2 + c^2 - (1 + m^2)a^2$ so that $c^2 = (1 + m^2)a^2$.

Text answers

T1 The quadratic equation becomes $x^2 + (3x + 5\sqrt{10})^2 = 5^2$ so that $10x^2 + 30\sqrt{10}x + 225 = 0$. The quadratic equation $Ax^2 + Bx + C = 0$ has equal roots if $B^2 - 4AC = 0$, and in this case $B^2 - 4AC = (30\sqrt{10})^2 - 40 \times 225 = 0$.

Also $c^2 - (1 + m^2) = 250 - (1 + 9) \times 25 = 0$ so that Equation 3 is satisfied.

The line $y = 3x + 5\sqrt{10}$ is a tangent to the circle $x^2 + y^2 = 25$.

T2 The general equation of a tangent at (x_1, y_1) is $x_1x + y_1y = 4$ while the normal is given by Equation 5.

(a) tangent: $\sqrt{2}x + \sqrt{2}y = 4$ or $x + y = 2\sqrt{2}$; normal: $y = x$

(b) tangent: $2y = 4$ i.e. $y = 2$; normal: $x = 0$

(c) tangent: $2x = 4$ i.e. $x = 2$; normal: $y = 0$

(d) tangent: $-x - \sqrt{3}y = 4$; normal: $y = \sqrt{3}x$

(e) tangent: $1.2x - 1.6y = 4$ i.e. $3x - 4y = 10$; normal: $y = -4x/3$

T3 The normal has equation

$$y = \left(\frac{4\sin\theta}{4\cos\theta}\right)x, \quad \text{that is } y = x\tan\theta$$

T4 (a) Here $4a = 1/8$ so that $a = \frac{1}{32}$ and the general equation of the tangent at (x_1, y_1) is given by Equation 13. The normal at this point has gradient $-(y_1/2a) = -16y_1$ and so its equation is $y - y_1 = 16y_1(x - x_1)$. At the point $(1/2, 1/4)$ these equations become:

tangent $y/4 = (x + 1/2)/16$, i.e. $8y = 2x + 1$

normal $y - 1/4 = -4(x - 1/2)$

i.e. $16x + 4y = 9$

At the point $(1/2, -1/4)$ these equations become:

tangent $-y/4 = (x + 1/2)/16$

i.e. $8y + 2x + 1 = 0$

normal $y + 1/4 = 4(x - 1/2)$

i.e. $16x - 4y = 9$

(b) Putting $x_1 = a/m^2$ and $y_1 = 2a/m$ we have $y_1^2 = 4a^2/m^2 = 4ax_1$ and therefore M lies on the parabola. From Equation 13 the tangent at M is $2ay/m = 2a(x + a/m^2)$ which simplifies to give Equation 14.

T5 The line through P and Q is

$$\frac{y - 2at_1}{2at_2 - 2at_1} = \frac{x - at_1^2}{at_2^2 - at_1^2}$$

which simplifies to

$$\frac{y - 2at_1}{2} = \frac{x - at_1^2}{t_2 + t_1}$$

Since the line passes through the focus $(a, 0)$ we have

$$-\frac{2at_1}{2} = \frac{a(1 - t_1^2)}{t_2 + t_1}$$

so that $-t_1(t_2 + t_1) = 1 - t_1^2$ or $t_2 t_1 = -1$.

T6 The parametric equation of a tangent is $y = x/t + at$ so the tangents at P and Q are $x - t_1 y + at_1^2 = 0$ and $x - t_2 y + at_2^2 = 0$, respectively. These meet where $(t_2 - t_1)y + a(t_1^2 - t_2^2) = 0$, i.e. if $y = a(t_1 + t_2)$; substituting into the equation of either tangent we have $x = at_1 t_2$. Hence R is $(at_1 t_2, a(t_1 + t_2))$

T7 The parabola passes through the points $x = 2$, $y = 0$ and $x = 0$, $y = 4$ and the only option that satisfies these requirements is option (e) $y^2 - 16 = -8x$

T8 From Equation 20, the equation of the tangent to the given ellipse at P (x_1, y_1) is $x_1 x/4 + y_1 y/16 = 1$. This can be written

$$y = -\left(\frac{4x_1}{y_1}\right)x + \frac{16}{y_1}$$

and so has gradient $-4x_1/y_1$. Thus the normal at P has gradient $y_1/4x_1$, and the equation of the normal at P is therefore

$$y - y_1 = \frac{y_1}{4x_1}(x - x_1)$$

which simplifies to

$$y = \frac{y_1 x}{4x_1} + \frac{3y_1}{4}.$$

(a) At $(1, 2\sqrt{3})$ the equations are:

tangent: $\dfrac{1}{4}x + \dfrac{2\sqrt{3}}{16}y = 1$ or $2x + \sqrt{3}\,y = 8$

normal: $y = \dfrac{\sqrt{3}}{2}x \times \dfrac{3}{2}\sqrt{3}$

(b) At $(1, -2\sqrt{3})$:

tangent: $2x - \sqrt{3}y = 8$

normal: $y = -\dfrac{\sqrt{3}}{2}x - \dfrac{3}{2}\sqrt{3}$

(c) At $(-1, 2\sqrt{3})$:

tangent: $-2x + \sqrt{3}y = 8$

normal: $y = -\dfrac{\sqrt{3}}{2}x + \dfrac{3}{2}\sqrt{3}$

(d) At $(2, 0)$: tangent $x = 2$, normal $y = 0$ (the x-axis).

(e) At $(0, -4)$: tangent: $y = -4$, normal $x = 0$ (the y-axis).

T9 The tangent at the point corresponding to $\theta = \theta_1$ is given by Equation 22

$$\frac{\cos\theta_1}{a}x + \frac{\sin\theta_1}{b}y = 1 \qquad\qquad\text{(i)}$$

While the tangent at the point corresponding to $\theta = \theta_1 + 90°$ is

$$\frac{\cos(\theta_1 + 90°)}{a} x + \frac{\sin(\theta_1 + 90°)}{b} y = 1$$

which becomes

$$\frac{-\sin\theta_1}{a} x + \frac{\cos\theta_1}{b} y = 1 \tag{ii}$$

The tangents meet where Equation (i) and Equation (ii) are satisfied simultaneously.

Multiplying (i) by $\cos\theta_1$ and (ii) by $\sin\theta_1$ and then subtracting gives

$$\frac{\cos^2\theta_1}{a} x + \frac{\sin^2\theta_1}{a} x = \cos\theta_1 - \sin\theta_1$$

so that $x = a(\cos\theta_1 - \sin\theta_1)$.

Multiplying (i) by $\sin\theta_1$ and (ii) by $\cos\theta_1$ and then adding them together gives

$$\frac{\sin^2\theta_1}{b} y + \frac{\cos^2\theta_1}{b} y = \cos\theta_1 + \sin\theta_1$$

so that $y = b(\cos\theta_1 + \sin\theta_1)$.

It follows that

$$\frac{x^2}{a^2} + \frac{y^2}{b^2} = (\cos^2\theta_1 - 2\cos\theta_1\sin\theta_1 + \sin^2\theta_1) +$$

$$(\cos^2\theta_1 + 2\cos\theta_1\sin\theta_1 + \sin^2\theta_1) = 2$$

T10 The major axis is along the x-axis.

T11 The hyperbola is $\dfrac{x^2}{4} - \dfrac{y^2}{9} = 1$

Then from Equation 26, the equation of the tangent at $(4, 3\sqrt{3})$ is $4x/4 - 3\sqrt{3}y/9 = 1$ which simplifies to $\sqrt{3}x - y = \sqrt{3}$.

The gradient of this tangent is $\sqrt{3}$ so that the gradient of the normal is

$$-\frac{1}{\sqrt{3}}$$

It has equation

$$y - 3\sqrt{3} = -\frac{1}{\sqrt{3}}(x - 4)$$

which simplifies to $x + \sqrt{3}y = 13$.

T12 From Equation 29 the tangent at the point on the hyperbola corresponding to the parameter θ is

$$\frac{\cosh\theta}{a} x - \frac{\sinh\theta}{b} y = 1$$

and this passes through the point $(a/2, 0)$ if

$$\frac{\cosh\theta}{a}\left(\frac{a}{2}\right) - \frac{\sinh\theta}{b}(0) = 1$$

i.e. if $\cosh\theta = 2$. Since $\sinh^2\theta = \cosh^2\theta - 1$ it follows that

$$\sinh\theta = \pm\sqrt{\cosh^2\theta - 1} = \pm\sqrt{4 - 1} = \pm\sqrt{3}$$

The equations of the tangent lines are therefore

$$\frac{2}{a} x \pm \frac{\sqrt{3}}{b} y = 1.$$

T13 (a) Here $A = 5$, $B = -4$, $H = 3$ and $H^2 = 9 > -20 = AB$. The curve is a hyperbola.

(b) $A = 2$, $B = 2$, $H = \frac{3}{2}$ and $H^2 = \frac{9}{4} < 4 = AB$. The curve is an ellipse.

T14 Ellipse: $\dfrac{(x + 1)^2}{a^2} + \dfrac{(y - 2)^2}{b^2} = 1$.

Hyperbola: $\dfrac{(x + 1)^2}{a^2} - \dfrac{(y - 2)^2}{b^2} = 1$.

Chapter 11

Fast track answers

F1 (a) The *velocity* of light is a vector quantity so its complete specification requires a magnitude *and* a direction. No direction is given in the statement, only the *speed* of light has been specified.

(b) Similarly, acceleration is a vector quantity so its complete specification requires a magnitude *and* a direction. It is conventional to refer to g as the 'acceleration due to gravity' but in strict vector terms it should be called the magnitude of the acceleration due to gravity.

(c) Mass is a scalar quantity, so it has no associated direction and cannot push down on anything. Forces exerted by your head (by virtue of gravity acting on its mass) do have direction and *can* 'push down' on your neck.

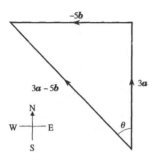

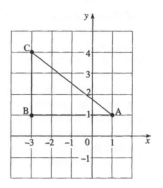

Figure 11.14 See Answer F2.

Figure 11.15 See Answer R1.

F2 (a) The displacement $3a - 5b$ is shown in Figure 11.14. Its magnitude can be determined graphically or algebraically by means of Pythagoras's theorem. Using this latter technique

$$| 3a - 5b |^2 = 9\, |\, a\, |^2 + 25\, |\, b\, |^2$$

i.e. $\quad | 3a - 5b |^2 = [9(4.32) + 5(2.40)^2]\ \text{km}^2$

Thus, $\quad | 3a - 5b | = 17.7\ \text{km}$

The displacement is at an angle θ west of north, where

$$\theta = \arctan\left(\frac{|\,15b\,|}{|\,3a\,|}\right) = \arctan\left(\frac{12.00}{12.96}\right) = 42.80°$$

(b) The orthogonal component vector in the given direction will be inclined at 22.80° (= 42.80° − 20°) to $3a - 5b$ so its magnitude will be

$$| 3a - 5b |\cos 22.80° = (17.7)0.922\ \text{km}$$

$$= 16.3\ \text{km}$$

Ready to study answers

R1 See Figure 11.15. (You are not expected to draw the complete grid.)

Length of side AB = 1 − (−3) = 4

Length of side BC = 4 − 1 = 3

Length of side CA = $\sqrt{(4)^2 + (3)^2} = \sqrt{25} = 5$.
This side is called the *hypotenuse*.

R2 It follows from the definitions of the *trigonometric ratios* that:

Length of side AB = $c \cos \theta$

Length of side BC = $c \sin \theta$

Thus,

$$\frac{\sin \theta}{\cos \theta} = \tan \theta = \frac{\text{length of side BC}}{\text{length of side AB}} = \frac{3}{4}$$

If follows that $\theta = \arctan\,(0.75) = 36.9°$.

R3 The *modulus* $|\, x\, |$ is defined by the requirement

$$\text{if } x \geqslant 0 \quad |\, x\, | = x \quad \text{and} \quad |\, -x\, | = x$$

Thus, $|\, x\, |$ is always a non-negative quantity, irrespective of the sign of x itself.

(a) $|\, 3\, | = 3$

(b) $|\, -3\, | = 3$

(c) $|\, (-2)(3.1)\, | = |\, -6.2\, | = 6.2$

(d) $|\, (-2.4)/2\, | = |\, -1.2\, | = 1.2$

Text answers

T1 (a) The displacement from Exeter to Belfast is 465 km, at 21° west of north.

(b) The displacement from Belfast to Edinburgh is 230 km, at 48° east of north. (You may have specified the directions differently, but your specifications should be equivalent to those given in the answer.)

T2 Only displacement (c) is equal to the displacement from Edinburgh to Belfast. Displacement (a) is in the exactly opposite direction and displacement (b) has a negative magnitude − which is meaningless. (*Note*: Magnitudes can *never* be negative. Negative magnitudes do not indicate a reversal of direction − they are simply meaningless.)

T3 Only rate of change of acceleration is a vector, since it alone requires a *direction* as part of its complete specification. Energy, distance and time can all be expressed as single numbers together with appropriate units, so they are all scalars. Electric charge is also a scalar, for the same reason, even though it may be positive or negative. Similarly, altitude above sea level is a scalar since there is no need to include any directional information (such as 'upwards') when specifying values of the altitude − it is enough to say 10 m, for example.

T4 (a)   It is the *velocity* of the car which is to be specified, not the car itself. Whereas the specification of the car includes information on its ownership, the velocity itself does not. The velocity so defined is completely specified by a magnitude and a direction.

(b) This claim is reasonable but misguided. It is indeed necessary to know about compass bearings (or something equivalent) but this is part of the specification of direction and not something separate. So, the statement that a displacement is defined by a magnitude and a direction is still true.

T5 (a)   You can find the magnitude of $\overrightarrow{PR}$ just by measuring its length with a ruler and then interpreting your measurement in terms of the scale indicated on the Cartesian axes. The magnitude of the displacement $\overrightarrow{PR}$ is 5 m. (Alternatively, you may treat the line PR as the hypotenuse of a right-angled triangle, as shown in Figure 11.16 and use Pythagoras's theorem to deduce that the magnitude of $\overrightarrow{PR}$ is

$$\sqrt{(3 \text{ m})^2 + (4 \text{ m})^2} = \sqrt{25} \text{ m} = 5 \text{ m}.)$$

(b) $\overrightarrow{PR}$ points in the direction from P to R. The direction may be described by saying that the angle θ in Figure 11.16 is

arctan (3/4) = arctan 0.75 = 36.9°

(c) Using Pythagoras's theorem, the magnitude of the displacement $\overrightarrow{OQ}$ is

$$\sqrt{(2 \text{ m})^2 + (2 \text{ m})^2} = \sqrt{8} \text{ m} = 2.83 \text{ m}$$

T6 See Figure 11.17. Make sure you have included the wavy underlines beneath the vectors.

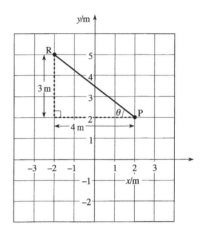

Figure 11.16   See Answer T5.

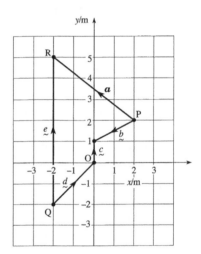

Figure 11.17   See Answer T6.

T7 (a)   Using Pythagoras's theorem

$$| \overrightarrow{RS} | = \sqrt{(2 \text{ m})^2 + (4 \text{ m})^2} = \sqrt{20} \text{ m} = 4.47 \text{ m}$$

(b) $| b | = \sqrt{(2 \text{ m})^2 + (1 \text{ m})^2} = \sqrt{5} \text{ m} = 2.24 \text{ m}$

$| c | = 1 \text{ m}$

(c) $| b | + | c | = (1 + 2.24) \text{ m} = 3.24 \text{ m}$

$| b | \, | c | = (2.24 \text{ m})(1 \text{ m}) = 2.24 \text{ m}^2$

(d) $| a | = 5 \text{ m}, | b | = 2.24 \text{ m}, | c | = 1 \text{ m}$

$| d | = \sqrt{(2 \text{ m})^2 + (2 \text{ m})^2} = \sqrt{8} = 2.83 \text{ m}$

$| e | = 7 \text{ m}$

So, $| a | + | b | + | c | + | d | + | e | = 18.07 \text{ m}$

(e) $2|\boldsymbol{a}| + \dfrac{|\boldsymbol{b}||\boldsymbol{c}|}{|\boldsymbol{d}|} = 2(5\text{ m}) + \dfrac{(2.24\text{ m})\,(1\text{ m})}{2.83\text{ m}}$

$= 10\text{ m} + 0.79\text{ m} = 10.79\text{ m}$

Note that the magnitudes of vectors are simply non-negative scalar quantities, so they can be added and subtracted, multiplied and divided just like any other scalar quantities.

T8 If $m = 6$ kg and $\boldsymbol{v} = 5$ m s^{-1} north-west, then

$m\boldsymbol{v} = 30$ kg m s^{-1} north-west.

The magnitude of this momentum is $|\,m\boldsymbol{v}\,|$ and its direction is north-west.

T9 (a) $\alpha\boldsymbol{a}$ is a vector, so $|\alpha\boldsymbol{a}|$ is the magnitude of a vector. Since any vector magnitude is always a *non-negative* scalar it follows that $|\alpha\boldsymbol{a}|$ cannot be less than zero, so the given statement is always true.

(b) Whether α is positive or negative, α^2 will always be positive so $\alpha^2\boldsymbol{a}$ is always in the same direction as $\boldsymbol{a}$ and the statement is always true.

(c) $-\alpha\boldsymbol{a}$ will be antiparallel to $\boldsymbol{a}$ if $\alpha > 0$, but it will be parallel to $\boldsymbol{a}$ if $\alpha < 0$. Thus the given statement is sometimes true and sometimes false.

(d) $\hat{\boldsymbol{a}} = \dfrac{\boldsymbol{a}}{|\boldsymbol{a}|}$ may be rewritten as

$\hat{\boldsymbol{a}} = \left(\dfrac{1}{|\boldsymbol{a}|}\right)\boldsymbol{a}$

so it simply represents a vector that has been scaled by the reciprocal of its own magnitude. Since $|\boldsymbol{a}|$ must be a positive quantity $\hat{\boldsymbol{a}}$ must point in the same direction as $\boldsymbol{a}$ and its magnitude must be

$\dfrac{1}{|\boldsymbol{a}|}|\boldsymbol{a}| = 1$

Thus, the given statement is always true. Vectors such as $\hat{\boldsymbol{a}}$ are called *unit vectors* and are of great use in more advanced work.

(e) This statement is always false since it equates a scalar (on the left-hand side) with a vector (on the right-hand side).

(f) This statement is always true. It expresses symbolically part of the definition of scaling.

T10 (a) $\overrightarrow{OS} + \overrightarrow{SP} = \overrightarrow{OP}$

(b) $\overrightarrow{PS} + \overrightarrow{SR} = \overrightarrow{PR}$

(c) $(\overrightarrow{PS} + \overrightarrow{SR}) + \overrightarrow{RQ} = \overrightarrow{PR} + \overrightarrow{RQ} = \overrightarrow{PQ}$

(d) $\overrightarrow{PS} + (\overrightarrow{SR} + \overrightarrow{RQ}) = \overrightarrow{PS} + \overrightarrow{SQ} = \overrightarrow{PQ}$

Comment Parts (c) and (d) of this answer show that we may add three (or more) vectors together and that the result does not depend on the order in which the additions are carried out. Thus, it is not at all ambiguous to write $\overrightarrow{PS} + \overrightarrow{SR} + \overrightarrow{RQ}$ without any brackets.

T11 See Figure 11.18. Note that in the case of the vectors $\boldsymbol{E}$ and $\boldsymbol{F}$ the 'triangle' has collapsed into a straight line. Nonetheless, the principles of the triangle rule still apply.

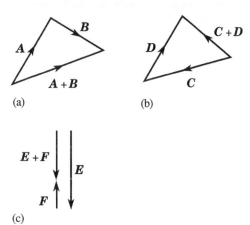

Figure 11.18 See Answer T11.

T12 See Figure 11.19.

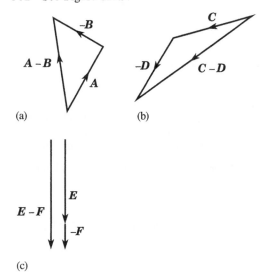

Figure 11.19 See Answer T12.

TI3 (a) The velocity of the duck relative to the angler will be the vector sum of the duck's velocity relative to the water and the water's velocity relative to the angler, i.e. $u + v$.

(b) The angler's velocity relative to the duck will be equal in magnitude, but opposite in direction to the duck's velocity relative to the angler, i.e. $-u - v$.

TI4 As you can see in Figure 11.20, F_1 is at 60° to W. (This is the crucial angle, not the 30° between the plane and the horizontal.) Thus

$$|F_1| = |W| \cos 60° = 2.50 \text{ N}$$

$$|F_2| = |W| \sin 60° = 4.33 \text{ N}$$

F_1 is parallel to the inclined plane in the 'downhill' direction.

F_2 is normal to the inclined plane, in the downward sense.

(Note that simply saying normal (or parallel) to the plane is ambiguous – you need to specify the upward or downward sense.)

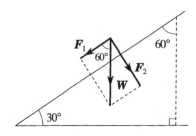

Figure 11.20 See Answer T14.

TI5 The north-west course of the plane means that its velocity is directed 45° to the north-south line. Thus, the orthogonal component vectors of the plane's velocity, parallel and normal to the line are:

$(500 \text{ km h}^{-1}) \cos 45°$, due west $= 354 \text{ km h}^{-1}$, due west

and

$(500 \text{ km h}^{-1}) \sin 45°$, due north $= 354 \text{ km h}^{-1}$, due north.

Chapter 12

Fast track answers

FI (a) If the particle is in equilibrium then $F_1 + F_2 + F_3 = 0$ and therefore

$$F_3 = -(F_1 + F_2)$$

so that

$$F_3 = -[(-i + 2j + 4k) + (-i - 5k)] \text{ N}$$

$$= -(-2i + 2j - k) \text{ N} = (2i - 2j + k) \text{ N}$$

(b) Now $\quad 2F_1 = (-2i + 4j + 8k) \text{ N},$

$$F_2 = (-i - 5k) \text{ N}$$

and $\quad 3F_3 = (6i - 6j + 3k) \text{ N}$

so that $\quad 2F_1 + F_2 + 3F_3 = (3i - 2j + 6k) \text{ N}$

Using Equation 8,

$$|2F_1 + F_2 + 3F_3| = [3^2 + (-2)^2 + 6^2]^{1/2} \text{ N}$$

$$= 7 \text{ N}$$

Using Equation 1, the required unit vector is obtained by dividing the vector $2F_1 + F_2 + 3F_3$ by $|2F_1 + F_2 + 3F_3|$, it is therefore

$$\left(\frac{3}{7}i - \frac{2}{7}j + \frac{6}{7}k\right) = 0.429i - 0.286j + 0.857k$$

F2 $v_p + v_c = (2, 2, -1) \text{ m s}^{-1} + (-1, 1, 3) \text{ m s}^{-1}$

$$= (1, 3, 2) \text{ m s}^{-1}$$

therefore $|v_p + v_c| = [1^2 + 3 + 2^2]^{1/2} \text{ m s}^{-1}$

$$= \sqrt{14} \text{ m s}^{-1}$$

therefore time $=$ distance/speed

$$= (5/\sqrt{14}) \text{ s} = 1.34 \text{ s}$$

Ready to study answers

RI From the definition of the tangent of an angle:

$$\tan \hat{BAC} \approx \left(\frac{5}{12}\right)$$

so that $\quad \hat{BAC} = \arctan\left(\frac{5}{12}\right) \approx 22.6°$

so $\quad \hat{ACB} \approx 90° - 22.6 \approx 67.4°$

From *Pythagoras's theorem*:

$$AC = (25 + 144)^{1/2} \text{ cm} = 13 \text{ cm}$$

R2 Let the angle of inclination be θ (i.e. the angle between the ladder and the ground), then from the definition of the *sine of an angle*:

$$\sin \theta = 4.5/5$$

so that $\quad \theta = \arcsin (4.5/5) \approx 64.2°$

From the definition of *cosine*, the distance of the foot of the ladder from the wall is $(5 \cos \theta)$ m $= (5 \cos 64.2°)$ m ≈ 2.18 m

R3 You may find it helpful to plot A and B on a *Cartesian coordinate system*.

From *Pythagoras's theorem*:

$$OA = (1^2 + 2^2)^{1/2} = 5^{1/2} \approx 2.24$$

Similarly: $\quad OB = (4^2 + 3^2)^{1/2} = 25^{1/2} = 5$

Also: $\quad AB = [(4 - 1)^2 + (3 - 2)^2]^{1/2}$

$$= 10^{1/2} \approx 3.16$$

The angle between OA and the x-axis is arctan (2), while the angle between OB and the x-axis is arctan (3/4), so that

$$A\hat{O}B = \arctan (2) - \arctan (3/4)$$

$$\approx 63.43° - 36.87° \approx 26.6°$$

Text answers

T1 If F is vertically downwards, then the required unit vector upwards $= -F/|F|$.

T2 (a) In Figure 12.3,

$$\overrightarrow{KU} = 2a + 2b, \quad \overrightarrow{WL} = a - 3b$$

and $\quad \overrightarrow{YK} = -2a - 3b$.

(b) $2a - 3b = \overrightarrow{WM} = \overrightarrow{VL}$.

T3 If $\hat{v}$ and $\hat{h}$ are unit vectors in the direction of the upward vertical, and a horizontal in the direction of steepest slope, respectively, then the horizontal and vertical component vectors are:

$$[(20 \cos 25°) \text{ N}] \, \hat{h} \approx (18.13 \text{ N}) \, \hat{h}$$

and $\quad [(20 \sin 25°) \text{ N}] \, \hat{v} \approx (8.45 \text{ N}) \, \hat{v}$

T4 The coordinate system is right-handed. The displacement from the origin to the mid-point of the north wall is $p = (1.5i + 5j + 1.25k)$ m. The displacement from the centre of the ceiling to the mid-point of the north wall is given by

$$s = -r + p$$

$$= - (1.5i + 2.5j + 2.5k) \text{ m}$$

$$+ (1.5i + 5j + 1.25k) \text{ m}$$

$$= (2.5j - 1.25k) \text{ m}$$

T5 The magnitude of the displacement is given by Equation 8:

$$|s| = [0^2 + (2.5)^2 + (1.25)^2]^{1/2} \text{ m} \approx 2.80 \text{ m}$$

T6 The combined momentum is

$$h + t = (3.0i + 75 \times 10^4 j) \text{ kg km h}^{-1}$$

and

$$|h + t| = \sqrt{(3.0)^2 + (75 \times 10^4)^2} \text{ kg km h}^{-1}$$

$$\approx 75 \times 10^4 \text{ kg km h}^{-1}$$

In the five minutes between 3a.m. and 3.05a.m. the hedgehog covered a distance of about 167 m, and since the total width of the M1 (including both carriageways, the central reservation and the emergency lanes) is about 40 m, the hedgehog was long gone by the time the truck passed. The answer for the momentum is the same whether they collide or not. (The hedgehog's name was Lucky!)

T7 (a) Since momentum $= mv$, the required momenta are obtained by scaling the given velocities by the appropriate masses. Hence:

momentum of A $= (2 \text{ kg}) \, v_A$

$$= 2(i - 2j + 4k) \text{ kg m s}^{-1}$$

$$= (2i - 4j + 8k) \text{ kg m s}^{-1}$$

and momentum of B $= (3 \text{ kg}) \, v_B$

$$= 3(3i + j - 2k) \text{ kg m s}^{-1}$$

$$= (9i + 3j - 6k) \text{ kg m s}^{-1}$$

(b) The total momentum

$$= (2 \text{ kg}) \, v_A + (3 \text{ kg}) \, v_B$$

$$= [(2i - 4j + 8k) + (9i + 3j - 6k)] \text{ kg m s}^{-1}$$

$$= (11i - j + 2k) \text{ kg m s}^{-1}$$

(c) The magnitudes of the initial momenta are obtained using Equation 8:

$$|(2 \text{ kg}) \, v_A| = [2^2 + (-4)^2 + 8^2]^{1/2} \text{ kg m s}^{-1}$$

$$= 84^{1/2} \text{ kg m s}^{-1} \approx 9.17 \text{ kg m s}^{-1}$$

and $|(3 \text{ kg}) \, v_B| = [9^2 + 3^2 + (-6)^2]^{1/2} \text{ kg m s}^{-1}$

$$= 126^{1/2} \text{ kg m s}^{-1} \approx 11.2 \text{ kg m s}^{-1}$$

(d) The final total momentum equals the total mass of the combined body multiplied by the final velocity, v_F, say. Therefore v_F is obtained by dividing the final momentum by the total mass, so that

$$v_F = \left(\frac{11i - j + 2k}{5}\right) \text{ m s}^{-1}$$

$$= (2.2i - 0.2j + 0.4k) \text{ m s}^{-1}$$

T8 Equation 8 gives the expression for the magnitude of the vector. Hence:

$$|(5, -2, -1)| = [5^2 + (-2)^2 + (-1)^2]^{1/2}$$

$$= 30^{1/2} \approx 5.48$$

T9 The resultant of $(2, -1, -2)$ and $(3, -2, 7)$ is given by $(5, -3, 5)$ in ordered triple notation. To find the required unit vector we need to find the magnitude of the resultant, and this is done using Equation 8. Therefore:

$$\text{magnitude} = [5^2 + (-3)^2 + 5^2]^{1/2} = 59^{1/2}$$

$$= 7.68$$

The required unit vector is oppositely directed to the resultant and can be obtained using Equation 1:

$$\text{required unit vector} = -\frac{(5, -3, 5)}{\sqrt{59}}$$

$$\approx (-0.651, 0.391, -0.651)$$

T10 The individual displacements s_1, s_2 and s_3, say, are obtained by scaling the velocities by the appropriate time intervals. Therefore

$$s_1 = t_1 v_1, \quad s_2 = t_2 v_2 \quad \text{and} \quad s_3 = t_3 v_3$$

so that

$$s_1 = 5(2, 1, -2) \text{ m} = (0, 5, -10) \text{ m}$$

$$s_2 = 2(-1, 2, 3) \text{ m} = (-2, 4, 6) \text{ m}$$

and $s_3 = 3(-2, -1, 2) \text{ m} = (-6, -3, 6) \text{ m}$

The resultant displacement is obtained by adding s_1, s_2 and s_3. Therefore

$$s_1 + s_2 + s_3 = (2, 6, 2) \text{ m}$$

The magnitude of this resultant is given by Equation 8:

$$|s_1 + s_2 + s_3| = [2^2 + 6^2 + 2^2]^{1/2} \text{ m}$$
$$= 44^{1/2} \text{ m} \approx 6.63 \text{ m}$$

The unit vector is given by Equation 1:

$$\text{unit vector} = \frac{s_1 + s_2 + s_3}{|s_1 + s_2 + s_3|} = \frac{(2, 6, 2)}{\sqrt{44}}$$

$$\approx (0.302, 0.905, 0.302)$$

T11 The displacement of the particle after 5 s is

$$\overrightarrow{PQ} = (5 \text{ s}) \, v = 5 \, (1, 1, 2) \text{ m}$$

so that $\overrightarrow{OQ} = \overrightarrow{OP} + \overrightarrow{PQ}$

$$= (2, 3, -4) \text{ m} + 5 \, (1, 1, 2) \text{ m}$$

$$= (7, 8, 6) \text{ m}$$

Chapter 13

Fast track answers

F1 The car starts from rest at time 0 moving in the positive direction (forward), increasing its velocity until time A, at which point it travels at constant velocity until time B. Then it slows down, coming momentarily to a halt at C. It then reverses direction with increasing speed (the velocity is becoming more negative) until time D when it slows down, coming to a halt at time E. It remains stationary until time F when it moves forward, increasing its velocity until time G when it starts to slow and it momentarily stops and then reverses direction at time H. It now increases the speed with which it reverses (again, the velocity is becoming more negative) until I when it starts to slow and comes to a halt at its start point at time J.

(a) The acceleration is negative between B and D and between G and I.

(b) The acceleration is zero between A and B and between E and F.

(c) The acceleration is positive in the periods from O to A, D to E, F to G, I to J.

F2 (a) dx/dt is the rate of change of position with respect to time, i.e. the velocity v_x. It is a function of time and at any particular time t its value is equal to the gradient of the tangent to the position–time graph at that time.

Its formal definition is

$$\frac{dx}{dt} = \lim_{\Delta t \to 0} \left(\frac{\Delta x}{\Delta t}\right) = \lim_{\Delta t \to 0} \left[\frac{x(y + \Delta t) - x(t)}{\Delta t}\right]$$

To apply this definition we calculate

$x(t + \Delta t = -p(t + \Delta t) + q(t + \Delta t)^2$

$\qquad = -pt - p(\Delta t) + qt^2 + 2qt(\Delta t) + q(\Delta t)^2$

So, $x(t + \Delta t) - x(t) = -p(\Delta t) + 2qt(\Delta t) + q(\Delta t)^2$

Dividing both sides by Δt gives

$$\left[\frac{x(t + \Delta t) - x(t)}{\Delta t} \right] = p + 2qt + q(\Delta t)$$

So $\dfrac{dx}{dt} = \lim\limits_{\Delta t \to 0} \left(\dfrac{\Delta x}{\Delta t} \right) = \lim\limits_{\Delta t \to 0} \left[\dfrac{x(y + \Delta t) - x(t)}{\Delta t} \right]$

$\qquad = -p + 2qt$

(b) The velocity is $v_x = -p + 2qt$ and is zero when $t = p/(2q) = (1/3)$ s.

When $t = \frac{1}{3}$ s,

$\qquad x = -2 \text{ m s}^{-1} \times \frac{1}{3} \text{ s} + 3 \text{ m s}^{-2} \times (\frac{1}{3} \text{ s})^3$

$\qquad = -\frac{1}{3}$ m

The acceleration at this time is

$\qquad a_x = \dfrac{dv_x}{dt} = 2q = 6 \text{ m s}^{-2}.$

Ready to study answers

R1 See Figure 13.20. The *gradients* of the lines are as follows:

$\qquad \text{gradient of AB} = \dfrac{\text{difference in } y\text{-coordinates}}{\text{difference in } x\text{-coordinates}}$

$$= \frac{3 - 1}{3 - 1} = 1$$

$\qquad \text{gradient of BC} = \dfrac{\text{difference in } y\text{-coordinates}}{\text{difference in } x\text{-coordinates}}$

$$= \frac{6 - 3}{2 - 3} = \frac{3}{-1} = -3$$

$\qquad \text{gradient of AC} = \dfrac{\text{difference in } y\text{-coordinates}}{\text{difference in } x\text{-coordinates}}$

$$= \frac{6 - 1}{2 - 1} = \frac{5}{1} = 5$$

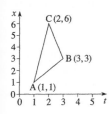

Figure 13.20 See Answer R1.

R2 The graph of $x(t) = 1.2 + 3t$ is given in Figure 13.21. The gradient is 3. This may be found from the coordinate differences between any two points on the line (as in Answer R1) or as the constant that multiplies t in the original function.

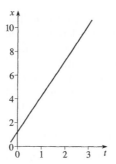

Figure 13.21 See Answer R2.

R3 The graph of $x(t) = 5 + t + 3t^2$ is given in Figure 13.22.

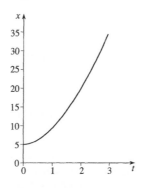

Figure 13.22 See Answer R3.

R4 (a) $x(0) = -4 + 0 + 0 = -4$

(b) $x(2) = -4 + 6 \times 2 + 2 \times 4 = 16$

(c) $|x(-1)| = |-4 - 6 + 2(-1)^2|$

$\qquad\qquad = |-8| = 8$

($|x|$ should be read as '*modulus of x*')

(d) $x(2.5) - x(2) = -4 + 15 + 2 \times 6.25 - 16$

$\qquad\qquad\qquad = 7.5$

(e) $x(2.1) - x(2) = -4 + 12.6 + 2 \times 4.41 - 16$

$\qquad\qquad\qquad = 1.42$

(f) $\dfrac{x(2 + \Delta t) - x(2.0)}{0.50} = \dfrac{x(2.5) - x(2.0)}{0.50}$

$\qquad = \dfrac{7.50}{0.50} = 15.0$

(g) $\dfrac{x(t + \Delta t) - x(t)}{\Delta t} = \dfrac{x(2.1) - x(2.0)}{0.10}$

$= \dfrac{1.42}{0.10} = 14.2$

R5 They all represent the same *function*. The letters are not important. Any other letters can be used provided they are always replaced consistently. So (b) can be written as $s(t) = (1 + t)^2$ replacing y by s and x by t. But $(1 + t)^2 = 1 + 2t + t^2$ which is (a). Similarly, expanding and simplifying the right-hand side of (c) while replacing v by s also gives (a). (It is assumed that all the functions have the same *codomain*.)

Comment *Functions* are reviewed briefly in this chapter, but if you are unsure about the answers to Questions R4 and R5 you should review Chapter 3.

Text answers

T1 (a) The data are given in Table 13.6 and plotted in Figure 13.23. Figure 13.23 is Figure 13.3 shifted downwards. Different choices of reference point (the origin) simply shift the position–time graph upwards or downwards without changing its

Table 13.6 See Answer T1.

Time t/s	Position coordinate x/m	Time t/s	Position coordinate x/m
0	−20.0	35	48.0
5	−18.3	40	64.0
10	−13.2	45	79.0
15	−5.0	50	95.0
20	6.0	55	110.0
25	19.0	60	126.0
30	33.0		

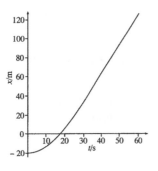

Figure 13.23 See Answer T1 (a).

shape. The new choice of origin has not affected the gradient, so the velocities found using this graph would be exactly the same as those from Figure 13.3.

(b) The data are given in Table 13.7 and plotted in Figure 13.24. Figure 13.24 is Figure 13.23 turned upside down. Reversing the choice of positive direction turns the graph upside down. The choice of positive direction affects the sign of the gradient and hence velocities found from this graph will have the opposite sign to those found using Figure 13.23 − the car is now travelling in the negative direction so its velocity is negative.

Table 13.7 See Answer T1 (b)

Time t/s	Position coordinate x/m	Time t/s	Position coordinate x/m
0	20.0	35	−48.0
5	18.3	40	−64.0
10	13.2	45	−79.0
15	5.0	50	−95.0
20	−6.0	55	−110.0
25	−19.0	60	−126.0
30	−33.0		

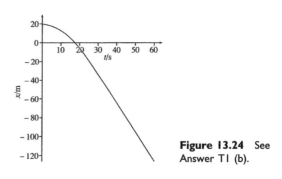

Figure 13.24 See Answer T1 (b).

T2 (a) The steeper the graph, the greater the change in position in a given interval of time, and therefore the greater the speed. So in terms of increasing speed the order is B, D, A, C.

(b) Graphs A and B have *positive* gradients (since *x increases* as *t* decreases). The gradient of A is greater (i.e. more positive) than B, therefore A has the greatest velocity of the four.

(c) Graphs C and D have *negative* gradients (since *x decreases* as *t* increases). The gradient of C is more negative than D, therefore C has the least velocity of the four.

T3 You should obtain the values given in Table 13.8, to within an uncertainty of about 5%, or so. Figure 13.25 shows the points from Table 13.8 added to Figure 13.7.

Table 13.8   See Answer T3

Interval	Average velocity
5.0 s to 10.0 s	$(7.0 \text{ m} - 1.8 \text{ m})/(10.0 \text{ s} - 5.0 \text{ s}) = 1.0 \text{ m s}^{-1}$
5.0 s to 9.0 s	$(5.5 \text{ m} - 1.8 \text{ m})/(9.0 \text{ s} - 5.0 \text{ s}) = 0.95 \text{ m s}^{-1}$
5.0 s to 8.0 s	$(4.5 \text{ m} - 1.8 \text{ m})/(8.0 \text{ s} - 5.0 \text{ s}) = 0.90 \text{ m s}^{-1}$

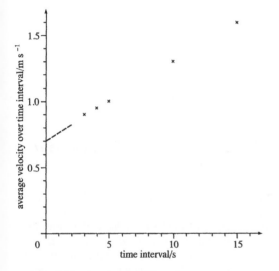

Figure 13.25   See Answer T3.

T4 By drawing tangents at 10 s, 40 s and 55 s and measuring their gradients the accelerations are found to be 0.66 m s^{-2}, 0 m s^{-2} and -1.3 m s^{-2}, respectively.

The car accelerates from rest to a steady velocity of 17.5 m s^{-1} in about 34 s, remains at that velocity for about 10 s, and then decelerates to rest in about a further 15 s.

T5 The rate of change of volume with respect to pressure is approximately $-4.5 \text{ litres atm}^{-1}$ and is given by the gradient of the tangent to the curve at $P = 1$ atm. As the pressure increases the rate of change of volume with respect to pressure increases (i.e. becomes less negative). To find the rate of change of pressure with respect to volume you really need a graph of P against V, though you might

be able to imagine what it looks like from Figure 13.13. In any event, the rate of change of pressure with respect to volume is also negative and it too increases (i.e. becomes less negative) as the volume increases since the gas becomes easier to compress and a given increase in volume corresponds to a smaller and smaller reduction in pressure.

T6 The graph of $x(t) = 4t + 7$ is shown in Figure 13.26.

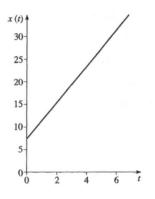

Figure 13.26   See Answer T6.

T7 Since the graph of $v_x(t)$ is a straight line the derivative is the gradient which is a_x, not only when $t = 2$ s but also at all other values of t. So a_x, as the notation suggests, is the acceleration. When $t = 0$, $v_x = u_x$, so u is the initial velocity.

T8 The completed Table 13.4 is given in Table 13.9. Recall that the gradient is $6 + \Delta t$.

Table 13.9   See Answer T8

$t + \Delta t$	Δt	Gradient
3.5	0.5	6.5
3.4	0.4	6.4
3.2	0.2	6.2
3.01	0.01	6.01

T9 (a) $x(t) = 14 - t$,

so $x(t + \Delta t) = 14 - (t + \Delta t)$.

Thus $\Delta x = x(t + \Delta t) - x(t)$

$= 14 - (t + \Delta t) - (14 - t)$

$= -t - \Delta t + t = -\Delta t$

so $\quad \dfrac{\Delta x}{\Delta t} = \dfrac{-\Delta t}{\Delta t} = -1$

hence $\quad \dfrac{dx}{dt} = \lim_{\Delta t \to 0} \left(\dfrac{\Delta x}{\Delta t} \right) = -1$

An alternative way of answering is to note that the position–time graph is a straight line with gradient -1.

(b) $\quad x(t) = 1 + t + t^2$

so $\quad x(t + \Delta t) = 1 + (t + \Delta t) + (t + \Delta t)^2$

Thus $\quad \Delta x = \Delta t + (t + \Delta t)^2 - t^2$

$= \Delta t + t^2 + 2t(\Delta t) + (\Delta t)^2 - t^2$

$= \Delta t + 2t(\Delta t) + (\Delta t)^2$

so $\quad \Delta x/\Delta t = 1 + 2t + (\Delta t)$

hence $\quad \dfrac{dx}{dt} = \lim_{\Delta t \to 0} \left(\dfrac{\Delta x}{\Delta t} \right) = 1 + 2t$

(c) $\quad x(t) = (1 - t)^2 = 1 - 2t + t^2$

so $\quad x(t + \Delta t) = 1 - 2(t + \Delta t) + (t + \Delta t)^2$

$= 1 - 2t - 2\Delta t + t^2 + 2t\Delta t + (\Delta t)^2$

Thus $\quad \Delta x = -2\Delta t + 2t\Delta t + (\Delta t)^2$

so $\quad \Delta x/\Delta t = -2 + 2t + \Delta t$

hence $\quad \dfrac{dx}{dt} = \lim_{\Delta t \to 0} \left(\dfrac{\Delta x}{\Delta t} \right) = -2 + 2t = -2(1 - t)$

T10 $\quad x(t) = p + qt + rt^2$, therefore

$\Delta x = x(t + \Delta t) - x(t)$

$= p + q(t + \Delta t) + r(t + \Delta t)^2 - p$

$- qt - rt^2$

$= q\Delta t + 2rt(\Delta t) + r(\Delta t)^2$

So $\quad \Delta x/\Delta t = q + 2rt + r\Delta t$

hence $\quad \dfrac{dx}{dt} = \lim_{\Delta t \to 0} \left(\dfrac{\Delta x}{\Delta t} \right) = q + 2rt$

This is the velocity v_x.

T11 From Equation 6, $v_x(t) = 0$ when $t = -q/2r = 1.5$ s. The initial position, p, does not affect v_x.

T12 Since $v_x = \dfrac{ds_x}{dt}$ and $s_x(t)$ is a quadratic function it follows from Equations 4 and 5 (or by differentiating from first principles as in Question T10) that $v_x = u_x + a_x t$ as required.

T13 The graph of temperature against altitude (Figure 13.27) is linear so $dT/dh = a$. Physically, a is the rate of change of temperature with respect to altitude, and b is the temperature at altitude $h = 0$. (The choice of where $h = 0$ is arbitrary, and would generally be taken to be sea level.) The SI units of a would be $\mathrm{K\,m^{-1}}$ or $\mathrm{°C\,m^{-1}}$ (since T decreases with increasing altitude, a would have a negative sign) and SI units of b would be K or °C. See Figure 13.27, which has a gradient of $a = -2.5 \times 10^{-3}\ \mathrm{°C\,m^{-1}}$.

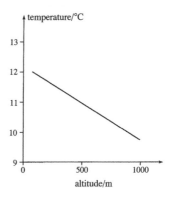

Figure 13.27
See Answer T13.

T14 The function $E_{\mathrm{kin}}(v) = \tfrac{1}{2} mv^2$ has the same form as the function $x(t) = \tfrac{1}{2} mt^2$ and comparison with Equations 4 and 5 shows that in that case (with $p = q = 0$ and $r = m/2$) $dx/dt = mt$. So $dE_{\mathrm{kin}}/dv = mv$. The expression dE_{kin}/dv represents the rate of change of kinetic energy with respect to speed.

Chapter 14

Fast track answers

F1 (a) $f'(x) = x^2 \cos x + 2x \sin x + (1/x)$.

(b) $f'(x) = (x^3 - 6x)(4x + 5) + (3x^2 - 6)$

$(2x^2 + 5x - 1)$

$= 10x^4 + 20x^3 - 39x^2 - 60x + 6$.

(c) $f(x) = 1/\sin x = \operatorname{cosec} x$,

so $f'(x) = -\operatorname{cosec} x \cot x$.

(d) $f'(x) = \dfrac{(2x + 4)(x^2 + 9) - (x^2 + 4x - 1)2x}{(x^2 + 9)^2}$

(e) $f'(x) = \dfrac{(x \sec^2 x + \tan x)}{(3x^2 - 2x + 1)^2} \times$

$\dfrac{(3x^2 - 2x + 1) - x \tan x \,(6x - 2)}{(3x^2 - 2x + 1)^2}$

(f) $f'(x) = (\log_e 2)\, 2^x + 2x$

Note that e^2 is a constant.

(g) $f(x) = (e^x + e^{-x})(e^x + e^{-x})$

so using the product rule we find

$f'(x) = (e^x + e^{-x})(e^x - e^{-x}) + (e^x - e^{-x})$

$\qquad (e^x + e^{-x})$

$\qquad = 2(e^{2x} - e^{-2x})$

F2 $s_x(t) = \exp(-\alpha t)[A \cos(\omega t) + B \sin(\omega t)]$

$v_x(t) = \dfrac{d}{dt} s_x(t)$

$\qquad = \exp(-\alpha t)[-A\omega \sin(\omega t) + B\omega \cos(\omega t)]$

$\qquad - \alpha \exp(-\alpha t)[A \cos(\omega t) + B \sin(\omega t)]$

So $v_x(t) =$

$\omega \exp(-\alpha t)[-A \sin(\omega t) + B \cos(\omega t)] - \alpha s_x(t)$

$a_x(t) = v_x'(t)$

$\qquad = \omega^2 \exp(-\alpha t)[-A \cos(\omega t) - B \sin(\omega t)]$

$\qquad - \alpha\omega \exp(-\alpha t)[-A \sin(\omega t) +$

$\qquad B \cos(\omega t)] - \alpha v_x(t)$

So $a_x(t) = -\omega^2 s_x(t) - \alpha\omega \exp(-\alpha t)[-A \sin(\omega t)$

$\qquad + B \cos(\omega t)] - \alpha v_x(t)$

$\qquad = -\omega^2 s_x(t) - \alpha v_x(t) - \alpha^2 s_x(t) - \alpha v_x(t)$

using the expression for $v_x(t)$, and therefore

$a_x(t) + 2\alpha\, v_x(t) + (\omega^2 + \alpha^2)s_x(t) = 0$

Ready to study answers

R1 (a) $\log_{10} xy$ (b) $(2 \times 2)^3 = 64$

(c) $\log_e(x/y)$ (d) e^{x-y} (e) e^{2x-2y}

(f) $3 \log_e x = \log_e x^3$.

R2 (a) $[f(x) + g(x)]^2 = (\cos x + \sin x)^2$

$\qquad = \cos^2 x + 2 \cos x \sin x + \sin^2 x$

Since $\sin^2 x + \cos^2 x = 1$

and $\sin(2x) = 2 \sin x \cos x$

for all x, we can write

$\qquad [f(x) + g(x)]^2 = 1 + \sin(2x)$.

(b) $f(x)h(x) + g(x)/h(x) = \cos x \tan x + \sin x/\tan x$

$\qquad = \cos x \times \dfrac{\sin x}{\cos x} + \sin x \times \dfrac{\cos x}{\sin x}$

$\qquad = \sin x + \cos x$

(c) $\dfrac{f(x) - g(x)}{1 + h(x)} = \dfrac{\cos x - \sin x}{1 + \tan x}$

which cannot be simplified.

R3 $f(1) = -0.75, f(0) = -1, f(-2) = 0,$
$|f(0)| = 1, |f(0) - f(-2)| = 1.$

The graph of $y = \dfrac{x^2}{4} - 1$ is shown in Figure 14.7.

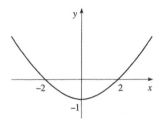

Figure 14.7 See Answer R3.

R4 (a) $2 \log_{10} x$, (b) 10^x, (c) 2.

Text answers

T1 (a) $f(\sqrt{x}) = (\sqrt{x})^2 + 1 = x + 1$, and $g(x/2)$
$= 2(x/2) = x$.

(b) We need $x^2 + 1 = 2x$, i.e. $x^2 - 2x + 1 = 0$. The solution of this is $x = 1$.

(c) $f(x + 0.1) - f(x) = (x + 0.1)^2 + 1 - (x^2 + 1)$

$\qquad = x^2 + 0.2x + 0.01 + 1 - x^2 - 1$

$\qquad = 0.2x + 0.01$

So, $\dfrac{f(x + 0.1) - f(x)}{0.1} = \dfrac{0.2x + 0.01}{0.1}$

$\qquad = 2x + 0.1$

If this equals 0.2 we must have $2x = 0.1$, which gives $x = 0.05$.

T2 (a) $\dfrac{f(x+h)-f(x)}{h}=\dfrac{1}{h}\left[\dfrac{1}{\omega(x+h)}-\dfrac{1}{\omega x}\right]$

$=\dfrac{1}{\omega}\times\dfrac{1}{h}\left(\dfrac{1}{x+h}-\dfrac{1}{x}\right)$

which is $1/\omega$ times the expression discussed earlier in this subsection, and so

$f'(x)=\dfrac{1}{\omega}\left(-\dfrac{1}{x^2}\right)=-\dfrac{1}{\omega x^2}$

(b) g is the same function as f and therefore

$g'(t)=-\dfrac{1}{\omega t^2}$

and $g'(2t)=-\dfrac{1}{\omega(2t)^2}=-\dfrac{1}{4\omega t^2}$

T3 (a) Trying various small values (either positive or negative) for h, you should have convinced yourself that

$\lim_{h\to 0}\left[\dfrac{\exp(h)-1}{h}\right]=1$, for example

$\dfrac{\exp(0.001)-\exp(0)}{0.001}=\dfrac{1.001\,000\,500\,17-1}{0.001}$

$=1.000\,500\,17$

(b) Part (a) should have convinced you that

$f'(0)=\lim_{h\to 0}\left[\dfrac{\exp(h)-\exp(0)}{h}\right]$

$=\lim_{h\to 0}\left[\dfrac{\exp(h)-1}{h}\right]=1$

(c) Keeping in mind the result of part (b)

$f'(x)=\lim_{h\to 0}\left[\dfrac{\exp(x+h)-\exp(x)}{h}\right]$

$=\lim_{h\to 0}\left[\dfrac{\exp(x)\exp(h)-\exp(x)}{h}\right]$

$=\lim_{h\to 0}\left[\exp(x)\dfrac{\exp(h)-1}{h}\right]$

$=\exp(x)\lim_{h\to 0}\left[\dfrac{\exp(h)-1}{h}\right]=\exp(x)$

T4 If $f(x)=x^n$ then, using the binomial theorem

$f(x+h)=(x+h)^n$

$=x^n+nx^{n-1}h+\tfrac{1}{2}n(n-1)x^{n-2}+\ldots+h^n$

So, $\dfrac{f(x+h)-f(x)}{h}=$

$nx^{n-1}+\tfrac{1}{2}n(n-1)x^{n-2}h+\ldots+h^{n-1}$

All the terms except the first involve a positive power of h, so the limit of each of them is zero, giving the limit of right-hand side as nx^{n-1} as required.

(This result holds true even when n is not a positive integer, but the above demonstration is not sufficiently general to prove that.)

T5 The gradient the graph of $y=\sin x$ at any value of x, is given by $dy/dx=\cos x$ (from Table 14.1).

(a) To find where the gradient is 0 we need to find the values of x for which $\cos x=0$, which are

$x=\pm\dfrac{\pi}{2},\pm\dfrac{3\pi}{2},\pm\dfrac{5\pi}{2}$, etc.

(b) In this case we need to solve $\cos x=-1$, which gives $x=\pm\pi,\pm3\pi,\pm5\pi$, etc.

(c) In this case we need to solve $\cos x=+1$, which gives $x=0,\pm2\pi,\pm4\pi,\pm6\pi$, etc.

T6 (a) Table 14.2 shows approximate values of

$\dfrac{\cos(h)-1}{h}$ and $\dfrac{\sin h}{h}$

for various values of h.

Table 14.2 See Answer T6

h	$[\cos(h)-1]/h$	$(\sin h)/h$
0.1	−0.05	0.998
−0.1	0.05	0.998
0.01	−0.005	0.999 98
−0.01	0.005	0.999 98
0.0001	−0.000 05	0.999 999 998
−0.0001	0.000 05	0.999 999 998

Thus it seems very likely that

$\lim_{h\to 0}\left[\dfrac{\cos(h)-1}{h}\right]=0$ and $\lim_{h\to 0}\left(\dfrac{\sin h}{h}\right)=1$

(These results are true, but we will not prove them rigorously.)

(b) $f(x) = \sin x$, so $f(x + h) = \sin(x + h)$

and (using the hint)

$$\frac{f(x+h) - f(x)}{h} = \frac{\sin(x+h) - \sin x}{h}$$

$$= \frac{\sin x \cos h + \cos x \sin h - \sin x}{h}$$

$$= \sin x \frac{\cos(h) - 1}{h} + \cos x \frac{\sin h}{h}$$

Since $\lim\limits_{h \to 0} \left[\dfrac{\cos(h) - 1}{h} \right] = 0$

and $\lim\limits_{h \to 0} \left(\dfrac{\sin h}{h} \right) = 1$

we obtain $\lim\limits_{h \to 0} \left[\dfrac{f(x+h) - f(x)}{h} \right] =$

$$= (\sin x) \times 0 + (\cos x) \times 1$$

$$= \cos x$$

In other words $\dfrac{d}{dx}(\sin x) = \cos x$.

T7 (a) The derivative of $\tan x$ is $\sec^2 x$ which is always positive because it is a square. Since e^x is always positive and is its own derivative, e^x always has a positive derivative.

The last function with a positive derivative is $\log_e x$. This involves a tricky point. Since $f(x) = \log_e x$ it is only defined for $x > 0$: so is its derivative, $f'(x)$. In its domain $f'(x)$ takes the same value as $1/x$, and on this domain $1/x$ is positive. (Many authors prefer to discuss the function $\log_e |x|$ since this function has $1/x$ as its derivative for all values of $x \neq 0$.)

(b) Since the derivative of $\cot x$ is $-\text{cosec}^2 x$ it is always negative.

(c) All the other derivatives in Table 14.1b can take both positive and negative values.

T8 (a) $y = 2 + \sqrt[3]{x} + e^x = 2 + x^{1/3} + e^x$ is the sum of three functions each belonging to Table 14.1.

So, $\dfrac{dy}{dx} = \dfrac{d}{dx}(2 + x^{1/3} + e^x) =$

$$\frac{d}{dx}(2) + \frac{d}{dx}(x^{1/3}) + \frac{d}{dx}(e^x)$$

$$= 0 + \frac{1}{3}x^{-2/3} + e^x = \frac{1}{3}x^{-2/3} + e^x$$

(b) Expand $(1 + \sqrt{x})^2$ to find

$$1 + 2\sqrt{x} + (\sqrt{x})^2 = 1 + \sqrt{x} + \sqrt{x} + x$$

(We have written $2\sqrt{x}$ as $\sqrt{x} + \sqrt{x}$ in order to express it as a sum of functions appearing in Table 14.1. In Subsection 14.3.3 another rule is developed which makes this unnecessary.) Hence

$$\frac{dy}{dx} = \frac{d}{dx}(1 + \sqrt{x})^2$$

$$= \frac{d}{dx}(1 + \sqrt{x} + \sqrt{x} + x)$$

$$= 0 + \frac{1}{2}x^{-1/2} + \frac{1}{2}x^{-1/2} + 1 = \frac{1}{\sqrt{x}} + 1$$

(c) $\log_e(xe^x) = \log_e x + \log_e(e^x) = \log_e x + x$

Hence

$$\frac{dy}{dx} = \frac{d}{dx}(\log_e xe^x) = \frac{d}{dx}(\log_e x + x)$$

$$= \frac{1}{x} + 1$$

(d) $\cot x \sin x + \tan x \cos x$

$$= \frac{\cos x}{\sin x} \times \sin x + \frac{\sin x}{\cos x} \times \cos x$$

$$= \cos x + \sin x$$

Hence

$$\frac{dy}{dx} = \frac{d}{dx}(\cot x \sin x + \tan x \cos x)$$

$$= \frac{d}{dx}(\cos x + \sin x)$$

$$= -\sin x + \cos x$$

T9 You would have to assume (correctly, though we will not prove it) that the limit of a sum equals the sum of the limits.

T10 (a) Expanding the bracket gives us

$$x^2 - 4x + 6 - 4/x + 1/x^2 =$$

$$x^2 - 4x + 6 - 4x^{-1} + x^{-2}$$

So, $\dfrac{d}{dx}\left[\left(\sqrt{x} - \dfrac{1}{\sqrt{x}}\right)^4\right]$

$$= 2x - 4 - 4(-1)x^{-2} + (-2)x^{-3}$$

$$= 2x - 4 + 4/x^2 - 2/x^3$$

(b) $\log_e \sqrt[3]{x} = \log_e x^{1/3} = \dfrac{1}{3} \log_e x$

So $\dfrac{d}{dx} \log_e \sqrt[3]{3} = \dfrac{d}{dx}\left(\dfrac{1}{3}\log_e x\right)$

$= \dfrac{1}{3}\dfrac{d}{dx}\log_e x = \dfrac{1}{3x}$

(c) $\left(\sin \dfrac{x}{2} + \cos \dfrac{x}{2}\right)^2$

$= \left(\sin \dfrac{x}{2}\right)^2 + 2 \sin \dfrac{x}{2} \cos \dfrac{x}{2} + \left(\cos \dfrac{x}{2}\right)^2$

Since $\sin^2 A + \cos^2 A = 1$, and $2 \sin A \cos A = \sin(2A)$ for any angle A, this becomes $1 + \sin x$ with the derivative

$\dfrac{d}{dx}(1 + \sin x) = 0 + \cos x = \cos x$

(d) Expanding the bracket gives us

$(\pi - x)^2 = \pi^2 - 2\pi x + x^2$

Now π^2 is just a constant and so has a derivative of 0.

Therefore

$\dfrac{d}{dx}[(\pi - x)^2] = 0 - 2\pi + 2x = -2(\pi - x)$

(e) $\sin(x + 2)$ does not appear in Table 14.1 so we need to expand it using $\sin(A + B) = \sin A \cos B + \cos A \sin B$ to find

$\sin(x + 2) = \sin x \cos 2 + \cos x \sin 2$

Note that $\cos 2$ and $\sin 2$ are just constants, so the derivative is

$\dfrac{d}{dx}[\sin(x + 2)]$

$= \cos 2 \dfrac{d}{dx}(\sin x) + \sin 2 \dfrac{d}{dx}(\cos x)$

$= \cos 2 \cos x - \sin 2 \sin x = \cos(x + 2)$

(since $\cos(A + B) = \cos A \cos B - \sin A \sin B$)

(f) Using $\tan(A + B) = \dfrac{\tan A + \tan B}{1 - \tan A \tan B}$

the function can be written as

$\dfrac{\tan x + \tan(\pi/4)}{1 - \tan x \tan(\pi/4)}(1 - \tan x)$

We can use $\tan(\pi/4) = 1$ to simplify this expression to $\tan x + \tan(\pi/4) = 1 + \tan x$ which has the derivative $\sec^2 x$.

T11 By definition

$f'(x) = \lim_{h \to 0}\left[\dfrac{f(x + h) - f(x)}{h}\right]$

$[kf(x)]' = \lim_{h \to 0}\left[\dfrac{kf(x + h) - kf(x)}{h}\right]$

Therefore we can write

$[kf(x)]' = \lim_{h \to 0} k\left[\dfrac{f(x + h) - f(x)}{h}\right]$

$= k \lim_{h \to 0}\left[\dfrac{f(x + h) - f(x)}{h}\right]$

$= kf'(x)$

We have assumed (correctly) that the limit of a constant multiple of an expression is the same as the constant multiple of the limit of the expression.

T12 (a) If $f(x) = x^2$ and $g(x) = \log_e x$, then

$f(x)g(x) = x^2 \log_e x$

$f'(x) = 2x$ and $g'(x) = 1/x$

So, $\dfrac{d}{dx}(x^2 \log_e x)$

$= 2x \log_e x + x^2 \times \dfrac{1}{x} = x(2\log_e x + 1)$

(b) Frequently there is more than one way to break down the given function in terms of simpler functions, the derivatives of which we know. In this exercise we could multiply out the brackets, but that would still involve using the product rule, so it is shorter to take

$f(x) = (\sin x - 2\cos x)$ and

$g(x) = (\sin x - 2\cos x)$

so that the given function is $f(x)g(x)$, then

$f'(x) = (\cos x + 2\sin x)$ and

$g'(x) = (\cos x + 2\sin x)$

The derivative of the function $f(x)g(x)$ is therefore

$(\cos x + 2\sin x)(\sin x - 2\cos x)$

$+ (\sin x - 2\cos x) \times (\cos x + 2\sin x)$

$= 2(\cos x + 2\sin x)(\sin x - 2\cos x)$

(This sort of problem is better treated by another method known as the *chain rule*. This rule is discussed in the next chapter.)

(c) This is probably a case where it is easier to expand the brackets and then differentiate, rather than to use the product rule directly. However, to use the product rule directly, put

$$f(x) = 1 - \sqrt{x} = 1 - x^{1/2} \text{ and}$$

$$g(x) = 1 + \sqrt[3]{x} = 1 + x^{1/3}$$

Therefore $f'(x) = -\frac{1}{2}x^{-1/2}$ and $g'(x) = \frac{1}{3}x^{-2/3}$

Therefore the derivative is

$$(-\tfrac{1}{2}x^{-1/2})(1 + x^{1/3}) + (1 - x^{1/2})\tfrac{1}{3}x^{-2/3}$$

$$= -\tfrac{1}{2}x^{-1/2} - \tfrac{5}{6}x^{-1/6} + \tfrac{1}{3}x^{-2/3}$$

(d) Take $f(x) = e^x$ and $g(x) = e^x$ then $f'(x) = e^x$ and $g'(x) = e^x$

So, $\dfrac{d}{dx}(e^{2x}) = e^x e^x + e^x e^x = 2e^{2x}$

as can be verified from Table 14.1.

(e) Using the product rule (and without thinking!) take $f(x) = \sec x$ and $g(x) = \cot x$

$$f'(x) = \sec x \tan x \quad \text{and} \quad g'(x) = -\csc^2 x$$

The derivative of the product of f and g is

$$(\sec x \tan x) \cot x + \sec x (-\csc^2 x)$$

$$= \sec x - \sec x \csc^2 x$$

(using $\cot x = 1/\tan x$)

$$= \sec x (1 - \csc^2 x)$$

$$= \frac{1}{\cos x}\left[1 - \frac{1}{(\sin x)^2}\right] = \frac{(\sin^2 x - 1)}{\cos x \sin^2 x}$$

$$= \frac{-\cos^2 x}{\cos x \sin x} = -\csc x \cot x$$

On the other hand you may have realised that

$$\sec x \cot x = \frac{1}{\cos x}\frac{\cos x}{\sin x} = \csc x$$

and from Table 14.1 the derivative of $\csc x$ is $-\csc x \cot x$. A little thought can often save a lot of hard work!

(f) Putting $f(x) = x^3$ and $g(x) = F(x)$ we have

$$\frac{d}{dx}[x^3 F(x)] = 3x^2 F(x) + x^3 F'(x)$$

T13 There are several ways to tackle this question, but the following is probably the easiest.

$$v_x(t) = \frac{d}{dt}[s_x(t)]$$

$$\frac{d}{dt}\{[\cos(\omega t) - 2\sin(\omega t)]\exp(\alpha t)\}$$

From the product rule we can write this as

$$\frac{d}{dt}\{[\cos(\omega t) - 2\sin(\omega t)]\exp(\alpha t)\}$$

$$= \frac{d}{dt}\{[\cos(\omega t) - 2\sin(\omega t)]\} \times \exp(\alpha t)$$
$$\quad + [\cos(\omega t) - 2\sin(\omega t)]\frac{d}{dt}\exp(\alpha t)$$

$$= [-\omega\sin(\omega t) - 2\omega\cos(\omega t)]\exp(\alpha t)$$
$$\quad + \alpha[\cos(\omega t) - 2\sin(\omega t)]\exp(\alpha t)$$

so, $v_x(t) = \omega[-\sin(\omega t) - 2\cos(\omega t)]\exp(\alpha t)$
$$\quad + \alpha s_x(t)$$

Differentiating both sides of the last equation with respect to t we find

$$a_x(t) = \frac{d}{dt}v_x(t)$$

$$= \frac{d}{dt}\{\omega[-\sin(\omega t) - 2\cos(\omega t)]\exp(\alpha t)\}$$

$$\quad + \alpha\frac{d}{dt}s_x(t)$$

$$= \omega^2[-\cos(\omega t) + 2\sin(\omega t)]\exp(\alpha t)$$

$$\quad + \omega\alpha[-\sin(\omega t) - 2\cos(\omega t)]\exp(\alpha t)$$

$$\quad + \alpha v_x(t)$$

We can now use the original specification for $s_x(t)$ to eliminate $[-\cos(\omega t) + 2\sin(\omega t)]\exp(\alpha t)$ from the above equation, and we can use the formula for $v_x(t)$ to eliminate $\omega[-\sin(\omega t) - 2\cos(\omega t)]\exp(\alpha t)$ from the same equation by writing

$$\omega[-\sin(\omega t) - 2\cos(\omega t)]\exp(\alpha t)$$

$$= v_x(t) - \alpha s_x(t)$$

Consequently, we obtain

$$a_x(t) = -\omega^2 s_x(t) + \alpha[v_x(t) - s_x(t)] + \alpha v_x(t)$$

$$= 2\alpha v_x(t) - (\omega^2 + \alpha^2)s_s(t)$$

T14   To make use of the product rule we need to write $f(x)g(x)h(x)$ as the product of the two functions $f(x)g(x)$ and $h(x)$. Then

$$\frac{d}{dx}[f(x)g(x)h(x)]$$

$$= \frac{d}{dx}[f(x)g(x)]h(x) + f(x)g(x)\frac{d}{dx}h(x)$$

$$= f'(x)g(x)h(x) + f(x)g'(x)h(x)$$

$$+ f(x)g(x)h'(x)$$

T15   Take $f(x) = e^x, g(x) = \log_e x$ and $h(x) = \sin x$

then $f'gh + fg'h + fgh'$

$$= e^x \log_e x \sin x + e^x (1/x) \sin x$$

$$+ e^x \log_e x \cos x$$

T16   (a)   Set $f(x) = x$ and $g(x) = e^x$, to give $y = f(x)/g(x)$.

Now, $f'(x) = 1$ and $g'(x) = e^x$, so that the quotient rule gives us

$$\frac{dy}{dx} = \frac{1 \times e^x - x \times e^x}{(e^x)^2}$$

$$= \frac{e^x(1 - x)}{e^{2x}} = \frac{1 - x}{e^x}$$

(b)   Set $f(x) = 1 - x^2$ and $g(x) = \cos x$, to give $y = f(x)/g(x)$.

Now, $f'(x) = -2x$ and $g'(x) = -\sin x$, so that the quotient rule gives us

$$\frac{dy}{dx} = \frac{(-2x)(\cos x) - (1 - x^2)(-\sin x)}{\cos^2 x}$$

$$= \frac{-2x \cos x + (1 - x^2) \sin x}{\cos^2 x}$$

(c)   Set $f(x) = x^5 - x^4 + x^3 - x^2$

and $g(x) = (x + 1)^2$.

To apply the quotient rule we need to differentiate $(x + 1)^2$ which is easiest done if we expand it to obtain $x^2 + 2x + 1$.

Thus $f'(x) = 5x^4 - 4x^3 + 3x^2 - 2x$ and

$$g'(x) = 2x + 2.$$

So, $\dfrac{dy}{dx} = [(5x^4 - 4x^3 + 3x^2 - 2x)(x + 1)^2$

$$- (x^5 - x^4 + x^3 - x^2)(2x + 2)]/(x + 1)^4$$

$$= [(5x^4 - 4x^3 + 3x^2 - 2x)(x + 1)$$

$$-2(x^5 - x^4 + x^3 - x^2)]/(x + 1)^3$$

$$= \frac{x(3x^4 + 3x^3 - 3x^2 + 3x - 2)}{(x = 1)^3}$$

(d)   With $f(x) = x \log_e x$, and $g(x) = \tan x$, we can apply the quotient rule. As an intermediate stage we differentiate $f(x)$ using the product rule:

$$f'(x) = x\frac{1}{x} + 1 \times \log_e x = \log_e x$$

$$g'(x) = \sec^2 x$$

$$\frac{dy}{dx} = \frac{(1 + \log_e x) \tan x - x \log_e x \sec^2 x}{\tan^2 x}$$

T17   (a) $f(x) - g(x) = (x - 1)^2 - xe^x$

$$= x^2 - 2x + 1 - xe^x$$

$$\frac{d}{dx}[f(x) - g(x)] = 2x - 2 - (xe^x + 1 \times e^x)$$

$$= 2x - 2 - (x + 1)e^x$$

(b) $\dfrac{d}{dx}[f(x)g(x)] = \dfrac{d}{dx}[(x - 1)^2 x e^x]$

$$= (2x - 2)xe^x + (x - 1)^2 e^x + (x - 1)^2 xe^x$$

$$= (x - 1)e^x (x^2 + 2x - 1)$$

(c) $\dfrac{f(x)}{g(x)} = \dfrac{(x - 1)^2}{xe^x}$

$$f'(x) = (2x - 2) = 2(x - 1)$$

$$g'(x) = xe^x + e^x = (x + 1) e^x$$

$$\frac{d}{dx}\left[\frac{f(x)}{g(x)}\right] = \frac{2(x - 1)xe^x - (x - 1)^2(x + 1)e^x}{x^2 e^{2x}}$$

$$= \frac{(x - 1)(1 + 2x - x^2)}{x^2 e^x}$$

(d)   This can certainly be done using the quotient rule, but we can also obtain the answer from the reciprocal rule (i.e. Equation 13) and the solution to part (c) of this question as follows. First we let

$$F(x) = \frac{f(x)}{g(x)}$$

so that, from the reciprocal rule

$$\frac{d}{dx}\left[\frac{g(x)}{f(x)}\right] = \frac{d}{dx}\left[\frac{1}{F(x)}\right] = -\frac{F'(x)}{[F(x)]^2}$$

$$= -\frac{(x-1)(1+2x-x^2)}{x^2\,e^x}\bigg/\left[\frac{f(x)}{g(x)}\right]^2$$

(from part (c))

$$= -\frac{(x-1)(1+2x-x^2)}{x^2 e^x} \times \left[\frac{xe^x}{(x-1)^2}\right]^2$$

$$= \frac{(x^2-2x-1)e^x}{(x-1)^3}$$

T18 (a) Differentiating both sides of the following equation

$$f(x)h(x) = 1$$

using the product rule, we obtain

$$f'(x)h(x) + f(x)h'(x) = 0$$

Using the fact that $h(x) = \dfrac{1}{f(x)}$,

this can be rearranged to give

$$h'(x) = \frac{f'(x)h(x)}{f(x)} = -\frac{f'(x)}{[f(x)]^2}$$

and, since

$$h'(x) = \frac{d}{dx}\left[\frac{1}{f(x)}\right]$$

we have obtained the required result.

(b) Differentiating $f(x) \times \dfrac{1}{g(x)}$ using the product rule, we have

$$\frac{d}{dx}\left[f(x) \times \frac{1}{g(x)}\right]$$

$$= f'(x) \times \frac{1}{g(x)} + f(x)\frac{d}{dx}\left[\frac{1}{g(x)}\right]$$

$$= \frac{f'(x)}{g(x)} + f(x) \times \left\{-\frac{g'(x)}{[g(x)]^2}\right\}$$

$$= \frac{f'(x)g(x) - f(x)g'(x)}{[g(x)]^2}$$

T19 (a) $f(x) = \sin x$ and $g(x) = \cos x$

so, $f'(x) = \cos x$ and $g'(x) = -\sin x$

thus $\dfrac{d}{dx}(\tan x) = \dfrac{d}{dx}\left(\dfrac{\sin x}{\cos x}\right)$

$$= \frac{(\cos x)(\cos x) - (\sin x)(-\sin x)}{\cos^2 x}$$

and, since $\cos^2 x + \sin^2 x = 1$, this gives

$$\frac{d}{dx}(\tan x) = \frac{\cos^2 x + \sin^2 x}{\cos^2 x}$$

$$= \frac{1}{\cos^2 x} = \sec^2 x$$

(b) To differentiate $\cot x = \dfrac{\cos x}{\sin x}$

let $f(x) = \cos x$ and $g(x) = \sin x$

so, $f'(x) = -\sin x$ and $g'(x) = \cos x$

Then

$$\frac{d}{dx}(\cot x) = \frac{(-\sin x)(\sin x) - (\cos x)(\cos x)}{\sin^2 x}$$

$$= \frac{-1}{\sin^2 x} = -\cosec^2 x$$

T20 (a) $f(x) = e^{-x} = \dfrac{1}{e^x} = \dfrac{1}{g(x)}$

where $g(x) = e^x$.

Since $g'(x) = e^x$ we have

$$f'(x) = -\frac{e^x}{(e^x)^2} = -\frac{1}{e^x} = -e^{-x}$$

(b) $f(x)$ is a constant multiple of a sum of two functions, so, using our rules and the result of part (a),

$$\frac{df}{dx} = \frac{1}{2}(e^x - e^{-x})$$

(The function $\dfrac{1}{2}(e^x - e^{-x})$ is known as the *hyperbolic sine* of x and is denoted by $\sinh x$.)

$$[f(x)]^2 - \left(\frac{df}{dx}\right)^2 = \frac{1}{4}(e^x + e^{-x})^2 - \frac{1}{4}(e^x - e^{-x})^2$$

$$= \frac{1}{4}[(e^{2x} + 2 + e^{-2x}) - (e^{2x} - 2 + e^{-2x})] = 1$$

thus establishing the useful result that:

$$\cosh^2 x - \sinh^2 x = 1$$

T21 $F(x)$ is proportional to the reciprocal of $(x - a)^2$ and

$$\frac{d}{dx}(x - a)^2 = \frac{d}{dx}(x^2 - 2ax + a^2)$$
$$= 2x - 2a = 2(x - a)$$

Hence, using the reciprocal rule,

$$F'(x) = -GMm\frac{2(x - a)}{(x - a)^4} = -\frac{2GMm}{(x - a)^3}$$

T22   (a) $\sec x = \dfrac{1}{\cos x}$

so $\dfrac{d}{dx}\left(\dfrac{1}{\cos x}\right) = -\dfrac{(-\sin x)}{\cos^2 x}$

$= \dfrac{1}{\cos x} \times \dfrac{\sin x}{\cos x} = \sec x \tan x$

(b) $\operatorname{cosec} x = \dfrac{1}{\sin x}$

so $\dfrac{d}{dx}\left(\dfrac{1}{\sin x}\right) = \dfrac{(-\cos x)}{\sin^2 x}$

$= -\dfrac{1}{\sin x} \times \dfrac{\cos x}{\sin x} = -\operatorname{cosec} x \cot x$

T23   You should have been able to establish this result using the quotient rule, but the following method, known as *implicit differentiation*, requires a little less algebra. (This method is explored more fully in the next chapter.)

First write $y(1 + Ce^x) = Ce^x$

Differentiate both sides, using the product rule on the left-hand side

$$y'(1 + Ce^x) + yCe^x = Ce^x$$

Now eliminate $(1 + Ce^x)$ by writing

$$(1 + Ce^x) = \frac{Ce^x}{y}$$

and divide the resulting equation by Ce^x to get the required result

$$\frac{y'}{y} + y = 1 \quad \text{so} \quad y' = y(1 - y)$$

(Notice that by using y' in this way we have avoided the need to differentiate y explicitly, hence the name *implicit differentiation*.)

T24 $\dfrac{d}{dx}[(e^x + e^{-x})(e^x - e^{-x})] = \dfrac{d}{dx}(e^{2x} - e^{-2x})$

$= 2e^{2x} + 2e^{-2x}$

T25   (a) $(\log_e 3)3^x$

(b) $(\log_e 3)3^{x-1}$ (if we write $3^{x-1} = 3^x/3$)

(c) $(\log_e 5)5^x$

(d) $\dfrac{d}{dx}(2^x e^x) = \dfrac{d}{dx}[(2e)^x] = [\log_e (2e)](2e)^x$

$= (1 + \log_e 2)2^x e^x$

T26   (a) $\dfrac{d}{dx}(\log_2 x) = \dfrac{1}{x \log_e 2}$

(b) $\dfrac{d}{dx}(\log_{10} x^2) = \dfrac{d}{dx}(2\log_{10} x) = \dfrac{2}{x \log_e 10}$

(c)   The answer is zero, which is seen most easily by putting

$$\frac{\log_2 x}{\log_{10} x} = \log_2 10$$

so we are simply required to differentiate a constant.

(d) $\dfrac{d}{dx}(a^x \log_a x)$

$= (a^x \log_e a) \times \log_a x + a^x \times \dfrac{1}{x \log_e a}$

$= a^x\left(\log_e x + \dfrac{1}{x \log_e a}\right)$

if we use $(\log_e a) \times (\log_a x) = \log_e x$.

T27   Since

$$y(x) = \frac{\log_e 2 - x}{\log_e 2 + x}$$

the quotient rule implies

$$\frac{d}{dx}[y(x)] = \frac{d}{dx}\left(\frac{\log_e 2 - x}{\log_e 2 + x}\right)$$

$$= \frac{-1 \times (\log_e 2 + x) - (\log_e 2 - x) \times 1}{(\log_e 2 + x)^2}$$

$$= \frac{-2\log_e 2}{(\log_e 2 + x)^2}$$

It follows that

$$2(\log_e 2)y'(x) + [1 + y(x)]^2$$

$$2\,(\log_e 2)\frac{-2\log_e 2}{(\log_e 2 + x)^2} + \left(1 + \frac{\log_e 2 - x}{\log_e 2 + x}\right)^2$$

which (eventually) simplifies to zero, so the expression $y(x)$ satisfies the equation.

Chapter 15

Fast track answers

F1 (a) $F'(x) = 3\cos x - 4\sin x$

$F''(x) = -3\sin x - 4\cos x$

$F'''(x) = -3\cos x + 4\sin x$

$F^{(iv)}(x) = 3\sin x + 4\cos x$

(b) $\dfrac{dy}{dx} = \dfrac{1}{x}$, $\dfrac{d^2y}{dx^2} = \dfrac{-1}{x^2}$, $\dfrac{d^3y}{dx^3} = \dfrac{2}{x^3}$, $\dfrac{d^4y}{dx^4} = \dfrac{-6}{x^4}$

Comment   The two notations used in parts (a) and (b) are equivalent, i.e. $F'(x)$ is equivalent to dy/dx.

F2 $\dfrac{dx}{d\theta} = a(1 - \cos \theta)$ and $\dfrac{dy}{d\theta} = a \sin \theta$

so $\dfrac{dy}{dx} = \dfrac{dy}{d\theta} \times \dfrac{d\theta}{dx} = \dfrac{dy}{d\theta} \Big/ \dfrac{dx}{d\theta}$

$$= \frac{a \sin \theta}{a\,(1 - \cos \theta)} = \frac{\sin \theta}{1 - \cos \theta}$$

i.e. $\dfrac{dy}{dx} = \dfrac{\sin \theta}{1 - \cos \theta}$

Also $\dfrac{d^2y}{dx^2} = \dfrac{d}{dx}\left(\dfrac{\sin \theta}{1 - \cos \theta}\right)$

$$= \frac{d}{d\theta}\left(\frac{\sin \theta}{1 - \cos \theta}\right) \times \frac{d\theta}{dx}$$

so $\dfrac{d^2y}{dx^2} = \left[\dfrac{(1 - \cos \theta)\cos \theta - \sin \theta\,(+\sin \theta)}{(1 - \cos \theta)^2}\right] \times \dfrac{d\theta}{dx}$

i.e. $\dfrac{d^2y}{dx^2} = \left[\dfrac{\cos \theta - \cos^2 \theta - \sin^2 \theta}{(1 - \cos \theta)^2}\right] \Big/ \dfrac{dx}{d\theta}$

$$\frac{d^2y}{dx^2} = \left[\frac{\cos \theta - 1}{(1 - \cos \theta)^2}\right] \times \frac{1}{a\,(1 - \cos \theta)}$$

$$= \frac{-1}{a\,(1 - \cos \theta)^2}$$

F3 (a)   Put $u = \alpha t + \sin(\omega t)$ and $F(u) = e^u$

then $\dfrac{du}{dt} = \alpha + \omega \cos(\omega t)$ and $F'(u) = e^u$

Hence, by the *chain rule*

$$\frac{dF}{dt} = \frac{dF}{du} \times \frac{du}{dt} = e^u \times [\alpha + \omega \cos(\omega t)]$$

$$= [\alpha + \omega \cos(\omega t)]\,e^u$$

$$= [a + \omega \cos(\omega t)] \times \exp[\alpha t + \sin(\omega t)]$$

i.e. $\dfrac{dF}{dt} = [a + \omega \cos(\omega t)] \times \exp[\alpha t + \omega \sin(\omega t)]$

By the *product rule*

$$\frac{d^2F}{dt^2} = [\alpha + \omega \cos(\omega t)] \times \frac{d}{dt}[\exp[\alpha t + \omega \sin(\omega t)]]$$

$$+ [-\omega^2 \sin(\omega t)] \times \exp[\alpha t + \omega \sin(\omega t)]$$

i.e. $\dfrac{d^2F}{dt^2} = [\alpha + \omega \cos(\omega t)] \times [\alpha + \omega^2 \cos(\omega t)]$

$$\times \exp[\alpha t + \omega \sin(\omega t)]$$

$$- \omega^2 \sin(\omega t) \times \exp[\alpha t + \omega \sin(\omega t)]$$

so $\dfrac{d^2F}{dt^2} = \{[\alpha + \omega \cos(\omega t)] \times [\alpha + \omega^2 \cos(\omega t)]$

$$- \omega^2 \sin(\omega t)]\} \times \exp[\alpha + \omega \sin(\omega t)]$$

(b) Put $u = \sin^2 x + 1$ and $F(u) = 5 \log_e u$ then, using either the product rule $(\sin^2 x = \sin x \times \sin x)$ or the chain rule, we find

$$\frac{du}{dx} = 2 \sin x \cos x \quad \text{and} \quad F'(u) = \frac{5}{u}$$

$$\frac{dF}{dx} = \frac{dF}{du} \times \frac{du}{dx} = \frac{5}{u} \times (2 \sin x \cos x)$$

$$= \frac{5 \times 2 \sin x \cos x}{\sin^2 x + 1} = \frac{10 \sin x \cos x}{\sin^2 x + 1}$$

Applying the quotient rule we find

$$\frac{d^2F}{dx^2} = [(\sin^2 x + 1) \times \frac{d}{dx}(10 \sin x \cos x)$$

$$-10 \sin x \cos x\,(2 \sin x \cos x)]/(\sin^2 x + 1)^2$$

$$= \{(\sin^2 x + 1)[10 \cos x \times \cos x + 10 \sin x$$

$$\times (-\sin x)] - 10 \sin x \cos x$$

$$\times (2 \sin x \cos x)\}/(\sin^2 x + 1)^2$$

$$= \frac{10(\sin^2 x + 1)(\cos^2 x - \sin^2 x)}{(\sin^2 x + 1)^2}$$

$$- \frac{20 \sin^2 x \cos^2 x}{(\sin^2 x + 1)^2}$$

(c) Let $y = \arcsin(5x)$ then $\sin y = 5x$.

Hence, differentiating implicitly with respect to x, we find

$$\cos y \times \frac{dy}{dx} = 5$$

so $$\frac{dy}{dx} = \frac{5}{\cos y} = \frac{5}{\sqrt{1 - 25x^2}}$$

Now put

$$\frac{dy}{dx} = \frac{5}{\sqrt{1 - 25x^2}} = 5(1 - 25x^2)^{-1/2} = G(x)$$

let $u = 1 - 25x^2$ and $g(u) = 5u^{-1/2}$

so $$\frac{du}{dx} = -50x \text{ and } \frac{dg}{du} = -\frac{1}{2} \times 5u^{-3/2} = \frac{-5}{2u^{3/2}}$$

Hence, by the chain rule

$$\frac{d^2F}{dx^2} = \frac{dG}{dx} = \frac{dg}{du} \times \frac{du}{dx}$$

$$= \frac{-5}{2u^{3/2}} \times (-50x) = \frac{125x}{(1 - 25x^2)^{3/2}}$$

Ready to study answers

R1 (a) $f(x)$ is an acceptable function of x and there are no restrictions on the value of x. Each value of x corresponds to a single value of y.

(b) y is an acceptable function of x, provided x is in the range $-1 \leqslant x \leqslant 1$ (this defines the *domain* of the function). There are many different values of y that could correspond to a given value of x, so to ensure that each admissible value of x corresponds to only a single value of y it is also necessary to impose restrictions on y in this case. Conventionally, y is required to be in the range $-\pi/2 \leqslant y \leqslant +\pi/2$, i.e. $-90° \leqslant y \leqslant +90°$ (this defines the *codomain* of the function).

(c) y is not a function of x since the given formula implies that $y = \pm 2x$, so it is *not* the case that a single value of y corresponds to each value of x.

R2 (a) $x = (y/9)^{1/3}$

(b) $x = \arctan y$

(c) $2x = \log_e y$ so, $x = (\log_e y)/2$

(d) $3x = e^y$, so $x = e^y/3$

R3 If the expression

$$\frac{f(a + h) - f(a)}{h}$$

approaches a unique limiting value as h approaches zero, then that limiting value is called the *derivative* of $f(x)$ at $x = a$ and is written $f'(a)$.

R4 (a) $f'(x) = 48x^5 + 18x^2 - 10x$

(b) $f'(x) = 3\cos x - 4 \sin x$

R5 (a) $\frac{dy}{dx} = 6e^x$

(b) $\frac{dy}{dx} = 2x - \frac{5}{x}$

R6 (a) $x^2 \times e^x + 2xe^x = (x^2 + 2x)e^x$ (*product rule*)

(b) $x^3 \cos x + 3x^2 \sin x$ (*product rule*)

(c) $e^x \times (-\sin x) + (e^x) \times \cos x$ (*product rule*)

$= e^x(-\sin x + \cos x)$

(d) $\dfrac{(3x^2 + x)\left(\dfrac{1}{x}\right) - \log_e x \times (6x + 1)}{(3x^2 + x)^2}$

(*quotient rule*)

$= \dfrac{(3x + 1) - \log_e x \times (6x + 1)}{(3x^2 + x)^2}$

(e) $\dfrac{(2 + \sin x) \times (-3 \sin x) - 3 \cos x \times (\cos x)}{(2 + \sin x)^2}$

$= \dfrac{-6 \sin x - 2 \sin^2 x - 3 \cos^2 x}{(2 + \sin x)^2}$

(*quotient rule*)

$= \dfrac{-3(2 \sin x + 1)}{(2 + \sin x)^2}$

(f) $\dfrac{(2 + \cos x) \times (1) - x \times (- \sin x)}{(2 + \cos x)^2}$

(*quotient rule*)

$= \dfrac{\cos x + 2 + x \sin x}{(2 + \cos x)^2}$

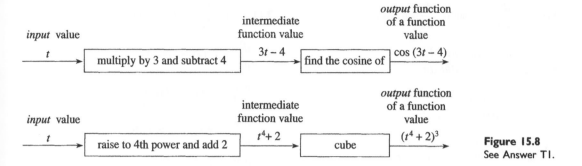

Figure 15.8
See Answer T1.

Text answers

T1 The box representations of the functions are shown in Figure 15.8.

T2 (a) Composite; $g(x) = u = 2x$, $f(u) = e^u$

(b) Not composite; F is a product.

(c) Composite; the log *of* x cubed.
$$u = g(x) = x^3, f(u) = \log_e u$$

(d) Composite; $u = g(x) = x^2 + 2x + 1$,
$$f(u) = \sin u$$

(e) Composite; the square *of* sin x.
$$u = g(x) = \sin x, f(u) = u^2.$$
(It is also a product.)

(f) Composite; $u = g(x) = \sin x$, $f(u) = u^3$.
(It is also a product.)

(g) Composite; $u = g(x) = \cos x$, $f(u) = u^{0.5}$.

T3 $f(x)$ is defined only for $x \geq -1$;
$g(x)$ is defined only for $x > 0$.

$f(g(x))$ is defined only for those x for which $\log_e x > -1$, i.e. $x > 1/e$.

$g(f(x))$ is always defined if $x > -1$, for then $f(x) > 0$.

T4 (a) $dq/dx = 2(2x - 3) \times 2 = 4(2x - 3)$

(b) $dq/dx = 2(3x^4 + 1) \times 12x^3 = 24x^3(3x^4 + 1)$

(c) $dq/dx = 2(x + \sin x) \times (1 + \cos x)$

T5 (a) $2e^{2x}$

(b) $2x \exp(x^2)$

(c) $5(x^2 + 1)^4 \times 2x = 10x(x^2 + 1)^4$

(d) $6 \cos(6x)$

(e) $-3 \sin(3x)$

(f) $\dfrac{1}{e^x} \times e^x = 1$

(g) $2x/(1 + x^2)$

(h) $-\omega \sin(\omega t + \phi)$

T6 $5y^4 \dfrac{dy}{dx} = \cos x$, so $\dfrac{dy}{dx} = \dfrac{\cos x}{5y^4}$

T7 (a) $y = \arcsin(x^2)$, so $\sin y = x^2$

so, differentiating both sides gives us
$$\cos y \, \frac{dy}{dx} = 2x$$
$$\frac{dy}{dx} = \frac{2x}{\cos y} = \frac{2x}{\sqrt{1 - \sin^2 y}} = \frac{2x}{\sqrt{1 - x^4}}$$

Note that we have taken the positive square root expression for cos y, as in the text, to ensure that dy/dx has the required sign.

(b) There are no (real) values of x for which $\arccos(x^2 + 4)$ is meaningful. The function $\arccos u$ is only defined when its argument u is in the range $-1 \leq u \leq 1$. No x will produce an argument in this range. *Moral*: don't just differentiate without thinking first!

(c) $y = \arctan(3 - x)$, $\tan y = 3 - x$

therefore $\sec^2 y \dfrac{dy}{dx} = -1$

therefore
$$\frac{dy}{dx} = \frac{-1}{\sec^2 y} = \frac{-1}{1 + \tan^2 y} = \frac{-1}{1 + (3 - x)^2}$$
$$= \frac{-1}{10 - 6x + x^2}$$

T8 (a) $\dfrac{dx}{d\theta} = -a \sin \theta \quad \dfrac{dy}{d\theta} = b \cos \theta$

so, $\dfrac{dy}{dx} = \dfrac{dy}{d\theta} \times \dfrac{d\theta}{dx} = \dfrac{dy}{d\theta} \Big/ \dfrac{dx}{d\theta}$

$= \dfrac{-b \cos \theta}{a \sin \theta} = \dfrac{-b}{a} \cot \theta$

(b) $\dfrac{dx}{dt} = 2at, \quad \dfrac{dy}{dt} = 2a, \quad \dfrac{dy}{dx} = \dfrac{2a}{2at} = \dfrac{1}{t}$

(c) $\dfrac{dx}{dt} = c, \quad \dfrac{dy}{dt} = \dfrac{-c}{t^2}, \quad \dfrac{dy}{dx} = \dfrac{-1}{t^2}$

T9 (a) $f'(x) = 6x^5 - 4x, \quad f''(x) = 30x^4 - 4$

(b) $f'(x) = a, \quad f''(x) = 0$

(c) $f'(x) = a \cos x - b \sin x$

$f''(x) = -a \sin x - b \cos x$

Note that $f''(x) = -f(x)$, in this example.

(d) $f'(x) = 3e^{3x}, \quad f''(x) = 9e^{3x}$

(e) $f'(x) = \dfrac{2x}{x^2 + 1}$

$f''(x) = \dfrac{(x^2 + 1) \times 2 - 2x \times (2x)}{(x^2 + 1)^2}$

$= \dfrac{2 - 2x^2}{(x^2 + 1)^2}$

(f) This is exactly the same as parts (a) to (e) except that $f'(x)$ is replaced by dy/dx and $f''(x)$ is replaced by d^2y/dx^2 in each case.

T10 (a) $v(t) = dx/dt = p + 2qt$, i.e. $\dot{x}(t)$

$= p + 2qt;$

$a(t) = d^2x/dt^2 = 2q$, i.e. $\ddot{x}(t) = 2q$.

(b) When $x(t) = A \sin (\omega t), \dot{x}(t) = A \omega \cos (\omega t)$ and

$\ddot{x}(t) = -A \omega^2 \sin (\omega t)$.

When $x(t) = B \cos (\omega t), \dot{x}(t) = -B \omega \sin (\omega t)$ and

$\ddot{x}(t) = -B \omega^2 \cos (\omega t)$

Note that both these expressions for x satisfy the equation

$\ddot{x} = -\omega^2 x.$

T11 (a) $\dfrac{dy}{dx} = f'(x) = 20e^{5x},$

$\dfrac{d^2y}{dx^2} = f''(x) = 100e^{5x}$

$\dfrac{d^3y}{dx^3} = f'''(x) = 500e^{5x}$

$\dfrac{d^4y}{dx^4} = f^{(iv)}(x) = 2500e^{5x}$

$\dfrac{d^ny}{dx^n} = f^{(n)}(x) = 4 \times 5^n \times e^{5x}$

(b) $\dfrac{dy}{dx} = 9x^2, \quad \dfrac{d^2y}{dx^2} = 18x, \quad \dfrac{d^3y}{dx^3} = 18$

$\dfrac{d^4}{dx^4} = 0, \quad \dfrac{d^ny}{dx^n} = 0 \quad \text{for } n \geq 4$

(c) $\dfrac{dy}{dx} = -2A \sin (2x) + 2 B \cos (2x)$

$\dfrac{d^2y}{dx^2} = -4A \cos (2x) - 4B \sin (2x)$

$\dfrac{d^3y}{dx^3} = 8a \sin (2x) - 8B \cos (2x)$

$\dfrac{d^4y}{dx^4} = 16A \cos (2x) + 16B \sin (2x)$

Note that $\dfrac{d^2y}{dx^2} = -2^2 y$

and that $\dfrac{d^4y}{dx^4} = 2^4 y$

If n is even $\dfrac{d^ny}{dx^n} = (-1)^{n/2} (2)^n y,$

if n is odd $\dfrac{d^ny}{dx^n} = (-1)^{(n-1)/2} (2)^{n-1} \dfrac{dy}{dx}$

(d) $\dfrac{dy}{dx} = \dfrac{1}{x}, \quad \dfrac{d^2y}{dx^2} = \dfrac{-1}{x^2}, \quad \dfrac{d^3y}{dx^3} = \dfrac{2}{x^3}$

$\dfrac{d^4y}{dx^4} = \dfrac{-6}{x^4}, \quad \dfrac{d^ny}{dx^n} = \dfrac{(-1)^{n-1}(n-1)(n-2)\dots 2}{x^n}$

T12 (You may wish to look back at Answer T8 for some of the results used in this answer.)

(a) $\dfrac{dx}{d\theta} = -\sin \theta, \quad \dfrac{dy}{d\theta} = \cos \theta$

therefore

$$\frac{dy}{dx} = -\frac{\cos\theta}{\sin\theta} = -\cot\theta$$

$$\frac{d^2y}{dx^2} = \frac{d}{d\theta}\frac{-\cot\theta}{-\sin\theta} = \frac{\csc^2\theta}{(-\sin\theta)}$$

$$= \frac{1}{\sin^2\theta} \times \left(-\frac{1}{\sin\theta}\right) = -\frac{1}{\sin^3\theta} = -\csc^3\theta$$

(b) $\dfrac{dx}{d\theta} = -a\sin\theta, \quad \dfrac{dy}{d\theta} = b\cos\theta$

therefore

$$\frac{dy}{dx} = \frac{-b\cos\theta}{a\sin\theta} = \frac{-b}{a}\cot\theta$$

$$\frac{d^2y}{dx^2} = \frac{d}{d\theta}\left(-\frac{b}{a}\cot\theta\right)\bigg/(-a\sin\theta)$$

$$= \frac{b}{a}\csc^2\theta\bigg/(-a\sin\theta)$$

$$= \frac{b}{a}\frac{1}{\sin^2\theta}\times\left(-\frac{1}{a\sin\theta}\right) = -\frac{b}{a^2\sin^3\theta}$$

$$= \frac{-b}{a^2}\csc^3\theta$$

(c) $\dfrac{dx}{dt} = c, \quad \dfrac{dy}{dt} = -\dfrac{c}{t^2}$

therefore $\quad \dfrac{dy}{dx} = -\dfrac{1}{t^2}$

$$\frac{d^2y}{dx^2} = \frac{d}{dt}\left(-\frac{1}{t^2}\right)\bigg/c = +\frac{2}{ct^3}$$

Chapter 16

Fast track answers

F1 Using the *quotient rule* we find

$$E'(r) = \frac{(r+1)^2 \times 4 - 2(r+1) \times 4r}{(r+1)^4}$$

So, $E'(r) = \dfrac{4 - 4r}{(r+1)^3} = \dfrac{4(1-r)}{(r+1)^3}$

When $r = 1$, $E'(r) = 0$.

For r just less than 1, $E'(r) > 0$

For r just greater than 1, $E'(r) < 0$

Hence there is a local maximum at $r = 1$ and when $r = 1$, $E(r) = 1$.

Furthermore, $E(0) = 0$ and $E(r) > 0$ for $r > 0$.

As $r \to \infty$, $E(r) \to 0$.

Figure 16.23 shows a sketch graph of

$$E(r) = \frac{4r}{(r+1)^2}$$

(There is also a point of inflection for $r > 1$, but this is not a stationary point.)

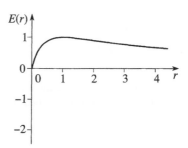

Figure 16.23 See Answer F1.

F2 The maximum value of $\sin(2\theta)$ is 1 and occurs when $2\theta = \pi/2$, i.e. $\theta = \pi/4$. So, $R_{\max} = u^2/g$.

To confirm that $\theta = \pi/4$ is a maximum note that

$$\frac{dR}{d\theta} = \frac{2u^2}{g}\cos(2\theta)$$

and that $dR/d\theta = 0$ when $\theta = \pi/4$. Also note that

$$\frac{d^2R}{d\theta^2} = \frac{-4u^2}{g}\sin(2\theta)$$

So $\dfrac{d^2R}{d\theta^2} < 0$ when $\theta = \pi/4$.

Thus, according to the second derivative test, $\theta = \pi/4$ is a local maximum.

F3 $f(x)$ is an even function. It is not defined for $x = 0$. There is no intercept on either axis. For $|x| \gg 1$, $f(x) \approx x^2$ while for $|x| \ll 1$, $f(x) \approx 1/x^2$.

$$f'(x) = 2x - \frac{2}{x^3} = \frac{2(x^4 - 1)}{x^3}$$

so that $f'(1) = 0$ and $f'(-1) = 0$, and $f(1) = 2$ and $f(-1) = 2$.

Also $\quad f''(x) = 2 + \dfrac{6}{x^4}$

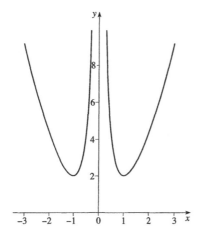

Figure 16.24 See Answer F3.

so that $f''(x) \geqslant 0$ for $x = \pm 1$ and so these points are local minima. Figure 16.24 shows a sketch graph of $y = f(x)$.

Ready to study answers

R1 (a) Using the *sum rule* and the *constant multiple rule* of differentiation, we find

$$f'(x) = \frac{df}{dx} = 4x^3 + 6x - 7$$

(b) Using the *chain rule* of differentiation, we find

$$f'(x) = \frac{df}{dx} = 2 \cos (2x)$$

(c) Using the *quotient rule* of differentiation, we find

$$f'(x) = \frac{df}{dx} = \frac{3(x^2 + 4) - 3x(2x)}{(x^2 + 4)^2} = \frac{3(4 - x^2)}{(x^2 + 4)^2}$$

(d) $f(x) = (2x - 5)(x^2 + a^2)^{-1/2}$

Using the *product rule* and the *chain rule*, we find

$$f'(x)\frac{df}{dx}$$

$$= 2(x^2 + a^2)^{-1/2} - \frac{1}{2}(2x - 5)(x^2 + a^2)^{-3/2}(2x)$$

i.e. $f'(x) = \dfrac{2(x^2 + a^2) - (2x^2 - 5x)}{(x^2 + a^2)^{3/2}} = \dfrac{2a^2 + 5x}{(x^2 + a^2)^{3/2}}$

(e) Using the *product rule* of differentiation, we find

$$f'(x) = \frac{df}{dx} = 2xe^x + x^2e^x = x(x + 2)e^x$$

R2 The *second derivatives* are as follows:

(a) $12x^2 + 6$,   (b) $-4 \sin (2x)$,

(c) $\dfrac{6x(x^2 - 12)}{(x^2 + 4)^3}$

R3 *Factorising* the left-hand side of $x^4 - x^3 = 0$ we obtain $x^3(x - 1) = 0$ and therefore the roots are $x = 0$ or $x = 1$.

Text answers

T1 (a) $df/dx = 12x - 3$. Therefore the function is increasing for $x > 1/4$, decreasing for $x < 1/4$, and stationary at $x = 1/4$.

(b) $dg/dx = 2x + 5$. Therefore the function is increasing for $x > -5/2$, decreasing for $x < -5/2$, and stationary at $x = -5/2$.

(c) $dh/dx = 1 - 6x$. Therefore the function is increasing for $x < 1/6$, decreasing for $x > 1/6$, and stationary at $x = 1/6$.

(d) $dk/dx = 3x^2 - 12x + 9 = 3(x - 3)(x - 1)$. Since $dk/dx < 0$ if $1 < x < 3$ and $dk/dx > 0$ if $x < 1$ or $x > 3$, it follows that the function is increasing for $x < 1$ or $x > 3$, decreasing for $1 < x < 3$ and stationary at $x = 1$ and $x = 3$.

(e) $dl/dx = 4 - x^2 = (2 - x)(2 + x)$. Therefore the function is increasing for $-2 < x < 2$, decreasing for $x < -2$ or $x > 2$, and stationary at $x = 2$ and $x = -2$.

(f) $dm/dx = 3x^2$ and it follows that $dm/dx > 0$ if $x \neq 0$ and therefore $m(x)$ is increasing if $x > 0$ or if $x < 0$, and it is stationary at $x = 0$.

T2 If $f(x) = x^4 - 4x^3$ then $f'(x) = 4x^3 - 12x^2 = 4x^2(x - 3)$. The stationary points occur when $x = 0$ or $x = 3$, and both points lie in the interval $-1 \leqslant x \leqslant 4$. We consider the values $f(-1) = 5$, $f(0) = 0, f(3) = -27, f(4) = 0$ and we see that the global maximum is 5 (occurring at $x = -1$) while the global minimum is -27 (occurring at $x = 3$).

T3 (a) $f'(x) = 4$ which is > 0 and thus the gradient is positive for all x, and $f(x)$ is increasing in any interval.

(i) Over $-1 \leqslant x \leqslant 4$ the global minimum value of $f(x)$ is -1 and is attained at $x = -1$ (the left-hand end of the interval) while the global maximum value, 19, is attained at $x = 4$ (the right-hand end of the interval).

(ii) Over $-3 \leqslant x \leqslant 2$, the global minimum occurs at $x = -3$ with $f(-3) = -9$; the global maximum occurs at $x = 2$ with $f(2) = 11$.

(b) $g'(x) = 2x$ so there is a stationary point at $x = 0$.

(i) Over $-2 \leqslant x \leqslant 1$ we examine the values of the function at the stationary point (which occurs in the interval) and at the ends of the interval to obtain $g(-2) = 8$, $g(0) = 4$ and $g(1) = 5$.

Therefore there is a global minimum at $x = 0$ with $g(0) = 4$; a global maximum at $x = -2$ with $g(-2) = 8$.

(ii) The derivative is positive over the interval $2 < x < 5$ so that $g(x)$ is increasing and therefore takes its global minimum value at the left-hand end of the interval and its global maximum value at the right-hand end.

Therefore there is a global minimum at $x = 2$ with $f(2) = 8$, and a global maximum at $x = 5$ with $f(5) = 29$.

(iii) Over $-3 \leqslant x \leqslant 2$ we examine the values of the function at the stationary point (which occurs in the interval) and at the ends of the interval to obtain $g(-3) = 13$, $g(0) = 4$ and $g(2) = 8$.

Therefore there is a global minimum at $x = 0$ with $g(0) = 4$, and a global maximum at $x = -3$ with $g(-3) = 13$.

(c) Since $h'(x) = 2(1 - x)$ there is a stationary point at $x = 1$.

(i) Over $0 \leqslant x \leqslant 3$ we evaluate the function at the stationary point (which occurs in the interval) and at the ends of the interval to obtain $h(0) = 3$, $h(1) = 4$, and $h(3) = 0$, so that the global maximum value in the interval is 4, attained at $x = 1$, while the global minimum value is 0 attained at $x = 3$.

(ii) In the interval $2 \leqslant x \leqslant 3$ we see that $h'(x) = 2(1 - x)$ which is < 0 and therefore the function is decreasing. The global maximum value is attained at the left-hand end of the interval value in the interval, i.e. $h(2) = 3$, while the global minimum value occurs at the right-hand end of the interval, i.e. $h(3) = 0$.

T4 (a) $f'(x) = 3x^2 - 6x = 3x(x - 2)$ so that $x = 0$ and $x = 2$ are stationary points. $x = 0$ is a turning point which is actually a local maximum since the derivative changes from positive to negative as x passes through 0; $x = 2$ is also a turning point and a local minimum since the derivative changes from negative to positive as x passes through 2.

(b) $g'(x) = x^2 + 4x + 3 = (x + 1)(x + 3)$ so that $x = -1$ and $x = -3$ are stationary points. $x = -3$ is a local maximum since the derivative changes from positive to negative as x passes through -3; $x = -1$ is a local minimum since the derivative changes from negative to positive as x passes through -1.

(c) $h'(x) = 4x^3 - 12x^2 + 16 = 4(x^3 - 3x^2 + 4)$

$\qquad = 4(x + 1)(x^2 - 4x + 4) = 4(x + 1)(x - 2)^2$

$h'(-1) = 0$, hence $x = -1$ is a stationary point and by considering $h'(x)$ either side of $x = -1$ we see that $x = -1$ is a local minimum.

There is also a stationary point at $x = 2$, but this point is not a turning point, and so cannot be an extremum, since $h'(x)$ is positive on each side of $x = 2$.

T5 $f(x) = \dfrac{1}{[(x - 2)^2 + 1]^2}$

As $x \to \pm\infty$, $f(x) \to 0$, but has no least value. The expression $[(x - 2)^2 + 1]$ is least when $x = 2$ so that $f(x)$ attains its greatest value of 1 when $x = 2$.

T6 (a) $f(x) = -x^2 + 7x$, $f'(x) = -2x + 7$, $f''(x) = -2$.

Since $f''(x) < 0$ the graph of the function is concave downwards everywhere.

(b) $f'(x) = -2(x + 2)$, $f''(x) = -2$.

Since $f''(x) < 0$ the graph of the function is concave downwards everywhere.

(c) $f''(x) = 6(2 - x)$

$f''(x) > 0$ if $x < 2$, therefore the graph of the function is concave upwards.

$f''(x) < 0$ if $x > 2$, therefore the graph of the function is concave downwards.

(d) $f''(x) = 6(x + 5)$

$f''(x) > 0$ if $x > -5$, therefore the graph of the function is concave upwards.

$f''(x) < 0$ if $x < -5$, therefore the graph of the function is concave downwards.

(e) $f''(x) = 12(x - 2)(x - 4)$

$f''(x) > 0$ if $x < 2$ or $x > 4$, therefore the graph of the function is concave upwards.

$f''(x) < 0$ if $2 < x < 4$, therefore the graph of the function is concave downwards.

(f) $f''(x) = (x + 2)e^x$

$f''(x) > 0$ if $x > -2$, therefore the graph of the function is concave upwards.

$f''(x) < 0$ if $x < -2$, therefore the graph of the function is concave downwards.

T7 (a) $f''(x) = 2(1 - 3x)$. Inflection at $x = 1/3$ since $f''(x)$ changes sign as x passes through $x = 1/3$.

(b) $f''(x) = 3x^2 - 4x + 1 = (3x - 1)(x - 1)$, so that $x = 1/3$ and $x = 1$ are points of inflection since $f''(x)$ changes sign as x passes through each of these points.

(c) $f''(x) = 6(2x - 3)$ and $x = 3/2$ is a point of inflection since $f''(x)$ changes sign as x passes through $x = 3/2$.

(d) $f''(x) = 20x^3 - 10x = 10x(2x^2 - 1)$. Points of inflection at $x = 0$, $x = \pm 1/\sqrt{2}$ as $f''(x)$ changes sign as x passes through each of these points.

(e) $f''(x) = -\sin x$. Points of inflection at $x = 0$, $\pm\pi$, $\pm 2\pi$, ... since at each of these points $f''(x)$ changes sign.

(f) $f''(x) = 20x^2(x - 1)$. Point of inflection at $x = 1$ because $f''(x)$ changes sign as x passes through the value 1. Also $f''(x) = 0$ if $x = 0$ but there is no point of inflection at this point because the second derivative does not change sign.

T8 (a) $f'(x) = 12x - 3$, $f''(x) = 12 > 0$. So $x = 1/4$ is a local minimum.

(b) $g'(x) = 2x + 5$, $g''(x) = 2 > 0$. So, $x = -5/2$ is a local minimum.

(c) $h'(x) = 1 - 6x$, $h''(x) = -6 < 0$. So, $x = 1/6$ is a local maximum.

(d) $k'(x) = 3x^2 - 12x + 9 = 3(x - 3)(x - 1)$, $k''(x) = 6x - 12$.

At $x = 1$ $k''(1) = -6 < 0$, a local maximum.

At $x = 3$, $k''(3) = 6 > 0$, a local minimum.

(e) $l'(x) = 4 - x^2 = (2 - x)(2 + x)$, $l''(x) = -2x$

At $x = 2$, $l''(2) = -4 < 0$, a local maximum.

At $x = -2$, $l''(-2) = 4 > 0$, a local minimum.

(f) $m(x) = x^3$, $m'(x) = 3x^2$, $m''(x) = 6x$

There is a point of inflection at $x = 0$.

T9 (a) Odd,

since $(-x) + \sin(-x) = -(x + \sin x)$

(b) Even, since $(-x)^2 - 2\cos(-x) = x^2 - 2\cos x$

(c) Neither, since $(-x)^2 + \sin(-x) = x^2 - \sin x$

(d) Even,

since $(-x)\sin(-x) = -(-x)\sin x = x\sin x$

(e) Odd, since $(-x)\cos(-x) = -(x\cos x)$

T10 (a) This is an odd function. The graph intercepts the x-axis when $x = \pm 1$. $f(x) \approx -1/x$ when $|x|$ is small and $f(x) \approx x$ when $|x|$ is large.

Also $f'(x) = 1 + \dfrac{1}{x^2} > 0$.

Hence there are no stationary points, and, since $f'(x) > 0$ for all x (except $x = 0$), $f(x)$ is increasing if $x < 0$ and also increasing if $x > 0$. The graph is sketched in Figure 16.25.

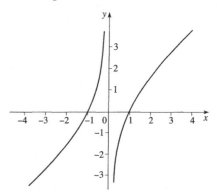

Figure 16.25   See Answer T10(a).

(b) $f'(x) = 4x^3 - 12x^2 = 4x^2(x - 3)$. By the first derivative test, $x = 0$ is a point of inflection, and $x = 3$ is a local minimum.

$f(0) = 10$, $f(3) = -17$. $f(x) = 0$ for $x \approx 1.6$ and 3.8. For large x, $f(x) \approx x^4$. The graph is sketched in Figure 16.26.

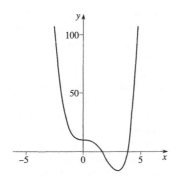

Figure 16.26   See Answer T10(b).

(c) $f(x)$ is an even function and $f(-1) = f(0)$ $= f(1) = 0$. $f'(x) = 2x(2x^2 - 1)$, $f''(x)$ $12x^2 - 2$, $f'(x) = 0$ at $x = 0, \pm 1/\sqrt{2}$. Using the second derivative test we find that $x = 0$ is a local maximum, $x = \pm 1/\sqrt{2}$ are local minima, and that there are points of inflection at $x = \pm 1/\sqrt{6}$ (since $f''(x)$ changes sign at these points). The graph is sketched in Figure 16.27.

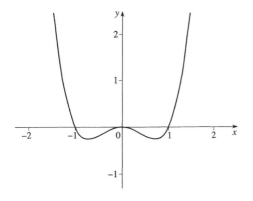

Figure 16.27   See Answer T10(c)

Chapter 17

Fast track answers

F1 (a) $\int (1 + 2x + 3x^2)dx = x + x^2 + x^3 + C$

where C is an arbitrary constant, since

$$\frac{d}{dx}(x + x^2 + x^3 + C) = 1 + 2x + 3x^2$$

(b) $\int [\sin(2x) + \cos(3x)]dx$

$$= -\tfrac{1}{2}\cos(2x) + \tfrac{1}{3}\sin(3x) + C$$

where C is an arbitrary constant, since

$$\frac{d}{dx}[-\tfrac{1}{2}\cos(2x) + \tfrac{1}{3}\sin(3x) + C]$$
$$= \sin(2x) + \cos(3x)$$

(c) $\int \left(e^t + \dfrac{1}{t^2}\right) dt = \int (e^t + t^{-2})\, dt$

$$= e^t - t^{-1} + C = e^t - \frac{1}{t} + C$$

where C is an arbitrary constant, since

$$\frac{d}{dt}\left(e^t - \frac{1}{t} + C\right) = e^t + \frac{1}{t^2}$$

(d) $\int dw = \int 1\, dw = w + C$

where C is an arbitrary constant, since

$$\frac{d}{dw}(w + C) = 1$$

F2 (a) $\displaystyle\int_{-1/4}^{1/4} \cos(2\pi x)\, dx = \left[\frac{\sin(2\pi x)}{2\pi}\right]_{-1/4}^{1/4}$

$$= \frac{1}{2\pi}\left[\sin\left(\frac{\pi}{2}\right) - \sin\left(-\frac{\pi}{2}\right)\right]$$

$$= \frac{1}{2\pi}[1 - (-1)] = \frac{1}{\pi}$$

(b) $\displaystyle\int_0^3 (2t - 1)^2\, dt = \int_0^3 (4t^2 - 4t + 1)\, dt$

$$= \left[\frac{4}{3}t^3 - 2t^2 + t\right]_0^3$$

$$= (\tfrac{4}{3} \times 27 - 18 + 3) - 0$$

$$= 36 - 18 + 3 = 21$$

(c) $\displaystyle\int_1^2 \frac{(1 + e^t)^2}{e^t}\, dt = \int_1^2 \frac{1 + 2e^t + e^{2t}}{e^t}\, dt$

$$= \int_1^2 (e^{-t} + 2 + e^t)\, dt$$

$$= [-e^{-t} + 2t + e^t]_1^2 \quad \left(\text{since } \frac{d}{dt}e^{kt} = ke^{kt}\right)$$

$$= (-e^{-2} + 4 + e^2) - (-e^{-1} + 2 + e)$$

$$= e^2 - e^{-2} + e^{-1} - e + 2$$

(d) $\int_4^9 \sqrt{x}\left(x - \dfrac{1}{x}\right) dx = \int_4^9 (x^{3/2} - x^{-1/2}) \, dx$

$= \left[\dfrac{2}{5} x^{5/2} - 2x^{1/2}\right]_4^9$

since $\dfrac{d}{dx}\left(\dfrac{x^n}{n}\right) = x^{n-1}$ if $n \neq 0$

$= \left(\dfrac{2}{5} \times 3^5 - 6\right) - \left(\dfrac{2}{5} \times 2^5 - 4\right)$

$= \dfrac{2}{5}(3^5 - 2^5) - 2 = \dfrac{412}{5} = 82.4$

F3 Since $a_x(t) = a - bt$, where $a = 5 \text{ m s}^{-2}$ and $b = 3 \text{ m s}^{-3}$, the velocity is

$$v_x(t) = \int a_x(t) \, dt = \int (a - bt) \, dt$$

$$= at - \dfrac{bt^2}{2} + C = \dfrac{t}{2}(2a - bt) + C$$

for some particular constant C. But $v_x = 0$ when $t = 0$, so $C = 0$ and

$$v_x(t) = \dfrac{t}{2}(2a - bt)$$

Similarly, the position is given by

$$x(t) = \int v_x(t) \, dt = \int \left(at - \dfrac{bt^2}{2}\right) dt$$

$$= \dfrac{at^2}{2} - \dfrac{bt^3}{6} + D = \dfrac{t^2}{6}(3a - bt) + D$$

where D is some particular constant.

(a) False. The velocity v_x is zero when $t = 0$ and again when $t = 2a/b = (10/3)$ s.

(b) True. The position x is D when $t = 0$ and again when $t = 3a/b = 5$ s. It follows that the displacement (i.e. the difference in position) from the initial position is zero when $t = 5$ s.

(c) True. Generally the area under the velocity–time graph between two values of t represents the change in position (which may be positive or negative) over that time interval, rather than the distance moved (which must be positive). However in this case the velocity is positive in the time interval under consideration so the two are identical.

(d) False. Speed (a positive quantity) is the magnitude of velocity (which may be positive or negative). The area under the acceleration–time

graph between two values of t represents the change in the velocity of the object over that time interval, so, since the velocity is zero when $t = 0$, the area from $t = 0$ to $t = (5/3)$ s represents the velocity when $t = (5/3)$ s. At this time the acceleration is zero, and the velocity attains a value of 4.167 m s^{-1} (which happens to be its positive maximum). However, this does *not* represent the maximum speed. The velocity becomes negative when t exceeds $(10/3)$ s, and subsequently stays negative, becoming ever larger in magnitude. The velocity when $t = 10$ s is -100 m s^{-1} so the speed at that time is 100 m s^{-1}, which clearly exceeds the speed at $t = (5/3)$ s.

Ready to study answers

RI (a) If $f(t) = t^{2/3}$

then $\dfrac{df}{dt} = f'(t) = \dfrac{2}{3} t^{-1/3}$

(b) If $f(t) = \sin\left(\dfrac{\pi t}{2}\right)$

then $\dfrac{df}{dt} = f'(t) = \dfrac{\pi}{2} \cos\left(\dfrac{\pi t}{2}\right)$

(c) If $f(t) = 5e^{at}$

then $\dfrac{df}{dt} = f'(t) = 5a e^{at}$

R2 (a) $\dfrac{(1 + e^x)^2}{e^x} = \dfrac{(1 + 2e^x + e^{2x})}{e^x}$

$= e^{-x} + 2 + e^x$

(b) $\sqrt{x}\left(x - \dfrac{1}{x}\right) = x^{3/2} - x^{-1/2}$

R3 $F(-2) = (-2)[\sin(-\pi) - 1] = 2$

(since $\sin(-\pi) = 0$)

$F(-1) = (-1)[\sin(-\pi/2) - 1]$

$= (-1)(-2) = 2$

(since $\sin(-\pi/2) = -1$)

$F(1) = 1[\sin(\pi/2) - 1] = 0$

(since $\sin(\pi/2) = 1$)

Therefore

(a) $F(-1) - F(-2) = 2 - 2 = 0$

(b) $F(1) - F(-1) = 0 - 2 = -2$

(c) $F(1) - F(-2) = -2$

Text answers

T1 $x(t) = A \sin(\omega t) + Ct + D$ where D is some particular constant.

Since $x = 0$ when $t = 0$, it follows that $D = 0$. Also, since $v_x = 0$ when $t = \pi/(2\omega)$, Equation 11 implies $C = 0$.

Thus $x(t) = A \sin(\omega t)$

T2 Inverse derivatives of x^2, x^3 and x^4 are, respectively,

$$\frac{x^3}{3} + C, \quad \frac{x^4}{4} + C, \quad \text{and} \quad \frac{x^5}{5} + C$$

where C is a constant.

The required rule for finding the inverse derivative of x^p where the power $p \neq -1$ is:

$$\frac{x^{p+1}}{p+1} + C$$

Informally, 'add one to the power and put it underneath, then add C'.

T3 (a) Any function of the form

$$F(x) = -e^{-x} + C$$

is a suitable answer.

(b) Any function of the form

$$F(x) = \frac{x^2}{2} + e^x + C$$

is a suitable answer.

T4 (a) $x(t)$ must be an inverse derivative of $v_x(t)$, so it must be of the form

$$x(t) = v_0 t - b\frac{t^3}{3} + C$$

so the required displacement is given by

$$x(2\,\text{s}) - x(0) = \left[v_0 t - b\frac{t^3}{3} + C \right]_0^{2\,\text{s}}$$

$$= \left(2\,\text{m s}^{-1} \times 2\,\text{s} - \frac{1\,\text{m s}^{-1} \times (2\,\text{s})^3}{3} \right)$$

$$= \frac{4}{3}\,\text{m} \approx 1.333\,\text{m}$$

Since $x(t)$ is one of the inverse derivatives of $v_x(t)$, it follows that the displacement $x(2\,\text{s}) - x(0)$ is given by the *area under the graph* of v_x against t, between $t = 0$ and $t = 2\,\text{s}$ (with the areas of regions below the t-axis being treated as negative quantities).

(b) The function $v_x(t)$ is positive for $0 \leqslant t \leqslant \sqrt{2}\,\text{s}$, so that for this time interval the displacement and the distance travelled are identical and equal to

$$x(\sqrt{2}\,\text{s}) - x(0) = \left[v_0 t - b\frac{t^3}{3} + C \right]_0^{1/2\,\text{s}}$$

$$-\left(2\sqrt{2} = \frac{2\sqrt{2}}{3} \right)\text{m} = \frac{4\sqrt{2}}{3}\,\text{m} \approx 1.886\,\text{m}$$

During the time interval $\sqrt{2}\,\text{s} < t < 2\,\text{s}$ the velocity is negative and the displacement is

$$x(2\text{s}) - x(\sqrt{2}\,\text{s}) = \left[v_0 t - b\frac{t^3}{3} \right]_{(2)^{1/2}\,\text{s}}^{2\,\text{s}}$$

$$= \left(4 - \frac{8}{3} \right)\text{m} - \left(2\sqrt{2} - \frac{2\sqrt{2}}{3} \right)\text{m}$$

$$= \left(\frac{4}{3} - \frac{4\sqrt{2}}{3} \right)\text{m} = \frac{4}{3}(1 - \sqrt{2})\,\text{m}$$

$$\approx -0.552\,\text{m}$$

However the distance travelled during this second period is given by the *magnitude* of the displacement, and is therefore 0.552 m. Thus, although the total displacement is $1.886\,\text{m} - 0.552\,\text{m} \approx 1.33\,\text{m}$, the distance travelled is $1.886\,\text{m} + 0.552\,\text{m} \approx 2.44\,\text{m}$.

In terms of a graph of v_x against t, the distance travelled is represented by the sum of the *magnitudes* of the areas of the various regions enclosed between the graph and the t-axis between $t = 0$ and $t = 2\,\text{s}$.

T5 (a) $\displaystyle\int_1^2 x^{1/2}\,dx = \left[\frac{2x^{3/2}}{3} \right]_1^2 = \frac{2}{3}(2^{3/2} - 1) \approx 1.219$

since $\displaystyle\frac{d}{dx}\left(\frac{2x^{3/2}}{3} \right) = x^{1/2}$

(b) $\displaystyle\int_{-2}^0 4x^2\,dx = \left[\frac{4x^3}{3} \right]_{-2}^0 = (0) - \left(\frac{-32}{3} \right) \approx 10.67$

(c) $\displaystyle\int_{-\pi/2}^{\pi/2} \sin x \, dx = [= \cos x]_{-\pi/2}^{\pi/2} = (0) - (0) = 0$

T6 (a) $\displaystyle\int t^3 \, dt = \frac{1}{4}t^4 + C$

since $\displaystyle\frac{d}{dt}\left(\frac{1}{4}t^4 + C\right) = t^3$

(b) $\displaystyle\int \cos x \, dx = \sin x + C$

since $\displaystyle\frac{d}{dx}(\sin x + C) = \cos x$

(c) $\displaystyle\int e^{2x} \, dx = \tfrac{1}{2}\,e^{2x} + C$

since $\displaystyle\frac{d}{dx}\left(\frac{1}{2}e^{2x} + C\right) = e^{2x}$

T7 (a) $\displaystyle\int a_x(t) \, dt = \frac{a_0}{k}(e^{kt} + e^{-kt}) + C$

where C is an arbitrary constant.

The velocity of the moving object $v_x(t)$ is of this form, though it includes a particular (even if unknown) value of C. Thus we can say that, apart from an additive constant, $\int a_x(t)\,dt$ represents the velocity of the moving object.

(b) From the previous part we know

$$v_x(t) = \frac{a_0}{k}(e^{kt} + e^{-kt}) + C$$

This is a velocity. We can therefore determine

$$x(t) = \int v_x(t) \, dt = \frac{a_0}{k}(e^{kt} - e^{-kt}) + Ct + D$$

We now have two unknown constants C and D, since we have integrated twice, but we are given two pieces of information to determine them.

First, $0 = x(0) = \dfrac{a_0}{k}(e^0 - e^0) + D$ giving $D = 0$.

Second, since $D = 0$ and $k = s^{-1}$

$$1 \text{ m} = x(1 \text{ s}) = a_0(e^1 - e^{-1}) + C$$

so that the required constant of integration is

$$C = 1 - a_0(e^1 + e^{-1}) = 1 - a_0\left(e - \frac{1}{e}\right)$$

T8 The mass of the bar between x and $x + \Delta x$ is approximately

$$\pi R^2 \rho(x)\Delta x$$

The exact total mass is

$$\int_{-L}^{L} \pi R^2 \rho(x) \, dx$$

Note that the limits are $-L$ and L since the mid-point of the bar is given as the origin for the density. Hence the total mass is

$$\int_{-L}^{L} \pi R^2 A\, e^{2Bx} \, dx = \left[\frac{\pi R^2}{2B} A\, e^{2Bx}\right]_{-L}^{L}$$

since $\displaystyle\frac{d}{dx}\left(\frac{\pi R^2}{2B} A\, e^{2Bx}\right) = \pi R^2 A\, e^{2Bx}$

i.e. $\displaystyle\int_{-L}^{L} \pi R^2 A\, e^{2Bx} \, dx = \frac{\pi R^2}{2B} A\, (e^{2BL} - e^{-2BL})$

T9 (a) $\displaystyle\langle T\rangle = \frac{1}{2L}\int_{-L}^{L} T(x) \, dx$

$$= \frac{1}{2L}\int_{-L}^{L} A \cos(Bx) \, dx = \frac{1}{2L}\left[\frac{A}{B}\sin(Bx)\right]_{-L}^{L}$$

$$= \frac{1}{2L}\left[\frac{A}{B}\sin(BL) + \frac{A}{B}\sin(BL)\right] = \frac{A\sin(BL)}{BL}$$

then $\displaystyle\langle T\rangle = \frac{(30 \text{ K})\sin(0.8)}{0.8} = 26.9 \text{ K}$

T10 (a) $\displaystyle\int (1 + x + x^2)dx = x + \frac{x^2}{2} + \frac{x^3}{3} + C$

(b) $\displaystyle\int (\sin x + \cos x) \, dx = \sin x - \cos x + C$

(c) $\displaystyle\int \left(e^{-x} + \frac{1}{x^3}\right) dx = -e^{-x} - \frac{x^{-2}}{2} + C$

(d) $\displaystyle\int 2dw = 2w + C$

T11 (a) $\displaystyle\int_{-1/2}^{1/2} \sin(\pi x) \, dx = \left[-\frac{\cos(\pi x)}{\pi}\right]_{-1/2}^{1/2} = 0$

You see why this *must* be zero just by sketching a graph of $y = \sin(\pi x)$.

(b) $\displaystyle\int_{0}^{3} (3t + 2)^2 \, dt = \int_{0}^{3} (9t^2 + 12t + 4) \, dt$

$$= [3t^3 + 6t^2 + 4t]_0^3 = 147$$

(c) $\int_0^1 \dfrac{(1 + e^x)}{e^x} \, dx = \int_0^1 (e^{-x} + 1) \, dx$

$$= \left[-e^{-x} + x \right]_0^1 = \left(-\dfrac{1}{e} + 1 \right) - (-1)$$

$$= 2 - \dfrac{1}{e}$$

(d) $\int_1^4 \sqrt{x} \, (x + 1) \, dx = \int_1^4 (x^{3/2} + x^{1/2}) \, dx$

$$= \left[\dfrac{2x^{5/2}}{5} + \dfrac{2x^{3/2}}{3} \right]_1^4 = \dfrac{256}{15} \approx 17.067$$

T12 We must evaluate the area in two parts, since the graph lies below the x-axis for $-1 \leqslant x < 0$ and above the x-axis for $0 < x \leqslant 1$ (see Figure 17.12).

$$\int_{-1}^0 (e^x - 1) dx = [e^x - x]_{-1}^0$$

$$= 1 - [e^{-1} - (-1)]$$

$$= -\dfrac{1}{e}$$

and $\int_0^1 (e^x - 1) \, dx = [e^x - x]_0^1$

$$= (e - 1) - 1 = e - 2$$

We would normally say that these answers represented the (signed) *areas under the graph* in the two regions, even though one of them is a negative quantity because it lies below the horizontal axis. However, in this problem we are specifically asked for the *magnitudes* of the areas *enclosed*, so the required answer is

$$|e - 2| + \left| -\dfrac{1}{e} \right| = e + \dfrac{1}{e} - 2$$

T13 $\int_{t_1}^{t_2} a_x(t) \, dt$ represents the area under the graph of $a_x(t)$ against t between the points $t = t_1$ and $t = t_2$, assuming that areas above the t-axis are assigned a positive value and areas below the t-axis are assigned a negative value. Physically, it represents the change in the velocity of the object from time t_1 to time t_2.

$\int_{t_1}^{t_2} v_x(t) \, dt$ represents the area under the graph of $v_x(t)$ against t between the points $t = t_1$ and $t = t_2$, assuming that areas above the t-axis are assigned a positive value and areas below the t-axis are

assigned a negative value. Physically, it represents the change in the position of the object from time t_1 to time t_2.

$\int_{t_1}^{t_2} x(t) \, dt$ represents the area under the graph of $x(t)$ against t between the points $t = t_1$ and $t = t_2$, assuming that areas above the t-axis are assigned a positive value and areas below the t-axis are assigned a negative value. There is no other particular physical significance for this integral.

Chapter 18

Fast track answers

F1 The area under the curve is divided into n thin strips of width Δx_i and height $f(x_i)$, as shown in Figure 18.5. The approximate area of such a strip is $f(x_i)\Delta x_i$, and the total area is

$$\left(\sum_{i=1}^n f(x_i) \, \Delta x_i \right)$$

where the summation is taken over all strips. As the number of strips increases and the width of each strip decreases, this summation tends to

$$\int_a^b f(x) \, dx$$

which is the exact area required.

When the graph crosses the axis at a point c where $a < c < b$, each region where $f(x)$ has a fixed sign must be considered separately. Thus the magnitudes of the enclosed areas in the case $f(x) = x^3$, where the graph crosses the horizontal axis at $x = 0$, will be

$$A_1 = \left| \int_{-2}^0 x^3 \, dx \right| = \left| \left[\dfrac{x^4}{4} \right]_{-2}^{-0} \right| = \left| 0 - \dfrac{16}{4} \right| = 4$$

and

$$A_2 = \left| \int_0^1 x^3 \, dx \right| = \left| \left[\dfrac{x^4}{4} \right]_0^1 \right| = \dfrac{1}{4}$$

F2 (a) $\int (4x^2 + 7x - 5) dx = \frac{4}{3}x^3 + \frac{7}{2}x^2 - 5x + C$

(b) $\int e^{-6x} \, dx = -\frac{1}{6}e^{-6x} + C$

(c) $\int [3 \cos (4x) - 5 \sin (4x)] dx$

$$= \frac{3}{4} \sin (4x) + \frac{5}{4} \cos (4x) + C$$

(d) $\int 3 \log_e (2x) dx = 3x \log_e (2x) - 3x + C$

F3 (a) $\int_{-2}^{3} (3 - 2x - x^2)\, dx = \left[3x - x^2 - \frac{x^3}{3} \right]_{-2}^{3}$

$= (9 - 9 - 9) - \left(-6 - 4 + \frac{8}{3} \right) = -\frac{5}{3}$

$= -1.6667$

(b) $\int_{\pi/4}^{\pi/3} [6 \cos (3x) - 10 \cos (2x)]dx$

$= [2 \sin (3x) - 5 \sin (2x)]_{\pi/4}^{\pi/3}$

$= \left(0 - \frac{5\sqrt{3}}{2} \right) - \left(2 \times \frac{1}{\sqrt{2}} - 5 \times 1 \right)$

$= \frac{-5\sqrt{3}}{2} - \sqrt{2} + 5 \approx -0.7443$

(c) $\int_{1}^{2} 4 \log_e x\, dx = 4[x \log_e x - x]_{1}^{2}$

$= 4(2 \log_e 2 - 1) \approx 1.5452$

Ready to study answers

R1 (a) $df/dx = 48x^5 + 18x^2 - 10x$

(b) $df/dx = 3 \cos (x) - 4 \sin (x)$

(c) $df/dx = 10e^{2x} + 4e^{-2x}$

(d) $\dfrac{df}{dx} = \dfrac{2}{x} + \dfrac{3}{x}$

R2 (a) $\dfrac{dy}{dx} = -5 \sin (x) + 4x - 7$

(b) $\dfrac{dy}{dx} = -18e^{-3x}$

(c) $\dfrac{dy}{dx} = 2x - \dfrac{5}{x}$

R3 (a) $\lim\limits_{x \to 1} (2 - x) = 1$

(b) $\sum\limits_{i=1}^{3} 2i = 2 + 4 + 6 = 12$

(c) $\arcsin (-1) = -\pi/2 = n\pi/2$, where $n = -3$, $-2, -1, 2, 3, 5 \ldots$

Text answers

T1 (a) This is a function, every value of x gives rise to a unique value for y.

(b) This is a function because we are taking $\sqrt{x}$ to be the positive square root of x. (Note that if we were following the alternative convention of allowing $\sqrt{x}$ to denote either of the square roots of x, so that $\sqrt{4} = \pm 2$, then $\sqrt{x}$ would *not* have defined a function.)

(c) This definitely does *not* define y as a function of x. For each positive value of x there are two possible values of y.

(d) Perhaps it is a surprise, but this *does* define y as a function of x: it is a constant function, $f(x) = 2$.

(e) In this case $y = \pm 2$, so y is not uniquely defined and so cannot be a function of x.

T2 Since $F(x)$ is an indefinite integral of $f(x)$ we know that

$$\frac{dF}{dx} = f(x)$$

Thus,

$$\frac{d}{dx} [aF(x)] = a\frac{d}{dx} [F(x)] = af(x)$$

Hence $aF(x)$ is an indefinite integral of $af(x)$.

$$\frac{d}{dx} [F(x) + G(x)] = \frac{d}{dx} [F(x)] + \frac{d}{dx} [G(x)]$$

$$= f(x) = g(x)$$

Hence $F(x) + G(x)$ is an indefinite integral of $f(x) + g(x)$.

An indefinite integral of $af(x) + bg(x)$ is $aF(x) + bG(x)$.

Comment Note that we can also add an arbitrary constant to any of these indefinite integrals to obtain another indefinite integral.

T3 (a) $3x + C$, (b) $\dfrac{x^6}{6} + C$, (c) $4x^{5/4} + C$.

Comment The indefinite integral of x^p where the power $p \neq -1$ is:

$$\frac{x^{p+1}}{p + 1} + C$$

T4 (a) $\dfrac{d}{dx}(\sqrt{2x+2}+C)=\dfrac{d}{dx}[(2x+2)^{1/2}+C]$

$=\dfrac{1}{2}(2x+2)^{-1/2}\times 2=\dfrac{1}{\sqrt{2x+2}}$

(b) $\dfrac{d}{dx}\left[\dfrac{1}{18}(x^2-2x)^9+C\right]$

$=\dfrac{1}{18}\times 9\,(x^2-2x)^8\times(2x-2)$

$=(x^2-2x)^8(x-1)$

(c) $\dfrac{d}{dx}[-(2+\sin x)^{-1}+C]=(2+\sin x)^{-2}(\cos x)$

$=\dfrac{\cos x}{(2+\sin x)^2}$

T5 (a) $\displaystyle\int_{-1}^{2}3\,dx=[3x]_{-1}^{2}=(6)-(-3)=9$

(b) $\displaystyle\int_{0}^{2}x^5\,dx=\left[\dfrac{x^6}{6}\right]_0^2=\dfrac{64}{6}=10.67$

(c) $\displaystyle\int_{16}^{81}5x^{1/4}\,dx=[4x^{5/4}]_{16}^{81}=[4(3)^5]-[4(2)^5]$

$=972-128=844$

T6 The torque has a magnitude Γ that is approximately given by

$\Gamma\approx\displaystyle\sum_{i=1}^{n}\lambda g x_i\,\Delta x_i$

So in the limit as $\Delta x_i\to 0$

$\Gamma=\lambda g\displaystyle\int_0^L x\,dx=\lambda g\left[\dfrac{x^2}{2}\right]_0^L=\dfrac{\lambda g L^2}{2}$

Thus the beam acts as though its entire mass (λL) was concentrated at its mid-point $(x=L/2)$.

T7 The function $f(x)=x^5$ changes sign at $x=0$ and therefore the required sum is

$\left|\displaystyle\int_{-1}^{0}bx^5\,dx\right|+\left|\displaystyle\int_{0}^{1}bx^5\,dx\right|$

$=\left|\left[\dfrac{bx^6}{6}\right]_{-1}^{0}\right|+\left|\left[\dfrac{bx^6}{6}\right]_{0}^{1}\right|$

$=\left[\dfrac{2\text{ m}^{-4})\times(1\text{ m}^6)}{6}\right]+\left[\dfrac{(2\text{ m}^{-4})\times(1\text{ m}^6)}{6}\right]$

$=\dfrac{2\text{ m}^2}{6}+\dfrac{2\text{ m}^2}{6}=\dfrac{2}{3}\text{ m}^2$

T8 (a) From Table 1 $\displaystyle\int kx^{-1}\,dx=k\log_e|x|+C$

So $\displaystyle\int 8x^{-1}\,dx=8\log_e|x|+C$

(b) $\displaystyle\int_1^3 8x^{-1}\,dx=8\,[\log_e|x|]_1^3$

$=8(\log_e|3|)-8(\log_e|1|)$

$=8[\log_e(3)]-8[\log_e(1)]$

$=8\log_e(3)=8.789$

(c) $\displaystyle\int\sec^2(4x)\,dx=\tfrac14\tan(4x)+C$

(d) $\displaystyle\int_0^{\pi/16}\sec^2(4x)\,dx=\left[\dfrac{1}{4}\tan(4x)\right]_0^{\pi/16}$

$=\left(\dfrac{1}{4}\tan\dfrac{\pi}{4}\right)-0=\dfrac{1}{4}$

T9 (a) $\displaystyle\int\sqrt{x}(x^2-3)\,dx=\int(x^{5/2}-3x^{1/2})\,dx$

$=\dfrac{2}{7}x^{7/2}-2x^{3/2}+C=\dfrac{2x\sqrt{x}}{7}(x^2-7)+C$

(b) $\displaystyle\int\left(\dfrac{5}{t^{2/3}}+\dfrac{2}{t^{1/3}}\right)dt=\int(5t^{-2/3}+2t^{-1/3})\,dt$

$=15t^{1/3}+3t^{2/3}+C$

(c) $\displaystyle\int_1^3\dfrac{(x+2)^2}{\sqrt{x}}\,dx=\int_1^3\left(\dfrac{x^2+4x+4}{\sqrt{x}}\right)dx$

$=\displaystyle\int(x^{3/2}+4x^{1/2}+4x^{-1/2})\,dx$

$=\left[\dfrac{2}{5}x^{5/2}+\dfrac{8}{3}x^{3/2}+8x^{1/2}\right]_1^3$

$\left(\dfrac{2}{5}\times 9\sqrt{3}+\dfrac{8}{3}\times 3\sqrt{3}+8\sqrt{3}\right)-\left(\dfrac{2}{5}+\dfrac{8}{3}+8\right)$

$=\dfrac{98}{5}\sqrt{3}-\dfrac{166}{15}\approx 22.882$

(d) $\int \dfrac{t^2 - 6t + 1}{t^4}\, dt = \int (t^{-2} - 6t^{-3} + t^{-4})\, dt$

$\qquad = -t^{-1} + 3t^{-2} - \dfrac{1}{3} t^{-3} + C$

$\qquad = -\dfrac{1}{t} + \dfrac{3}{t^2} - \dfrac{1}{3t^3} + C$

(e) $\int [-3\cos(x) + 2\sec^2(x)]dx$

$\qquad = -3\sin(x) + 2\tan(x) + C$

(f) $\int \tan^2(\theta)\, d\theta = \int (\sec^2(\theta) - 1)d\theta$

$\qquad = \int \sec^2(\theta)d\theta - \int 1 d\theta$

$\qquad = \tan(\theta) - \theta + C$

therefore $\quad \displaystyle\int_0^{\pi/4} \tan^2(\theta)d\theta = [\tan(\theta) - \theta]_0^{\pi/4}$

$\qquad = \left(1 - \dfrac{\pi}{4}\right) - 0 = 1 - \dfrac{\pi}{4} \approx 0.215$

T10 $\displaystyle\int_1^2 [\sin(\pi x) + \pi \log_e(3x)]\, dx$

$= \left[-\dfrac{1}{\pi}\cos(\pi x) + x\pi \log_e(3x) = x\pi\right]_1^2$

$= \left(-\dfrac{1}{\pi}\cos(2\pi) + 2\pi \log_e(6) - 2\pi\right)$

$\quad - \left(-\dfrac{1}{\pi}\cos(\pi) + \pi \log_e(3) - \pi\right)$

$= \left(-\dfrac{1}{\pi} + 2\pi \log_e(6) - 2\pi\right)$

$\quad - \left(\dfrac{1}{\pi} + \pi \log_e(3) - \pi\right)$

$= -\dfrac{2}{\pi} + \pi \log_e 12 - \pi \approx 4.028\,35$

T11 (a) $\displaystyle\int_0^2 e^{3-2x}\, dx = \left[-\dfrac{1}{2}e^{3-2x}\right]_0^2$

$= \left(-\dfrac{1}{2}e^{-1}\right) - \left(-\dfrac{1}{2}e^3\right) = \dfrac{1}{2}\left(e^3 - \dfrac{1}{e}\right)$

(b) $\displaystyle\int \left(\dfrac{3x+2}{x+4}\right)dx = \int \left[\dfrac{3(x+4) - 10}{x+4}\right]dx$

$= \displaystyle\int \left(3 - \dfrac{10}{x+4}\right)dx$

$= 3x - \log_e |x+4| + C$

(c) $\displaystyle\int_2^4 \dfrac{1}{5x^2 + 20}\, dx = \dfrac{1}{5}\int_2^4 \dfrac{1}{x^2 + 4}\, dx$

$= \left[\dfrac{1}{5} \times \dfrac{1}{2} \arctan\left(\dfrac{x}{2}\right)\right]_2^4$

$= \left[\dfrac{1}{10}\arctan(2) = \dfrac{1}{10}\arctan(1)\right]$

$= \dfrac{1}{10}[\arctan(2) - (\pi/4)]$

(d) $\displaystyle\int_0^{1/3} \dfrac{3}{(8 - 9x^2)^{1/2}}\, dx = \left[\arcsin\left(\dfrac{3x}{2\sqrt{2}}\right)\right]_0^{1/3}$

$= \arcsin\left(\dfrac{1}{2\sqrt{2}}\right)$

T12 $f(x) - g(x) \geq 0$ for $a \leq x \leq b$. By Property 5,

$$\int_a^b [f(x) - g(x)]dx \geq 0$$

i.e. $\displaystyle\int_a^b f(x)\, dx - \int_a^b g(x)\, dx \geq 0$

Hence the result

$$\int_a^b f(x)\, dx \geq \int_a^b g(x)\, dx$$

T13 The work done is given by

$$W = \lim_{\Delta x \to 0}\left(\sum_{i=1}^n - F(x_i)\,\Delta x_i\right)$$

where $x_1 = 0$, $x_{n+1} = a/b$, and Δx represents the largest of the Δx_i which are themselves defined by $\Delta x_i = x_{i+1} - x_i$ where $x_1 < x_2 < x_3 \ldots < x_{n-1} < x_n < x_{n+1}$.

It follows from this definition that

$$W = \int_0^{a/b} -F(x)\, dx = -\int_0^{a/b} F(x)\, dx$$

Since the initial kinetic energy is *minus* the work done by the resistive force in bringing the bullet to rest.

The initial kinetic energies:

$$= \int_0^{a/b} F(x)\,dx = \int_0^{a/b} \sqrt[3]{(a^3 - b^3 x^3)}\, m\, dx$$

The greatest value of

$$\sqrt[3]{(a^3 - b^3 x^3)}\, m$$

in the interval $0 \leqslant x \leqslant a/b$ occurs at $x = 0$, so that by Property 6

$$\int_0^{a/b} \sqrt[3]{(a^3 - b^3 x^3)}\, m\, dx \leqslant \sqrt[3]{a^3}\, m \times \frac{a}{b} = \frac{ma^2}{b}$$

T14 (a) $\displaystyle\int_{-5}^5 3x\,dx = 0$ because the integrand is an odd function but the range of integration is symmetric about the origin.

(b) $\displaystyle\int_{-\pi}^{\pi} \sin^2(x/2)\,dx$ is non-zero because the integrand is never negative and is non-zero over at least part of the range of integration.

(c) $\displaystyle\int_{3\pi}^{7\pi} [3\sin(3x) + \sin^3(x)]\,dx = 0$ because the integrand is an odd periodic function (of period 2π) that is being integrated over two full periods. Due to the periodic nature of the integrand we can take the range of integration to be symmetric about the origin; the oddness then ensures that the integral is zero.

T15 $\displaystyle\int_{-1}^1 \frac{1}{x}\,dx = \lim_{\delta \to 0}\left(\int_{-1}^{-\delta}\frac{1}{x}\,dx\right) + \lim_{\varepsilon \to 0}\left(\int_{\varepsilon}^1 \frac{1}{x}\,dx\right)$

where δ and ε are positive quantities. (Note that δ and ε are independent quantities, it would be wrong to use the same variable in both limits.) Thus

$$\int_{-1}^1 \frac{1}{x}\,dx = \lim_{\delta \to 0}\left([\log_e|x|]_{-1}^{-\delta}\right) + \lim_{\varepsilon \to 0}\left([\log_e|x|]_{\varepsilon}^1\right)$$

$$= \lim_{\delta \to 0}(\log_e \delta) + \lim_{\varepsilon \to 0}(-\log_e \varepsilon)$$

but neither of these limits exists, so the integral is divergent. If we consider the following joint limit

$$\lim_{\varepsilon \to 0}\left(\int_{-1}^{-\varepsilon}\frac{1}{x}\,dx + \int_{\varepsilon}^1 \frac{1}{x}\,dx\right)$$

$$= \lim_{\varepsilon \to 0}\left([\log_e|x|]_{-1}^{-\varepsilon} + [\log_e|x|]_{\varepsilon}^1\right)$$

$$= \lim_{\varepsilon \to 0}(\log_e \varepsilon - \log_e \varepsilon) = 0$$

we can see that it is possible to associate a finite value with this particular improper integral, even though it is formally divergent. The value obtained by this particular kind of limiting process is called the *Cauchy principal value* of the integral. Such principal values are sometimes of interest, but their consideration is beyond the scope of this chapter.

Chapter 19

Fast track answers

Note In the answers that follow, C denotes a constant of integration.

F1 Both these integrals can be found using the formula for integration by parts,

$$\int f(x)\,g(x)\,dx = F(x)\,g(x) - \int F(x)\frac{dg}{dx}\,dx$$

where $\dfrac{dF}{dx} = f(x)$.

(a) Let $f(x) = e^{-x}$ and $g(x) = x^2$.

Then $F(x) = -e^{-x}$ and $\dfrac{dg}{dx} = 2x$.

$$\int x^2 e^{-x}\,dx = -x^2 e^{-x} + 2\int x\, e^{-x}\,dx \qquad \text{(i)}$$

Now consider $\int x\, e^{-x}\,dx$.

Let $f(x) = e^{-x}$ and $g(x) = x$.

Then $F(x) = -e^{-x}$ and $\dfrac{dg}{dx} = 1$.

So $\displaystyle\int x e^{-x}\,dx = x e^{-x} + \int e^{-x}\,dx$

$$= -x e^{-x} - e^{-x} + C$$

Substituting this into Equation (i) gives us

$$\int x^2 e^{-x}\,dx = -x^2 e^{-x} - 2x e^{-x} - 2e^{-x} + C$$

(b) Let $f(x) = e^{-x}$ and $g(x) = \sin(3x)$.

Then $F(x) = -e^{-x}$ and $\dfrac{dg}{dx} = 3\cos(3x)$.

Hence

$$\int e^{-x} \sin(3x)\,dx = -e^{-x} \sin(3x)$$
$$+ 3\int e^{-x} \cos(3x)\,dx \qquad \text{(ii)}$$

Now apply integration by parts to $\int e^{-x}\cos(3x)\,dx$. Take $f(x) = e^{-x}$ and $g(x) = \cos(3x)$.

Then $F(x) = -e^{-x}$ and $\dfrac{dg}{dx} = -3\sin(3x)$. Hence

$$\int e^{-x} \cos(3x)\,dx = -e^{-x} \cos(3x)$$
$$- 3\int e^{-x} \sin(3x)\,dx$$

Substituting this result into Equation (ii) gives us

$$\int e^{-x}\sin(3x)\,dx = -e^{-x}\sin(3x) - 3e^{-x}\cos(3x)$$
$$-9\int e^{-x}\sin(3x)\,dx$$

Collecting the $\int e^{-x}\sin(3x)\,dx$ terms on the left-hand side gives us

$$10\int e^{-x}\sin(3x)\,dx$$
$$= -e^{-x}\sin(3x) - 3e^{-x}\cos(3x) + C$$

i.e.

$$\int e^{-x}\sin(3x)\,dx$$
$$= -\tfrac{1}{10}e^{-x}\sin(3x) - \tfrac{3}{10}e^{-x}\cos(3x) + C$$

F2 Make the substitution $y = 1 + \cos x$, $dy = -\sin x\,dx$. Then

$$\int \sin x\,(1 + \cos x)^4\,dx = -\int y^4\,dy = -\tfrac{1}{5}y^5 + C$$
$$= -\tfrac{1}{5}(1 + \cos x)^5 + C$$

F3 Make the substitution $y = x + 1$, so that $x = y - 1$, and $dx = dy$. Then the integrand becomes

$$\frac{y - 1}{\sqrt{y}} = \sqrt{y} - \frac{1}{\sqrt{y}}$$

When $x = 0$, $y = 1$; and when $x = 1$, $y = 2$.

Therefore

$$\int_0^1 \frac{x}{\sqrt{1+x}}\,dx = \int_1^2 \left(\sqrt{y} - \frac{1}{\sqrt{y}}\right)dy$$

$$= \int_1^2 (y^{1/2} - y^{-1/2})\,dy$$

$$= \left[\frac{2}{3}y^{3/2} - 2y^{1/2}\right]_1^2 = 0.391$$

Ready to study answers

R1 Here we use the *product rule*:

$$\frac{d}{dx}[f(x)\,g(x)] = \frac{df}{dx}g(x) + f(x)\frac{dg}{dx}.$$

(a) Taking $f(x) = x^6$, $g(x) = \log_e x$, we have

$$\frac{df}{dx} = 6x^5 \quad \text{and} \quad \frac{dg}{dx} = \frac{1}{x}.$$

So $\dfrac{d}{dx}(x^6\log_e x) = (6x^5 \times \log_e x) + \left(x^6 \times \dfrac{1}{x}\right)$

$$= x^5\,(1 + 6\log_e x)$$

(b) Taking $f(x) = e^x$, $g(x) = \cos(2x)$, we have

$$\frac{df}{dx} = e^x \quad \text{and} \quad \frac{dg}{dx} = -2\sin 2x\,.$$

So $\dfrac{d}{dx}[e^x\cos(2x)] = e^x\cos(2x) + e^x[-2\sin(2x)]$

$$= e^x\,[\cos(2x) - 2\sin(2x)]$$

R2 Here we use the *chain rule*:

$$\frac{d}{dx}\{f[y(x)]\} = \frac{df}{dy}\frac{dy}{dx}$$

(a) Taking $f(y) = \cos y$, $y(x) = x^4$, we have

$$\frac{df}{dy} = -\sin y\,, \quad \frac{dy}{dx} = 4x^3$$

So $\dfrac{d}{dx}(\cos(x^4)) = -4x^3\sin y = -4\,x^3\sin(x^4)$

(b) Taking $f(y) = \log_e y$, $y(x) = 1 + x^2$, we have

$$\frac{df}{dy} = \frac{1}{y} \quad \text{and} \quad \frac{dy}{dx} = 2x$$

So $\dfrac{d}{dx}(\log_e(1 + x^2)) = \dfrac{2x}{y} = \dfrac{2x}{1 + x^2}$

Note In the answers to Question R3 and R4, C denotes an arbitrary constant of integration.

R3 To differentiate $\arcsin(2x)$, write $y = \arcsin(2x)$. Since arcsin is the *inverse function* of sin, it follows that $\sin y = 2x$. Now differentiate both sides of this equation with respect to x, using the *chain rule* to differentiate $\sin y$. This gives

$$\cos y\,\frac{dy}{dx} = 2 \quad \text{or} \quad \frac{dy}{dx} = \frac{2}{\cos y}$$

To express dy/dx in terms of x, use the identity $\cos^2 y + \sin^2 y = 1$ or

$$\cos y = \sqrt{1 - \sin^2 y} = \sqrt{1 - 4x^2}$$

Then $\dfrac{d}{dx}[\arcsin(2x)] = \dfrac{2}{\sqrt{1 - 4x^2}}$

This shows that $\dfrac{2}{\sqrt{1 = 4x^2}}$ has the *inverse derivative* $\arcsin(2x) + C$, and so (dividing by 2), the inverse derivative of $\dfrac{1}{\sqrt{1 - 4x^2}}$ is $\dfrac{1}{2}\arcsin(2x) + C$.

R4 These integrals may all be found by making an intelligent guess at the form of the answer, then checking by differentiation. The answers are

(a) $-3x^{-1/3} + C$

(b) $-\frac{1}{2}\cos(2x) + 3\sin x + C$ (here, we also use the fact that an integral of a sum of terms is equal to the sum of the integrals of each term).

(c) $4\exp(x/4) + C$.

R5 $\int_{1}^{2} x^2\, dx = \left[\frac{1}{3}x^3\right]_{1}^{2} = \frac{8}{3} - \frac{1}{3} = \frac{7}{3} \ (\approx 2.333)$

R6 (a) Here we use the *trigonometric identity*

$$\cos^2 x + \sin^2 x = 1.$$

Thus if $\sin x = 1/3$ then

$$\cos^2 x = 1 - \left(\frac{1}{3}\right)^2 = \frac{8}{9}.$$

So $\cos x = \pm\dfrac{2\sqrt{2}}{3}$

(b) Here we use the *trigonometric identity*

$$1 + \tan^2 x = \sec^2 x.$$

Thus if $\sec x = \sqrt{5}$ then $\tan^2 x = 5 - 1 = 4$,

so $\tan x = \pm 2$

R7 If $y = x + 3$, then $x + 1 = y - 2$.

So $\dfrac{(x+1)^2}{x+3} = \dfrac{(y-2)^2}{y}$

Text answers

T1 $\dfrac{d}{dx}(xe^x) = e^x + xe^x$ using the product rule.

So

$$xe^x = \int (e^x + xe^x)\, dx = \int e^x\, dx + \int xe^x\, dx$$

and therefore,

$$\int xe^x\, dx = xe^x - \int e^x\, dx = xe^x - e^x + C$$

T2 Take $f(x) = x^2, g(x) = \log_e x$. Then $F(x) = \frac{1}{3}x^3$ and $dg/dx = 1/x$. Substituting these expressions in Equation 2 we obtain

$$\int x^2 \log_e x\, dx = \frac{1}{3}x^3 \log_e x - \frac{1}{3}\int x^2\, dx$$

$$= \frac{1}{3}x^3 \log_e x - \frac{1}{9}x^3 + C$$

T3 Take $f(x) = e^{3x}, g(x) = x$. Then $F(x) = \frac{1}{3}e^{3x}$ and $dg/dx = 1$. Substituting these expressions in Equation 5 we obtain

$$\int_{0}^{1} xe^{3x}\, dx = \left[\frac{1}{3}xe^{3x}\right]_{0}^{1} - \frac{1}{3}\int_{0}^{1} e^{3x}\, dx$$

$$= \left[\frac{1}{3}xe^{3x}\right]_{0}^{1} - \left[\frac{1}{9}e^{3x}\right]_{0}^{1} = 4.5746$$

T4 Take $f(x) = e^{2x}, g(x) = x^2$. Then $F(x) = \frac{1}{2}e^{2x}$ and $dg/dx = 2x$. Substituting these expressions into Equation 2 gives us

$$\int x^2 e^{2x}\, dx = \frac{1}{2}x^2 e^{2x} - \int xe^{2x}\, dx \qquad \text{(iii)}$$

Now apply integration by parts to the integral $\int xe^{2x}\, dx$ taking $f(x) = e^{2x}$, $g(x) = x$, giving $F(x) = \frac{1}{2}e^{2x}$ and $dg/dx = 1$. Thus

$$\int xe^{2x}\, dx = \frac{1}{2}xe^{2x} - \frac{1}{2}\int e^{2x}\, dx$$

$$= \frac{1}{2}xe^{2x} - \frac{1}{2}e^{2x} + C$$

Substituting this result in Equation (iii) gives us

$$\int x^2 e^{2x}\, dx = \frac{1}{2}x^2 e^{2x} - \frac{1}{2}xe^{2x} + \frac{1}{4}e^{2x} + C$$

T5 Take $f(x) = 1, g(x) = \arctan x$. Then $F(x) = x$ and

$$\frac{dg}{dx} = \frac{1}{1 + x^2}$$

So Equation 2 gives us

$$\int \arctan x\, dx = x \arctan x - \int \frac{x}{1 + x^2}\, dx$$

and, on using the result given in the question,

$$\int \arctan x\, dx =$$

$$x \arctan x - \frac{1}{2}\log_e(1 + x^2) + C$$

T6 Taking $f(x) = \cos x$ and $g(x) = e^{2x}$ we have $F(x) = \sin x$ and $dg/dx = 2e^{2x}$.

Then Equation 2 gives us

$$J = \int e^{2x} \cos x \, dx$$

$$= e^{2x} \sin x - 2\int e^{2x} \sin x \, dx \qquad \text{(iv)}$$

Now apply integration by parts to $\int e^{2x} \sin x \, dx$ taking $f(x) = \sin x$ and $g(x) = e^{2x}$. Then $F(x) = -\cos x$ and $dg/dx = 2e^{2x}$.

Hence $\int e^{2x} \sin x \, dx$

$$= -e^{2x} \cos x + 2\int e^{2x} \cos x \, d$$
$$x$$
$$= -e^{2x} \cos x + 2J$$

Substituting this result into Equation (iv) gives us

$$J = e^{2x} \sin x - 2(-e^{2x} \cos x + 2J)$$

and collecting all terms in $\int e^{2x} \cos x \, dx$ on the left-hand side gives us

$$5\int e^{2x} \cos x \, dx = e^{2x} \sin x + 2e^{2x} \cos x + C$$

so that $\quad J = \frac{1}{5} e^{2x} \sin x + \frac{2}{5} e^{2x} \cos x + C$

T7 With $f(x) = \cos x$, $g(x) = \cos^3 x$ we have $F(x) = \sin x$ and $dg/dx = -3 \sin x \cos^2 x$. So Equation 2 gives

$$\int \cos^4 x \, dx = \sin x \cos^3 x$$
$$+ 3\int \sin^2 x \cos^2 x \, dx \qquad \text{(v)}$$

Writing $\sin^2 x = 1 - \cos^2 x$ in the integral on the right-hand side of Equation (v) gives

$$\int \cos^4 x \, dx = \sin x \cos^3 x + 3\int \cos^2 x(1 - \cos^2 x) \, dx$$

$$= \sin x \cos^3 x + 3\int \cos^2 x \, dx - 3\int \cos^4 x \, dx$$

Collecting together all terms in $\int \cos^4 x \, dx$ on the left-hand side gives

$$4\int \cos^4 x \, dx = \sin x \cos^3 x + 3\int \cos^2 x \, dx$$

or $\quad \int \cos^4 x \, dx = \frac{1}{4} \sin x \cos^3 x + \frac{3}{4} \int \cos^2 x \, dx$

T8 In this case, we make the substitution $y = x - 2$, and, since $dy/dx = 1$, we also write $dy = dx$. The integral then becomes

$$\int (x - 2)^{4/3} \, dx = \int y^{4/3} \, dy$$

We now evaluate this integral with respect to y

$$\int y^{4/3} \, dy = \frac{3}{7} y^{7/3} + C$$

and finally write $y = x - 2$ in this result, to obtain an answer in terms of x:

$$\int (x - 2)^{4/3} \, dx = \frac{3}{7} (x - 2)^{7/3} + C$$

To differentiate $\frac{3}{7} (x - 2)^{7/3}$ we use the chain rule. This function can be written as $f[y(x)]$, where

$f(y) = \frac{3}{7} y^{7/3}$ and $y = x - 2$. So $df/dy = y^{4/3}$ and $dy/dx = 1$. Substituting these results in the chain rule,

$$\frac{d}{dx} \{f[y(x)]\} = \frac{df}{dy} \frac{dy}{dx}$$

we find $\dfrac{d}{dx} [\frac{3}{7} (x - 2)^{7/3}] = y^{4/3} \times 1 = (x - 2)^{4/3}$

which confirms the result obtained for the integral $\int (x - 2)^{4/3} \, dx$.

T9 (a) *Step 1* We try the substitution $y = g(x) = 5 + 2x^2$.

Step 2 Then $dg/dx = 4x$, so we write $dy = 4x \, dx$ or, since the product $x \, dx$ appears in the integral, $x \, dx = \frac{1}{4} \, dy$.

Step 3 We now put $(5 + 2x^2)^{16} = y^{16}$ and $x \, dx = \frac{1}{4} \, dy$ in the integral, to obtain

$$\int x (5 + 2x^2)^{16} \, dx = \frac{1}{4} \int y^{16} \, dy$$

Step 4 Now we integrate to obtain

$$\frac{1}{4} \int y^{16} \, dy = \frac{1}{4} \times \frac{1}{17} y^{17} + C = \frac{1}{68} y^{17} + C$$

Step 5 Finally, we substitute $y = 5 + 2x^2$ into the answer obtained in Step 4, to arrive at an answer which is a function of x:

$$\int x(5 + 2x^2)^{16} \, dx = \frac{1}{68} (5 + 2x^2)^{17} + C$$

(b) *Step 1* We try the substitution $y = g(x) = \cos x$.

Step 2 Then $dg/dx = -\sin x$, so we write $dy = -\sin x \, dx$ or, since the product $\sin x \, dx$ appears in the integral, $\sin x \, dx = - dy$.

Step 3 We now put $y = \cos x$ and $\sin x \, dx = -dy$ in the integral, to obtain

$$\int \cos^4 x \sin x \, dx = - \int y^4 \, dy$$

Step 4 Now we integrate to obtain $\int y^4 \, dy = -\frac{1}{5} y^5 + C$.

Step 5 Finally we substitute $y = \cos x$ into the answer obtained in Step 4, to arrive at an answer which is a function of x:

$$\int \cos^4 x \sin x \, dx = -\frac{1}{5} \cos^5 x + C$$

T10 (a) *Step 1* We use the suggested substitution $y = g(x) = \sin x$.

Step 2 Then $dg/dx = \cos x$ so we write $dy = \cos(x)\,dx$

Step 3 We now put $e^{\sin x} = e^y$ and $dy = \cos(x)\,dx$ in the integral, to obtain

$$\int (\cos x)\, e^{\sin x}\, dx = \int y^4\, dy$$

Step 4 Now we integrate: $\int e^y\, dy = e^y + C$

Step 5 Finally, we write $y = \sin x$ in the answer obtained in Step 4, to arrive at an answer which is a function of x:

$$\int (\cos x) e^{\sin x}\, dx = e^{\sin x} + C$$

T11 (a) *Step 1* Note first that the derivative of $(2 + \cos x)$ is $(-\sin x)$. So we can write the integrand in the form '$p[g(x)] \times g'(x)$', where $g(x) = 2 + \cos x$. This suggests the substitution $y = 2 + \cos x$.

Step 2 Then we write $dy = -\sin x\, dx$

Step 3 Substituting $(2 + \cos x)^7$ and $-\sin x\, dx = dy$ in the integral, we obtain

$$\int \sin x (2 + \cos x)^7\, dx = -\int y^7\, dy$$

Step 4 Now we integrate $\int - y^7\, dy = -\frac{1}{8} y^8 + C$

Step 5 Finally, we write $y = 2 + \cos x$, to arrive at an answer which is a function of x:

$$\int \sin x (2 + \cos x)^7\, dx = -\frac{1}{8} (2 + \cos x)^8 + C$$

(b) *Step 1* Note first that the derivative of $(x^2 + 1)$ is $2x$. So we can write the integrand in the form $p[g(x)] \times g'(x)$, where $g(x) = x^2 + 1$. This suggests the substitution $y = x^2 + 1$.

Step 2 Then $dy = 2x\, dx$, or $x\, dx = \frac{1}{2}\, dy$.

Step 3 Substituting

$$\frac{1}{x^2 + 1} = \frac{1}{y} \quad \text{and} \quad x\, dx = \frac{1}{2}\, dy$$

in the integral, we obtain

$$\int \frac{x}{x^2 + 1}\, dx = \frac{1}{2} \int \frac{1}{y}\, dy$$

Step 4 Now we integrate: $\frac{1}{2} \int \frac{1}{y}\, dy = \frac{1}{2} \log_e y + C$

Step 5 Finally, we write $y = x^2 + 1$ to give an answer which is a function of x:

$$\int \frac{x}{x^2 + 1}\, dx = \frac{1}{2} \log_e (x^2 + 1) + C$$

T12 *Step 1* We take $y = 1 + x$.

Step 2 Then $dy = dx$.

Step 3 To express the integrand as a function of y, we invert the relation $y = 1 + x$ to obtain $x = y - 1$. Then $x(1 + x)^{5/2} = (y - 1)y^{5/2}$, so we have

$$\int x(1 + x)^{5/2}\, dx = \int (y - 1)y^{5/2}\, dy$$

Step 4 Integrate with respect to y:

$$\int (y - 1)y^{5/2}\, dy = \int y^{7/2}\, dy - \int y^{5/2}\, dy$$

$$= \frac{2}{9} y^{9/2} = \frac{2}{7} y^{7/2} + C$$

Step 5 Express the result in terms of x:

$$\int x(1 + x)^{5/2}\, dx = \frac{2}{9} (1 + x)^{9/2} - \frac{2}{7}$$
$$(1 + x)^{7/2} + C$$

T13 *Step 1* We want to make a substitution of the form $x = a \sin y$, which will ensure that $\sqrt{1 - 16x^2}$ is a multiple of $\cos y$. However $1 - 16a^2 \sin^2 y = \cos^2 y$ if $a = \frac{1}{4}$.

So we take $x = \frac{1}{4} \sin y$.

Step 2 Then $dx = \frac{1}{4} \cos y\, dy$.

Step 3 With $\sqrt{1 - 16x^2} = \cos y$ and $dx = \frac{1}{4} \cos y\, dy$ the integral becomes

$$\int \frac{1}{\sqrt{1 - 16x^2}}\, dx = \int \frac{1}{4} \frac{\cos y}{\cos y}\, dy = \int \frac{1}{4}\, dy$$

Step 4 Integrate with respect to y:

$$\int \frac{1}{4}\, dy = \frac{1}{4} y + C$$

Step 5 If $x = \frac{1}{4} \sin y$, $y = \arcsin (4x)$. So expressing the result in terms of x gives us

$$\int \frac{1}{\sqrt{1 - 16x^2}}\, dx = \frac{1}{4} \arcsin (4x) + C$$

T14 *Step 1* Take $x = \sin y$.

Step 2 Then $dx = \cos y\, dy$.

Step 3 The integrand

$$\sqrt{1 - x^2} = \cos y$$

(using $\cos^2 y + \sin^2 y = 1$). So the integral becomes

$$\int \sqrt{1-x^2}\, dx = \int \cos^2 y \, dy$$

Step 4 The integral with respect to y was found in Example 7:

$$\int \cos^2 y \, dy = \tfrac{1}{2} \sin y \cos y + \tfrac{1}{2} y + C$$

Step 5 To express this answer in terms of x, we put $\sin y = x$, so $y = \arcsin x$. We must also express $\cos y$ in terms of x; and since $\cos^2 y + \sin^2 y = 1$, we have

$$\cos y = \sqrt{1 - \sin^2 y} = \sqrt{1 - x^2}$$

So $\int \sqrt{1-x^2}\, dx = \tfrac{1}{2} x \sqrt{1-x^2} + \tfrac{1}{2} \arcsin x + C$

T15 *Step 1* We want a substitution of the form $x = a \tan y$, such that $(5 + 4x^2)$ is a multiple of $\sec^2 y$. We see that

$$5 + 4a^2 \tan^2 y = 5 \sec^2 y \text{ if } 4a^2 = 5,$$

i.e. if $a = \tfrac{1}{2} \sqrt{5}$. So we take $x = \tfrac{1}{2} \sqrt{5} \tan y$.

Step 2 Then $dx = \tfrac{1}{2} \sqrt{5} \sec^2 y \, dy$.

Step 3 Since $5 + 4x^2 = 5 \sec^2 y$ the integral becomes

$$\int \frac{1}{5+4x^2}\, dx = \int \frac{1}{2} \frac{\sqrt{5} \sec^2 y}{5 \sec^2 y} \, dy = \int \frac{1}{2\sqrt{5}} \, dy$$

Step 4 Integrating with respect to y gives us

$$\int \frac{1}{2\sqrt{5}}\, dy = \frac{1}{2\sqrt{5}} y + C$$

Step 5 Expressing this result in terms of x, using

$$y = \arctan\left(\frac{2x}{\sqrt{5}}\right)$$

gives us $\int \dfrac{1}{5+4x^2}\, dx = \dfrac{1}{2\sqrt{5}} \arctan\left(\dfrac{2x}{\sqrt{5}}\right) + C$

T16 *Step 1* We take $y = 4 - x$.

Step 2 Then $dy = -dx$.

Step 3 If $y = 4 - x$, then $x = 4 - y$ so the integrand $x \sqrt{4-x}$ becomes $(4 - y) \sqrt{y}$.

When $x = 0$, $y = 4$, and when $x = 4$, $y = 0$. Thus the integral becomes

$$\int_0^4 x \sqrt{4-x}\, dx = \int_4^0 (4 - y) \sqrt{y}\, (-dy)$$

$$= -\int_4^0 4 y^{1/2}\, dy + \int_4^0 y^{3/2}\, dy$$

Step 4 We now evaluate the definite integrals with respect to y:

$$-\int_4^0 4 \sqrt{y}\, dy + \int_4^0 y^{3/2}\, dy$$

$$= -\left[\frac{8}{3} y^{3/2}\right]_4^0 + \left[\frac{2}{5} y^{5/2}\right]_4^0$$

$$= -\frac{8}{3} (0 - 4^{3/2}) + \frac{2}{5} (0 - 4^{5/2}) = 8.5333$$

T17 (a) This integral can be found by a suitable substitution.

Step 1 Since $(x + 5)$ appears in the denominator of the integral, the best substitution to try is $y = x + 5$.

Step 2 Then $dx = dy$.

Step 3 If $y = x + 5$, then $x + 3 = y - 2$. So the integral becomes

$$\int \frac{(x+3)^2}{(x+5)^4}\, dx = \int \frac{(y-2)^2}{y^4}\, dy$$

$$= \int \left(\frac{1}{y^2} - \frac{4}{y^3} + \frac{4}{y^4}\right) dy$$

Step 4 Integrating with respect to y gives us

$$\int \left(\frac{1}{y^2} - \frac{4}{y^3} + \frac{4}{y^4}\right) dy = -\frac{1}{y} + \frac{2}{y^2} - \frac{4}{3y^3} + C$$

Step 5 Expressing the result in terms of x gives us

$$\int \frac{(x+3)^2}{(x+5)^4}\, dx = -\frac{1}{x+5} + \frac{2}{(x+5)^2}$$

$$- \frac{4}{3(x+5)^3} + C$$

(b) This integral can be found by *integrating by parts*. We take $f(x) = \cos(x/2)$, $g(x) = 3x$; then $F(x) = 2 \sin(x/2)$ and $dg/dx = 3$. Substituting into Equation 2 gives us

$$\int 3x \cos\frac{x}{2}\, dx = 6x \sin\frac{x}{2} - 6 \int \sin\frac{x}{2}\, dx$$

$$= 6x \sin\frac{x}{2} + 12 \cos\frac{x}{2} + C$$

(c) This integral can be found by means of a suitable substitution.

Step 1 We want a substitution of the form $x = a \sin y$ that will convert $\sqrt{2 - x^2}$ into a multiple of $\cos y$. This can be achieved if we take $x = \sqrt{2} \sin y$.

Step 2 Then $dx = \sqrt{2} \cos y \, dy$.

Step 3 Since $\sqrt{2 - x^2} = \sqrt{2} \cos y$ and $dx = \sqrt{2} \cos y \, dy$ the integral becomes simply $\int 1 \, dy$.

Step 4 Integrate with respect to y:

$$\int 1 \, dy = y + C.$$

Step 5 Expressing the result in terms of x, using

$$y = \arcsin\left(\frac{x}{\sqrt{2}}\right)$$

gives us

$$\int \frac{1}{\sqrt{2 - x^2}} \, dx = \arcsin\left(\frac{x}{\sqrt{2}}\right) + C$$

(d) This integral can be found by means of a suitable substitution.

Step 1 The integrand can be written as a product of a function of $(2 - x^2)$ and the derivative of $(2 - x^2)$, namely $-2x$. So we make the substitution $y = (2 - x^2)$.

Step 2 Then $dy = -2x \, dx$, or $x \, dx = -\frac{1}{2} \, dy$.

Step 3 As $\sqrt{2 - x^2} = \sqrt{y}$ the integral becomes

$$\int \frac{x}{\sqrt{2 - x^2}} \, dx = -\int \frac{1}{2\sqrt{y}} \, dy$$

Step 4 Integrating with respect to y gives us

$$-\int \frac{1}{2\sqrt{y}} \, dy = -\sqrt{y} + C$$

Step 5 Expressing the result in terms of x gives us

$$\int \frac{x}{\sqrt{2 - x^2}} \, dx = -\sqrt{2 - x^2} + C$$

Comment In part (d), you may have reasoned that as the integrand contains $\sqrt{2 - x^2}$ then the substitution $x = \sqrt{2} \sin y$ would be suitable (as in part (c)). This will indeed work; the integral becomes $\sqrt{2} \int \sin y \, dy = -\sqrt{2} \cos y + C = -\sqrt{2 - x^2} + C$. Quite often, there is more than one substitution which will enable you to find a given integral.

(e) The integral $\int \sqrt{x} \, (1 + x) \, dx$ can be found by quite simple methods. Expanding the integrand

$$\int \sqrt{x} \, (1 + x) dx = \int (x^{-1/2} + x^{3/2}) dx$$
$$= \tfrac{1}{2} x^{3/2} + \tfrac{2}{5} x^{5/2} + C.$$

(The moral is to try the simplest method first.)

Index

The terms printed in **bold** are also emboldened at the corresponding point in the main text and usually represent major references. Numerical entries indicate location in terms of chapter, section and (where appropriate) subsection. Thus, **angle, 5.2.1** refers to an emboldened appearance of the word 'angle' in Chapter 5, Subsection 2.1. *Italic* entries refer to figures and tables (e.g. *F6.1*, *T4.2*), and MH indicates an entry in the *Maths handbook*.

Printed and bound by CPI Group (UK) Ltd, Croydon, CR0 4YY

27/10/2024

14580378-0005